Essential Mathematics for NMR and MRI Spectroscopists

Essential Mathematics for NMR and MRI Spectroscopists

Keith C. Brown

University of Saskatchewan, Saskatoon, Canada
Email: brownk@chem4823.usask.ca

ROYAL SOCIETY OF CHEMISTRY

THE QUEEN'S AWARDS FOR ENTERPRISE: INTERNATIONAL TRADE 2013

Print ISBN: 978-1-78262-797-5

A catalogue record for this book is available from the British Library

Published by The Royal Society of Chemistry,
Thomas Graham House, Science Park,
Milton Road, Cambridge CB4 0WF, UK
Registered Charity Number 207890

Visit our website at www.rsc.org/books

Printed in the United Kingdom by CPI Group (UK) Ltd, Croydon, CR0 4YY, UK

Preface

Philosophy is written in this grand book, the universe which stands continually open to our gaze. But the book cannot be understood unless one first learns to comprehend the language and interpret the alphabet in which it is composed. It is written in the language of mathematics.

Galileo Galilei

The genesis of this book was a desire on my part to pass on what I had learned about NMR spectroscopy to the students passing through my laboratory. The original manuscript was a short document giving very brief outlines of the vector model of NMR spectroscopy, the product operator formalism and various minor topics, all with very little mathematics. I felt the need, however, to be able to explain in detail how, for example, the Fourier transform works, which led to the first draft of what is now Chapter 8 in this book. Gradually other chapters were added, each with a leaning towards the mathematics involved in understanding NMR spectroscopy. In order to write each chapter much reading was required and several new (to me at least) concepts were encountered. It was, however, very tough going, especially when reading the classic books such as those by Abragam and Schlicter. In effect, I was translating their concepts into something more readable by the newcomer to NMR spectroscopy. It was this realisation that led me to consider writing a "real" book along these lines for everyone, not just my students. I have therefore produced a book that is hopefully intermediate between the very excellent introductory NMR/MRI spectroscopy texts that are now available and the very high-level texts.

Essential Mathematics for NMR and MRI Spectroscopists
By Keith C. Brown

Published by the Royal Society of Chemistry, www.rsc.org

As an undergraduate and as a postgraduate student I found the mathematical explanations of various physical phenomena given in texts and lectures to be lacking in critical details for understanding the material. Frequently the text would declare "it can be shown that ..." or "it is easily shown that ..." and then leave it to the reader to figure out how it can be shown. This is undoubtedly a Good Thing as it spurs the curious reader on to actually dig into the specifics of the problem at hand but it is also sometimes very frustrating. If one cannot "show that ..." then how can one proceed in a logical progression leading to understanding? The physicist Leonard Susskind in his book "The Black Hole War" says "I fear that I will have to take the scoundrel's way out and say "It can be shown that ..."; the implication being that it is as distasteful to him as to me. Others feel this way about scientific writing, as well. The tongue-in-cheek editorial in *Analytical Chemistry*, 2011, **83**, 633, describing how *not* to write a scientific article has a particular resonance with respect to this text. Therefore, I will try to painstakingly explain each concept from beginning to end as simply as possible and not skip any (major) steps, hopefully leaving the reader with an understanding and an appreciation for the beauty and the utility of the mathematical description of NMR spectroscopy. There may be those who will condemn this approach as "hand-holding" ... this is entirely the intention of this text; readability and comprehension are the ultimate goal, rigour and obtuse or confusing prose are *not*. I expect, however, that despite my best efforts there will be some who will not be able to grasp some of the concepts presented here for one reason or another. My experience is that in this situation another viewpoint is frequently of value. To assist the reader who finds herself/himself in this state I have included at the end of each chapter several sources for further reading that I have found particularly valuable in understanding the ideas presented here.

The subject of this book is the mathematical description of nuclear magnetic resonance spectroscopy. Many, if not all, aspects of NMR spectroscopy have a mathematical underpinning and that are not, as far as I can tell, all treated in one intermediate level book although there are certainly several excellent modern texts available. In this book I describe angular momentum in both classical and quantum mechanical terms as they apply to NMR spectroscopy, Fourier series and transforms, linear algebra as it applies to NMR spectroscopy and other pertinent topics. Also, some mathematics as it applies to the more practical aspects of NMR spectroscopy such as high-frequency electronics is included. It is also my belief that the history of the development of these areas of mathematics is helpful in understanding the subject, not to mention interesting. To that end, I include some brief descriptions of the history of the development of some of the various areas of mathematics and physics that are discussed here. Although I would not in any way describe myself an NMR imaging spectroscopist I have included a short chapter on that type of spectroscopy for the interested reader. A couple of very much expanded treatises on the subject are given in the excellent books by Brown *et al.* and Bernstein, listed in the further reading section of Chapter 22 for those who wish to delve more deeply into the subject. There are NMR and MRI topics that are not addressed in the text such as chemical exchange, metabolomics data analysis techniques and MRI *k*-space trajectories. The potential range of topics for a book such as this is *extremely* large and much as I would like to include them all, there comes a point when one must just simply say "enough".

My hope is that this text will be seen as a vehicle for further understanding and perhaps as a first approach to in-depth expertise in NMR or MRI spectroscopy. I do require that the reader be proficient in basic high-school algebra and at least somewhat familiar with simple introductory differential and integral calculus. Some common derivatives and integrals and trigonometric identities are included in the appendices for those who are rusty in these areas. Many may not be familiar with vector calculus so the basics are included in the chapter on vector calculus.

The text is aimed at NMR and MRI spectroscopists such as laboratory managers, academics and postgraduate students wishing to grasp the fundamentals of NMR/MRI spectroscopy at a deeper level than is currently available in beginner texts. I can promise that "It can be shown that..." does not appear anywhere without my actually showing it or, at the *very least*, actually saying where it is explicitly shown.

Of course, it goes without saying that there are errors and inconsistencies in the text of which I am not aware. I have tried to make it as clear and error-free as possible but perfection is unfortunately not within my grasp just yet. Until that time I would be very grateful to know of any errors that appear in the text.

The figures in the book were prepared using the SciDraw scientific figure preparation system (M. A. Caprio, *Comput. Phys. Commun.*, 2005, **107**, 171; http://scidraw.nd.edu) and the MaTeX software for LaTeX typesetting in Mathematica®. Symbolic algebra calculations were done using wxMaxima. Product operator calculations were performed using my own software, wxProdOp, freely available at https://sourceforge.net/projects/wxProdOp.

I would like to thank all of those involved in assisting in the preparation of this document. I am especially indebted to Mr Hridaynath Bhattacharjee for very kindly producing the cover photograph of the Math Mug. The Math Mug photo is used with the very kind permission of The Unemployed Philosophers Guild: www.philosophersguild.com. I would also like to thank Prof. K. Rossberg of Oklahoma City University for very kind assistance in the understanding of simple analytical mechanics and Dr G. Schatte, of Queens University, for kindly pointing out the occasional short-circuit between my ears in dealing with practical day-to-day problems in the NMR laboratory (in between writing sessions). Also, I am grateful to Dr Sergiy Nokhrin, University of Toronto, for his assistance in unravelling the mysteries of spin-one density matrices. Dr Bob Berno at McMaster University and Dr Glenn Facey at the University of Ottawa were very helpful in spotting errors and omissions from early draft documents and in providing valuable suggestions for corrections and improvements to the document. A most special tribute is due to my late friend and colleague of 27 years, Prof. J. A. Weil, University of Saskatchewan, for many discussions (and the odd argument) about magnetic resonance spectroscopy. Last, but certainly not least, I thank my lovely wife, Janice, for her patience and understanding and for putting up with equations day-in and day-out during the writing of this text.

Keith Brown
University of Saskatchewan

Contents

Essential Mathematics for NMR and MRI Spectroscopists
By Keith C. Brown

Published by the Royal Society of Chemistry, www.rsc.org

1 Complex Numbers

It is India that gave us the ingenious method of expressing all numbers by means of ten symbols, each symbol receiving a value of position as well as an absolute value; a profound and important idea which appears so simple to us now that we ignore its true merit.

Pierre-Simon Laplace

1.1 The History of Numbers

The first humans presumably had no need for numbers and would therefore probably not have understood the concept. Eventually, of course, numbers were invented for use in counting … taking stock of animals and items of trade. The first set of numbers was undoubtedly simply a set of integers beginning at 1 and increasing from there … we are all familiar with them as we learn them at a very early age. This is very obvious and intuitive to us; anyone that counts uses these numbers. It turned out, however, that this number system was to be altered to include zero. Why? It was originally used as a placeholder for larger numbers. Thus, in the number 1053, the zero is in the "hundreds" column and indicates that are no hundreds to be added into the number. It consists of 3 ones, 5 tens, 0 hundreds and 1 thousands. Used at first in India for practical calculations it was imported into ancient Babylonian mathematics to replace a cumbersome placeholder system of slanted wedges. Its use was only as a placeholder and as it was not used alone it was not really considered to be a number. It did, however, come to be used as a number and it is

Essential Mathematics for NMR and MRI Spectroscopists
By Keith C. Brown

Published by the Royal Society of Chemistry, www.rsc.org

now an integral part of our number systems. So, the simple number system was altered out of necessity to produce what we now refer to as the natural numbers.

Negative numbers were introduced to represent debts, again in India and again the number system was altered to include negative numbers into what we now refer to as the integers. There were those in ancient Greece who considered negative solutions to equations to be false and "absurd" so the use of negative numbers has not been without its critics.

Next, in order to express fractional quantities another set of numbers was needed. Using fractions of integers we come up with the *rational* numbers. Thus:

$$\frac{1}{2} = 0.5$$

$$\frac{1}{3} = 0.333...$$

but we still do not have a complete set of numbers. The Greek Pythagoreans considered the integers to be "perfect". Since the rational numbers are expressible using integers, they too were considered to be perfect. However, for them, a major problem existed, ironically from the results of what we now call the Pythagorean theory. The theory says the sum of the squares of the two sides of a right-angle triangle is equal to the square of the hypotenuse and has been proven many times in many different ways. If the two sides are each one unit in length then the length of the hypotenuse must be equal to the square root of two. The square root of two cannot be expressed as a ratio of integers which made the Pythagoreans very uncomfortable since it could not be constructed from "perfect numbers". It is now referred to as an *irrational number*. Another example of an irrational number is π. So yet again, the number set was added to with rational and irrational numbers. Together they make up the set of real numbers.

1.2 Why Complex Numbers?

As NMR spectroscopists we deal with nuclei that possess the property of intrinsic angular momentum or "spin" and in order to model this property we use the mathematics of classical mechanical rotations, suitably modified for quantum mechanics. It turns out that complex numbers are ideally suited to this task and, indeed, it is well-nigh impossible to do without them, as we shall see in later chapters. It is amazing that such a fundamental property of matter that was totally unknown 100 years ago can be modelled with relatively simple mathematics. Complex numbers are used in many areas of physics and engineering and the modern practitioner of these sciences must be thoroughly familiar with them. It is often puzzling though to the uninitiated (and perhaps the initiated as well) as to why one would want to use a number that is defined as the square root of a negative number. How are we to picture such a number? Where does it fit in with our "regular" numbers? Hopefully we can draw aside the veil of mystery surrounding these numbers and reveal them to be really no different in many respects from ordinary numbers.[†]

A complex number consists of a real part and an imaginary part and is generally represented as:

$$A = x + iy \tag{1.1}$$

Since $\sqrt{-1}$ arises so frequently it is given the symbol i in the sciences (the symbol j is used extensively in engineering as i is used for electric current there). The *complex conjugate* of A is generally represented as A^* and is defined as:

$$A^* = x - iy \tag{1.2}$$

We see that the difference between a complex number and its conjugate is the sign in front of the term containing i.

Where do complex numbers come from? Under what circumstances would anyone want to contemplate using such strange numbers? Essentially, the necessity for complex numbers arises from a consideration of the roots of certain polynomials. A very simple example would be the roots of $x^2 + 4 = 0$:

$$\begin{aligned} x^2 &= -4 \\ x &= \pm\sqrt{-4} = \pm 2\sqrt{-1} \\ &= \pm 2i \end{aligned}$$

[†]In writing this text I have come to think of numbers as objects with their own sets of addition, subtraction, multiplication and division rules and that complex numbers are simply a type of number object with their own rules for addition and multiplication. This idea is easily extended to matrices, vectors and tensors and, I feel, makes understanding them that much simpler. As it turns out, this is the basic idea behind group theory (see Appendix A).

In this case the solution of the equation is an imaginary number[‡] but in general a complex number of the form introduced above in eqn (1.1) can be expected.

As mentioned above, the solutions to some equations require the use of the root of a negative number. Of course this idea would have caused the ancient Greeks much trouble! How can one take the root of a negative number? Nonetheless, it is necessary to use this concept to solve these equations so we again alter our number set to include the root of −1 in the form of complex numbers. Each time that a new type of number was needed in the past it was added to our set of numbers.

If you look at the general solution of polynomials you find that for a polynomial of degree n there are n factors from which we can extract n roots of the equation.[§] For example:

$$x^3 - 3x^2 + 6x - 4 = 0$$

This polynomial is of order three and has the three roots:

$$x = 1,\ x = 1 + \sqrt{-3},\ x = 1 - \sqrt{-3}$$

You can see from the above example equation that there are two complex roots as part of the solution to the equation. Furthermore, they are conjugates of each other. Thus, the use of $x + iy$ to represent these types of numbers. In general for a polynomial:

$$x^n + Ax^{n-1} + Bx^{n-2} + Cx^{n-3} + \ldots + N = 0$$

there are n factors:

$$(x - \alpha)\,(x - \beta)\,(x - \gamma)\,(x - \delta)\ldots(x - \nu) = 0$$

such that:

$A =$ the sum of $-\alpha,\ -\beta,\ -\gamma\ \ldots\ -\nu$
$B =$ the sum of all their products two at at time
$C =$ the sum of all their products three at a time
⋮
$N =$ the product of all taken together

[‡]Note that there is a difference between the terms "imaginary number" and "complex number". An imaginary number is just some multiple of $\sqrt{-1}$ whereas a complex number consists of an imaginary number and a real number combined, as we saw in eqn (1.1) and (1.2).

[§]There is a fascinating and very readable translation of an article published by Leonhard Euler in 1751 on complex numbers and their origin. Please see the Further Reading list at the end of the chapter.

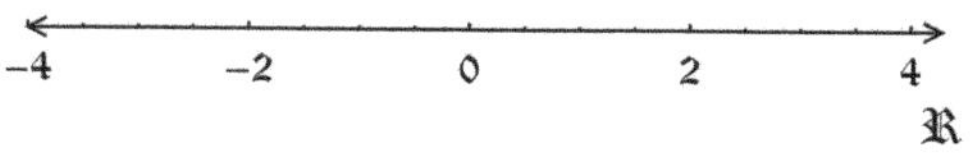

Figure 1.1 The simple number line containing the integers and real numbers, both rational and irrational.

As humans we have a built-in need to connect ideas with pictures and this is the case with numbers. In the case of integers one sometimes encounters numbers arrayed along a number line in early grade school as we have seen in Figure 1.1. How are we to picture the inclusion of imaginary numbers in this picture? There is a problem in this respect with imaginary numbers that is not immediately apparent. Suppose that we take a number that lies on the number line ... it can be an integer or a rational or an irrational number, it does not matter. If we square that number we will generate a new number that is *also* on the number line. Furthermore, the new number will be positive since, as we all learn in our early schooling, a positive number times a positive number gives us a positive result and a negative number times a negative number also gives us a positive result. $\sqrt{-1}$ or i does not do this and therefore i *cannot* lie on the number line. Thus, $3.77i$, $-5.018i$ and indeed any multiple of i cannot lie on the number line ... with one important exception, $0i$. $0i = 0$, which *does* lie on the number line. Thus, we can view the imaginary numbers as lying on their own distinct number line that intersects the real number line at 0. The two number lines need not be orthogonal but for convenience this is how they are generally represented in the form of an Argand diagram (see Figure 1.2) where $\Re$ and $\Im$ indicate the real and imaginary axes, respectively.

1.3 Properties of Complex Numbers

As shown in eqn (1.1) and (1.2), the conjugate of a complex number differs from the complex number in that the opposite sign is used in front of the term containing i so that, for example, $x + iy$ and $x - iy$ are complex conjugates of each other. Later in the text when we are speaking of mathematical objects that have this type of conjugate relationship we will simply use a superscripted asterisk to indicate the conjugate ... A^* is the conjugate of A.

The inverse of i equals $-i$:

$$\frac{1}{i} = \frac{1}{i} \cdot \frac{i}{i} = \frac{i}{-1} = -i \tag{1.3}$$

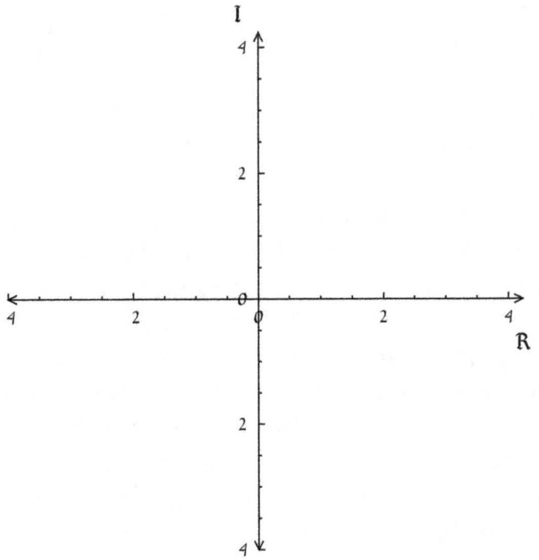

Figure 1.2 The Argand diagram, which is just two simple number lines drawn at right angles to each other and crossing at zero on each line. By convention, the horizontal line contains the real numbers and the vertical line contains the imaginary numbers.

The square of i is -1:¶

$$i^2 = -1$$

$$-i^2 = +1$$

There is an interesting pattern that emerges as one increases the exponent of i:

$$i^0 = 1$$

$$i^1 = i$$

$$i^2 = i \cdot i = -1$$

$$i^3 = i \cdot i^2 = -i$$

$$i^4 = i \cdot i^3 = +1$$

$$i^5 = i \cdot i^4 = i$$

$$\vdots$$

¶Indeed, this is the very definition of i. Since a complex number consists of a real portion and imaginary portion we may visualise it in an Argand diagram as a line running from the origin (0,0) to coordinate (a, ib). See Figure 1.3.

Even powers of i result in either $+1$ or -1 and odd powers of i give either i or $-i$. and a repeating pattern is observed; $1, i, -1, -i$. If plotted on the Argand number-line diagram these appears as four points on the lines that describe a counter-clockwise circular movement around the origin.

The sums and products of conjugate complex numbers are real:

$$\begin{aligned}(x+iy)+(x-iy)&=2x\\(x+iy)\cdot(x-iy)&=x^2+y^2\end{aligned}\tag{1.4}$$

and the difference between complex conjugates is imaginary:

$$(x+iy)-(x-iy)=2iy$$

If complex conjugates are equal then the number is real:

$$\begin{aligned}&\text{if: } x+iy=x-iy\\&\text{solving for y:}\\&2iy=0\rightarrow y=0\end{aligned}\tag{1.5}$$

From the Argand diagram in Figure 1.3, the length of the line r is called the *modulus* of complex number A:

$$r=\sqrt{a^2+b^2}=|A|$$

Some properties of the modulus are:

$$\begin{aligned}&|A|=|A^*|\\&|A||A^*|=|A^2|\\&|A_1A_2|=|A_1||A_2|\end{aligned}\tag{1.6}$$

The modulus is also called the *absolute value* of the complex number.

We can extract the real and imaginary parts of the complex number using the appropriate functions:

$$\begin{aligned}&\text{if: } z=x+iy\\&\mathrm{Re}(z)=x\\&\mathrm{Im}(z)=y\end{aligned}$$

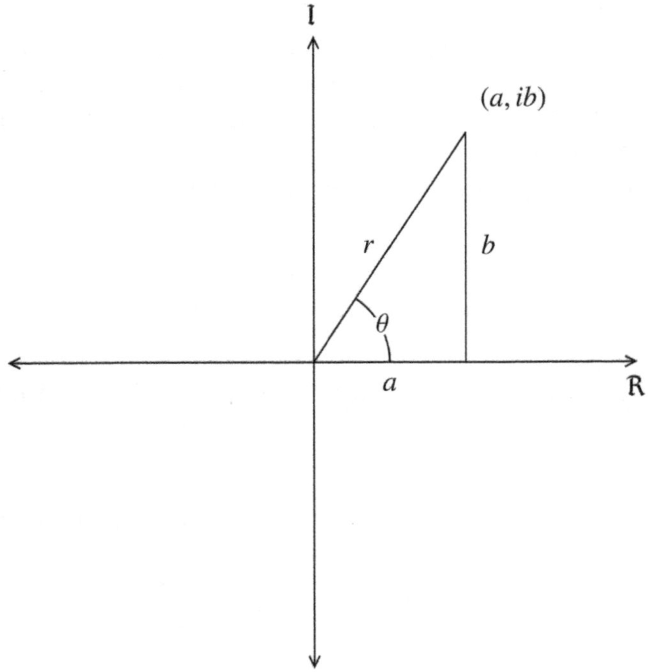

Figure 1.3 An Argand diagram showing the line representing the complex number $a+ib$ and the angle θ that the line makes with the real axis.

One can see an angle, θ, in the Argand diagram in Figure 1.3. Thus, the complex number can also be represented in polar coordinate form:

$$A = re^{i\theta} \tag{1.7}$$

Euler's theorem (see Appendix A, eqn (A.2)) is stated as:

$$e^{i\theta} = \cos(\theta) + i\ \sin\ (\theta)$$

so that:

$$A = re^{i\theta} = r[\cos(\theta) + i\ \sin(\theta)]$$

Also from Euler's theorem:

$$e^{i\frac{\pi}{2}} = -i, \quad e^{i\pi} = -1$$

or

$$e^{i\pi} + 1 = 0$$

From the Argand diagram (Figure 1.3):

$$\cos(\theta) = \frac{a}{r}, \sin(\theta) = \frac{b}{r}$$

$$\theta = \arccos\left(\frac{a}{r}\right) = \cos^{-1}\left(\frac{a}{r}\right)$$

$$= \arcsin\left(\frac{b}{r}\right) = \sin^{-1}\left(\frac{b}{r}\right)$$

and so:

$$A = r\left[\frac{a}{r} + i\frac{b}{r}\right] = a + ib$$

hence the intimate association of the complex number with a phase angle, θ.

The product of two complex numbers can be done in two ways; either using the formal complex notation definition of the number (eqn (1.1)) or using the polar coordinate form (eqn (1.7)). Thus, for two complex numbers:

$$A_1 = 4 + 5i$$

$$A_2 = 3 + i$$

or in polar notation:

$$A_1 = 6.40e^{0.8960i}$$

$$A_2 = 3.16e^{0.3218i}$$

the product of A_1 and A_2 using the polar notation is:

$$A_1 \cdot A_2 = 6.40e^{0.8960i} \cdot 3.16e^{0.3218i} = 20.25e^{1.2178i}$$

(the phase angles here are in radians). We can show that these two results are equivalent by converting one into the other. Let us convert the complex notation answer into polar notation:

$$r = \sqrt{a^2 + b^2} = \sqrt{7^2 + 19^2} = 20.25$$

$$\theta = \sin^{-1}\left(\frac{b}{r}\right) = \sin^{-1}\left(\frac{19}{20.25}\right) = 1.2178$$

$$r = 20.25e^{1.2178i}$$

The product of a complex number and its conjugate is real and positive:

$$A \cdot A^* = re^{i\phi} \cdot re^{-i\phi} = r^2 = a^2 + b^2 = |A|^2 \tag{1.8}$$

The inverse of a complex number is:

$$\frac{1}{A} = \frac{1}{a + ib}$$

however, this is not in standard form (see eqn (1.1)) so:

$$\frac{1}{a+ib} = \frac{1}{a+ib} \cdot \frac{a-ib}{a-ib} = \frac{a-ib}{a^2+b^2} = \frac{a}{a^2+b^2} + i\frac{b}{a^2+b^2}$$

Thus, we can divide two complex numbers:

$$\text{if } A = x_A + iy_A \text{ and } B = x_B + iy_B$$

$$\frac{A}{B} = \frac{x_A + iy_A}{x_B + iy_B} = (x_A + iy_A)\left[\frac{x_B}{x_B^2 + y_B^2} - i\frac{y_B}{x_B^2 + y_B^2}\right] \tag{1.9}$$

$$= \frac{x_A x_B + y_A y_B}{x_B^2 + y_B^2} + i\frac{x_B y_A - x_A y_B}{x_B^2 + y_B^2}$$

We can, of course, also do division using the polar form:

$$A = r_A e^{i\theta_A} \quad B = r_B e^{i\theta_B}$$

$$\frac{A}{B} = \frac{r_A}{r_B} e^{i(\theta_A - \theta_B)}$$

As with the product of two complex numbers in polar form, this is equivalent to the standard form quotient:

$$\frac{A}{B} = \frac{r_A}{r_B} e^{i(\theta_A - \theta_B)}$$

$$= \frac{r_A}{r_B}\left[\frac{x_A x_B + y_A y_B}{x_A^2 + y_B^2} + i\frac{x_B y_A - x_A y_B}{x_A^2 + y_B^2}\right]$$

which is of the same form as eqn (1.9).

Finally, from the Euler theorem in Appendix A, sine and cosine can be expressed in terms of exponentials:

$$\cos(\theta) = \frac{e^{i\theta} + e^{-i\theta}}{2}$$

$$\sin(\theta) = \frac{e^{i\theta} - e^{-i\theta}}{2i}$$

which will be of use to us in later chapters.

Further Reading

1. T. Andreescu and D. Andrica, *Complex Numbers from A to ... Z*, Springer, New York, 2nd edn, 2014.
2. L. Euler, Recherches sur les racines imaginaires des equations, http://eulerarchive.maa.org/.
3. D. W. Jordan and P. Smith, *Mathematical Techniques: an introduction for the engineering, physical, and mathematical sciences*, Oxford University Press, London, 1994.
4. J. Stilwell, *Mathematics and Its History*, Springer, New York, 3rd edn, 2010.

2 Matrices

It is utterly implausible that a mathematical formula should make the future known to us, and those who think it can would once have believed in witchcraft.

Jacob Bernoulli

2.1 Matrix Algebra

Matrices[†] are arrays of (usually) scalars arranged in rows and columns. For example:

$$\hat{\mathbf{A}} = \begin{bmatrix} 1 & 2 & 3 \\ 4 & 5 & 6 \\ 7 & 8 & 9 \end{bmatrix} \quad \hat{\mathbf{B}} = \begin{bmatrix} 2 & 3 & 4 \\ 5 & 6 & 7 \\ 8 & 9 & 1 \end{bmatrix}$$

Matrices of equivalent dimensions are added or subtracted by adding or subtracting their corresponding individual elements:

$$\hat{\mathbf{A}} + \hat{\mathbf{B}} = \begin{bmatrix} 1 & 2 & 3 \\ 4 & 5 & 6 \\ 7 & 8 & 9 \end{bmatrix} + \begin{bmatrix} 2 & 3 & 4 \\ 5 & 6 & 7 \\ 8 & 9 & 1 \end{bmatrix} = \begin{bmatrix} 1+2 & 2+3 & 3+4 \\ 4+5 & 5+6 & 6+7 \\ 7+8 & 8+9 & 9+1 \end{bmatrix} = \begin{bmatrix} 3 & 5 & 7 \\ 9 & 11 & 13 \\ 15 & 17 & 10 \end{bmatrix}$$

$$\hat{\mathbf{A}} + \hat{\mathbf{B}} = \hat{\mathbf{B}} + \hat{\mathbf{A}} \text{ (commutative)} \tag{2.1}$$

[†]Matrices tend to intimidate the beginner because there are so many numbers packed into such a small space. If you feel this way, please try to set aside the feeling and judge for yourself whether or not it is justified, at the end of the chapter. These are exceedingly powerful mathematical objects and will be of great use to us. Also, I hope that you will come to appreciate their natural beauty.

Essential Mathematics for NMR and MRI Spectroscopists
By Keith C. Brown

Published by the Royal Society of Chemistry, www.rsc.org

$$\hat{\mathbf{A}} + (\hat{\mathbf{B}} + \hat{\mathbf{C}}) = (\hat{\mathbf{A}} + \hat{\mathbf{B}}) + \hat{\mathbf{C}} \text{ (associative)} \tag{2.2}$$

$$\hat{\mathbf{A}} + \hat{\mathbf{0}} = \hat{\mathbf{A}} \text{ (identity)} \tag{2.3}$$

$$\hat{\mathbf{A}} + (-\hat{\mathbf{A}}) = \hat{\mathbf{0}} \text{ (additive inverse)} \tag{2.4}$$

Normal matrix multiplication proceeds row by column:

$$\hat{\mathbf{A}}\hat{\mathbf{B}} = \begin{bmatrix} 1 & 2 & 3 \\ 4 & 5 & 6 \\ 7 & 8 & 9 \end{bmatrix} \times \begin{bmatrix} 2 & 3 & 4 \\ 5 & 6 & 7 \\ 8 & 9 & 1 \end{bmatrix}$$

$$= \begin{bmatrix} 1\times2+2\times5+3\times8 & 1\times3+2\times6+3\times9 & 1\times4+2\times7+3\times1 \\ 4\times2+5\times5+6\times8 & 4\times3+5\times6+6\times9 & 4\times4+5\times7+6\times1 \\ 7\times2+8\times5+9\times8 & 7\times3+8\times6+9\times9 & 7\times4+8\times7+9\times1 \end{bmatrix}$$

$$= \begin{bmatrix} 36 & 42 & 63 \\ 85 & 96 & 57 \\ 126 & 150 & 93 \end{bmatrix}$$

Note that matrix multiplication is not necessarily commutative. By this is meant that $\hat{\mathbf{A}}\hat{\mathbf{B}}$ does not necessarily equal $\hat{\mathbf{B}}\hat{\mathbf{A}}$.[‡] Also, the matrices to not have to be square but the number of rows of $\hat{\mathbf{A}}$ must be the *same* as the number of columns in $\hat{\mathbf{B}}$ in order to do a normal multiplication; we say that they are *conformable* when this is so. In this text all of our operator matrices[§] will be square and of the same dimension and therefore conformable to normal matrix multiplication, as required.

Matrices may be multiplied or divided by a scalar:

$$6\times\begin{bmatrix} 1 & 2 \\ 3 & 4 \end{bmatrix} = \begin{bmatrix} 6\times1 & 6\times2 \\ 6\times3 & 6\times4 \end{bmatrix} = \begin{bmatrix} 6 & 12 \\ 18 & 24 \end{bmatrix}$$

$$\alpha(\hat{\mathbf{A}} + \hat{\mathbf{B}}) = \alpha\hat{\mathbf{A}} + \alpha\hat{\mathbf{B}} \text{ (scalar distributive)} \tag{2.5}$$

$$(\alpha + \beta)\hat{\mathbf{A}} = \alpha\hat{\mathbf{A}} + \beta\hat{\mathbf{A}} \text{ (matrix distributive)} \tag{2.6}$$

$$(\alpha\beta)\hat{\mathbf{A}} = \alpha(\beta\hat{\mathbf{A}}) \text{ (associative law for multiplication)} \tag{2.7}$$

These definitions for addition and scalar multiplication are of importance in the field of vector analysis. We will return to this later.

[‡]This is a *very* important idea for us as NMR spectroscopists. We will return to it several times throughout the text.

[§]We will elaborate on the operator-as-matrix concept in Section 3.5 of Chapter 3.

The direct product (sometimes called the Kronecker product or tensor product) proceeds as follows:

$$\hat{\mathbf{A}} \otimes \hat{\mathbf{B}} = \begin{bmatrix} 1 & 2 \\ 3 & 4 \end{bmatrix} \begin{bmatrix} 5 & 6 \\ 7 & 8 \end{bmatrix} = \begin{bmatrix} 1\begin{bmatrix} 5 & 6 \\ 7 & 8 \end{bmatrix} & 2\begin{bmatrix} 5 & 6 \\ 7 & 8 \end{bmatrix} \\ 3\begin{bmatrix} 5 & 6 \\ 7 & 8 \end{bmatrix} & 4\begin{bmatrix} 5 & 6 \\ 7 & 8 \end{bmatrix} \end{bmatrix} = \begin{bmatrix} 5 & 6 & 10 & 12 \\ 7 & 8 & 14 & 16 \\ 15 & 18 & 20 & 24 \\ 21 & 24 & 28 & 32 \end{bmatrix} \tag{2.8}$$

As with normal matrix multiplication, the direct product is not generally commutative. This type of matrix multiplication will be of importance to us when we encounter product operators of multiple spins.

The *zero matrix*, referred to in eqn (2.4), is (not surprisingly) full of zeros:

$$\begin{bmatrix} 0 & 0 & 0 & \cdots \\ 0 & 0 & 0 & \cdots \\ 0 & 0 & 0 & \cdots \\ \vdots & \vdots & \vdots & \ddots \end{bmatrix} = \hat{\mathbf{0}} \tag{2.9}$$

The unit or *identity matrix* is:

$$\begin{bmatrix} 1 & 0 & 0 & \cdots \\ 0 & 1 & 0 & \cdots \\ 0 & 0 & 1 & \cdots \\ \vdots & \vdots & \vdots & \ddots \end{bmatrix} = \hat{\mathbf{1}} \tag{2.10}$$

with 1s on the diagonal[¶] so that:

$$\begin{aligned} \hat{\mathbf{A}}\hat{\mathbf{0}} = \hat{\mathbf{0}} \quad \hat{\mathbf{0}}\hat{\mathbf{A}} = \hat{\mathbf{0}} \\ \hat{\mathbf{A}}\hat{\mathbf{1}} = \hat{\mathbf{A}} \quad \hat{\mathbf{1}}\hat{\mathbf{A}} = \hat{\mathbf{A}} \end{aligned} \tag{2.11}$$

If $\hat{\mathbf{A}}\hat{\mathbf{B}} = \hat{\mathbf{1}}$ where $\hat{\mathbf{1}}$ is the identity matrix then $\hat{\mathbf{A}}$ and $\hat{\mathbf{B}}$ are the *inverse* of each other ... $\hat{\mathbf{A}}^{-1} = \hat{\mathbf{B}}$ and $\hat{\mathbf{A}} = \hat{\mathbf{B}}^{-1}$. The inverse of the product of two matrices is equal to the product of the inverses of the two matrices *in reverse order*:

$$(\hat{\mathbf{A}}\hat{\mathbf{B}})^{-1} = \hat{\mathbf{B}}^{-1}\hat{\mathbf{A}}^{-1} \tag{2.12}$$

[¶]The diagonal of a matrix runs from the upper left to the lower right.

That this must be so can be seen from the following:

$$\begin{aligned}\hat{\mathbf{A}}\hat{\mathbf{B}}(\hat{\mathbf{A}}\hat{\mathbf{B}})^{-1} &= \hat{\mathbf{A}}\hat{\mathbf{B}}\hat{\mathbf{B}}^{-1}\hat{\mathbf{A}}^{-1} \\ &= \hat{\mathbf{A}}\hat{\mathbf{1}}\hat{\mathbf{A}}^{-1} \\ &= \hat{\mathbf{A}}\hat{\mathbf{A}}^{-1} \\ &= \hat{\mathbf{1}}\end{aligned}$$

Not all matrices have an inverse. If the determinant of the matrix is equal to zero then there will be no inverse matrix.

The *transpose matrix* $\hat{\mathbf{A}}^T$ is one in which the rows and columns have been interchanged. Thus, if $\hat{\mathbf{A}}$ is a $n \times m$ matrix and $\hat{\mathbf{A}}^T$ is the transpose of $\hat{\mathbf{A}}$ then $\hat{\mathbf{A}}^T$ will be a $m \times n$ matrix.[‖]

$$\hat{\mathbf{A}} = \begin{bmatrix} 0 & 1 & 2 \\ 3 & 4 & 5 \\ 6 & 7 & 8 \end{bmatrix} \hat{\mathbf{A}}^T = \begin{bmatrix} 0 & 3 & 6 \\ 1 & 4 & 7 \\ 2 & 5 & 8 \end{bmatrix} \tag{2.13}$$

If we write a general matrix with subscripts on the matrix components then the transpose of $\hat{\mathbf{A}}$ will have its subscripts *exchanged*:

$$\hat{\mathbf{A}} = \begin{bmatrix} a_{11} & a_{12} & a_{13} \\ a_{21} & a_{22} & a_{23} \\ a_{31} & a_{32} & a_{33} \end{bmatrix} \hat{\mathbf{A}}^T = \begin{bmatrix} a_{11} & a_{21} & a_{31} \\ a_{12} & a_{22} & a_{32} \\ a_{13} & a_{23} & a_{33} \end{bmatrix}$$

Subscript ij becomes subscript ji. With this in mind it is obvious that the transpose of the transpose of a matrix equals the original matrix:

$$(\hat{\mathbf{A}}^T)^T = \hat{\mathbf{A}}$$

The transpose of the sum of two matrices is the same as the sum of their transposes:

$$\begin{aligned}&(\hat{\mathbf{A}} + \hat{\mathbf{B}})^T = \hat{\mathbf{A}}^T + \hat{\mathbf{B}}^T \\ &\text{proof:} \\ &\text{let } \hat{\mathbf{C}} = (\hat{\mathbf{A}} + \hat{\mathbf{B}})^T \\ &\text{then } c_{ij} = (a_{ij} + b_{ij})^T = a_{ji} + b_{ji}\end{aligned} \tag{2.14}$$

[‖]We will be dealing in later chapters almost exclusively with square matrices; that is, matrices of dimensions $n \times n$.

The transpose of the product of two matrices is equal to the *reverse* product of the transposes of the matrices:

$$(\hat{\mathbf{A}}\hat{\mathbf{B}})^T = \hat{\mathbf{B}}^T\hat{\mathbf{A}}^T \tag{2.15}$$

If the product of $\hat{\mathbf{A}}$ and $\hat{\mathbf{A}}^T$ is equal to the identity matrix they are *orthogonal*:

$$\hat{\mathbf{A}}\hat{\mathbf{A}}^T = \hat{\mathbf{A}}^T\hat{\mathbf{A}} = \hat{\mathbf{1}} \tag{2.16}$$

If a matrix is equal to its transpose it is *symmetric*:

$$\hat{\mathbf{A}} = \hat{\mathbf{A}}^T \text{ (symmetric)} \tag{2.17}$$

If $\hat{\mathbf{A}}$ and $\hat{\mathbf{B}}$ are both symmetric then their sum is another symmetric matrix (using eqn (2.14)):

$$\hat{\mathrm{C}} = \hat{\mathbf{A}} + \hat{\mathbf{B}} = \hat{\mathbf{A}}^T + \hat{\mathbf{B}}^T = (\hat{\mathbf{A}} + \hat{\mathbf{B}})^T = \hat{\mathbf{C}}^T \tag{2.18}$$

In order for the product of two symmetric matrices to be symmetric, the two matrices *must* commute:

$$\hat{\mathbf{C}}^T = (\hat{\mathbf{A}}\hat{\mathbf{B}})^T = \hat{\mathbf{B}}^T\hat{\mathbf{A}}^T = \hat{\mathbf{B}}\hat{\mathbf{A}} = \hat{\mathbf{A}}\hat{\mathbf{B}} = \hat{\mathbf{C}}$$

The *conjugate* matrix $\hat{\mathbf{A}}^*$ is one in which the elements are the complex conjugates of the original matrix elements.

$$\hat{\mathbf{A}} = \begin{bmatrix} 0 & 1 & -2i \\ 3 & 4 & 5 \\ 6 & 7i & 8 \end{bmatrix} \hat{\mathbf{A}}^* = \begin{bmatrix} 0 & 1 & 2i \\ 3 & 4 & 5 \\ 6 & -7i & 8 \end{bmatrix} \tag{2.19}$$

The matrix $(\hat{\mathbf{A}}^*)^T$ is the transpose of the complex conjugate of matrix $\hat{\mathbf{A}}$. If $(\hat{\mathbf{A}}^*)^T = \hat{\mathbf{A}}$ it is *Hermitian*. Obviously, a Hermitian matrix must be square. If $(\hat{\mathbf{A}}^*)^T = \hat{\mathbf{A}}^{-1}$ it is *unitary* (and also square). Thus, we can write:

$$\begin{aligned} &\text{If } (\hat{\mathbf{A}}^*)^T = \hat{\mathbf{A}} \text{ (Hermitian)} \\ &(\hat{\mathbf{A}}^*)^T \cdot \hat{\mathbf{A}}^{-1} = \hat{\mathbf{A}} \cdot \hat{\mathbf{A}}^{-1} = \hat{\mathbf{1}} \\ \\ &\text{If } (\hat{\mathbf{A}}^*)^T = \hat{\mathbf{A}}^{-1} \text{ (unitary)} \\ &(\hat{\mathbf{A}}^*)^T \cdot \hat{\mathbf{A}} = \hat{\mathbf{A}}^{-1}\hat{\mathbf{A}} = \hat{\mathbf{1}} \end{aligned} \tag{2.20}$$

Some examples of unitary matrices are:

$$\hat{\mathbf{A}} = \begin{bmatrix} 0 & -i \\ i & 0 \end{bmatrix} \hat{\mathbf{A}}^{-1} = \begin{bmatrix} 0 & -i \\ i & 0 \end{bmatrix} (\hat{\mathbf{A}}^*)^T = \begin{bmatrix} 0 & -i \\ i & 0 \end{bmatrix}$$

$$(\hat{\mathbf{A}}^*)^T\hat{\mathbf{A}} = \begin{bmatrix} 0 & -i \\ i & 0 \end{bmatrix} \times \begin{bmatrix} 0 & -i \\ i & 0 \end{bmatrix} = \begin{bmatrix} 1 & 0 \\ 0 & 1 \end{bmatrix} = \hat{\mathbf{1}}$$

and an example of Hermitian matrices is:

$$\hat{\mathbf{A}} = \begin{bmatrix} 0 & -i \\ i & 0 \end{bmatrix} \quad \hat{\mathbf{A}}^{-1} = \begin{bmatrix} 0 & -i \\ i & 0 \end{bmatrix} \quad (\hat{\mathbf{A}}^*)^T = \begin{bmatrix} 0 & -i \\ i & 0 \end{bmatrix}$$

$$(\hat{\mathbf{A}}^*)^T \hat{\mathbf{A}}^{-1} = \begin{bmatrix} 0 & -i \\ i & 0 \end{bmatrix} \times \begin{bmatrix} 0 & -i \\ i & 0 \end{bmatrix} = \begin{bmatrix} 1 & 0 \\ 0 & 1 \end{bmatrix} = \hat{\mathbf{1}}$$

In this case you can see that $\hat{\mathbf{A}}$ is *both* Hermitian and unitary.

A small amount of effort on the reader's part will reveal that the diagonal elements of the Hermitian matrix must, from the definition (eqn (2.20)), be real.** Hermitian matrices play a central role in quantum mechanics and will figure prominently in our discussion of the quantum mechanics of spins.

The *trace* of a matrix is the sum of the diagonal elements:

$$\mathrm{Tr}(\hat{\mathbf{A}}) = \sum_i a_{ii}$$

$$\text{If } \hat{\mathbf{A}} = \begin{bmatrix} 1 & 2 \\ 3 & 4 \end{bmatrix} \text{ then } \mathrm{Tr}(\hat{\mathbf{A}}) = 1 + 4 = 5 \tag{2.21}$$

and is equal to the sum of the eigenvalues of the matrix (see below for a discussion of eigenvalues). Note that the trace is a scalar; a simple number in other words. The trace of the product of two matrices is (or rather, will be) of interest. We start with 2×2 matrices, $\hat{\mathbf{A}}$ and $\hat{\mathbf{B}}$:

$$\hat{\mathbf{A}} = \begin{bmatrix} a_{11} & a_{12} \\ a_{21} & a_{22} \end{bmatrix} \quad \hat{\mathbf{B}} = \begin{bmatrix} b_{11} & b_{12} \\ b_{21} & b_{22} \end{bmatrix}$$

The product of these matrices, $\hat{\mathbf{C}} = \hat{\mathbf{A}}\hat{\mathbf{B}}$ is:

$$\hat{\mathbf{C}} = \begin{bmatrix} a_{11} & a_{12} \\ a_{21} & a_{22} \end{bmatrix} \begin{bmatrix} b_{11} & b_{12} \\ b_{21} & b_{22} \end{bmatrix} = \begin{bmatrix} a_{11}b_{11} + a_{12}b_{21} & a_{11}b_{12} + a_{12}b_{22} \\ a_{21}b_{11} + a_{22}b_{21} & a_{21}b_{12} + a_{22}b_{22} \end{bmatrix}$$

and the trace of $\hat{\mathbf{C}}$ is:

$$\mathrm{Tr}\,\hat{\mathbf{C}} = a_{11}b_{11} + a_{12}b_{21} + a_{21}b_{12} + a_{22}b_{22}$$

**Hint: to verify this, apply the definition of the Hermitian matrix to the diagonal elements and use eqn (1.5).

and in general, for $n \times n$ matrices the trace is easily calculated:

$$\mathrm{Tr}(\hat{\mathbf{A}}\hat{\mathbf{B}}) = \sum_i^n \sum_j^n a_{ij} b_{ji} \tag{2.22}$$

Note that the subscripts i and j are reversed from matrix components a to b. What about the trace of $\hat{\mathbf{B}}$ times $\hat{\mathbf{A}}$? We know that reversing the order of multiplication of matrices does not necessarily give the same result as $\hat{\mathbf{A}}$ times $\hat{\mathbf{B}}$ but how does this reversal affect the trace? Again we start with our 2×2 matrices and compute $\hat{\mathbf{D}} = \hat{\mathbf{B}}\hat{\mathbf{A}}$:

$$\hat{\mathbf{D}} = \begin{bmatrix} b_{11} & b_{12} \\ b_{21} & b_{22} \end{bmatrix} \times \begin{bmatrix} a_{11} & a_{12} \\ a_{21} & a_{22} \end{bmatrix}$$

$$= \begin{bmatrix} b_{11}a_{11} + b_{12}a_{21} & b_{11}a_{12} + b_{12}a_{22} \\ b_{21}a_{11} + b_{22}a_{21} & b_{21}a_{12} + b_{22}a_{22} \end{bmatrix}$$

The trace of $\hat{\mathbf{D}}$ is:

$$\mathrm{Tr}\ \hat{\mathbf{D}} = b_{11}a_{11} + b_{12}a_{21} + b_{21}a_{12} + b_{22}a_{22}$$

which is the same as the trace of $\hat{\mathbf{C}}$. In general the trace of the product of two matrices does *not* rely on the order of multiplication of the matrices:

$$\mathrm{Tr}(\hat{\mathbf{A}}\hat{\mathbf{B}}) = \sum_i^n \sum_j^n a_{ij} b_{ji} = \sum_i^n \sum_j^n b_{ij} a_{ji} = \mathrm{Tr}(\hat{\mathbf{B}}\hat{\mathbf{A}}) \tag{2.23}$$

Even more generally, the trace of multiple matrices is invariant under *cyclic permutation*:

$$\mathrm{Tr}(\hat{\mathbf{A}}\hat{\mathbf{B}}\hat{\mathbf{C}}\hat{\mathbf{D}}) = \mathrm{Tr}(\hat{\mathbf{B}}\hat{\mathbf{C}}\hat{\mathbf{D}}\hat{\mathbf{A}})$$

but *not* under arbitrary permutation:

$$\mathrm{Tr}(\hat{\mathbf{A}}\hat{\mathbf{B}}\hat{\mathbf{C}}\hat{\mathbf{D}}) \neq \mathrm{Tr}(\hat{\mathbf{B}}\hat{\mathbf{A}}\hat{\mathbf{C}}\hat{\mathbf{D}})$$

The sum of traces of **AB** and **AC** is the same as the trace of $\mathbf{A}(\mathbf{B}+\mathbf{C})$:

$$\mathrm{Tr}(\mathbf{AB}) = \sum_i^n \sum_j^n a_{ij} b_{ji} \quad \text{and} \quad \mathrm{Tr}(\mathbf{AC}) = \sum_i^n \sum_j^n a_{ij} c_{ji}$$

$$\mathrm{Tr}(\mathbf{AB}) + \mathrm{Tr}(\mathbf{AC}) = \sum_i^n \sum_j^n a_{ij}(b_{ji} + c_{ji}) = \mathrm{Tr}(\mathbf{A}(\mathbf{B}+\mathbf{C})) \tag{2.24}$$

If the trace of the product of two matrices is zero the matrices in question are said to be *trace orthogonal*:

$$\mathrm{Tr}(\hat{\mathbf{A}}\hat{\mathbf{B}}) = 0 \tag{2.25}$$

2.2 Determinants

A determinant is a form of matrix that can be evaluated to a scalar. The procedure is to multiply each member of a row or column by an element of the matrix that is *not* in the same row and column.

Note that there a couple of ways to denote determinants:

$$\det \hat{\mathbf{A}} \text{ or } |\hat{\mathbf{A}}|$$

We will use both of these interchangeably.

For our example matrix, $\hat{\mathbf{A}}$, the determinant is:

$$\begin{aligned}
\det \begin{bmatrix} 1 & 2 & 3 \\ 4 & 5 & 6 \\ 7 & 8 & 9 \end{bmatrix}
&= 1\cdot5\cdot9 - 1\cdot8\cdot6 - 2\cdot4\cdot9 + 2\cdot7\cdot6 + 3\cdot4\cdot8 - 3\cdot7\cdot5 \\
&= 45 - 48 - 72 + 84 + 96 - 105 \\
&= 0
\end{aligned}$$

A more useful evaluation technique for the general matrix:

$$\hat{\mathbf{E}} = \begin{bmatrix} e_{11} & e_{12} & e_{13} \\ e_{21} & e_{22} & e_{23} \\ e_{31} & e_{32} & e_{33} \end{bmatrix} \tag{2.26}$$

is to evaluate by taking a row (or a column) and multiplying each member of that row (column) by its *cofactor*. What is the cofactor? In our general matrix, $\hat{\mathbf{E}}$, let us choose the first (top) row. Then, for e_{11} the cofactor is:

$$E_{11} = \begin{vmatrix} e_{22} & e_{23} \\ e_{32} & e_{33} \end{vmatrix}$$

where E_{11} refers to the cofactor of element e_{11}. Generally, here we will refer to cofactors in subscripted uppercase. Note that the minor submatrix includes all terms from the original matrix that are *not in the same row or column* as e_{11}. It is particularly simple to evaluate a 3×3 matrix since at this point we evaluate the 2×2 *cofactor* of the minor matrix by doing:

$$E_{11} = e_{22} \cdot e_{33} - e_{32} \cdot e_{23}$$

which is itself a determinant calculation of a 2×2 matrix. We then multiply by e_{11} to complete the calculation:

$$e_{11}\left(e_{22} \cdot e_{33} - e_{32} \cdot e_{23}\right)$$

We do the same for the other terms in the first row with one very minor complication ... for the second term we multiply by −1 and by +1 for the third term:

$$-e_{12}(e_{21} \cdot e_{33} - e_{31} \cdot e_{23}) + e_{13}(e_{21} \cdot e_{32} - e_{31} \cdot e_{22})$$

Cofactors are signed either +1 or −1 simply by calculating the sum of their subscripts. An even result leads to +1 and an odd result leads to −1. Thus, the sign of the cofactor is determined by:

$$(-1)^{i+j}$$

Thus, cofactor E_{11} has sign +1 and cofactor E_{21} has sign −1. One can visualise required signs nicely for a 3×3 determinant:

$$\begin{vmatrix} + & - & + \\ - & + & - \\ + & - & + \end{vmatrix}$$

So the whole determinant evaluation looks like:

$$\begin{aligned}
\det\hat{\mathbf{E}} &= \begin{vmatrix} e_{11} & e_{12} & e_{13} \\ e_{21} & e_{22} & e_{23} \\ e_{31} & e_{32} & e_{33} \end{vmatrix} \\
&= e_{11}\begin{vmatrix} e_{22} & e_{23} \\ e_{32} & e_{33} \end{vmatrix} - e_{12}\begin{vmatrix} e_{21} & e_{23} \\ e_{31} & e_{33} \end{vmatrix} + e_{13}\begin{vmatrix} e_{21} & e_{22} \\ e_{31} & e_{32} \end{vmatrix} \qquad (2.27) \\
&= e_{11}\left(e_{22}e_{33} - e_{32}e_{23}\right) - e_{12}\left(e_{21}e_{33} - e_{31}e_{23}\right) + e_{13}\left(e_{21}e_{32} - e_{31}e_{22}\right) \\
&= e_{11}e_{22}e_{33} - e_{11}e_{32}e_{23} - e_{12}e_{21}e_{33} + e_{12}e_{31}e_{23} + e_{13}e_{21}e_{32} - e_{13}e_{31}e_{22}
\end{aligned}$$

This is a very simplified explanation and there is much more to it than this but this is all we will need in this text. For more detailed information regarding determinants please see any competent text on matrices. Most of our determinant evaluations will be for 2×2 and 3×3 matrices.

The *cofactor matrix* will be of use later in this chapter. Using our general matrix, $\hat{\mathbf{E}}$, we can define the ijth component of the cofactor matrix as $(-1)^{ij}|E_{ij}|$ and our cofactor matrix would be:

$$\hat{\mathbf{C}} = \begin{bmatrix} |E_{11}| & -|E_{12}| & |E_{13}| \\ -|E_{21}| & |E_{22}| & -|E_{23}| \\ |E_{31}| & -|E_{32}| & |E_{33}| \end{bmatrix} \tag{2.28}$$

A concrete example will illustrate this idea. Suppose we have the matrix:

$$\hat{\mathbf{A}} = \begin{bmatrix} 1 & 2 & 3 \\ 7 & 8 & 9 \\ 4 & 5 & 6 \end{bmatrix}$$

then its cofactor matrix is:

$$\hat{\mathbf{C}} = \begin{bmatrix} \begin{vmatrix} 8 & 9 \\ 5 & 6 \end{vmatrix} & -\begin{vmatrix} 7 & 9 \\ 4 & 6 \end{vmatrix} & \begin{vmatrix} 7 & 8 \\ 4 & 5 \end{vmatrix} \\ -\begin{vmatrix} 2 & 3 \\ 5 & 6 \end{vmatrix} & \begin{vmatrix} 1 & 3 \\ 4 & 6 \end{vmatrix} & -\begin{vmatrix} 1 & 2 \\ 4 & 5 \end{vmatrix} \\ \begin{vmatrix} 2 & 3 \\ 8 & 9 \end{vmatrix} & -\begin{vmatrix} 1 & 3 \\ 7 & 9 \end{vmatrix} & \begin{vmatrix} 1 & 2 \\ 7 & 8 \end{vmatrix} \end{bmatrix}$$
$$= \begin{bmatrix} 3 & -6 & 3 \\ 3 & -6 & 3 \\ -6 & 12 & -6 \end{bmatrix}$$

Determinants have several interesting properties that may be taken advantage of when calculating their value. First, interchanging rows or columns in the determinant changes the sign but not the value of the scalar result. An even number of interchanges produces no sign change and an odd number of interchanges produces a sign change. We can show this with our example determinant; we will exchange rows 1 and 2 to produce a new determinant, $\det\hat{\mathbf{E}}'$:

$$\det\hat{\mathbf{E}}' = \begin{vmatrix} e_{21} & e_{22} & e_{23} \\ e_{11} & e_{12} & e_{13} \\ e_{31} & e_{32} & e_{33} \end{vmatrix} = e_{21}E'_{21} + e_{22}E'_{22} + e_{23}E'_{23}$$

$$= e_{21}(e_{12}e_{33} - e_{32}e_{13}) - e_{22}(e_{11}e_{33} - e_{31}e_{13}) + e_{23}(e_{11}e_{32} - e_{31}e_{12})$$

$$= e_{21}e_{12}e_{33} - e_{21}e_{32}e_{13} - e_{22}e_{11}e_{33} + e_{22}e_{31}e_{13} + e_{23}e_{11}e_{32} - e_{23}e_{31}e_{12} \tag{2.29}$$

This is equal to but of opposite sign to the result in eqn (2.27); that is, each term in eqn (2.29) is the same as in eqn (2.27) but is of opposite sign.

Multiplication of each of the elements in a row or column in the determinant by a scalar results in the value of the determinant being multiplied by the scalar.

$$\det\hat{\mathbf{E}}'' = \begin{vmatrix} \lambda e_{11} & \lambda e_{12} & \lambda e_{13} \\ e_{21} & e_{22} & e_{23} \\ e_{31} & e_{32} & e_{33} \end{vmatrix}$$

$$\begin{aligned} &= \lambda e_{11}E_{11} + \lambda e_{12}E_{12} + \lambda e_{13}E_{13} \\ &= \lambda e_{11}(e_{22}e_{33} - e_{32}e_{23}) \\ &\quad - \lambda e_{12}(e_{21}e_{33} - e_{31}e_{23}) \\ &\quad + \lambda e_{13}(e_{21}e_{32} - e_{31}e_{22}) \\ &= \lambda\det\hat{\mathbf{E}} \end{aligned} \tag{2.30}$$

If k times the value of the elements of a row (or column) are added to another row (or column) the value of the determinant is unchanged.

$$\text{if} \begin{vmatrix} e_{11} & e_{12} & e_{13} \\ e_{21} & e_{22} & e_{23} \\ e_{31} & e_{32} & e_{33} \end{vmatrix} = e_{11}E_{11} + e_{12}E_{12} + e_{13}E_{13}$$

$$= e_{11}(e_{22}e_{33} - e_{32}e_{23}) - e_{12}(e_{21}e_{33} - e_{31}e_{23}) + e_{13}(e_{21}e_{32} - e_{31}e_{22})$$

$$= e_{11}e_{22}e_{33} - e_{11}e_{32}e_{23} - e_{12}e_{21}e_{33} + e_{12}e_{31}e_{23} + e_{13}e_{21}e_{32} - e_{13}e_{31}e_{22}$$

$$\begin{vmatrix} e_{11} & e_{12}+ke_{11} & e_{13} \\ e_{21} & e_{22}+ke_{21} & e_{23} \\ e_{31} & e_{32}+ke_{31} & e_{33} \end{vmatrix}$$

$$= e_{11}E_{11} + (e_{12}+ke_{11})E_{12} + e_{13}E_{13}$$

$$= e_{11}([e_{22}+ke_{21}]e_{33} - [e_{32}+ke_{31}\}]e_{23})$$

$$- [e_{12}+ke_{11}](e_{21}e_{33} - e_{31}e_{23})$$

$$+ e_{13}(e_{21}[e_{32}+ke_{31}] - e_{31}[e_{22}+ke_{21}])$$

$$= e_{11}e_{22}e_{33} + ke_{11}e_{21}e_{33} - e_{11}e_{32}e_{23} - ke_{11}e_{31}e_{23}$$

$$- e_{12}e_{21}e_{33} - ke_{11}e_{21}e_{33} + e_{12}e_{31}e_{22} + ke_{11}e_{31}e_{23}$$

$$e_{13}e_{21}e_{32} + ke_{13}e_{21}e_{31} - e_{13}e_{31}e_{22} - ke_{13}e_{31}e_{21}$$

$$= e_{11}e_{22}e_{33} - e_{11}e_{32}e_{23} - e_{12}e_{21}e_{33} + e_{12}e_{31}e_{23} + e_{13}e_{21}e_{32} - e_{13}e_{31}e_{22}$$

$$= \begin{vmatrix} e_{11} & e_{12} & e_{13} \\ e_{21} & e_{22} & e_{23} \\ e_{31} & e_{32} & e_{33} \end{vmatrix}$$

$$\therefore \ \det \hat{\mathbf{E}} = \begin{vmatrix} e_{11} & e_{12} & e_{13} \\ e_{21} & e_{22} & e_{23} \\ e_{31} & e_{32} & e_{33} \end{vmatrix} = \begin{vmatrix} e_{11} & e_{12}+ke_{11} & e_{13} \\ e_{21} & e_{22}+ke_{21} & e_{23} \\ e_{31} & e_{32}+ke_{31} & e_{33} \end{vmatrix} \tag{2.31}$$

If every element of a row or column is equal to zero then the value of the determinant is zero.

$$\det \hat{\mathbf{E}} = \begin{vmatrix} 0 & 0 & 0 \\ e_{21} & e_{22} & e_{23} \\ e_{31} & e_{32} & e_{33} \end{vmatrix} = 0E_{11} + 0E_{12} + 0E_{13} \tag{2.32}$$

$$= 0(e_{22}e_{33} - e_{32}e_{23}) - 0(e_{21}e_{33} - e_{31}e_{23}) + 0(e_{21}e_{32} - e_{31}e_{22})$$

$$= 0$$

The value of the determinant of the transpose of matrix $\hat{\mathbf{E}}$ is equal to the value of the determinant of matrix $\hat{\mathbf{E}}$.

$$\det \hat{\mathbf{E}}^T = \begin{vmatrix} e_{11} & e_{21} & e_{31} \\ e_{12} & e_{22} & e_{32} \\ e_{13} & e_{23} & e_{33} \end{vmatrix}$$

$$= e_{11}(e_{22}e_{33} - e_{32}e_{23}) - e_{12}(e_{21}e_{33} - e_{31}e_{23}) + e_{13}(e_{21}e_{32} - e_{31}e_{22})$$

$$= e_{11}e_{22}e_{33} - e_{11}e_{32}e_{23} - e_{12}e_{21}e_{33} + e_{12}e_{31}e_{23} + e_{13}e_{21}e_{32} - e_{13}e_{31}e_{22}$$

$$= \det \hat{\mathbf{E}} \tag{2.33}$$

Note that in eqn (2.33) we have used the cofactors of column 1 rather than row 1, as we have been doing up until now. If the reader looks at any competent introductory text on matrices she/he will find that one may use any row or column in the matrix and produce cofactors from it.

If two rows or columns are identical then the value of the determinant is zero.

$$\det \hat{\mathbf{E}}''' = \begin{vmatrix} e_{11} & e_{12} & e_{13} \\ e_{11} & e_{12} & e_{13} \\ e_{31} & e_{32} & e_{33} \end{vmatrix} = e_{11}E'''_{11} + e_{12}E'''_{12} + e_{13}E'''_{13}$$

$$= e_{11}(e_{12}e_{33} - e_{32}e_{13}) - e_{12}(e_{11}e_{33} - e_{31}e_{13}) + e_{13}(e_{11}e_{32} - e_{31}e_{12})$$

$$= e_{11}e_{12}e_{33} - e_{11}e_{32}e_{13} - e_{12}e_{11}e_{33} + e_{12}e_{31}e_{13} + e_{13}e_{11}e_{32} - e_{13}e_{31}e_{12}$$

$$= 0 \tag{2.34}$$

If a row or column in the determinant can be obtained by a linear combination of the other rows or columns in the matrix then the value of the determinant is zero.

$$\det \hat{\mathbf{E}} = \begin{vmatrix} \alpha e_{21} + \beta e_{31} & \alpha e_{22} + \beta e_{32} & \alpha e_{23} + \beta e_{33} \\ e_{21} & e_{22} & e_{23} \\ e_{31} & e_{32} & e_{33} \end{vmatrix}$$

$$\begin{aligned}
&= (\alpha e_{21} + \beta e_{31})E_{11} + (\alpha e_{22} + \beta e_{32})E_{22} + (\alpha e_{23} + \beta e_{33})E_{33} \\
&= (\alpha e_{21} + \beta e_{31})(e_{22}e_{33} - e_{32}e_{23}) \\
&\quad - (\alpha e_{22} + \beta e_{32})(e_{21}e_{33} - e_{31}e_{23}) \\
&\quad + (\alpha e_{23} + \beta e_{33})(e_{21}e_{32} - e_{31}e_{22}) \\
&= \alpha e_{21}e_{22}e_{33} + \beta e_{31}e_{22}e_{33} - \alpha e_{21}e_{32}e_{23} - \beta e_{31}e_{32}e_{23} \\
&\quad - \alpha e_{22}e_{21}e_{33} - \beta e_{32}e_{21}e_{33} + \alpha e_{22}e_{31}e_{23} + \beta e_{32}e_{31}e_{23} \\
&\quad + \alpha e_{23}e_{21}e_{32} + \beta e_{33}e_{21}e_{32} - \alpha e_{23}e_{31}e_{22} - \beta e_{33}e_{31}e_{22} \\
&= 0
\end{aligned} \tag{2.35}$$

Close inspection of the resulting terms in eqn (2.35) reveals that each one has its opposite and thus the entire expression collapses to zero. This result will be of use to us in our explorations of vectors. The determinant of a product of matrices is:[††]

$$\det(\hat{\mathbf{A}}\hat{\mathbf{B}}) = \det \hat{\mathbf{A}} \det \hat{\mathbf{B}} \tag{2.36}$$

We can show this by using a pair of 2×2 matrices:

$$\hat{\mathbf{A}} = \begin{bmatrix} a_{11} & a_{12} \\ a_{21} & a_{22} \end{bmatrix} \quad \hat{\mathbf{B}} = \begin{bmatrix} b_{11} & b_{12} \\ b_{21} & b_{22} \end{bmatrix}$$

$$\det\hat{\mathbf{A}} = a_{11}a_{22} - a_{21}a_{12}$$

$$\det\hat{\mathbf{B}} = b_{11}b_{22} - b_{21}b_{12}$$

$$\begin{aligned}
\det\hat{\mathbf{A}}\det\hat{\mathbf{B}} &= (a_{11}a_{22} - a_{21}a_{12})(b_{11}b_{22} - b_{21}b_{12}) \\
&= a_{11}a_{22}b_{11}b_{22} - a_{11}a_{22}b_{21}b_{12} - a_{21}a_{12}b_{11}b_{22} + a_{21}a_{12}b_{21}b_{12}
\end{aligned}$$

[††]You could also do this with higher-order matrices but it would be rather tedious. You could still do the calculations reasonably quickly, however, using a symbolic algebra program of your choice. There are several excellent proprietary programs available, however, my choice for this task was wxMaxima, a very capable and *freely* downloadable symbolic algebra program.

Now, using this result and multiplying $\hat{\mathbf{A}}$ and $\hat{\mathbf{B}}$:

$$\hat{\mathbf{A}}\hat{\mathbf{B}} = \begin{bmatrix} a_{11} & a_{12} \\ a_{21} & a_{22} \end{bmatrix} \times \begin{bmatrix} b_{11} & b_{12} \\ b_{21} & b_{22} \end{bmatrix}$$

$$= \begin{bmatrix} a_{11}b_{11} + a_{12}b_{21} & a_{11}b_{12} + a_{12}b_{22} \\ a_{21}b_{11} + a_{22}b_{21} & a_{21}b_{12} + a_{22}b_{22} \end{bmatrix}$$

$$\begin{aligned}
\det(\hat{\mathbf{A}}\hat{\mathbf{B}}) &= (a_{11}b_{11} + a_{12}b_{21})(a_{21}b_{12} + a_{22}b_{22}) - (a_{21}b_{11} + a_{22}b_{21})(a_{11}b_{12} + a_{12}b_{22}) \\
&= a_{11}b_{11}a_{21}b_{12} + a_{11}b_{11}a_{22}b_{22} + a_{12}b_{21}a_{21}b_{12} + a_{12}b_{21}a_{22}b_{22} \\
&\quad - a_{21}b_{11}a_{11}b_{12} - a_{21}b_{11}a_{12}b_{22} - a_{22}b_{21}a_{11}b_{12} - a_{22}b_{21}a_{12}b_{22} \\
&= a_{11}b_{11}a_{22}b_{22} + a_{12}b_{21}a_{21}b_{12} - a_{21}b_{11}a_{12}b_{22} - a_{22}b_{21}a_{11}b_{12} \\
&= \det\hat{\mathbf{A}}\det\hat{\mathbf{B}}
\end{aligned} \tag{2.37}$$

The determinant of the identity matrix is (not unexpectedly) equal to one:

$$\begin{vmatrix} 1 & 0 & 0 & \cdots \\ 0 & 1 & 0 & \cdots \\ 0 & 0 & 1 & \cdots \\ \vdots & \vdots & \vdots & \ddots \end{vmatrix} = 1 \times \begin{vmatrix} 1 & 0 & \cdots \\ 0 & 1 & \cdots \\ \vdots & \vdots & \ddots \end{vmatrix} + (n-1) \times 0 \times \begin{vmatrix} 1 & 0 & \cdots \\ 0 & 1 & \cdots \\ \vdots & \vdots & \ddots \end{vmatrix}$$

$$= 1 \times \begin{vmatrix} 1 & 0 & \cdots \\ 0 & 1 & \cdots \\ \vdots & \vdots & \ddots \end{vmatrix}$$

This may need some clarification. Since this is the $(n \times n)$ identity matrix the first row of its determinant will consist of a 1 in the first position and $n-1$ zeros in all the other positions. The 1 will multiply into its cofactors, as in our above examples. The zeros will multiply into their cofactors but will, of course, produce a zero result leaving only 1 times its cofactors. The same procedure is applied to the cofactors until we arrive at a 2×2 cofactor:

$$1\times1\times1\ldots\begin{vmatrix}1 & 0\\ 0 & 1\end{vmatrix}=1\times1\times1\ldots\times(1\times1-0\times0)$$
$$=\prod_{1}^{n}(1\times1)=1$$

The determinant of an inverse matrix is simply related to the original matrix:

$$\begin{aligned}&\hat{\mathbf{A}}^{-1}\hat{\mathbf{A}}=\hat{\mathbf{1}}\\&\det(\hat{\mathbf{A}}^{-1}\hat{\mathbf{A}})=\det\hat{\mathbf{1}}=1\\&\det\hat{\mathbf{A}}^{-1}\det\hat{\mathbf{A}}=1\\&\det\hat{\mathbf{A}}^{-1}=\frac{1}{\det\hat{\mathbf{A}}}\end{aligned}\tag{2.38}$$

We are in a position to define the *adjoint matrix* now. For our general matrix, $\hat{\mathbf{E}}$, the adjoint is the *transpose* of the cofactor matrix of eqn (2.28):

$$\text{adj}\hat{\mathbf{E}}=\mathbf{C}^T=\begin{bmatrix}|E_{11}| & -|E_{21}| & |E_{31}|\\ -|E_{12}| & |E_{22}| & -|E_{23}|\\ |E_{13}| & -|E_{23}| & |E_{33}|\end{bmatrix}\tag{2.39}$$

where, as above, E_{ij} is the cofactor of element e_{ij} in $\hat{\mathbf{E}}$. This will shortly be of use in solving matrix equations.

2.3 Solving Matrix Equations

Let us suppose that we have three equations:

$$\begin{aligned}6x-4y+z&=3\\x-2y+5z&=1\\3x+2y+z&=0\end{aligned}$$

We can represent these equations using three matrices:‡‡

$$\hat{\mathbf{A}}=\begin{bmatrix}6 & -4 & 1\\ 1 & -2 & 5\\ 3 & 2 & 1\end{bmatrix},\quad \hat{\mathbf{x}}=\begin{bmatrix}x\\ y\\ z\end{bmatrix},\quad \hat{\mathbf{d}}=\begin{bmatrix}3\\ 1\\ 0\end{bmatrix}$$

Then we may write:

$$\hat{\mathbf{A}}\hat{\mathbf{x}}=\hat{\mathbf{d}}$$

$$\begin{bmatrix}6 & -4 & 1\\ 1 & -2 & 5\\ 3 & 2 & 1\end{bmatrix}\begin{bmatrix}x\\ y\\ z\end{bmatrix}=\begin{bmatrix}3\\ 1\\ 0\end{bmatrix}$$

which regenerates our original equations. Now, what we usually want to do is solve for x, y and z. What if we were to multiply both sides of the matrix equation by the inverse of $\hat{\mathbf{A}}$?

$$\hat{\mathbf{A}}^{-1}\hat{\mathbf{A}}\hat{\mathbf{x}}=\hat{\mathbf{A}}^{-1}\hat{\mathbf{d}}$$

and since $\hat{\mathbf{A}}^{-1}\hat{\mathbf{A}}=\hat{\mathbf{1}}$

$$\hat{\mathbf{x}}=\hat{\mathbf{A}}^{-1}\hat{\mathbf{d}}$$

Matrix $\hat{\mathbf{x}}$ now holds the solutions that we are looking for. The problem now is "how do we find the inverse matrix?". We will proceed to find out how using our general matrix, $\hat{\mathbf{E}}$:

$$\hat{\mathbf{E}}\hat{\mathbf{x}}=\hat{\mathbf{d}}$$

$$\begin{bmatrix}e_{11} & e_{12} & e_{13}\\ e_{21} & e_{22} & e_{23}\\ e_{31} & e_{32} & e_{33}\end{bmatrix}\begin{bmatrix}x\\ y\\ z\end{bmatrix}=\begin{bmatrix}d_1\\ d_2\\ d_3\end{bmatrix}$$

‡‡We will see in the next chapter that vectors are conveniently represented by column and row matrices and we will at that point change our notation from $\hat{x}$ to $\vec{x}$ for these special types of matrices.

In equation form this looks like:

$$e_{11}x + e_{12}y + e_{13}z = d_1$$
$$e_{21}x + e_{22}y + e_{23}z = d_2$$
$$e_{31}x + e_{32}y + e_{33}z = d_3$$

The inverse of matrix $\hat{\mathbf{E}}$ is:

$$\hat{\mathbf{E}}^{-1} = \frac{\text{adj}\hat{\mathbf{E}}}{\det\hat{\mathbf{E}}} \tag{2.40}$$

Note that since this depends on the value of det$\hat{\mathbf{E}}$; if det$\hat{\mathbf{E}}$ equals zero the inverse of $\hat{\mathbf{E}}$ is undefined. We can show that eqn (2.40) is valid (assuming det$\hat{\mathbf{E}}$ does not equal zero):

$$\hat{\mathbf{E}} \times \frac{\text{adj}\hat{\mathbf{E}}}{\det\hat{\mathbf{E}}}$$

$$= \frac{\begin{bmatrix} e_{11} & e_{12} & e_{13} \\ e_{21} & e_{22} & e_{23} \\ e_{31} & e_{32} & e_{33} \end{bmatrix} \begin{bmatrix} |E_{11}| & -|E_{21}| & |E_{31}| \\ -|E_{12}| & |E_{22}| & -|E_{23}| \\ |E_{13}| & -|E_{23}| & |E_{33}| \end{bmatrix}}{\det\hat{\mathbf{E}}}$$

$$= \frac{1}{\det\hat{\mathbf{E}}} \begin{bmatrix} e_{11}E_{11} - e_{12}E_{12} + e_{13}E_{13} & -e_{11}E_{21} + e_{12}E_{22} - e_{13}E_{23} & e_{11}E_{31} - e_{12}E_{32} + e_{13}E_{33} \\ e_{21}E_{11} - e_{22}E_{12} + e_{23}E_{13} & -e_{21}E_{21} + e_{22}E_{22} - e_{23}E_{23} & e_{21}E_{31} - e_{22}E_{32} + e_{23}E_{33} \\ e_{31}E_{11} - e_{32}E_{12} + e_{33}E_{13} & e_{31}E_{21} - e_{32}E_{22} + e_{33}E_{23} & e_{31}E_{31} - e_{32}E_{32} + e_{33}E_{33} \end{bmatrix}$$

$$= \frac{1}{\det\hat{\mathbf{E}}} \begin{bmatrix} \det\hat{\mathbf{E}} & 0 & 0 \\ 0 & \det\hat{\mathbf{E}} & 0 \\ 0 & 0 & \det\hat{\mathbf{E}} \end{bmatrix}$$

$$= \begin{bmatrix} 1 & 0 & 0 \\ 0 & 1 & 0 \\ 0 & 0 & 1 \end{bmatrix}$$

$$= \hat{\mathbf{1}}$$

The diagonal terms in the second line are the definitions of det$\hat{\mathbf{E}}$ and of course, in line three, det$\hat{\mathbf{E}}$ divided by det$\hat{\mathbf{E}}$ equals 1 but why are the off-diagonal terms zero? Let us look at one of them:

$$
\begin{aligned}
&e_{21}E_{11} - e_{22}E_{12} + e_{23}E_{13} \\
&= e_{21}\begin{vmatrix} e_{22} & e_{23} \\ e_{32} & e_{33} \end{vmatrix} - e_{22}\begin{vmatrix} e_{21} & e_{23} \\ e_{31} & e_{33} \end{vmatrix} + e_{23}\begin{vmatrix} e_{21} & e_{22} \\ e_{31} & e_{32} \end{vmatrix} \\
&= e_{21}(e_{22}e_{33} - e_{32}e_{23}) - e_{22}(e_{21}e_{33} - e_{31}e_{23}) + e_{23}(e_{21}e_{32} - e_{31}e_{22}) \\
&= e_{21}e_{22}e_{33} - e_{21}e_{32}e_{23} - e_{22}e_{31}e_{33} + e_{22}e_{31}e_{23} + e_{23}e_{21}e_{32} - e_{23}e_{31}e_{22} \\
&= 0
\end{aligned}
$$

The same result will be obtained with any of the other off-diagonal terms.

We now can write:

$$
\begin{aligned}
\hat{\mathbf{x}} &= \hat{\mathbf{E}}^{-1}\hat{\mathbf{d}} \\
&= \frac{\mathrm{adj}\hat{\mathbf{E}}}{\det \hat{\mathbf{E}}}\hat{\mathbf{d}}
\end{aligned}
$$

Thus, if we can determine the value of det$\hat{\mathbf{E}}$ and what the adjoint matrix of $\hat{\mathbf{E}}$ is, we can solve for the variables in the equations. So, using our example matrix, $\hat{\mathbf{A}}$ from the beginning of this section:

$$
\hat{\mathbf{A}} = \begin{bmatrix} 6 & -4 & 1 \\ 1 & -2 & 5 \\ 3 & 2 & 1 \end{bmatrix}
$$

$$
\begin{aligned}
\det \hat{\mathbf{A}} &= 6((-2)\cdot 1 - 2\cdot 5) - (-4)(1\cdot 1 - 3\cdot 5) + 1(1\cdot 2 - 3\cdot(-2)) \\
&= -72 - 56 + 8 \\
&= -120
\end{aligned}
$$

$$
\begin{aligned}
\mathrm{adj}\hat{\mathbf{A}} &= \begin{bmatrix} \det A_{11} & -\det A_{21} & \det A_{31} \\ -\det A_{12} & \det A_{22} & -\det A_{32} \\ \det A_{13} & -\det A_{23} & \det A_{33} \end{bmatrix} \\
&= \begin{bmatrix} -12 & 6 & -18 \\ 14 & 2 & -29 \\ 8 & -24 & -8 \end{bmatrix}
\end{aligned}
$$

and we may now set up the equations:

$$\hat{\mathbf{x}} = \hat{\mathbf{A}}^{-1}\hat{\mathbf{d}}$$

$$= \frac{\operatorname{adj}\hat{\mathbf{A}}}{\det\hat{\mathbf{A}}}\hat{\mathbf{d}}$$

$$= \frac{\begin{bmatrix} -12 & 6 & -18 \\ 14 & 3 & -29 \\ 8 & -24 & -8 \end{bmatrix}}{-120}\begin{bmatrix} 3 \\ 1 \\ 0 \end{bmatrix} = \begin{bmatrix} \frac{-12}{-120} & \frac{6}{-120} & \frac{-18}{-120} \\ \frac{14}{-120} & \frac{3}{-120} & \frac{-29}{-120} \\ \frac{8}{-120} & \frac{-24}{-120} & \frac{-8}{-120} \end{bmatrix}\begin{bmatrix} 3 \\ 1 \\ 0 \end{bmatrix} = \begin{bmatrix} \frac{-30}{-120} \\ \frac{45}{-120} \\ 0 \end{bmatrix}$$

and:

$$x = \frac{-30}{-120} = 0.25$$

$$y = \frac{45}{-120} = -0.375$$

$$z = 0$$

We can check our arithmetic by producing the inverse matrix:

$$\hat{\mathbf{A}} \times \hat{\mathbf{A}}^{-1} = \hat{\mathbf{1}}$$

$$\hat{\mathbf{A}} \times \operatorname{adj}\hat{\mathbf{A}} \times \frac{1}{\det\hat{\mathbf{A}}}$$

$$= \begin{bmatrix} 6 & -4 & 1 \\ 1 & -2 & 5 \\ 3 & 2 & 1 \end{bmatrix} \times \begin{bmatrix} -12 & 6 & -18 \\ 14 & 2 & -29 \\ 8 & -24 & -8 \end{bmatrix}\frac{1}{-120}$$

$$= \begin{bmatrix} 1 & 0 & 0 \\ 0 & 1 & 0 \\ 0 & 0 & 1 \end{bmatrix} = \hat{\mathbf{1}}$$

Also widely used to solve matrix equations is *Cramer's Rule*. This involves a couple of the properties developed earlier for determinants in Section 2.2. Our matrix equation is:

$$\hat{\mathbf{E}}\hat{\mathbf{x}} = \hat{\mathbf{d}}$$

$$\begin{bmatrix} e_{11} & e_{12} & e_{13} \\ e_{21} & e_{22} & e_{23} \\ e_{31} & e_{32} & e_{33} \end{bmatrix} \begin{bmatrix} x \\ y \\ z \end{bmatrix} = \begin{bmatrix} d_1 \\ d_2 \\ d_3 \end{bmatrix}$$

According to eqn (2.30) we may multiply det$\hat{\mathbf{E}}$ by a factor ... let us choose the variable x:

$$x \cdot \det \hat{\mathbf{E}} = x \cdot \begin{vmatrix} e_{11} & e_{12} & e_{13} \\ e_{21} & e_{22} & e_{23} \\ e_{31} & e_{32} & e_{33} \end{vmatrix} = \begin{vmatrix} xe_{11} & e_{12} & e_{13} \\ xe_{21} & e_{22} & e_{23} \\ xe_{31} & e_{32} & e_{33} \end{vmatrix}$$

Also, according to eqn (2.31), we may add the elements of one row times a factor to another row without changing the value of the determinant.

$$x \cdot \det \hat{\mathbf{E}} = \begin{bmatrix} xe_{11} + ye_{12} & e_{12} & e_{13} \\ xe_{21} + ye_{22} & e_{22} & e_{23} \\ xe_{31} + ye_{32} & e_{32} & e_{33} \end{bmatrix} = \begin{bmatrix} d_1 & e_{12} & e_{13} \\ d_2 & e_{22} & e_{23} \\ d_3 & e_{32} & e_{33} \end{bmatrix}$$

and:

$$x = \frac{\begin{bmatrix} d_1 & e_{12} & e_{13} \\ d_2 & e_{22} & e_{23} \\ d_3 & e_{32} & e_{33} \end{bmatrix}}{\det \hat{\mathbf{E}}} \tag{2.41}$$

This is Cramer's Rule. The corresponding solutions for y and z are:

$$y = \frac{\begin{bmatrix} e_{11} & d_1 & e_{13} \\ e_{21} & d_2 & e_{23} \\ e_{31} & d_3 & e_{33} \end{bmatrix}}{\det \hat{\mathbf{E}}} \quad z = \frac{\begin{bmatrix} e_{11} & e_{12} & d_1 \\ e_{21} & e_{22} & d_2 \\ e_{31} & e_{32} & d_3 \end{bmatrix}}{\det \hat{\mathbf{E}}}$$

There is yet another way to solve for the xs by a technique known as *Gaussian elimination*. We use a simple set of elementary row operations that follow directly from the algebra of equations:

1. Any equation can be multiplied by a non-zero constant. This follows from the fact that doing the same operation on each side of an equation leaves each side equal to each other.
2. Any two equations can be interchanged. It does not matter what order the equations are presented to you ... any order of presentation is surely arbitrary.
3. Any equation can be replaced by the sum of itself and a multiple of another equation. This is really just rule 1 in disguise.

Let us use our example equations from the beginning of this section:

$$1\colon 6x - 4y + z = 3$$

$$2\colon x - 2y + 5z = 1$$

$$3\colon 3x + 2y + z = 0$$

From rule 1 we can, say, multiply eqn 3 by 6:

$$1\colon 6x - 4y + z = 3$$

$$2\colon x - 2y + 5z = 1$$

$$3\colon 18x + 12y + 6z = 0$$

and from rule 3 we can add it to eqn 1:

$$1 + 3\colon 24x - 8y + 7z = 3$$

$$2\colon x - 2y + 5z = 1$$

$$3\colon 18x + 12y + 6z = 0$$

and so on.

How does this help us? We can use these rules to solve the equations by Gaussian elimination. The technique is to manipulate the equations using the elementary operations so that in one of the equations the coefficients of x and y are zero. The value of z can then be solved for and back-substituted to solve for x and y. Also, there is no need to write the entire equation but only the coefficients and the right-hand side of the equations in an *augmented matrix*:

$$\begin{bmatrix} 6 & -4 & 1 & 3 \\ 1 & -2 & 5 & 1 \\ 3 & 2 & 1 & 0 \end{bmatrix}$$

We can now apply our elementary row operations to get the matrix into *echelon* form in which there are all zeros below the diagonal elements of the variable coefficients starting in the upper left. Let us proceed:

$$\begin{bmatrix} 6 & -4 & 1 & 3 \\ 1 & -2 & 5 & 1 \\ 3 & 2 & 1 & 0 \end{bmatrix} \xrightarrow{2\times row3} \begin{bmatrix} 6 & -4 & 1 & 3 \\ 1 & -2 & 5 & 1 \\ 6 & 4 & 2 & 0 \end{bmatrix} \xrightarrow{row3-row1} \begin{bmatrix} 6 & -4 & 1 & 3 \\ 1 & -2 & 5 & 1 \\ 0 & 8 & 1 & -3 \end{bmatrix}$$

Here we have multiplied row 3 by 2 and then subtracted row 1 from row 3. Next:

$$\begin{bmatrix} 6 & -4 & 1 & 3 \\ 1 & -2 & 5 & 1 \\ 0 & 8 & 1 & -3 \end{bmatrix} \xrightarrow{6\times row2} \begin{bmatrix} 6 & -4 & 1 & 3 \\ 6 & -12 & 30 & 6 \\ 0 & 8 & 1 & -3 \end{bmatrix} \xrightarrow{row2-row1} \begin{bmatrix} 6 & -4 & 1 & 3 \\ 0 & -8 & 29 & 3 \\ 0 & 8 & 1 & -3 \end{bmatrix}$$

$$\xrightarrow{row2+row3} \begin{bmatrix} 6 & -4 & 1 & 3 \\ 0 & -8 & 29 & 3 \\ 0 & 0 & 30 & 0 \end{bmatrix}$$

The final matrix is now in echelon form and written out as a set of equations is:

$$1:\ 6x - 4y + z = 3$$

$$2:\ 0x - 8y + 29z = 1$$

$$3:\ 0x + 0y + 30z = 0$$

We solve for z in the last equation to get $z = 0$. Substituting this into eqn 2 we solve for y to get $y = -8/3$. Then substituting these values in eqn 1 we get $x = -21/18$.

It is quite possible that there will be either no solution or an infinite number of solutions to a set of equations. For example:

$$1:\ x + y - z = 3$$

$$2:\ 3x - y + 3z = 5$$

$$3:\ x - y + 2z = 2$$

These equations cannot be solved using the inverse matrix method and the Gaussian elimination gives the absurd conclusion that $0 = 1$. Try it for yourself.

If we alter the above set of equations to produce another example:

$$1:\ x + y - z = 1$$

$$2:\ 3x - y + 3z = 5$$

$$3:\ x - y + 2z = 2$$

Gaussian elimination using the augmented matrix gives us:

$$\begin{bmatrix} 1 & 1 & -1 & 1 \\ 3 & -1 & 3 & 5 \\ 1 & -1 & 2 & 2 \end{bmatrix} \xrightarrow[row3-row1]{row2-3\times row1} \begin{bmatrix} 1 & 1 & -1 & 1 \\ 0 & -4 & 6 & 2 \\ 0 & -2 & 3 & 1 \end{bmatrix} \xrightarrow{row3-\frac{1}{2}row1} \begin{bmatrix} 1 & 1 & -1 & 1 \\ 0 & -4 & 6 & 2 \\ 0 & 0 & 0 & 0 \end{bmatrix}$$

in which the last matrix is in echelon form. When we look at solving for z we see that *any* value of z will work:

$$z = \eta$$

where η is any number. Back-substituting, we get for x and y:

$$x = \frac{3}{2} - \frac{\eta}{2}$$

$$y = -\frac{1}{2} + \frac{3}{2}\eta$$

Thus, for this particular set of equations there exist an infinite number of solutions.

Under some circumstances the matrix equation is said to be *homogeneous*:

$$\hat{\mathbf{A}}\hat{\mathbf{x}} = \hat{0} \tag{2.42}$$

Obviously, in this situation we cannot use an inverse matrix, $\hat{\mathbf{A}}^{-1}$, to get anything other than a solution of zero for $\hat{\mathbf{x}}$:

$$\hat{\mathbf{A}}^{-1}\hat{\mathbf{A}}\hat{\mathbf{x}} = \hat{\mathbf{A}}^{-1}\hat{\mathbf{0}}$$

$$\hat{\mathbf{x}} = \hat{\mathbf{A}}^{-1}\hat{\mathbf{0}} = \hat{\mathbf{0}}$$

This type of solution is called a *trivial* solution to the equation. There may well be *non-trivial* solutions, however. If $\det\hat{\mathbf{A}} = \mathbf{0}$ there will be an infinite number of non-trivial solutions and if $\det\hat{\mathbf{A}} \neq \mathbf{0}$ the only solution is $\hat{\mathbf{x}} = \hat{\mathbf{0}}$.

The usefulness of this is in the search for *eigenvalues* of a matrix. Let us suppose that we have the matrix equation:

$$\hat{\mathbf{A}}\hat{\mathbf{x}} = \lambda\hat{\mathbf{x}}$$

or

$$\hat{\mathbf{A}}\hat{\mathbf{x}} - \lambda\hat{\mathbf{x}} = \hat{\mathbf{0}} \tag{2.43}$$

or

$$(\hat{\mathbf{A}} - \lambda)\hat{\mathbf{x}} = \hat{\mathbf{0}}$$

This is called the *characteristic equation*. The λs are the scalar eigenvalues and $\hat{\mathrm{x}}$ is the *eigenvector* or more accurately, the right eigenvector. If:

$$\hat{\mathrm{A}}^T\hat{\mathrm{x}} = \lambda\hat{\mathrm{x}}$$

then:

$$\lambda\hat{\mathbf{x}}^T = (\lambda\hat{\mathbf{x}})^T = (\hat{\mathbf{A}}^T\hat{\mathbf{x}})^T = \hat{\mathbf{x}}^T\hat{\mathbf{A}} \tag{2.44}$$

where $\hat{\mathbf{x}}^T$ is now referred to as the *left eigenvector*.

There will be non-trivial solutions for the characteristic equation only if:

$$\det(\hat{\mathbf{A}} - \lambda\hat{\mathbf{1}}) = 0 \tag{2.45}$$

or:

$$\begin{vmatrix} a_{11} - \lambda & a_{12} & \cdots & a_{12} \\ a_{21} & a_{22} - \lambda & \cdots & a_{12} \\ \vdots & \vdots & \ddots & a_{12} \\ a_{n1} & a_{n2} & \cdots & a_{nn} - \lambda \end{vmatrix} = 0$$

which will evaluate to a polynomial called the *characteristic polynomial.*

An example will help to illustrate this:

$$\hat{\mathbf{A}} = \begin{bmatrix} 2 & 1 & 0 \\ 1 & 4 & 0 \\ 2 & 5 & 2 \end{bmatrix}$$

The characteristic equation is:

$$\begin{vmatrix} 2-\lambda & 1 & 0 \\ 1 & 2-\lambda & 0 \\ 2 & 5 & 2-\lambda \end{vmatrix} = 0$$

Evaluation of this determinant gives us:

$$\begin{aligned} &(2-\lambda)\begin{vmatrix} 4-\lambda & 0 \\ 5 & 2-\lambda \end{vmatrix} - \begin{vmatrix} 1 & 0 \\ 2 & 2-\lambda \end{vmatrix} \\ &= (2-\lambda)(4-\lambda)(2-\lambda) - (2-\lambda) \\ &= (2-\lambda)[\lambda^2 - 6\lambda + 7] \\ &= (2-\lambda)(3+\sqrt{2}-\lambda)(3-\sqrt{2}-\lambda) = 0 \end{aligned}$$

Thus, λ can equal 2, $3+\sqrt{2}$ or $3-\sqrt{2}$ in order to satisfy the characteristic equation. Note that since we have solved a polynomial for its roots it is quite conceivable that those roots will be complex numbers and we must be prepared for that possibility. A quick check of these eigenvalues is to calculate the trace of the matrix and compare it to the sum of the eigenvalues (see eqn (2.21)). They should be equal:

$$\begin{aligned} &\mathrm{Tr}\hat{\mathbf{A}} = 2 + 4 + 2 = 8 \\ &\sum_i^n \lambda_i = 2 + (3+\sqrt{2}) + (3-\sqrt{2}) = 8 \\ &\mathrm{Tr}\hat{\mathbf{A}} = \sum_i^n \lambda_i \end{aligned} \tag{2.46}$$

We ask the reader's permission to defer proof of this until a bit later in the chapter.

Knowing the eigenvalues, we can now proceed to find possible eigenvectors from our eigenvalue equation, starting with $\lambda = 2$:

$$\hat{\mathbf{A}}\hat{\mathbf{x}} = \lambda\hat{\mathbf{x}}$$

$$\begin{bmatrix} 2 & 1 & 0 \\ 1 & 4 & 0 \\ 2 & 5 & 2 \end{bmatrix} \begin{bmatrix} x_1 \\ x_2 \\ x_3 \end{bmatrix} = 2 \begin{bmatrix} x_1 \\ x_2 \\ x_3 \end{bmatrix} = \begin{bmatrix} 2x_1 \\ 2x_2 \\ 2x_3 \end{bmatrix}$$

Our simultaneous equations are:

$$2x_1 + x_2 = 2x_1$$

$$x_1 + 4x_2 = 2x_2$$

$$2x_1 + 5x_2 + 2x_3 = 2x_3$$

Solution of these simultaneous equations shows that $x_1 = x_2 = 0$. A possible value for x_3 could be, for example, 1. This being the case, a possible eigenvector is:

$$\begin{bmatrix} 0 \\ 0 \\ 1 \end{bmatrix}$$

This is, of course, one of an infinite number of possible eigenvectors, all of which have different values of x_3. For the other values of λ:

$$\lambda = 3 + \sqrt{2}$$

$$\begin{bmatrix} 2 & 1 & 0 \\ 1 & 4 & 0 \\ 2 & 5 & 2 \end{bmatrix} \begin{bmatrix} x_1 \\ x_2 \\ x_3 \end{bmatrix} = (3 + \sqrt{2}) \begin{bmatrix} x_1 \\ x_2 \\ x_3 \end{bmatrix} = \begin{bmatrix} (3 + \sqrt{2})x_1 \\ (3 + \sqrt{2})x_2 \\ (3 + \sqrt{2})x_3 \end{bmatrix}$$

$$2x_1 + x_2 = (3 + \sqrt{2})x_1$$

$$x_1 + 4x_2 = (3 + \sqrt{2})x_2$$

$$2x_1 + 5x_2 + 2x_3 = (3 + \sqrt{2})x_3$$

$$x_1 = x_1$$

$$x_2 = (1 + \sqrt{2})x_1$$

$$x_3 = (3 + 2\sqrt{2})x_1$$

and a possible eigenvector is:

$$\begin{bmatrix} 1 \\ (1+\sqrt{2}) \\ (3+2\sqrt{2}) \end{bmatrix}$$

Similar analysis using the third eigenvalue gives a third possible eigenvector:

$$\begin{bmatrix} 1 \\ (1-\sqrt{2}) \\ (3-2\sqrt{2}) \end{bmatrix}$$

These results are easily checked by doing the matrix multiplication:

$$\hat{\mathbf{A}}\hat{\mathbf{x}} = \lambda\hat{\mathbf{x}}$$

and substituting the appropriate eigenvector. Let us check using the last eigenvector:

$$\hat{\mathbf{A}}\hat{\mathbf{x}} = \lambda\hat{\mathbf{x}}$$

$$\begin{bmatrix} 2 & 1 & 0 \\ 1 & 4 & 0 \\ 2 & 5 & 2 \end{bmatrix} \begin{bmatrix} 1 \\ (1-\sqrt{2}) \\ (3-2\sqrt{2}) \end{bmatrix} = \begin{bmatrix} 3-\sqrt{2} \\ 5-4\sqrt{2} \\ 13-9\sqrt{2} \end{bmatrix} = (3-\sqrt{2}) \begin{bmatrix} 1 \\ (1-\sqrt{2}) \\ (3-2\sqrt{2}) \end{bmatrix} \tag{2.47}$$

You can check the other two eigenvectors for yourself.

A step further is to combine the separate column eigenvectors, $\hat{\mathbf{x}}$, into one matrix, $\hat{\mathbf{X}}$, and to put the eigenvalues into a diagonal matrix, $\hat{\Lambda}$, so that we have a characteristic equation with all eigenvalues and eigenvectors:

$$\hat{\mathbf{A}}\hat{\mathbf{X}} = \hat{\mathbf{X}}\hat{\Lambda} \tag{2.48}$$

Why put $\hat{\mathbf{X}}$ first and $\hat{\Lambda}$ second on the right side of the equation instead of the other way around? We have constructed $\hat{\mathbf{X}}$ from our column vectors, $\hat{\mathbf{x}}_i$, each of which is associated with an eigenvalue, λ_i.

You can see an example of this relationship in eqn (2.47). In order to maintain this association we must multiply $\hat{\Lambda}$ from the *left* by $\hat{\mathbf{X}}$:

$$\hat{\mathbf{X}}\hat{\Lambda}$$

$$\begin{bmatrix} X_{11} & X_{12} & X_{13} \\ X_{21} & X_{22} & X_{23} \\ X_{31} & X_{32} & X_{33} \end{bmatrix} \begin{bmatrix} \lambda_1 & 0 & 0 \\ 0 & \lambda_2 & 0 \\ 0 & 0 & \lambda_3 \end{bmatrix}$$

$$= \begin{bmatrix} \lambda_1 X_{11} & \lambda_2 X_{12} & \lambda_3 X_{13} \\ \lambda_1 X_{21} & \lambda_2 X_{22} & \lambda_3 X_{23} \\ \lambda_1 X_{31} & \lambda_2 X_{32} & \lambda_3 X_{33} \end{bmatrix}$$

Thus, λ_1 is associated with column 1, which is one of its eigenvectors. Similarly, λ_2 and λ_3 are in columns 2 and 3, respectively. We can also explicitly show that eqn (2.48) is valid by writing:

$$\hat{\mathbf{X}}^{-1}\hat{\mathbf{A}}\hat{\mathbf{X}} = \hat{\Lambda}$$

$$\hat{\mathbf{X}}\hat{\mathbf{X}}^{-1}\hat{\mathbf{A}}\hat{\mathbf{X}} = \hat{\mathbf{X}}\hat{\Lambda}$$

$$\hat{\mathbf{A}}\hat{\mathbf{X}} = \hat{\mathbf{X}}\hat{\Lambda}$$

Using eqn (2.23) and (2.48), we can now offer a proof of eqn (2.46). Multiplication of both sides of eqn (2.48) by $\hat{\mathbf{X}}^{-1}$ gives:

$$\begin{aligned}
&\hat{\mathbf{X}}^{-1}\hat{\mathbf{A}}\hat{\mathbf{X}} = \hat{\mathbf{X}}^{-1}\hat{\mathbf{X}}\hat{\Lambda} = \hat{\Lambda} \\
&\mathrm{Tr}(\hat{\mathbf{X}}^{-1}\hat{\mathbf{A}}\hat{\mathbf{X}}) = \mathrm{Tr}(\hat{\Lambda}) \\
&\mathrm{Tr}(\hat{\mathbf{X}}\hat{\mathbf{X}}^{-1}\hat{\mathbf{A}}) = \mathrm{Tr}(\hat{\Lambda})\,(\text{by cyclic permutation}) \\
&\mathrm{Tr}(\hat{\mathbf{A}}) = \mathrm{Tr}(\hat{\Lambda}) \\
&\mathrm{Tr}(\hat{\mathbf{A}}) = \sum_i \lambda_i
\end{aligned} \tag{2.49}$$

2.4 Transforms

Two $n \times n$ matrices $\hat{\mathbf{A}}$ and are said to be *similar matrices* if there exists an invertible $n \times n$ matrix, $\hat{\mathbf{T}}$ such that:

$$\hat{\mathbf{B}} = \hat{\mathbf{T}}^{-1}\hat{\mathbf{A}}\hat{\mathbf{T}} \tag{2.50}$$

and the transformation from $\hat{\mathbf{A}}$ to $\hat{\mathbf{B}}$ is called a *similarity transform*. We can back-transform $\hat{\mathbf{B}}$ to $\hat{\mathbf{A}}$ again:

$$\begin{aligned}\hat{\mathbf{T}}\hat{\mathbf{B}}\hat{\mathbf{T}}^{-1} &= \hat{\mathbf{T}}\hat{\mathbf{T}}^{-1}\hat{\mathbf{A}}\hat{\mathbf{T}}\hat{\mathbf{T}}^{-1} \\ &= \hat{\mathbf{A}}\end{aligned}$$

Note that reversing the order of the $\hat{\mathbf{T}}$ matrices gives a different matrix, $\hat{\mathrm{C}}$, but is still, of course, a similarity transformation:

$$\hat{\mathbf{C}} = \hat{\mathbf{T}}\hat{\mathbf{A}}\hat{\mathbf{T}}^{-1}$$

Note, also, that $\hat{\mathbf{C}}$ is similar to $\hat{\mathbf{A}}$ *and* to $\hat{\mathbf{B}}$:

$$\begin{aligned}\hat{\mathbf{C}} &= \hat{\mathbf{T}}\hat{\mathbf{A}}\hat{\mathbf{T}}^{-1} \\ \hat{\mathbf{A}} &= \hat{\mathbf{T}}^{-1}\hat{\mathbf{T}}\hat{\mathbf{A}}\hat{\mathbf{T}}^{-1}\hat{\mathbf{T}} = \hat{\mathbf{T}}^{-1}\hat{\mathbf{C}}\hat{\mathbf{T}} \\ \hat{\mathbf{B}} &= \hat{\mathbf{T}}^{-1}\hat{\mathbf{A}}\hat{\mathbf{T}} = \hat{\mathbf{T}}^{-1}\hat{\mathbf{T}}^{-1}\hat{\mathbf{C}}\hat{\mathbf{T}}\hat{\mathbf{T}} = (\hat{\mathbf{T}}^{-1})^2\hat{\mathbf{C}}(\hat{\mathbf{T}})^2\end{aligned}$$

The reason that $\hat{\mathbf{A}}$ and $\hat{\mathbf{B}}$ (and $\hat{\mathrm{C}}$) are called similar is that they have several properties in common. First, the determinants of $\hat{\mathbf{A}}$ and $\hat{\mathbf{B}}$ are the same. We can show this quite easily using eqn (2.37) and (2.38):

$$\begin{aligned}\hat{\mathbf{B}} &= \hat{\mathbf{T}}^{-1}\hat{\mathbf{A}}\hat{\mathbf{T}} \\ \det\hat{\mathbf{B}} &= \det(\hat{\mathbf{T}}^{-1}\hat{\mathbf{A}}\hat{\mathbf{T}}) \\ &= \det\hat{\mathbf{T}}^{-1}\det\hat{\mathbf{A}}\det\hat{\mathbf{T}} \\ &= \det\hat{\mathbf{A}}\end{aligned} \tag{2.51}$$

Next, using eqn (2.23), the traces of $\hat{\mathbf{A}}$ and $\hat{\mathbf{B}}$ are equal:

$$\begin{aligned}\hat{\mathbf{B}} &= \hat{\mathbf{T}}^{-1}\hat{\mathbf{A}}\hat{\mathbf{T}} \\ \mathrm{Tr}(\hat{\mathbf{B}}) &= \mathrm{Tr}(\hat{\mathbf{T}}^{-1}\hat{\mathbf{A}}\hat{\mathbf{T}}) \\ &= \mathrm{Tr}(\hat{\mathbf{T}}^{-1}\hat{\mathbf{T}}\hat{\mathbf{A}}) \\ &= \mathrm{Tr}(\hat{\mathbf{1}}\hat{\mathbf{A}}) \\ &= \mathrm{Tr}(\hat{\mathbf{A}})\end{aligned} \tag{2.52}$$

Two matrices related by a similarity transformation have the same eigenvalues. If we have matrices $\hat{\mathbf{A}}$ and $\hat{\mathbf{B}}$ related by a similarity transformation and eigenvector matrix $\hat{\mathbf{X}}$ then:

$$\hat{\mathbf{B}} = \hat{\mathbf{T}}\hat{\mathbf{A}}\hat{\mathbf{T}}^{-1}$$

$$\hat{\mathbf{A}}\hat{\mathbf{X}} = \lambda\hat{\mathbf{X}}$$

Now, using the transformation matrix $\hat{\mathbf{T}}$, we write:

$$\begin{aligned} \hat{\mathbf{B}}\hat{\mathbf{T}}\hat{\mathbf{X}} &= (\hat{\mathbf{T}}\hat{\mathbf{A}}\hat{\mathbf{T}}^{-1})\hat{\mathbf{T}}\hat{\mathbf{X}} \\ &= \hat{\mathbf{T}}(\hat{\mathbf{A}}\hat{\mathbf{X}}) \\ &= \hat{\mathbf{T}}(\lambda\hat{\mathbf{X}}) \\ &= \lambda\hat{\mathbf{T}}\hat{\mathbf{X}} \end{aligned} \tag{2.53}$$

Thus, if $\hat{\mathbf{X}}$ is an eigenvector matrix of $\hat{\mathbf{A}}$ then $\hat{\mathbf{T}}\hat{\mathbf{X}}$ is an eigenvector matrix of $\hat{\mathbf{B}}$, and therefore $\hat{\mathbf{B}}$ has the same eigenvalues as $\hat{\mathbf{A}}$.

From a practical point of view, the order in which the (left-to-right) multiplication of the matrices is done is irrelevant. That is, for the similarity transform:

$$\hat{\mathbf{B}} = \hat{\mathbf{T}}^{-1}\hat{\mathbf{A}}\hat{\mathbf{T}}$$

if we make the definitions:

$$\hat{\mathbf{M}} = \hat{\mathbf{T}}^{-1}\hat{\mathbf{A}}$$

and

$$\hat{\mathbf{N}} = \hat{\mathbf{A}}\hat{\mathbf{T}}$$

then we can write our similarity transform in two equivalent ways:

$$\hat{\mathbf{B}} = \hat{\mathbf{M}}\hat{\mathbf{T}}$$

or

$$\hat{\mathbf{B}} = \hat{\mathbf{T}}^{-1}\hat{\mathbf{N}}$$

each representing a different order of multiplication of matrices $\hat{\mathbf{T}}^{-1}$, $\hat{\mathbf{A}}$ and $\hat{\mathbf{T}}$.

Another type of transform is the *orthogonal* transformation that uses the transpose matrix:

$$\hat{\mathbf{B}} = \hat{\mathbf{T}}^T \hat{\mathbf{A}} \hat{\mathbf{T}} \tag{2.54}$$

This only works, however, if:

$$\hat{\mathbf{T}}^T = \hat{\mathbf{T}}^{-1}$$

and, this being the case, all of the properties of the similar matrices discussed above apply here as well.

The last type of transformation that we will be interested in is the *unitary* transformation:

$$\hat{\mathbf{B}} = (\hat{\mathbf{T}}^*)^T \hat{\mathbf{A}} \hat{\mathbf{T}} \tag{2.55}$$

where:

$$(\hat{\mathbf{T}}^*)^T = \hat{\mathbf{T}}^{-1}$$

In other words, the matrix $\hat{\mathbf{T}}$ is unitary (see eqn (2.20)). If all the elements of $\hat{\mathbf{T}}$ are real then it is also an orthogonal transformation. Again, there is much that is parallel between these and the similarity and orthogonal transformations. The determinant, trace and eigenvalues of the transformed matrix are invariant under the transformation.

Very closely related to the unitary transformation is the *Hermitian* transformation:

$$\hat{\mathbf{B}} = (\hat{\mathbf{T}})^T \hat{\mathbf{A}} \hat{\mathbf{T}}^{-1} \tag{2.56}$$

where now:

$$(\hat{\mathbf{T}}^*)^T = \hat{\mathbf{T}}$$

Note that the inverse matrix is used explicitly in this transformation, unlike the previous ones. As before, we can show that the trace, determinant and eigenvalues are invariant with this transformation.

We have already used a similarity transformation to advantage in eqn (2.49). We can use these ideas in our investigation of eigenvalue equations. In fact, eqn (2.49) is referred to as the *diagonalisation* of $\hat{\mathbf{A}}$ since it transforms $\hat{\mathbf{A}}$ into $\hat{\Lambda}$, which is, of course, a diagonal matrix of eigenvalues. The diagonalisation problem is essentially the problem of finding the eigenvalues for a given matrix and is therefore an important one in the physical sciences.

If $\hat{\mathbf{A}}$ is Hermitian then the matrices that diagonalise it will be unitary. An example of a unitary matrix is:

$$\hat{\mathbf{T}} = \begin{bmatrix} \cos(\varphi) & \sin(\varphi) \\ -\sin(\varphi) & \cos(\varphi) \end{bmatrix}$$

$$(\hat{\mathbf{T}}^{*})^{T} = \begin{bmatrix} \cos(\varphi) & -\sin(\varphi) \\ \sin(\varphi) & \cos(\varphi) \end{bmatrix}$$

Simple multiplication will confirm that

$$\hat{\mathbf{T}}(\hat{\mathbf{T}}^{*})^{T} = \hat{\mathbf{1}}$$

Let us take an example matrix and diagonalise it:

$$\hat{\mathbf{A}} = \begin{bmatrix} 0 & b \\ d & 0 \end{bmatrix}$$

This is obviously not diagonal but application of the unitary transformation will give us the diagonal matrix:

$$\begin{aligned} \hat{\mathbf{A}}_{D} &= (\hat{\mathbf{T}}^{*})^{T}\hat{\mathbf{A}}\hat{T} \\ &= \begin{bmatrix} \cos(\varphi) & -\sin(\varphi) \\ \sin(\varphi) & \cos(\varphi) \end{bmatrix} \begin{bmatrix} 0 & b \\ d & 0 \end{bmatrix} \begin{bmatrix} \cos(\varphi) & \sin(\varphi) \\ -\sin(\varphi) & \cos(\varphi) \end{bmatrix} \\ &= \begin{bmatrix} -(d+b)\cos(\varphi)\sin(\varphi) & b\cos^{2}(\varphi) - d\sin^{2}(\varphi) \\ d\cos^{2}(\varphi) - b\sin^{2}(\varphi) & (d+b)\cos(\varphi)\sin(\varphi) \end{bmatrix} \end{aligned}$$

If this matrix is to be diagonal then:

$$b\cos^{2}(\varphi) - d\sin^{2}(\varphi) = 0$$

and

$$d\cos^{2}(\varphi) - b\sin^{2}(\varphi) = 0$$

or:

$$b\cos^{2}(\varphi) = d\sin^{2}(\varphi)$$

and

$$d\cos^{2}(\varphi) = b\sin^{2}(\varphi)$$

For this to be true, $b = d$ and $\cos(\phi) = \sin(\phi)$ and therefore, $\phi = \pi/4$ and our diagonal matrix now looks like:

$$\hat{\mathbf{A}}_D = \begin{bmatrix} -2b\cos(\varphi)\sin(\varphi) & 0 \\ 0 & 2b\cos(\varphi)\sin(\varphi) \end{bmatrix}$$

Using trigonometric identity A.14:

$$\hat{\mathbf{A}}_D = \begin{bmatrix} b\sin(2\varphi) & 0 \\ 0 & b\sin(2\varphi) \end{bmatrix}$$

Since $\phi = \pi/4$ we can write:

$$\hat{\mathbf{A}}_D = \begin{bmatrix} b & 0 \\ 0 & b \end{bmatrix}$$

This was a somewhat contrived example and a more generally useful demonstration of a unitary transformation resulting in a diagonalised matrix would start with a general matrix, $\hat{\mathbf{A}}$:

$$\hat{\mathbf{A}} = \begin{bmatrix} a & b \\ d & e \end{bmatrix}$$

Subjecting $\hat{\mathbf{A}}$ to a unitary matrix transformation using $\hat{\mathbf{T}}$ and $(\hat{\mathbf{T}}^*)^T$ from above we get:[§§]

$$\hat{\mathbf{A}}_D = \begin{bmatrix} es^2(\varphi) + ac^2(\varphi) - (d+b)c(\varphi)s(\varphi) & (a-e)c(\varphi)s(\varphi) + bc^2(\varphi) - ds^2(\varphi) \\ (a-e)c(\varphi)s(\varphi) + dc^2(\varphi) - bs^2(\varphi) & ec^2(\varphi) + as^2(\varphi) + (d+b)c(\varphi)s(\varphi) \end{bmatrix}$$

where we have substituted $s(\phi) = \sin(\phi)$ and $c(\phi) = \cos(\phi)$. For $\hat{\mathbf{A}}$ to be diagonal it must be true that:

$$1. (a-e)c(\varphi)s(\varphi) + bc^2(\varphi) - ds^2(\varphi) = 0$$

$$2. (a-e)c(\varphi)s(\varphi) + dc^2(\varphi) - bs^2(\varphi) = 0$$

Subtraction of 2 from 1 gives:

$$\begin{gathered} (b-d)(c^2(\varphi) + s^2(\varphi)) = 0 \\ \text{and therefore} \\ b = d \end{gathered} \tag{2.57}$$

[§§]Matrix multiplications can be very tedious and are prone to human error. For this reason alone, symbolic mathematics software is very useful. I recommend wxMaxima, a freely available program that does much in addition to matrix manipulations.

which tells us that the matrix $\hat{\mathbf{A}}$ must be symmetric in order to be diagonalised. This gives us:

$$(a-e)c(\varphi)s(\varphi)+b(c^2(\varphi)-s^2(\varphi))=0=(a-e)\frac{\sin(2\varphi)}{2}+b\cos(2\varphi)$$
$$\tan(2\varphi)=\frac{-2b}{a-e} \tag{2.58}$$

where trigonometric identities (A.14) and (A.15) have been used. We see that the diagonalisation angle, 2ϕ, is dependent on the values of the elements in the matrix and that if $e=a$ then the value of 2ϕ must be $\pi/2$.

The form of our diagonalised matrix is:

$$\hat{\mathbf{A}}_D=\begin{bmatrix} es^2(\varphi)+ac^2(\varphi)-bs(2\varphi) & 0 \\ 0 & ec^2(\varphi)+as^2(\varphi)-bs(2\varphi) \end{bmatrix}$$

We know from our above discussion (see eqn (2.49)) that the trace of the matrix is invariant to diagonalisation and can use this fact as a test of our procedure. Taking the trace of $\hat{\mathbf{A}}_D$:

$$\begin{aligned}\mathrm{Tr}(\hat{\mathbf{A}}_D)&=es^2(\varphi)+ac^2(\varphi)-bs(2\varphi)+ec^2(\varphi)+as^2(\varphi)+bs(2\varphi)\\&=e(s^2(\varphi)+c^2(\varphi))+a(c^2(\varphi)+s^2(\varphi))-bs(2\varphi)+bs(2\varphi)\\&=e+a\\&=\mathrm{Tr}(\hat{\mathbf{A}})\end{aligned}$$

Using the already discussed method involving eqn (2.45) we can find a general expression for the eigenvalues of our symmetric matrix:

$$\begin{aligned}\begin{vmatrix} a-\lambda & b \\ b & e-\lambda \end{vmatrix}&=0\\ \lambda^2-(a+e)\lambda+(ae-b^2)&=0\\ \lambda=\frac{(a+e)\pm\sqrt{(a+e)^2-4(ae-b^2)}}{2}\\ =\frac{(a+e)\pm\sqrt{4b^2+(a-e)^2}}{2}\end{aligned} \tag{2.59}$$

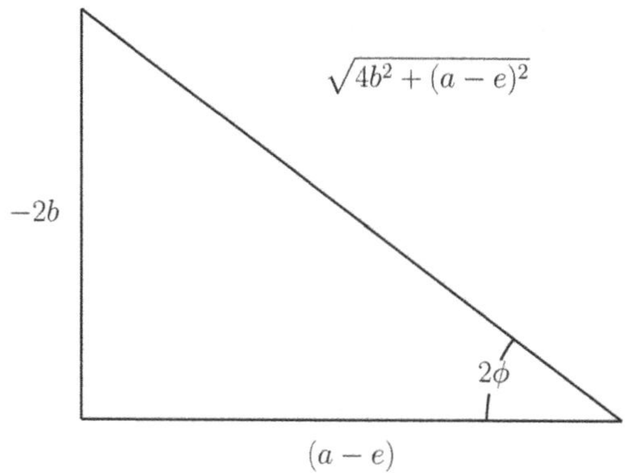

Figure 2.1 Matrix diagonalisation quantities visualised as components of a right-angle triangle.

Equations (2.58) and (2.59) point to a geometrical interpretation of the diagonalisation procedure. eqn (2.58) includes an angle, 2ϕ, and eqn (2.59) includes $\sqrt{4b^2 + (a-e)^2}$ that looks suspiciously Pythagorean. From these two equations a picture can be drawn that will help (see Figure 2.1).

What is this angle? At the risk of getting ahead of ourselves, one may view this as a vector rotation about the z-axis. Diagonalisation is accomplished by rotation through an angle of 2ϕ. We will have much more to say about vectors and vector rotations in the next chapter.

2.5 Exponentials of Matrices

We will have occasion in later chapters to want the exponentials of matrices. For diagonal matrices this is particularly easy. Thus, for diagonal matrix $\hat{\mathbf{A}}$:

$$\hat{\mathbf{A}} = \begin{bmatrix} a & 0 \\ 0 & b \end{bmatrix}$$

we will want to be able to write the matrix corresponding to $e^{\hat{\mathbf{A}}}$. To be able to write this we start with the powers of $\hat{\mathbf{A}}$:

$$\hat{\mathbf{A}}^2 = \begin{bmatrix} a & 0 \\ 0 & b \end{bmatrix}\begin{bmatrix} a & 0 \\ 0 & b \end{bmatrix} = \begin{bmatrix} a^2 & 0 \\ 0 & b^2 \end{bmatrix}$$

$$\hat{\mathbf{A}}^3 = \begin{bmatrix} a^2 & 0 \\ 0 & b^2 \end{bmatrix}\begin{bmatrix} a & 0 \\ 0 & b \end{bmatrix} = \begin{bmatrix} a^3 & 0 \\ 0 & b^3 \end{bmatrix}$$

etc.

The Euler expansion (see Appendix A) of $e^{\hat{\mathbf{A}}}$ is:

$$\begin{aligned} e^{\hat{\mathbf{A}}} &= \hat{\mathbf{1}} + \hat{\mathbf{A}} + \frac{\hat{\mathbf{A}}^2}{2!} + \frac{\hat{\mathbf{A}}^3}{3!} \cdots = \sum_{k=0}^{\infty} \frac{1}{k!} \mathbf{A}^k \\ &= \begin{bmatrix} 1 & 0 \\ 0 & 1 \end{bmatrix} + \begin{bmatrix} a & 0 \\ 0 & b \end{bmatrix} + \frac{\begin{bmatrix} a^2 & 0 \\ 0 & b^2 \end{bmatrix}}{2!} + \frac{\begin{bmatrix} a^3 & 0 \\ 0 & b^3 \end{bmatrix}}{3!} \cdots \\ &= \begin{bmatrix} e^a & 0 \\ 0 & e^b \end{bmatrix} \end{aligned} \tag{2.60}$$

So, given a diagonal matrix the exponential of that matrix is simply another diagonal matrix whose diagonal elements are the exponentials of the original elements of the matrix. The exponential of non-diagonal matrices may also be calculated but the procedure is less trivial. We will explore this in a later chapter.

There is a connection here to diagonalised matrices that will prove useful. The fundamental diagonalisation equation (see eqn (2.48)) is:

$$\hat{\mathrm{D}}\hat{\mathrm{X}} = \hat{\mathrm{X}}\hat{\Lambda}$$

or:

$$\hat{\mathbf{D}}\hat{\mathbf{X}}\hat{\mathbf{X}}^{-1} = \hat{\mathbf{X}}\hat{\Lambda}\hat{\mathbf{X}}^{-1}$$

$$\hat{\mathbf{D}} = \hat{\mathbf{X}}\hat{\Lambda}\hat{\mathbf{X}}^{-1}$$

in which the eigenvalue matrix $\hat{\Lambda}$ is back-transformed into $\hat{\mathbf{D}}$. Now, matrix $\hat{\mathbf{D}}$ is not necessarily diagonal and therefore finding its exponential matrix is not necessarily trivial. However, matrix $\hat{\Lambda}$ *is* diagonal and is related to $\hat{\mathbf{D}}$ by the unitary transformation using the eigenvector matrices $\hat{\mathbf{X}}$ and $\hat{\mathbf{X}}^{-1}$. Its exponential matrix is also diagonal as we have just seen. If we use the same eigenvector matrices to back transform it we will arrive at the exponential of matrix $\hat{\mathbf{D}}$:

$$e^{\hat{\mathbf{D}}} = \mathbf{X} e^{\hat{\Lambda}} \mathbf{X}^{-1} \tag{2.61}$$

We can explicitly show that this is so by again invoking the Euler expansion formula. If

$$
\begin{aligned}
e^{\hat{\mathbf{D}}} &= \mathbf{X}e^{\hat{\Lambda}}\mathbf{X}^{-1} \\
&= \hat{\mathbf{1}} + \mathbf{X}\hat{\Lambda}\mathbf{X}^{-1} + \frac{(\mathbf{X}\hat{\Lambda}\mathbf{X}^{-1})(\mathbf{X}\hat{\Lambda}\mathbf{X}^{-1})}{2!} + \frac{(\mathbf{X}\hat{\Lambda}\mathbf{X}^{-1})(\mathbf{X}\hat{\Lambda}\mathbf{X}^{-1})(\mathbf{X}\hat{\Lambda}\mathbf{X}^{-1})}{3!}\cdots \\
&= \hat{\mathbf{1}} + \mathbf{X}\hat{\Lambda}\mathbf{X}^{-1} + \frac{(\mathbf{X}\hat{\Lambda}^{2}\mathbf{X}^{-1})}{2!} + \frac{(\mathbf{X}\hat{\Lambda}^{3}\mathbf{X}^{-1})}{3!}\cdots \\
&= \mathbf{X}\left(\hat{\mathbf{1}} + \hat{\Lambda} + \frac{\hat{\Lambda}^{2}}{2!} + \frac{\hat{\Lambda}^{3}}{3!}\cdots\right)\mathbf{X}^{-1} \\
&= \mathbf{X}e^{\hat{\Lambda}}\mathbf{X}^{-1}
\end{aligned}
$$

A question that naturally comes to mind is this: If matrices $\hat{\mathbf{A}}$ and $\hat{\mathbf{B}}$ commute, do the exponentials of these matrices also commute? They do. To see that this is so consider:

$$e^{\mathrm{a}+\mathrm{b}} = e^{\mathrm{a}}e^{\mathrm{b}} = e^{\mathrm{b}}e^{\mathrm{a}}$$

For integer and real numbers *a* and *b* that always commute, this relationship is always true. However, this is not always true for exponentials of matrices $\hat{\mathbf{A}}$ and $\hat{\mathbf{B}}$ since they are not guaranteed to commute as are integers and real numbers. If they do commute though, the corresponding exponential matrix equation is also true. We show this here by first assuming that:

$$\hat{\mathrm{A}}\hat{\mathrm{B}} = \hat{\mathrm{B}}\hat{\mathrm{A}}$$

If this is the case then any power of $\hat{\mathbf{A}}$ will commute with any power of $\hat{\mathbf{B}}$. This is easily shown by a simple example:

$$\text{let } \hat{\mathbf{A}}^{3} = \hat{\mathbf{A}}\hat{\mathbf{A}}\hat{\mathbf{A}} \text{ and } \hat{\mathbf{B}}^{2} = \hat{\mathbf{B}}\hat{\mathbf{B}}$$

$$
\begin{aligned}
\hat{\mathbf{A}}^{3}\hat{\mathbf{B}}^{2} &= \hat{\mathbf{A}}\hat{\mathbf{A}}\hat{\mathbf{A}}\hat{\mathbf{B}}\hat{\mathbf{B}} \\
&= \hat{\mathbf{A}}\hat{\mathbf{A}}\hat{\mathbf{B}}\hat{\mathbf{A}}\hat{\mathbf{B}} = \hat{\mathbf{A}}\hat{\mathbf{A}}\hat{\mathbf{B}}\hat{\mathbf{B}}\hat{\mathbf{A}} \\
&= \hat{\mathbf{A}}\hat{\mathbf{B}}\hat{\mathbf{B}}\hat{\mathbf{A}}\hat{\mathbf{A}} \\
&= \hat{\mathbf{B}}\hat{\mathbf{B}}\hat{\mathbf{A}}\hat{\mathbf{A}}\hat{\mathbf{A}} = \hat{\mathbf{B}}^{2}\hat{\mathbf{A}}^{3}
\end{aligned}
$$

This being the case, we must be able multiply matrices $\hat{\mathbf{A}}$ and $\hat{\mathbf{B}}$ in just the same way that we would for integers and real numbers. Specifically, we must be able to invoke the binomial theorem using matrices $\hat{\mathbf{A}}$ and $\hat{\mathbf{B}}$:

$$(\hat{\mathbf{A}}+\hat{\mathbf{B}})^n=\sum_{m=0}^{n}\frac{n!}{m!(n-m)!}(\hat{\mathbf{A}}^m\hat{\mathbf{B}}^{(n-m)})$$

Writing out the product of the exponentials of matrices $\hat{\mathbf{A}}$ and $\hat{\mathbf{B}}$ and grouping in sums of powers:

$$\begin{aligned}
e^{\hat{\mathbf{A}}}e^{\hat{\mathbf{B}}} &= \left(\hat{\mathbf{1}}+\frac{\hat{\mathbf{A}}}{1!}+\frac{\hat{\mathbf{A}}^2}{2!}+\frac{\hat{\mathbf{A}}^3}{3!}\cdots\right)\left(\hat{\mathbf{1}}+\frac{\hat{\mathbf{B}}}{1!}+\frac{\hat{\mathbf{B}}^2}{2!}+\frac{\hat{\mathbf{B}}^3}{3!}\cdots\right)\\
&=\hat{\mathbf{1}}+\left(\frac{\hat{\mathbf{A}}}{1!}+\frac{\hat{\mathbf{B}}}{1!}\right)+\left(\frac{\hat{\mathbf{A}}^2}{2!}+\frac{\hat{\mathbf{A}}\hat{\mathbf{B}}}{1!1!}+\frac{\hat{\mathbf{B}}^2}{2!}\right)\cdots\\
&\quad+\left(\frac{\hat{\mathbf{A}}^m}{m!}+\frac{\hat{\mathbf{A}}^{(m-1)}\hat{\mathbf{B}}}{1!(m-1)!}\cdots\frac{\hat{\mathbf{A}}\hat{\mathbf{B}}^{(m-1)}}{1!(m-1)!}+\frac{\hat{\mathbf{B}}^m}{m!}\right)\\
&=\hat{\mathbf{1}}+\frac{1}{1!}(\hat{\mathbf{A}}+\hat{\mathbf{B}})+\frac{1}{2!}\left(\hat{\mathbf{A}}^2+2\hat{\mathbf{A}}\hat{\mathbf{B}}+\hat{\mathbf{B}}^2\right)\cdots\frac{1}{n!}\sum_{m=0}^{n}\frac{n!}{m!(n-m)!}\hat{\mathbf{A}}^m\hat{\mathbf{B}}^{(n-m)}\\
&=\hat{\mathbf{1}}+\frac{1}{1!}(\hat{\mathbf{A}}+\hat{\mathbf{B}})+\frac{1}{2!}(\hat{\mathbf{A}}+\hat{\mathbf{B}})^2\cdots\frac{1}{n!}(\hat{\mathbf{A}}+\hat{\mathbf{B}})^n\\
&=e^{(\hat{\mathbf{A}}+\hat{\mathbf{B}})}
\end{aligned}$$

Furthermore, since commuting matrices may be reordered in strings of multiplications, as we saw above, it must be so that $e^{\hat{\mathbf{A}}}e^{\hat{\mathbf{B}}}=e^{\hat{\mathbf{B}}}e^{\hat{\mathbf{A}}}$. We see then, if matrices $\hat{\mathbf{A}}$ and $\hat{\mathbf{B}}$ commute, that the exponentials of $\hat{\mathbf{A}}$ and $\hat{\mathbf{B}}$ also commute.

Further Reading

1. J. S. Golan, *The Linear Algebra a Beginning Graduate Student Ought to Know*, Kluwer Academic Publishers, Dordrecht, Boston, 2004.
2. M. O'nan, *Linear Algebra*, Harcourt, Brace, Jovanovich, New York, 1976.
3. F. L. Pilar, *Elementary Quantum Chemistry*, McGraw-Hill Book Company, New York, 1968.
4. G. Strang, *Linear Algebra and its Applications*, Harcourt, Brace, Jovanovich, San Diego, 1988.
5. K. A. Stroud, *Linear Algebra*, Industrial Press Inc., New York, 2008.

3 Vectors

But, in my opinion, all things in nature occur mathematically.

Rene Descartes

3.1 A Preliminary Definition of Vectors

We wish to provide a very loose general definition of vectors that we will "tighten" in Section 3.2. We consider sets of mathematical objects that follow a very specific set of rules for their computational properties. The term "objects" is purposely vague here. The object might be a number or a matrix or a function. For clarity one might think "real numbers" here in order to follow the definitions and rules. Note the use of boldface characters with an arrow above to denote a vector as distinct from a "normal" number or scalar. Our purpose here is to put forward the mechanism for identifying a vector and being able to say "yes that is a vector" or "no that is not a vector". It is difficult to give examples that are easy to visualise at this stage so we will defer examples until later sections. We will represent vectors here in boldface with an arrow above. Thus:

$$\vec{\mathbf{a}} \in V$$

represents a vector $\vec{\mathbf{a}}$ belonging to the set of vectors, V.

Vectors are number objects that belong to a set of similar objects called a *vector space*. The words "vector" and "space" are inextricably linked and when someone mentions the word "space" the mathematician thinks immediately "vectors" and *vice versa*. Thus, the

Essential Mathematics for NMR and MRI Spectroscopists
By Keith C. Brown

Published by the Royal Society of Chemistry, www.rsc.org

complete set of vectors, V, comprises a vector space. In order to be classified as vectors, these objects must satisfy some very basic algebraic criteria.

The basic vector algebra rules are as follows:

1. Addition is commutative: $\vec{\mathbf{a}}+\vec{\mathbf{b}}=\vec{\mathbf{b}}+\vec{\mathbf{a}}$
2. Addition is associative: $\vec{\mathbf{a}}+(\vec{\mathbf{b}}+\vec{\mathbf{c}})=(\vec{\mathbf{a}}+\vec{\mathbf{b}})+\vec{\mathbf{c}}$
3. There is an additive identity element, $\vec{\mathbf{0}}:\vec{\mathbf{a}}+\vec{\mathbf{0}}=\vec{\mathbf{a}}$
4. Every element of the set has an additive inverse: $\vec{\mathbf{a}}+(-\vec{\mathbf{a}})=\vec{\mathbf{0}}$

Scalar multiplication is also included in the algebra of vectors:

1. $\alpha(\vec{\mathbf{a}}+\vec{\mathbf{b}})=\alpha\vec{\mathbf{a}}+\alpha\vec{\mathbf{b}}$ (distributive)
2. $(\alpha+\beta)\,\vec{\mathbf{a}}=\alpha\vec{\mathbf{a}}+\beta\vec{\mathbf{a}}$ (distributive)
3. $(\alpha\beta)\vec{\mathbf{a}}=\alpha(\beta\vec{\mathbf{a}})$ (associative law for multiplication)

The new vector(s) produced by these additions and scalar multiplications *must* also be an element of the vector space, V. This being the case, we say that vectors are *closed* under addition and scalar multiplication, which simply means that addition of two vectors or multiplication of a vector by a scalar produces a new vector that is in the *same* vector space.

A *subspace* is a set of vectors that constitutes a subset of a full vector space. The subspace will have the properties:

1. When $\vec{\mathbf{a}}$ and $\vec{\mathbf{b}}$ belong to the same subspace then $\vec{\mathbf{a}}+\vec{\mathbf{b}}$ also belongs to the subspace.
2. If $\vec{\mathbf{a}}$ belongs to the subspace and α is a scalar then $\alpha\vec{\mathbf{a}}$ also belongs to the subspace.

One may write a linear combination of vectors:

$$\sum_{1}^{n} c_i\vec{\mathbf{e}}_i = c_1\vec{\mathbf{e}}_1 + c_2\vec{\mathbf{e}}_2 + c_3\vec{\mathbf{e}}_3 + \ldots \tag{3.1}$$

If we do this and are able to find a set of cs (other than all zeros) such that:

$$c_1\vec{\mathbf{e}}_1 + c_2\vec{\mathbf{e}}_2 + c_3\vec{\mathbf{e}}_3 + \ldots = 0 \tag{3.2}$$

then the set of vectors is said to be *linearly dependent*. This simply means that one or more of the vectors in the set can be constructed by a linear combination of other vectors in the space. To see that this is so, consider the case when c_1 is non-zero. If we divide through by c_1 and rearrange then we have:

$$\vec{\mathbf{e}}_1 = +\frac{c_2}{c_1}\vec{\mathbf{e}}_2 + \frac{c_3}{c_1}\vec{\mathbf{e}}_3 + \ldots$$

$\vec{\mathbf{e}}_1$ is shown to be constructed from a linear combination of one or more of the other vectors in the set. If, otherwise, in order to satisfy eqn (3.2) we must set all cs to zero, the set is *linearly independent*. In this case, it is not possible to construct any of the vectors in the set from any of the other vectors in the set. A set such as either of these is called a *basis* set of vectors. We will be primarily interested in linearly independent basis sets.

A set of vectors is said to *span* a vector space if every element of the vector space can be expressed as a linear combination of members of the set of vectors:

$$\vec{\mathbf{a}} = a_1\vec{\mathbf{e}}_1 + a_2\vec{\mathbf{e}}_2 + a_3\vec{\mathbf{e}}_3 \ldots \tag{3.3}$$

In other words, if any vector in a vector space can be written as a linear combination of a set of vectors that set of vectors spans the vector space. What is really of interest here is that if this set of spanning vectors is *also* linearly independent it is said to form a *basis* set for the vector space. The total number of vectors in the basis set is the *dimension* of the vector space. Also of interest is that if we have a basis set of vectors, $(\vec{\mathbf{e}}_1, \vec{\mathbf{e}}_2, \vec{\mathbf{e}}_3 \ldots)$, *any* vector in the full set of vectors of which the basis set is a part can be constructed as a linear combination of these basis vectors:

$$\vec{\mathbf{a}} = a_1\vec{\mathbf{e}}_1 + a_2\vec{\mathbf{e}}_2 + a_3\vec{\mathbf{e}}_3 \ldots \tag{3.4}$$

The a_is are called *components* of the vector, $\vec{\mathbf{a}}$.

In order to qualify as a vector space the algebraic rules set out at the beginning of this section must be satisfied for all objects in the vector space. It must also be possible to produce the linear combinations of objects that have been discussed.

Having been suitably general (or perhaps vague from your point of view) let us look at some examples of vector spaces. Remember, to qualify as a vector space the vectors in the set must satisfy the basic algebra rules. We will start with a set of 2×2 matrices of real numbers (ℝ is the set of all real numbers):

$$\begin{bmatrix}1&0\\0&0\end{bmatrix}\begin{bmatrix}0&1\\0&0\end{bmatrix}\begin{bmatrix}0&0\\1&0\end{bmatrix}\begin{bmatrix}0&0\\0&1\end{bmatrix}$$

$$\begin{bmatrix}1&2\\3&4\end{bmatrix}\begin{bmatrix}12&0\\0&0\end{bmatrix}\begin{bmatrix}3&6\\9&12\end{bmatrix}\begin{bmatrix}6&2\\32&9\end{bmatrix}$$

$$\begin{bmatrix}a_{11}&a_{12}\\a_{21}&a_{22}\end{bmatrix}\mathrm{a}_{ij}\in\mathbb{R}$$

These are but a few of an infinite number of 2×2 matrices that we could cite. The first row comprises a basis set from which all other 2×2 matrices can be constructed from:

$$\hat{\mathbf{A}}=\begin{bmatrix}a_{11}&a_{12}\\a_{21}&a_{22}\end{bmatrix}=a_{11}\begin{bmatrix}1&0\\0&0\end{bmatrix}+a_{12}\begin{bmatrix}0&1\\0&0\end{bmatrix}+a_{21}\begin{bmatrix}0&0\\1&0\end{bmatrix}+a_{22}\begin{bmatrix}0&0\\0&1\end{bmatrix}$$

$$\hat{\mathbf{B}}=\begin{bmatrix}b_{11}&b_{12}\\b_{21}&b_{22}\end{bmatrix}=b_{11}\begin{bmatrix}1&0\\0&0\end{bmatrix}+b_{12}\begin{bmatrix}0&1\\0&0\end{bmatrix}+b_{21}\begin{bmatrix}0&0\\1&0\end{bmatrix}+b_{22}\begin{bmatrix}0&0\\0&1\end{bmatrix}$$

How do we know that it is a basis set? The test is "is it linearly independent?". It is:

$$x_1\begin{bmatrix}1&0\\0&0\end{bmatrix}+x_2\begin{bmatrix}0&1\\0&0\end{bmatrix}+x_3\begin{bmatrix}0&0\\1&0\end{bmatrix}+x_4\begin{bmatrix}0&0\\0&1\end{bmatrix}=\begin{bmatrix}0&0\\0&0\end{bmatrix}$$

The only way this equation can be true is if all xs are equal to zero, which means the matrices are linearly independent.

As we saw in Chapter 2, matrices $\hat{\mathbf{A}}$ and $\hat{\mathbf{B}}$ can be added to give matrix $\hat{\mathbf{C}}$:

$$\hat{\mathbf{C}}=\hat{\mathbf{A}}+\hat{\mathbf{B}}=\begin{bmatrix}a_{11}&a_{12}\\a_{21}&a_{22}\end{bmatrix}+\begin{bmatrix}b_{11}&b_{12}\\b_{21}&b_{22}\end{bmatrix}=\begin{bmatrix}a_{11}+b_{11}&a_{12}+b_{12}\\a_{21}+b_{21}&a_{22}+b_{22}\end{bmatrix}$$

which is another matrix in the vector space. We can multiply matrix $\hat{\mathbf{A}}$ (or $\hat{\mathbf{B}}$ or $\hat{\mathbf{C}}$) by a scalar to produce yet another matrix in the vector space:

$$\alpha\hat{\mathbf{A}}=\alpha\begin{bmatrix}a_{11}&a_{12}\\a_{21}&a_{22}\end{bmatrix}=\begin{bmatrix}\alpha a_{11}&\alpha a_{12}\\\alpha a_{21}&\alpha a_{22}\end{bmatrix}$$

In fact, if you look at it carefully, you will see that these 2×2 matrices satisfy each of the rules of algebra for vectors. Hence, we can say that they constitute a set of vectors.

Another example is the set of all polynomial functions of real numbers. For example:

$$f(x) = x^4 - 2x$$

$$g(x) = x^{12} + 3x^5 - 2$$

We may add these functions to produce a new function of real numbers:

$$k(x) = x^{12} + 3x^5 + x^4 - 2x - 2$$

We may multiply any of these functions by a scalar to produce a new function of real numbers:

$$m(x) = 6f(x) = 6x^4 - 12x$$

Thus, the set of all functions of real numbers is a set of vectors by our current definition. We can try to find a basis set for these vectors. Let us think in terms of constructing a vector such as $f(x)$ from the basis set:

$$f(x) = 1 \times (x)^4 + (-2) \times (x)$$

What we have done here is to construct $f(x)$ from a linear combination of powers of x so a polynomial basis set must consist of the individual powers of x:

$$\sum_{i=0}^{\infty} b_i x^i$$

These are linearly independent since no one x^i can be constructed from a linear combination of the other x^i. Also, the number of members of the basis set is infinite, so the dimension of the polynomial space is infinite. Infinite dimensional spaces appear in the Schrodinger formulation of quantum mechanics.

3.2 Euclidean Vector Algebra and Inner Product Spaces

We have, so far, talked about addition and multiplication by scalars but *not* about multiplication of vectors by vectors. We now do so.

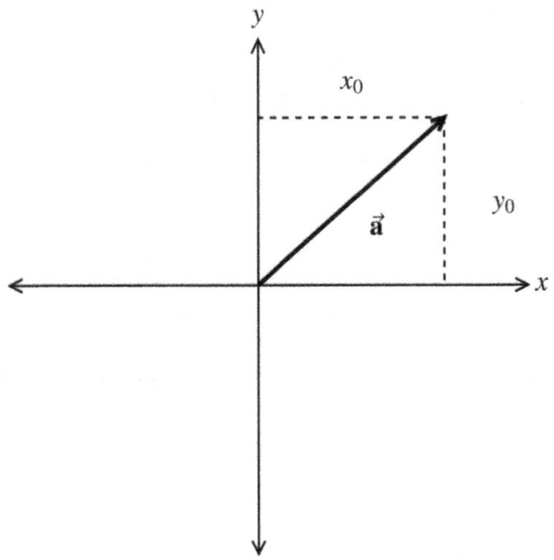

Figure 3.1 A two-dimensional Euclidean vector.

The most familiar vector spaces are two and three-dimensional Euclidean spaces, which we will denote using $\mathbb{R}^2$ and $\mathbb{R}^3$, respectively, to emphasise that we are using real numbers in our vectors (for the moment).† Vectors in these spaces can be conveniently represented by ordered pairs or triples of real numbers and are frequently depicted graphically as arrows beginning at the origin (see Figure 3.1). These vectors may be added together using the *parallelogram rule* (see Figure 3.2) or multiplied by real numbers (scalar multiplication) and in general, satisfy the criteria for vectors given above. We leave it to the reader to verify this. In addition we define the concepts of length and angles between vectors.

A vector in the Euclidean sense is a quantity that has, in addition to the above properties of vectors, both a scalar value (or a magnitude) and a direction. It is thus easy to represent two-dimensional vectors as arrows on a plane. In Figure 3.1 we have represented a vector $\vec{\mathbf{a}}$ beginning at the origin as is the general convention and pointing (in this case) into the upper right quadrant. The end point of the vector can be represented by its x-y coordinates … this is an often-used way to compactly talk about vectors. It can of course be expanded to higher dimensions … as high as desired. The length or *norm* of the vector is $||\vec{\mathbf{a}}|| = \sqrt{x^2 + y^2}$ in two dimensions and $||\vec{\mathbf{a}}|| = \sqrt{x^2 + y^2 + z^2}$

†Do not confuse these superscripts with powers. They are meant instead to indicate the number of dimensions of the space.

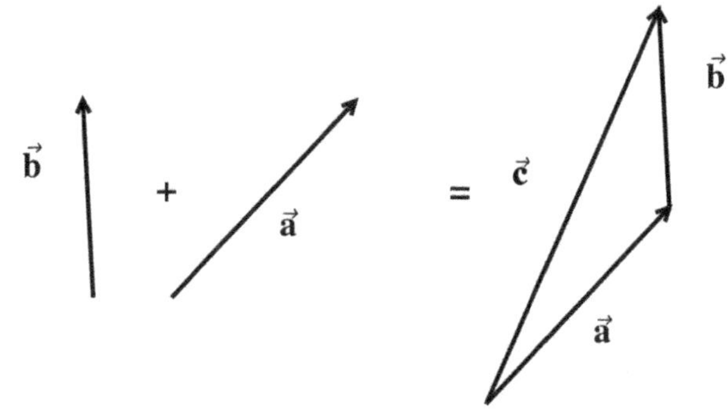

Figure 3.2 The parallelogram method of vector addition.

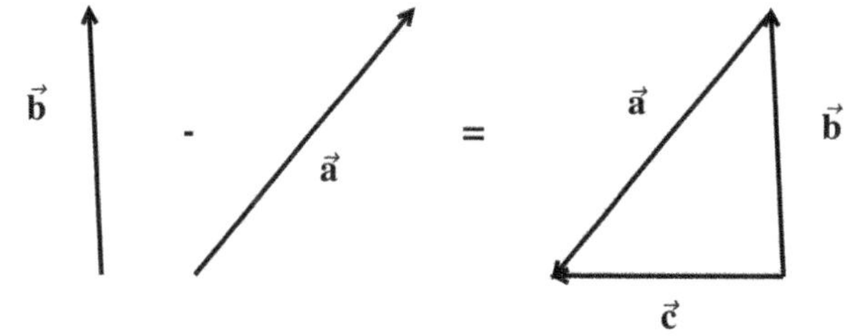

Figure 3.3 The parallelogram method of vector subtraction.

in three dimensions, easily calculated from the Pythagorean theorem.[‡]

Algebra on these types of vectors can be done in several ways. The simplest is to use the parallelogram method. Thus, addition is done by lining the vectors up head-to-tail (the head is the end with the arrow head) and drawing a resultant vector from the tail of the first to the head of the last (Figure 3.2).

Vectors $\vec{\mathbf{a}}$ and $\vec{\mathbf{b}}$ are placed head-to-tail to produce the resultant vector $\vec{\mathbf{c}}$. Subtraction is similar except that negative vectors are reversed (Figure 3.3). So, $\vec{\mathbf{a}}-\vec{\mathbf{b}}$ is similar to $\vec{\mathbf{a}}+\vec{\mathbf{b}}$ except that the direction of $\vec{\mathbf{b}}$ is reversed. After reversal, the procedure is identical to addition.

A common use of vectors is to denote the spatial position of an object. The *position vector*, often given the symbol $\vec{\mathbf{r}}$, is anchored at the origin and points to the position in space where the object resides (Figure 3.4). Position vectors can be used to calculate the vector between two points in space by simply doing a vector difference calculation (Figure 3.5).

[‡]Surely the best-known mathematical theorem and certainly one of the most used.

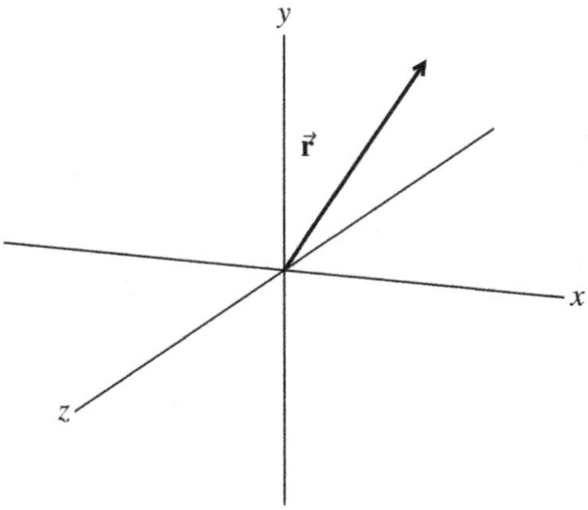

Figure 3.4 A position vector, $\vec{\mathbf{r}}$.

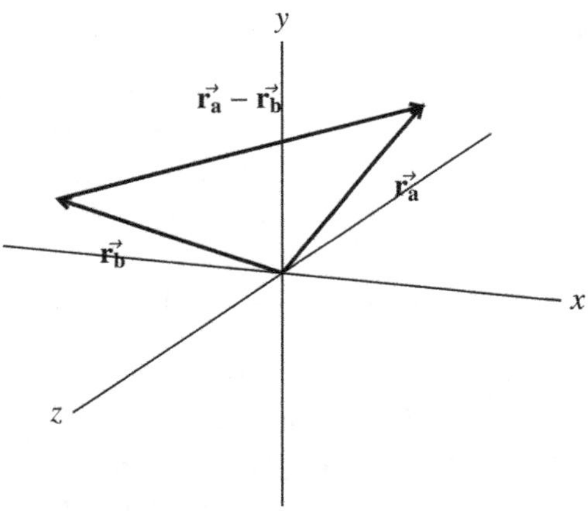

Figure 3.5 A position vector difference calculation.

Very frequently we use *unit vector*s multiplied by a scalar number to represent vectors. Thus, we might show the three-dimensional vector $\vec{\mathbf{a}}$ as the linear combination:

$$\vec{\mathbf{a}} = 3\vec{\mathbf{u}}_x + 6\vec{\mathbf{u}}_y + \vec{\mathbf{u}}_z$$

and vector $\vec{\mathbf{b}}$ as:

$$\vec{\mathbf{b}} = \vec{\mathbf{u}}_x - 5\vec{\mathbf{u}}_y + 2\vec{\mathbf{u}}_z$$

$\vec{\mathbf{u}}_x$, $\vec{\mathbf{u}}_y$ and $\vec{\mathbf{u}}_z$ are, of course, a basis set for three-dimensional vector space, as discussed in Section 3.1. That this is so can be seen from the definition of a basis set. First, we can construct *any* vector in $\mathbb{R}^3$ space from a linear combination of these three vectors, as indeed we have

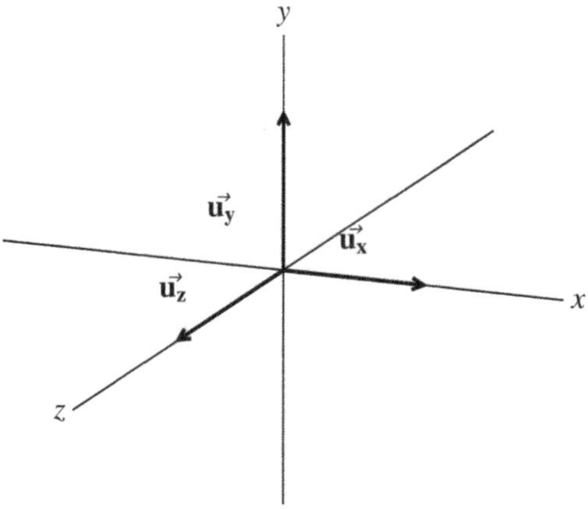

Figure 3.6 Three-dimensional unit vectors.

just done with $\vec{\mathbf{a}}$ and $\vec{\mathbf{b}}$. Secondly, these three unit vectors are linearly independent. That is, *via* eqn (3.2):

$$c_x\vec{\mathbf{u}}_x + c_y\vec{\mathbf{u}}_y + c_z\vec{\mathbf{u}}_z = 0$$

and the *only* way to do this is to have each c be equal to zero. Visually, for Euclidean vectors this means that these three unit vectors are aligned along the three Cartesian coordinate axes (Figure 3.6) and, as we will see, have some very special and useful properties.

Addition and subtraction are quite easy. One simply adds or subtracts the scalars multiplying into the unit vectors:

$$\begin{aligned}\vec{\mathbf{a}} + \vec{\mathbf{b}} &= 3\vec{\mathbf{u}}_x + 6\vec{\mathbf{u}}_y + \vec{\mathbf{u}}_z + \vec{\mathbf{u}}_x - 5\vec{\mathbf{u}}_y + 2\vec{\mathbf{u}}_z \\ &= (3+1)\vec{\mathbf{u}}_x + (6-5)\vec{\mathbf{u}}_y + (1+2)\vec{\mathbf{u}}_z \\ &= 4\underset{\rightarrow}{\mathbf{u}}_x + \vec{\mathbf{u}}_y + 3\vec{\mathbf{u}}_z\end{aligned}$$

The scalars in front of the unit vectors are the components or co-ordinates of the vector ... in this case in three dimensions.

Any one of the familiar planes in $\mathbb{R}^3$ space is a vector subspace of $\mathbb{R}^3$ space as defined above. Thus, if we consider the *x*,*y*-plane and two vectors, $\vec{\mathbf{a}}$ and $\vec{\mathbf{b}}$ within the plane (Figure 3.7), vector $\vec{\mathbf{a}}$ plus vector $\vec{\mathbf{b}}$ gives a new vector $\vec{\mathbf{c}}$ that is *still* in the *x*,*y*-plane. Also, any scalar multiplied into $\vec{\mathbf{a}}$ or $\vec{\mathbf{b}}$ will result in a vector that is also still in the *x*,*y*-plane. As you will recall from the above discussion on subspaces, this fits the definition of a subspace and therefore the *x*,*y*-plane is a subspace of $\mathbb{R}^3$. There are two unit vectors, $\vec{\mathbf{u}}_x$ and $\vec{\mathbf{u}}_y$ directed along

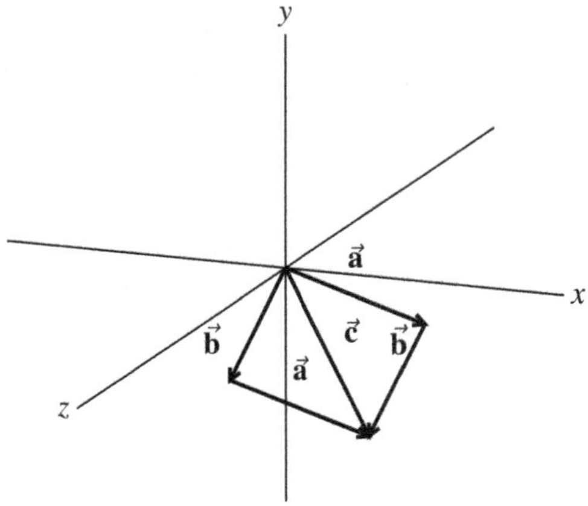

Figure 3.7 Two-dimensional vector subspace of 3D space.

the x and y axes, respectively, that form a basis for this particular subspace so this is therefore a $\mathbb{R}^2$.

Also, from Figure 3.7, we note that vector $\vec{\mathbf{c}}$ is a result of a linear combination of $\vec{\mathbf{a}}$ and $\vec{\mathbf{b}}$, which are themselves linear combinations of the unit vectors. This being the case we can say that $\vec{\mathbf{a}}$ and $\vec{\mathbf{b}}$ span the subspace that is the x,y-plane:

$$\begin{aligned}\vec{\mathbf{c}} &= \vec{\mathbf{a}} + \vec{\mathbf{b}} \\ &= a_x\vec{\mathbf{u}}_x + a_y\vec{\mathbf{u}}_y + a_z\vec{\mathbf{u}}_z + b_x\vec{\mathbf{u}}_x + b_y\vec{\mathbf{u}}_y + b_z\vec{\mathbf{u}}_z \\ &= (a_x + b_x)\vec{\mathbf{u}}_x + (a_y + b_y)\vec{\mathbf{u}}_y + (a_z + b_z)\vec{\mathbf{u}}_z\end{aligned} \tag{3.5}$$

3.3 Vector Multiplication

We can multiply a vector by a scalar number that simply scales the vector. Thus:

$$\begin{aligned}2\vec{\mathbf{a}} &= 2(3\vec{\mathbf{u}}_x + 6\vec{\mathbf{u}}_y + \vec{\mathbf{u}}_z) \\ &= 6\vec{\mathbf{u}}_x + 12\vec{\mathbf{u}}_y + 2\vec{\mathbf{u}}_z\end{aligned}$$

Note that $\mathbb{R}^2$ and $\mathbb{R}^3$ vectors satisfy the addition and scalar multiplication criteria set out in Section 3.1.

It is intuitive that there must be a length to a Euclidean vector. One can see it in the graphical picture of a two-dimensional vector and calculate it using the Pythagorean theorem:

$$\vec{\mathbf{a}} = 3\vec{\mathbf{u}}_x + 6\vec{\mathbf{u}}_y$$

The length of $\vec{\mathbf{a}}$ (represented here by $\|\vec{\mathbf{a}}\|$) is:

$$\|\vec{\mathbf{a}}\| = \sqrt{3^2 + 6^2}$$
$$= 6.708$$

To do this calculation we have to do some multiplication. In fact, it is as though we have taken *two* copies of $\vec{\mathbf{a}}$ and multiplied their x and y components and then taken the square root of the sum of the two. At first glance, it may seem a bit silly to multiply a number and then take its square root, however, since a length is a positive real quantity we do this so that any negative numbers involved result in positive lengths. So, for our vector $\vec{\mathbf{a}}$:

$$\vec{\mathbf{a}} = 3\vec{\mathbf{u}}_x + 6\vec{\mathbf{u}}_y$$
$$\|\vec{\mathbf{a}}\| = \sqrt{\vec{\mathbf{a}} \cdot \vec{\mathbf{a}}}$$
$$= \sqrt{3^2 + 6^2}$$

and generally for a $\mathbb{R}^3$ vector $\vec{\mathbf{a}}$:

$$\begin{aligned} \vec{\mathbf{a}} &= a_x\mathbf{u}_x + a_y\mathbf{u}_y + a_z\mathbf{u}_z s \\ \|\vec{\mathbf{a}}\| &= \sqrt{\vec{\mathbf{a}} \cdot \vec{\mathbf{a}}} \\ &= \sqrt{a_x a_x + a_y a_y + a_z a_z} \\ &\sqrt{{a_x}^2 + {a_y}^2 + {a_z}^2} \end{aligned} \tag{3.6}$$

The length of vector $\vec{\mathbf{a}}$ is called the *norm*, which we will represent in this text as $\|\vec{\mathbf{a}}\|$. We will produce a slightly more general definition of the norm below.

Might we not be able to do something similar with two *different* two- or three-dimensional vectors? That is, instead of using $\vec{\mathbf{a}} \cdot \vec{\mathbf{a}}$, can we use $\vec{\mathbf{a}} \cdot \vec{\mathbf{b}}$? Indeed, we can do so. We define the multiplication of two different vectors, $\vec{\mathbf{a}}$ and $\vec{\mathbf{b}}$, much as we did with two of copies of $\vec{\mathbf{a}}$:

$$\begin{aligned} \vec{\mathbf{a}} &= a_x\vec{\mathbf{u}}_x + a_y\vec{\mathbf{u}}_y + a_z\vec{\mathbf{u}}_z \\ \vec{\mathbf{b}} &= b_x\vec{\mathbf{u}}_x + b_y\vec{\mathbf{u}}_y + b_z\vec{\mathbf{u}}_z \\ \vec{\mathbf{a}} \cdot \vec{\mathbf{b}} &= a_x b_x + a_y b_y + a_z b_z \end{aligned} \tag{3.7}$$

Note that, just as with the length calculation, this quantity is a scalar or a simple number. Also, note that the product of $\vec{\mathbf{a}}$ and $\vec{\mathbf{b}}$ may be either negative or positive. This quantity is called by several different names including *inner product* and *dot product*. The inner product is the more general term, while the dot product is mostly used to refer to Euclidean vectors in $\mathbb{R}^2$ and $\mathbb{R}^3$. There are also several different ways that one may see this product denoted:

$$\vec{\mathbf{a}} \cdot \vec{\mathbf{b}} \text{ or } (\vec{\mathbf{a}},\vec{\mathbf{b}}) \text{ or } \langle\vec{\mathbf{a}}|\vec{\mathbf{b}}\rangle \tag{3.8}$$

We will speak more of the last of these notations later. A vector space for which we can do inner product calculations is called an *inner product space*. $\mathbb{R}^2$ and $\mathbb{R}^3$ are two such spaces.

There are some properties that follow from our definition of the inner product for $\mathbb{R}^2$ and $\mathbb{R}^3$:

1. $\vec{\mathbf{a}} \cdot \vec{\mathbf{a}} = \|\vec{\mathbf{a}}\|^2 \geq 0$ – the length of a vector is zero or greater
2. $\vec{\mathbf{a}} \cdot \vec{\mathbf{b}} = \vec{\mathbf{b}} \cdot \vec{\mathbf{a}}$ – the inner product is commutative (however, see below in eqn (3.50))
3. $(\alpha\vec{\mathbf{a}} \cdot \vec{\mathbf{b}}) = \alpha(\vec{\mathbf{a}} \cdot \vec{\mathbf{b}})$ – scalar multiplication
4. $\vec{\mathbf{a}} \cdot (\vec{\mathbf{b}} + \vec{\mathbf{c}}) = (\vec{\mathbf{a}} \cdot \vec{\mathbf{b}}) + (\vec{\mathbf{a}} \cdot \vec{\mathbf{c}})$ – the inner product is distributive

The inner product for two different vectors in a two- or three-dimensional space must include the two ideas of length and angle between vectors. We have investigated the length idea but not the angle. Let us do so now using the law of cosines (see appendix A, eqn (A.21)) to get at it. Imagine vectors $\vec{\mathbf{a}}$ and $\vec{\mathbf{b}}$ in a two-dimensional space and vector $\vec{\mathbf{c}} = \vec{\mathbf{a}} - \vec{\mathbf{b}}$ (see Figure 3.8). From the law of cosines:

$$\|\vec{\mathbf{a}} - \vec{\mathbf{b}}\| = \|\vec{\mathbf{c}}\|^2 = \|\vec{\mathbf{a}}\|^2 + \|\vec{\mathbf{b}}\|^2 - 2\|\vec{\mathbf{a}}\|\,\|\vec{\mathbf{b}}\|\,\cos(\theta)$$

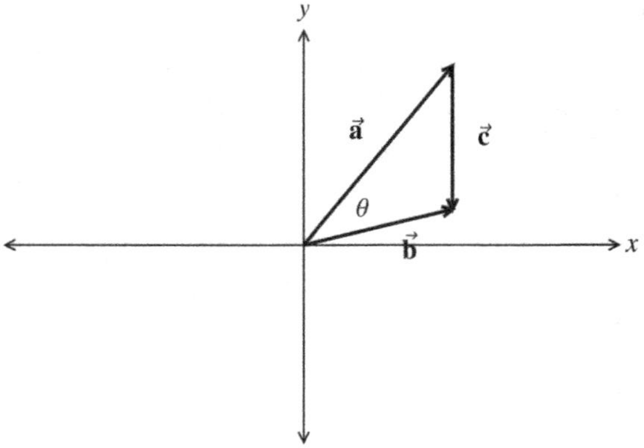

Figure 3.8 The law of cosines.

and from our definitions above:

$$\|\vec{\mathbf{a}}-\vec{\mathbf{b}}\|^2=\|\vec{\mathbf{c}}\|^2=(\vec{\mathbf{a}}-\vec{\mathbf{b}})\cdot(\vec{\mathbf{a}}-\vec{\mathbf{b}})$$

$$=\vec{\mathbf{a}}\cdot(\vec{\mathbf{a}}-\vec{\mathbf{b}})-\vec{\mathbf{b}}\cdot(\vec{\mathbf{a}}-\vec{\mathbf{b}})\quad\text{(distributive property)}$$

$$=\vec{\mathbf{a}}\cdot\vec{\mathbf{a}}-\vec{\mathbf{a}}\cdot\vec{\mathbf{b}}-\vec{\mathbf{b}}\cdot\vec{\mathbf{a}}+\vec{\mathbf{b}}\cdot\vec{\mathbf{b}}$$

$$=\vec{\mathbf{a}}\cdot\vec{\mathbf{a}}+\vec{\mathbf{b}}\cdot\vec{\mathbf{b}}-2(\vec{\mathbf{a}}\cdot\vec{\mathbf{b}})\quad\text{(commutative property)}$$

$$\|\vec{\mathbf{a}}\|^2=\vec{\mathbf{a}}\cdot\vec{\mathbf{a}}\ \|\vec{\mathbf{b}}\|^2=\vec{\mathbf{b}}\cdot\vec{\mathbf{b}}$$

and we can then substitute into the law of cosines:

$$(\vec{\mathbf{a}}-\vec{\mathbf{b}})\cdot(\vec{\mathbf{a}}-\vec{\mathbf{b}})=\vec{\mathbf{a}}\cdot\vec{\mathbf{a}}+\vec{\mathbf{b}}\cdot\vec{\mathbf{b}}-2\|\vec{\mathbf{a}}\|\|\vec{\mathbf{b}}\|\cos(\theta)$$

$$\vec{\mathbf{a}}\cdot\vec{\mathbf{a}}+\vec{\mathbf{b}}\cdot\vec{\mathbf{b}}-2(\vec{\mathbf{a}}\cdot\vec{\mathbf{b}})=\vec{\mathbf{a}}\cdot\vec{\mathbf{a}}+\vec{\mathbf{b}}\cdot\vec{\mathbf{b}}-2\|\vec{\mathbf{a}}\|\|\vec{\mathbf{b}}\|\cos(\theta)$$

We note that $\vec{\mathbf{a}}\cdot\vec{\mathbf{a}}$ and $\vec{\mathbf{b}}\cdot\vec{\mathbf{b}}$ and (-2) cancel out on both sides so we have:

$$\vec{\mathbf{a}}\cdot\vec{\mathbf{b}}=\|\vec{\mathbf{a}}\|\ \|\vec{\mathbf{b}}\|\ \cos(\theta) \tag{3.9}$$

Thus, we see that the inner product is the product of the lengths of vectors $\vec{\mathbf{a}}$ and $\vec{\mathbf{b}}$ and the cosine of the angle between them. Note that eqn (3.9) applies only to vector spaces comprised of real numbers. Complex numbers complicate things ... in particular, the commutative property of vectors is altered, as we will see.

Because there is generally an angle between two vectors $\vec{\mathbf{a}}$ and $\vec{\mathbf{b}}$, the maximum value that the inner product can take occurs when that angle is zero radians. Any non-zero angle means a cosine value less than or equal to 1 and therefore:

$$\vec{\mathbf{a}}\cdot\vec{\mathbf{b}}\le\|\vec{\mathbf{a}}\|\|\vec{\mathbf{b}}\|$$

or

$$(\vec{\mathbf{a}}\cdot\vec{\mathbf{b}})^2\le\|\vec{\mathbf{a}}\|^2\|\vec{\mathbf{b}}\|^2 \tag{3.10}$$

or

$$(\vec{\mathbf{a}}\cdot\vec{\mathbf{b}})^2\le(\vec{\mathbf{a}}\cdot\vec{\mathbf{a}})(\vec{\mathbf{b}}\cdot\vec{\mathbf{b}})$$

This is known as the Cauchy–Schwartz inequality and will be of use to us later on in our look at quantum mechanics.

If the angle between the two vectors in question is 90° then from eqn (3.9) the cosine of the angle equals zero. Under these circumstances we say that they are *orthogonal* vectors:

$$\vec{\mathbf{a}} \cdot \vec{\mathbf{b}} = \|\vec{\mathbf{a}}\| \|\vec{\mathbf{b}}\| \cos\left(\frac{\pi}{2}\right) = 0 \tag{3.11}$$

As far as a basis set in Euclidean space is concerned it is convenient under many circumstances to use one in which each element is orthogonal to the others and in which each element is normalised. Given the above definition of the dimension of a vector space it is obvious that for $\mathbb{R}^2$ and $\mathbb{R}^3$ spaces there are two and three basis vectors, respectively, which we represent here using $\vec{\mathbf{u}}_x$, $\vec{\mathbf{u}}_y$ and $\vec{\mathbf{u}}_z$. Since these are both orthogonal and normalised we call this an *orthonormal* basis set.

We say that a vector is *normalised* when:

$$\vec{\mathbf{a}} \cdot \vec{\mathbf{a}} = 1 \tag{3.12}$$

A vector may always be normalised using the definition of the inner product and the norm:

$$\begin{aligned} \vec{\mathbf{a}} \cdot \vec{\mathbf{a}} &= \|\vec{\mathbf{a}}\|^2 \\ \frac{\vec{\mathbf{a}} \cdot \vec{\mathbf{a}}}{\|\vec{\mathbf{a}}\|^2} &= 1 \\ \text{let } \vec{\mathbf{a}}' &= \frac{\vec{\mathbf{a}}}{\|\vec{\mathbf{a}}\|} \\ \text{then } \vec{\mathbf{a}}' \cdot \vec{\mathbf{a}}' &= 1 \end{aligned} \tag{3.13}$$

We can do an inner product calculation for each pair of our orthogonal and normalised unit vectors:

$$\begin{gathered} \vec{\mathbf{u}}_x \cdot \vec{\mathbf{u}}_x = 1 \quad \vec{\mathbf{u}}_x \cdot \vec{\mathbf{u}}_y = 0 \quad \vec{\mathbf{u}}_x \cdot \vec{\mathbf{u}}_z = 0 \\ \vec{\mathbf{u}}_y \cdot \vec{\mathbf{u}}_x = 0 \quad \vec{\mathbf{u}}_y \cdot \vec{\mathbf{u}}_y = 1 \quad \vec{\mathbf{u}}_y \cdot \vec{\mathbf{u}}_z = 0 \\ \vec{\mathbf{u}}_z \cdot \vec{\mathbf{u}}_x = 0 \quad \vec{\mathbf{u}}_z \cdot \vec{\mathbf{u}}_y = 0 \quad \vec{\mathbf{u}}_z \cdot \vec{\mathbf{u}}_z = 1 \\ \text{or} \\ \vec{\mathbf{u}}_i \cdot \vec{\mathbf{u}}_j = \delta_{ij} \begin{Bmatrix} \delta_{ij} = 1 \text{ if } i = j \\ \delta_{ij} = 0 \text{ if } i \neq j \end{Bmatrix} \end{gathered} \tag{3.14}$$

δ_{ij} is called the Kronecker delta. We can use this to produce a more general definition of the norm and the results of the inner product calculation (which we merely stated in eqn (3.7)). Suppose that we have a pair of vectors, $\vec{\mathbf{a}}$ and $\vec{\mathbf{b}}$:

$$\vec{\mathbf{a}} = a_x\vec{\mathbf{u}}_x + a_y\vec{\mathbf{u}}_y + a_z\vec{\mathbf{u}}_z$$

$$\vec{\mathbf{b}} = b_x\vec{\mathbf{u}}_x + b_y\vec{\mathbf{u}}_y + b_z\vec{\mathbf{u}}_z$$

Calculation of the norm of $\vec{\mathbf{a}}$ involves doing the inner product of $\vec{\mathbf{a}}$ with itself:

$$\begin{aligned} \|\vec{\mathbf{a}}\| &= \sqrt{\vec{\mathbf{a}} \cdot \vec{\mathbf{a}}} \\ &= \sqrt{(a_x\vec{\mathbf{u}}_x + a_y\vec{\mathbf{u}}_y + a_z\vec{\mathbf{u}}_z) \cdot (a_x\vec{\mathbf{u}}_x + a_y\vec{\mathbf{u}}_y + a_z\vec{\mathbf{u}}_z)} \end{aligned}$$

Since, as we have seen in eqn (3.14), the inner product of orthonormal basis vectors is ruled by a Kronecker delta, only those components with identical subscripts will survive:

$$\begin{aligned} \|\vec{\mathbf{a}}\| &= \sqrt{a_x^{\,2} + a_y^{\,2} + a_z^{\,2}} \\ &= \sqrt{\sum a_i a_j \delta_{ij}} \end{aligned} \tag{3.15}$$

Similarly, the inner product of $\vec{\mathbf{a}}$ and $\vec{\mathbf{b}}$ results in the survival of only those components of each vector with identical subscripts:

$$\begin{aligned} \vec{\mathbf{a}} \cdot \vec{\mathbf{b}} &= (a_x\vec{\mathbf{u}}_x + a_y\vec{\mathbf{u}}_y + a_z\vec{\mathbf{u}}_z) \cdot (b_x\vec{\mathbf{u}}_x + b_y\vec{\mathbf{u}}_y + b_z\vec{\mathbf{u}}_z) \\ &= a_x b_x + a_y b_y + a_z b_z \\ &= \sum a_i b_j \delta_{ij} \end{aligned} \tag{3.16}$$

There are two main types of vector multiplication, the inner or dot product that we have just seen and the *cross product*, so named because of the notation used to represent it:

$$\vec{\mathbf{c}} = \vec{\mathbf{a}} \times \vec{\mathbf{b}} \tag{3.17}$$

The cross product is heavily used in all areas of the physical sciences and is crucial to understanding many aspects of NMR spectroscopy. The cross product also produces a *new* vector, $\vec{\mathbf{c}}$ rather than a scalar as the dot product does. However, as we will see, this new vector is at right angles to the plane defined by $\vec{\mathbf{a}}$ and $\vec{\mathbf{b}}$ and its magnitude depends on the angle, θ, between $\vec{\mathbf{a}}$ and $\vec{\mathbf{b}}$, as does the inner product. We define the cross product multiplication for two $\mathbb{R}^3$ vectors $\vec{\mathbf{a}}$ and $\vec{\mathbf{b}}$ as follows:

$$\vec{\mathbf{a}} = a_x\vec{\mathbf{u}}_x + a_y\vec{\mathbf{u}}_y + a_z\vec{\mathbf{u}}_z$$

$$\vec{\mathbf{b}} = b_x\vec{\mathbf{u}}_x + b_y\vec{\mathbf{u}}_y + b_z\vec{\mathbf{u}}_z$$

$$\vec{\mathbf{a}} \times \vec{\mathbf{b}} = \begin{vmatrix} a_y & a_z \\ b_y & b_z \end{vmatrix} \vec{\mathbf{u}}_x + \begin{vmatrix} a_x & a_z \\ b_x & b_z \end{vmatrix} \vec{\mathbf{u}}_y + \begin{vmatrix} a_x & a_y \\ b_x & b_y \end{vmatrix} \vec{\mathbf{u}}_z \tag{3.18}$$

$$= \begin{vmatrix} \vec{\mathbf{u}}_x & \vec{\mathbf{u}}_y & \vec{\mathbf{u}}_z \\ a_x & a_y & a_z \\ b_x & b_y & b_z \end{vmatrix}$$

As with the inner product, given this definition there are properties that follow from it:

1. $\vec{\mathbf{a}}\times\vec{\mathbf{b}} = -\vec{\mathbf{b}}\times\vec{\mathbf{a}}$ – swapping rows (see the properties of determinants, Chapter 2)
2. $\vec{\mathbf{a}}\times\gamma\vec{\mathbf{b}} = \gamma\vec{\mathbf{a}}\times\vec{\mathbf{b}}$ – scalar multiplication (see eqn (2.31))
3. $\vec{\mathbf{a}}\times(\vec{\mathbf{b}}+\vec{\mathbf{c}}) = \vec{\mathbf{a}}\times\vec{\mathbf{b}}+\vec{\mathbf{a}}\times\vec{\mathbf{c}}$ – distributive property
4. $\vec{\mathbf{a}}\times\vec{\mathbf{a}} = \vec{0}$ – the identity vector

Property 4 follows from property 1:

$$\vec{\mathbf{a}} \times \vec{\mathbf{a}} = -(\vec{\mathbf{a}} \times \vec{\mathbf{a}})$$

therefore

$$\vec{\mathbf{a}} \times \vec{\mathbf{a}} = \vec{\mathbf{0}}$$

Note the "zero" vector in the last result. Since the result of the cross product is always a vector we define the zero vector in this fashion. *Every* vector space has a zero vector. We see from the definition of the cross product that the result is a new vector. What is the direction of this new vector? We can examine this using:

$$\vec{\mathbf{a}} \cdot (\vec{\mathbf{b}} \times \vec{\mathbf{c}})$$

that is, the inner product between $\vec{\mathbf{a}}$ and the result of cross product of $\vec{\mathbf{b}}$ and $\vec{\mathbf{c}}$. We further stipulate that $\vec{\mathbf{a}}$ is in the same plane as that defined by $\vec{\mathbf{b}}$ and $\vec{\mathbf{c}}$. Proceeding, we get:

$$\begin{aligned}
&\vec{\mathbf{a}}\cdot(\vec{\mathbf{b}}\times\vec{\mathbf{c}})\\
&=(a_x\vec{\mathbf{u}}_x+a_y\vec{\mathbf{u}}_y+a_z\vec{\mathbf{u}}_z)\cdot\left(\begin{vmatrix} b_y & b_z\\ c_y & c_z\end{vmatrix}\vec{\mathbf{u}}_x+\begin{vmatrix} b_x & b_z\\ c_x & c_z\end{vmatrix}\vec{\mathbf{u}}_y+\begin{vmatrix} b_x & b_y\\ c_x & c_y\end{vmatrix}\vec{\mathbf{u}}_z\right)\\
&=a_x\begin{vmatrix} b_y & b_z\\ c_y & c_z\end{vmatrix}\vec{\mathbf{u}}_x+a_y\begin{vmatrix} b_x & b_z\\ c_x & c_z\end{vmatrix}\vec{\mathbf{u}}_y+a_z\begin{vmatrix} b_x & b_y\\ c_x & c_y\end{vmatrix}\vec{\mathbf{u}}_z\\
&=\begin{vmatrix} a_x & a_y & a_z\\ b_x & b_y & b_z\\ c_x & c_y & c_z\end{vmatrix}
\end{aligned}$$

Now, if $\vec{\mathbf{a}}$ is in the same plane as that defined by $\vec{\mathbf{b}}$ and $\vec{\mathbf{c}}$, then it must be possible to write $\vec{\mathbf{a}}$ as a linear combination of $\vec{\mathbf{b}}$ and $\vec{\mathbf{c}}$ since $\vec{\mathbf{b}}$ and $\vec{\mathbf{c}}$ span the subspace of the plane that they define. This being the case, the first row in the determinant is a linear combination of the second and third rows. From eqn (2.36) this means that the result will equal zero, which means in turn that $\vec{\mathbf{b}}\times\vec{\mathbf{c}}$ must be orthogonal to the plane defined by vectors $\vec{\mathbf{b}}$ and $\vec{\mathbf{c}}$:

$$\text{let }(\vec{\mathbf{b}}\times\vec{\mathbf{c}})=\vec{\mathbf{d}}$$

$$\vec{\mathbf{a}}\cdot\vec{\mathbf{d}}=\|\vec{\mathbf{a}}\|\,\|\vec{\mathbf{d}}\|\cos(\theta)$$

If $\vec{\mathbf{a}}\cdot\vec{\mathbf{d}}=0$ then either $\vec{\mathbf{a}}$ and/or $\vec{\mathbf{d}}$ equals zero or $\cos(\theta)=0$. The first is trivial and the second indicates that θ must be some multiple of 90°. In other words, $\vec{\mathbf{d}}$ (= $\vec{\mathbf{b}}\times\vec{\mathbf{c}}$) is orthogonal to the $\vec{\mathbf{b}}$, $\vec{\mathbf{c}}$-plane.

As with the dot product, we ask ourselves "what is the magnitude of $\vec{\mathbf{b}}\times\vec{\mathbf{c}}$?":

$$\begin{aligned}
\vec{\mathbf{d}}=\vec{\mathbf{b}}\times\vec{\mathbf{c}}&=\begin{vmatrix} \vec{\mathbf{u}}_x & \vec{\mathbf{u}}_y & \vec{\mathbf{u}}_z\\ b_x & b_y & b_z\\ c_x & c_y & c_z\end{vmatrix}\\
&=\begin{vmatrix} b_y & b_z\\ c_y & c_z\end{vmatrix}\vec{\mathbf{u}}_x+\begin{vmatrix} b_x & b_z\\ c_x & c_z\end{vmatrix}\vec{\mathbf{u}}_y+\begin{vmatrix} b_x & b_y\\ c_x & c_y\end{vmatrix}\vec{\mathbf{u}}_z\\
&=d_x\vec{\mathbf{u}}_x+d_y\vec{\mathbf{u}}_y+d_z\vec{\mathbf{u}}_z
\end{aligned}$$

From eqn (3.15) the length (or the norm) of $\vec{\mathbf{d}}$:

$$\|\vec{\mathbf{d}}\| = \sqrt{d_x^{\,2} + d_y^{\,2} + d_z^{\,2}}$$

$$\begin{aligned}
\|\vec{\mathbf{d}}\|^2 &= d_x^{\,2} + d_y^{\,2} + d_z^{\,2} \\
&= \begin{vmatrix} b_y & b_z \\ c_y & c_z \end{vmatrix}^2 + \begin{vmatrix} b_x & b_z \\ c_x & c_z \end{vmatrix}^2 + \begin{vmatrix} b_x & b_y \\ c_x & c_y \end{vmatrix}^2 \\
&= (b_y c_z - c_y b_z)^2 + (b_x c_z - c_x b_z)^2 + (b_x c_y - c_x b_y)^2 \\
&= b_y^{\,2} c_z^{\,2} + c_y^{\,2} b_z^{\,2} - 2 b_y b_z c_y c_z \\
&\quad + b_x^{\,2} c_z^{\,2} + c_x^{\,2} b_z^{\,2} - 2 b_x b_z c_x c_z \\
&\quad + b_x^{\,2} c_y^{\,2} + c_x^{\,2} b_y^{\,2} - 2 b_x b_y c_x c_y
\end{aligned}$$

At this point we add some terms into the expression:

$$\begin{aligned}
\|\vec{\mathbf{d}}\|^2 &= (b_y c_z - c_y b_z)^2 + (b_x c_z - c_x b_z)^2 + (b_x c_y - c_x b_y)^2 \\
&= b_y^{\,2} c_z^{\,2} + c_y^{\,2} b_z^{\,2} - 2 b_y b_z c_y c_z \\
&\quad + b_x^{\,2} c_z^{\,2} + c_x^{\,2} b_z^{\,2} - 2 b_x b_z c_x c_z \\
&\quad + b_x^{\,2} c_y^{\,2} + c_x^{\,2} b_y^{\,2} - 2 b_x b_y c_x c_y \\
&\quad + b_x^{\,2} c_x^{\,2} - b_x^{\,2} c_x^{\,2} + b_y^{\,2} c_y^{\,2} - b_y^{\,2} c_y^{\,2} + b_z^{\,2} b_z^{\,2} - b_z^{\,2} b_z^{\,2}
\end{aligned}$$

Since each term in the last line has its negative counterpart we have not changed the value of the whole expression. However, it does allow us to proceed. We can now gather all of the terms and simplify things using our formula for the length of a vector, eqn (3.15), our formulas for the inner product, and and a trigonometric identity (see Appendix A, eqn (A.7)):

$$\begin{aligned}
\|\vec{\mathbf{d}}\|^2 &= (b_x^{\,2} + b_y^{\,2} + b_z^{\,2})(c_x^{\,2} + c_y^{\,2} + c_z^{\,2}) - (b_x c_x + b_y c_y + b_z c_z)^2 \\
&= \|\vec{\mathbf{b}}\|^2 \|\vec{\mathbf{c}}\|^2 - \|\vec{\mathbf{b}}\|^2 \|\vec{\mathbf{c}}\|^2 \cos^2(\theta) \\
&= \|\vec{\mathbf{b}}\|^2 \|\vec{\mathbf{c}}\|^2 (1 - \cos^2(\theta)) \\
&= \|\vec{\mathbf{b}}\|^2 \|\vec{\mathbf{c}}\|^2 \sin^2(\theta)
\end{aligned}$$

and

$$\|\vec{\mathbf{d}}\|^2 = \|\vec{\mathbf{b}} \times \vec{\mathbf{c}}\|^2 = \|\vec{\mathbf{b}}\|^2 \|\vec{\mathbf{c}}\|^2 \sin^2(\theta)$$

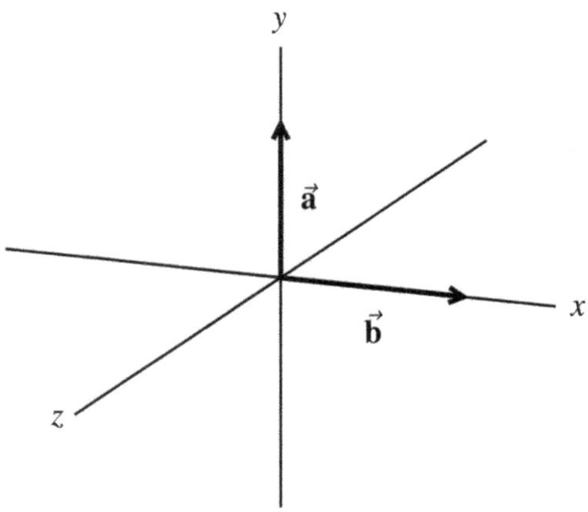

Figure 3.9 Vectors $\vec{\mathbf{a}}$ and $\vec{\mathbf{b}}$.

So, finally we reach our goal. The equation for the length of the new vector arising from the cross product of two vectors is:

$$\|\vec{\mathbf{c}}\| = \|\vec{\mathbf{a}}\times\vec{\mathbf{b}}\| = \|\vec{\mathbf{a}}\|\,\|\vec{\mathbf{b}}\|\ \sin(\theta) \tag{3.19}$$

Thus, if $\vec{\mathbf{a}}$ and $\vec{\mathbf{b}}$ are parallel the magnitude of $\vec{\mathbf{c}}$ will be zero.

We know that the new vector, $\vec{\mathbf{c}}$, will be perpendicular to the plane defined by $\vec{\mathbf{a}}$ and $\vec{\mathbf{b}}$ but what direction will the new vector, $\vec{\mathbf{c}}$, be in after $\vec{\mathbf{a}}\times\vec{\mathbf{b}}$? For simplicity let us suppose that $\vec{\mathbf{a}}$ lies along the $+y$-axis and $\vec{\mathbf{b}}$ lies along the $+x$-axis as in Figure 3.9.

Using our definition of the cross product, eqn (3.18), we write:

$$\vec{\mathbf{a}} = 0\vec{\mathbf{u}}_x + a_y\vec{\mathbf{u}}_y + 0\vec{\mathbf{u}}_z$$

$$\vec{\mathbf{b}} = b_x\vec{\mathbf{u}}_x + 0\vec{\mathbf{u}}_y + 0\vec{\mathbf{u}}_z$$

$$\begin{aligned}\vec{\mathbf{a}}\times\vec{\mathbf{b}} &= \begin{vmatrix} a_y & 0 \\ 0 & 0 \end{vmatrix}\vec{\mathbf{u}}_x + \begin{vmatrix} 0 & 0 \\ b_x & 0 \end{vmatrix}\vec{\mathbf{u}}_y + \begin{vmatrix} 0 & a_y \\ b_x & 0 \end{vmatrix}\vec{\mathbf{u}}_z \\ &= -a_y b_x \vec{\mathbf{u}}_z \\ &= \vec{\mathbf{c}}\end{aligned}$$

Evidently, in this case $\vec{\mathbf{c}}$ lies along the z-axis (Figure 3.10). In order to know where the resultant vector will be we use the *right-hand rule* to determine this. Imagine vector $\vec{\mathbf{a}}$ rotating about the z-axis towards vector $\vec{\mathbf{b}}$. Imagine also, that this rotation corresponds to the rotational direction that a standard wood screw is turned in order to drive it into the wood. The inward motion of the screw is in the direction of the

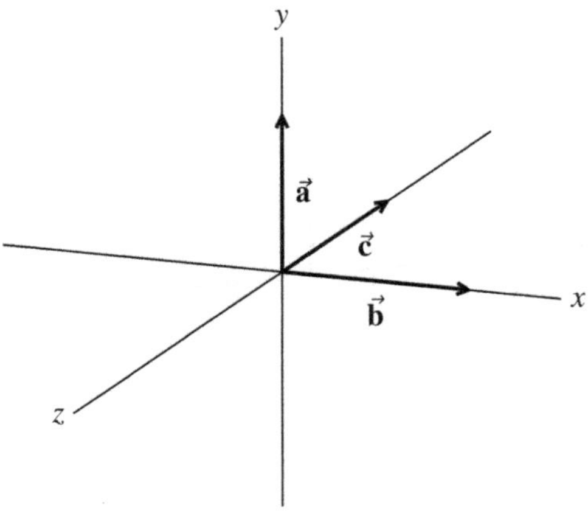

Figure 3.10 The cross product of vectors $\vec{\mathbf{a}}$ and $\vec{\mathbf{b}}$.

new vector, $\vec{\mathbf{c}}$. It should now be obvious that if we write $\vec{\mathbf{b}}\times\vec{\mathbf{a}}$ that the resultant vector $\vec{\mathbf{c}}$ will be along the $+z$-axis.

Of course, the cross product applies to our unit basis vectors as well:

$$
\begin{aligned}
&\vec{\mathbf{u}}_x \times \vec{\mathbf{u}}_x = 0 \quad \vec{\mathbf{u}}_x \times \vec{\mathbf{u}}_y = \vec{\mathbf{u}}_z \quad \vec{\mathbf{u}}_x \times \vec{\mathbf{u}}_z = -\vec{\mathbf{u}}_y \\
&\vec{\mathbf{u}}_y \times \vec{\mathbf{u}}_x = -\vec{\mathbf{u}}_z \quad \vec{\mathbf{u}}_y \times \vec{\mathbf{u}}_y = 0 \quad \vec{\mathbf{u}}_y \times \vec{\mathbf{u}}_z = \vec{\mathbf{u}}_x \\
&\vec{\mathbf{u}}_z \times \vec{\mathbf{u}}_x = \vec{\mathbf{u}}_y \quad \vec{\mathbf{u}}_z \times \vec{\mathbf{u}}_y = -\vec{\mathbf{u}}_x \quad \vec{\mathbf{u}}_z \times \vec{\mathbf{u}}_z = 0
\end{aligned}
$$

$$\text{or} \tag{3.20}$$

$$
\vec{\mathbf{u}}_i \times \vec{\mathbf{u}}_j = \epsilon_{ijk}\vec{\mathbf{u}}_k \begin{Bmatrix} 1 & \text{if i, j, k are an even permutation of } 1, 2, 3 \\ -1 & \text{if i, j, k are an odd permutation of } 1, 2, 3 \\ 0 & \text{if i, j, k are not a permutation of } 1, 2, 3 \end{Bmatrix}
$$

The symbol "ϵ_{ijk}" is the Lev–Civita permutation symbol. It is useful to note the cyclic permutation of $\vec{\mathbf{u}}_x$, $\vec{\mathbf{u}}_y$ and $\vec{\mathbf{u}}_z$ in these calculations. Thus, if you go from left to right and cycle back to the beginning and go from $\vec{\mathbf{u}}_x$ to $\vec{\mathbf{u}}_y$ to $\vec{\mathbf{u}}_z$ then the result of the calculation will be positive. Otherwise, it will be negative. For example, take $\vec{\mathbf{u}}_x\times\vec{\mathbf{u}}_y=\vec{\mathbf{u}}_z$ then permute by cycling each unit vector one position to the right, $\vec{\mathbf{u}}_z\times\vec{\mathbf{u}}_x=\vec{\mathbf{u}}_y$ and once again, $\vec{\mathbf{u}}_y\times\vec{\mathbf{u}}_z=\vec{\mathbf{u}}_x$. Any combination that is out of order will result in a negative cross product, *i.e.* $\vec{\mathbf{u}}_y\times\vec{\mathbf{u}}_x=-\vec{\mathbf{u}}_z$.

Using our example vectors and eqn (3.14) and (3.20) we can do dot and cross products:

$$\vec{\mathbf{a}} = 3\vec{\mathbf{u}}_x \times 6\vec{\mathbf{u}}_y + \vec{\mathbf{u}}_z \quad \vec{\mathbf{b}} = \vec{\mathbf{u}}_x - 5\vec{\mathbf{u}}_y + 2\vec{\mathbf{u}}_z$$

The dot (or scalar or inner) product would be:

$$\begin{aligned}
\vec{\mathbf{a}} \cdot \vec{\mathbf{b}} &= (3\vec{\mathbf{u}}_x + 6\vec{\mathbf{u}}_y + \vec{\mathbf{u}}_z) \cdot (\vec{\mathbf{u}}_x - 5\vec{\mathbf{u}}_y + 2\vec{\mathbf{u}}_z) \\
&= 3\vec{\mathbf{u}}_x \cdot \vec{\mathbf{u}}_x - 15\vec{\mathbf{u}}_x \cdot \vec{\mathbf{u}}_y + 6\vec{\mathbf{u}}_x \cdot \vec{\mathbf{u}}_z \\
&\quad - 30\vec{\mathbf{u}}_y \cdot \vec{\mathbf{u}}_y + 12\vec{\mathbf{u}}_y \cdot \vec{\mathbf{u}}_z + \vec{\mathbf{u}}_z \cdot \vec{\mathbf{u}}_x - 5\vec{\mathbf{u}}_z \cdot \vec{\mathbf{u}}_y + 2\vec{\mathbf{u}}_z \cdot \vec{\mathbf{u}}_z \\
&= 3 - 0 + 0 + 0 - 30 + 0 + 0 - 0 + 2 \\
&= -25
\end{aligned}$$

and the cross product is:

$$\begin{aligned}
\vec{\mathbf{a}} \times \vec{\mathbf{b}} &= (3\vec{\mathbf{u}}_x + 6\vec{\mathbf{u}}_y + \vec{\mathbf{u}}_z) \times (\vec{\mathbf{u}}_x - 5\vec{\mathbf{u}}_y + 2\vec{\mathbf{u}}_z) \\
&= 3\vec{\mathbf{u}}_x \times \vec{\mathbf{u}}_x - 15\vec{\mathbf{u}}_x \times \vec{\mathbf{u}}_y + 6\vec{\mathbf{u}}_x \times \vec{\mathbf{u}}_z - 30\vec{\mathbf{u}}_y \times \vec{\mathbf{u}}_y \\
&\quad + 12\vec{\mathbf{u}}_y \times \vec{\mathbf{u}}_z + \vec{\mathbf{u}}_z \times \vec{\mathbf{u}}_x - 5\vec{\mathbf{u}}_z \times \vec{\mathbf{u}}_y + 2\vec{\mathbf{u}}_z \times \vec{\mathbf{u}}_z \\
&= 0 - 15\vec{\mathbf{u}}_z - 6\vec{\mathbf{u}}_y - 6\vec{\mathbf{u}}_z - 0 + 12\vec{\mathbf{u}}_x + \vec{\mathbf{u}}_y + 5\vec{\mathbf{u}}_x + 0 \\
&= 17\vec{\mathbf{u}}_x - 5\vec{\mathbf{u}}_y - 21\vec{\mathbf{u}}_z \\
&= \vec{\mathbf{c}}
\end{aligned}$$

As you can see, the dot product results in an ordinary number (or a scalar) and the cross product results in a new vector. The new vector, $\vec{\mathbf{c}}$, will be at right angles to the plane defined by $\vec{\mathbf{a}}$ and $\vec{\mathbf{b}}$, as we have seen. How would one show this formally? The definition of orthogonality is that the dot product of two vectors must equal zero for the angle between them to be 90°.

$$\begin{aligned}
\vec{\mathbf{a}} \cdot \vec{\mathbf{c}} &= (3\vec{\mathbf{u}}_x + 6\vec{\mathbf{u}}_y + \vec{\mathbf{u}}_z) \cdot (17\vec{\mathbf{u}}_x - 5\vec{\mathbf{u}}_y - 21\vec{\mathbf{u}}_z) \\
&= 3 \cdot 17 + 6 \cdot (-5) + 1 \cdot (-21) \\
&= 51 - 30 - 21 \\
&= 0
\end{aligned}$$

Therefore, $\vec{\mathbf{a}}$ and $\vec{\mathbf{c}}$ are orthogonal. We will leave it to the reader to confirm that $\vec{\mathbf{b}}$ and $\vec{\mathbf{c}}$ are also orthogonal.

The cross product may also be calculated using the determinant method (eqn (3.18)):

$$\vec{\mathbf{a}} \times \vec{\mathbf{b}} = \begin{vmatrix} \vec{\mathbf{u}}_x & \vec{\mathbf{u}}_y & \vec{\mathbf{u}}_z \\ 3 & 6 & 1 \\ 1 & -5 & 2 \end{vmatrix}$$

$$= (6 \cdot 2 - (-5) \cdot 1)\vec{\mathbf{u}}_x - (3 \cdot 2 - 1 \cdot 1)\vec{\mathbf{u}}_y + (3 \cdot (-5) - 1 \cdot 6)\vec{\mathbf{u}}_z$$

$$= 17\vec{\mathbf{u}}_x - 5\vec{\mathbf{u}}_y - 21\vec{\mathbf{u}}_z$$

The norm of vector $\vec{\mathbf{a}}$ is:

$$\begin{aligned} \vec{\mathbf{a}} \cdot \vec{\mathbf{a}} &= (3\vec{\mathbf{u}}_x + 6\vec{\mathbf{u}}_y + \vec{\mathbf{u}}_z) \cdot (3\vec{\mathbf{u}}_x + 6\vec{\mathbf{u}}_y + \vec{\mathbf{u}}_z) \\ &= (3\vec{\mathbf{u}}_x \cdot 3\vec{\mathbf{u}}_x) + (3\vec{\mathbf{u}}_x \cdot 6\vec{\mathbf{u}}_y) + (3\vec{\mathbf{u}}_x \cdot \vec{\mathbf{u}}_z) \\ &\quad + (6\vec{\mathbf{u}}_y \cdot 3\vec{\mathbf{u}}_x) + (6\vec{\mathbf{u}}_y \cdot 6\vec{\mathbf{u}}_y) + (6\vec{\mathbf{u}}_y \cdot \vec{\mathbf{u}}_z) \\ &\quad + (\vec{\mathbf{u}}_z \cdot 3\vec{\mathbf{u}}_x) + (\vec{\mathbf{u}}_z \cdot 6\vec{\mathbf{u}}_y) + (\vec{\mathbf{u}}_z \cdot \vec{\mathbf{u}}_z) \\ &= 9 + 0 + 0 + 0 + 36 + 0 + 0 + 0 + 1 \\ &= 46 \end{aligned}$$

There is yet another type of vector multiplication that produces an object called a *dyad* from two vectors using the "normal" distributive rules of multiplication from grade school:

$$\begin{aligned} \vec{\mathbf{a}}\vec{\mathbf{b}} &= (3\vec{\mathbf{u}}_x + 6\vec{\mathbf{u}}_y + \vec{\mathbf{u}}_z)(\vec{\mathbf{u}}_x - 5\vec{\mathbf{u}}_y + 2\vec{\mathbf{u}}_z) \\ &= 3\vec{\mathbf{u}}_x\vec{\mathbf{u}}_x - 15\vec{\mathbf{u}}_x\vec{\mathbf{u}}_y + 6\vec{\mathbf{u}}_x\vec{\mathbf{u}}_z \\ &\quad + 6\vec{\mathbf{u}}_y\vec{\mathbf{u}}_x - 30\vec{\mathbf{u}}_y\vec{\mathbf{u}}_y + 12\vec{\mathbf{u}}_y\vec{\mathbf{u}}_z \\ &\quad + \vec{\mathbf{u}}_z\vec{\mathbf{u}}_x - 5\vec{\mathbf{u}}_z\vec{\mathbf{u}}_y + 2\vec{\mathbf{u}}_z\vec{\mathbf{u}}_z \end{aligned}$$

We will have more to say about this in Chapter 5.

There are two useful extensions of inner and cross products, the *scalar triple product* and the *vector triple product*. The scalar triple product is:

$$\vec{\mathbf{a}} \cdot (\vec{\mathbf{b}} \times \vec{\mathbf{c}})$$

in which the three vectors are non-coplanar. This is the formula for the volume of a parallelepiped whose sides are the three non-coplanar vectors, $\vec{\mathbf{a}}$, $\vec{\mathbf{b}}$ and $\vec{\mathbf{c}}$:

$$V = \vec{\mathbf{a}} \cdot (\vec{\mathbf{b}} \times \vec{\mathbf{c}})$$

That this is so is evident from Figure 3.11 and the following argument. The volume of the parallelepiped is the area of the parallelogram defined by vectors $\vec{\mathbf{b}}$ and $\vec{\mathbf{c}}$ times the height, h, of the parallelepiped. The area of the base of the parallelepiped is:

$$\|\vec{\mathbf{b}}\times\vec{\mathbf{c}}\| = \|\vec{\mathbf{b}}\|\,\|\vec{\mathbf{c}}\|\,\sin(\theta)$$

that is, the magnitude of the cross product (see Appendix A). The vector defined by the cross product is the *area vector* and is orthogonal to the plane described by vectors $\vec{\mathbf{b}}$ and $\vec{\mathbf{c}}$ (see Figure 3.12).

We now are in a position to show that our assertion concerning the volume of the parallelepiped is true:

$$\begin{aligned} V &= \vec{\mathbf{a}}\cdot(\vec{\mathbf{b}}\times\vec{\mathbf{c}}) \\ &= \vec{\mathbf{a}}\cdot\vec{\mathbf{d}} \\ &= \|\vec{\mathbf{a}}\|\|\vec{\mathbf{b}}\times\vec{\mathbf{c}}\|\cos(\alpha) \\ &= h\|\vec{\mathbf{b}}\times\vec{\mathbf{c}}\| \end{aligned}$$

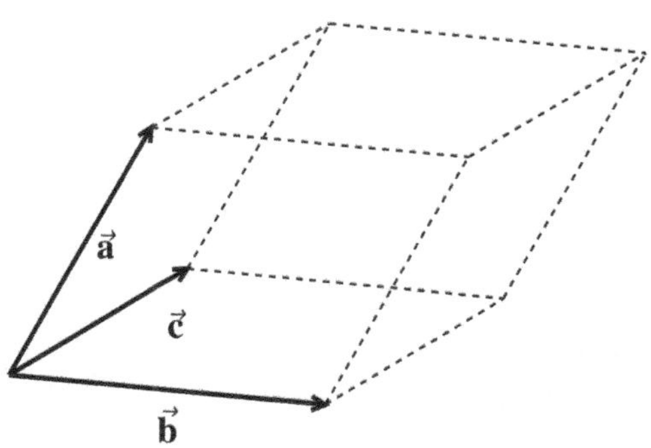

Figure 3.11 A parallelepiped defined by vectors $\vec{\mathbf{a}}$, $\vec{\mathbf{b}}$ and $\vec{\mathbf{c}}$.

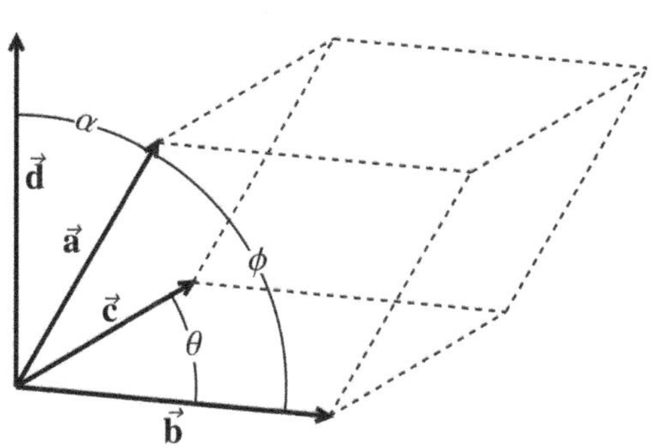

Figure 3.12 The area vector $\vec{\mathbf{d}} = \vec{\mathbf{b}}\times\vec{\mathbf{c}}$.

In other words, the height, h, of the parallelepiped times the area defined by the base (not basis!) vectors of the parallelepiped. We may represent this result in determinant form:

$$\vec{\mathbf{b}} \times \vec{\mathbf{c}} = \begin{vmatrix} \mathbf{u}_x & \mathbf{u}_y & \mathbf{u}_z \\ b_x & b_y & b_z \\ c_x & c_y & c_z \end{vmatrix}$$

$$\vec{\mathbf{a}} \cdot (\vec{\mathbf{b}} \times \vec{\mathbf{c}}) = \vec{\mathbf{a}} \cdot \vec{\mathbf{d}} = \vec{\mathbf{a}} \cdot \begin{vmatrix} \mathbf{u}_x & \mathbf{u}_y & \mathbf{u}_z \\ b_x & b_y & b_z \\ c_x & c_y & c_z \end{vmatrix}$$

$$= \begin{vmatrix} a_x & a_y & a_z \\ b_x & b_y & b_z \\ c_x & c_y & c_z \end{vmatrix}$$

The proof of this uses the matrix notation for the inner product operations in the next section.

We may also write the scalar triple product as:

$$(\vec{\mathbf{a}} \times \vec{\mathbf{b}}) \cdot \vec{\mathbf{c}}$$

which is equivalent to our first scalar triple product:

$$\vec{\mathbf{a}} \cdot (\vec{\mathbf{b}} \times \vec{\mathbf{c}}) = (\vec{\mathbf{a}} \times \vec{\mathbf{b}}) \cdot \vec{\mathbf{c}}$$

That this is so follows from our geometrical arguments. If we define the base of the parallelepiped using vectors $\vec{\mathbf{a}}$ and $\vec{\mathbf{b}}$ then, as above, $\vec{\mathbf{a}} \times \vec{\mathbf{b}}$ is the area of the base and then the height of the parallelepiped is given by $\vec{\mathbf{c}} \cdot \vec{\mathbf{d}}$.

What if the result of the scalar triple product is zero? This could be so if one or more of the vectors is the zero vector, however, this is trivial and of more interest is the situation in which all three vectors are coplanar and are therefore dependent vectors. Given our above geometrical arguments this would naturally be expected to give a zero volume for the parallelepiped.

$$\vec{\mathbf{a}} \cdot (\vec{\mathbf{b}} \times \vec{\mathbf{c}}) = \begin{vmatrix} a_x & a_y & a_z \\ b_x & b_y & b_z \\ c_x & c_y & c_z \end{vmatrix} = 0$$

Next, if in our expression $(\vec{\mathbf{a}}\times\vec{\mathbf{b}})\cdot\vec{\mathbf{c}}$ the vector $\vec{\mathbf{c}}$ is a unit vector with a norm of 1, then the expression corresponds to the area of the parallelepiped defined by $\vec{\mathbf{a}}\times\vec{\mathbf{b}}$:

$$
\begin{aligned}
(\vec{\mathbf{a}}\times\vec{\mathbf{b}})\cdot\vec{\mathbf{c}} &= \|\vec{\mathbf{a}}\times\vec{\mathbf{b}}\|\|\vec{\mathbf{c}}\|\cos(\theta)\\
&= \|\vec{\mathbf{a}}\times\vec{\mathbf{b}}\|\cos(\theta)\\
&= \|\vec{\mathbf{a}}\|\|\vec{\mathbf{b}}\|\sin(\theta)\cos(\theta)
\end{aligned}
$$

As in the appendix, $\vec{\mathbf{a}}\times\vec{\mathbf{b}}$ is the area of a parallelogram.

Since the scalar triple product corresponds to the calculation of the volume of a parallelpiped we might be tempted to think that the order of multiplication is not of importance. This is almost true; we can change the order of multiplication but only by shifting each vector left or right. Thus:

$$\vec{\mathbf{a}}\cdot(\vec{\mathbf{b}}\times\vec{\mathbf{c}})=\vec{\mathbf{c}}\cdot(\vec{\mathbf{a}}\times\vec{\mathbf{b}})=\vec{\mathbf{b}}\cdot(\vec{\mathbf{c}}\times\vec{\mathbf{a}}) \tag{3.21}$$

Finally, we can use the commutative property of vectors in real space to show that:

$$
\begin{aligned}
\vec{\mathbf{a}}\cdot(\vec{\mathbf{b}}\times\vec{\mathbf{c}}) &= (\vec{\mathbf{b}}\times\vec{\mathbf{c}})\cdot\vec{\mathbf{a}}\\
\vec{\mathbf{a}}\cdot\vec{\mathbf{d}} &= \vec{\mathbf{d}}\cdot\vec{\mathbf{a}}
\end{aligned}
$$

The *vector triple product* is:

$$\vec{\mathbf{a}}\times(\vec{\mathbf{b}}\times\vec{\mathbf{c}})=\vec{\mathbf{b}}\,(\vec{\mathbf{a}}\cdot\vec{\mathbf{c}})-\vec{\mathbf{c}}\,(\vec{\mathbf{a}}\cdot\vec{\mathbf{b}}) \tag{3.22}$$

The proof for this is straightforward, if somewhat tedious. Consider first the operation $\vec{\mathbf{b}}\times\vec{\mathbf{c}}$. Geometrically, the resultant vector, $\vec{\mathbf{d}}$, will be situated at right angles to the plane defined by $\vec{\mathbf{b}}$ and $\vec{\mathbf{c}}$ (see Figure 3.13).

If we add in vector $\vec{\mathbf{a}}$ to the picture then the plane defined by vectors $\vec{\mathbf{a}}$ and $\vec{\mathbf{d}}$ is orthogonal to the plane defined by vectors $\vec{\mathbf{b}}$ and $\vec{\mathbf{c}}$ by virtue of the fact that vector $\vec{\mathbf{d}}$ is orthogonal to the $\vec{\mathbf{b}}$, $\vec{\mathbf{c}}$-plane (see Figure 3.14).

Now, the operation $\vec{\mathbf{a}}\times\vec{\mathbf{d}}$ will produce a new vector, $\vec{\mathbf{e}}$, which will be orthogonal to the plane defined by vectors $\vec{\mathbf{a}}$ and $\vec{\mathbf{d}}$, which means that it *must* be in the same plane as vectors $\vec{\mathbf{b}}$ and $\vec{\mathbf{d}}$ (Figure 3.15). This being the case, it must be possible to express vector $\vec{\mathbf{e}}$ as a linear combination of vectors $\vec{\mathbf{b}}$ and $\vec{\mathbf{c}}$:

$$\vec{\mathbf{e}}=a\vec{\mathbf{b}}+b\vec{\mathbf{c}}$$

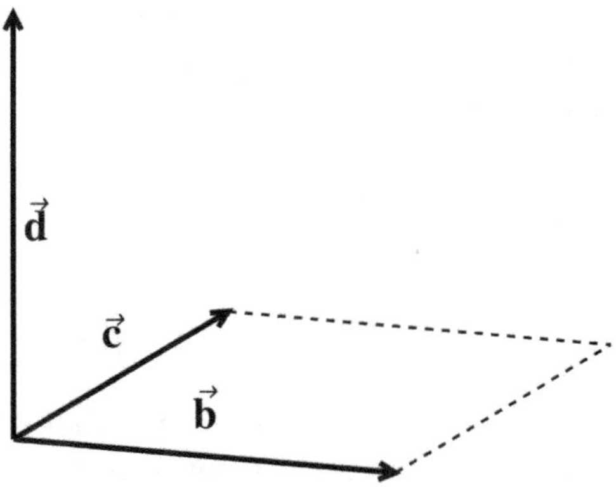

Figure 3.13 The cross product of vectors $\vec{b}$ and $\vec{c}$.

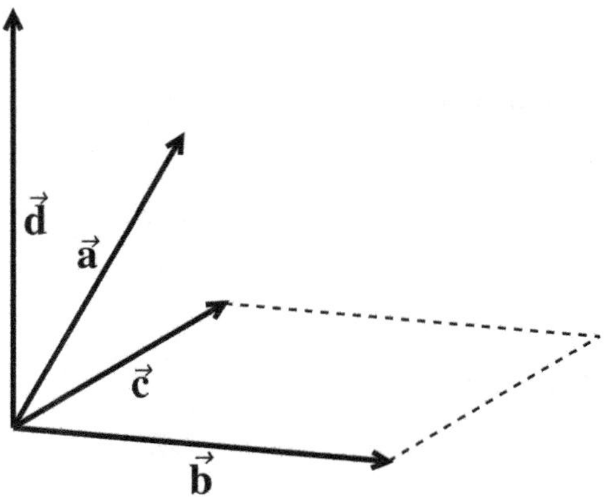

Figure 3.14 The cross product of vectors $\vec{b}$ and $\vec{c}$ with vector $\vec{a}$ added in.

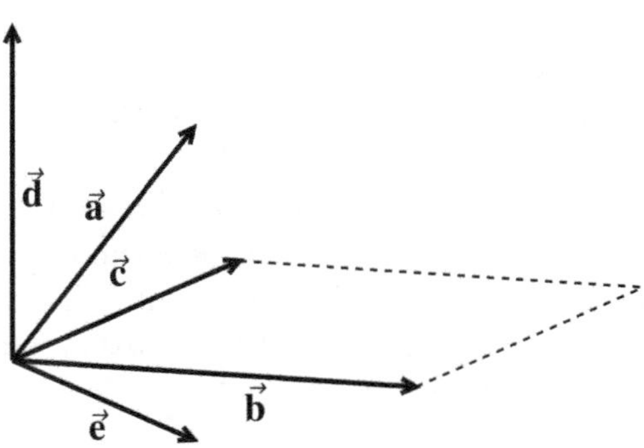

Figure 3.15 Vector $\vec{e}$ as a linear combination of $\vec{b}$ and $\vec{c}$.

Because, as we have seen, all vectors in a plane belong to the same vector space and the appropriate combination of $\vec{\mathbf{b}}$ and $\vec{\mathbf{c}}$ will produce $\vec{\mathbf{e}}$.

Now, again consider the operation $\vec{\mathbf{a}}\times\vec{\mathbf{d}}=\vec{\mathbf{e}}$ but now from the point of view of the determinant representation:

$$\vec{\mathbf{e}}=\begin{vmatrix} \mathbf{u}_x & \mathbf{u}_y & \mathbf{u}_z \\ a_x & a_y & a_z \\ d_x & d_y & d_z \end{vmatrix}$$

$$=(a_yd_z-a_zd_y)\mathbf{u}_x-(a_xd_z-a_zd_x)\mathbf{u}_y+(a_xd_y-a_zd_x)\mathbf{u}_z$$

$$=(a_yd_z-a_zd_y)\mathbf{u}_x+(a_zd_x-a_xd_z)\mathbf{u}_y+(a_xd_y-a_zd_x)\mathbf{u}_z$$

$$=e_x\mathbf{u}_x+e_y\mathbf{u}_y+e_z\mathbf{u}_z$$

Let us look closely at e_x:

$$e_x=a_yd_z-a_zd_y$$

The components d_y and d_z result from the first cross product, $\vec{\mathbf{b}}\times\vec{\mathbf{c}}$, and have the values:

$$d_y=b_zc_x-b_xc_z$$

$$d_z=b_xc_y-b_yc_x$$

Substituting these for d_y and d_z in our expression for e_x:

$$e_x=a_y(b_xc_y-b_yc_x)-a_z(b_zc_x-b_xc_z)$$

$$=a_yb_xc_y-a_yb_yc_x-a_zb_zc_x+a_zb_xc_z$$

Now, we add and subtract $a_xb_xc_x$:

$$e_x=a_yb_xc_y-a_yb_yc_x-a_zb_zc_x+a_zb_xc_z+a_xb_xc_x-a_xb_xc_x$$

and collect terms in b_x:

$$e_x=b_x(a_xc_x+a_yc_y+a_zc_z)-c_x(a_xb_x+a_yb_y+a_zb_z)$$

$$=b_x(\vec{\mathbf{a}}\cdot\vec{\mathbf{c}})-c_x(\vec{\mathbf{a}}\cdot\vec{\mathbf{b}})$$

Similarly:

$$e_y=b_y(\vec{\mathbf{a}}\cdot\vec{\mathbf{c}})-c_y(\vec{\mathbf{a}}\cdot\vec{\mathbf{b}})$$

$$e_z=b_z(\vec{\mathbf{a}}\cdot\vec{\mathbf{c}})-c_z(\vec{\mathbf{a}}\cdot\vec{\mathbf{b}})$$

and from our original linear combination of $\vec{\mathbf{b}}$ and $\vec{\mathbf{c}}$:

$$\begin{aligned}\vec{\mathbf{a}} \times (\vec{\mathbf{b}} \times \vec{\mathbf{c}}) &= \vec{\mathbf{e}} \\ &= a\vec{\mathbf{b}} + b\vec{\mathbf{c}} \\ &= \vec{\mathbf{b}}(\vec{\mathbf{a}} \cdot \vec{\mathbf{c}}) - \vec{\mathbf{c}}(\vec{\mathbf{a}} \cdot \vec{\mathbf{b}})\end{aligned}$$

as was originally asserted for the vector triple product.

3.4 Matrix Notation

We can use matrix notation to compactly represent our vectors. Rather than using:

$$\vec{\mathbf{a}} = 3\vec{\mathbf{u}}_x + 6\vec{\mathbf{u}}_y + \vec{\mathbf{u}}_z$$

we can simply use:

$$\vec{\mathbf{a}} = \begin{bmatrix} 3 \\ 6 \\ 1 \end{bmatrix} \tag{3.23}$$

The inner product of two vectors:

$$\begin{aligned}\vec{\mathbf{a}} \cdot \vec{\mathbf{b}} &= (3\vec{\mathbf{u}}_x + 6\vec{\mathbf{u}}_y + \vec{\mathbf{u}}_z) \cdot (\vec{\mathbf{u}}_x - 5\vec{\mathbf{u}}_y + 2\vec{\mathbf{u}}_z) \\ &= 3\vec{\mathbf{u}}_x \cdot \vec{\mathbf{u}}_x - 15\vec{\mathbf{u}}_x \cdot \vec{\mathbf{u}}_y + 6\vec{\mathbf{u}}_x \cdot \vec{\mathbf{u}}_z \\ &\quad - 30\vec{\mathbf{u}}_y \cdot \vec{\mathbf{u}}_y + 12\vec{\mathbf{u}}_y \cdot \vec{\mathbf{u}}_z + \vec{\mathbf{u}}_z \cdot \vec{\mathbf{u}}_x - 5\vec{\mathbf{u}}_z \cdot \vec{\mathbf{u}}_y + 2\vec{\mathbf{u}}_z \cdot \vec{\mathbf{u}}_z \\ &= 3 - 0 + 0 + 0 - 30 + 0 + 0 - 0 + 2 \\ &= -25\end{aligned}$$

is represented much more compactly in matrix notation by:

$$\vec{\mathbf{a}} \cdot \vec{\mathbf{b}} = \vec{\mathbf{a}}^T \vec{\mathbf{b}} = [3 \quad 6 \quad 1] \begin{bmatrix} 1 \\ -5 \\ 2 \end{bmatrix} = -25 \tag{3.24}$$

Here, the use of a superscript "T" indicates the transpose of the corresponding column vector, as was done in Chapter 2. Many texts leave this out but if we are thinking "matrices" when looking at vectors this notation serves as a reminder that this operation is necessary in order to do the inner product calculation. We will use the superscript "T" when emphasizing the matrix nature of the calculation, however, the reader should bear in mind that even when the superscript is absent the matrix calculation should be "lurking" in the background of her/his mind. Note that in order to accomplish the dot product multiplication the first is a single-row matrix and the second a single-column matrix and that transforming from a column matrix to a row matrix is the equivalent of taking the transpose of the column matrix. We will continue to use this notation throughout the text where it serves to remove any ambiguity, however, it is understood that the dot product notation implies the transpose of the first of the two vectors.

Orthogonality can be easily demonstrated from the dot product of two vectors at right angles to each other:

$$\begin{bmatrix} 0 & 6 & 0 \end{bmatrix} \begin{bmatrix} 1 \\ 0 \\ 0 \end{bmatrix} = 0$$

This is the matrix equivalent of writing:

$$6\vec{\mathbf{u}}_y \cdot \vec{\mathbf{u}}_x = 0$$

What happens if, instead of multiplying a row vector by a column vector, we reverse the order of multiplication? Let us give it a try:

$$\vec{\mathbf{a}}\vec{\mathbf{b}}^T = \begin{bmatrix} 3 \\ 6 \\ 1 \end{bmatrix} \begin{bmatrix} 1 & -5 & 2 \end{bmatrix} = \begin{bmatrix} 3 & -15 & 6 \\ 6 & -30 & 12 \\ 1 & -5 & 2 \end{bmatrix} \tag{3.25}$$

This is obviously *not* the scalar that we are used to by now but rather, a new matrix. This operation is called the *outer product* and is considered to be a special case of the Kronecker product (see eqn (2.8)). If we do the outer product of a $m \times 1$ vector and a $1 \times n$ vector we will obtain an $m \times n$ matrix. In the situation in which neither of the vectors is labelled as being transposed we use the "$\otimes$" operator here to clearly distinguish between the inner product and the outer product when which vector is the transposed one is not obvious:

$$\vec{\mathbf{a}}\vec{\mathbf{b}}^T = \vec{\mathbf{a}} \otimes \vec{\mathbf{b}} \tag{3.26}$$

As with our inner product, it is now understood from this notation that the *second* of the two vectors is transposed. For our familiar $\mathbb{R}^2$ and $\mathbb{R}^3$ spaces the outer product is the matrix product of a column matrix with 2 or three rows and a row matrix with 2 or

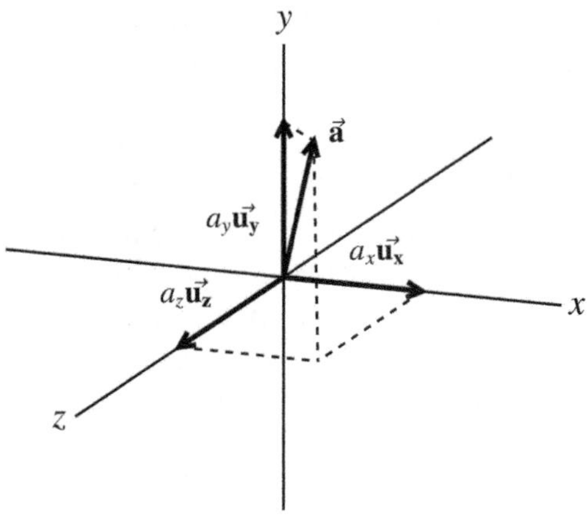

Figure 3.16 The components of a vector.

3 columns: The resulting matrix is a 2×2 or 3×3 matrix, which is not a normal two- or three-dimensional Euclidean vector.

The outer product will be of some considerable interest later, so we will investigate it a bit further. Consider a vector $\vec{\mathbf{a}}$ in $\mathbb{R}^3$, where $\vec{\mathbf{a}}$ is, as usual, composed of a linear combination of the basis vectors $\vec{\mathbf{u}}_x$, $\vec{\mathbf{u}}_y$ and $\vec{\mathbf{u}}_z$ (see Figure 3.16). Now, consider what we get when we do an inner product calculation using one of our basis vectors, $\vec{\mathbf{u}}_x$ for example, and $\vec{\mathbf{a}}$:

$$\begin{aligned}\vec{\mathbf{u}}_x^T\vec{\mathbf{a}} &= \vec{\mathbf{u}}_x \cdot \vec{\mathbf{a}} \\ &= \vec{\mathbf{u}}_x^T(a_x\vec{\mathbf{u}}_x + a_y\vec{\mathbf{u}}_y + a_z\vec{\mathbf{u}}_z) \\ &= a_x\vec{\mathbf{u}}_x^T\vec{\mathbf{u}}_x + a_y\vec{\mathbf{u}}_x^T\vec{\mathbf{u}}_y + a_z\vec{\mathbf{u}}_x^T\vec{\mathbf{u}}_z \\ &= a_x\vec{\mathbf{u}}_x \cdot \vec{\mathbf{u}}_x + a_y\vec{\mathbf{u}}_x \cdot \vec{\mathbf{u}}_y + a_z\vec{\mathbf{u}}_x \cdot \vec{\mathbf{u}}_z \\ &= a_x\end{aligned}$$

Similarly, for $\vec{\mathbf{u}}_y$ and $\vec{\mathbf{u}}_z$:

$$\begin{aligned}\vec{\mathbf{u}}_y^T\vec{\mathbf{a}} &= a_y \\ \vec{\mathbf{u}}_z^T\vec{\mathbf{a}} &= a_z\end{aligned}$$

Now, consider:

$$\vec{\mathbf{u}}_x\,(\vec{\mathbf{u}}_x^T\vec{\mathbf{a}}) = \vec{\mathbf{u}}_x\,(\vec{\mathbf{u}}_x \cdot \vec{\mathbf{a}}) = a_x\vec{\mathbf{u}}_x \tag{3.27}$$

With reference to Figure 3.16 you can see that what we have done is generate the *projection* of $\vec{\mathbf{a}}$ onto the x-axis and have generated a vector that is proportional to the unit vector along the x-axis.

If we define the outer products as operators, $\hat{\mathrm{P}}_i$:

$$\begin{aligned}\hat{\mathbf{P}}_x &\equiv \vec{\mathbf{u}}_x\vec{\mathbf{u}}_x^T = \vec{\mathbf{u}}_x \otimes \vec{\mathbf{u}}_x \\ \hat{\mathbf{P}}_y &\equiv \vec{\mathbf{u}}_y\vec{\mathbf{u}}_y^T = \vec{\mathbf{u}}_y \otimes \vec{\mathbf{u}}_y \\ \hat{\mathbf{P}}_z &\equiv \vec{\mathbf{u}}_z\vec{\mathbf{u}}_z^T = \vec{\mathbf{u}}_z \otimes \vec{\mathbf{u}}_z\end{aligned} \tag{3.28}$$

then eqn (3.27) becomes:

$$\hat{\mathbf{P}}_x\vec{\mathbf{a}} = a_x\vec{\mathbf{u}}_x \tag{3.29}$$

Similarly:

$$\begin{aligned}\hat{\mathbf{P}}_y\vec{\mathbf{a}} &= a_y\vec{\mathbf{u}}_y \\ \hat{\mathbf{P}}_y\vec{\mathbf{a}} &= a_y\vec{\mathbf{u}}_y\end{aligned} \tag{3.30}$$

$\hat{\mathbf{P}}_x$, $\hat{\mathbf{P}}_y$ and $\hat{\mathbf{P}}_z$ generate the projection of a vector onto the corresponding unit basis vector's axis and are called *projection operators*. Note that they are *not* written here as vectors because they are not vectors but rather matrix operators. Using our unit vectors in $\mathbb{R}^3$ these matrices are:

$$\hat{\mathbf{P}}_x = \vec{\mathbf{u}}_x\vec{\mathbf{u}}_x^T = \begin{bmatrix}1\\0\\0\end{bmatrix}\begin{bmatrix}1 & 0 & 0\end{bmatrix} = \begin{bmatrix}1 & 0 & 0\\0 & 0 & 0\\0 & 0 & 0\end{bmatrix} \quad \hat{\mathbf{P}}_y = \vec{\mathbf{u}}_y\vec{\mathbf{u}}_y^T = \begin{bmatrix}0\\1\\0\end{bmatrix}\begin{bmatrix}0 & 1 & 0\end{bmatrix} = \begin{bmatrix}0 & 0 & 0\\0 & 1 & 0\\0 & 0 & 0\end{bmatrix}$$

$$\hat{\mathbf{P}}_z = \vec{\mathbf{u}}_z\vec{\mathbf{u}}_z^T = \begin{bmatrix}0\\0\\1\end{bmatrix}\begin{bmatrix}0 & 0 & 1\end{bmatrix} = \begin{bmatrix}0 & 0 & 0\\0 & 0 & 0\\0 & 0 & 1\end{bmatrix}$$

Notice that:

$$\hat{\mathbf{P}}_x + \hat{\mathbf{P}}_y + \hat{\mathbf{P}}_z = \hat{\mathbf{1}} \tag{3.31}$$

This is generally the case with projection operators constructed from orthonormal basis sets and is referred to as the *completeness theorem* (which allows us to determine whether or not we have a complete set of basis vectors ... hence the name). We will encounter this property later. In fact, you could think of eqn (3.31) as saying that all of the possible projection operators for a particular space put together project a vector onto itself:

$$(\hat{\mathbf{P}}_x + \hat{\mathbf{P}}_y + \hat{\mathbf{P}}_z)\vec{\mathbf{a}} = \hat{\mathbf{1}}\vec{\mathbf{a}}$$

From our visual picture of the projection operator and what it does it should be no surprise that the projection of a unit basis vector onto itself is:

$$\hat{\mathbf{P}}_i\,\vec{\mathbf{u}}_i = \vec{\mathbf{u}}_i$$

With this in mind and using eqn (3.29):

$$\begin{aligned}\hat{\mathbf{P}}_i^2\vec{\mathbf{a}} &= \hat{\mathbf{P}}_i(\hat{\mathbf{P}}_i\vec{\mathbf{a}}) \\ &= \hat{\mathbf{P}}_i(a_i\vec{\mathbf{u}}_i) \\ &= a_i\hat{\mathbf{P}}_i\vec{\mathbf{u}}_i \\ &= a_i\vec{\mathbf{u}}_i\end{aligned}$$

from which we infer that:

$$\hat{\mathbf{P}}_i^2 = \hat{\mathbf{P}}_i$$

and we say that $\hat{\mathbf{P}}_i$ is *idempotent*.

Again, from our visual picture of what the projection operator does, we can predict that $\hat{\mathbf{P}}_i\hat{\mathbf{P}}_j\vec{\mathbf{a}}$ (where i and j refer to orthogonal basis vectors) should be zero:

$$\begin{aligned}\hat{\mathbf{P}}_i\hat{\mathbf{P}}_j\vec{\mathbf{a}} &= \vec{\mathbf{u}}_i\vec{\mathbf{u}}_i^{\,T}\vec{\mathbf{u}}_j\vec{\mathbf{u}}_j^{\,T}\vec{\mathbf{a}} \\ &= \vec{\mathbf{u}}_i(\vec{\mathbf{u}}_i^{\,T}\vec{\mathbf{u}}_j)\vec{\mathbf{u}}_j^{\,T}\vec{\mathbf{a}} \\ &= \vec{\mathbf{u}}_i(0)\vec{\mathbf{u}}_j^{\,T}\vec{\mathbf{a}} \\ &= 0\end{aligned}$$

We need not restrict ourselves to using basis set vectors to define projection operators. We can be completely general and say that:

$$\hat{\mathbf{P}}_b = \vec{\mathbf{b}}\vec{\mathbf{b}}^T \tag{3.32}$$

and this, when operating on a vector $\vec{\mathbf{a}}$ will constitute a projection of vector $\vec{\mathbf{a}}$ onto vector $\vec{\mathbf{b}}$:

$$\vec{\mathbf{b}}^T\vec{\mathbf{a}} = \vec{\mathbf{b}} \cdot \vec{\mathbf{a}} = k$$

$$\hat{\mathbf{P}}_b\vec{\mathbf{a}} = \vec{\mathbf{b}}(\vec{\mathbf{b}}^T\vec{\mathbf{a}})$$

$$= k\vec{\mathbf{b}}$$

The projection of $\vec{\mathbf{a}}$ onto $\vec{\mathbf{b}}$ is a scalar times $\vec{\mathbf{b}}$ and is obviously parallel to $\vec{\mathbf{b}}$. What is the value of k and how can we use it? Consider vectors $\vec{\mathbf{v}}$ and $\vec{\mathbf{w}}$ (see Figure 3.17). Vector $\vec{\mathbf{p}}$ is the projection of $\vec{\mathbf{v}}$ on $\vec{\mathbf{w}}$ and $\vec{\mathbf{q}} = \vec{\mathbf{v}} - \vec{\mathbf{p}}$ is at right angles to $\vec{\mathbf{w}}$ and $\vec{\mathbf{p}}$ (which of course means that they are orthogonal). Vector $\vec{\mathbf{p}}$ is also proportional to $\vec{\mathbf{w}}$.

$$\vec{\mathbf{p}} = k\vec{\mathbf{w}}$$

Since $\vec{\mathbf{q}}$ is orthogonal to $\vec{\mathbf{w}}$ we can write:

$$\vec{\mathbf{w}} \cdot \vec{\mathbf{q}} = 0$$

$$= \vec{\mathbf{w}} \cdot (\vec{\mathbf{v}} - \vec{\mathbf{p}})$$

$$= \vec{\mathbf{w}} \cdot \vec{\mathbf{v}} - \vec{\mathbf{w}} \cdot \vec{\mathbf{p}}$$

$$= \vec{\mathbf{w}} \cdot \vec{\mathbf{v}} - k\vec{\mathbf{w}} \cdot \vec{\mathbf{w}}$$

Adding $k\vec{\mathbf{w}} \cdot \vec{\mathbf{w}}$ to each side:

$$k\vec{\mathbf{w}} \cdot \vec{\mathbf{w}} = \vec{\mathbf{w}} \cdot \vec{\mathrm{v}}$$

and:

$$k = \frac{\vec{\mathbf{w}} \cdot \vec{\mathbf{v}}}{\vec{\mathbf{w}} \cdot \vec{\mathbf{w}}}$$

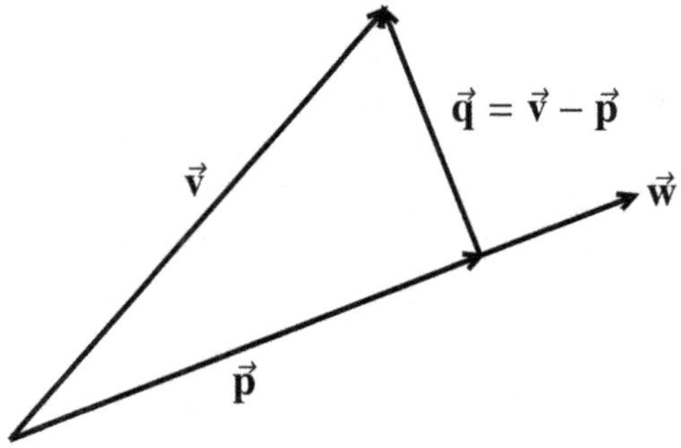

Figure 3.17 The projection of $\vec{\mathrm{u}}$ on $\vec{\mathrm{v}}$.

Therefore:

$$\vec{\mathbf{p}} = k\vec{\mathbf{w}} = \frac{\vec{\mathbf{w}} \cdot \vec{\mathbf{v}}}{\vec{\mathbf{w}} \cdot \vec{\mathbf{w}}}\vec{\mathbf{w}} \tag{3.33}$$

Equation (3.33) gives us a tool for finding an orthogonal basis set *via* the Gram–Schmidt orthogonalisation procedure. For example, a two-dimensional vector set in $\mathbb{R}^2$ is, in matrix notation:

$$\vec{\mathbf{e}}_1 = \begin{bmatrix} 1 \\ 1 \end{bmatrix}, \vec{\mathbf{e}}_2 = \begin{bmatrix} 1 \\ 0 \end{bmatrix}$$

To test whether or not this is a basis set it is necessary to demonstrate that they span $\mathbb{R}^2$ and that they are linearly independent. To demonstrate the first property we imagine a vector, $\vec{\mathbf{a}}$, in $\mathbb{R}^2$ that is a linear combination of the two:

$$\vec{\mathbf{a}} = \alpha_1\vec{\mathbf{e}}_1 + \alpha_2\vec{\mathbf{e}}_2$$

or

$$\begin{bmatrix} a_1 \\ a_2 \end{bmatrix} = \alpha_1\begin{bmatrix} 1 \\ 1 \end{bmatrix} + \alpha_2\begin{bmatrix} 1 \\ 0 \end{bmatrix} = \begin{bmatrix} \alpha_1 + \alpha_2 \\ \alpha_1 + 0 \end{bmatrix}$$

When we solve for the *a*s we get:

$$\alpha_1 = a_2$$

$$\alpha_2 = a_2 - a_1$$

That these vectors are also linearly independent is shown by:

$$\beta_1\vec{\mathbf{e}}_1 + \beta_2\vec{\mathbf{e}}_2 = \vec{\mathbf{0}}$$

or:

$$\beta_1\begin{bmatrix} 1 \\ 1 \end{bmatrix} + \beta_2\begin{bmatrix} 1 \\ 0 \end{bmatrix} = \vec{\mathbf{0}}$$

which can *only* be so if β_1 and β_2 are both equal to zero. Hence, vectors $\vec{\mathbf{e}}_1$ and $\vec{\mathbf{e}}_2$ are linearly independent and form a spanning set and therefore form a basis set. They are *not* however, orthogonal. The test for this is, of course, the inner product, which as we have seen must be equal to zero for orthogonal vectors:

$$\vec{\mathbf{e}}_1 \cdot \vec{\mathbf{e}}_2 = \vec{\mathbf{e}}_1^{\,T}\vec{\mathbf{e}}_2 = [1 \quad 1]\begin{bmatrix} 1 \\ 0 \end{bmatrix} = 1$$

We can, however, construct an orthogonal basis set using the Gram–Schmidt orthogonalisation method. We first normalise $\vec{\mathbf{e}}_1$:

$$\vec{\mathbf{e}}_1 \cdot \vec{\mathbf{e}}_1 = \|\vec{\mathbf{e}}_1\|^2 = 2$$

$$\vec{\mathbf{e}}_1' = \frac{\vec{\mathbf{e}}_1}{\|\vec{\mathbf{e}}_1\|} = \frac{\vec{\mathbf{e}}_1}{\sqrt{2}}$$

so that:

$$\vec{\mathbf{e}}_1' \cdot \vec{\mathbf{e}}_1' = 1$$

Now, since we are looking for an orthogonal basis set we project $\vec{\mathbf{e}}_2$ onto $\vec{\mathbf{e}}_1'$ using eqn (3.33):

$$\vec{\mathbf{p}} = \frac{\vec{\mathbf{e}}_1' \cdot \vec{\mathbf{e}}_2}{\vec{\mathbf{e}}_1' \cdot \vec{\mathbf{e}}_1'} \vec{\mathbf{e}}_1'$$

$$\vec{\mathbf{e}}_1' = \frac{1}{\sqrt{2}} \begin{bmatrix} 1 \\ 1 \end{bmatrix}$$

$$\begin{aligned} \vec{\mathbf{e}}_1' \cdot \vec{\mathbf{e}}_2 &= \frac{1}{\sqrt{2}} [1 \quad 1] \begin{bmatrix} 1 \\ 0 \end{bmatrix} \\ &= \frac{1}{\sqrt{2}} \end{aligned}$$

and:

$$\vec{\mathbf{p}} = \frac{1}{\sqrt{2}} \vec{\mathbf{e}}_1'$$

Obviously, $\vec{\mathbf{p}}$ is not orthogonal to $\vec{\mathbf{e}}_1'$ but rather, is parallel. $\vec{\mathbf{e}}_2 - \vec{\mathbf{p}}$, however, *is* orthogonal to $\vec{\mathbf{e}}_1'$:

$$\begin{aligned} \vec{\mathbf{e}}_1' \cdot (\vec{\mathbf{e}}_2 - \vec{\mathbf{p}}) &= \vec{\mathbf{e}}_1' \cdot \vec{\mathbf{e}}_2 - \vec{\mathbf{e}}_1' \cdot \vec{\mathbf{p}} \\ &= \vec{\mathbf{e}}_1' \cdot \vec{\mathbf{e}}_2 - \frac{1}{\sqrt{2}} \vec{\mathbf{e}}_1' \cdot \vec{\mathbf{e}}_1' \\ &= \frac{1}{\sqrt{2}} - \frac{1}{\sqrt{2}} \\ &= 0 \end{aligned}$$

So, our second orthogonal basis vector is:

$$\begin{aligned}\vec{\mathbf{e}}_2^0 &= \vec{\mathbf{e}}_2 - \vec{\mathbf{p}} \\ &= \vec{\mathbf{e}}_2 - \frac{\vec{\mathbf{e}}_1' \cdot \vec{\mathbf{e}}_2}{\vec{\mathbf{e}}_1' \cdot \vec{\mathbf{e}}_1'} \vec{\mathbf{e}}_1' \\ &= \vec{\mathbf{e}}_2 - \frac{1}{\sqrt{2}} \vec{\mathbf{e}}_1' \\ &= \vec{\mathbf{e}}_2 - \frac{1}{2} \vec{\mathbf{e}}_1 \\ &= \begin{bmatrix} 1 \\ 0 \end{bmatrix} - \frac{1}{2} \begin{bmatrix} 1 \\ 1 \end{bmatrix} = \begin{bmatrix} \frac{1}{2} \\ \frac{-1}{2} \end{bmatrix}\end{aligned}$$

where $\vec{\mathbf{e}}_2^0$ is orthogonal to $\vec{\mathbf{e}}_1$ but *not* yet normalised. We can easily normalise $\vec{\mathbf{e}}_2^0$:

$$\begin{aligned}\vec{\mathbf{e}}_2' &= \frac{\vec{\mathbf{e}}_2^0}{\|\vec{\mathbf{e}}_2^0\|} \\ &= \sqrt{2} \vec{\mathbf{e}}_2^0 \\ &= \sqrt{2} \begin{bmatrix} \frac{1}{\sqrt{2}} \\ \frac{-1}{\sqrt{2}} \end{bmatrix}\end{aligned}$$

and our (now) orthonormal basis set is:

$$\vec{\mathbf{e}}_1' = \begin{bmatrix} \frac{1}{\sqrt{2}} \\ \frac{1}{\sqrt{2}} \end{bmatrix} \qquad \vec{\mathbf{e}}_2' = \begin{bmatrix} \frac{1}{\sqrt{2}} \\ \frac{-1}{\sqrt{2}} \end{bmatrix}$$

You can verify for yourself that they are indeed now both orthogonal and normalised.

Thus, we can always construct an orthogonal basis set from a non-orthogonal one *via* projection operations.

3.5 Vector Transformations

A given vector $\vec{\mathbf{a}}$ can be transformed to a new vector $\vec{\mathbf{b}}$ with the application of an *operator*:

$$\vec{\mathbf{b}} = \hat{\mathbf{O}} \vec{\mathbf{a}}$$

We have already seen this in the case of the projection operator. Generally, the operator $\hat{\mathbf{O}}$ can be represented by an *nxn* matrix, "*n*" being the dimension of the vector space. Again, we have seen this in the case of the projection operator above.

There are two things an operation on a vector can do.[§] The length of the vector can be changed, accomplished by simple multiplication by a scalar and the direction of the vector can be changed by rotating it. These two fundamental operations can be done simultaneously; the projection operator does this by creating a new vector that is parallel to another vector (a rotation) and that generally has a different length from the starting vector.

An important (for us) operation that can be performed on a vector is a rotation. We apply the operator to vector $\vec{\mathbf{a}}$ to produce vector $\vec{\mathbf{b}}$ that is rotated in the vector space by the specified angle. This, of course requires a Euclidean space since angles and vector lengths are not defined in our more basic definition of vectors. Let us look at a simple example in two-dimensional $\mathbb{R}^2$ space. We start with a vector $\vec{\mathbf{a}}$ that is for simplicity, along the *y*-axis (Figure 3.18).

$$\vec{\mathbf{a}} = a_y \vec{\mathbf{u}}_y$$

We now apply a rotation in the form of an operator, $\hat{\mathbf{R}}(90)$, so that the new vector $\vec{\mathbf{b}}$ (Figure 3.19) lies along the *x*-axis:

$$\vec{\mathbf{b}} = \hat{\mathbf{R}}(90)\vec{\mathbf{a}} \tag{3.34}$$

The new vector $\vec{\mathbf{b}}$ is of the form:

$$\vec{\mathbf{b}} = a_y \vec{\mathbf{u}}_x$$

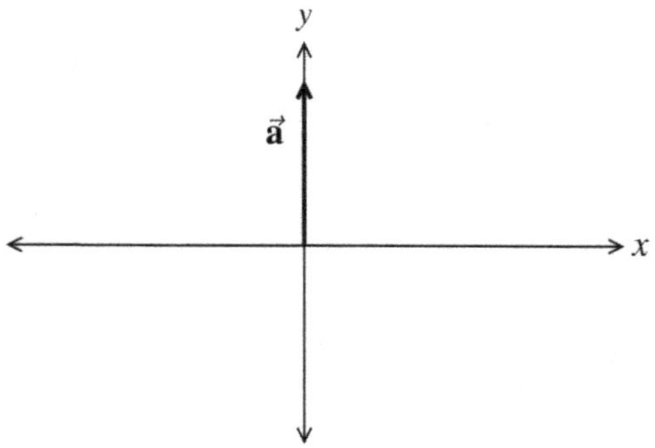

Figure 3.18 Vector $\vec{\mathbf{a}}$ before transformation.

[§]Three things, actually. The third is that the vector may be moved within its coordinate system. However, this is rarely done and certainly we will not be doing it here in this text.

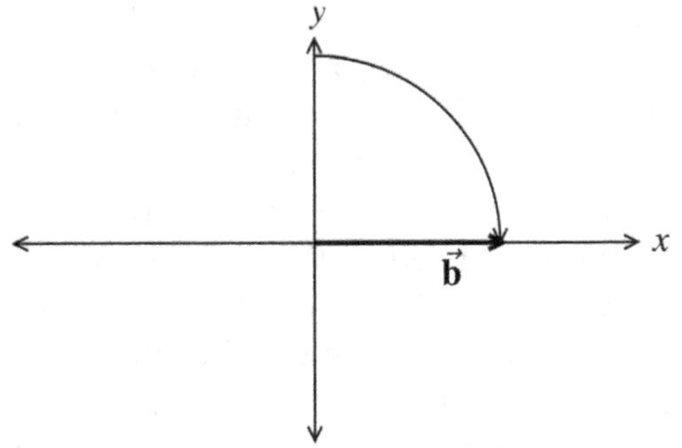

Figure 3.19 Vector $\vec{\mathbf{a}}$ after transformation to $\vec{\mathbf{b}}$.

Note that the new vector, $\vec{\mathbf{b}}$, must have the same length (or norm) as the old vector, $\vec{\mathbf{a}}$, as we would expect from a simple rotation of a vector. This is a very important property of vectors and can be viewed as one of the defining properties of vectors. This is referred to as *invariance of the norm* and we will explore this in more detail later.

What will the matrix form of the rotation operator look like? The operator must be such that it converts the matrix representation of $\vec{\mathbf{a}}$ to the matrix representation of $\vec{\mathbf{b}}$. A rotation matrix that does this is:

$$\hat{\mathbf{R}}(90) = \begin{bmatrix} 0 & 1 \\ -1 & 0 \end{bmatrix} \tag{3.35}$$

We can now write eqn (3.34) in matrix form:

$$\begin{bmatrix} a_y \\ 0 \end{bmatrix} = \begin{bmatrix} 0 & 1 \\ -1 & 0 \end{bmatrix} \begin{bmatrix} 0 \\ a_y \end{bmatrix}$$

$\hat{\mathbf{R}}(-90)$ will, of course, effect a rotation by 90° in the opposite direction and its matrix will be:

$$\hat{\mathbf{R}}(-90) = \begin{bmatrix} 0 & -1 \\ 1 & 0 \end{bmatrix} \tag{3.36}$$

which is simply the transpose of $\hat{\mathbf{R}}(90)$. Thus, the application of $\hat{\mathbf{R}}(90)$ followed by $\hat{\mathbf{R}}(-90)$ will bring a vector back to its original place. In other words, sequential application of these two operators is equivalent to using the identity operator:

$$\hat{\mathbf{R}}(90)\hat{\mathbf{R}}(-90) = \hat{\mathbf{1}}$$

Evidently, $\hat{\mathbf{R}}(90)$ and $\hat{\mathbf{R}}(-90)$ are *inverses* of each other (see eqn (2.26)).

A more general clockwise rotation operator in $\mathbb{R}^2$ space is:

$$\hat{\mathbf{R}}(\varphi) = \begin{bmatrix} \cos(\varphi) & \sin(\varphi) \\ -\sin(\varphi) & \cos(\varphi) \end{bmatrix} \tag{3.37}$$

We can show this by a geometric argument. We see from Figure 3.20 that:

$$A = a_x \cos(\varphi)$$

$$B = a_y \sin(\varphi)$$

$$b_x = A + B$$

and thus:

$$b_x = a_x \cos(\varphi) + a_y \sin(\varphi)$$

Also:

$$C = C_1 + C_2 = a_y \cos(\varphi)$$

$$D = a_x \sin(\varphi)$$

$$b_y = C - D$$

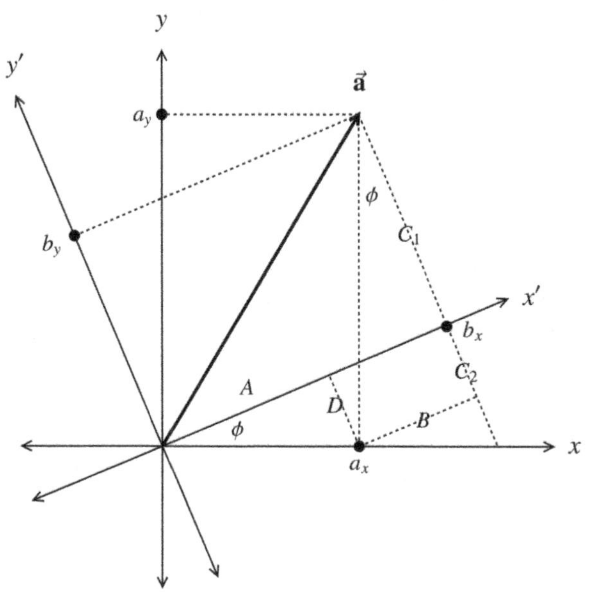

Figure 3.20 Rotation of coordinate axes x and y into x' and y'.

and:

$$b_y = a_y \cos(\varphi) + a_x \sin(\varphi)$$

Putting them together:

$$\begin{gathered} b_x = a_x \cos(\varphi) + a_y \sin(\varphi) \\ b_y = a_y \cos(\varphi) + a_x \sin(\varphi) \\ \text{or} \\ \begin{bmatrix} b_x \\ b_y \end{bmatrix} = \begin{bmatrix} \cos(\varphi) & \sin(\varphi) \\ -\sin(\varphi) & \cos(\varphi) \end{bmatrix} \begin{bmatrix} a_x \\ a_y \end{bmatrix} \end{gathered} \tag{3.38}$$

Using this we can show that the norm of a vector is invariant to rotation. Let the un-rotated vector be $\vec{\mathbf{a}}$ and the rotated vector be $\vec{\mathbf{b}}$ with components a_x, a_y and b_x, b_y, respectively. The square of the length of $\vec{\mathbf{a}}$ is:

$$\|\vec{\mathbf{a}}\|^2 = l^2 = a_x^2 + a_y^2$$

Using eqn (3.38):

$$\begin{aligned} \|\vec{\mathbf{b}}\|^2 &= b_x^2 + b_y^2 \\ &= a_x^2 \cos^2(\varphi) + a_x a_y \sin(\varphi)\cos(\varphi) + a_y^2 \sin^2(\varphi) \\ &\quad + a_x^2 \sin^2(\varphi) - a_x a_y \sin(\varphi)\cos(\varphi) + a_y^2 \cos^2(\varphi) \\ &= a_x^2(\cos^2(\varphi) + \sin^2(\varphi)) + a_y^2(\cos^2(\varphi) + \sin^2(\varphi)) \\ &= a_x^2 + a_y^2 \end{aligned} \tag{3.39}$$

This is a most fundamental property of vectors and as we have indicated above, is part of their definition: the norm of a vector is *invariant* to rotation. Note also that, in essence, when calculating the norm we have converted a vector into a scalar. This will be a useful concept to keep in mind when we tackle the potentially confusing concept of tensors in Chapter 5.

Rotation in the opposite direction, counter-clockwise, would involve the matrix operator:

$$\hat{\mathbf{R}}(-\varphi) = \begin{bmatrix} \cos(\theta) & -\sin(\varphi) \\ \sin(\varphi) & \cos(\varphi) \end{bmatrix}$$

A moment's reflection will reveal that this is simply the transpose of the matrix operator for the clockwise vector rotation (eqn (3.38)). It is generally the case that the transpose of the rotation operator results in an equal and opposite rotation. Of course, this requires that the rotation operator matrix and its transpose also be the inverse of each other:

$$\begin{aligned}\hat{\mathbf{R}}(\varphi)\hat{\mathbf{R}}(-\varphi) &= \hat{\mathbf{1}} \\ \hat{\mathbf{R}}(-\varphi) &= \hat{\mathbf{R}}^{T}(\varphi) = \hat{\mathbf{R}}^{-1}(\varphi)\end{aligned} \tag{3.40}$$

From Chapter 2, eqn (2.55) we can see that the vector rotation operator is orthogonal.

It should be noted that one can view these operations as either rotations of vectors *or* as rotations of the coordinate system (in the opposite direction). It is often the case that the latter is the point of view taken and is referred to as a *change of basis*. We will examine this again shortly.

A slightly different way to write the rotation is to use the cosines of angles between the new coordinate axes with the old axes.¶ Thus, we would write for the transformation from vector $\vec{\mathbf{a}}$ to $\vec{\mathbf{b}}$:

$$\vec{\mathbf{a}} = a_x\vec{\mathbf{u}}_x + a_y\vec{\mathbf{u}}_y + a_z\vec{\mathbf{u}}_z$$

$$\vec{\mathbf{b}} = b_x\vec{\mathbf{u}}_x + b_y\vec{\mathbf{u}}_y + b_z\vec{\mathbf{u}}_z$$

$$b_x = \cos(\vec{\mathbf{u}}'_x, \vec{\mathbf{u}}_x)a_x + \cos(\vec{\mathbf{u}}'_x, \vec{\mathbf{u}}_y)a_y + \cos(\vec{\mathbf{u}}'_x, \vec{\mathbf{u}}_z)a_z$$

where $\cos(\vec{\mathbf{u}}'_i, \vec{\mathbf{u}}_i)$ is the cosine of the angle between the new axis unit vectors $\vec{\mathbf{u}}'_i$ and the old axis unit vectors $\vec{\mathbf{u}}_i$ and is referred to as the *direction cosine* (see Appendix A). Similarly:

$$b_y = \cos(\vec{\mathbf{u}}'_y, \vec{\mathbf{u}}_x)a_x + \cos(\vec{\mathbf{u}}'_y, \vec{\mathbf{u}}_y)a_y + \cos(\vec{\mathbf{u}}'_z, \vec{\mathbf{u}}_z)a_z$$

$$b_z = \cos(\vec{\mathbf{u}}'_z, \vec{\mathbf{u}}_x)a_x + \cos(\vec{\mathbf{u}}'_z, \vec{\mathbf{u}}_y)a_y + \cos(\vec{\mathbf{u}}'_z, \vec{\mathbf{u}}_z)a_z$$

or more succinctly:

$$b_i = \sum_{j}^{3} \cos(\vec{\mathbf{u}}'_i, \vec{\mathbf{u}}_j)a_j \tag{3.41}$$

¶Or perhaps more accurately, the angle between the unit vectors along the old and new coordinate axes.

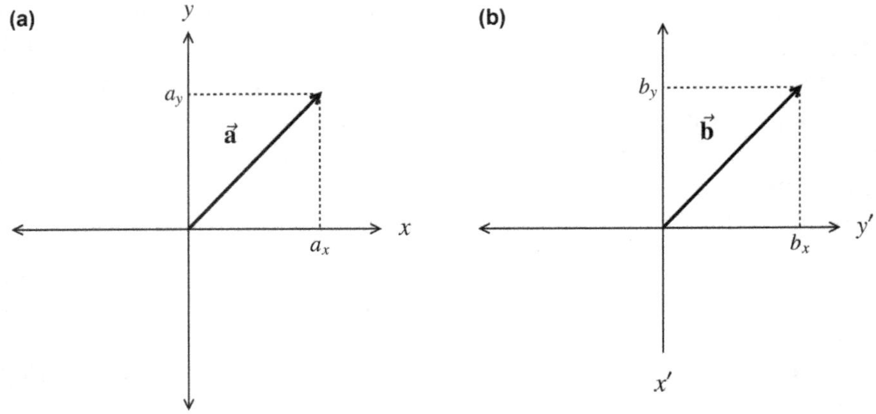

Figure 3.21 (a) Before 90° clockwise rotation of the axes; (b) after 90° clockwise rotation of the axes.

Note that the subscripts x, y and z have been replaced with numbers, which will be more convenient in later chapters. From this point we will use numbers or letters, as is convenient.

This can be understood easily by using a two-dimensional example. Let us suppose that we have a vector $\vec{\mathbf{a}}$ in coordinate system A and that we transform the basis set such that the axes rotate by 90° in the clockwise direction (see Figure 3.21).

The vector of course remains the same vector;[||] only the coordinate axes have changed. We will have more to say about this when we discuss tensors. Our transformation equation indicates that:

$$b_1 = \cos(\vec{\mathbf{u}}'_1, \vec{\mathbf{u}}_1)a_1 + \cos(\vec{\mathbf{u}}'_1, \vec{\mathbf{u}}_2)a_2$$

If the new coordinate axes are superimposed on top of the old ones the angle between the new x-axis and the old x-axis is 90 degrees and its cosine value is zero. The angle between the new x-axis and the old y-axis is 180° with a cosine value of -1:

$$b_1 = 0 \cdot a_1 - 1 \cdot a_2$$

Similarly:

$$b_2 = 1 \cdot a_1 - 0 \cdot a_2$$

[||]A vector is a vector is a vector, regardless of changes in the coordinate system. Silly as this might sound it is really part of the fundamental definition of vectors and, as we will see later, tensors in general.

Note that we can express this as a matrix equation:

$$\begin{bmatrix} b_1 \\ b_2 \end{bmatrix} = \begin{bmatrix} 0 & -1 \\ 1 & 0 \end{bmatrix} \begin{bmatrix} a_1 \\ a_2 \end{bmatrix}$$

where the rotation matrix is identical to eqn (3.36) in which we rotated the vector by $-90°$. Of course, a counter-clockwise rotation of a vector is the equivalent of a clockwise rotation of the axes.

The general two-dimensional rotation matrix, eqn (3.38), consists of sines and cosines. How does this fit with what we have just said about direction cosines? In other words, using direction cosines the transformation from vector $\vec{\mathbf{a}}$ to $\vec{\mathbf{b}}$ as above is:

$$\begin{bmatrix} b_1 \\ b_2 \end{bmatrix} = \begin{bmatrix} \cos(\vec{\mathbf{u}}'_1, \vec{\mathbf{u}}_1) & \cos(\vec{\mathbf{u}}'_1, \vec{\mathbf{u}}_2) \\ \cos(\vec{\mathbf{u}}'_2, \vec{\mathbf{u}}_1) & \cos(\vec{\mathbf{u}}'_2, \vec{\mathbf{u}}_2) \end{bmatrix} \begin{bmatrix} a_1 \\ a_2 \end{bmatrix} \tag{3.42}$$

but using eqn (3.38) it is:

$$\begin{bmatrix} b_1 \\ b_2 \end{bmatrix} = \begin{bmatrix} \cos(\varphi) & \sin(\varphi) \\ -\sin(\varphi) & \cos(\varphi) \end{bmatrix} \begin{bmatrix} a_1 \\ a_2 \end{bmatrix}$$

Using eqn (A.9) and (A.10) from Appendix A:

$$\begin{bmatrix} b_1 \\ b_2 \end{bmatrix} = \begin{bmatrix} \cos(\varphi) & \cos\left(\frac{\pi}{2} - \varphi\right) \\ \cos\left(\frac{\pi}{2} + \varphi\right) & \cos(\varphi) \end{bmatrix} \begin{bmatrix} a_1 \\ a_2 \end{bmatrix}$$

$$= \begin{bmatrix} \cos(\varphi) & \sin(\varphi) \\ -\sin(\varphi) & \cos(\varphi) \end{bmatrix} \begin{bmatrix} a_1 \\ a_2 \end{bmatrix}$$

We see that in the case of the two-dimensional rotation the angle between $\vec{\mathbf{u}}'_1$ and $\vec{\mathbf{u}}_1$ is the same as that between $\vec{\mathbf{u}}'_2$ and $\vec{\mathbf{u}}_2$. This would not necessarily be the case for a three-dimensional rotation.

For convenience, let us replace the cosine function in eqn (3.41) with γs:

$$b_i = \sum_j^3 \gamma_{ij} a_j$$

and to shorten this even further we can use Einstein notation:**

$$b_i = \gamma_{ij} a_j \tag{3.43}$$

**This is sometimes referred to as indicial notation.

In this notation where there are two (or more) occurrences of an index in a term that is then the summation index. A short set of rules for the application of this notation is:

1. In any term a letter index may occur no more than twice. In eqn (3.43) this index is j.
2. When an index appears twice in a term that index is taken as the summation index. Thus, in eqn (3.43) the summation index is j and the summation is $b_i = \gamma_{i1}a_1 + \gamma_{i2}a_2 + \gamma_{i3}a_3$.
3. Unrepeated indices are called "free indices" and may take any value in the range of 1 to N, where N is the dimension of the space. Each term in an equation must have the *same* free indices. Equation uses i as the free index and can be equal to 1, 2 or 3 (for a three-dimensional space).

We will find this notation convenient in our discussion of tensors.

There is one final thing to note about the transformation matrix that will be of use later. We may differentiate eqn (3.43):

$$\frac{\partial b_i}{\partial a_j} = \gamma_{ij} = \cos(\vec{\mathbf{u}}_i, \vec{\mathbf{u}}_j) \tag{3.44}$$

and using this, the transformation eqn (3.42) may be written as:

$$\begin{bmatrix} b_1 \\ b_2 \end{bmatrix} = \begin{bmatrix} \gamma_{11} & \gamma_{12} \\ \gamma_{21} & \gamma_{22} \end{bmatrix} \begin{bmatrix} a_1 \\ a_2 \end{bmatrix}$$

From the discussion of matrices in Chapter 2 there is another type of operator that we can consider; the eigenvalue operator. That is, the operator that when applied to a vector produces a new version of the same vector scaled by some value called the eigenvalue. The vector is called the *eigenvector*.

$$\hat{\mathbf{O}}\vec{\mathbf{a}} = \lambda\vec{\mathbf{a}} \tag{3.45}$$

Everything that was discussed about eigenvalue equations with respect to matrices applies here to vectors. The astute reader will by now be beginning to suspect that vectors and matrices are connected in many subtle and useful ways. This is so ... it is very useful to think of vectors in their matrix forms when studying their properties.

In our use of operators with vectors we assume linearity. That is:

$$\hat{\mathbf{O}}(\vec{\mathbf{a}}+\vec{\mathbf{b}})=\hat{\mathbf{O}}\vec{\mathbf{a}}+\hat{\mathbf{O}}\vec{\mathbf{b}}$$

$$(\hat{\mathbf{O}}+\hat{\mathbf{P}})\vec{\mathbf{a}}=\hat{\mathbf{O}}\vec{\mathbf{a}}+\hat{\mathbf{P}}\vec{\mathbf{a}}$$

$$\hat{\mathbf{O}}\hat{\mathbf{P}}\vec{\mathbf{a}}=\hat{\mathbf{O}}(\hat{\mathbf{P}}\vec{\mathbf{a}})$$

$$\hat{\mathbf{O}}(\hat{\mathbf{P}}\hat{\mathbf{Q}})=(\hat{\mathbf{O}}\hat{\mathbf{P}})\hat{\mathbf{Q}}$$

We will see that a *very* important aspect of operators is their commutative properties:

$$[\hat{\mathbf{O}},\hat{\mathbf{P}}]=\hat{\mathbf{O}}\hat{\mathbf{P}}-\hat{\mathbf{P}}\hat{\mathbf{O}} \qquad (3.46)$$

If $\hat{\mathbf{O}}\hat{\mathbf{P}}=\hat{\mathbf{P}}\hat{\mathbf{O}}$ and $[\hat{\mathbf{O}},\hat{\mathbf{P}}]=\hat{\mathbf{0}}$ then we say that the operators *commute*. Since we have learned that matrices in general do not commute we expect to find the same behaviour in our vector operators.

3.6 Dirac Notation and Operator Matrices

We continue our exploration of operators but with a new notation in hand; the Dirac notation. This is a shorthand way of representing vectors introduced by Paul Dirac in 1947 and is very frequently used today.

We represent a vector $\vec{\mathbf{a}}$ as:

$$\vec{\mathbf{a}}\equiv|a\rangle \qquad (3.47)$$

This is called a "ket" and in this particular case we would say "ket a". Note that we are not using boldface vector notation inside the ket since this would seem to be redundant. If it is inside the ket then, by definition, it is a vector and we need not explicitly indicate it as we have been doing so far. Thus, our eigenvalue equation, for example, becomes:

$$\hat{\mathbf{O}}|a\rangle=\lambda|a\rangle$$

What about the inner product? Recall that the inner product of vector $\vec{\mathbf{a}}$ with itself is:

$$\vec{\mathbf{a}}^{T}\cdot\hat{\mathbf{a}}=\|a\|^{2}$$

where we multiply the transpose (thinking of the matrix version of vectors) of $\vec{\mathbf{a}}$ by $\vec{\mathbf{a}}$ to get a scalar quantity. It is defined in this way since we want to get a real, non-negative value for $\|a\|$. In Dirac notation the transpose of $\vec{\mathbf{a}}$ is:

$$\vec{\mathbf{a}}^T \equiv \langle a|$$

which is called the "bra" and we would say "bra *a*". Using the bra and the ket the inner product is written as:

$$\langle a|a\rangle = \|a\|^2$$

which is, of course, the square of the length or norm of the vector. How do we say this? "braket a". Surely, the largely taciturn professor Dirac had a bit of a sense of humour.[††]

From the above definitions:

$$|a\rangle = \langle a|^T$$

An operator can also operate on the bra:

$$\langle a'| = \langle a|\hat{\mathbf{O}}$$

Note that in the case of the ket, the operator is placed to the left of the vector and operates left-to-right. For the bra, the operator is placed to the right and operates right-to-left (or at least, appears to operate right-to-left ... see below).

The Dirac notation is useful because it is very compact and general. It applies to *any* vectors for which an inner product is defined, not just our familiar two- and three-dimensional Euclidean ones. Also useful is that in order to understand a concept in an abstract vector space we need only ask ourselves what would be the case in simple two- or three-dimensional space.

There is another generalisation that needs to be introduced. Up until now we have been using numbers from the set of real numbers for our coefficients of the unit vectors:

$$|a\rangle = a_x|u_x\rangle + a_y|u_y\rangle + a_z|u_z\rangle$$

$$a_i \in \mathbb{R}$$

[††]To some this must have been somewhat surprising. Professor Dirac had a reputation for being a very quiet, non-verbose person who simply worked away at his work. However, he did love to read comics and must therefore have enjoyed humour.

We wish to now generalise this so that the coefficients are in the set of complex numbers:

$$|a\rangle = a_x|u_x\rangle + a_y|u_y\rangle + a_z|u_z\rangle$$

$$a_i \in \mathbb{C}$$

From here on we will assume complex numbers in order to be completely general and that the vector spaces are defined as $\mathbb{C}^n$ indicating n-dimensional vector spaces using complex numbers.

This change requires some minor alterations in how we do our inner product calculations. Since we are still looking for a real, non-negative number the bra portion of the product now represents the transpose of the complex conjugate:

$$|a\rangle = \begin{bmatrix} a_1 \\ a_2 \\ \vdots \\ a_n \end{bmatrix}$$

$$\langle a| = [a_1{}^* \quad a_2{}^* \quad \cdots \quad a_n{}^*]$$

and

$$\langle a|a\rangle = [a_1{}^* \quad a_2{}^* \quad \cdots \quad a_n{}^*] \begin{bmatrix} a_1 \\ a_2 \\ \vdots \\ a_n \end{bmatrix}$$

We can multiply our bra and ket by a scalar, however, we must remember the complex conjugate nature of the bra:

$$\begin{aligned} \alpha|a\rangle &= |\alpha a\rangle \\ \alpha\langle a| &= \langle a\alpha^*| \\ \langle a\alpha| &= \alpha^*\langle a| \end{aligned} \tag{3.48}$$

Also:

$$|a\rangle = \langle a|^* \tag{3.49}$$

and:

$$\langle a|b\rangle = \langle b|a\rangle^* \tag{3.50}$$

Note that "*" in the context of the Dirac notation refers to the transpose of the complex conjugate. We can best illustrate this with a 2×2 matrix example:

$$|a\rangle = \begin{bmatrix} \alpha_1 + \beta_1 i \\ \alpha_2 + \beta_2 i \end{bmatrix} \quad |b\rangle = \begin{bmatrix} \alpha_3 + \beta_3 i \\ \alpha_4 + \beta_4 i \end{bmatrix}$$

$$\begin{aligned}
\langle a|b\rangle &= (\alpha_1 - \beta_1 i)(\alpha_3 + \beta_3 i) + (\alpha_2 - \beta_2 i)(\alpha_4 + \beta_4 i) \\
&= \alpha_1\alpha_3 - \beta_1\beta_3 i^2 - \alpha_3\beta_1 i + \alpha_1\beta_3 i + \alpha_2\alpha_4 - \beta_2\beta_4 i^2 + \alpha_2\beta_4 i - \alpha_4\beta_2 i \\
&= (\alpha_1\alpha_3 + \beta_1\beta_3 + \alpha_2\alpha_4 + \beta_2\beta_4) + (\alpha_1\beta_3 - \alpha_3\beta_1 + \alpha_2\beta_4 - \alpha_4\beta_2)i \\
\langle b|a\rangle &= (\alpha_3 - \beta_3 i)(\alpha_1 + \beta_1 i) + (\alpha_4 - \beta_4 i)(\alpha_2 + \beta_2 i) \\
&= \alpha_3\alpha_1 - \beta_3\beta_1 i^2 - \alpha_1\beta_3 i + \alpha_3\beta_1 i + \alpha_2\alpha_4 - \beta_2\beta_4 i^2 - \alpha_2\beta_4 i + \alpha_4\beta_2 i \\
&= (\alpha_3\alpha_1 + \beta_3\beta_1 + \alpha_2\alpha_4 + \beta_2\beta_4) - (\alpha_1\beta_3 - \alpha_3\beta_1 + \alpha_2\beta_4 - \alpha_4\beta_2)i \\
&= \langle a|b\rangle^*
\end{aligned}$$

This is different from the normal two- and three-dimensional vector dot product and is referred to as *skew-symmetry* and results in some unusual properties of inner products involving complex numbers. For example, if the ket is a linear combination of vectors and we take an inner product, all is as we suspect it should be:

$$\begin{aligned}
&\langle a\,|\alpha b + \beta c\rangle \\
&= \langle a|\alpha b\rangle + \langle a|\beta c\rangle \\
&= \alpha\langle a|b\rangle + \beta\langle a|c\rangle
\end{aligned}$$

However, if the linear combination of vectors is in the bra then the result is not quite as expected:

$$\begin{aligned}
\langle \alpha b + \beta c|a\rangle &= \langle a|\alpha b + \beta c\rangle^* \\
&= (\alpha\langle b|a\rangle + \beta\langle a|c\rangle)^* \\
&= \alpha^*\langle b|a\rangle + \beta^*\langle c|a\rangle
\end{aligned}$$

using eqn (3.50) in the first and last lines and eqn (3.48) in the last line. So, the skew-symmetry property results in an *anti-symmetry* property with respect to the bra. Also, from eqn (3.50) (and remembering eqn (1.5) from Chapter 1):

$$\langle a|a\rangle = \langle a|a\rangle^*$$

which means that the inner product of $\vec{\mathbf{a}}$ with itself is real, as we know it should be.

We now turn our attention back to operators; specifically we want to produce a general way of writing the matrix form of an operator. To do so we work with our unit vectors since any vector in our vector space can be built up from a linear combination of them and they are thus the "lowest common denominator", so to speak. We assume that the operator produces a new vector that is in the same vector space as the old vector. Consider then, the effect of an operator on a unit vector:

$$|u'_i\rangle = \hat{\mathbf{O}}|u_i\rangle$$

Now let us form an inner product of this result with another unit vector:

$$\langle u_j|u'_i\rangle = \langle u_j|\hat{\mathbf{O}}|u_i\rangle = O_{ji} \tag{3.51}$$

We use this to find our matrix representation of operator $\hat{\mathbf{O}}$:

$$\hat{\mathbf{O}} = \begin{bmatrix} O_{11} & O_{12} & O_{13} & \cdots \\ O_{21} & O_{22} & O_{23} & \cdots \\ O_{31} & O_{32} & O_{33} & \cdots \\ \vdots & \vdots & \vdots & \ddots \end{bmatrix} \tag{3.52}$$

Note that we could just as well have done this 'backwards' with the operator acting on a bra unit vector instead:

$$\langle u_i|\hat{\mathbf{O}} = \langle u'_i|$$

$$\langle\langle u'_i|u_j\rangle = \langle u_i|\hat{\mathbf{O}}|u_j\rangle$$

Thus, the expression:

$$\langle u_i|\hat{\mathbf{O}}|u_j\rangle$$

can refer to the operator acting in either direction on the bra or the ket.

We can use eqn (3.51) to construct our $\hat{\mathbf{R}}(90)$ operator from above. Recall that $\hat{\mathbf{R}}(90)$ means rotate by 90° clockwise; thus unit vector $\vec{\mathbf{u}}_x$, for example, becomes $-\vec{\mathbf{u}}_y$, *etc.*:

$$\hat{\mathbf{R}}(90) = \begin{bmatrix} R_{11} & R_{12} \\ R_{21} & R_{22} \end{bmatrix}$$

$$= \begin{bmatrix} \langle u_x \mid \hat{\mathbf{R}} \mid u_x \rangle & \langle u_x \mid \hat{\mathbf{R}} \mid u_y \rangle \\ \langle u_y \mid \hat{\mathbf{R}} \mid u_x \rangle & \langle u_y \mid \hat{\mathbf{R}} \mid u_y \rangle \end{bmatrix}$$

$$= \begin{bmatrix} 0 & 1 \\ -1 & 0 \end{bmatrix}$$

Now that we know how to construct the matrix form of the operator it must be said that there is another way to look at the rotation operator (and by extension, rotation operators in general). We have been looking at it as though it causes a rotation in the vector, however, it is also possible to look at it as causing a rotation of the basis vectors, the vector itself remaining motionless (see Figure 3.22).

Thus, vector $\vec{\mathbf{a}}$ remains motionless but attains a new set of coordinates in the new basis set:

$$\vec{\mathbf{a}}' = \hat{\mathbf{R}}(\varphi)\vec{\mathbf{a}}$$

$$\begin{bmatrix} a'_x \\ a'_y \end{bmatrix} = \begin{bmatrix} \cos(\varphi) & \sin(\varphi) \\ -\sin(\varphi) & \cos(\varphi) \end{bmatrix} \begin{bmatrix} a_x \\ a_y \end{bmatrix}$$

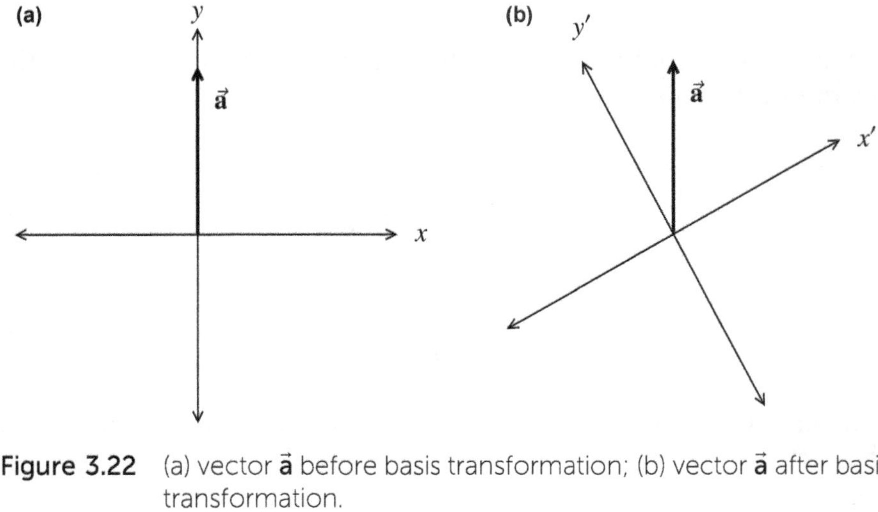

Figure 3.22 (a) vector $\vec{\mathbf{a}}$ before basis transformation; (b) vector $\vec{\mathbf{a}}$ after basis transformation.

So, what we viewed earlier as a clockwise motion of the vector can also be viewed as a counter-clockwise motion of the basis vectors. This is referred to as a *change of basis* and is generalised as follows:

$$
\begin{aligned}
e'_1 &= a_{11}e_1 + a_{12}e_2 + a_{13}e_1 \ldots a_{1n}e_n \\
e'_2 &= a_{21}e_1 + a_{22}e_2 + a_{23}e_1 \ldots a_{2n}e_n \\
e'_3 &= a_{31}e_1 + a_{32}e_2 + a_{33}e_1 \ldots a_{3n}e_n \\
&\vdots \\
e'_n &= a_{n1}e_1 + a_{n2}e_2 + a_{n3}e_1 \ldots a_{nn}e_n
\end{aligned}
$$

or in matrix notation:

$$
\begin{bmatrix} e'_1 \\ e'_2 \\ e'_3 \\ \vdots \\ e'_n \end{bmatrix} = \begin{bmatrix} a_{11} & a_{12} & a_{13} & a_{1n} \\ a_{21} & a_{22} & a_{23} & a_{2n} \\ a_{31} & a_{32} & a_{33} & a_{3n} \\ & & & \\ a_{n1} & a_{n2} & a_{n3} & a_{nn} \end{bmatrix} \begin{bmatrix} e_1 \\ e_2 \\ e_3 \\ \vdots \\ e_n \end{bmatrix}
$$

where the e_i are the old basis set and the e'_i are the new basis set. One can see that for any given vector there can be an infinite number of basis sets corresponding to all possible rotations.

A very natural question to ask would be "are the new basis vectors orthogonal if the old ones were?" for our rotation example. Visually, from Figure 3.22 you can see that they are but this really is not proof. We can prove it though. We must show using eqn (3.14) that:

$$\vec{\mathbf{u}}'_i \cdot \vec{\mathbf{u}}'_j = \delta_{ij}$$

The transformation from the old basis to the new basis is:

$$
\begin{bmatrix} u'_x \\ u'_y \end{bmatrix} = \begin{bmatrix} \cos(\varphi) & \sin(\varphi) \\ -\sin(\varphi) & \cos(\varphi) \end{bmatrix} \begin{bmatrix} u_x \\ u_y \end{bmatrix}
$$

or:

$$
\begin{aligned}
\vec{\mathbf{u}}'_x &= \cos(\varphi)\vec{\mathbf{u}}_x + \sin(\varphi)\vec{\mathbf{u}}_y \\
\vec{\mathbf{u}}'_y &= -\sin(\varphi)\vec{\mathbf{u}}_x + \cos(\varphi)\vec{\mathbf{u}}_y
\end{aligned}
$$

each new basis vector being a linear combination of the old basis vector set. Writing the inner product:

$$\begin{aligned}\vec{\mathbf{u}}'_x \cdot \vec{\mathbf{u}}'_y &= (\cos(\varphi)\vec{\mathbf{u}}_x + \sin(\varphi)\vec{\mathbf{u}}_y) \cdot (-\sin(\varphi)\vec{\mathbf{u}}_x + \cos(\varphi)\vec{\mathbf{u}}_y)\\
&= -\cos(\varphi)\sin(\varphi)\vec{\mathbf{u}}_x \cdot \vec{\mathbf{u}}_x + \cos(\varphi)\cos(\varphi)\vec{\mathbf{u}}_x \cdot \vec{\mathbf{u}}_y\\
&\quad -\sin(\varphi)\sin(\varphi)\vec{\mathbf{u}}_y\vec{\mathbf{u}}_x + \sin(\varphi)\cos(\varphi)\vec{\mathbf{u}}_y \cdot \vec{\mathbf{u}}_y\\
&= -\cos(\varphi)\sin(\varphi) + 0 + 0 + \sin(\varphi)\cos(\varphi)\\
&= 0\end{aligned}$$

Thus, the new basis set is orthogonal as was the old one, which means that the angle between basis vectors is invariant. We can similarly show that the new basis vectors are normalised if the old ones were:

$$\begin{aligned}\vec{\mathbf{u}}'_x \cdot \vec{\mathbf{u}}'_x &= (\cos(\varphi)\vec{\mathbf{u}}_x + \sin(\varphi)\vec{\mathbf{u}}_y) \cdot (\cos(\varphi)\vec{\mathbf{u}}_x + \sin(\varphi)\vec{\mathbf{u}}_y)\\
&= \cos(\varphi)\cos(\varphi)\vec{\mathbf{u}}_x \cdot \vec{\mathbf{u}}_x + 2\sin(\varphi)\cos(\varphi)\vec{\mathbf{u}}_x \cdot \vec{\mathbf{u}}_y + \sin(\varphi)\sin(\varphi)\vec{\mathbf{u}}_y \cdot \vec{\mathbf{u}}_y\\
&= 1\end{aligned}$$

using the orthonormal properties of the original basis set.

With reference to eqn (3.48) we draw the analogy between scalar multipliers and operators:

$$\begin{gathered}\hat{\mathbf{O}}\,|\mathbf{a}\rangle = \left|\hat{\mathbf{O}}\mathbf{a}\right\rangle\\
\langle a|\hat{\mathbf{O}} = \left\langle (\hat{\mathbf{O}}^T)^* a\right|\\
\text{or}\\
\left|\hat{\mathbf{O}}\mathbf{a}\right\rangle = \hat{\mathbf{O}}|a\rangle\\
\left\langle \hat{\mathbf{O}}a\right| = \langle a|(\hat{\mathbf{O}}^T)^*\end{gathered} \tag{3.53}$$

There is a non-intuitive subtlety of notation in eqn (3.53) that should be mentioned. As far as the ket goes, whether the operator is inside or outside does not affect anything much. The bra is a different story. To understand this think "matrices" again. The expression:

$$\langle \hat{\mathbf{O}}a|$$

has $\hat{\mathbf{O}}$ inside the bra. In terms of matrices this means multiply the column vector $\vec{\mathbf{a}}$ by the square matrix $\hat{\mathbf{O}}$ *inside the bra*. We could not write it as $\vec{\mathbf{a}}\hat{\mathbf{O}}$ outside of the bra and have it make sense since the corresponding matrix multiplication does not work. However, because they are enclosed in the bra we take the complex transpose and use it for any further calculations. If we take $\hat{\mathbf{O}}$ and $\vec{\mathbf{a}}$ out of the bra altogether and use our original vector notation we must take the complex transpose (and reverse the order, see eqn (2.15):

$$\langle \hat{\mathbf{O}}a| = \vec{\mathbf{a}}^T(\hat{\mathbf{O}}^T)^*$$

or, if we remove only the operator:

$$\langle \hat{\mathbf{O}}a| = \langle a|(\hat{\mathbf{O}}^T)^*$$

In a sense one could view the bra as itself being an operator meaning "take the complex transpose of whatever is inside me". To convince yourself that this is so you need only set up a simple two-dimensional vector/operator pair and see how it works.

There are special types of operators linked very closely to the special matrices mentioned in Chapter 2. When dealing with real numbers only, an *orthogonal* operator is one whose transpose is equal to its inverse:

$$\hat{\mathbf{O}}^T = \hat{\mathbf{O}}^{-1} \tag{3.54}$$

In this case the product of the operator and its transpose equals the unit operator:

$$\hat{\mathbf{O}}\hat{\mathbf{O}}^T = \hat{\mathbf{O}}^T\hat{\mathbf{O}} = \hat{1}$$

We have already seen an example of this in our two-dimensional rotation operator.

The next type of operator is the *unitary* operator and is associated with complex numbers in the vector. This is one in which the complex conjugate of the transpose of the operator equals the inverse of the operator:

$$(\hat{\mathbf{O}}^T)^* = \hat{\mathbf{O}}^{-1} \tag{3.55}$$

and

$$\hat{\mathbf{O}}(\hat{\mathbf{O}}^T)^* = (\hat{\mathbf{O}}^T)^*\hat{\mathbf{O}} = \hat{\mathbf{1}}$$

A bit of thought will reveal that if unitary operator matrix consists of all real numbers then it is also an orthogonal operator. These two operators are ones that leave the vector norms (lengths) and angles between vectors invariant. In equation form, using the unitary operator, this is:

$$\langle \hat{\mathbf{O}}a|\hat{\mathbf{O}}b\rangle = \langle a|b\rangle$$

This is easily shown to be so:

$$\langle \hat{\mathbf{O}}a\,|\,\hat{\mathbf{O}}b\rangle = \langle a|(\vec{\mathbf{O}}^{T})^{*}\vec{\mathbf{O}}|b\rangle$$
$$= \langle a|\hat{\mathbf{1}}|b\rangle$$
$$= \langle a\,|\,b\rangle$$

The unitary operator has an interesting feature in that its eigenvalues are all equal to 1. As usual, we can show this. We start with:

$$\lambda^{*}\lambda\langle a|a\rangle$$

where $|a\rangle$ is a non-zero vector and λ is the eigenvalue of operator $\hat{\mathbf{O}}$. Since the λs are scalars we can move them into the bra and ket:

$$\langle \lambda a|\lambda a\rangle$$

making use of eqn (3.48) in the bra. Now we substitute our unitary operator:

$$\langle \hat{\mathbf{O}}a\,|\,\hat{\mathbf{O}}a\rangle$$
$$= \langle a|(\hat{\mathbf{O}}^{T})^{*}\hat{\mathbf{O}}|a\rangle$$
$$= \langle a|\hat{\mathbf{1}}|a\rangle$$
$$= \langle a\,|\,a\rangle$$

So

$$\lambda^{*}\lambda\langle a|a\rangle = \langle a|a\rangle$$

and since $\langle a|a\rangle$ is not equal to zero, $\lambda^{*}\lambda$ must be equal to one.

Finally, we consider the *Hermitian* operator:

$$(\hat{\mathbf{O}}^{T})^{*} = \hat{\mathbf{O}} \tag{3.56}$$

and

$$\hat{\mathbf{O}}^{-1}(\hat{\mathbf{O}}^{T})^{*} = (\hat{\mathbf{O}}^{T})^{*}\hat{\mathbf{O}}^{-1} = \hat{\mathbf{1}}$$

We can use the Hermitian operator in an inner product calculation:

$$\langle a|\vec{\mathbf{O}}|b\rangle = \langle a|(\vec{\mathbf{O}}^T)^*|b\rangle$$

Let us take the complex conjugate of this:

$$\begin{aligned}
\langle a|\hat{\mathbf{O}}|b\rangle^* &= \langle a|(\hat{\mathbf{O}}^T)^*|b\rangle^* \\
&= \langle \hat{\mathbf{O}}a \,|\, b\rangle^* \\
&= (\langle \hat{\mathbf{O}}a|)^*(|b\rangle)^* \\
&= \langle b \,|\, \hat{\mathbf{O}}a\rangle \\
&= \langle b|\hat{\mathbf{O}}|a\rangle
\end{aligned} \tag{3.57}$$

Products of operators are worth a mention here. We treat a product of two or more operators acting on a vector $\vec{\mathbf{a}}$ as follows:

$$\hat{\mathbf{O}}_1\hat{\mathbf{O}}_2|a\rangle = \hat{\mathbf{O}}_1(\hat{\mathbf{O}}_2|a\rangle) = |\hat{\mathbf{O}}_1(\hat{\mathbf{O}}_2 a)\rangle$$

That is, operator $\hat{\mathbf{O}}_2$ acts first on ket a followed by $\hat{\mathbf{O}}_1$. Recall that the order of operations is generally important since we do not expect that operators will commute (see eqn (3.46)). The situation with the bra is not quite the same, as a result of anti-linearity:

$$\langle a|\hat{\mathbf{O}}_1\hat{\mathbf{O}}_2 = \langle(\hat{\mathbf{O}}_1\hat{\mathbf{O}}_2)^{T*}a| = \langle(\hat{\mathbf{O}}_2)^{T*}(\hat{\mathbf{O}}_1)^{T*}a|$$

From our look at matrices we saw that taking the transpose of a product of matrices is the same as the product of the transposes of each matrix in reverse order (eqn (2.15)). The tight relationship between operators and matrices suggests the same behaviour for operators.

There are some interesting and, as we will see later, useful properties of these types of operators. First, we show that commuting operators have simultaneously defined eigenvectors. This may seem a bit backwards but we begin by letting $\hat{\mathbf{A}}$ and $\hat{\mathbf{B}}$ be two linear operators that have a common set of eigenvectors:

$$\hat{\mathbf{A}}|f\rangle = a|f\rangle \quad \hat{\mathbf{B}}|f\rangle = b|f\rangle$$

and then show that $[\hat{\mathbf{A}}, \hat{\mathbf{B}}]$ is always equal to 0 for any arbitrary vector, $\vec{\mathbf{f}}$, or, what is equivalent, that $(\hat{\mathbf{A}}\hat{\mathbf{B}} - \hat{\mathbf{B}}\hat{\mathbf{A}})\vec{\mathbf{f}} = 0$. We start by defining $\vec{\mathbf{f}}$ in terms of the complete set of basis vectors:

$$\vec{\mathbf{f}} = |f\rangle = \sum_i c_i \vec{\mathbf{u}}_i$$

then:

$$\begin{aligned}(\hat{\mathbf{A}}\hat{\mathbf{B}} - \hat{\mathbf{B}}\hat{\mathbf{A}})|f\rangle &= (\hat{\mathbf{A}}\hat{\mathbf{B}} - \hat{\mathbf{B}}\hat{\mathbf{A}})\left|\sum_i c_i u_i\right\rangle \\ &= \sum_i c_i(\hat{\mathbf{A}}\hat{\mathbf{B}} - \hat{\mathbf{B}}\hat{\mathbf{A}})|u_i\rangle = \sum_i c_i\big(\hat{\mathbf{A}}(\hat{\mathbf{B}}|u_i\rangle) - \hat{\mathbf{B}}(\hat{\mathbf{A}}|u_i\rangle)\big) \\ &= \sum_i c_i\big(\hat{\mathbf{A}}B_i|u_i\rangle - \hat{\mathbf{B}}A_i|u_i\rangle\big) = \sum_i c_i(A_iB_i|u_i\rangle - B_iA_i|u_i\rangle) \\ &= 0\end{aligned} \tag{3.58}$$

where A_i and B_i are, of course, the eigenvalues associated with their respective operators. $\hat{\mathbf{A}}$ and $\hat{\mathbf{B}}$ are therefore commuting operators and have the common eigenvector, $|f\rangle$. Furthermore, eqn (3.58) shows us that two operators with a common eigenvector will *always* commute.

Next, we look at the Hermitian operator and assert that the eigenvalues of Hermitian operators are real. That is, for:

$$\hat{\mathrm{O}}|a\rangle = \lambda|a\rangle$$

where it is understood that $\hat{\mathrm{O}}$ is Hermitian as defined above and that $|a\rangle$ is normalised, the eigenvalue, λ, is a real number. We construct an inner product and do the eigenvalue calculation:

$$\langle a|\hat{\mathbf{O}}|a\rangle = \langle a|\lambda|a\rangle = \lambda\langle a|a\rangle = \lambda$$

Now, we do the same for the complex conjugate:

$$\langle a|\hat{\mathbf{O}}|a\rangle^* = \langle a|\lambda|a\rangle^* = \lambda^*\langle a|a\rangle^* = \lambda^*$$

Looking at eqn (3.57) we realise that, if $\hat{\mathbf{O}}$ is Hermitian:

$$\langle a|\hat{\mathbf{O}}|a\rangle = \langle a|\hat{\mathbf{O}}|a\rangle^*$$

and that therefore:

$$\lambda = \lambda^*$$

Assuming that λ is not equal to zero then, since $\lambda = \lambda^*$, λ *must* be real.

We also assert that the eigenvectors of a Hermitian operator are orthogonal. Consider the eigenvalue equations on two different eigenvectors:

$$\hat{\mathbf{O}}|a\rangle = \lambda_a|a\rangle \quad \hat{\mathbf{O}}|b\rangle = \lambda_b|b\rangle$$

where, as before, $\hat{\mathbf{O}}$ is Hermitian. We form the inner products and do the eigenvalue calculation:

$$\langle b|\hat{\mathbf{O}}|a\rangle = \lambda_a\langle b|a\rangle \quad \langle a|\hat{\mathbf{O}}|b\rangle = \lambda_b\langle a|b\rangle$$

Now we take the complex conjugate of one of them (it does not matter which for our purposes) and, again, do the eigenvalue calculation:

$$\langle a|\hat{\mathbf{O}}|b\rangle^* = \lambda_b{}^*\langle a|b\rangle^* = \lambda_b{}^*\langle b|a\rangle$$

using eqn (3.50). Now we subtract the complex conjugate calculation from our first eigenvalue calculation:

$$\langle b|\hat{\mathbf{O}}|a\rangle - \langle a|\hat{\mathbf{O}}|b\rangle^* = (\lambda_a - \lambda_b{}^*)\langle b|a\rangle \tag{3.59}$$

and, using eqn (3.57) we can say:

$$\langle b|\hat{\mathbf{O}}|a\rangle = \langle a|\hat{\mathbf{O}}|b\rangle^*$$

Since they are equal because $\hat{\mathbf{O}}$ is Hermitian, their difference is zero:

$$\langle b|\hat{\mathbf{O}}|a\rangle - \langle a|\hat{\mathbf{O}}|b\rangle^* = 0$$

which means that in eqn (3.59):

$$(\lambda_a - \lambda_b{}^*)\langle b|a\rangle = 0$$

If $|a\rangle$ and $\langle b|$ are not degenerate so that λ_a and $\lambda_b{}^*$ are not equal and their difference is non-zero, then:

$$\langle b|a\rangle = 0$$

which means that $|a\rangle$ and $\langle b|$ are orthogonal.

We can write our operators in terms of outer products. Specifically, if we use an orthonormal basis set for our outer products we write:

$$\hat{\mathbf{O}} = \sum_{ij}^{n} O_{ij}|u_i\rangle\langle u_j| \tag{3.60}$$

As an example let us consider two-dimensional space $\mathbb{R}^2$, and the standard orthonormal basis set:

$$|u_x\rangle = \begin{bmatrix} 1 \\ 0 \end{bmatrix} \quad |u_y\rangle = \begin{bmatrix} 0 \\ 1 \end{bmatrix}$$

The four possible outer products of these two vectors are:

$$|u_x\rangle\langle u_x| = \begin{bmatrix} 1 & 0 \\ 0 & 0 \end{bmatrix}$$

$$|u_x\rangle\langle u_y| = \begin{bmatrix} 0 & 1 \\ 0 & 0 \end{bmatrix}$$

$$|u_y\rangle\langle u_x| = \begin{bmatrix} 0 & 0 \\ 1 & 0 \end{bmatrix}$$

$$|u_y\rangle\langle u_y| = \begin{bmatrix} 0 & 0 \\ 0 & 1 \end{bmatrix}$$

That the first and last of these are also projection operators follows from our definition of the projection operator in eqn (3.28) and (3.29). Recall that our 90° rotation operator in two-dimensional space is:

$$\hat{\mathbf{R}}(90) = \begin{bmatrix} 0 & 1 \\ -1 & 0 \end{bmatrix}$$

In terms of the outer product matrices this is:

$$\hat{\mathbf{R}}(90) = R_{11}|u_x\rangle\langle u_x| + R_{12}|u_x\rangle\langle u_y| + R_{21}|u_y\rangle\langle u_x| + R_{22}|u_y\rangle\langle u_y|$$

$$\hat{\mathbf{R}}(90) = |u_x\rangle\langle u_y| - |u_y\rangle\langle u_x|$$

Note that in this special case the projection operators, $|u_x\rangle\langle u_x|$ and $|u_y\rangle\langle u_y|$, are not part of the operator or rather, since this is a 90° rotation their respective positions in the rotation matrix are equal to zero *via* the cosine of 90°. Had we been more general and specified a rotation angle of ϕ then the projection operators would have been part of the rotation operator:

$$\hat{\mathbf{R}}(\varphi) = \begin{bmatrix} \cos(\varphi) & \sin(\varphi) \\ -\sin(\varphi) & \cos(\varphi) \end{bmatrix}$$

and:

$$\hat{\mathbf{R}}(\varphi) = R_{11}|u_x\rangle\langle u_x| + R_{12}|u_x\rangle\langle u_y| + R_{21}|u_y\rangle\langle u_x| + R_{22}|u_y\rangle\langle u_y|$$

$$= \cos(\varphi)|u_x\rangle\langle u_x| + \sin(\varphi)|u_x\rangle\langle u_y| + \cos(\varphi)|u_y\rangle\langle u_x| + \cos(\varphi)|u_y\rangle\langle u_y|$$

Another place where outer products appear is in *spectral decomposition*. The complete set of eigenvalues of an operator is called the *spectrum*.

This time only the outer products corresponding to projection operators are used. Suppose we have the eigenvalue equation:

$$\hat{\mathbf{O}}|a_i\rangle = \lambda|a_i\rangle$$

where there are several eigenvectors and associated eigenvalues. We can also write the projection operator on vector $|a_i\rangle$ as:

$$\hat{\mathbf{P}} = |a_i\rangle\langle a_i|$$

and a typical matrix representation of a projection operator is:

$$\hat{\mathbf{P}}_i = \begin{bmatrix} 0 & \cdots & \cdots & 0 \\ \vdots & \ddots & \cdots & 0 \\ 0 & \cdots & 1 & \vdots \\ \vdots & \vdots & \vdots & \vdots \end{bmatrix}$$

Recall that the closure relation indicates that:

$$\sum_i \hat{\mathbf{P}}_i = \begin{bmatrix} 1 & \cdots & \cdots & 0 \\ \vdots & 1 & \cdots & 0 \\ 0 & \cdots & \ddots & \vdots \\ \vdots & \vdots & \vdots & 1 \end{bmatrix}$$

This is of some importance in quantum mechanics so we will briefly explore the Dirac notation for the closure relation. From the outer product definition of $\hat{\mathbf{P}}$ (see eqn (3.28)) written in Dirac notation:

$$\hat{\mathbf{P}} = |u_i\rangle\langle u_i|$$

we have seen that the sum of all projections operators is the unity operator:

$$\hat{\mathbf{P}} = \sum_i \hat{\mathbf{P}}_i = \sum_i |u_i\rangle\langle u_i| = \hat{\mathbf{1}}$$

This means that it would be acceptable to do something like:

$$\langle u_j|\hat{\mathbf{A}}\hat{B}|u_k\rangle = \sum_i \langle u_j|\hat{\mathbf{A}}|u_i\rangle\langle u_i|\hat{\mathbf{B}}|u_k\rangle \tag{3.61}$$

In essence, we are simply inserting a unity operator into the expression that, of course, changes nothing. This type of manipulation happens frequently in quantum mechanics and we will have occasion to use it.

The eigenvalue matrix looks like:

$$\hat{\Lambda} = \begin{bmatrix} \lambda_1 & 0 & \cdots & 0 \\ 0 & \lambda_2 & \cdots & 0 \\ 0 & 0 & \ddots & \vdots \\ \vdots & \vdots & \vdots & \lambda_n \end{bmatrix}$$

Looking carefully at the last two matrices one can write:

$$\begin{aligned} \hat{\Lambda} &= \sum_i \lambda_i \hat{\mathbf{1}} \\ &= \sum_i \lambda_i \mathbf{P}_i \\ &= \sum_i \lambda_i |a_i\rangle\langle a_i| \end{aligned}$$

What, if anything, does $\hat{\mathbf{O}}^2|a\rangle$ mean? What we do in this situation is successively apply the operator to the vector:

$$\hat{\mathbf{O}}^2|a\rangle = \hat{\mathbf{O}}(\hat{\mathbf{O}}|a\rangle)$$

If $|a\rangle$ is an eigenfunction of $\hat{\mathbf{O}}$ then we simply get the square of the eigenvalue:

$$\begin{aligned} \hat{\mathbf{O}}^2|a\rangle &= \hat{\mathbf{O}}(\hat{\mathbf{O}}|a\rangle) \\ &= \hat{\mathbf{O}}(\lambda|a\rangle) \\ &= \lambda(\hat{\mathbf{O}}|a\rangle) \\ &= \lambda^2|a\rangle \end{aligned}$$

More generally:

$$\hat{\mathbf{O}}^N|a\rangle = \lambda^N|a\rangle$$

Alternately, if $\hat{\mathbf{O}}$ is a rotation operator we simply get N successive rotations by an angle, φ.

What is the eigenvalue of the exponential of an operator?

$$\begin{aligned}\exp(\hat{\mathbf{O}})|a\rangle &= (\hat{\mathbf{1}} + \hat{\mathbf{O}} + \frac{\hat{\mathbf{O}}^2}{2!} + \frac{\hat{\mathbf{O}}^3}{3!} + \cdots)|a\rangle \\ &= (1 + \lambda + \lambda^2 + \lambda^3 + \cdots)|a\rangle \\ &= \exp(\lambda)|a\rangle\end{aligned} \tag{3.62}$$

3.7 Hilbert Space

The mathematical concept of a Hilbert space, named after the mathematician David Hilbert, generalises the notion of a vector space and is referred to frequently in quantum mechanics. It extends the methods of vector algebra from the two-dimensional plane, R^2, and three-dimensional space R^3 to include infinite dimensional spaces in R^{∞} and C^{∞}, where R, as usual, refers to real numbers and C refers to complex numbers. In fact, a Hilbert space can be comprised of vectors that are functions rather than numbers!

The characteristics of a Hilbert space that are important to us are:

1. It has a scalar or inner product, as have most of the vector spaces that we have looked at. This scalar product is not necessarily commutative but:

$$\langle a|b\rangle = \langle b|a\rangle^* \tag{3.63}$$

where, of course , the asterisk indicates the complex conjugate. The inner product of a vector with itself is:

$$\langle a|a\rangle = \langle a|a\rangle^*$$

which we have seen is necessarily real.
2. The real value of the inner product, $\langle a|a\rangle$, is positive and its square root is called the norm of the vector, $|a\rangle$.

For a Hilbert space there will always be a set of basis vectors of unit norm such that:

$$\langle e_i|e_j\rangle = \delta_{ij} \tag{3.64}$$

These vectors constitute an orthonormal basis set from which we can construct any other arbitrary vector as a linear combination of them:

$$\vec{\mathbf{A}} = \sum_i a_i e_i$$

in exactly the same way we have done above for a vector in 3D Euclidean space using $\vec{\mathbf{u}}_x$, $\vec{\mathbf{u}}_y$ and $\vec{\mathbf{u}}_z$. With this, orthonormality and the inner product we can write:

$$\langle e_i | A \rangle = a_i$$

$$\langle A | e_i \rangle = a_i{}^*$$

The complication that arises here is the inclusion of infinite-dimensional spaces. In order that we have something that can be managed mathematically in terms of our inner products we must add to the definition of a Hilbert space that the sum of the squares of the components of a vector be a finite number. Otherwise we could possibly end up with a vector of infinite length! Recall that the definition of the inner product is:

$$\begin{aligned}\langle a | a \rangle &= \langle e_1 | a_1{}^* a_1 | e_1 \rangle + \langle e_2 | a_2{}^* a_2 | e_2 \rangle + \langle e_3 | a_3{}^* a_3 | e_3 \rangle \; \ldots \; \langle e_n | a_n{}^* a_n | e_n \rangle \\ &= a_1{}^* a_1 \langle e_1 | e_1 \rangle + a_2{}^* a_2 \langle e_2 | e_2 \rangle + a_3{}^* a_3 \langle e_3 | e_3 \rangle \; \ldots \; a_n{}^* a_n \langle e_n | e_n \rangle \\ &= a_1{}^* a_1 + a_2{}^* a_2 + a_3{}^* a_3 \; \ldots \; a_n{}^* a_n \\ &= |a_1|^2 + |a_2|^2 + |a_3|^2 \ldots \; |a_n|^2 \\ &= \sum_i^n |a_i|^2\end{aligned}$$

It is this sum (a Cauchy sequence) that we cannot allow to be infinite. In other words, it must converge to a finite number in order to be of use to us, even if n goes to infinity. That this sequence converges is the third requirement of a Hilbert space, the first two being the normal requirements of an inner product vector space that we have discussed so far.

This is not really so very different from the Euclidean vector spaces that we have already seen and, in fact, Euclidean vector spaces are Hilbert spaces. They satisfy the regular requirements of a vector space and the sums of the squares of their vector components are finite. We will deal almost exclusively with finite-dimensional spaces in our exploration of nuclear magnetic resonance spectroscopy.

3.8 Merging Vector Spaces

There will come a point in our mathematical explorations when we will be considering two spaces that we will want to combine into one.

The way that we will do this is by using the direct product of the two spaces to produce the new, larger space. Let us consider two spaces, each of dimension two and for simplicity let us assume that each uses the same simple orthonormal basis set:

$$\text{space A}: \quad \begin{bmatrix} 1 \\ 0 \end{bmatrix} \quad \text{and} \quad \begin{bmatrix} 0 \\ 1 \end{bmatrix} \quad \text{or} \quad \begin{bmatrix} 1 & 0 \\ 0 & 1 \end{bmatrix}$$

$$\text{space B}: \quad \begin{bmatrix} 1 \\ 0 \end{bmatrix} \quad \text{and} \quad \begin{bmatrix} 0 \\ 1 \end{bmatrix} \quad \text{or} \quad \begin{bmatrix} 1 & 0 \\ 0 & 1 \end{bmatrix}$$

The direct product (see eqn [2.8]) of these basis sets gives:

$$\begin{bmatrix} 1 & 0 \\ 0 & 1 \end{bmatrix} \otimes \begin{bmatrix} 1 & 0 \\ 0 & 1 \end{bmatrix} = \begin{bmatrix} 1 & 0 & 0 & 0 \\ 0 & 1 & 0 & 0 \\ 0 & 0 & 1 & 0 \\ 0 & 0 & 0 & 1 \end{bmatrix}$$

which is the orthonormal basis set in the new, combined space. We started with two spaces of dimension 2 each and ended up with a new space of dimension 4. Generally, if space A is of dimension n and space B is of dimension m then the direct product matrix will be of dimension $n \times m$:

$$\begin{bmatrix} A_1 \\ A_2 \\ \vdots \\ A_n \end{bmatrix} \otimes \begin{bmatrix} B_1 \\ B_2 \\ \vdots \\ B_m \end{bmatrix} = \begin{bmatrix} A_1B_1 \\ A_1B_2 \\ \vdots \\ A_2B_1 \\ \vdots \\ A_nB_m \end{bmatrix} \tag{3.65}$$

In eqn (3.65) we begin with a vector of dimension n and a vector of dimension m and end up with a vector of dimension $n \times m$.[‡‡] The situation is similar with respect to operators. If the matrix representation of operator $\hat{\mathbf{M}}$ is of dimension $n \times n$ and operator $\hat{\mathbf{N}}$ is of dimension $m \times m$ then the direct product matrix will be of dimension $n \times m$.

In the literature one sees the following notation used to indicate the combining of two spaces:

$$|\varphi\rangle \otimes |\eta\rangle$$

‡‡By using $n \times m$ with respect to a vector I mean that the number of rows in the column matrix is $n \times m$. With respect to a multicolumn matrix, $n \times m$ means n rows by m columns.

or more simply (if somewhat confusingly to the uninitiated):

$$|\varphi\rangle|\eta\rangle \text{ or } |\varphi\eta\rangle$$

We will encounter this concept of merging vector spaces in our exploration of density matrices and product operators.

There are a few things to consider with respect to the direct product. First, for $|\Psi_A\rangle$ and $|\Psi_B\rangle$ in spaces A and B, respectively:

$$a(|\Psi_A\rangle \otimes |\Psi_B\rangle) = (a|\Psi_A\rangle \otimes |\Psi_B\rangle) = |\Psi_A\rangle \otimes (a|\Psi_B\rangle)\ (a \in \mathbb{C}) \tag{3.66}$$

This is easily shown to be so *via* matrix notation for the direct product of vectors $|\Psi_A\rangle$ and $|\Psi_B\rangle$:

$$a \times \begin{bmatrix} A_1 \\ A_2 \\ \vdots \end{bmatrix} \otimes \begin{bmatrix} B_1 \\ B_2 \\ \vdots \end{bmatrix} = \begin{bmatrix} aA_1 \\ aA_2 \\ \vdots \end{bmatrix} \otimes \begin{bmatrix} B_1 \\ B_2 \\ \vdots \end{bmatrix} = \begin{bmatrix} aA_1 \begin{bmatrix} B_1 \\ B_2 \\ \vdots \end{bmatrix} \\ aA_2 \begin{bmatrix} B_1 \\ B_2 \\ \vdots \end{bmatrix} \\ \vdots \end{bmatrix} = \begin{bmatrix} aA_1B_1 \\ aA_1B_2 \\ \vdots \\ aA_2B_1 \\ \vdots \end{bmatrix}$$

Secondly, for vectors vectors $|\Psi_A\rangle$ and $|\Psi_B\rangle$, both in space A, and $|\Psi_C\rangle$ in space B:

$$(|\Psi_A\rangle + |\Psi_B\rangle) \otimes |\Psi_C\rangle = |\Psi_A\rangle \otimes |\Psi_C\rangle + |\Psi_B\rangle \otimes |\Psi_C\rangle \tag{3.67}$$

which, again, is easily shown using matrix notation for vectors $|\Psi_A\rangle$, $|\Psi_B\rangle$ and $|\Psi_C\rangle$ with basis coefficients A, B and C, respectively:

$$\left(\begin{bmatrix} A_1 \\ A_2 \\ \vdots \end{bmatrix} + \begin{bmatrix} B_1 \\ B_2 \\ \vdots \end{bmatrix}\right) \otimes \begin{bmatrix} C_1 \\ C_2 \\ \vdots \end{bmatrix} = \begin{bmatrix} A_1 + B_1 \\ A_2 + b_2 \\ \vdots \end{bmatrix} \otimes \begin{bmatrix} C_1 \\ C_2 \\ \vdots \end{bmatrix} = \begin{bmatrix} (A_1 + B_1) \begin{bmatrix} C_1 \\ C_2 \\ \vdots \end{bmatrix} \\ (A_2 + B_2) \begin{bmatrix} C_1 \\ C_2 \\ \vdots \end{bmatrix} \\ \vdots \end{bmatrix}$$

$$= \begin{bmatrix} A_1C_1 + B_1C_1 \\ A_1C_2 + B_1C_2 \\ \vdots \\ A_2C_1 + B_2C_1 \\ \vdots \end{bmatrix} = \begin{bmatrix} A_1C_1 \\ A_1C_2 \\ \vdots \\ A_2C_1 \\ \vdots \end{bmatrix} + \begin{bmatrix} B_1C_1 \\ B_1C_2 \\ \vdots \\ B_2C_1 \\ \vdots \end{bmatrix} = \begin{bmatrix} A_1 \\ A_2 \\ \vdots \end{bmatrix} \otimes \begin{bmatrix} C_1 \\ C_2 \\ \vdots \end{bmatrix} + \begin{bmatrix} B_1 \\ B_2 \\ \vdots \end{bmatrix} \otimes \begin{bmatrix} C_1 \\ C_2 \\ \vdots \end{bmatrix}$$

Thirdly, for $|\Psi_A\rangle$ and $|\Psi_B\rangle$ in spaces A and B, respectively, and operators $\hat{\mathbf{M}}$ and $\hat{\mathbf{N}}$, also of spaces A and B, respectively we define:

$$(\hat{\mathbf{M}}\otimes\hat{\mathbf{N}})\,(|\Psi_A\rangle)\otimes|\Psi_B\rangle)\equiv\hat{\mathbf{M}}|\Psi_A\rangle\otimes\hat{\mathbf{N}}|\Psi_B\rangle \tag{3.68}$$

We can, of course, show that this is a reasonable definition ... again by using matrices (it is a bit tedious so you are allowed to take our word for it if you wish):

$$\hat{\mathbf{M}}\otimes\hat{\mathbf{N}}=\begin{bmatrix} M_{11} & M_{12} & \cdots \\ M_{21} & M_{22} & \cdots \\ \vdots & \vdots & \ddots \end{bmatrix}\otimes\begin{bmatrix} N_{11} & N_{12} & \cdots \\ N_{21} & N_{22} & \cdots \\ \vdots & \vdots & \ddots \end{bmatrix}$$

$$=\begin{bmatrix} M_{11}N_{11} & M_{11}N_{12} & \cdots & M_{12}N_{11} & \cdots \\ M_{11}N_{21} & M_{11}N_{22} & \cdots & M_{12}N_{21} & \cdots \\ \vdots & \vdots & \ddots & \cdots & \cdots \\ M_{21}N_{11} & M_{21}N_{11} & \vdots & \ddots & \cdots \\ \vdots & \vdots & \vdots & \vdots & M_{nn}N_{mm} \end{bmatrix}$$

$$|\Psi_A\rangle\otimes|\Psi_B\rangle=\begin{bmatrix} A_1 \\ A_2 \\ \vdots \\ A_n \end{bmatrix}\otimes\begin{bmatrix} B_1 \\ B_2 \\ \vdots \\ B_m \end{bmatrix}=\begin{bmatrix} A_1B_1 \\ A_1B_2 \\ \vdots \\ A_2B_1 \\ \vdots \\ A_nB_m \end{bmatrix}$$

$$(\hat{\mathbf{M}}\otimes\hat{\mathbf{N}})(|\Psi_A\rangle\otimes|\Psi_B\rangle)=\begin{bmatrix} M_{11}N_{11} & M_{11}N_{12} & \cdots & M_{12}N_{11} & \cdots \\ M_{11}N_{21} & M_{11}N_{22} & \cdots & M_{12}N_{21} & \cdots \\ \vdots & \vdots & \ddots & \cdots & \cdots \\ M_{21}N_{11} & M_{21}N_{11} & \vdots & \ddots & \cdots \\ \vdots & \vdots & \vdots & \vdots & M_{nn}N_{mm} \end{bmatrix}\begin{bmatrix} A_1B_1 \\ A_1B_2 \\ \vdots \\ A_2B_1 \\ \vdots \\ A_nB_m \end{bmatrix}$$

$$=\begin{bmatrix} \sum_j^n\sum_i^m M_{1j}N_{1i}A_jB_i \\ \sum_j^n\sum_i^m M_{1j}N_{2i}A_jB_i \\ \vdots \\ \sum_j^n\sum_i^m M_{2j}N_{1i}A_jB_i \\ \vdots \\ \sum_j^n\sum_i^m M_{nj}N_{mi}A_jB_i \end{bmatrix}$$

That is the left-hand side of eqn (3.68). Now for the right side:

$$\hat{\mathbf{M}} \mid \Psi_A\rangle = \begin{bmatrix} M_{11} & M_{12} & \cdots & \cdots \\ M_{21} & M_{22} & \cdots & \cdots \\ \vdots & \vdots & \ddots & \cdots \\ \vdots & \vdots & \vdots & M_{nn} \end{bmatrix} \begin{bmatrix} A_1 \\ A_2 \\ \vdots \\ A_n \end{bmatrix} = \begin{bmatrix} \sum_j^n M_{1j}A_j \\ \sum_j^n M_{2j}A_j \\ \vdots \\ \sum_j^n M_{nj}A_j \end{bmatrix}$$

$$\hat{\mathbf{N}}|\Psi_B\rangle = \begin{bmatrix} N_{11} & N_{12} & \cdots & \cdots \\ N_{21} & N_{22} & \cdots & \cdots \\ \vdots & \vdots & \ddots & \cdots \\ \vdots & \vdots & \vdots & N_{mm} \end{bmatrix} \begin{bmatrix} B_1 \\ B_2 \\ \vdots \\ B_m \end{bmatrix} = \begin{bmatrix} \sum_i^m N_{1i}B_i \\ \sum_i^m N_{2i}B_i \\ \vdots \\ \sum_i^m N_{mi}B_i \end{bmatrix}$$

$$\hat{\mathbf{M}}|\Psi_A\rangle \otimes \hat{\mathbf{N}}|\Psi_B\rangle = \begin{bmatrix} \sum_j^n M_{1j}A_j \\ \sum_j^n M_{2j}A_j \\ \vdots \\ \sum_j^n M_{nj}A_j \end{bmatrix} \otimes \begin{bmatrix} \sum_i^m N_{1i}B_i \\ \sum_i^m N_{2i}B_i \\ \vdots \\ \sum_i^m N_{mi}B_i \end{bmatrix}$$

$$= \begin{bmatrix} \sum_j^n M_{1j}A_j \sum_i^m N_{1i}B_i \\ \sum_j^n M_{1j}A_j \sum_i^m N_{2i}B_i \\ \vdots \\ \sum_j^n M_{2j}A_j \sum_i^m N_{1i}B_i \\ \vdots \\ \sum_j^n M_{nj}A_j \sum_i^m N_{mi}B_i \end{bmatrix} = \begin{bmatrix} \sum_j^n \sum_i^m M_{1j}N_{1i}A_jB_i \\ \sum_j^n \sum_i^m M_{1j}N_{2i}A_jB_i \\ \vdots \\ \sum_j^n \sum_i^m M_{2j}N_{1i}A_jB_i \\ \vdots \\ \sum_j^n \sum_i^m M_{nj}N_{mi}A_jB_i \end{bmatrix}$$

As you can see, this is the same as the left side of the equation.

Finally, if the basis set of each vector is orthonormal then the basis set of the direct product will be orthonormal as well.

$$\langle\Psi_A\Psi_B|\Psi_A\Psi_B\rangle = 1 \text{ if } \langle\Psi_A|\Psi_A\rangle = \langle\Psi_B|\Psi_B\rangle = 1 \tag{3.69}$$

We assume here that $\langle\Psi_A|\Psi_A\rangle$ and $\langle\Psi_B|\Psi_B\rangle$ are normalised since, as we have seen above, they can always be made to be so. As usual, we invoke matrices to show this:

$$|\Psi_A\rangle = \sum_i^n A_i|\psi_i\rangle$$

$$\langle\Psi_A|\Psi_A\rangle = \sum_i^n A_i{}^*A_i\langle\psi_i|\psi_i\rangle = \sum_i^n A_i{}^*A_i = 1$$

similarly

$$\langle\Psi_B|\Psi_B\rangle = \sum_j^n B_j{}^*B_j\left\langle\psi_j|\psi_j\right\rangle = \sum_j^n B_j{}^*B_j = 1$$

This is so, of course, only if the basis sets are orthonormal. Proceeding:

$$|\Psi_A\rangle \otimes |\Psi_B\rangle = |\Psi_A\Psi_B\rangle$$

$$= \begin{bmatrix} A_1 \\ A_2 \\ \vdots \\ A_n \end{bmatrix} \otimes \begin{bmatrix} B_1 \\ B_2 \\ \vdots \\ B_m \end{bmatrix} = \begin{bmatrix} A_1B_1 \\ A_1B_2 \\ \vdots \\ A_2B_1 \\ \vdots \\ A_nB_m \end{bmatrix}$$

$$\langle\Psi_A\Psi_B|\Psi_A\Psi_B\rangle = \begin{bmatrix} A_1{}^*B_1{}^* & A_1{}^*B_2{}^* & \cdots & A_2{}^*B_1{}^* & \cdots & A_n{}^*B_m{}^* \end{bmatrix} \begin{bmatrix} A_1B_1 \\ A_1B_2 \\ \vdots \\ A_2B_1 \\ \vdots \\ A_nB_m \end{bmatrix}$$

$$= \sum_i^n A_i{}^*A_i \sum_j^n B_j{}^*B_j$$

$$= \langle\Psi_A|\Psi_A\rangle\langle\Psi_B|\Psi_B\rangle = 1$$

Evidently, if the basis sets of the separate vectors are orthonormal then so is the basis set of the new vector.

Further Reading

1. J. S. Faires and B. T. Faires, *Calculus*, Random House, New York, 1988.
2. J. S. Golan, *The Linear Algebra a Beginning Graduate Student Ought to Know*, Kluwer Academic Publishers, Dordrecht, Boston, 2004.
3. R. I. G. Hughes, *The Structure and Interpretation of Quantum Mechanics*, Harvard University Press, Cambridge, Mass, 1989.
4. J. C. Kolecki, *Foundations of Tensor Analysis for Students of Physics and engineering With an Introduction to the Theory of Relativity*, Report TP-2005-213115, NASA Science and Technical Information, 2005.
5. M. O'nan, *Linear Algebra*, Harcourt, Brace Jovanovich, New York, 1976.
6. R. Shankar, *Principles of Quantum Mechanics*, Kluwer Academic Publishers, New York, 1994.
7. K. A. Stroud, *Linear Algebra*, Industrial Press Inc., New York, 2008.

4 Vector Calculus

Mathematics is the queen of the sciences and arithmetic is the queen of mathematics. She often condescends to render service to astronomy and other natural sciences, but in all relations, she is entitled to first rank.

Carl Friedrich Gauss

4.1 Vector Differentiation

In order to introduce electric and magnetic field theories and, ultimately, the rationale for various parts of the NMR Hamiltonian we must review our vector calculus.†

Vector differentiation (with respect to time, which is what we will ultimately be interested in) is defined for a static coordinate system by:

$$\frac{\mathrm{d}\vec{\mathbf{a}}}{\mathrm{d}t} = \frac{\mathrm{d}a_x}{\mathrm{d}t}\vec{\mathbf{u}}_x + \frac{\mathrm{d}a_y}{\mathrm{d}t}\vec{\mathbf{u}}_y + \frac{\mathrm{d}a_z}{\mathrm{d}t}\vec{\mathbf{u}}_z \tag{4.1}$$

and for a dynamic coordinate system by:

$$\begin{aligned}\frac{\mathrm{d}\vec{\mathbf{a}}}{\mathrm{d}t} &= \frac{\mathrm{d}a_x}{\mathrm{d}t}\vec{\mathbf{u}}_x + \frac{\mathrm{d}a_y}{\mathrm{d}t}\vec{\mathbf{u}}_y + \frac{\mathrm{d}a_z}{\mathrm{d}t}\vec{\mathbf{u}}_z \\ &\quad + a_x\frac{\mathrm{d}\vec{\mathbf{u}}_x}{\mathrm{d}t} + a_y\frac{\mathrm{d}\vec{\mathbf{u}}_y}{\mathrm{d}t} + a_z\frac{\mathrm{d}\vec{\mathbf{u}}_z}{\mathrm{d}t}\end{aligned} \tag{4.2}$$

†I did say in the preface that I would presume an understanding of calculus on the part of the reader but these topics are of sufficient importance to at least treat this chapter as a review if not a completely new topic for some readers. There is much to cover here and it may at times seem overwhelming but I have tried to make things as clear as possible. To the reader I say "stay the course" and you will be that much richer intellectually at the end, to say nothing about being prepared for the chapter on magnetic and electric fields.

Essential Mathematics for NMR and MRI Spectroscopists
By Keith C. Brown

Published by the Royal Society of Chemistry, www.rsc.org

for a rotating coordinate system in which the unit vectors are changing direction with time.

Vector differentiation follows much the same rules for normal function differentiation:

1. $\frac{\mathrm{d}}{\mathrm{d}t}(\vec{\mathbf{a}}+\vec{\mathbf{b}})=\frac{\mathrm{d}\vec{\mathbf{a}}}{\mathrm{d}t}+\frac{\mathrm{d}\vec{\mathbf{b}}}{\mathrm{d}t}$ (distributive)
2. $\frac{\mathrm{d}}{\mathrm{d}t}(\vec{\mathbf{a}}\cdot\vec{\mathbf{b}})=\vec{\mathbf{b}}\cdot\frac{\mathrm{d}\vec{\mathbf{a}}}{\mathrm{d}t}+\vec{\mathbf{a}}\cdot\frac{\mathrm{d}\vec{\mathbf{b}}}{\mathrm{d}t}$ (product rule)
3. $\frac{\mathrm{d}}{\mathrm{d}t}(\vec{\mathbf{a}}\times\vec{\mathbf{b}})=\vec{\mathbf{b}}\times\frac{\mathrm{d}\vec{\mathbf{a}}}{\mathrm{d}t}+\vec{\mathbf{a}}\times\frac{\mathrm{d}\vec{\mathbf{b}}}{\mathrm{d}t}$ (product rule)

For the orthogonal Cartesian unit vectors:

$$\vec{\mathbf{u}}_x\cdot\vec{\mathbf{u}}_y=0$$

$$\frac{\mathrm{d}}{\mathrm{d}t}(\vec{\mathbf{u}}_x\cdot\vec{\mathbf{u}}_y)=\frac{\mathrm{d}}{\mathrm{d}t}0$$

$$\vec{\mathbf{u}}_x\cdot\frac{\mathrm{d}\vec{\mathbf{u}}_y}{\mathrm{d}t}+\frac{\mathrm{d}\vec{\mathbf{u}}_x}{\mathrm{d}t}\cdot\vec{\mathbf{u}}_y=0$$

What direction is $\mathrm{d}\vec{\mathbf{u}}_x/\mathrm{d}t$ in? Since $\vec{\mathbf{u}}_x$ is a unit vector, $\mathrm{d}\vec{\mathbf{u}}_x/\mathrm{d}t$ cannot change in magnitude. The only other change that is possible is a change in direction … a rotation in other words. It is easiest to picture in the situation in which the rotation is about the z-axis (see Figure 4.1).

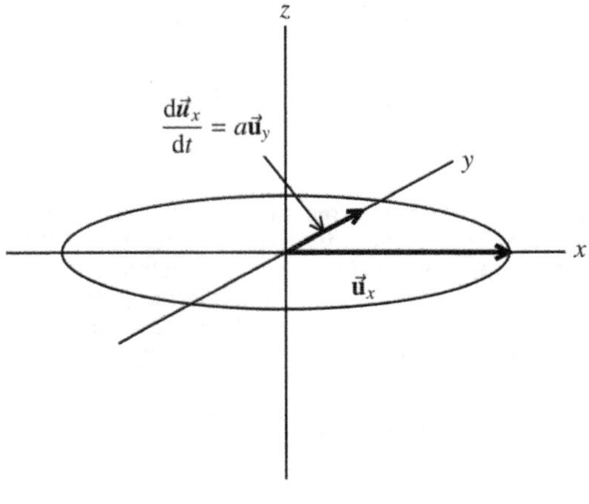

Figure 4.1 The rotating unit vector, $\vec{\mathbf{u}}$.

So, $d\vec{\mathbf{u}}_x/dt$ is at right angles to $\vec{\mathbf{u}}_x$ along the y-axis and we can write:

$$\vec{\mathbf{u}}_x \cdot \frac{d\vec{\mathbf{u}}_y}{dt} + \frac{d\vec{\mathbf{u}}_x}{dt} \cdot \vec{\mathbf{u}}_y = 0$$

$$\vec{\mathbf{u}}_x \cdot b\vec{\mathbf{u}}_x + \cdot\vec{\mathbf{u}}_y \cdot a\vec{\mathbf{u}}_y = 0$$

$$= b + a = 0$$

Why $\vec{\mathbf{u}}_x \cdot b\vec{\mathbf{u}}_x$? Since $d\vec{\mathbf{u}}_y/dt$ must be at right angles to $\vec{\mathbf{u}}_y$, it must be parallel and proportional to $\vec{\mathbf{u}}_x$. The same reasoning applies to $\vec{\mathbf{u}}_y \cdot a\vec{\mathbf{u}}_y$.

The vector operator, $\vec{\nabla}$, is often referred to as *del* or *nabla* and is defined as:

$$\vec{\nabla} = \frac{\partial}{\partial x}\vec{\mathbf{u}}_x + \frac{\partial}{\partial y}\vec{\mathbf{u}}_y + \frac{\partial}{\partial z}\vec{\mathbf{u}}_z \tag{4.3}$$

The *gradient* is defined in terms of del and a scalar function, $f(x, y, z)$:

$$\vec{\nabla} f(x,y,z) \ (\text{or}\,\text{grad} f(x,y,z)) = \left(\frac{\partial}{\partial x}\vec{\mathbf{u}}_x + \frac{\partial}{\partial y}\vec{\mathbf{u}}_y + \frac{\partial}{\partial z}\vec{\mathbf{u}}_z\right) f(x,y,z)$$

$$= \frac{\partial f}{\partial x}\vec{\mathbf{u}}_x + \frac{\partial f}{\partial y}\vec{\mathbf{u}}_y + \frac{\partial f}{\partial z}\vec{\mathbf{u}}_z \tag{4.4}$$

$$= P(x,y,z)\vec{\mathbf{u}}_x + Q(x,y,z)\vec{\mathbf{u}}_y + R(x,y,z)\vec{\mathbf{u}}_z$$

$f(x, y, z)$ is a scalar function that defines a *scalar field* and the action of $\vec{\nabla}$ on $f(x, y, z)$ produces a vector function (sometimes referred to as a vector-valued function). eqn (4.4) contains three components of the vector function, P, Q and R. A scalar field is a region of space in which there is a scalar value associated with each point in the space. For example, one might define a scalar field of temperatures within a region of the atmosphere. Similarly, a *vector field* is a region of space where there is a vector defined for each point in the space. Atmospheric wind velocity would constitute a tangible example of a vector field.

The *divergence* is the inner product of del and a vector and, as you might suspect if you have read Chapter 3, produces a scalar:

$$\vec{\nabla} \cdot \vec{\mathbf{a}} \ (\text{or}\,\text{div}\,\vec{\mathbf{a}}) = \left(\frac{\partial}{\partial x}\vec{\mathbf{u}}_x + \frac{\partial}{\partial y}\vec{\mathbf{u}}_y + \frac{\partial}{\partial z}\vec{\mathbf{u}}_z\right) \cdot (a_x\vec{\mathbf{u}}_x + a_y\vec{\mathbf{u}}_y + a_z\vec{\mathbf{u}}_z)$$

$$= \frac{\partial a_x}{\partial x} + \frac{\partial a_y}{\partial y} + \frac{\partial a_{xz}}{\partial z} \tag{4.5}$$

The third important operation is the *curl* and is defined as the vector cross product of del and a vector:

$$\vec{\nabla} \times \vec{\mathbf{a}}\ (\text{or curl}\ \vec{\mathbf{a}}) = \left(\frac{\partial}{\partial x}\vec{\mathbf{u}}_x + \frac{\partial}{\partial y}\vec{\mathbf{u}}_y + \frac{\partial}{\partial z}\vec{\mathbf{u}}_z\right) \times (a_x\vec{\mathbf{u}}_x + a_y\vec{\mathbf{u}}_y + a_z\vec{\mathbf{u}}_z)$$

$$= \begin{vmatrix} \vec{\mathbf{u}}_x & \vec{\mathbf{u}}_y & \vec{\mathbf{u}}_z \\ \frac{\partial}{\partial x} & \frac{\partial}{\partial y} & \frac{\partial}{\partial z} \\ a_x & a_y & a_z \end{vmatrix} \tag{4.6}$$

$$= \left(\frac{\partial a_z}{\partial y} - \frac{\partial a_y}{\partial z}\right)\vec{\mathbf{u}}_x - \left(\frac{\partial a_z}{\partial x} - \frac{\partial a_x}{\partial z}\right)\vec{\mathbf{u}}_y + \left(\frac{\partial a_y}{\partial x} - \frac{\partial a_x}{\partial y}\right)\vec{\mathbf{u}}_z$$

The divergence is analogous to the scalar product discussed in Chapter 3 and is a measure of, well . . . the divergence of a vector field. The curl is a measure of the "vorticity" of the vector field. In a sense these are opposites; a function that is purely divergent has no vorticity and a purely vortex function has no divergence. To gain a better understanding of the meaning of these operations we will use the following pictures. We start with the picture of a two-dimensional vector field, $\vec{\mathbf{F}} = x\vec{\mathbf{u}}_x + y\vec{\mathbf{u}}_y$ (Figure 4.2).

We imagine that these are, for example, force vectors and that a particle's motion will follow the direction of the force vector where it finds itself in the field. Next, we imagine groups of particles each of which is free to follow its own path that is determined by its particular force vector. We place a box around the group and then follow the group as each particle moves and, in particular, we are interested in what happens to the box around the particles. For our current example we show the state of the "boxes" at two different times (see Figures 4.3 and 4.4).

We see that in this case with this particular vector field the particle "box" increases in size, which indicates that the particles within the box area are moving apart from each other . . . they are diverging from each other. A divergence calculation should therefore give a positive number for this vector field (we are using the two-dimensional version of del here):

$$\vec{\nabla} \cdot \vec{\mathbf{F}} = \left(\frac{\partial}{\partial x}\vec{\mathbf{u}}_x + \frac{\partial}{\partial y}\vec{\mathbf{u}}_y\right) \cdot (x\vec{\mathbf{u}}_x + y\vec{\mathbf{u}}_y)$$

$$= \frac{\partial x}{\partial x} + \frac{\partial y}{\partial y}$$

$$= 1 + 1 = 2$$

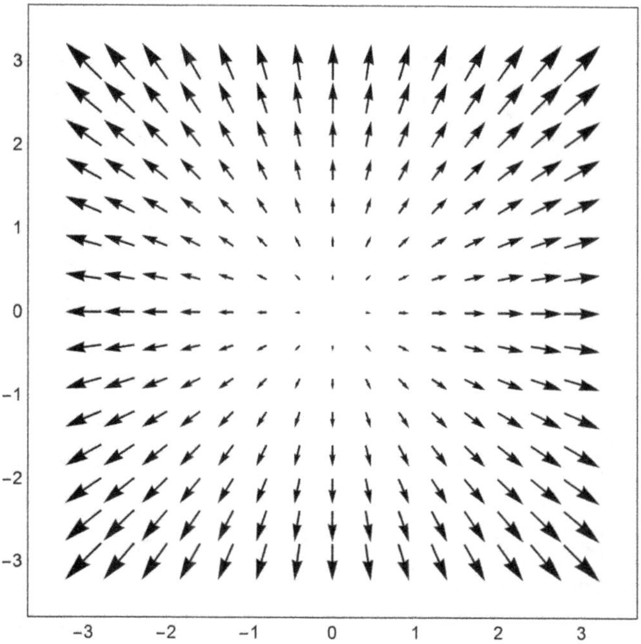

Figure 4.2 Vector field, $\vec{\mathbf{F}} = x\vec{\mathbf{u}}_x + y\vec{\mathbf{u}}_y$.

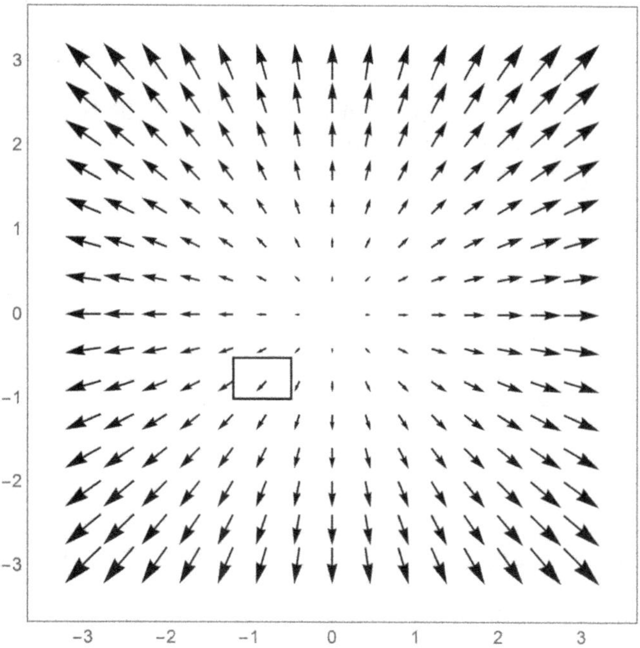

Figure 4.3 Particle group box in field $\vec{\mathbf{F}} = x\vec{\mathbf{u}}_x + y\vec{\mathbf{u}}_y$ at time t.

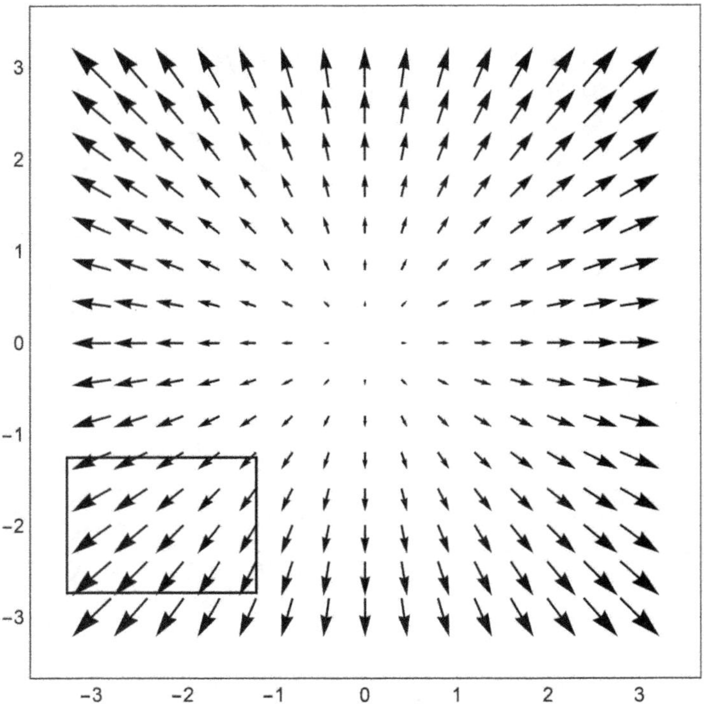

Figure 4.4 Particle group box in field $\vec{\mathbf{F}} = x\vec{\mathbf{u}}_x + y\vec{\mathbf{u}}_y$ at time $t + t'$.

Furthermore, we do not expect any "vorticity" or rotational motion associated with any points in the vector field that should be reflected in a zero value for the curl:

$$\begin{aligned}
\vec{\nabla} \times \vec{\mathbf{F}} &= \left(\frac{\partial}{\partial x}\vec{\mathbf{u}}_x + \frac{\partial}{\partial y}\vec{\mathbf{u}}_y + \frac{\partial}{\partial z}\vec{\mathbf{u}}_z\right) \times (x\vec{\mathbf{u}}_x + y\vec{\mathbf{u}}_y) \\
&= \begin{vmatrix} \vec{\mathbf{u}}_x & \vec{\mathbf{u}}_y & \vec{\mathbf{u}}_z \\ \dfrac{\partial}{\partial x} & \dfrac{\partial}{\partial y} & \dfrac{\partial}{\partial z} \\ x & y & 0 \end{vmatrix} \\
&= \left(\frac{\partial 0}{\partial y} - \frac{\partial y}{\partial z}\right)\vec{\mathbf{u}}_x - \left(\frac{\partial 0}{\partial x} - \frac{\partial x}{\partial z}\right)\vec{\mathbf{u}}_y + \left(\frac{\partial y}{\partial x} - \frac{\partial x}{\partial y}\right)\vec{\mathbf{u}}_z \\
&= 0 - 0 + 0 \\
&= 0
\end{aligned}$$

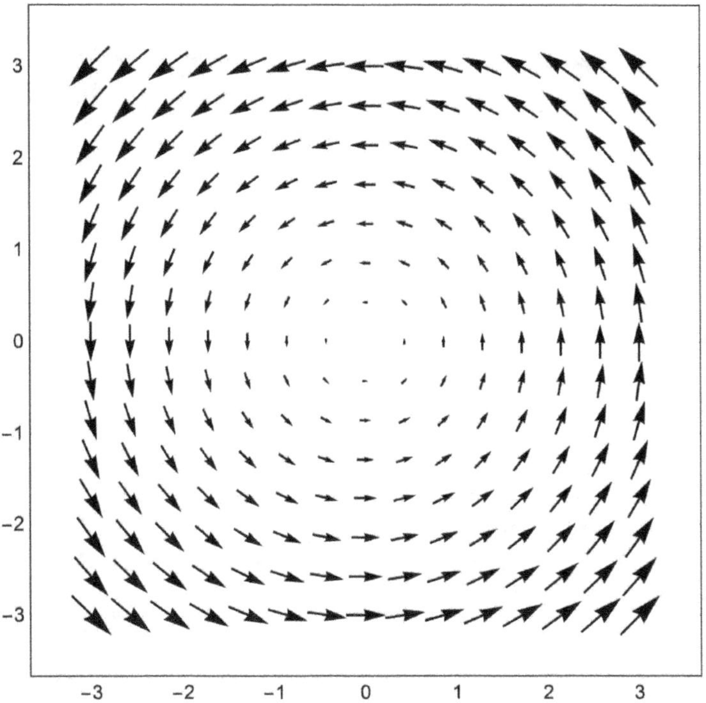

Figure 4.5 Vector field, $\vec{\mathbf{F}} = y\vec{\mathbf{u}}_x + x\vec{\mathbf{u}}_y$.

We can go to the other extreme and consider an example of a vector-field with zero divergence and positive vorticity, $\vec{\mathbf{F}} = -y\vec{\mathbf{u}}_x + x\vec{\mathbf{u}}_y$ (Figure 4.5).

We see here that rather than pointing outwards from the origin, the vectors are circulating about the origin. Now, as before, we imagine a group of particles that follow the "flow" of the vectors in the field and enclose the group in a box (Figure 4.6).

All of the particles, in this case, will flow such that the box rotates counter-clockwise but does not change in size (Figure 4.7). This means that there is a positive curl and a zero divergence.

Again, we may verify this. For the divergence:

$$\begin{aligned}
\vec{\nabla} \cdot \vec{\mathbf{F}} &= \left(\frac{\partial}{\partial x}\vec{\mathbf{u}}_x + \frac{\partial}{\partial y}\vec{\mathbf{u}}_y \right) \cdot (-y\vec{\mathbf{u}}_x + x\vec{\mathbf{u}}_y) \\
&= -\frac{\partial y}{\partial x} + \frac{\partial x}{\partial y} \\
&= 0
\end{aligned}$$

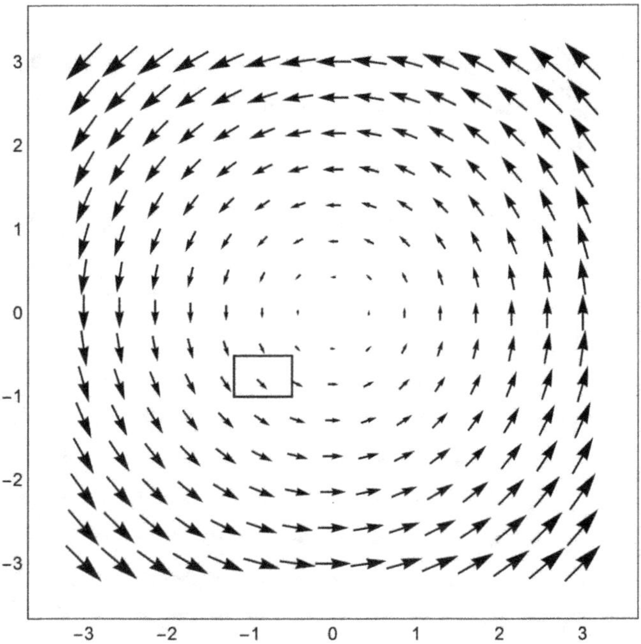

Figure 4.6 Particle group box in field $\vec{\mathbf{F}} = y\vec{\mathbf{u}}_x + x\vec{\mathbf{u}}_y$ at time t.

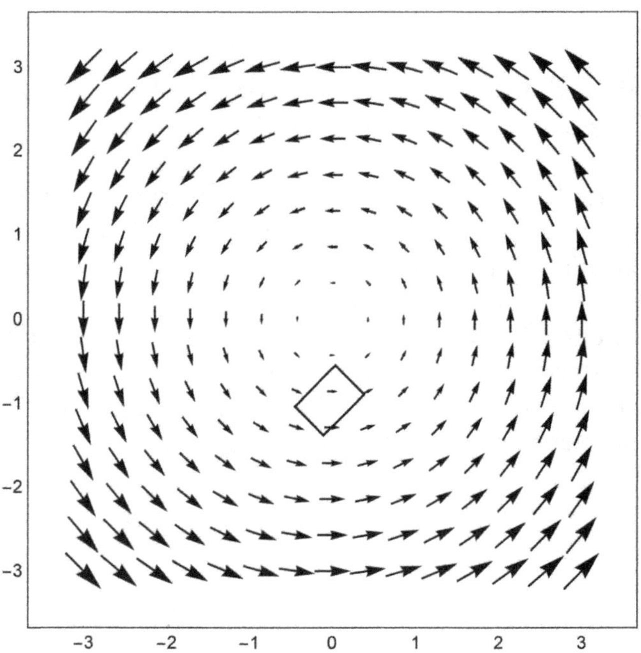

Figure 4.7 Particle group box in field $\vec{\mathbf{F}} = y\vec{\mathbf{u}}_x + x\vec{\mathbf{u}}_y$ at time $t + t'$.

and for the curl:

$$\vec{\nabla} \times \vec{\mathbf{F}} = \left(\frac{\partial}{\partial x}\vec{\mathbf{u}}_x + \frac{\partial}{\partial y}\vec{\mathbf{u}}_y + \frac{\partial}{\partial z}\vec{\mathbf{u}}_z\right) \times (-y\vec{\mathbf{u}}_x + x\vec{\mathbf{u}}_y)$$

$$= \begin{vmatrix} \vec{\mathbf{u}}_x & \vec{\mathbf{u}}_y & \vec{\mathbf{u}}_z \\ \frac{\partial}{\partial x} & \frac{\partial}{\partial y} & \frac{\partial}{\partial z} \\ -y & x & 0 \end{vmatrix}$$

$$= \left(\frac{\partial 0}{\partial y} - \frac{\partial x}{\partial z}\right)\vec{\mathbf{u}}_x - \left(\frac{\partial 0}{\partial x} - \frac{\partial y}{\partial z}\right)\vec{\mathbf{u}}_y + \left(\frac{\partial x}{\partial x} + \frac{\partial y}{\partial y}\right)\vec{\mathbf{u}}_z$$

$$= 0 - 0 + 2\vec{\mathbf{u}}_z$$

Based on these three operations, del, div and curl, we can list some identities involving the del operator:

1. $\vec{\nabla}\,(f+g) = \vec{\nabla}f + \vec{\nabla}g$ (distributive)
2. $\vec{\nabla}\cdot(\vec{\mathbf{a}}+\vec{\mathbf{b}}) = \vec{\nabla}\cdot\vec{\mathbf{a}} + \vec{\nabla}\cdot\vec{\mathbf{b}}$ (distributive)
3. $\vec{\nabla}\times(\vec{\mathbf{a}}+\vec{\mathbf{b}}) = \vec{\nabla}\times\vec{\mathbf{a}} + \vec{\nabla}\times\vec{\mathbf{b}}$ (distributive)
4. $\vec{\nabla}\cdot(f\vec{\mathbf{a}}) = (\vec{\nabla}\cdot f)\,\vec{\mathbf{a}} + f(\vec{\nabla}\cdot\vec{\mathbf{a}}$ (via the product rule)
5. $\vec{\nabla}\times(f\vec{\mathbf{a}}) = (\vec{\nabla}f\times\vec{\mathbf{a}} + f(\vec{\nabla}\times\vec{\mathbf{a}}$ (via the product rule)
6. $\vec{\nabla}\cdot(\vec{\nabla}f) = \nabla^2 f$ (the Laplacian operator)
7. $\vec{\nabla}\times(\vec{\nabla}\times\vec{\mathbf{a}}) = \vec{\nabla}(\vec{\nabla}\cdot\vec{\mathbf{a}}) + \nabla^2\,\vec{\mathbf{a}}$
8. $\vec{\nabla}\times\vec{\mathbf{u}}_x = \vec{\nabla}\times\vec{\mathbf{u}}_y = \vec{\nabla}\times\vec{\mathbf{u}}_z = 0$

We have not yet said what will be the effect of using the del operator directly on a vector. This operation produces an object called a dyad, the discussion of which we will defer to Chapter 5.

We will use these ideas and what follows in our discussion of electric and magnetic fields, which both turn out to be vector fields.

4.2 Vector Integration

The integral of a vector $\vec{\mathbf{V}}(t)$, which is a vector function of a scalar variable or a vector field, is defined in terms of its components:

$$\begin{aligned}\vec{\mathbf{V}}(t) &= x(t)\vec{\mathbf{u}}_x + y(t)\vec{\mathbf{u}}_y + z(t)\vec{\mathbf{u}}_z \\ \int \vec{\mathbf{V}}(t)\mathrm{d}t &= \int (x(t) + y(t) + z(t))\mathrm{d}t \\ &= \int x(t)\mathrm{d}t + \int y(t)\mathrm{d}t + \int z(t)\mathrm{d}t\end{aligned} \tag{4.7}$$

$x(t)$, $y(t)$ and $z(t)$ are the x, y and z components of the vector expressed as functions of t. This an example of a parametric equation from first calculus courses, with t as the parameter in this case (see Appendix A).

Alternatively, if we know that there is a vector, $\vec{\mathbf{S}}(t)$, such that:

$$\vec{\mathbf{V}}(t) = \frac{\mathrm{d}}{\mathrm{d}t}\vec{\mathbf{S}}(t)$$

then the indefinite integral of this expression is:

$$\int \vec{\mathbf{V}}(t)\mathrm{d}t = \int \frac{\mathrm{d}}{\mathrm{d}t}\vec{\mathbf{S}}(t)\mathrm{d}t$$

Of more interest to us in the context of magnetic fields and NMR spectroscopy is the *line integral*. Let us suppose that we have a parametric vector function $\vec{\mathbf{r}}(t)$, which is a position vector:

$$\vec{\mathbf{r}}(t) = x(t)\vec{\mathbf{u}}_x + y(t)\vec{\mathbf{u}}_y + z(t)\vec{\mathbf{u}}_z$$

$\vec{\mathbf{r}}(t)$ defines a curve in three-dimensional space between two points, P_1 and P_2 such that $\vec{\mathbf{r}}(t_1)$ defines P_1 and $\vec{\mathbf{r}}(t_2)$ defines P_2 (see Figure 4.8). The vector between P_1 and P_2 (not shown in Figure 4.8) is just the vector difference between $\vec{\mathbf{r}}(t_2)$ and $\vec{\mathbf{r}}(t_1)$, $\vec{\mathbf{r}}(t_2) - \vec{\mathbf{r}}(t_1)$.

The derivative of $\vec{\mathbf{r}}(t)$ using eqn (4.1) is:

$$\frac{\mathrm{d}}{\mathrm{d}t}\vec{\mathbf{r}}(t) = \frac{\mathrm{d}x(t)}{\mathrm{d}t}\vec{\mathbf{u}}_x + \frac{\mathrm{d}y(t)}{\mathrm{d}t}\vec{\mathbf{u}}_y + \frac{\mathrm{d}z(t)}{\mathrm{d}t}\vec{\mathbf{u}}_z$$

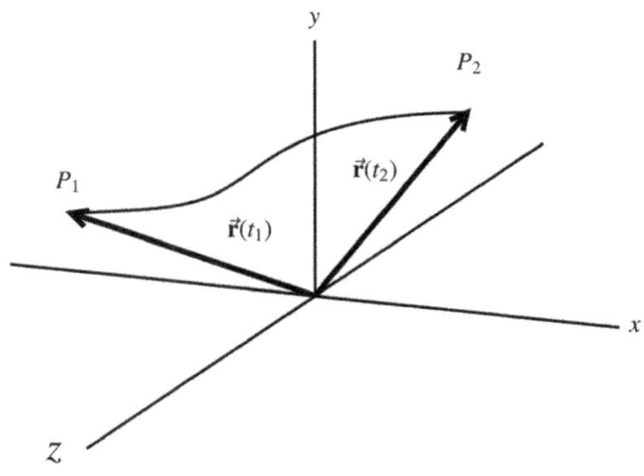

Figure 4.8 Curve defined by $\vec{\mathbf{r}}(t)$ between points P_1 and P_2.

$\frac{\mathrm{d}}{\mathrm{d}t}\vec{\mathbf{r}}(t_1)$ is also a vector for which we can do all of the by now familiar types of vector algebra on, including the calculation of its norm (using eqn (3.6)):

$$\left\|\frac{\mathrm{d}}{\mathrm{d}t}\vec{\mathbf{r}}(t)\right\| = \sqrt{\frac{\mathrm{d}}{\mathrm{d}t}\vec{\mathbf{r}}(t)\cdot\frac{\mathrm{d}}{\mathrm{d}t}\vec{\mathbf{r}}(t)}$$
$$= \sqrt{\left(\frac{\mathrm{d}x(t)}{\mathrm{d}t}\right)^2 + \left(\frac{\mathrm{d}y(t)}{\mathrm{d}t}\right)^2 + \left(\frac{\mathrm{d}z(t)}{\mathrm{d}t}\right)^2} \tag{4.8}$$
$$= \sqrt{(x'(t))^2 + (y'(t))^2 + (z'(t))^2}$$

We now ask "What is the length of the curve defined by $\vec{\mathbf{r}}(t)$ between points $P_1$and P_2?". If it were a straight line the answer would be simply the norm of the vector $\vec{\mathbf{r}}(t_2) - \vec{\mathbf{r}}(t_1)$ but this will obviously not work for a curved line. We *can* use straight line secant segments to approximate the length of the curve, however. Thus, a series of straight lines summed approximates the length of the curve. If we break the line up into n points and join them with secant segments, $\Delta\vec{\mathbf{r}}(t_i)$, as in Figure 4.9 we might represent this approximation as follows:

$$s \approx \sum_i^n \|\vec{\mathbf{r}}(t_i) - \vec{\mathbf{r}}(t_{i-1})\| = \sum_i^n \sqrt{(\vec{\mathbf{r}}(t_i) - \vec{\mathbf{r}}(t_{i-1}))\cdot(\vec{\mathbf{r}}(t_i) - \vec{\mathbf{r}}(t_{i-1}))}$$
$$= \sum_i^n \sqrt{\Delta\vec{\mathbf{r}}(t_i)\cdot\Delta\vec{\mathbf{r}}(t_i)}$$

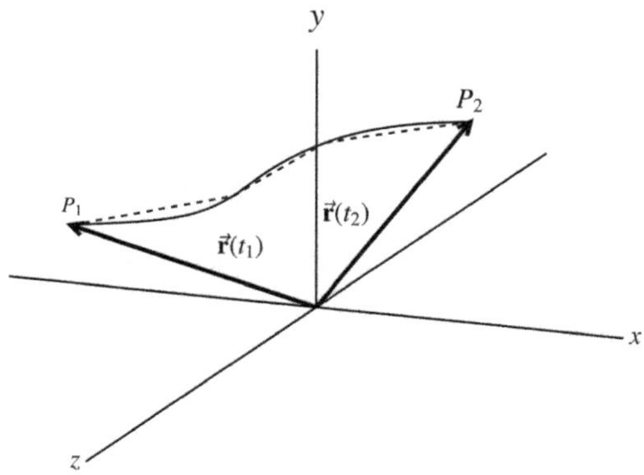

Figure 4.9 Position vector difference calculation.

$\vec{\mathbf{r}}(t)$ is written as a parametric equation, so we can write this in terms of the derivatives of the components of $\vec{\mathbf{r}}(t)$:

$$s \approx \sum_i^n \sqrt{\Delta\vec{\mathbf{r}}(t_i) \cdot \Delta\vec{\mathbf{r}}(t_i)} = \sum_i^n \|\Delta\vec{\mathbf{r}}(t_i)\|$$

$$\approx \sum_i^n \sqrt{(x(t_i) - x(t_{i-1}))^2 + (y(t_i) - y(t_{i-1}))^2 + (z(t_i) - z(t_{i-1}))^2}$$

(see equation (3.6))

$$\approx \sum_i^n \sqrt{\frac{(x(t_i) - x(t_{i-1}))^2}{t_i - t_{i-1}} + \frac{(y(t_i) - y(t_{i-1}))^2}{t_i - t_{i-1}} + \frac{(z(t_i) - z(t_{i-1}))^2}{t_i - t_{i-1}}}\Delta t$$

$$\approx \sum_i^n \sqrt{x'(w)^2 + y'(v)^2 + z'(q)^2}\Delta t$$

The last line is arrived at *via* the Mean Value theorem of calculus that states (very briefly) that at least once in the interval $[t_{i-1}, t_i]$ there will be a point on the line at which the tangent slope will be equal to the secant line slope between points P_{i-1} and P_{i}. In terms of equations:

$$\frac{(x(t_i) - x(t_{i-1}))^2}{t_i - t_{i-1}} = x'(w)$$

where w is some point between t_{i-1} and t_i at which the tangent slope is equal to the slope of the secant line. Any decent calculus text will have more details regarding this theorem; it really is a very simple idea. In the limit as n approaches infinity the secants, $\Delta\vec{\mathbf{r}}(t_{\text{i}})$, become tangents to the curve and at any point along the curve $d\vec{\mathbf{r}}(t_i)$ is the infinitesimal at that point. Summing them will then give us the exact arc length of the curve:

$$s=\int_{t_1}^{t_2}\sqrt{x'(w)^2+y'(v)^2+z'(q)^2}\,dt \tag{4.9}$$

and from the first part of the Fundamental Theorem of Calculus (see Appendix A) it follows that:

$$\frac{ds}{dt}=\sqrt{x'(t)^2+y'(t)^2+z'(t)^2}=\left\|\frac{d}{dt}\vec{\mathbf{r}}(t)\right\|$$

or (4.10)

$$ds=\sqrt{x'(t)^2+y'(t)^2+z'(t)^2}\,dt=\left\|\frac{d}{dt}\vec{\mathbf{r}}(t)\right\|dt=\|\vec{\mathbf{r}}'(t)\|dt$$

(see also eqn (4.8)). $\frac{d}{dt}\vec{\mathbf{r}}(t)$is tangent to the curve at $\vec{\mathbf{r}}(t)$ and from this we can find the *unit tangent vector* (using eqn (3.13)):

$$\frac{d}{dt}\vec{\mathbf{r}}(t)=\vec{\mathbf{r}}'(t)$$

$$\frac{\vec{\mathbf{r}}'(t)}{\|\vec{\mathbf{r}}'(t)\|}=\frac{\vec{\mathbf{r}}'(t)}{\frac{ds}{dt}}=\frac{d}{dt}\vec{\mathbf{r}}(t)\frac{dt}{ds}=\frac{d}{ds}\vec{\mathbf{r}}(t)\equiv\vec{\mathbf{T}}(t) \tag{4.11}$$

or in differential form:

$$d\vec{\mathbf{r}}(t)=\vec{\mathbf{T}}(t)d\,s \tag{4.12}$$

$\vec{\mathbf{T}}(t)$ is necessarily collinear with $\vec{\mathbf{r}}'(t)$ from eqn (4.11) and is therefore also tangent to the curve in question. We will use this idea momentarily. Squaring the norm expression for $\vec{\mathbf{T}}(t)$:

$$\|\vec{\mathbf{T}}(t)\|^2=1=\vec{\mathbf{T}}(t)\cdot\vec{\mathbf{T}}(t)$$

and differentiating both sides of the equation gives:

$$0=(\vec{\mathbf{T}}(t)\cdot\vec{\mathbf{T}}(t))'=\vec{\mathbf{T}}'(t)\cdot\vec{\mathbf{T}}(t)+\vec{\mathbf{T}}(t)\cdot\vec{\mathbf{T}}'(t)\quad\text{(product rule)}$$

$$=2\vec{\mathbf{T}}'(t)\cdot\vec{\mathbf{T}}(t)$$

and therefore

$$\vec{\mathbf{T}}'(\mathrm{t})\perp\vec{\mathbf{T}}(\mathrm{t})$$

"Why is this so?" you ask. The answer is that we see here that $\vec{\mathbf{T}}'(t)\cdot\vec{\mathbf{T}}(t)$ must equal 0 that, as you will recall from Chapter 3, means that the angle between the two vectors is some multiple of $\frac{\pi}{2}$. From this result we define the *unit normal vector*, $\vec{\mathbf{n}}$ (using eqn (3.13)):

$$\vec{\mathbf{n}} \equiv \frac{\vec{\mathbf{T}}'(t)}{\|\vec{\mathbf{T}}'(t)\|} \tag{4.13}$$

We now have a unit tangent vector and a unit normal vector that we may also express in component form. From the above analysis (eqn (4.11)) we know:

$$\vec{\mathbf{T}}(t) = \frac{\mathrm{d}}{\mathrm{d}s}\vec{\mathbf{r}}(t)$$

and if we express $\frac{\mathrm{d}}{\mathrm{d}s}\vec{\mathbf{r}}(t)$ in vector component form:‡

$$\vec{\mathbf{r}}(t) = x(t)\vec{\mathbf{u}}_x + y(t)\vec{\mathbf{u}}_y$$

$$\vec{\mathbf{T}}(t) = \frac{\mathrm{d}}{\mathrm{d}s}\vec{\mathbf{r}}(t) = \frac{\mathrm{d}x(t)}{\mathrm{d}s}\vec{\mathbf{u}}_x + \frac{\mathrm{d}y(t)}{\mathrm{d}s}\vec{\mathbf{u}}_y$$

A 90° clockwise rotation (see eqn (3.35)) of this vector will preserve its unit length to produce the unit normal vector in component form:

$$\hat{\mathbf{R}}(90)\vec{\mathbf{T}}(t) = \begin{bmatrix} 0 & 1 \\ -1 & 0 \end{bmatrix} \begin{bmatrix} \frac{\mathrm{d}x(t)}{\mathrm{d}s} \\ \frac{\mathrm{d}y(t)}{\mathrm{d}s} \end{bmatrix} = \begin{bmatrix} \frac{\mathrm{d}y(t)}{\mathrm{d}s} \\ \frac{-\mathrm{d}x(t)}{\mathrm{d}s} \end{bmatrix} \Rightarrow \frac{\mathrm{d}y(t)}{\mathrm{d}s}\vec{\mathbf{u}}_x - \frac{\mathrm{d}x(t)}{\mathrm{d}s}\vec{\mathbf{u}}_y = \vec{\mathbf{n}} \tag{4.14}$$

Why not a counter-clockwise rotation? At the risk of getting ahead of ourselves . . . by convention, for a closed curve the positive direction is defined as a counter-clockwise movement around the curve in order to fit in with Green's theorem (*vide infra*). This being the case we also consider that the normal is "positive" when pointing out of the surface enclosed by a curve, also for compatibility with Green's theorem. This is accomplished with a clockwise rotation of $\vec{\mathbf{r}}(t)$ towards n (Figure 4.10)§.

‡We will temporarily revert to two dimensions for clarity.
§Note that in Figure 4.10, since $\vec{\mathbf{T}}$ is pointed to the right, a 90° clockwise rotation from our point of view should result in a vector that is pointed downwards. However, if one's point of view was on the *other* side of the curve then a clockwise rotation would be as shown.

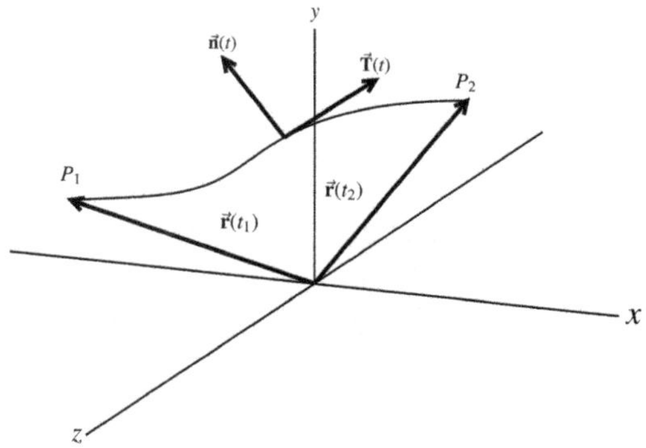

Figure 4.10 Unit tangent and unit normal vectors on a curve at point $P(t)$.

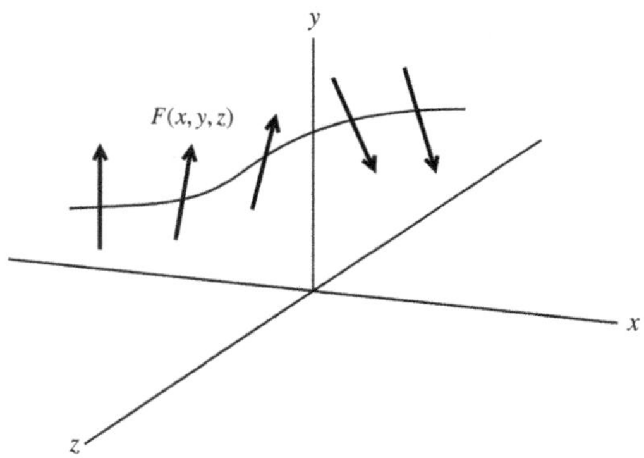

Figure 4.11 A vector field, $\vec{F}(x, y, z)$, along a curve.

Now, let us turn our attention to how we might calculate the work done on a particle that moves along a curve rather than a straight line. The straight line calculation is, well . . . straightforward. Assuming the force vector and the direction of change of motion to be parallel, as it is usually presented in first physics courses, the vector notation for the work done collapses to a scalar equation:

$$W = \vec{\mathbf{F}} \cdot \vec{\mathrm{r}}' = \|\vec{\mathbf{F}}\| \, \|\vec{\mathbf{r}}'\| \cos(\theta) = \|\vec{\mathbf{F}}\| \, \|\vec{\mathbf{r}}'\| \;\; (\theta = 0)$$

On a curve, however, the force vector will not necessarily be collinear with $\vec{\mathbf{r}}'$ and may very well have different directions and magnitudes at different places along the curve (see Figure 4.11). Let us try to visualise what is happening here as a particle moves along a curve such as we have been looking at. As it moves from point 1 to point 2 it passes through a vector field defined by $\vec{\mathbf{F}}(x, y, z)$ where each of the components of $\vec{\mathbf{F}}(x, y, z)$ is a function of x, y and z:

$$\vec{\mathbf{F}}(x, y, z) = P(x, y, z)\vec{\mathbf{u}}_x + Q(x, y, z)\vec{\mathbf{u}}_y + R(x, y, z)\vec{\mathbf{u}}_z$$

At a point $P(t)$ where the particle is on the curve, the force acts on the particle as it moves along the curve. So far, so good.

Now, if we want to calculate the infinitesimal work done on the particle at point $P(t)$ on the curve we apply the work equation:

$$\mathrm{d}W = \vec{\mathbf{F}}(x, y, z) \cdot \mathrm{d}\vec{\mathbf{r}}'$$

in which we cannot assume that the force and direction vectors are collinear (Figure 4.12) and must use the full inner product calculation. In order to complete the calculation we must compute the component of $\vec{\mathbf{F}}(x, y, z)$ that is collinear with $\mathrm{d}\vec{\mathbf{r}}'$:

$$\begin{aligned}\mathrm{d}W &= \vec{\mathbf{F}}(x,y,z) \cdot \frac{\mathrm{d}\vec{\mathbf{r}}'}{\mathrm{d}s}\mathrm{d}s \\ &= \vec{\mathbf{F}}(x,y,z) \cdot \vec{\mathbf{T}}(t)\mathrm{d}s\end{aligned}$$

That this is the component of $\vec{\mathbf{F}}(x, y, z)$ in the direction of $\vec{\mathbf{T}}(t)$ we can see from the definition of the scalar value of the inner product of $\vec{\mathbf{F}}(x, y, z)$ and $\vec{\mathbf{T}}(t)$ (Figure 4.13):

$$\begin{aligned}\vec{\mathbf{F}}(x,y,z) \cdot \vec{\mathbf{T}}(t) &= \left\|\vec{\mathbf{F}}(x,y,z)\right\| \left\|\vec{\mathbf{T}}(t)\right\| \cos(\theta) \\ &= \left\|\vec{\mathbf{F}}(x,y,z)\right\| \cos(\theta)\end{aligned}$$

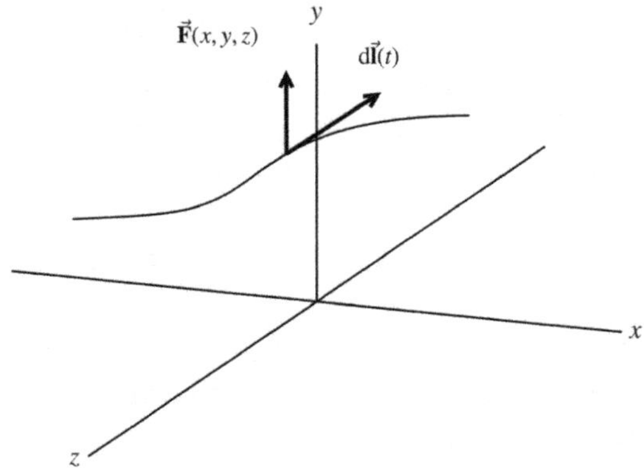

Figure 4.12 Force and direction vectors on a curve.

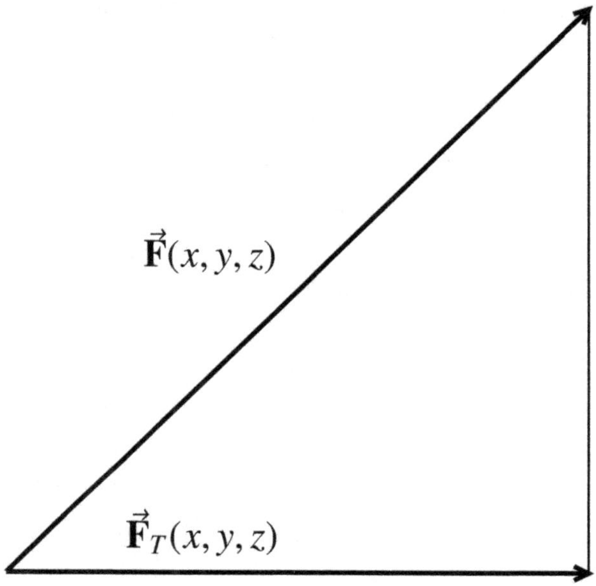

Figure 4.13 Component of $\vec{\mathbf{F}}(x,y,z)$, in the direction of $\vec{\mathbf{T}}$.

Next, we sum up all the infinitesimals along the curve:

$$W = \int_C \vec{\mathbf{F}}(x, y, z) \cdot \vec{\mathbf{T}}(t) \mathrm{d}s$$

where $\vec{\mathbf{T}}(t)$ is the unit tangent vector and the symbol $\int_C$ refers to a special kind of integral called a *line integral*, alluded to earlier. This is an integral with respect to arc length instead of with respect to coordinate axes as we usually learn in our first calculus courses. Making a few substitutions using our previous results:

$$\begin{aligned} W &= \int_C \vec{\mathbf{F}}(x, y, z) \cdot \vec{\mathbf{T}}(t) \mathrm{d}s \\ &= \int_C \vec{\mathbf{F}}(x, y, z) \cdot \frac{\vec{\mathbf{r}}'(t)}{\|\vec{\mathbf{r}}'(t)\|} \mathrm{d}s \\ &= \int_C \vec{\mathbf{F}}(x, y, z) \cdot \frac{\vec{\mathbf{r}}'(t)}{\|\vec{\mathbf{r}}'(t)\|} \|\vec{\mathbf{r}}'(t)\| \mathrm{d}t \\ &= \int_C \vec{\mathbf{F}}(x, y, z) \cdot \vec{\mathbf{r}}'(t) \mathrm{d}t \end{aligned}$$

To understand how to get from line 2 to line 3 please see eqn (4.10). Frequently, the last line is written in a bit more compact form:

$$\begin{aligned} W &= \int_C \vec{\mathbf{F}}(x,y,z) \cdot \vec{\mathbf{r}}'(t)\mathrm{d}t \\ &= \int_C \vec{\mathbf{F}}(x,y,z) \cdot \mathrm{d}\vec{\mathbf{r}}' \end{aligned}$$

We may also write this as follows:

$$\begin{aligned} \vec{\mathbf{F}} &= P\vec{\mathbf{u}}_x + Q\vec{\mathbf{u}}_y + R\vec{\mathbf{u}}_z \\ \mathrm{d}\vec{\mathbf{r}}' &= \mathrm{d}x\vec{\mathbf{u}}_x + \mathrm{d}y\vec{\mathbf{u}}_y + \mathrm{d}z\vec{\mathbf{u}}_z \\ \vec{\mathbf{F}} \cdot \mathrm{d}\vec{\mathbf{r}}' &= (P\vec{\mathbf{u}}_x + Q\vec{\mathbf{u}}_y + R\vec{\mathbf{u}}_z) \cdot (\mathrm{d}x\vec{\mathbf{u}}_x + \mathrm{d}y\vec{\mathbf{u}}_y + \mathrm{d}z\vec{\mathbf{u}}_z) \\ &= P\mathrm{d}x + Q\mathrm{d}y + R\mathrm{d}z \\ \int_C \vec{\mathbf{F}}(x,y,z) \cdot \mathrm{d}\vec{\mathbf{r}}' &= \int_C P\mathrm{d}x + Q\mathrm{d}y + R\mathrm{d}z \end{aligned} \tag{4.15}$$

In the above analysis, the component of force collinear with the unit tangent vector $\vec{\mathbf{T}}(t)$ was used to calculate the work done in moving the particle along the line. What about the other component of the force vector? This is the component that is collinear with the unit normal vector $\vec{\mathbf{n}}$, mentioned earlier and illustrated in Figure 4.10. The force in this direction is called the *flux* of the force crossing the line and is an important concept in connection with Green's theorem and others that we will look at a bit later. Figure 4.14 shows the flux vector, $\vec{\mathbf{F}}_n(x, y, z)$ normal to a point along the curve. The word "flux" comes from Latin and means "flow". The physical meaning of the flux is a bit clearer if we consider that the function $\vec{\mathbf{F}}_n(x, y, z)$ represents the direction of flow of a fluid rather than a force function. In this case we see that the normal component of the fluid flow at point P is the flux across the line. The concept, however, applies to any vector field function that crosses a curve (or a surface as we will see) but does not actually "flow" in the fluid sense. The calculation is very similar to the line integral:

$$\begin{aligned} W &= \int_C \vec{\mathbf{F}}(x,y,z) \cdot \vec{\mathbf{n}}\mathrm{d}s \\ &= \int_C \vec{\mathbf{F}}(x,y,z) \cdot \frac{\vec{\mathbf{T}}'(t)}{\left\|\vec{\mathbf{T}}'(t)\right\|}\mathrm{d}s \end{aligned}$$

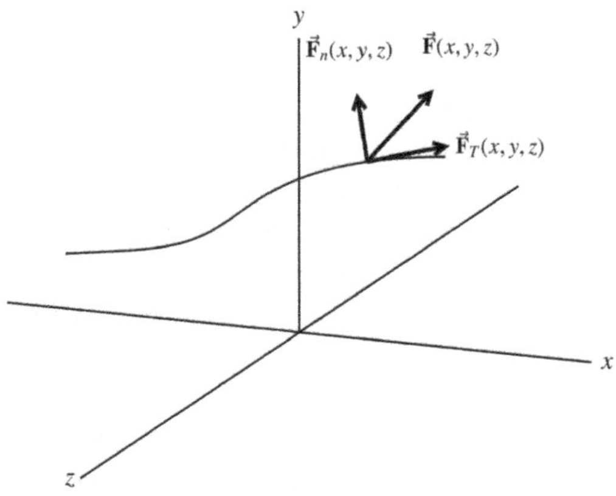

Figure 4.14 $\vec{F}(x,y,z)$, and its components tangent and normal to the curve at point P.

In which we have used eqn (4.13). More succinctly:

$$W = \int_C \vec{\mathbf{F}} \cdot \vec{\mathbf{n}} \mathrm{d}s \tag{4.16}$$

We now invoke the del operator. Specifically, we want to use it on a scalar function, $f(x, y, z)$ to produce, as we have seen above, a vector function, $\vec{\mathbf{F}}(x, y, z)$. If this is our vector function for the force on the particle on the line then $f(x, y, z)$ is referred to as the *scalar potential function*. There is a happy consequence of this in that the value of the line integral for the work done does not matter what path is taken. In other words, between points P_1 and P_2, it does not matter what curve is used to move the particle … the work done will be the same regardless of which path is used. Using the definition for del and our line integral for work done, we can show this:

$$\vec{\mathbf{F}}(x,y,z) = \vec{\nabla} f(x,y,z)$$

$$\begin{aligned}
W &= \int_C \vec{\mathbf{F}} \cdot \mathrm{d}\vec{\mathbf{r}}' \\
&= \int_C \vec{\nabla} f(x,y,z) \cdot \vec{\mathbf{r}}'(t) \mathrm{d}t \\
&= \int_a^b \left(\frac{\mathrm{d}f}{\mathrm{d}x} \vec{\mathbf{u}}_x + \frac{\mathrm{d}f}{\mathrm{d}y} \vec{\mathbf{u}}_y + \frac{\mathrm{d}f}{\mathrm{d}z} \vec{\mathbf{u}}_z \right) \cdot \left(\frac{\mathrm{d}x}{\mathrm{d}t} \vec{\mathbf{u}}_x + \frac{\mathrm{d}y}{\mathrm{d}t} \vec{\mathbf{u}}_y + \frac{\mathrm{d}z}{\mathrm{d}t} \vec{\mathbf{u}}_z \right) \mathrm{d}t \\
&= \int_a^b \left(\frac{\mathrm{d}f}{\mathrm{d}x} \frac{\mathrm{d}x}{\mathrm{d}t} + \frac{\mathrm{d}f}{\mathrm{d}y} \frac{\mathrm{d}y}{\mathrm{d}t} + \frac{\mathrm{d}f}{\mathrm{d}z} \frac{\mathrm{d}z}{\mathrm{d}t} \right) \mathrm{d}t \\
&= \int_a^b \frac{\mathrm{d}f}{\mathrm{d}t} \mathrm{d}t = \int_a^b \mathrm{d}f = f(x_b, y_b, z_b) - f(x_a, y_a, z_a)
\end{aligned} \tag{4.17}$$

Thus, given the dependence of $\vec{\mathbf{F}}(x, y, z)$ on the gradient of a scalar potential function, $f(x, y, z)$, the value of the line integral depends only on the values of the potential function at the beginning and end of the curve and not on the path taken between them.¶ The vector function $\vec{\mathbf{F}}(x, y, z)$ so generated is said to be *conservative*. This result has another very interesting facet that we can emphasise... this is the line integral version of the fundamental theorem of calculus in much the same form as expressed in the fundamental theorem of calculus as shown in Appendix A. The scalar function $f(x, y, z)$ is an anti-derivative of $\frac{df}{dt}$. Its importance lies in the idea that we do not need to know what the curve is between points a and b, just the difference between the values of the scalar function at the starting and ending points of the curve, $f(x_a, y_a, z_a)$ and $f(x_b, y_b, z_b)$. The end points are called the "boundaries". This is a common theme in vector calculus, as we will see.

What happens if our curve loops back to its original starting point? In other words, what about a closed curve? If our vector field function, $\vec{\mathbf{F}}(x, y, z)$, is conservative then evaluation of the integral will give a zero result:

$$\begin{aligned}\vec{\mathbf{F}}(x,y,z) &= \vec{\nabla}f(x,y,z)\\ W &= \oint \vec{\mathbf{F}}\cdot d\vec{\mathbf{r}}'\\ &= \oint \vec{\nabla}f(x,y,z)\cdot\vec{\mathbf{r}}'(t)dt\\ &= \oint df = f(x_a,y_a,z_a) - f(x_a,y_a,z_a) = 0\end{aligned} \tag{4.18}$$

Note the use of the symbol $\oint$ to denote a line integral on a closed curve. This can also work in reverse; evaluation of a line integral over a closed curve to give a zero result indicates that the vector field is conservative.

We diverge momentarily in order to prepare for the next topic. Let us consider a surface, $z = f(x, y) = x^2 + y^2$ (see Figure 4.15). We can slice the surface with a plane that is parallel to the x, y-plane to produce a curve on the surface that is defined by the height of the plane above the x, y-plane (Figure 4.16).

In our case, with the plane located at c units above the x,y-plane, we find that the curve so outlined has the equation $f(x,y) = x^2 + y^2 = c$.

¶Note that the scalar function $f(x, y, z)$ is not unique. Any constant, K added to $f(x, y, z)$ will also serve to generate $\vec{\mathbf{F}}(x,y,z)$ since the derivative of a constant is zero, *i.e.* $\vec{\nabla}(f+K) = \vec{\nabla}f$.

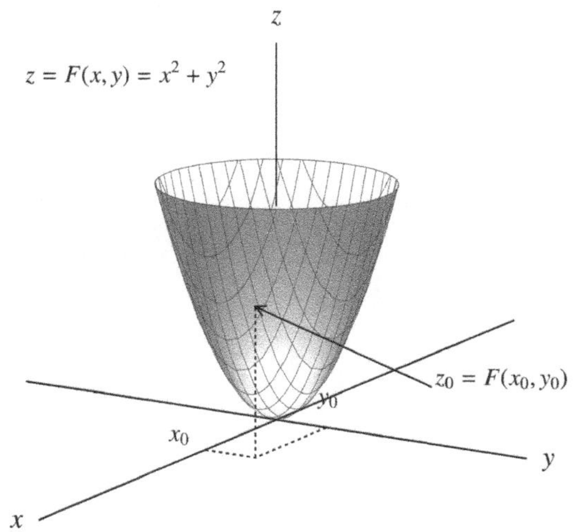

Figure 4.15 Three-dimensional shape defined by $z = F(x, y) = x^2 + y^2$.

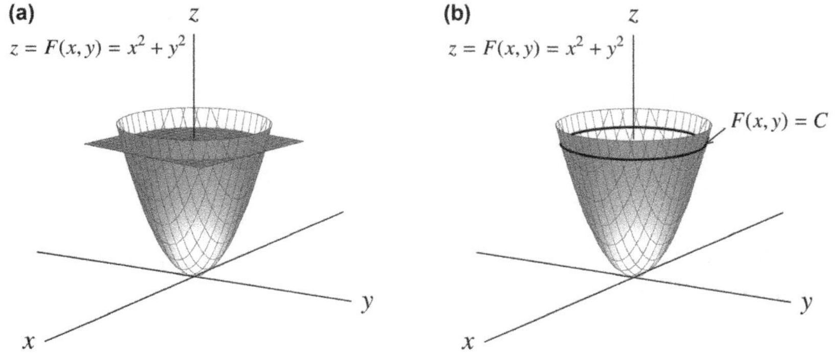

Figure 4.16 (a) Three-dimensional shape defined by $z = F(x, y) = x^2 + y^2$ sliced by a plane parallel to the x, y-plane; (b) Portion of the shape sliced by the plane.

Let us pick a point on the curve, say x_0, y_0, and calculate the slope of the surface at that point. There are of course, an infinite number of slopes to be calculated at this point but the one that we are currently interested in lies in a plane that is perpendicular to the x, y-plane (but not necessarily parallel to either the x or y axis), cuts the surface and contains a unit vector, $\vec{\mathbf{u}} = u_1\vec{\mathbf{u}}_x + u_2\vec{\mathbf{u}}_y$ that is parallel to the x, y-plane (see Figure 4.17) where the plane cuts the surface at curve C as described.

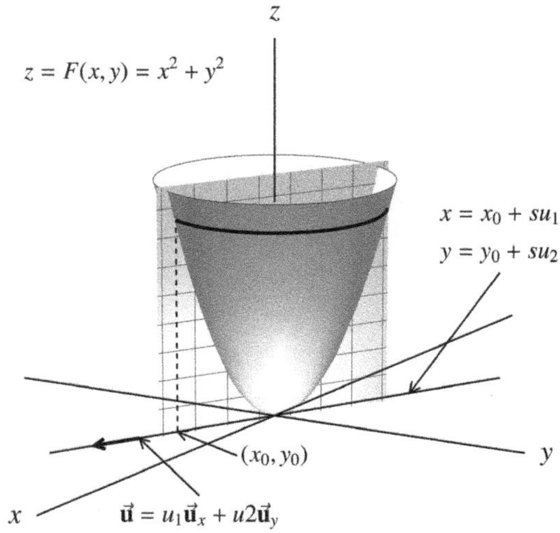

Figure 4.17 Vertical plane intersecting $z = F(x, y) = x^2 + y^2$.

The projection of curve C on the surface onto the x, y-plane can be expressed in a set of parametric equations (see Appendix A):

$$x = x_0 + su_1 \Rightarrow \frac{dx}{ds} = u_1$$

$$y = y_0 + su_2 \Rightarrow \frac{dy}{ds} = u_2$$

We wish to know the rate of change of $f(x, y)$ at (x_0, y_0) *in the direction of* $\vec{\mathbf{u}}$. By the chain rule:

$$\frac{df}{ds} = \frac{\partial f}{\partial x}\frac{dx}{ds} + \frac{\partial f}{\partial y}\frac{dy}{ds}$$

$$= \frac{\partial f}{\partial x}u_1 + \frac{\partial f}{\partial y}u_2$$

This is equivalent to $\vec{\nabla} f(x, y) \cdot \vec{\mathbf{u}}$ (write it out longhand to convince yourself) and is called the *directional derivative*. Note that here we are using a two-dimensional version of the gradient operator, $\vec{\nabla}$. The directional derivative gives the rate of change in the direction of the vector $\vec{\mathbf{u}}$. That we must have such an entity follows from the consideration of the question "What is the derivative of $z = f(x, y)$ at the point (x_0, y_0)?". Since $z = f(x, y)$ is a three-dimensional surface, the derivative or slope at (x_0, y_0) will depend on the direction of interest and there will be an infinite number of choices for the direction. The directional derivative specifies the direction of interest. One often encounters this written as $D_u\vec{\mathbf{F}} = \vec{\nabla}f \cdot \vec{\mathbf{u}}$. $\vec{\nabla}f$ and $\vec{\mathbf{u}}$ are both vectors and their inner product will involve a cosine term. With this in mind, we note that the directional derivative will have its largest value when $\theta = 0$ or, in other words, when $\vec{\mathbf{u}}$ is in the same direction as $\vec{\nabla}f$:

$$\begin{aligned} D_u\vec{\mathbf{F}} &= \vec{\nabla}f \cdot \vec{\mathbf{u}} \\ &= \left\|\vec{\nabla}f\right\| \|\vec{\mathbf{u}}\| \cos(\theta) \\ &= \left\|\vec{\nabla}f\right\| \cos(\theta) \end{aligned}$$

Also, when $\vec{\mathbf{u}}$ is perpendicular to the gradient, $\vec{\nabla}f$, $D_u\vec{\mathbf{F}} = 0$. In other words, in this direction there is zero change in the function. In our pictorial example above this would involve having $\vec{\mathbf{u}}$ tangent to the circle $f(x, y) = x^2 + y^2 = c$. This means that the gradient is directed perpendicularly to the curve in our example or, for a level curve, perpendicular to the surface … in the direction of largest change (Figure 4.18).

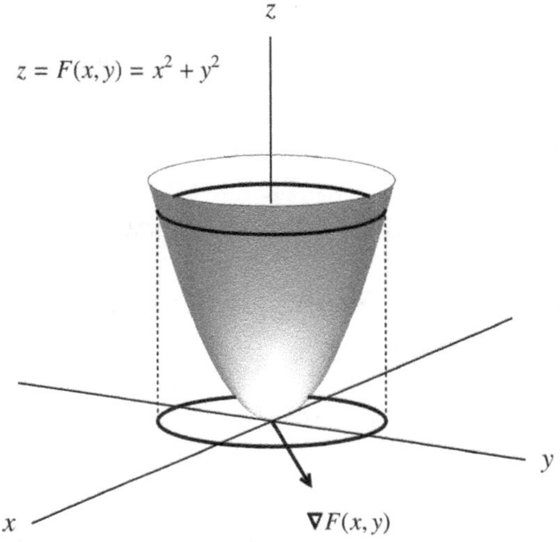

Figure 4.18 The directional derivative of $F(x, y) = x^2 + y^2 = c$.

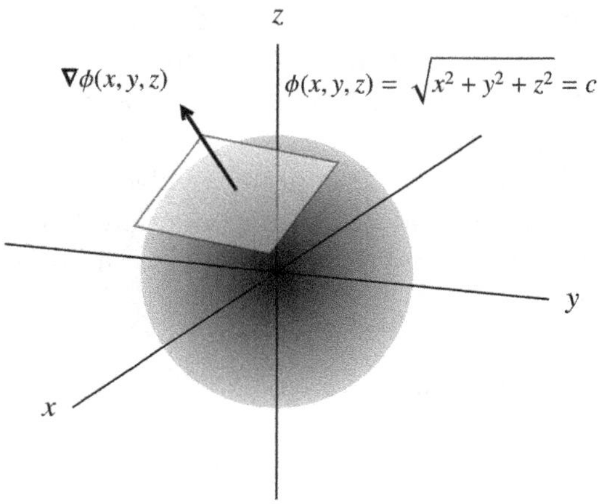

Figure 4.19 The directional derivative of $F(x, y, z) = c$.

We can extend this idea to another dimension using the function $f(x, y, z)$ (see Figure 4.19). We cannot plot $f(x, y, z)$ since it is now a four-dimensional object. We can, however, plot "slices", which are surfaces, in analogy to the curve we just looked at. The term *level surface* is used to describe these objects. Therefore, we consider a level surface $f(x, y, z) = c$ and calculate the directional derivative as for the previous case, except that we now use a three-dimensional version of $\vec{\nabla}$.

In the situation with an added dimension where we are looking at level surfaces rather than curves, the gradient is normal to the tangent plane at a point on the surface. For example, if our function is $f(x,y,z) = \sqrt{x^2 + y^2 + z^2} = c$ this level surface is in the shape of a sphere. A tangent plane at a point x_0, y_0, z_0 will tell us what direction $\vec{\nabla} f$ is in.

Now that we have some feeling for the meanings of $\vec{\nabla} f$, $\vec{\nabla} \cdot \vec{\mathbf{a}}$ and $\vec{\nabla} \times \vec{\mathbf{a}}$ (or grad, div and curl) we will see if we can consolidate our understanding a little. If the gradient is normal to a boundary of some kind and the curl is the rotational behaviour or "vorticity" of whatever it is that is "flowing" then the curl of the gradient should be zero since the gradient is not rotating but pointing directly across the boundary. Likewise the divergence of the curl should be zero for similar reasons. Let us see if this is really so. First, curl-grad:

$$\text{curl}(\text{grad} f) = \vec{\nabla} \times (\vec{\nabla} f)$$

$$\vec{\nabla} f = \vec{\mathbf{F}} = \left(\frac{\partial}{\partial x}\vec{\mathbf{u}}_x + \frac{\partial}{\partial y}\vec{\mathbf{u}}_y + \frac{\partial}{\partial z}\vec{\mathbf{u}}_z\right) f(x, y, z)$$

$$= \frac{\partial f}{\partial x}\vec{\mathbf{u}}_x + \frac{\partial f}{\partial y}\vec{\mathbf{u}}_y + \frac{\partial f}{\partial z}\vec{\mathbf{u}}_z$$

$$= \vec{\nabla} \times \vec{\mathbf{F}} = \begin{vmatrix} \vec{\mathbf{u}}_x & \vec{\mathbf{u}}_y & \vec{\mathbf{u}}_z \\ \dfrac{\partial}{\partial x} & \dfrac{\partial}{\partial y} & \dfrac{\partial}{\partial z} \\ \dfrac{\partial f}{\partial x} & \dfrac{\partial f}{\partial y} & \dfrac{\partial f}{\partial z} \end{vmatrix}$$

$$= \left(\frac{\partial^2 f}{\partial y \partial z} - \frac{\partial^2 f}{\partial z \partial y}\right)\vec{\mathbf{u}}_x - \left(\frac{\partial^2 f}{\partial x \partial z} - \frac{\partial^2 f}{\partial z \partial x}\right)\vec{\mathbf{u}}_y + \left(\frac{\partial^2 f}{\partial x \partial y} - \frac{\partial^2 f}{\partial y \partial x}\right)\vec{\mathbf{u}}_z$$

$$= 0 \tag{4.19}$$

Now let us try div-curl:

$$\text{div}(\text{curl} f) = \vec{\nabla} \cdot (\vec{\nabla} \times \vec{\mathbf{F}})$$

$$\vec{\nabla} \times \vec{\mathbf{F}} = \vec{\mathbf{G}} = \begin{vmatrix} \vec{\mathbf{u}}_x & \vec{\mathbf{u}}_y & \vec{\mathbf{u}}_z \\ \dfrac{\partial}{\partial x} & \dfrac{\partial}{\partial y} & \dfrac{\partial}{\partial z} \\ P(x, y, z) & Q(x, y, z) & R(x, y, z) \end{vmatrix}$$

$$= \left(\frac{\partial R}{\partial y} - \frac{\partial Q}{\partial z}\right)\vec{\mathbf{u}}_x - \left(\frac{\partial R}{\partial x} - \frac{\partial P}{\partial z}\right)\vec{\mathbf{u}}_y + \left(\frac{\partial Q}{\partial x} - \frac{\partial P}{\partial y}\right)\vec{\mathbf{u}}_z \tag{4.20}$$

$$= \vec{\nabla} \cdot \vec{\mathbf{G}} = \left(\frac{\partial}{\partial x}\vec{\mathbf{u}}_x + \frac{\partial}{\partial y}\vec{\mathbf{u}}_y + \frac{\partial}{\partial z}\vec{\mathbf{u}}_z\right) \cdot$$

$$\left[\left(\frac{\partial R}{\partial y} - \frac{\partial Q}{\partial z}\right)\vec{\mathbf{u}}_x - \left(\frac{\partial R}{\partial x} - \frac{\partial P}{\partial z}\right)\vec{\mathbf{u}}_y + \left(\frac{\partial Q}{\partial x} - \frac{\partial P}{\partial y}\right)\vec{\mathbf{u}}_z\right]$$

$$= \frac{\partial^2 R}{\partial y \partial x} - \frac{\partial^2 Q}{\partial z \partial x} - \frac{\partial^2 R}{\partial x \partial y} + \frac{\partial^2 P}{\partial z \partial y} + \frac{\partial^2 Q}{\partial x \partial z} - \frac{\partial^2 P}{\partial y \partial z}$$

$$= 0$$

These results are of importance in electrodynamics so we summarise them:

$$\vec{\nabla} \times (\vec{\nabla} f) = 0$$
$$\vec{\nabla} \cdot (\vec{\nabla} \times \vec{\mathbf{F}}) = 0 \tag{4.21}$$

Generally when the curl of a vector-field is zero we say that it is *irrotational* and when the divergence of a vector-field is zero we call it *solenoidal*. Thus, a vector produced only from the gradient of a scalar potential function is irrotational and a vector produced only from the curl of another vector is solenoidal. The "other" vector here is called a *vector potential*and in eqn (4.21) the vector $\vec{\mathbf{F}}$ is therefore a vector potential.‖ The Helmholtz theorem** pulls the two of these ideas together into one and says that any vector-field may be decomposed into two parts; one generated from a scalar potential and the other from a vector potential:

$$\vec{\mathbf{F}} = \vec{\nabla} f + \vec{\nabla} \times \vec{\mathbf{A}} \tag{4.22}$$

where here, f is the scalar potential function and $\vec{\mathbf{A}}$ is the vector potential function. The irrotational and solenoidal vector-fields are at the extremes of this definition where for the former $\vec{\nabla} \times \vec{\mathbf{A}} = 0$ and for the latter $\vec{\nabla} f = 0$. Let us apply eqn (4.21) to eqn (4.22) and see what results. First, the curl:

$$\begin{aligned}\vec{\nabla} \times \vec{\mathbf{F}} &= \vec{\nabla} \times (\vec{\nabla} f + \vec{\nabla} \times \vec{\mathbf{A}})\\ &= \vec{\nabla} \times (\vec{\nabla} f) + \vec{\nabla} \times (\vec{\nabla} \times \vec{\mathbf{A}})\\ &= \vec{\nabla} \times (\vec{\nabla} \times \vec{\mathbf{A}})\end{aligned}$$

and then the divergence:

$$\begin{aligned}\vec{\nabla} \cdot \vec{\mathbf{F}} &= \vec{\nabla} \cdot (\vec{\nabla} f + \vec{\nabla} \times \vec{\mathbf{A}})\\ &= \vec{\nabla} \cdot (\vec{\nabla} f) + \vec{\nabla} \cdot (\vec{\nabla} \times \vec{\mathbf{A}})\\ &= \vec{\nabla} \cdot (\vec{\nabla} f)\\ &= \nabla^2 f\end{aligned}$$

‖As with the scalar potential, the vector potential that generates a field vector is not unique. One can add the gradient of an arbitrary scalar function and the gradient of the scalar function will vanish according to eqn (4.21).

**This is peripheral to our interests here and not necessary to our progress. I mention it only out of interest. Please see the Further Reading at the end of the chapter for more information on the Helmholz theorem.

It is time for a small leap. Let us consider that, rather than doing a line integral over a line in three-dimensional space, we are doing instead, a *surface* integral (or area integral) to calculate the flux. A calculation of this type is completely analogous to the line integral except that we are now integrating over an area:

$$\iint \vec{\mathbf{F}} \cdot \mathrm{d}\vec{\sigma} = \iint \vec{\mathbf{F}} \cdot \vec{\mathbf{n}} \mathrm{d}\vec{\sigma} \tag{4.23}$$

Here we are considering an area vector, $\mathrm{d}\vec{\sigma}$, or equivalently, $\vec{\mathbf{n}}\ \mathrm{d}\sigma$, where $\vec{\mathbf{n}}$ is the unit vector normal to $\mathrm{d}\sigma$. How can an area have a vector associated with it? Appendix A shows that the area of a parallelogram defined by vectors $\vec{\mathbf{a}}$ and $\vec{\mathbf{b}}$ can be calculated from the cross product of the two. Of course, the cross product produces another vector, $\vec{\mathbf{c}}$ whose magnitude is the area of the vector. The vector $\vec{\mathbf{c}}$ is then the associated area vector of the parallelogram and is necessarily normal to the plane of the parallelogram. As with the line integral, the normal vector, $\vec{\mathbf{n}}$, is in the direction of the flux of the vector field associated with the surface.

Why the double integrals? Since we are now considering areas we are now summing over two variables ... the length and width. Let us look closely at it for a bit. First, consider an area ΔP and its projection or its "shadow" on a plane below it.

ΔQ is the area in the plane and $\vec{\mathbf{u}}$ is a unit vector normal to the area represented by ΔQ. Now consider the vectors $\vec{\mathbf{AB}}$ and $\vec{\mathbf{CD}}$. With reference to Figure 4.20, by simple vector addition we can say:

$$\begin{aligned}\vec{\mathbf{a}} &= \vec{\mathbf{AB}} + \vec{\mathbf{v}} + \vec{\mathbf{CD}} \\ &= \vec{\mathbf{AB}} + \vec{\mathbf{v}} - \vec{\mathbf{DC}} \\ &= \vec{\mathbf{v}} + \alpha\vec{\mathbf{u}}\end{aligned}$$

The last line follows from the fact that the unit normal vector, $\vec{\mathbf{u}}$, is parallel and proportional to the sum of vectors $\vec{\mathbf{AB}}$ and $\vec{\mathbf{DC}}$. A similar analysis will show that:

$$\vec{\mathbf{b}} = \vec{\mathbf{w}} + \beta\vec{\mathbf{u}}$$

The cross product of $\vec{\mathbf{a}}$ and $\vec{\mathbf{b}}$ gives us:

$$\begin{aligned}\vec{\mathbf{a}} \times \vec{\mathbf{b}} &= (\vec{\mathbf{v}} + \alpha\vec{\mathbf{u}}) \times (\vec{\mathbf{w}} + \beta\vec{\mathbf{u}}) \\ &= \vec{\mathbf{v}} \times \vec{\mathbf{w}} + \vec{\mathbf{v}} \times \beta\vec{\mathbf{u}} + \alpha\vec{\mathbf{u}} \times \vec{\mathbf{w}} + \alpha\vec{\mathbf{u}} \times \beta\vec{\mathbf{u}} \\ &= \vec{\mathbf{v}} \times \vec{\mathbf{w}} + \vec{\mathbf{v}} \times \beta\vec{\mathbf{u}} + \alpha\vec{\mathbf{u}} \times \vec{\mathbf{w}}\end{aligned}$$

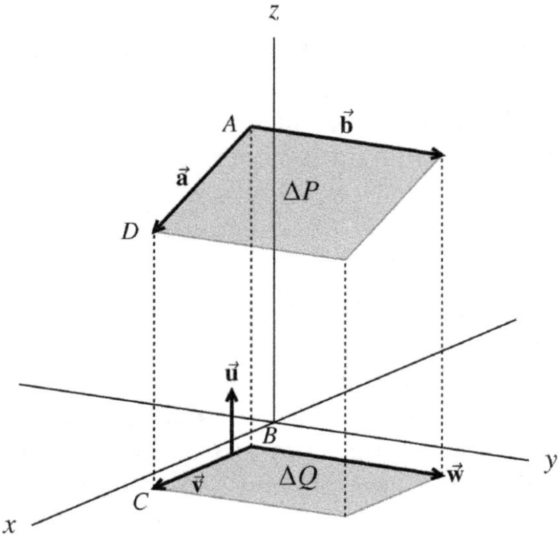

Figure 4.20 Projection of surface ΔP onto the x, y-plane.

What happened to the last term, $\alpha\vec{\mathbf{u}}\times\beta\vec{\mathbf{u}}$? Recall that the cross product of a vector with itself must be zero; *ie.* $\alpha\vec{\mathbf{u}}\times\beta\vec{\mathbf{u}}=\alpha\beta(\vec{\mathbf{u}}\times\vec{\mathbf{u}})=0$.

Now we do the scalar triple product of this with the unit vector $\vec{\mathbf{u}}$:

$$(\vec{\mathbf{a}}\times\vec{\mathbf{b}})\cdot\vec{\mathbf{u}}=(\vec{\mathbf{v}}\times\vec{\mathbf{w}}+\vec{\mathbf{v}}\times\beta\vec{\mathbf{u}}+\alpha\vec{\mathbf{u}}\times\vec{\mathbf{w}})\cdot\vec{\mathbf{u}}$$

The last two terms on the right side are orthogonal to $\vec{\mathbf{u}}$ so that their inner product with $\vec{\mathbf{u}}$ will be zero, so we can write:

$$(\vec{\mathbf{a}}\times\vec{\mathbf{b}})\cdot\vec{\mathbf{u}}=(\vec{\mathbf{v}}\times\vec{\mathbf{w}})\cdot\vec{\mathbf{u}}$$

$$\text{or} \tag{4.24}$$

$$(\vec{\mathbf{a}}\times\vec{\mathbf{b}})\cdot\vec{\mathbf{u}}=\|\vec{\mathbf{v}}\times\vec{\mathbf{w}}\|$$

The last line follows from the facts that a) $\vec{\mathbf{u}}$ is orthogonal to the plane defined by $\vec{\mathbf{v}}$ and $\vec{\mathbf{w}}$ and b) the cross product of $\vec{\mathbf{v}}$ with $\vec{\mathbf{w}}$ is orthogonal to the plane defined by them. Therefore, $\vec{\mathbf{v}}\times\vec{\mathbf{w}}$ and $\vec{\mathbf{u}}$ are collinear, the angle between them being 0° and the cosine of the angle being therefore equal to one.

$$\begin{aligned}(\vec{\mathbf{v}}\times\vec{\mathbf{w}})\cdot\vec{\mathbf{u}}&=\|\vec{\mathbf{v}}\times\vec{\mathbf{w}}\|\|\vec{\mathbf{u}}\|\cos(0)\\&=\|\vec{\mathbf{v}}\times\vec{\mathbf{w}}\|\end{aligned}$$

Another way to look at it is to recall from Chapter 3 that the scalar triple product corresponds to the volume of a parallelepiped. In this case the cross product is "dotted" with a unit vector so that the scalar triple product is simply an area.

The next step in our explorations is to add a level surface, $f(x, y, z) = c$, to our diagram such that ΔP is a tangent plane at some point on the surface (see Figure 4.21).

$\vec{\nabla} f$ is the gradient of the level surface at the tangent point and makes an angle θ with the unit normal $\vec{\mathbf{u}}$. As we saw above (eqn (4.24)), the area ΔQ can be related to the area ΔP:

$$\Delta P = \left\| \vec{\mathbf{a}} \times \vec{\mathbf{b}} \right\|$$

$$\Delta Q = \left\| (\vec{\mathbf{a}} \times \vec{\mathbf{b}}) \cdot \vec{\mathbf{u}} \right\|$$

$$= \Delta P \cos(\theta)$$

$$\Delta P = \frac{\Delta Q}{|\cos(\theta)|}$$

The $\cos(\theta)$ term comes from the fact that $\vec{\mathbf{a}} \times \vec{\mathbf{b}}$ is parallel to $\vec{\nabla} f$ and will therefore make the same angle θ with $\vec{\mathbf{u}}$ as does $\vec{\nabla} f$. We take the absolute values of course, since we do not want a negative area. Signs would be of importance if we were discussing area vectors.

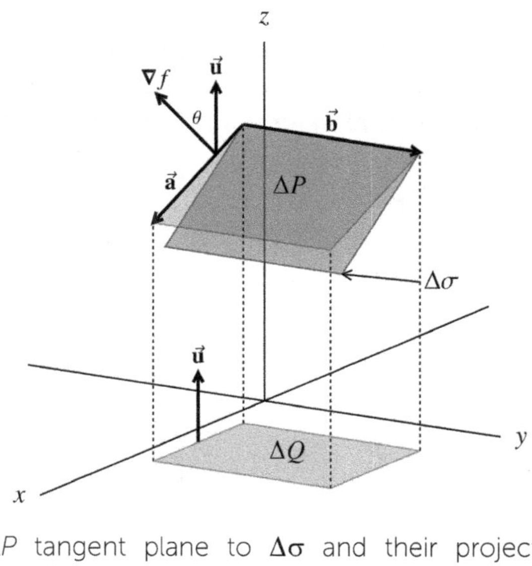

Figure 4.21 ΔP tangent plane to $\Delta\sigma$ and their projections onto the x, y-plane.

What is the area of ΔQ? Since it is in the x,y-plane it will be $\Delta x\Delta y$, so that ΔP can be written as:

$$\Delta P = \frac{\Delta x \Delta y}{|\cos(\theta)|} \approx \Delta \sigma$$

To get the area of the surface we integrate over x and y. Thus, using our formula for ΔP and summing over all x and y of interest for the surface:

$$\text{surface area} = \iint \frac{\mathrm{d}Q}{\cos(\theta)} = \iint \frac{\mathrm{d}x\mathrm{d}y}{|\cos(\theta)|} = \iint \mathrm{d}\sigma \tag{4.25}$$

We would not generally use this form of the surface area integral but rather convert the cosine term into a ratio of vector lengths:

$$\left\|\vec{\nabla} f \cdot \vec{\mathbf{u}}\right\| = \left\|\vec{\nabla} f\right\| \|\vec{\mathbf{u}}\| \cos(\theta)$$

$$= \left\|\vec{\nabla} f\right\| \cos(\theta)$$

$$\left|\frac{1}{\cos(\theta)}\right| = \frac{\left\|\vec{\nabla} f\right\|}{\left\|\vec{\nabla} f \cdot \vec{\mathbf{u}}\right\|}$$

and:

$$\text{surface area} = \iint \frac{\mathrm{d}Q}{\cos(\theta)} = \iint \frac{\left\|\vec{\nabla} f\right\|}{\left\|\vec{\nabla} f \cdot \vec{\mathbf{u}}\right\|} \mathrm{d}Q$$

Back to the surface integral. Let us suppose we have a function, $g(x, y, z)$, which represents the charge per unit area on a surface. We can use our surface integral expression to calculate the total charge residing on the surface. The charge on one small area of the surface, say ΔP in our above diagram, is:

$$g(x, y, z)\Delta P$$

Adding up all infinitesimal areas over the surface:

$$\text{total charge} = \iint \frac{g(x,y,z)\mathrm{d}Q}{|\cos(\theta)|}$$

$$= \iint g(x,y,z) \frac{\left\|\vec{\nabla} f\right\|}{\left\|\vec{\nabla} f \cdot \vec{\mathbf{u}}\right\|} \mathrm{d}Q$$

$$= \iint g\mathrm{d}\vec{\sigma} \text{ (short form)}$$

Note that there are two functions to keep track of here, $f(x, y, z)=c$ the surface function and $g(x, y, z)$ the charge density function. Also, note that if the function $g(x, y, z)$ is equal to a constant then the double integral evaluates to a surface area times the constant.

Analogous to the flux line integral we can do a flux surface (or area) integral:

$$\iint \vec{\mathbf{F}} \cdot \vec{\mathbf{n}} \mathrm{d}\sigma \tag{4.26}$$

and recalling that $\vec{\mathbf{n}}$ is the normal to the surface involved and that it is collinear with $\vec{\nabla} f$ we can then say (using eqn (3.13)):

$$\vec{\mathbf{n}} = \frac{\vec{\nabla} f}{\left\| \vec{\nabla} f \right\|}$$

which can then be substituted into the flux surface integral:

$$\iint \vec{\mathbf{F}} \cdot \vec{\mathbf{n}} \mathrm{d}\sigma = \iint \vec{\mathbf{F}} \cdot \frac{\vec{\nabla} f}{\left\| \vec{\nabla} f \right\|} \mathrm{d}\sigma$$

4.3 Curvilinear Coordinates and Grad, Div and Curl

Most of the time it is convenient to use basis sets that are orthogonal and normalised and most of us are used to the standard Cartesian axes that are defined by the now familiar orthogonal basis vectors $\vec{\mathbf{u}}_x$, $\vec{\mathbf{u}}_y$ and $\vec{\mathbf{u}}_z$. However, it is not strictly necessary to do so and in fact it may be advantageous to use a different orthogonal coordinate system such as cylindrical coordinates or spherical coordinates. We will be especially interested in spherical coordinates in the chapter on electrostatics.

Let us suppose that we may express Cartesian coordinates in terms of the coordinates of another coordinate system:

$$x = x(q_1, q_2, q_3)$$

$$y = y(q_1, q_2, q_3)$$

$$z = z(q_1, q_2, q_3)$$

and that we may also go the other way and express the coordinates of the other system in terms of the Cartesian coordinates:

$$q_1 = q_1(x, y, z)$$
$$q_2 = q_2(x, y, z)$$
$$q_3 = q_3(x, y, z)$$

In differential form we may write:

$$\begin{aligned} \mathrm{d}x &= \frac{\partial x}{\partial q_1}\mathrm{d}q_1 + \frac{\partial x}{\partial q_2}\mathrm{d}q_2 + \frac{\partial x}{\partial q_3}\mathrm{d}q_3 \\ \mathrm{d}y &= \frac{\partial y}{\partial q_1}\mathrm{d}q_1 + \frac{\partial y}{\partial q_2}\mathrm{d}q_2 + \frac{\partial y}{\partial q_3}\mathrm{d}q_3 \\ \mathrm{d}z &= \frac{\partial z}{\partial q_1}\mathrm{d}q_1 + \frac{\partial z}{\partial q_2}\mathrm{d}q_2 + \frac{\partial z}{\partial q_3}\mathrm{d}q_3 \end{aligned} \tag{4.27}$$

or expressed using matrices (see Appendix A, Section A.8):

$$\begin{bmatrix} \mathrm{d}x \\ \mathrm{d}y \\ \mathrm{d}z \end{bmatrix} = \begin{bmatrix} \frac{\mathrm{d}x}{\mathrm{d}q_1} & \frac{\mathrm{d}x}{\mathrm{d}q_2} & \frac{\mathrm{d}x}{\mathrm{d}q_3} \\ \frac{\mathrm{d}y}{\mathrm{d}q_1} & \frac{\mathrm{d}y}{\mathrm{d}q_2} & \frac{\mathrm{d}y}{\mathrm{d}q_3} \\ \frac{\mathrm{d}z}{\mathrm{d}q_1} & \frac{\mathrm{d}z}{\mathrm{d}q_2} & \frac{\mathrm{d}z}{\mathrm{d}q_3} \end{bmatrix} \begin{bmatrix} \mathrm{d}q_1 \\ \mathrm{d}q_2 \\ \mathrm{d}q_3 \end{bmatrix}$$

The square of the element of arc length along a curve in Cartesian notation is:

$$(\mathrm{d}s)^2 = (\mathrm{d}x)^2 + (\mathrm{d}y)^2 + (\mathrm{d}z)^2$$

which may be written in our alternate coordinate notation using the above:

$$\begin{aligned} (\mathrm{d}s)^2 = {} & \left(\frac{\partial x}{\partial q_1}\mathrm{d}q_1 + \frac{\partial x}{\partial q_2}\mathrm{d}q_2 + \frac{\partial x}{\partial q_3}\mathrm{d}q_3\right)^2 \\ & + \left(\frac{\partial y}{\partial q_1}\mathrm{d}q_1 + \frac{\partial y}{\partial q_2}\mathrm{d}q_2 + \frac{\partial y}{\partial q_3}\mathrm{d}q_3\right)^2 \\ & + \left(\frac{\partial z}{\partial q_1}\mathrm{d}q_1 + \frac{\partial z}{\partial q_2}\mathrm{d}q_2 + \frac{\partial z}{\partial q_3}\mathrm{d}q_3\right)^2 \end{aligned}$$

The (very tedious) expansion and collection of terms gives us:

$$(\mathrm{d}s)^2 = h_1{}^2\mathrm{d}q_1{}^2 + h_2{}^2\mathrm{d}q_2{}^2 + h_3{}^2\mathrm{d}q_3$$
$$+ 2h_{12}\mathrm{d}q_1\mathrm{d}q_2 + 2h_{13}\mathrm{d}q_1\mathrm{d}q_3 + 2h_{23}\mathrm{d}q_2\mathrm{d}q_3$$

where for brevity we have substituted:

$$h_{ij} = \frac{\partial x}{\partial q_i}\frac{\partial x}{\partial q_j} + \frac{\partial y}{\partial q_i}\frac{\partial y}{\partial q_j} + \frac{\partial z}{\partial q_i}\frac{\partial z}{\partial q_j} \quad (i=1,2,3;\ j>i)$$

and (4.28)

$$h_i^2 = \left(\frac{\partial x}{\partial q_i}\right)^2 + \left(\frac{\partial y}{\partial q_i}\right)^2 + \left(\frac{\partial z}{\partial q_i}\right)^2 \quad (i=1,2,3)$$

We refer to the h_i factors as the *scale factors*. In orthogonal coordinate systems[††] the h_{ij} factors are all zero leaving only the h_i factors, as we will see shortly.

Now, let us consider a line element in our general coordinate system. This is just the distance between two adjacent points on a coordinate axis in the coordinate system and is equivalent to an element of arc length along the axis. Since we are considering only one axis and are holding the others constant then from our above expression for $(\mathrm{d}s)^2$ we may write:

$$\mathrm{d}s_i = h_i\mathrm{d}q_i \tag{4.29}$$

The discussion to this point has been quite abstract and it is probably time to consider a practical example. Centering our attention on spherical coordinates and their transformation to rectangular coordinates we find (Appendix A, Section A.8):

$$x = r\sin(\theta)\cos(\varphi)$$
$$y = r\sin(\theta)\sin(\varphi)$$
$$z = r\cos(\theta)$$

and the corresponding differential equations (using eqn (4.27)) are:

$$\mathrm{d}x = \sin(\theta)\cos(\varphi)\mathrm{d}r + r\cos(\theta)\cos(\varphi)\mathrm{d}\theta - r\sin(\theta)\sin(\varphi)\mathrm{d}\varphi$$
$$\mathrm{d}y = \sin(\theta)\sin(\varphi)\mathrm{d}r + r\cos(\theta)\sin(\varphi)\mathrm{d}\theta + r\sin(\theta)\cos(\varphi)\mathrm{d}\varphi$$
$$\mathrm{d}z = \cos(\theta)\mathrm{d}r - r\sin(\theta)\mathrm{d}\theta$$

[††]This implies that one may consider non-orthogonal coordinate systems … intriguing, and probably messy!

We now will write an expression for the element of the square of the arc length:

$$(\mathrm{d}s)^2 = (\mathrm{d}x)^2 + (\mathrm{d}y)^2 + (\mathrm{d}z)^2$$

For spherical coordinates we substitute from above:

$$\begin{aligned}(\mathrm{d}s)^2 = &[\sin(\theta)\cos(\varphi)\mathrm{d}r + r\cos(\theta)\cos(\varphi)\mathrm{d}\theta - r\sin(\theta)\sin(\varphi)\mathrm{d}\varphi]^2 \\ &+ [\sin(\theta)\sin(\varphi)\mathrm{d}r + r\cos(\theta)\sin(\varphi)\mathrm{d}\theta + r\sin(\theta)\cos(\varphi)\mathrm{d}\varphi]^2 \\ &+ [\cos(\theta)\mathrm{d}r - r\sin(\theta)\mathrm{d}\theta]^2\end{aligned}$$

The (again, tedious) expansion of this expression is:

$$(\mathrm{d}s)^2 = (\mathrm{d}r)^2 + r^2\,(\mathrm{d}\theta)^2 + r^2\,\sin^2(\theta)\,(\mathrm{d}\phi)^2$$

Notice that the cross-terms, the h_{ij} terms from eqn (4.28), all vanish. This is a happy consequence of spherical coordinates being orthogonal as we noted earlier. We see here that:

$$\begin{aligned} h_1^2 &= 1 \\ h_2^2 &= r^2 \\ h_3^2 &= r^2\sin^2(\theta) \end{aligned} \tag{4.30}$$

The line elements are:

$$\begin{aligned} \mathrm{d}s_r &= \mathrm{d}r \\ \mathrm{d}s_\theta &= r\mathrm{d}\theta \\ \mathrm{d}s_\phi &= r\sin(\theta)\mathrm{d}\varphi \end{aligned}$$

as is shown in Appendix A, in the figures in Section A.8. After short reflection we see that in Cartesian coordinates:

$$h_1{}^2 = 1,\ h_2{}^2 = 1,\ h_3{}^2 = 1$$

and

$$s_x = 1,\ s_y = 1,\ s_z = 1$$

Back to our consideration of general coordinates. We now ask "What will the form of the del operator be in general coordinates?" as we want to be able to do so in whatever coordinate system we choose.

To answer this we will review our understanding of $\vec{\nabla}f$. We saw that this vector operator produces a vector from a scalar function, f, and that $\vec{\nabla}f$ is in the direction of greatest change of the function. From eqn (4.4) this is written (in Cartesian coordinates) as:

$$\vec{\nabla}f(x,y,z) = \frac{\partial f}{\partial x}\vec{\mathbf{u}}_x + \frac{\partial f}{\partial y}\vec{\mathbf{u}}_y + \frac{\partial f}{\partial z}\vec{\mathbf{u}}_z$$

Now consider the components of this vector along the general coordinate axes using the general line elements from above in eqn (4.29):

$$\frac{\mathrm{d}f}{\mathrm{d}s_i} = \frac{1}{h_i}\frac{\mathrm{d}\varphi}{\mathrm{d}q_i}$$

This can be written in terms of unit vectors for the general coordinate system to produce an expression for the del operator in an arbitrary coordinate system:

$$\vec{\nabla} = \frac{\vec{\mathbf{u}}_1}{h_1}\frac{\partial}{\partial q_1} + \frac{\vec{\mathbf{u}}_2}{h_2}\frac{\partial}{\partial q_2} + \frac{\vec{\mathbf{u}}_3}{h_3}\frac{\partial}{\partial q_3}$$

Since the h_i terms for Cartesian coordinates are all equal to one this becomes:

$$\vec{\nabla} = \vec{\mathbf{u}}_x\frac{\partial}{\partial x} + \vec{\mathbf{u}}_y\frac{\partial}{\partial y} + \vec{\mathbf{u}}_z\frac{\partial}{\partial z}$$

in accord with eqn (4.3). For spherical coordinates, using eqn (4.30) we can write del as:

$$\vec{\nabla} = \vec{\mathbf{u}}_r\frac{\partial}{\partial r} + \frac{\vec{\mathbf{u}}_\theta}{r}\frac{\partial}{\partial \theta} + \frac{\vec{\mathbf{u}}_\phi}{r\sin(\theta)}\frac{\partial}{\partial \varphi} \tag{4.31}$$

and the gradient $\vec{\nabla}f$ is therefore:

$$\vec{\nabla}f = \vec{\mathbf{u}}_r\frac{\partial f}{\partial r} + \frac{\vec{\mathbf{u}}_\theta}{r}\frac{\partial f}{\partial \theta} + \frac{\vec{\mathbf{u}}_\phi}{r\sin(\theta)}\frac{\partial f}{\partial \varphi}$$

and in general coordinates:

$$\vec{\nabla}f = \frac{\vec{\mathbf{u}}_1}{h_1}\frac{\partial f}{\partial q_1} + \frac{\vec{\mathbf{u}}_2}{h_2}\frac{\partial f}{\partial q_2} + \frac{\vec{\mathbf{u}}_3}{h_3}\frac{\partial f}{\partial q_3}$$

The divergence and curl can be derived using our new knowledge about general coordinates. First, let us calculate $\vec{\nabla} f$ when $f = q_i$:

$$\begin{aligned}\vec{\nabla} q_1 &= \frac{\vec{\mathbf{u}}_1}{h_1}\frac{\partial q_1}{\partial q_1} + \frac{\vec{\mathbf{u}}_2}{h_2}\frac{\partial q_1}{\partial q_2} + \frac{\vec{\mathbf{u}}_3}{h_3}\frac{\partial q_1}{\partial q_3} \\ &= \frac{\vec{\mathbf{u}}_1}{h_1}1 + \frac{\vec{\mathbf{u}}_2}{h_2}0 + \frac{\vec{\mathbf{u}}_3}{h_3}0 \\ &= \frac{\vec{\mathbf{u}}_1}{h_1} \\ \vec{\nabla} q_2 &= \frac{\vec{\mathbf{u}}_2}{h_2},\ \vec{\nabla} q_3 = \frac{\vec{\mathbf{u}}_3}{h_3}\end{aligned} \tag{4.32}$$

and the cross products of these (assuming an orthogonal coordinate system):

$$\begin{aligned}\vec{\nabla} q_1 \times \vec{\nabla} q_2 &= \frac{\vec{\mathbf{u}}_1 \times \vec{\mathbf{u}}_2}{h_1 h_2} = \frac{\vec{\mathbf{u}}_3}{h_1 h_2} \Rightarrow \vec{\mathbf{u}}_3 = h_1 h_2 \vec{\nabla} q_1 \times \vec{\nabla} q_2 \\ \vec{\nabla} q_2 \times \vec{\nabla} q_3 &= \frac{\vec{\mathbf{u}}_2 \times \vec{\mathbf{u}}_3}{h_2 h_3} = \frac{\vec{\mathbf{u}}_1}{h_2 h_3} \Rightarrow \vec{\mathbf{u}}_1 = h_2 h_3 \vec{\nabla} q_2 \times \vec{\nabla} q_3 \\ \vec{\nabla} q_3 \times \vec{\nabla} q_1 &= \frac{\vec{\mathbf{u}}_3 \times \vec{\mathbf{u}}_1}{h_3 h_1} = \frac{\vec{\mathbf{u}}_2}{h_3 h_1} \Rightarrow \vec{\mathbf{u}}_2 = h_3 h_1 \vec{\nabla} q_3 \times \vec{\nabla} q_1\end{aligned} \tag{4.33}$$

The divergence then is:

$$\begin{aligned}\vec{\nabla} \cdot \vec{\mathbf{F}} &= \vec{\nabla} \cdot (F_1 \vec{\mathbf{u}}_1 + F_2 \vec{\mathbf{u}}_2 + F_3 \vec{\mathbf{u}}_3) \\ &= F_1 \vec{\nabla} \cdot \vec{\mathbf{u}}_1 + F_2 \vec{\nabla} \cdot \vec{\mathbf{u}}_2 + F_3 \vec{\nabla} \cdot \vec{\mathbf{u}}_3\end{aligned}$$

and for convenience we will look only at coordinate q_1:

$$\vec{\nabla} \cdot F_1 \vec{\mathbf{u}}_1 = \vec{\nabla} \cdot (F_1 h_2 h_3\ \vec{\nabla} q_2 \times \vec{\nabla} q_3)$$

From the identities for del above in Section 4.1 (specifically, identity 4):

$$\vec{\nabla} \cdot F_1 \vec{\mathbf{u}}_1 = \vec{\nabla}(F_1 h_2 h_3) \cdot (\vec{\nabla} q_2 \times \vec{\nabla} q_3) + F_1 h_2 h_3 \vec{\nabla} \cdot (\vec{\nabla} q_2 \times \vec{\nabla} q_3)$$

The second term on the right is the divergence of a curl and, as we have seen above in eqn (4.21), is zero. Thus:

$$
\begin{aligned}
\vec{\nabla}\cdot F_1\vec{\mathbf{u}}_1 &= \vec{\nabla}(F_1h_2h_3)\cdot(\vec{\nabla}q_2\times\vec{\nabla}q_3)\\
&= \vec{\nabla}(F_1h_2h_3)\cdot\left(\frac{\vec{\mathbf{u}}_2\times\vec{\mathbf{u}}_3}{h_2h_3}\right)\\
&= \vec{\nabla}(F_1h_2h_3)\cdot\left(\frac{\vec{\mathbf{u}}_1}{h_2h_3}\right)\quad\text{(from equation(4.33))}\\
&= \left(\frac{\vec{\mathbf{u}}_1}{h_1}\frac{\partial(F_1h_2h_3)}{\partial q_1}+\frac{\vec{\mathbf{u}}_2}{h_2}\frac{\partial(F_2h_3h_1)}{\partial q_2}+\frac{\vec{\mathbf{u}}_3}{h_3}\frac{\partial(F_3h_1h_2)}{\partial q_3}\right)\cdot\left(\frac{\vec{\mathbf{u}}_1}{h_2h_3}\right)\\
&= \frac{1}{h_1h_2h_3}\frac{\partial(F_1h_2h_3)}{\partial q_1}
\end{aligned}
$$

We arrive at similar expressions for q_2 and q_3 so that the full expression for the divergence in general coordinates is:

$$
\vec{\nabla}\cdot\vec{\mathbf{F}} = \frac{1}{h_1h_2h_3}\left(\frac{\partial(F_1h_2h_3)}{\partial q_1}+\frac{\partial(F_2h_3h_1)}{\partial q_2}+\frac{\partial(F_3h_1h_2)}{\partial q_3}\right) \tag{4.34}
$$

What about the curl? Again, for convenience we focus on only coordinate q_1:

$$
\vec{\nabla}\times(F_1\,\vec{\mathbf{u}}_1) = \vec{\nabla}\times(F_1h_1\vec{\nabla}q_1)
$$

Again using our list of identities for del (identity 5 this time):

$$
\vec{\nabla}\times(F_1\vec{\mathbf{u}}_1) = \vec{\nabla}(F_1h_1)\times\vec{\nabla}q_1 + (F_1h_1)\,\vec{\nabla}\times\vec{\nabla}q_1
$$

and also, we recall that since the curl of a gradient is zero, the right-hand term vanishes:

$$
\begin{aligned}
\vec{\nabla}\times(F_1\vec{\mathbf{u}}_1) &= \vec{\nabla}(F_1h_1)\times\vec{\nabla}q_1\\
&= \vec{\nabla}(F_1h_1)\times\frac{\vec{\mathbf{u}}_1}{h_1}\\
&= \left(\frac{\vec{\mathbf{u}}_1}{h_1}\frac{\partial(F_1h_1)}{\partial q_1}+\frac{\vec{\mathbf{u}}_2}{h_2}\frac{\partial(F_1h_1)}{\partial q_2}+\frac{\vec{\mathbf{u}}_3}{h_3}\frac{\partial(F_1h_1)}{\partial q_3}\right)\times\frac{\vec{\mathbf{u}}_1}{h_1}\\
&= \frac{\vec{\mathbf{u}}_2}{h_3h_1}\frac{\partial(F_1h_1)}{\partial q_3}-\frac{\vec{\mathbf{u}}_3}{h_2h_1}\frac{\partial(F_1h_1)}{\partial q_2}
\end{aligned}
$$

The full equation for all three coordinates is:

$$
\begin{aligned}
\vec{\nabla}\cdot\vec{\mathbf{F}} &= \frac{\vec{\mathbf{u}}_2}{h_3h_1}\frac{\partial(F_1h_1)}{\partial q_3} - \frac{\vec{\mathbf{u}}_3}{h_2h_1}\frac{\partial(F_1h_1)}{\partial q_2} \\
&+ \frac{\vec{\mathbf{u}}_3}{h_1h_2}\frac{\partial(F_2h_2)}{\partial q_1} - \frac{\vec{\mathbf{u}}_1}{h_2h_3}\frac{\partial(F_2h_2)}{\partial q_3} \\
&+ \frac{\vec{\mathbf{u}}_1}{h_2h_3}\frac{\partial(F_3h_3)}{\partial q_2} - \frac{\vec{\mathbf{u}}_2}{h_3h_1}\frac{\partial(F_3h_3)}{\partial q_1} \\
&= \frac{\vec{\mathbf{u}}_1}{h_2h_3}\left(\frac{\partial(F_3h_3)}{\partial q_2} - \frac{\partial(F_2h_2)}{\partial q_3}\right) \\
&+ \frac{\vec{\mathbf{u}}_2}{h_3h_1}\left(\frac{\partial(F_1h_1)}{\partial q_3} - \frac{\partial(F_3h_3)}{\partial q_1}\right) \\
&+ \frac{\vec{\mathbf{u}}_3}{h_1h_2}\left(\frac{\partial(F_2h_2)}{\partial q_1} - \frac{\partial(F_1h_1)}{\partial q_2}\right)
\end{aligned} \tag{4.35}
$$

Once again we have strayed far into the abstract and we should probably come back to earth for a moment. The reason that we have been deriving these expressions is so that we can construct expressions for the divergence and the curl of a vector-field in spherical coordinates; they can now easily be written using eqn (4.34) and (4.35) and the values for the scale factors from above (see eqn (4.30)):

$$
\begin{aligned}
\vec{\nabla}\cdot\vec{\mathbf{F}} &= \frac{1}{r^2\sin(\theta)}\left(\frac{\partial(r^2\sin(\theta)F_r)}{\partial r} + \frac{\partial(r\sin(\theta)F_\theta)}{\partial\theta} + \frac{\partial(rF_\phi)}{\partial\varphi}\right) \\
&= \frac{1}{r^2}\frac{\partial(r^2F_r)}{\partial r} + \frac{1}{r\sin(\theta)}\frac{\partial(\sin(\theta)F_\theta)}{\partial\theta} + \frac{1}{r\sin}\frac{\partial(F_\phi)}{\partial\varphi}
\end{aligned} \tag{4.36}
$$

$$
\begin{aligned}
\vec{\nabla}\times\vec{\mathbf{F}} &= \frac{\vec{\mathbf{u}}_r}{r\sin(\theta)}\left(\frac{\partial\sin(\theta)F_\phi}{\partial\theta} - \frac{\partial F_\theta}{\partial\varphi}\right) \\
&+ \frac{\vec{\mathbf{u}}_\theta}{r}\left(\frac{1}{\sin(\theta)}\frac{\partial F_r}{\partial\varphi} - \frac{\partial rF_\phi}{\partial r}\right) \\
&+ \frac{\vec{\mathbf{u}}_\phi}{r}\left(\frac{\partial rF_\theta}{\partial r} - \frac{\partial F_r}{\partial\theta}\right)
\end{aligned} \tag{4.37}
$$

4.4 Green's Theorem, Gauss's Theorem and Stokes' Theorem

Green's theorem in the *x*,*y*-plane is stated as follows:

$$\iint_S \left(\frac{\partial Q}{\partial x} - \frac{\partial P}{\partial y}\right) \mathrm{d}x\mathrm{d}y = \oint_C (P\mathrm{d}x + Q\mathrm{d}y) \tag{4.38}$$

where S is a surface. We show that this is true using a diagram of the surface (Figure 4.22).

This particular surface is termed an x-simple and y-simple surface, which just means that a vertical line drawn from anywhere between a and b or a horizontal line drawn from anywhere between f and g will cut the boundary of the surface, at most, twice. On our surface boundary we have four points A, G, B and F along the curve surrounding the area corresponding to the extrema of the surface with respect to the axes.

To prove Green's theorem we start by considering two sections of the curve, AGB, which we will call C_1 and BFA, which we will call C_2. We will say that the function of x for the lower curve is $AGB(x)$ and $BFA(x)$ for the upper curve. Now, for any x between a and b we

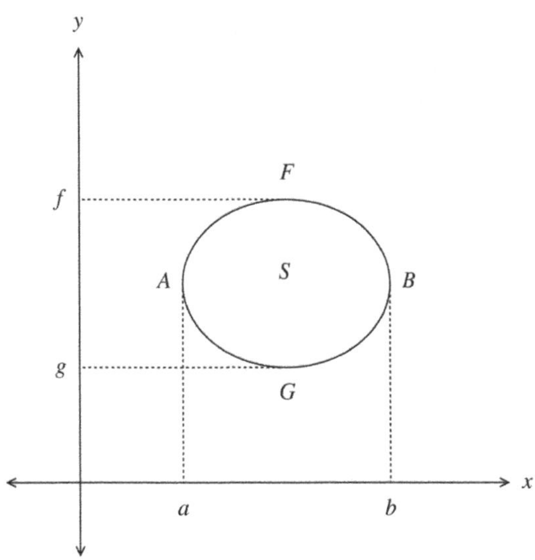

Figure 4.22 Gauss's theorem surface.

consider the integral of $\partial P/\partial y$ along the line between $AGB(x)$ and $BFA(x)$:

$$\int_{AGB(x)}^{BFA(x)} \frac{\partial P}{\partial y} \mathrm{d}y$$

$$= P(x, y) \mid_{AGB(x)}^{BFA(x)}$$

$$= P(x, BFA(x)) - P(x, AGB(x))$$

This is for a single vertical line through the area, S. Now we want to add up all of the other possible vertical lines within the surface so we integrate again over x:

$$\int_a^b \int_{AGB(x)}^{BFA(x)} \frac{\partial P}{\partial y} \mathrm{d}y\mathrm{d}x$$

$$= \int_a^b [P(x, BFA(x)) - P(x, AGB(x))]\mathrm{d}x$$

$$= \int_a^b P(x, BFA(x))\mathrm{d}x - \int_a^b P(x, AGB(x))\mathrm{d}x$$

$$= -\int_b^a P(x, BFA(x))\mathrm{d}x - \int_a^b P(x, AGB(x))\mathrm{d}x$$

$$= -\int_{C_2} P\mathrm{d}x - \int_{C_1} P\mathrm{d}x$$

The last line is just the by-now-familiar line integrals along the curves C_1 and C_2. Note that we are moving *counter-clockwise* around the curve. Furthermore, the sum of the two terms in the last line corresponds to the closed curve line integral arriving back at the starting point, B in our case:

$$\oint_C P\mathrm{d}x = \iint_S -\frac{\partial P}{\partial y} \mathrm{d}x\mathrm{d}y$$

If we do a very similar analysis for Q, this time using the portions of the curve "FAG" and "GBF" and doing our calculations between $y = f$ and $y = g$. First, a single horizontal line through the shape:

$$\int_{FAG(y)}^{GBF(y)} \frac{\partial Q}{\partial x} \mathrm{d}x$$

$$= Q(x, y)\Big|_{FAG(y)}^{GBF(y)}$$

$$= Q(GBF(y), y) - Q(FAG(y), y)$$

Next, add up all of the horizontal lines through the shape:

$$\begin{aligned}
&\int_f^g \int_{FAG(y)}^{GBF(y)} \frac{\partial Q}{\partial x} \mathrm{d}x\mathrm{d}y \\
&= \int_f^g [Q(GBF(y), y) - Q(FAG(y), y)]\mathrm{d}y \\
&= \int_f^g Q(GBF(y), y)\mathrm{d}y - \int_f^g Q(FAG(y), y)\mathrm{d}y \\
&= \int_f^g Q(GBF(y), y)\mathrm{d}y + \int_g^f Q(FAG(y), y)\mathrm{d}y \\
&= \int_{C_2} Q\mathrm{d}y + \int_{C_1} Q\mathrm{d}y
\end{aligned}$$

Also, we have that:

$$\oint_C Q\mathrm{d}y = \iint_S \frac{\partial Q}{\partial x} \mathrm{d}x\mathrm{d}y$$

Combining the two results gives us Green's theorem in the x,y-plane:

$$\oint_C P\mathrm{d}x + Q\mathrm{d}y = \iint_S \left(\frac{\partial Q}{\partial x} - \frac{\partial P}{\partial y}\right) \mathrm{d}x\mathrm{d}y \tag{4.39}$$

You will probably be thinking that each side of this equation looks a bit familiar. From our above discussion of the line integral recall that the left side may be written using eqn (4.15):

$$\oint_C P\mathrm{d}x + Q\mathrm{d}y = \int_C \vec{\mathbf{F}} \cdot \mathrm{d}\vec{\mathbf{r}}$$

Also, recall that the curl of vector $\vec{\mathbf{F}}$ is:

$$\begin{aligned}
\text{curl } \vec{\mathbf{F}} &= \vec{\nabla} \times \vec{\mathbf{F}} \\
&= \begin{vmatrix} \vec{\mathbf{u}}_x & \vec{\mathbf{u}}_y & \vec{\mathbf{u}}_z \\ \frac{\partial}{\partial x} & \frac{\partial}{\partial y} & \frac{\partial}{\partial z} \\ P & Q & R \end{vmatrix} \\
&= \left(\frac{\partial R}{\partial y} - \frac{\partial Q}{\partial z}\right)\vec{\mathbf{u}}_x + \left(\frac{\partial R}{\partial x} - \frac{\partial P}{\partial z}\right)\vec{\mathbf{u}}_y + \left(\frac{\partial Q}{\partial x} - \frac{\partial P}{\partial y}\right)\vec{\mathbf{u}}_z
\end{aligned}$$

If we take the last line and "dot" it with $\vec{\mathbf{u}}_z$ we get:

$$\left[\left(\frac{\partial R}{\partial y}-\frac{\partial Q}{\partial z}\right)\vec{\mathbf{u}}_x+\left(\frac{\partial R}{\partial x}-\frac{\partial P}{\partial z}\right)\vec{\mathbf{u}}_y+\left(\frac{\partial Q}{\partial x}-\frac{\partial P}{\partial y}\right)\vec{\mathbf{u}}_z\right]\cdot\vec{\mathbf{u}}_z$$

$$=\left(\frac{\partial Q}{\partial x}-\frac{\partial P}{\partial y}\right)$$

$$=(\vec{\nabla}\times\vec{\mathbf{F}})\cdot\vec{\mathbf{u}}_z$$

So we can write our double integral as:

$$\iint_S\left(\frac{\partial Q}{\partial x}-\frac{\partial P}{\partial y}\right)\mathrm{d}x\mathrm{d}y=\iint_S(\vec{\nabla}\times\vec{\mathbf{F}})\cdot\vec{\mathbf{u}}_z\mathrm{d}x\mathrm{d}y=\iint_S(\vec{\nabla}\times\vec{\mathbf{F}})\cdot\mathrm{d}\vec{\mathbf{A}}$$

Recall that we may calculate an area vector ... $\vec{\mathbf{u}}_z\,\mathrm{d}x\mathrm{d}y$ represents an area vector infinitesimal. Finally, we can write a version of Green's theorem in vector form:

$$\oint_C\vec{\mathbf{F}}\cdot\mathrm{d}\vec{\mathbf{r}}=\iint_S(\vec{\nabla}\times\vec{\mathbf{F}})\cdot\mathrm{d}\vec{\mathbf{A}}$$

$$\text{or}\qquad(4.40)$$

$$\oint_C\vec{\mathbf{F}}\cdot\mathrm{d}\vec{\mathbf{r}}=\iint_S(\vec{\nabla}\times\vec{\mathbf{F}})\cdot\vec{\mathbf{u}}_z\mathrm{d}x\mathrm{d}y$$

One can find Green's theorem written in various ways in texts, however, the most common forms appear to be the two written here.

What is the physical picture that we are to associate with Green's theorem? As with the line integrals above let us imagine an area, S, across which a vector field "flows".

If we calculate the line integral, $\oint_C\vec{\mathbf{F}}\cdot\mathrm{d}\vec{\mathbf{r}}$, we are actually adding up all of the infinitesimals $\vec{\mathbf{F}}\cdot\vec{\mathbf{T}}\mathrm{d}s$ on the curve bounding the area and calculating a "circulation" around the boundary of the area. Note that in Figure 4.23 the "circulation" around the boundary is depicted as being in a counter-clockwise direction. This is the positive direction by convention and is a consequence of using a right-hand coordinate system.

If we calculate the curl of the vector field at all points within the boundary of the area we will be also calculating the same thing as the circulation from the line integral. To see this we look at Figure 4.24.

We imagine two infinitesimal adjacent "chunks" of the area for which a curl calculation is done on each. The curl vectors on the sides of the chunks that are touching cancel each other out. Ultimately,

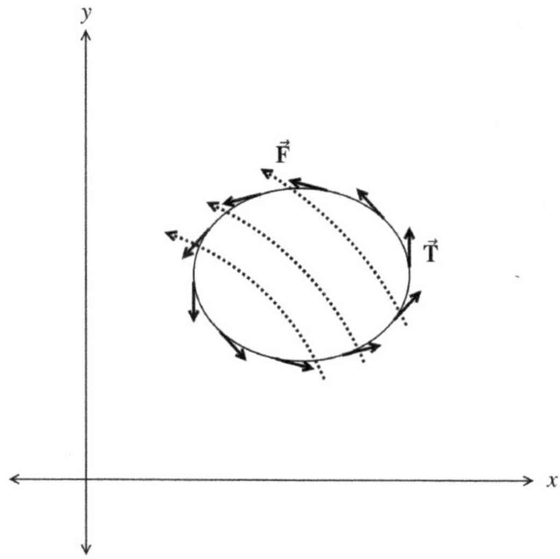

Figure 4.23 Tangent vectors, $\vec{T}$, along a closed curve and a field of force vectors crossing the curve.

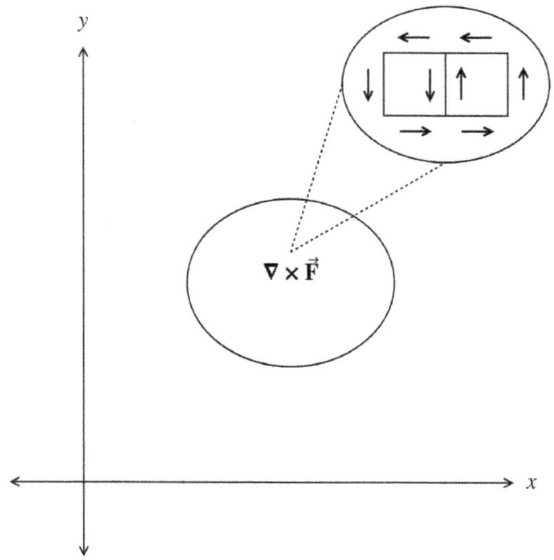

Figure 4.24 Equivalence of the line integral and the curl calculations.

when all chunks in the area S are added up the only parts of the curl vectors left are the ones on the boundary of the area . . . all the internal curl vectors have cancelled each other. Thus, we are left with a set of vectors that circulate on the boundary of the surface. As we have seen, this set of vectors is equal to the line integral over the boundary curve.

There are actually two parts to Green's theorem; we have seen the curl version and there is a divergence version as well. Recall that we can do two different types of line integrals; one calculating the sum of all the infinitesimals *tangent* to the curve and one calculating the sum of all of the infinitesimals *normal* to the curve. The second is what we referred to as the "flux" line integral and the second of the two Green's theorems is similar in that rather than doing a "circulation" line integral we do a "normal-to" line integral and relate that to a divergence calculation for the area:

$$\vec{\mathbf{F}} = P\vec{\mathbf{u}}_x + Q\vec{\mathbf{u}}_y$$

$$\vec{\mathbf{n}} = \frac{\mathrm{d}y}{\mathrm{d}s}\vec{\mathbf{u}}_x + \frac{\mathrm{d}x}{\mathrm{d}s}\vec{\mathbf{u}}_y$$

(from equation (4.14))

$$\begin{aligned}\oint_C \vec{\mathbf{F}}\cdot\vec{\mathbf{n}}\mathrm{d}s &= \oint_C (P\vec{\mathbf{u}}_x + Q\vec{\mathbf{u}}_y)\cdot\left(\frac{\mathrm{d}y}{\mathrm{d}s}\vec{\mathbf{u}}_x + \frac{\mathrm{d}x}{\mathrm{d}s}\vec{\mathbf{u}}_y\right)\mathrm{d}s \\ &= \oint_C -Q\mathrm{d}x + P\mathrm{d}y \\ &= \iint_S \left(\frac{\partial P}{\partial x} + \frac{\partial Q}{\partial y}\right)\mathrm{d}x\mathrm{d}y\end{aligned} \tag{4.41}$$

How is the last line arrived at? Note that in the second line P and Q have been swapped with respect to the first version of Green's theorem that we looked at. Also, the sign on Q has been changed. Having done this and then applied the first theorem (eqn (4.39)) we get the last line. Try it for yourself.

The second and third lines express a Cartesian version of the theorem but, as with the first theorem, we may write a vector version. We start by noting:

$$\begin{aligned}\vec{\nabla}\cdot\vec{\mathbf{F}} &= \left(\frac{\partial}{\partial x}\vec{\mathbf{u}}_x + \frac{\partial}{\partial y}\vec{\mathbf{u}}_y\right)\cdot(P\vec{\mathbf{u}}_x + Q\vec{\mathbf{u}}_y) \\ &= \frac{\partial P}{\partial x} + \frac{\partial Q}{\partial y}\end{aligned}$$

and substituting:

$$\iint_S \left(\frac{\partial P}{\partial x} + \frac{\partial Q}{\partial y}\right)\mathrm{d}x\mathrm{d}y = \iint_S \vec{\nabla}\cdot\vec{\mathbf{F}}\mathrm{d}a$$

So we may write this second version of Green's theorem in either Cartesian form or vector form:

$$\oint_C -Q\mathrm{d}x + P\mathrm{d}y = \iint_S \left(\frac{\partial P}{\partial x} + \frac{\partial Q}{\partial y}\right)\mathrm{d}x\mathrm{d}y$$

or (4.42)

$$\oint_C \vec{\mathbf{F}} \cdot \vec{\mathbf{n}}\mathrm{d}s = \iint_S \vec{\nabla} \cdot \vec{\mathbf{F}}\mathrm{d}a$$

The physical picture that we associate with this second version is one of net "flow" out of an area. This is the reason for doing our line integrals in a counter-clockwise direction. The line integral on the left of the vector version of the theorem measures the flux across the boundary of the area, as we have seen in our discussion of line integrals. The right-hand side of the vector version of the theorem gives the divergence of "fluid" within the area, which as we can see is equal to the flux of "fluid" across the area boundary. This second version of Green's theorem is occasionally referred to as Gauss's Divergence theorem in the plane.

Green's theorems are really just specific examples of two more general theorems, Stokes' theorem and Gauss's theorem. Green's theorems apply only in two-dimensional situations whereas Stokes' and Gauss's theorems apply to three-dimensional problems. We look first at Gauss's Divergence theorem. Since we are now working in three-space we suspect that rather than doing a line integral we will be doing a surface integral and this is exactly so. Gauss's theorem (also referred to as the *divergence theorem*) is stated in vector form as:

$$\oiint_S \vec{\mathbf{F}} \cdot \vec{\mathbf{n}}\mathrm{d}\sigma = \iiint_V \vec{\nabla} \cdot \vec{\mathbf{F}}\mathrm{d}V \tag{4.43}$$

or in Cartesian form as:

$$\oiint_S (P\cos(\alpha) + Q\cos(\beta) = R\cos(\gamma))\mathrm{d}\sigma = \iiint_V \left(\frac{\partial P}{\partial x} + \frac{\partial Q}{\partial y} + \frac{\partial R}{\partial z}\right)\mathrm{d}V \tag{4.44}$$

where the cosine terms are the *direction cosines* (see Appendix A) of the unit normal to the surface. The Cartesian form can easily be arrived at from the vector form using our by now familiar rules for calculating the inner product:

$$\vec{\mathbf{F}} = P\vec{\mathbf{u}}_x + Q\vec{\mathbf{u}}_y + R\vec{\mathbf{u}}_z$$

$$\vec{\mathbf{n}} = \cos(\alpha)\vec{\mathbf{u}}_x + \cos(\beta)\vec{\mathbf{u}}_y + \cos(\gamma)\vec{\mathbf{u}}_z$$

$$\begin{aligned}\vec{\mathbf{F}} \cdot \vec{\mathbf{n}} &= (P\vec{\mathbf{u}}_x + Q\vec{\mathbf{u}}_y + R\vec{\mathbf{u}}_z) \cdot (\cos(\alpha)\vec{\mathbf{u}}_x + \cos(\beta)\vec{\mathbf{u}}_y + \cos(\gamma)\vec{\mathbf{u}}_z) \\ &= P\cos(\alpha) + Q\cos(\beta) + R\cos(\gamma)\end{aligned}$$

and

$$\begin{aligned}\vec{\nabla} \cdot \vec{\mathbf{F}} &= \left(\frac{\partial}{\partial x}\vec{\mathbf{u}}_x + \frac{\partial}{\partial y}\vec{\mathbf{u}}_y + \frac{\partial}{\partial z}\vec{\mathbf{u}}_z\right) \cdot (P\vec{\mathbf{u}}_x + Q\vec{\mathbf{u}}_y + R\vec{\mathbf{u}}_z) \\ &= \frac{\partial P}{\partial x} + \frac{\partial Q}{\partial y} + \frac{\partial R}{\partial z}\end{aligned}$$

The physical meaning of this is very similar to Green's second theorem above. The normal version of the surface integral is equivalent to the flux of the vector $\vec{\mathbf{F}}$ across the boundary of the surface, S. The proof of this theorem follows much the same path as for Green's theorem. We start by looking at one term of the surface integral:

$$\oiint_S R\cos(\gamma)\mathrm{d}\sigma$$

γ is the direction cosine angle between the z-axis and the normal to the surface. As with Green's theorem we look at a surface split into sections (Figure 4.25).

This is an x-simple, y-simple and z-simple surface with a projected area, M, in the x, y-plane. For extreme simplicity we have chosen a box shape with side S_1 on the bottom, S_2 on the top and sides S_3 (only three are labelled in Figure 4.25) with the unit normal vectors shown on all but the back side. One could, however, choose any x-, y-, z-simple shape for the following analysis.

On each surface S_3, the value of the direction cosine, $\cos(\gamma)$, must be zero since γ is the (90°) angle between the unit normal vectors pointing outwards from surfaces S_3 and the z-axis. The complete surface integral is:

$$\begin{aligned}\oiint_S R\cos(\gamma)\mathrm{d}\sigma &= \iint_{S_1} R\cos(\gamma)\mathrm{d}\sigma + \oiint_{S_2} R\cos(\gamma)\mathrm{d}\sigma + \oiint_{S_3} R\cos(\gamma)\mathrm{d}\sigma \\ &= \oiint_{S_1} R\cos(\gamma)\mathrm{d}\sigma + \oiint_{S_2} R\cos(\gamma)\mathrm{d}\sigma\end{aligned}$$

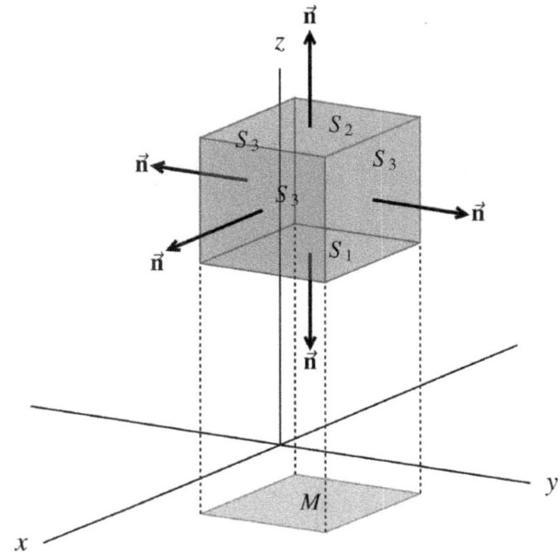

Figure 4.25 Gauss's theorem surface.

The last term over S_3 vanishes because of the zero value for $\cos(\gamma)$ mentioned above. From our previous discussion of surface area we found that $\mathrm{d}\sigma$ is equivalent to $\mathrm{d}x\mathrm{d}y/\cos(\gamma)$ or $\mathrm{d}\sigma \cos(\gamma) = \mathrm{d}x\mathrm{d}y$ (see eqn (4.25)). Substituting we get:

$$\oiint_S R\cos(\gamma)\mathrm{d}\sigma = -\oiint_{M_{xy}} R(x,y,f_1(x,y))\mathrm{d}x\mathrm{d}y + \oiint_{M_{xy}} R(x,y,f_2(x,y))\mathrm{d}x\mathrm{d}y$$

$$= \oiint_{M_{xy}} R(x,y,f_2(x,y))\mathrm{d}x\mathrm{d}y - \oiint_{M_{xy}} R(x,y,f_1(x,y))\mathrm{d}x\mathrm{d}y$$

$z = f_1(x,y)$ is the function corresponding to surface S_1 and $z = f_2(x, y)$ is for S_2 in Figure 4.25. Where does the minus sign come from? From the figure, the unit normal vector on surface S_1 has a direction cosine angle of 180 with respect to the z-axis and therefore a direction cosine value of -1. The terms in the brackets in the last line should look a bit familiar. They are the definite integral antiderivative terms of an integration over z from $f_1(x, y)$ to $f_2(x, y)$. We can therefore write:

$$\oiint_S R\cos(\gamma)d\sigma = \oiint_{M_{xy}} R(x,y,f_2(x,y))\mathrm{d}x\mathrm{d}y - \oiint_{M_{xy}} R(x,y,f_1(x,y))\mathrm{d}x\mathrm{d}y$$

$$= \iint_{M_{xy}} \left[\int_{f_1(x,y)}^{f_2(x,y)} \frac{\partial R}{\partial z} dz\right] \mathrm{d}x\mathrm{d}y$$

$$= \iiint_V \frac{\partial R}{\partial z} \mathrm{d}x\mathrm{d}y\mathrm{d}z$$

Similarly:

$$\oiint_S P\cos(\alpha)\mathrm{d}\sigma = \iiint_V \frac{\partial P}{\partial x}\mathrm{d}x\mathrm{d}y\mathrm{d}z$$

$$\oiint_S Q\cos(\beta)\mathrm{d}\sigma = \iiint_V \frac{\partial Q}{\partial y}\mathrm{d}x\mathrm{d}y\mathrm{d}z$$

and summing:

$$\oiint_S (P\cos(\alpha) + Q\cos(\beta) + R\cos(\gamma))\mathrm{d}\sigma = \iiint_V \left(\frac{\partial P}{\partial x} + \frac{\partial Q}{\partial y} + \frac{\partial R}{\partial z}\right)\mathrm{d}V$$

Stokes' theorem is stated as:

$$\oint_C \vec{\mathbf{F}} \cdot \mathrm{d}\vec{\mathbf{r}} = \iint_S (\vec{\nabla} \times \vec{\mathbf{F}}) \cdot \vec{\mathbf{n}}\mathrm{d}\sigma \tag{4.45}$$

This is very similar to Green's first theorem except that the curl term on the right is "dotted" with the unit normal to a three-dimensional surface rather than the unit normal to a planar surface. Thus, we are now considering a three-dimensional surface instead of a two-dimensional surface. Picture a child's bubble-blowing toy. The two-dimensional surface made by the soap solution on the bubble wand is blown out into a three-dimensional surface that is connected to the wand.

In Figure 4.26, S is the surface, A is the projection of the surface onto the x,y-plane and C is the boundary line around A.

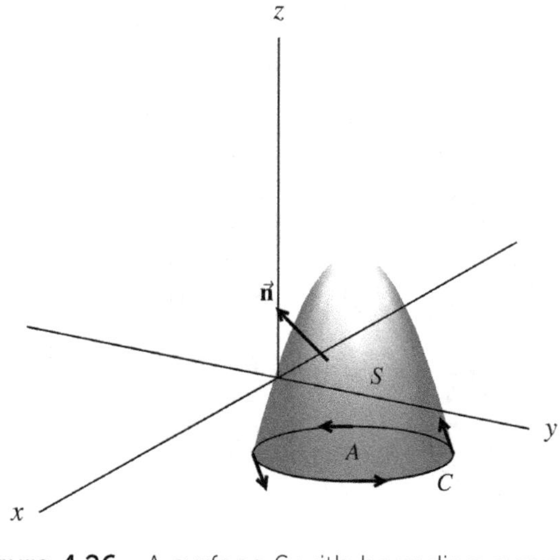

Figure 4.26 A surface S with bounding curve C.

We pause for a moment and consider the implications of eqn (4.45). The relationship says that the integral over the surface is equivalent to the line integral over the bounding curve. Another way to look at this is that the value of the surface integral is independent of the surface! That is, for a given boundary curve C, no matter what the connected surface is its surface integral will always be the same.

We can of course write Stokes' theorem in Cartesian form:

$$\vec{\mathbf{F}} = P\vec{\mathbf{u}}_x + Q\vec{\mathbf{u}}_y + R\vec{\mathbf{u}}_z$$

$$\vec{\mathbf{n}} = \cos(\alpha)\vec{\mathbf{u}}_x + \cos(\beta)\vec{\mathbf{u}}_y + \cos(\gamma)\vec{\mathbf{u}}_z$$

$$\mathrm{d}\vec{\mathbf{r}} = \mathrm{d}x\vec{\mathbf{u}}_x + \mathrm{d}y\vec{\mathbf{u}}_y + \mathrm{d}z\vec{\mathbf{u}}_z$$

$$\begin{aligned}\vec{\nabla} \times \vec{\mathbf{F}} &= \left(\frac{\partial R}{\partial y} - \frac{\partial Q}{\partial z}\right)\vec{\mathbf{u}}_x - \left(\frac{\partial R}{\partial x} - \frac{\partial P}{\partial z}\right)\vec{\mathbf{u}}_y + \left(\frac{\partial Q}{\partial x} - \frac{\partial P}{\partial y}\right)\vec{\mathbf{u}}_z \\ &= \left(\frac{\partial R}{\partial y} - \frac{\partial Q}{\partial z}\right)\vec{\mathbf{u}}_x + \left(\frac{\partial P}{\partial x} - \frac{\partial R}{\partial z}\right)\vec{\mathbf{u}}_y + \left(\frac{\partial Q}{\partial x} - \frac{\partial P}{\partial y}\right)\vec{\mathbf{u}}_z\end{aligned}$$

$$\vec{\mathbf{F}} \cdot \mathrm{d}\vec{\mathbf{r}} = P\mathrm{d}x + Q\mathrm{d}y + R\mathrm{d}z$$

$$\left(\vec{\nabla} \times \vec{\mathbf{F}}\right) \cdot \vec{\mathbf{n}} = \left(\frac{\partial R}{\partial y} - \frac{\partial Q}{\partial z}\right)\cos(\alpha) + \left(\frac{\partial P}{\partial x} - \frac{\partial R}{\partial z}\right)\cos(\beta) + \left(\frac{\partial Q}{\partial x} - \frac{\partial P}{\partial y}\right)\cos(\gamma)$$

and:

$$\begin{aligned}&\oint_C P\mathrm{d}x + Q\mathrm{d}y + R\mathrm{d}z \\ &= \iint_S \left[\left(\frac{\partial R}{\partial y} - \frac{\partial Q}{\partial z}\right)\cos(\alpha) + \left(\frac{\partial P}{\partial x} - \frac{\partial R}{\partial z}\right)\cos(\beta) + \left(\frac{\partial Q}{\partial x} - \frac{\partial P}{\partial y}\right)\cos(\gamma)\right]\mathrm{d}A\end{aligned} \tag{4.46}$$

One immediately appreciates that this becomes Green's first theorem if, for example $R(x, y, z) = 0$ and the surface is contained in the x, y-plane. In this case the unit normal vector is aligned parallel to the z-axis and $\cos(\gamma)$ equals one, while both $\cos(\alpha)$ and $\cos(\beta)$ are zero. This being the case:

$$\oint_C P\mathrm{d}x + Q\mathrm{d}y = \iint_S \left(\frac{\partial Q}{\partial x} - \frac{\partial P}{\partial y}\right)\mathrm{d}A$$

which is just the Cartesian version of Green's first theorem as we have seen it.

The proof is as follows. Consider:

$$\iint_S (\vec{\nabla} \times \vec{\mathbf{F}}) \cdot \vec{\mathbf{n}} d\sigma = \iint_S \vec{\nabla} \times (P\vec{\mathbf{u}}_x + Q\vec{\mathbf{u}}_y + R\vec{\mathbf{u}}_z) \cdot \vec{\mathbf{n}} d\sigma$$

We will look at the first term on the right-hand side:

$$\vec{\nabla} \times P\vec{\mathbf{u}}_x = \begin{vmatrix} \vec{\mathbf{u}}_x & \vec{\mathbf{u}}_y & \vec{\mathbf{u}}_z \\ \frac{\partial}{\partial x} & \frac{\partial}{\partial y} & \frac{\partial}{\partial z} \\ P & 0 & 0 \end{vmatrix} = \frac{\partial P}{\partial z}\vec{\mathbf{u}}_y - \frac{\partial P}{\partial y}\vec{\mathbf{u}}_z$$
$$\vec{\nabla} \times P\vec{\mathbf{u}}_x \cdot \vec{\mathbf{n}} d\sigma = \left(\frac{\partial P}{\partial z}\vec{\mathbf{u}}_y - \frac{\partial P}{\partial y}\vec{\mathbf{u}}_z\right) \cdot \vec{\mathbf{n}} d\sigma \tag{4.47}$$

If we understand that $z = f(x, y)$ and that the position vector for the surface S is $\vec{\mathbf{r}} = x\vec{\mathbf{u}}_x + y\vec{\mathbf{u}}_y + z\vec{\mathbf{u}}_z$ then we can write:

$$\begin{aligned} \vec{\mathbf{r}} &= x\vec{\mathbf{u}}_x + y\vec{\mathbf{u}}_y + z\vec{\mathbf{u}}_z \\ &= x\vec{\mathbf{u}}_x + y\vec{\mathbf{u}}_y + f(x, y)\vec{\mathbf{u}}_z \\ \frac{\partial}{\partial y}\vec{\mathbf{r}} &= \frac{\partial y}{\partial y}\vec{\mathbf{u}}_y + \frac{\partial f(x, y)}{\partial y}\vec{\mathbf{u}}_z \\ &= \vec{\mathbf{u}}_y + \frac{\partial z}{\partial y}\vec{\mathbf{u}}_z \end{aligned}$$

Recall from above that $\frac{\partial}{\partial y}\vec{\mathbf{r}}$ is in the tangent plane to the surface so that "dotting" it with the surface unit vector gives:

$$\frac{\partial}{\partial y}\vec{\mathbf{r}} \cdot \vec{\mathbf{n}} = 0 = \left(\vec{\mathbf{u}}_y + \frac{\partial z}{\partial y}\vec{\mathbf{u}}_z\right) \cdot \vec{\mathbf{n}} = \vec{\mathbf{u}}_y \cdot \vec{\mathbf{n}} + \frac{\partial z}{\partial y}\vec{\mathbf{u}}_z \cdot \vec{\mathbf{n}}$$

and rearranging

$$\vec{\mathbf{u}}_y \cdot \vec{\mathbf{n}} = -\frac{\partial z}{\partial y}\vec{\mathbf{u}}_z \cdot \vec{\mathbf{n}}$$

We now substitute into eqn (4.47):

$$\vec{\nabla} \times P\vec{\mathbf{u}}_x \cdot \vec{\mathbf{n}}\mathrm{d}\sigma = \left(\frac{\partial P}{\partial z}\vec{\mathbf{u}}_y - \frac{\partial P}{\partial y}\vec{\mathbf{u}}_z\right) \cdot \vec{\mathbf{n}}\mathrm{d}\sigma$$

$$= \left(\frac{\partial P}{\partial z}\vec{\mathbf{u}}_y \cdot \vec{\mathbf{n}} - \frac{\partial P}{\partial y}\vec{\mathbf{u}}_z \cdot \vec{\mathbf{n}}\right)\mathrm{d}\sigma$$

$$= \left(-\frac{\partial P}{\partial z}\frac{\partial z}{\partial y}\vec{\mathbf{u}}_z \cdot \vec{\mathbf{n}} - \frac{\partial P}{\partial y}\vec{\mathbf{u}}_z \cdot \vec{\mathbf{n}}\right)\mathrm{d}\sigma$$

$$= \left(-\frac{\partial P}{\partial z}\frac{\partial z}{\partial y} - \frac{\partial P}{\partial y}\right)\vec{\mathbf{u}}_z \cdot \vec{\mathbf{n}}\mathrm{d}\sigma$$

Now, P is formally a function of x, y and z, however, if $z = f(x, y)$ it can be considered to be a function of only two variables, x and y:

$$P(x,y,z) = P(x,y,f(x,y)) = G(x,y)$$

and

$$\frac{\partial P}{\partial y} + \frac{\partial P}{\partial z}\frac{\partial z}{\partial y} = \frac{\partial G}{\partial y}$$

and we can write

$$\nabla \times P\vec{\mathbf{u}}_x \cdot \vec{\mathbf{n}}\mathrm{d}\sigma = -\frac{\partial G}{\partial y}\vec{\mathbf{u}}_z \cdot \vec{\mathbf{n}}\mathrm{d}\sigma$$

$$= -\frac{\partial G}{\partial y}\cos(\gamma)\frac{\mathrm{d}x\mathrm{d}y}{\cos(\gamma)}$$

$$= -\frac{\partial G}{\partial y}\mathrm{d}x\mathrm{d}y$$

Therefore:

$$\iint_S \vec{\nabla} \times P\vec{\mathbf{u}}_x \cdot \vec{\mathbf{n}}\mathrm{d}\sigma = \iint_A -\frac{\partial G}{\partial y}\mathrm{d}x\mathrm{d}y$$

The right-hand side is part of what we have seen in Green's first theorem and so, using that theorem:

$$\iint_A -\frac{\partial G}{\partial y}\mathrm{d}x\mathrm{d}y = \oint_C G\mathrm{d}x$$

or

$$\iint_A -\frac{\partial P}{\partial y}\mathrm{d}x\mathrm{d}y = \oint_C P\mathrm{d}x = \iint_S \vec{\nabla} \times P\vec{\mathbf{u}}_x \cdot \vec{\mathbf{n}}\mathrm{d}\sigma$$

Similarly:

$$\oint_C Q dy = \iint_S \vec{\nabla} \times P\vec{\mathbf{u}}_y \cdot \vec{\mathbf{n}} d\sigma$$

$$\oint_C R dz = \iint_S \vec{\nabla} \times P\vec{\mathbf{u}}_z \cdot \vec{\mathbf{n}} d\sigma$$

Combining these results gives us:

$$\iint_S \vec{\nabla} \times (P\vec{\mathbf{u}}_x + Q\vec{\mathbf{u}}_y + R\vec{\mathbf{u}}_z) \cdot \vec{\mathbf{n}} d\sigma = \oint_C P\mathrm{d}x + Q\mathrm{d}y + R\mathrm{d}z \tag{4.48}$$

or, in vector form, Stokes' theorem is:

$$\iint_S (\vec{\nabla} \times \vec{\mathbf{F}}) \cdot \vec{\mathbf{n}} d\sigma = \oint_C \vec{\mathbf{F}} \cdot \mathrm{d}\vec{\mathbf{r}} \tag{4.49}$$

Stokes' theorem is of fundamental importance in electromagnetic theory and we will make much use of it later in that context.

We have seen some complicated and tedious calculations on our road to discovery of the differential and integral calculus of vectors. We have also seen some wonderful simplifications that we will summarise here.

First, we list the irrotational field relationships:

1. $\vec{\mathbf{F}} = \vec{\nabla} f$
2. $\vec{\nabla} \times \vec{\mathbf{F}} = 0$
3. $\int_a^b \vec{\mathbf{F}} \cdot \vec{\mathbf{r}}'$ (independent of path)
4. $\oint \vec{\mathbf{F}} \cdot \vec{\mathbf{r}}' = 0$ (independent of path)

The really interesting thing about these relationships is that if one of them is true then all of them are true. That is, if we generate a field from a scalar potential function only, as in point 1, then all of the other items are also satisfied. Given point 1, we have shown in eqn (4.19) that point 2 follows. Also, points 3 and 4 were shown to be so in eqns (4.17) and eqn (4.18), respectively, given point 1.

The solenoidal field relationships are:

1. $\vec{\mathbf{F}} = \vec{\nabla} \times \vec{\mathbf{A}}$
2. $\vec{\nabla} \cdot \vec{\mathbf{F}} = 0$

3. $\int_S \vec{\mathbf{F}} \cdot d\vec{\sigma}$ is independent of surface
4. $\int_S \vec{\mathbf{F}} \cdot d\vec{\sigma} = 0$ (for a closed surface)

and again, if one is true they are all true. The first point describes a field generated from a vector potential function. Point 2 follows from point 1, according to eqn (4.20). Point 3 is rationalised from Stokes' theorem:

$$\iint_S (\vec{\nabla} \times \vec{\mathbf{F}}) \cdot \vec{\mathbf{n}} d\sigma = \oint_C \vec{\mathbf{F}} \cdot d\vec{\mathbf{r}}$$

As discussed above with reference to Figure 4.26, the surface integral does not depend on the surface but rather on the bounding curve, hence point 3 when $\vec{\mathbf{F}}$ is a solenoidal vector field. Point 4 follows from point 3 if one imagines the boundary curve in Figure 4.26 shrinking down to nothing. The line integral in Stokes' theorem shinks to zero as well which means so does the surface integral.

We show now, *via* Stokes' theorem, a couple of relationships that will be of use later. First, letting $\vec{\mathbf{F}} = \vec{\mathbf{k}}T$ and stipulating that $\vec{\mathbf{k}}$ is a constant vector, we may use identity 5 from Section4.1 to write:

$$\begin{aligned} \vec{\nabla} \times \vec{\mathbf{k}}T &= \vec{\nabla}T \times \vec{\mathbf{k}} - \vec{\mathbf{k}} \times \vec{\nabla}T \\ &= -\vec{\mathbf{k}} \times \vec{\nabla}T \text{ (since } \vec{\mathbf{k}} \text{ is constant)} \end{aligned}$$

and inserting this into Stokes' theorem:

$$-\int_S \vec{\mathbf{k}} \times \vec{\nabla}T \cdot d\vec{\mathbf{a}} = \oint_C \vec{\mathbf{k}}T \cdot d\vec{\mathbf{l}}$$

Using the triple product relationship eqn (3.22), we may rearrange things a bit:

$$\vec{\mathbf{k}} \times (\vec{\nabla}T) \cdot d\vec{\mathbf{a}} = \vec{\mathbf{k}} \cdot (\vec{\nabla}T \times d\vec{\mathbf{a}})$$

and:

$$-\int_S \vec{\mathbf{k}} \cdot (\vec{\nabla}T \times d\vec{\mathbf{a}}) = \oint_C \vec{\mathbf{k}}T \cdot d\vec{\mathbf{l}} \tag{4.50}$$

Since $\vec{\mathbf{k}}$ is a constant vector we can pull it out of the integral:

$$-\vec{\mathbf{k}}\cdot\int_S(\vec{\nabla}T\times \mathrm{d}\vec{\mathbf{a}})=\vec{\mathbf{k}}\cdot\oint_C T\mathrm{d}\vec{\mathbf{l}}$$

$$\Downarrow \tag{4.51}$$

$$-\int_S(\vec{\nabla}T\times \mathrm{d}\vec{\mathbf{a}})=\oint_C T\mathrm{d}\vec{\mathbf{l}}$$

Now, letting $T=\vec{\kappa}\cdot\vec{\mathbf{r}}$, where both vectors are constants:

$$\begin{aligned}\vec{\nabla}T&=\vec{\nabla}(\vec{\kappa}\cdot\vec{\mathbf{r}})\\&=\vec{\kappa}\times(\vec{\nabla}\times\vec{\mathbf{r}})+(\vec{\kappa}\cdot\vec{\nabla})\vec{\mathbf{r}}\\&=(\vec{\kappa}\cdot\vec{\nabla})\vec{\mathbf{r}}\end{aligned}$$

We arrive at the last line by realising that $\vec{\nabla}\times\vec{\mathbf{r}}=0$ since $\vec{\mathbf{r}}$ is a constant vector. Now:

$$\begin{aligned}(\vec{\kappa}\cdot\vec{\nabla})\vec{\mathbf{r}}&=(\kappa_x\vec{\mathbf{u}}_x+\kappa_y\vec{\mathbf{u}}_y+\kappa_z\vec{\mathbf{u}}_z)\cdot\left(\frac{\partial}{\partial x}\vec{\mathbf{u}}_x+\frac{\partial}{\partial y}\vec{\mathbf{u}}_y+\frac{\partial}{\partial z}\vec{\mathbf{u}}_z\right)(x\vec{\mathbf{u}}_x+y\vec{\mathbf{u}}_y+z\vec{\mathbf{u}}_z)\\&=\left(\kappa_x\frac{\partial}{\partial x}+\kappa_y\frac{\partial}{\partial y}+\kappa_z\frac{\partial}{\partial z}\right)(x\vec{\mathbf{u}}_x+y\vec{\mathbf{u}}_y+z\vec{\mathbf{u}}_z)\\&=(\kappa_x\vec{\mathbf{u}}_x+\kappa_y\vec{\mathbf{u}}_y+\kappa_z\vec{\mathbf{u}}_z)\\&=\vec{\nabla}T\end{aligned}$$

and:

$$\oint T\mathrm{d}\vec{\mathbf{l}}=\oint(\vec{\kappa}\cdot\vec{\mathbf{r}})\mathrm{d}\vec{\mathbf{l}}=-\int_S\vec{\nabla}T\times\mathrm{d}\vec{\mathbf{a}}$$

from eqn (4.51). Furthermore:

$$\begin{aligned}-\int_S\vec{\nabla}T\times\mathrm{d}\vec{\mathbf{a}}&=-\int\vec{\kappa}\times\mathrm{d}\vec{\mathbf{a}}=-\vec{\kappa}\times\int\mathrm{d}\vec{\mathbf{a}}\\&=-\vec{\kappa}\times\vec{\mathbf{a}}\\&=\vec{\mathbf{a}}\times\vec{\kappa}\end{aligned}$$

and therefore:

$$\oint(\vec{\kappa}\cdot\vec{\mathbf{r}})\mathrm{d}\vec{\mathbf{l}}=-\vec{\kappa}\times\int\mathrm{d}\vec{\mathbf{a}}=\vec{\mathbf{a}}\times\vec{\kappa} \tag{4.52}$$

4.5 Vector Rotation in a Rotating Frame of Reference

We will see later that the application of an rf field, $\vec{\mathbf{B}}_1$, at right angles to the external field, $\vec{\mathbf{B}}_0$, results in changes to the macroscopic magnetisation that are difficult to picture clearly unless one invokes the *rotating frame* of reference. You will need a bit of vector algebra and calculus from the previous sections to understand the explanation.

Consider the vector $\vec{\mathbf{F}}$ in the transverse plane (the x,y-plane) broken up into its component vectors:

$$\vec{\mathbf{F}} = F_x\mathbf{u}_x + F_y\mathbf{u}_y + F_z\mathbf{u}_z$$

Now we write the rate of change of $\vec{\mathbf{F}}$ (using the product rule of differentiation):

$$\begin{aligned}\frac{\mathrm{d}\vec{\mathbf{F}}}{\mathrm{d}t} &= \frac{\mathrm{d}F_x}{\mathrm{d}t}\vec{\mathbf{u}}_x + F_x\frac{\mathrm{d}\vec{\mathbf{u}}_x}{\mathrm{d}t} + \frac{\mathrm{d}F_y}{\mathrm{d}t}\vec{\mathbf{u}}_y + F_y\frac{\mathrm{d}\vec{\mathbf{u}}_y}{\mathrm{d}t} \\ &= \left(\frac{\mathrm{d}F_x}{\mathrm{d}t}\vec{\mathbf{u}}_x + \frac{\mathrm{d}F_y}{\mathrm{d}t}\vec{\mathbf{u}}_y\right) + \left(F_x\frac{\mathrm{d}\vec{\mathbf{u}}_x}{\mathrm{d}t} + F_y\frac{\mathrm{d}\vec{\mathbf{u}}_y}{\mathrm{d}t}\right)\end{aligned} \tag{4.53}$$

The first term in brackets is the rate of change of the vector $\vec{\mathbf{F}}$ with respect to the x and y axes and is therefore the rate of change in the current frame of reference.

The second term in brackets is a bit more obtuse. Since it involves changes in unit vectors whose norms or lengths must remain constant, these must refer to changes in direction of the axes ... in other words, rotation. The x and y axes are rotating instead of being fixed!

Since $\mathrm{d}\vec{\mathbf{u}}_x/\mathrm{d}t$ must be perpendicular to $\vec{\mathbf{u}}_x$ as we saw above in Section 4.1 and since we are talking about a vector in the x,y-plane we can write:

$$\frac{\mathrm{d}\vec{\mathbf{u}}_x}{\mathrm{d}t} = \omega_x\vec{\mathbf{u}}_y \quad \text{and} \quad \frac{\mathrm{d}\vec{\mathbf{u}}_y}{\mathrm{d}t} = \omega_y\vec{\mathbf{u}}_x$$

From the their orthogonality property, $\vec{\mathbf{u}}_x \cdot \vec{\mathbf{u}}_y = 0$:

$$\vec{\mathbf{u}}_x \cdot \frac{\mathrm{d}\vec{\mathbf{u}}_y}{\mathrm{d}t} + \vec{\mathbf{u}}_y \cdot \frac{\mathrm{d}\vec{\mathbf{u}}_x}{\mathrm{d}t} = 0$$

$$\omega_x\vec{\mathbf{u}}_x \cdot \vec{\mathbf{u}}_x + \omega_y\vec{\mathbf{u}}_y \cdot \vec{\mathbf{u}}_y = 0$$

$$\omega_x + \omega_y = 0$$

This can only be true if either both ω_x and ω_y are zero or if they are equal in magnitude and opposite in sign. Since we are dealing with a rotating system we choose the latter and write:

$$\frac{\mathrm{d}\vec{\mathbf{u}}_x}{\mathrm{d}t} = -\omega\vec{\mathbf{u}}_y \quad \text{and} \quad \frac{\mathrm{d}\vec{\mathbf{u}}_y}{\mathrm{d}t} = \omega\vec{\mathbf{u}}_x$$

We use these to rewrite the second bracket in eqn (4.53):

$$F_x\omega\mathbf{u}_y - F_y\omega\mathbf{u}_x$$

This can also be expressed succinctly in a cross product:

$$\begin{aligned} \vec{\omega}\times\vec{\mathbf{F}} &= \begin{vmatrix} \vec{\mathbf{u}}_x & \vec{\mathbf{u}}_y & \vec{\mathbf{u}}_z \\ 0 & 0 & \omega \\ F_x & F_y & F_z \end{vmatrix} \\ &= (0F_z - \omega F_y)\vec{\mathbf{u}}_x - (0F_z - \omega F_x)\vec{\mathbf{u}}_y + (0F_y - 0F_x)\vec{\mathbf{u}}_z \\ &= \omega F_x\mathbf{u}_y - \omega F_y\mathbf{u}_x \end{aligned}$$

This might be a bit confusing at first glance. Why are the zeros there? Remember that we are talking about a vector rotating in the transverse plane and that the vector frequency is aligned along the z-axis. It must be so since for a system that is rotating at constant rotational velocity the vector frequency must be constant and this can only be so if it is aligned along the axis of rotation. If it is aligned along the z-axis it will have no components along the x and y axes ... hence the zeros. Also, the cross product of $\vec{\omega}$ and $\vec{\mathbf{F}}$ is at right angles to these two vectors and represents the change in $\vec{\mathbf{F}}$, which is at right angles to $\vec{\mathbf{F}}$ and is in the x,y-plane (see Figure 4.27).

So, we summarise eqn (4.53) in compact form:

$$\left(\frac{\mathrm{d}\vec{\mathbf{F}}}{\mathrm{d}t}\right)_{LAB} = \left(\frac{\mathrm{d}\vec{\mathbf{F}}}{\mathrm{d}t}\right)_{ROT} + (\vec{\omega}\times\vec{\mathbf{F}}) \tag{4.54}$$

where the subscripts *LAB* and *ROT* refer to the laboratory frame of reference and the rotating frame of reference respectively. $(\mathrm{d}\vec{\mathbf{F}}/\mathrm{d}t)_{ROT}$ refers to the rotation of vector $\vec{\mathbf{F}}$ with respect to the (possibly rotating) x, y and z axes and $\vec{\omega}\times\vec{\mathbf{F}}$ refers to the rotational movement of the x, y and z axes themselves.

Experience with trying to picture what eqn (4.54) refers to means that we should probably try to visualise it here. Picture yourself

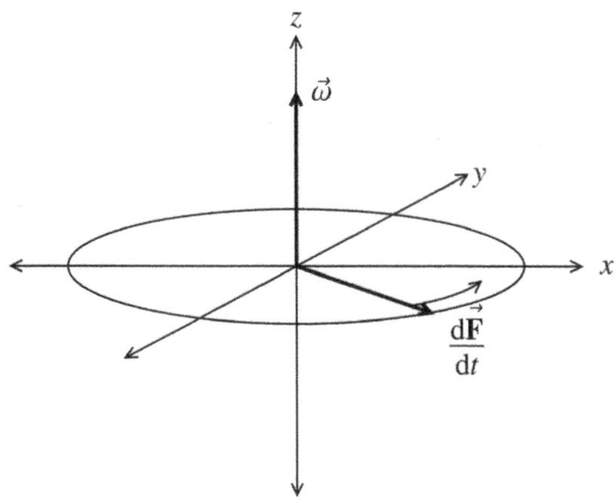

Figure 4.27 Direction of angular velocity vector, $\vec{\omega}$.

standing and watching a rotating vector. If we say for the moment that the frame of reference is fixed then $\vec{\omega} \times \vec{\mathbf{F}}$ in eqn (4.54) equals zero and:

$$\left(\frac{d\vec{\mathbf{F}}}{dt}\right)_{LAB} = \left(\frac{d\vec{\mathbf{F}}}{dt}\right)_{ROT}$$

The vector, $\vec{\mathbf{F}}$, rotates with angular velocity ω_{LAB} with respect to the static coordinate frame. Our current point of view is called the *laboratory frame* in NMR language. Obviously, it is measured relative to the laboratory frame where we consider the x, y and z axes to be fixed.

Now suppose that we change our point of view by taking the static frame of reference and making it dynamic, rotating in fact, as in Figure 4.28. Furthermore let us choose to make the x and y axes rotate about the z-axis in the same direction as $\vec{\mathbf{F}}$ with the same angular velocity. What would we see? From eqn (4.54):

$$\left(\frac{d\vec{\mathbf{F}}}{dt}\right)_{LAB} = (\vec{\omega} \times \vec{\mathbf{F}})$$

Because $(d\vec{\mathbf{F}}/dt)_{ROT}$ is now zero. The vector $\vec{\mathbf{F}}$ would appear to have stopped its rotation! We have changed our frame of reference to the rotating frame of reference from the laboratory or fixed frame of reference.

It is what you would see if you could jump onto a merry-go-round and watch someone . . . they move at this rate. On the merry-go-round this might be a person walking around the merry-go-round platform as it is turning. If $\vec{\mathbf{F}}$ is a simple rotating vector this term is zero and would correspond to a person on the merry-go-round standing still.

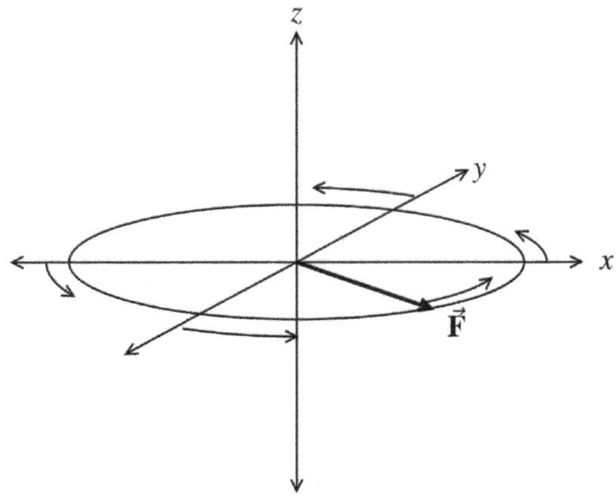

Figure 4.28 The motion of vector $\vec{\mathbf{F}}$ in a rotating frame of reference.

Further Reading

1. M. L. Boas, *Mathematical Methods in the Physical Sciences*, Wiley and Sons, New York, 2005.
2. J. D. Faires and B. T. Faires, *Calculus*, Random House, New York, 1988.
3. D. J. Griffiths, *Introduction to Electrodynamics*, Pearson Prentice Hall, Upper Saddle River, New Jersey, 1999.
4. G. B. Thomas and R.L. Finney, *Calculus and Analytic Geometry*, Addison-Wesley Publishing Company, New York, 1992.

5 Introductory Tensor Analysis

There is geometry in the humming of the strings, there is music in the spacing of the spheres.

Pythagoras

5.1 Dyadic Algebra

Consider two vectors, $\vec{\mathbf{a}}$ and $\vec{\mathbf{b}}$. As we saw in Chapter 3 we can write them as follows:

$$\vec{\mathbf{a}} = a_x\vec{\mathbf{u}}_x + a_y\vec{\mathbf{u}}_y + a_z\vec{\mathbf{u}}_z$$

$$\vec{\mathbf{b}} = b_x\vec{\mathbf{u}}_x + b_y\vec{\mathbf{u}}_y + b_z\vec{\mathbf{u}}_z$$

where each vector has three components in our Cartesian space. It will be of use to recast this using numbered subscripts rather than x, y and z:

$$\vec{\mathbf{a}} = a_1\vec{\mathbf{u}}_1 + a_2\vec{\mathbf{u}}_2 + a_3\vec{\mathbf{u}}_3$$

$$\vec{\mathbf{b}} = b_1\vec{\mathbf{u}}_1 + b_2\vec{\mathbf{u}}_2 + b_3\vec{\mathbf{u}}_3$$

If we multiply them in the "normal" distributive fashion:

$$\begin{aligned}\vec{\mathbf{a}} \otimes \vec{\mathbf{b}} &= (a_1\vec{\mathbf{u}}_1 + a_2\vec{\mathbf{u}}_2 + a_3\vec{\mathbf{u}}_3) \otimes (b_1\vec{\mathbf{u}}_1 + b_2\vec{\mathbf{u}}_2 + b_3\vec{\mathbf{u}}_3) \\ &= a_1b_1\vec{\mathbf{u}}_1 \otimes \vec{\mathbf{u}}_1 + a_1b_2\vec{\mathbf{u}}_1 \otimes \vec{\mathbf{u}}_2 + a_1b_3\vec{\mathbf{u}}_1 \otimes \vec{\mathbf{u}}_3 \\ &\quad + a_2b_1\vec{\mathbf{u}}_2 \otimes \vec{\mathbf{u}}_1 + a_2b_2\vec{\mathbf{u}}_2 \otimes \vec{\mathbf{u}}_2 + a_2b_3\vec{\mathbf{u}}_2 \otimes \vec{\mathbf{u}}_3 \\ &\quad + a_3b_1\vec{\mathbf{u}}_3 \otimes \vec{\mathbf{u}}_1 + a_3b_2\vec{\mathbf{u}}_3 \otimes \vec{\mathbf{u}}_2 + a_3b_3\vec{\mathbf{u}}_3 \otimes \vec{\mathbf{u}}_3\end{aligned}$$

Essential Mathematics for NMR and MRI Spectroscopists
By Keith C. Brown

Published by the Royal Society of Chemistry, www.rsc.org

This is the *direct* or *outer* product of $\vec{\mathbf{a}}$ and $\vec{\mathbf{b}}$, discussed in Chapters 2 and 3, and the resulting object is called a *dyad* or a *Cartesian tensor.* Note that, in our Cartesian space, there are now nine scalar coefficients, a_ib_j, that is, 3×3 from the vectors $\vec{\mathbf{a}}$ and $\vec{\mathbf{b}}$. One represents this compactly as:[†]

$$\tilde{\mathbf{D}} = \vec{\mathbf{a}} \otimes \vec{\mathbf{b}}$$

or

$$D_{ij} = a_i b_j \vec{\mathbf{u}}_i \otimes \vec{\mathbf{u}}_j$$

Just as $\vec{\mathbf{u}}_x$ is termed a unit vector, $\vec{\mathbf{u}}_1\vec{\mathbf{u}}_1$ is called a *unit dyad.* Is this product commutative, you ask? Let us see:

$$\begin{aligned}\vec{\mathbf{b}} \otimes \vec{\mathbf{a}} &= (b_1\vec{\mathbf{u}}_1 + b_2\vec{\mathbf{u}}_2 + b_3\vec{\mathbf{u}}_3) \otimes (a_1\vec{\mathbf{u}}_1 + a_2\vec{\mathbf{u}}_2 + a_3\vec{\mathbf{u}}_3)\\ &= b_1a_1\vec{\mathbf{u}}_1 \otimes \vec{\mathbf{u}}_1 + b_1a_2\vec{\mathbf{u}}_1 \otimes \vec{\mathbf{u}}_2 + b_1a_3\vec{\mathbf{u}}_1 \otimes \vec{\mathbf{u}}_3\\ &\quad + b_2a_1\vec{\mathbf{u}}_2 \otimes \vec{\mathbf{u}}_1 + b_2a_2\vec{\mathbf{u}}_2 \otimes \vec{\mathbf{u}}_2 + b_2a_3\vec{\mathbf{u}}_2 \otimes \vec{\mathbf{u}}_3\\ &\quad + b_3a_1\vec{\mathbf{u}}_3 \otimes \vec{\mathbf{u}}_1 + b_3a_2\vec{\mathbf{u}}_3 \otimes \vec{\mathbf{u}}_2 + b_3a_3\vec{\mathbf{u}}_3 \otimes \vec{\mathbf{u}}_3\end{aligned}$$

Now we subtract the dyadic products:

$$\begin{aligned}&\vec{\mathbf{a}} \otimes \vec{\mathbf{b}} - \vec{\mathbf{b}} \otimes \vec{\mathbf{a}}\\ &\quad = (a_1b_1 - b_1a_1)\vec{\mathbf{u}}_1 \otimes \vec{\mathbf{u}}_1 + (a_1b_2 - b_1a_2)\vec{\mathbf{u}}_1 \otimes \vec{\mathbf{u}}_2 + (a_1b_3 - b_1a_3)\vec{\mathbf{u}}_1 \otimes \vec{\mathbf{u}}_3\\ &\quad + (a_2b_1 - b_2a_1)\vec{\mathbf{u}}_2 \otimes \vec{\mathbf{u}}_1 + (a_2b_2 - b_2a_2)\vec{\mathbf{u}}_2 \otimes \vec{\mathbf{u}}_2 + (a_2b_3 - b_2a_3)\vec{\mathbf{u}}_2 \otimes \vec{\mathbf{u}}_3\\ &\quad + (a_3b_1 - b_3a_1)\vec{\mathbf{u}}_3 \otimes \vec{\mathbf{u}}_1 + (a_3b_2 - b_3a_2)\vec{\mathbf{u}}_3 \otimes \vec{\mathbf{u}}_2 + (a_3b_3 - b_3a_3)\vec{\mathbf{u}}_3 \otimes \vec{\mathbf{u}}_3\end{aligned}$$

The terms with the same subscripts are all zero, however, the non-identical subscript terms are not necessarily equal. The only dyads that are guaranteed to commute are those in which the two vectors making up the dyad are the same such as the projection operator in eqn (3.31). In general, however, we cannot say that the dyadic product is commutative.

Now, what would be the result of, say, the inner product of vector $\vec{\mathbf{c}}$ with dyad $\tilde{\mathbf{D}}$? We define this operation by "associating" the vector $\vec{\mathbf{c}}$ with the vector "beside" it in $\tilde{\mathbf{D}}$. Thus, if we pre-multiply by $\vec{\mathbf{c}}$:

$$\vec{\mathbf{c}} \cdot \tilde{\mathbf{D}} = (\vec{\mathbf{c}} \cdot \vec{\mathbf{a}})\,\vec{\mathbf{b}} = \lambda\vec{\mathbf{b}}$$

[†]We shall use the compact representation $\tilde{\mathbf{D}}$ here for dyads and all other tensors. Also, it will be useful to use Einstein notation as and where needed to reduce the number of written terms.

Post-multiplication gives:

$$\tilde{\mathbf{D}} \cdot \vec{\mathbf{c}} = \vec{\mathbf{a}}(\vec{\mathbf{b}} \cdot \vec{\mathbf{c}}) = \mu \vec{\mathbf{a}}$$

In effect, these algebraic operations have produced a new vector. We shall elaborate on this idea shortly. Note that this type of inner product is not generally commutative:

$$\vec{\mathbf{c}} \cdot \tilde{\mathbf{D}} \neq \tilde{\mathbf{D}} \cdot \vec{\mathbf{c}}$$

unlike the inner product of two vectors. We see that the inner product of a vector with a dyad gives back one of the vectors that make up the dyad multiplied by a constant. We will use this property to construct a set of very general Hamiltonian operators in Chapter 14.

The astute reader will by now be saying to him/herself "What is this?". To rationalise this in terms of previous discussions of vectors let us switch to our matrix representations and construct our dyad again:

$$\vec{\mathbf{a}} = \begin{bmatrix} a_1 \\ a_2 \\ a_3 \end{bmatrix} \vec{\mathbf{b}} = \begin{bmatrix} b_1 \\ b_2 \\ b_3 \end{bmatrix}$$

In order to construct the dyad such that doing an inner product of the dyad with a vector makes sense in terms of matrices we must use the transpose of $\vec{\mathbf{b}}$:

$$\vec{\mathbf{a}} = \begin{bmatrix} a_1 \\ a_2 \\ a_3 \end{bmatrix} \vec{\mathbf{b}}^T = [\, b_1 \quad b_2 \quad b_3 \,]$$

Now, if we do an inner product of $\tilde{\mathbf{D}}$ with vector $\vec{\mathbf{c}}$ explicitly in terms of matrices:

$$\begin{aligned} \tilde{\mathbf{D}} &= (\vec{\mathbf{a}} \otimes \vec{\mathbf{b}}) \cdot \vec{\mathbf{c}} \\ &= \begin{bmatrix} a_1 \\ a_2 \\ a_3 \end{bmatrix} [\, b_1 \quad b_2 \quad b_3 \,] \cdot \begin{bmatrix} c_1 \\ c_2 \\ c_3 \end{bmatrix} \\ &= \lambda \begin{bmatrix} a_1 \\ a_2 \\ a_3 \end{bmatrix} \end{aligned}$$

we observe that the inner product of $\vec{\mathbf{b}}$ with $\vec{\mathbf{c}}$ is, in terms of matrices, a scalar, as we have already seen in the Chapter 3. We might perhaps make this a bit clearer using Dirac notation:

$$\tilde{\mathbf{D}} = |a\rangle\langle b|$$

$$\tilde{\mathbf{D}} \cdot \vec{\mathbf{c}} = |a\rangle\langle b\,|\,c\rangle = \mu|a\rangle$$

What about pre-multiplication? A little thought will show that this must require the use of the transpose of vector $\vec{\mathbf{c}}$. Again, in the Dirac notation:

$$\tilde{\mathbf{D}} = |a\rangle\langle b|$$

$$\vec{\mathbf{c}}^T \cdot \tilde{\mathbf{D}} = \langle c\,|\,a\rangle\langle b| = \lambda\langle b|$$

We can do the same type of analysis with cross products:

$$\vec{\mathbf{c}} \times \tilde{\mathbf{D}} = (\vec{\mathbf{c}} \times \vec{\mathbf{a}}) \otimes \vec{\mathbf{b}} = \vec{\mathbf{d}} \otimes \vec{\mathbf{b}} = \tilde{\mathbf{N}}$$

$$\tilde{\mathbf{D}} \times \vec{\mathbf{c}} = \vec{\mathbf{a}} \otimes (\vec{\mathbf{b}} \times \vec{\mathbf{c}}) = \vec{\mathbf{a}} \otimes \vec{\mathbf{f}} = \tilde{\mathbf{O}}$$

and again we find that the product is not generally commutative:

$$\vec{\mathbf{c}} \times \tilde{\mathbf{D}} \neq \tilde{\mathbf{D}} \times \vec{\mathbf{c}}$$

but this time the result of the product of a dyad with a vector is a new dyad much as the cross product of two vectors is a new vector.

The third type of product that we will consider is the same type that we started with ... the normal distributive multiplication. Thus, we will multiply dyad $\tilde{\mathbf{D}}$ by vector $\vec{\mathbf{c}}$:

$$\vec{\mathbf{c}} \otimes \tilde{\mathbf{D}} = \vec{\mathbf{c}} \otimes \vec{\mathbf{a}} \otimes \vec{\mathbf{b}}$$

Long multiplication will produce:

$$\begin{aligned}\tilde{\mathbf{E}} &= \vec{\mathbf{c}} \otimes \vec{\mathbf{a}} \otimes \vec{\mathbf{b}} \\ &= (c_1\vec{\mathbf{u}}_1 + c_2\vec{\mathbf{u}}_2 + c_3\vec{\mathbf{u}}_3) \otimes (a_1\vec{\mathbf{u}}_1 + a_2\vec{\mathbf{u}}_2 + a_3\vec{\mathbf{u}}_3) \otimes (b_1\vec{\mathbf{u}}_1 + b_2\vec{\mathbf{u}}_2 + b_3\vec{\mathbf{u}}_3) \\ &= c_1a_1b_1\vec{\mathbf{u}}_1 \otimes \vec{\mathbf{u}}_1 \otimes \vec{\mathbf{u}}_1 + c_1a_1b_2\vec{\mathbf{u}}_1 \otimes \vec{\mathbf{u}}_1 \otimes \vec{\mathbf{u}}_2 + c_1a_1b_3\vec{\mathbf{u}}_1 \otimes \vec{\mathbf{u}}_1 \otimes \vec{\mathbf{u}}_3 \cdots \\ &\quad \cdots c_3a_3b_3\vec{\mathbf{u}}_3 \otimes \vec{\mathbf{u}}_3 \otimes \vec{\mathbf{u}}_3\end{aligned}$$

or

$$E_{ijk} = c_i a_j b_k \vec{\mathbf{u}}_i \otimes \vec{\mathbf{u}}_j \otimes \vec{\mathbf{u}}_k$$

in which there are now 27 or 3×3×3 coefficients. This product of three vectors is called a *triad*. Hopefully, you can see that we can take this as far as we wish to produce tetrads, pentads, *etc.* We can consider a way to calculate the number of terms or coefficients in each of these objects. Our vectors have three terms, our dyads have 9 terms and our triads have 27 terms. Let us say that a vector has a rank of 1, a dyad a rank of 2 and a triad has a rank of 3. Using these numbers we can now say that the number of coefficients in one of these objects is:

$$n_{\text{coefficients}} = 3^{\text{rank}}$$

Now, consider the inner product of a vector with a dyad as we have just discussed. The same operation on a triad will produce a dyad times a constant (try it for yourself). Thus, the inner product reduces the rank by one to produce a lower ranked object. With this in mind we can see that a scalar will be an object of rank 0 since the inner product of a rank 1 object with a vector (or simply the inner product of a vector with a vector) is a scalar as we have seen in Chapter 3.

Our determination of the number of coefficients is a little artificial since we have been using three-dimensional Cartesian space for our deliberations. To be completely general we would write:

$$n_{\text{coefficients}} = d^{\text{rank}}$$

where "d" represents the number of dimensions of the space under consideration. For clarity, however, we will continue to work with three-dimensional Cartesian space.

Let us take a closer look at the dyad, $\tilde{\mathbf{D}}$. Our longhand representation of it is:

$$\begin{aligned}
\tilde{\mathbf{D}} &= \vec{\mathbf{a}} \otimes \vec{\mathbf{b}} \\
&= (a_1\vec{\mathbf{u}}_1 + a_2\vec{\mathbf{u}}_2 + a_3\vec{\mathbf{u}}_3) \otimes (b_1\vec{\mathbf{u}}_1 + b_2\vec{\mathbf{u}}_2 + b_3\vec{\mathbf{u}}_3) \\
&= a_1b_1\vec{\mathbf{u}}_1 \otimes \vec{\mathbf{u}}_1 + a_1b_2\vec{\mathbf{u}}_1 \otimes \vec{\mathbf{u}}_2 + a_1b_3\vec{\mathbf{u}}_1 \otimes \vec{\mathbf{u}}_3 \\
&\quad + a_2b_1\vec{\mathbf{u}}_2 \otimes \vec{\mathbf{u}}_1 + a_2b_2\vec{\mathbf{u}}_2 \otimes \vec{\mathbf{u}}_2 + a_2b_3\vec{\mathbf{u}}_2 \otimes \vec{\mathbf{u}}_3 \\
&\quad + a_3b_1\vec{\mathbf{u}}_3 \otimes \vec{\mathbf{u}}_1 + a_3b_2\vec{\mathbf{u}}_3 \otimes \vec{\mathbf{u}}_2 + a_3b_3\vec{\mathbf{u}}_3 \otimes \vec{\mathbf{u}}_3
\end{aligned}$$

with coefficients and unit dyads, very similar to the longhand representation of vectors. Recall from Chapter 3 that a simple Euclidean vector can be represented using a 1×3 column matrix. Thus, vector $\vec{\mathbf{a}}$ has the components a_1, a_2 and a_3, which we include in a matrix representing the vector:

$$\vec{\mathbf{a}} = \begin{bmatrix} a_1 \\ a_2 \\ a_3 \end{bmatrix}$$

In a very similar fashion we may represent the dyad as a square matrix using the coefficients of the unit dyads:

$$\tilde{\mathbf{D}} = \begin{bmatrix} a_1b_1 & a_1b_2 & a_1b_3 \\ a_2b_1 & a_2b_2 & a_2b_3 \\ a_3b_1 & a_3b_2 & a_3b_3 \end{bmatrix}$$

This makes sense, considering our earlier discussion of the formation of a dyad from two vectors using the Dirac notation. This involves an outer product that, as we have seen in Chapter 3, results in a matrix. Since we can represent the dyad as a square matrix we expect that the algebra of the dyad will be much the same as that of matrices.

So, all dyads can be represented by matrices. How about the reverse? Are all square matrices representations of dyads? From our previous discussion this would require that we are able to factor a dyad into two vectors. In matrix notation this is:

$$\tilde{\mathbf{D}} = \vec{\mathbf{a}} \otimes \vec{\mathbf{b}}$$

or

$$\begin{bmatrix} a_1b_1 & a_1b_2 & a_1b_3 \\ a_2b_1 & a_2b_2 & a_2b_3 \\ a_3b_1 & a_3b_2 & a_3b_3 \end{bmatrix} = \begin{bmatrix} a_1 \\ a_2 \\ a_3 \end{bmatrix} \begin{bmatrix} b_1 & b_2 & b_3 \end{bmatrix}$$

Any matrix can have any values that we want to put into it, so if we have the matrix:

$$\begin{bmatrix} a_1b_1 & a_2b_2 & a_3b_3 \\ a_2b_1 & a_2b_2 & a_2b_3 \\ a_3b_1 & a_3b_2 & a_3b_3 \end{bmatrix}$$

in which the first row differs from the previous matrix, we cannot construct this matrix from the direct product of vectors $\vec{\mathbf{a}}$ and $\vec{\mathbf{b}}$ (except in the trivial cases of $\vec{0}$ and unit vectors) nor can it be factored into $\vec{\mathbf{a}}$ and $\vec{\mathbf{b}}$. Therefore we cannot say that in general, all square matrices are dyads.

There is an operation that we can do on a dyad called a *contraction*. As we have learned, the dyad can be constructed from the direct product of two vectors. The dyad is said to be contracted if the inner product is taken of the two component vectors (using Dirac notation):

$$\tilde{\mathbf{D}} = \vec{\mathbf{a}} \otimes \vec{\mathbf{b}}$$
$$\tilde{\mathbf{D}}_C = \langle a|b\rangle = \eta$$

We use in this text a subscript "C" to indicate the contraction operation. This reduces dyad, $\tilde{\mathbf{D}}$, to a scalar. Of course for higher rank objects there are multiple contractions ... in general, for a given rank, there will be $f(rank) = f(rank-1) + rank - 1$ possible ways to do a contraction. Thus, for a triad constructed from the direct product of three vectors, $\vec{\mathbf{a}}$, $\vec{\mathbf{b}}$ and $\vec{\mathbf{c}}$ there are three possible contractions:

$$\tilde{\mathbf{C}} = \vec{\mathbf{a}} \otimes \vec{\mathbf{b}} \otimes \vec{\mathbf{c}}$$

$$\tilde{\mathbf{C}}_C = \vec{\mathbf{a}} \cdot \vec{\mathbf{b}} \otimes \vec{\mathbf{c}} = \alpha \vec{\mathbf{c}}$$

or

$$\tilde{\mathbf{C}}_C = \vec{\mathbf{a}} \cdot \vec{\mathbf{c}} \otimes \vec{\mathbf{b}} = \beta \vec{\mathbf{c}}$$

or

$$\tilde{\mathbf{C}}_C = \vec{\mathbf{a}} \otimes \vec{\mathbf{b}} \cdot \vec{\mathbf{c}} = \eta \vec{\mathbf{c}}$$

Also, note that the contraction operation reduces the rank by *two*. The contraction of a rank-2 dyad is:

$$\tilde{\mathbf{D}}_C = \vec{\mathbf{a}} \cdot \vec{\mathbf{b}} = \sum a_i b_i \tag{5.1}$$

This is simply the trace of the dyad matrix.

We must point out some potential notational problems before proceeding. First, in our discussion of matrices we distinguished between the matrix product of two matrices and the direct product of two matrices (see eqn (2.8)). We must also be careful to do so here for dyads. "Regular" multiplication is the same as matrix multiplication:

$$\tilde{\mathbf{A}}\tilde{\mathbf{B}} = \begin{bmatrix} a_{11} & a_{12} & a_{13} \\ a_{21} & a_{22} & a_{23} \\ a_{31} & a_{32} & a_{33} \end{bmatrix} \begin{bmatrix} b_{11} & b_{12} & b_{13} \\ b_{21} & b_{22} & b_{23} \\ b_{31} & b_{32} & b_{33} \end{bmatrix}$$

$$= \begin{bmatrix} (a_{11}b_{11} + a_{12}b_{21} + a_{13}b_{31}) & (a_{11}b_{12} + a_{12}b_{22} + a_{13}b_{32}) & (a_{11}b_{13} + a_{12}b_{23} + a_{13}b_{33}) \\ (a_{21}b_{11} + a_{22}b_{21} + a_{23}b_{31}) & (a_{21}b_{12} + a_{22}b_{22} + a_{23}b_{32}) & (a_{21}b_{13} + a_{22}b_{23} + a_{23}b_{33}) \\ (a_{31}b_{11} + a_{32}b_{21} + a_{33}b_{31}) & (a_{31}b_{12} + a_{32}b_{22} + a_{33}b_{32}) & (a_{31}b_{13} + a_{32}b_{23} + a_{33}b_{33}) \end{bmatrix}$$

$$= \tilde{\mathbf{C}}$$

The direct product of two dyads is:

$$\tilde{\mathbf{A}} \otimes \tilde{\mathbf{B}} = \begin{bmatrix} a_{11} & a_{12} & a_{13} \\ a_{21} & a_{22} & a_{23} \\ a_{31} & a_{32} & a_{33} \end{bmatrix} \begin{bmatrix} b_{11} & b_{12} & b_{13} \\ b_{21} & b_{22} & b_{23} \\ b_{31} & b_{32} & b_{33} \end{bmatrix}$$

$$= \begin{bmatrix} a_{11}\begin{bmatrix} b_{11} & b_{12} & b_{13} \\ b_{21} & b_{22} & b_{23} \\ b_{31} & b_{32} & b_{33} \end{bmatrix} & a_{12}\begin{bmatrix} b_{11} & b_{12} & b_{13} \\ b_{21} & b_{22} & b_{23} \\ b_{31} & b_{32} & b_{33} \end{bmatrix} & a_{13}\begin{bmatrix} b_{11} & b_{12} & b_{13} \\ b_{21} & b_{22} & b_{23} \\ b_{31} & b_{32} & b_{33} \end{bmatrix} \\ a_{21}\begin{bmatrix} b_{11} & b_{12} & b_{13} \\ b_{21} & b_{22} & b_{23} \\ b_{31} & b_{32} & b_{33} \end{bmatrix} & a_{22}\begin{bmatrix} b_{11} & b_{12} & b_{13} \\ b_{21} & b_{22} & b_{23} \\ b_{31} & b_{32} & b_{33} \end{bmatrix} & a_{23}\begin{bmatrix} b_{11} & b_{12} & b_{13} \\ b_{21} & b_{22} & b_{23} \\ b_{31} & b_{32} & b_{33} \end{bmatrix} \\ a_{31}\begin{bmatrix} b_{11} & b_{12} & b_{13} \\ b_{21} & b_{22} & b_{23} \\ b_{31} & b_{32} & b_{33} \end{bmatrix} & a_{32}\begin{bmatrix} b_{11} & b_{12} & b_{13} \\ b_{21} & b_{22} & b_{23} \\ b_{31} & b_{32} & b_{33} \end{bmatrix} & a_{33}\begin{bmatrix} b_{11} & b_{12} & b_{13} \\ b_{21} & b_{22} & b_{23} \\ b_{31} & b_{32} & b_{33} \end{bmatrix} \end{bmatrix}$$

Also, in linear algebra it is common to write the pre-multiplication of a vector by a matrix as:

$$\hat{\mathbf{M}}\vec{\mathbf{x}} = \vec{\mathbf{y}}$$

the result of which is a new vector. However, in the context of dyads this would produce a triad, increasing the rank of the object:

$$\hat{\mathbf{M}}\vec{\mathbf{x}} = \tilde{\mathbf{O}}$$

or

$$\hat{\mathbf{M}} \otimes \vec{\mathbf{x}} = \tilde{\mathbf{O}}$$

To produce a vector we must use the inner product notation:[‡]

$$\hat{\mathbf{M}} \cdot \vec{\mathbf{x}} = \vec{\mathbf{y}}$$

We must take care not to confuse the two. Our notation so far has been to use $\hat{\mathbf{M}}$ to denote a matrix and $\tilde{\mathbf{M}}$ to denote a dyad (or a higher-rank tensor). Usually the context will tell us which is which; however, in other texts the distinction may not be so clear. Thus, for clarity, $\hat{\mathbf{A}}\hat{\mathbf{B}}$ is used in this text for standard matrix multiplication and $\tilde{\mathbf{A}} \otimes \tilde{\mathbf{B}}$ for dyad (or triad, tetrad, *etc.*) direct product multiplication.

So, to summarise, we have some new mathematical objects to play with that were developed from the application of the direct product of

[‡]This is actually the most important operation for our purposes. The general Hamiltonian of Chapter 14 uses this extensively.

vectors with each other. Each of these objects has a rank associated with it that is equivalent to the power that the dimensionality of the space of the vector(s) is raised to in order to generate the number of coefficients of the object. Thus, the dyad results from the direct product of two three-dimensional vectors and has rank 2 or the 3 to the power of 2. Three vectors give a triad of rank 3 and four vectors give a tetrad of rank 4. Scalars are ranked 0 since they consist of 3° components. The algebra of these mathematical objects is much the same as that for matrices with the exception of the contraction operation.

We have also seen that, from the tensor point of view, a scalar number is a tensor of rank 0, a vector is a tensor of rank 1 and a dyad is a tensor of rank 2.

5.2 The Gradient of a Vector

In Chapter 4 we alluded to the gradient calculation:

$$\vec{\nabla}\vec{\mathbf{a}} \text{ or grad } \vec{\mathbf{a}}$$

and made the assertion that the result is a dyad. We now show that this is so *via* the direct product of $\vec{\nabla}$ and $\vec{\mathbf{a}}$:

$$\begin{aligned}
\vec{\nabla}\otimes\vec{\mathbf{a}} &= \left(\frac{\partial}{\partial x}\vec{\mathbf{u}}_x+\frac{\partial}{\partial y}\vec{\mathbf{u}}_y+\frac{\partial}{\partial z}\vec{\mathbf{u}}_z\right)\otimes\left(a_x\vec{\mathbf{u}}_x+a_y\vec{\mathbf{u}}_y+a_z\vec{\mathbf{u}}_z\right)\\
&= \frac{\partial a_x}{\partial x}\vec{\mathbf{u}}_x\otimes\vec{\mathbf{u}}_x+\frac{\partial a_y}{\partial x}\vec{\mathbf{u}}_x\otimes\vec{\mathbf{u}}_y+\frac{\partial a_z}{\partial x}\vec{\mathbf{u}}_x\otimes\vec{\mathbf{u}}_z\\
&\quad+\frac{\partial a_x}{\partial x}\vec{\mathbf{u}}_y\otimes\vec{\mathbf{u}}_x+\frac{\partial a_y}{\partial x}\vec{\mathbf{u}}_y\otimes\vec{\mathbf{u}}_y+\frac{\partial a_z}{\partial x}\vec{\mathbf{u}}_y\otimes\vec{\mathbf{u}}_z\\
&\quad+\frac{\partial a_x}{\partial x}\vec{\mathbf{u}}_z\otimes\vec{\mathbf{u}}_x+\frac{\partial a_y}{\partial x}\vec{\mathbf{u}}_z\otimes\vec{\mathbf{u}}_y+\frac{\partial a_z}{\partial x}\vec{\mathbf{u}}_z\otimes\vec{\mathbf{u}}_z
\end{aligned}$$

The $\vec{\mathbf{u}}_i\otimes\vec{\mathbf{u}}_j$ are the unit dyads as above and the partial derivatives are the components of the dyad. We can compact this a bit using the matrix notation of Chapter 3, Section 3.3:

$$\vec{\nabla}\otimes\vec{\mathbf{a}}=\tilde{\nabla}=\begin{bmatrix}\dfrac{\partial a_x}{\partial x} & \dfrac{\partial a_y}{\partial x} & \dfrac{\partial a_z}{\partial x}\\[2ex] \dfrac{\partial a_x}{\partial y} & \dfrac{\partial a_y}{\partial y} & \dfrac{\partial a_z}{\partial y}\\[2ex] \dfrac{\partial a_x}{\partial z} & \dfrac{\partial a_y}{\partial z} & \dfrac{\partial a_z}{\partial z}\end{bmatrix}$$

5.3 Transformations

We have seen in Chapter 3, eqn (3.39) that the norm of a vector, or more generally the inner product of a pair of vectors, is invariant to rotations. Indeed, this constitutes a major part of the very definition of vectors. Rotation operators are orthogonal, which in visual terms means that an operator and its transpose rotate in opposite directions by the same amount (eqn (3.40)). Also, we have seen that the rotation operation may be considered a rotation of coordinates with a consequent change of basis set. In magnetic resonance spectroscopy we are primarily concerned with rotations. Intuitively, we expect that the norm of the vector will remain the same in the new coordinate system as in the old one.

Thus, a vector in coordinate system A is considered to be the same vector in coordinate system A', assuming the beginning and ending points of the vector do not move with the coordinate system change. The components of the vector in each coordinate system will generally not be equal, however, we expect the norm (the length) to remain constant. Let us suppose, then, that we have a two-dimensional vector, $\vec{\mathbf{a}} = a_1\overleftarrow{\mathbf{u}}_1 + a_2\overleftarrow{\mathbf{u}}_2$, in coordinate system A. To transform to coordinate system A' we use a function of some type:

$$a_1' = f_1(a_1, a_2)$$

$$a_2' = f_2(a_1, a_2)$$

and our vector is transformed to become $\vec{\mathbf{a}}' = a_1'\overleftarrow{\mathbf{u}}_1' + a_2'\overleftarrow{\mathbf{u}}_2'$. However, we have just said that we expect the norm of the new vector to be the same as the old vector since the vector itself does not change as a result of the transformation. An observer in coordinate system A must see the same vector that an observer in A' will see. To indicate that these are the same vector we shall write in this text:

$$\{\vec{\mathbf{a}} = \vec{\mathbf{a}}'\} \Rightarrow \langle a|a\rangle = \langle a'|a'\rangle = \tau$$

which is meant to indicate that (although their components are different) their norms are the same and that they are in reality the same vector.

Chapter 3 showed us that the coordinate transformation of a vector may be written as (using Einstein notation):

$$a_i' = \gamma_{ij}a_j \tag{5.2}$$

where γ_{ij} is the transformation matrix. Might we not apply something similar to our dyadic tensor? We assert that in transforming from coordinate system A to coordinate system A', the dyad in question will remain the same dyad in both coordinate systems much the same as is the case with vectors. Just as an observer in coordinate system A sees vector $\vec{\mathbf{a}}$ and an observer in coordinate system A' sees vector $\vec{\mathbf{a}}'$, an observer in A sees dyad $\tilde{\mathbf{D}}$ and an observer in A' sees dyad $\tilde{\mathbf{D}}'$. It should therefore be possible to construct a transformation function that when applied to $\tilde{\mathbf{D}}$ produces $\tilde{\mathbf{D}}'$. Let us investigate this idea.

Since we may think of the dyadic tensor as being composed of two vectors it is reasonable to expect that, in order to do a coordinate transformation on the dyad, we do a coordinate transformation on each vector. This would have the effect of producing two new vectors that when put together as a direct product produce the new dyad, $\tilde{\mathbf{D}}$ composed of vectors $\vec{\mathbf{c}}$ and $\vec{\mathbf{d}}$. With this in mind we write:

$$\vec{\mathbf{u}} = \tilde{\mathbf{D}} \cdot \vec{\mathbf{v}}$$
$$\vec{\mathbf{u}}' = \tilde{\mathbf{D}}' \cdot \vec{\mathbf{v}}' \tag{5.3}$$

This is just the inner product operation that we discussed above. It reduces the rank of tensor $\tilde{\mathbf{D}}$ and $\tilde{\mathbf{D}}'$ by one to produce a vector, as we saw. Now, using eqn (5.2) in matrix form, we can relate vectors $\vec{\mathbf{u}}$ and $\vec{\mathbf{u}}'$:

$$\vec{\mathbf{u}}' = \hat{\Gamma}\vec{\mathbf{u}}$$

The transformation matrix $\hat{\Gamma}$ is composed of the components γ_{ij} from eqn (5.2). Using this to substitute in eqn (5.3):

$$\hat{\Gamma}\vec{\mathbf{u}} = \hat{\Gamma}\tilde{\mathbf{D}} \cdot \vec{\mathbf{v}}$$

and

$$\vec{\mathbf{u}}' = \hat{\Gamma}\vec{\mathbf{u}} = \tilde{\mathbf{D}}' \cdot \vec{\mathbf{v}}' = \tilde{\mathbf{D}}' \cdot \hat{\Gamma}\vec{\mathbf{v}}$$

therefore

$$\hat{\Gamma}\tilde{\mathbf{D}} \cdot \vec{\mathbf{v}} = \tilde{\mathbf{D}}' \cdot \hat{\Gamma}\vec{\mathbf{v}}$$

This must mean that:

$$\hat{\Gamma}\tilde{\mathbf{D}} = \tilde{\mathbf{D}}' \cdot \hat{\Gamma}$$

or (5.4)

$$\tilde{\mathbf{D}}'\hat{\Gamma}\hat{\Gamma}^T = \tilde{\mathbf{D}}' = \hat{\Gamma}\tilde{\mathbf{D}}\hat{\Gamma}^T$$

This is just the orthogonal matrix transformation from Chapter 2, eqn (2.55). The Einstein notation version of eqn (5.4) is:

$$D'_{ij} = \gamma_{ik}\gamma_{jl}D_{kl} \tag{5.5}$$

Are dyads $\tilde{\mathbf{D}}$ and $\tilde{\mathbf{D}}'$ the same in the sense that vectors $\vec{\mathbf{a}}$ and $\vec{\mathbf{a}}'$ are the same in different coordinate systems? In the case of the vectors we used the inner product operation to reduce them to scalars that were invariant with respect to coordinate changes to show that their norms are equivalent. It is particularly simple to show that the same thing applies to dyads. Our tool for doing so is the dyad contraction that we have discussed above. As noted above in eqn (5.1), the scalar resulting from the dyad contraction is simply the trace of the dyadic matrix. However, we have shown in Chapter 2, eqn (2.53) that the trace of a matrix is invariant to transformations and therefore the contractions of dyads $\tilde{\mathbf{D}}$ and $\tilde{\mathbf{D}}'$ must be equal and these are then equivalent tensors.

5.4 A Tensor Definition

We can now define a tensor. A loose definition of a rank-2 tensor is as an operator that, when it operates on a vector, produces a new vector. This is what we have seen as the result of the inner product of a dyad with a vector and thus, using this definition, a dyad may be considered as a rank-2 tensor. We have also seen this type of behaviour in rotation operators (Section 3.4) and eigenfunction operators (Section 3.5). These operators are indeed, tensors. This definition is actually all that we will need for most of our purposes, however, we provide a more formal definition in the following paragraphs.

We may define a tensor as a mathematical object that is invariant to transformation of basis. We have already seen that the scalar object is invariant to a change of basis as are vectors and dyads. In other words a scalar such as the temperature of a cup of tea is the same whether the coordinate system's origin is on the earth or on the moon. Formally, if the temperature in coordinate system A is T and in coordinate system A' it is T' then the transformation from T to T' is:

$$T' = \hat{\mathbf{A}}T$$

where the transformation operator $\hat{\mathbf{A}}$ is *always* unity and T' is therefore equal to T and is said to be a tensor of rank 0.

The vector object is also a tensor if it too can be said to be the same in any basis. In terms of coefficients of a three-dimensional vector the transformation from coordinate basis $\vec{\mathbf{u}}$ to $\vec{\mathbf{u}}'$ is, for vector $\vec{\mathbf{a}}$:

$$\begin{aligned}\vec{\mathbf{a}} &= a_1\vec{\mathbf{u}}_1 + a_2\vec{\mathbf{u}}_2 + a_3\vec{\mathbf{u}}_3\\ \vec{\mathbf{a}}' &= a'_1\vec{\mathbf{u}}'_1 + a'_2\vec{\mathbf{u}}'_2 + a'_3\vec{\mathbf{u}}'_3\\ a'_i &= \sum_1^3 \cos(\vec{\mathbf{u}}'_i, \vec{\mathbf{u}}_j)a_j\end{aligned} \tag{5.6}$$

using eqn (3.41). Intuitively, we know that the vector itself does not change even though its coefficients may do so. Thus, a calculation of the norm of the vector will be the same in the new basis as in the old basis (eqn (3.39)). So, if two vectors are equivalent after a basis transformation then they are said to be tensors of rank 1.

We have seen that the dyad, as well, can be invariant to transformation such that dyad $\tilde{\mathbf{D}}$ is equivalent to dyad $\tilde{\mathbf{D}}'$ if $\tilde{\mathbf{D}}'$ is produced by a change of basis from $\tilde{\mathbf{D}}$. The dyad $\tilde{\mathbf{D}}$ produced from a pair of three-dimensional vectors can conveniently be represented by a 3×3 matrix with 9 components. In this sense $\tilde{\mathbf{D}}$ is similar to a vector in that it has components associated with it, the primary difference being that instead of three for the vector there are nine for the dyad.

In order to transform $\tilde{\mathbf{D}}$ to $\tilde{\mathbf{D}}'$ we must perform a set of operations that is very similar to the vector transformation. In the case of the vector (or rank-1 tensor in our present context) each new coefficient of the vector is a linear combination of all of the old coefficients. The case with rank-2 tensors is much the same; each coefficient of the new rank-2 tensor is a linear combination of all of the coefficients of the old tensor (see eqn (5.5)) and is expressed in a similar fashion to vectors:

$$\tilde{\mathbf{D}} = \begin{bmatrix} d_{11} & d_{12} & d_{13}\\ d_{21} & d_{22} & d_{23}\\ d_{31} & d_{32} & d_{33}\end{bmatrix} \quad \tilde{\mathbf{D}}' = \begin{bmatrix} d'_{11} & d'_{12} & d'_{13}\\ d'_{21} & d'_{22} & d'_{23}\\ d'_{31} & d'_{32} & d'_{33}\end{bmatrix}$$

Note that each of the coefficients now has *two* subscripts as opposed to the single subscript used for vector coefficients. This is because the vector matrix is represented by a $1\times n$ matrix, where n is the dimension of the space that the vector is in, whereas the rank-2 tensor is represented by a $n\times n$ matrix. To transform the vector we need only sum over one row of coefficients (as in eqn (5.2)) but for the rank-2 tensor we must sum over all coefficients of the $n \times n$ matrix. So, to transform from $\tilde{\mathbf{D}}$ to $\tilde{\mathbf{D}}'$:[§]

$$d'_{ij} = \sum_{1}^{3} \cos(\vec{\mathbf{u}}'_i i, \vec{\mathbf{u}}_k) \sum_{1}^{3} \cos(\vec{\mathbf{u}}'_j, \vec{\mathbf{u}}_l) d_{kl}$$

where, as in the vector case, $\cos(\vec{\mathbf{u}}'_i, \vec{\mathbf{u}}_k)$ is the direction cosine of the angle between axes i in the new set of coordinates and k in the old coordinates (see appendix A, Section A.4). As we have seen, we can simplify the equation using the compact Einstein notation:

$$d'_{ij} = \gamma_{ik}\gamma_{jl} d_{kl}$$

where the cosine terms have been replaced with γ_{ij} for brevity.

5.5 Isotropic Tensors

The unit tensor or identity tensor is defined as:

$$\tilde{\mathbf{1}} = \sum_i \sum_j \delta_{ij}$$

and we might write this tensor as a matrix:

$$\tilde{\mathbf{1}} = \begin{bmatrix} 1 & 0 & 0 \\ 0 & 1 & 0 \\ 0 & 0 & 1 \end{bmatrix}$$

This is identical to the unit matrix of Chapter 2 with the same properties. Specifically, we are interested in making use of the commutation property:

$$\tilde{\mathbf{1}}\tilde{\mathbf{A}} = \tilde{\mathbf{A}}\tilde{\mathbf{1}} \tag{5.7}$$

as implied in eqn (2.11).

[§]It begins to confuse the mind to try to keep track of these transformations as one goes up in rank with tensors. Fortunately, we will only be considering up to rank-2 tensors. A thorough understanding of the vector transformation will help greatly.

If the unit tensor is multiplied by a constant it is called a *constant tensor*:

$$\alpha\tilde{\mathbf{1}} = \alpha\sum_i\sum_j \delta_{ij}$$

$$\alpha\tilde{\mathbf{1}} = \tilde{\alpha} = \begin{bmatrix} \alpha & 0 & 0 \\ 0 & \alpha & 0 \\ 0 & 0 & \alpha \end{bmatrix}$$

and eqn (5.7) implies that:

$$\tilde{\alpha}\tilde{\mathbf{A}} = \tilde{\mathbf{A}}\tilde{\alpha}$$

Now, let us try to imagine a tensor that remains the same under all transformations. By this we mean that not only must the contractions of the initial and final tensors be the same but the actual components of the tensors must be equal. Let us suppose that tensor $\tilde{\mathbf{D}}$ is one such object and that when transformed by rotation operator $\hat{\mathbf{R}}$ becomes tensor $\tilde{\mathbf{D}}'$ that is equal, component for component, to tensor $\tilde{\mathbf{D}}$. Tensor $\tilde{\mathbf{D}}$ is said to be *isotropic* since its spatial orientation does not affect its components.¶ Let us look more closely at this:

$$\tilde{\mathbf{D}} = \begin{bmatrix} d_{11} & d_{12} & d_{13} \\ d_{21} & d_{22} & d_{23} \\ d_{31} & d_{32} & d_{33} \end{bmatrix} \qquad \tilde{\mathbf{D}}' = \begin{bmatrix} d'_{11} & d'_{12} & d'_{13} \\ d'_{21} & d'_{22} & d'_{23} \\ d'_{31} & d'_{32} & d'_{33} \end{bmatrix}$$

and

$$\tilde{\mathbf{D}} = \tilde{\mathbf{D}}'$$

For simplicity let us assume all Euler rotation angles are 90° rotation angles (see Chapter 16, Section 16.2 for more on Euler rotations). The first Euler rotation matrix, $\hat{\mathbf{R}}_z(\alpha)$ is:

$$\hat{\mathbf{R}}_z(\alpha) = \begin{bmatrix} 0 & 1 & 0 \\ -1 & 0 & 0 \\ 0 & 0 & 1 \end{bmatrix}$$

Rotation operator $\hat{\mathbf{R}}_z(\alpha)$ operates on tensor $\tilde{\mathbf{D}}$ to produce tensor $\tilde{\mathbf{D}}'$. Since we are assuming tensors $\tilde{\mathbf{D}}$ and $\tilde{\mathbf{D}}'$ to be isotropic then they

¶Or if you prefer, the basis of the tensor does not affect the components of the tensor.

must be equivalent, component by component. The transformation equation from $\tilde{\mathbf{D}}$ to $\tilde{\mathbf{D}}'$ is:

$$\tilde{\mathbf{D}}' = \hat{\mathbf{R}}_z(\alpha)\tilde{\mathbf{D}}\hat{\mathbf{R}}_z^{-1}$$

The explicit calculation using the tensor matrices is:

$$\hat{\mathbf{R}}_z(\alpha)\tilde{\mathbf{D}}\hat{\mathbf{R}}^T z(\alpha) = \begin{bmatrix} 0 & 1 & 0 \\ -1 & 0 & 0 \\ 0 & 0 & 1 \end{bmatrix} \begin{bmatrix} d_{11} & d_{12} & d_{13} \\ d_{21} & d_{22} & d_{23} \\ d_{31} & d_{32} & d_{33} \end{bmatrix} \begin{bmatrix} 0 & -1 & 0 \\ 1 & 0 & 0 \\ 0 & 0 & 1 \end{bmatrix}$$

$$= \begin{bmatrix} d_{22} & -d_{21} & -d_{23} \\ -d_{12} & d_{11} & d_{13} \\ -d_{32} & d_{31} & d_{33} \end{bmatrix}$$

and

$$\begin{bmatrix} d_{11} & d_{12} & d_{13} \\ d_{21} & d_{22} & d_{23} \\ d_{31} & d_{32} & d_{33} \end{bmatrix} = \begin{bmatrix} d_{22} & -d_{21} & -d_{23} \\ -d_{12} & d_{11} & d_{13} \\ -d_{32} & d_{31} & d_{33} \end{bmatrix}$$

We see here that:

$$d_{11} = d_{22}$$

$$d_{32} = d_{31} = -\ d_{32} = 0$$

$$d_{13} = -\ d_{23} = d_{23} = 0$$

and we can rewrite our transformed tensor matrix as:

$$\tilde{\mathbf{D}} = \begin{bmatrix} d_{11} & d_{12} & 0 \\ d_{21} & d_{11} & 0 \\ 0 & 0 & d_{33} \end{bmatrix}$$

Now let us do the next Euler rotation, $\hat{\mathbf{R}}_{y'}(\beta)$ on our updated tensor $\tilde{\mathbf{D}}$:

$$\hat{\mathbf{R}}_{y'}(\beta) = \begin{bmatrix} 0 & 0 & 1 \\ 0 & 1 & 0 \\ -1 & 0 & 0 \end{bmatrix}$$

$$\hat{\mathbf{R}}_{y'}(\beta)\tilde{\mathbf{D}}\hat{\mathbf{R}}^T y'(\beta) = \begin{bmatrix} 0 & 0 & 1 \\ 0 & 1 & 0 \\ -1 & 0 & 0 \end{bmatrix} \begin{bmatrix} d_{11} & d_{12} & 0 \\ d_{21} & d_{11} & 0 \\ 0 & 0 & d_{33} \end{bmatrix} \begin{bmatrix} 0 & 0 & -1 \\ 0 & 1 & 0 \\ 1 & 0 & 0 \end{bmatrix}$$

$$= \begin{bmatrix} d_{33} & 0 & 0 \\ 0 & d_{11} & d_{21} \\ 0 & d_{12} & d_{11} \end{bmatrix}$$

and

$$\begin{bmatrix} d_{11} & d_{12} & 0 \\ d_{21} & d_{11} & 0 \\ 0 & 0 & d_{33} \end{bmatrix} = \begin{bmatrix} d_{33} & 0 & 0 \\ 0 & d_{11} & d_{21} \\ 0 & d_{12} & d_{11} \end{bmatrix}$$

Clearly, $d_{12} = 0$, $d_{21} = 0$ and $d_{33} = d_{11}$, so isotropic tensor $\tilde{\mathbf{D}}$ must be such that all of its off-diagonal components are zero and its diagonal components are all equal:

$$\tilde{\mathbf{D}} = \begin{bmatrix} \tau & 0 & 0 \\ 0 & \tau & 0 \\ 0 & 0 & \tau \end{bmatrix} = \tau \begin{bmatrix} 1 & 0 & 0 \\ 0 & 1 & 0 \\ 0 & 0 & 1 \end{bmatrix} = \tau\tilde{\mathbf{1}} \tag{5.8}$$

The last rotation about the z''-axis must also preserve $\tilde{\mathbf{D}}$:

$$\hat{\mathbf{R}}_{z''}(\gamma) = \begin{bmatrix} 0 & 1 & 0 \\ -1 & 0 & 0 \\ 0 & 0 & 1 \end{bmatrix} \quad \tilde{\mathbf{D}} = \begin{bmatrix} \tau & 0 & 0 \\ 0 & \tau & 0 \\ 0 & 0 & \tau \end{bmatrix}$$

$$\begin{aligned}\hat{\mathbf{R}}_{z''}(\gamma)\tilde{\mathbf{D}}\hat{\mathbf{R}}^{T}_{z''}(\gamma) &= \hat{\mathbf{R}}_{z''}(\gamma)\tau\tilde{\mathbf{1}}\hat{\mathbf{R}}^{T}_{z''}(\gamma)\\ &= \tau\hat{\mathbf{R}}_{z''}(\gamma)\hat{\mathbf{R}}^{T}_{z''}(\gamma)\\ &= \tau\tilde{\mathbf{1}}\\ &= \tilde{\mathbf{D}}\end{aligned}$$

We see then, that for a tensor to be isotropic it must be a scalar multiple of the unit tensor.

5.6 Symmetric, Antisymmetric and Irreducible Tensors

As we saw in eqn (2.17) the symmetric matrix is equal to its transpose. This then, is also the case for rank-2 tensors:

$$\tilde{\mathbf{D}} = \tilde{\mathbf{D}}^T$$

or alternately:

$$d_{ij} = d_{ji}$$

An *antisymmetric* tensor is equal to the negative of its transpose:

$$\tilde{\mathbf{D}} = -\tilde{\mathbf{D}}^T$$

A little thought will guide you to the conclusion that the trace of the antisymmetric tensor is zero and that indeed, the diagonal elements of the tensor matrix are all equal to zero.

A normal tensor, $\tilde{\mathbf{D}}$, which is neither symmetric or antisymmetric can be decomposed into symmetric and antisymmetric tensors:

$$\tilde{\mathbf{D}}_S = \frac{1}{2}(\tilde{\mathbf{D}} + \tilde{\mathbf{D}}^T)$$

$$\tilde{\mathbf{D}}_{\mathrm{Anti}} = \frac{1}{2}(\tilde{\mathbf{D}} - \tilde{\mathbf{D}}^T)$$

$$\tilde{\mathbf{D}} = \tilde{\mathbf{D}}_S + \tilde{\mathbf{D}}_{\mathrm{Anti}}$$

A tensor that can undergo an orthogonal transformation may be decomposed a bit further into *irreducible tensors*. There are three sub-tensors that a tensor, $\tilde{\mathbf{D}}$, can be decomposed into, an isotropic tensor, an antisymmetric tensor and a symmetric tensor:

$$\tilde{\mathbf{D}} = \tilde{\mathbf{D}}_{\mathrm{Iso}} + \tilde{\mathbf{D}}_{\mathrm{Anti}} + \tilde{\mathbf{D}}_{\mathrm{Sym}}$$

$$= D_{\mathrm{iso}} \begin{bmatrix} 1 & 0 & 0 \\ 0 & 1 & 0 \\ 0 & 0 & 1 \end{bmatrix} + \begin{bmatrix} 0 & a_{12} & a_{13} \\ a_{21} & 0 & a_{23} \\ a_{31} & a_{32} & 0 \end{bmatrix} + \begin{bmatrix} (s_{11} - D_{\mathrm{Iso}}) & s_{12} & s_{13} \\ s_{12} & (s_{22} - D_{\mathrm{Iso}}) & s_{23} \\ s_{13} & s_{23} & (s_{33} - D_{\mathrm{Iso}}) \end{bmatrix} \quad (5.9)$$

D_{Iso} is $(1/3)\mathrm{Tr}(\tilde{\mathbf{D}})$. The symmetric tensor here is the symmetric tensor from above minus the isotropic tensor. The symmetric tensor here is the symmetric tensor from above minus the isotropic tensor:

$$\tilde{\mathbf{D}}_{\mathrm{sym}} = \tilde{\mathbf{D}}_s - \tilde{\mathbf{D}}_{\mathrm{iso}}$$

$$= \frac{1}{2}(\tilde{\mathbf{D}} + \tilde{\mathbf{D}}^T) - \tilde{\mathbf{D}}_{\mathrm{iso}}$$

The antisymmetric tensor is obviously traceless, its diagonal components all being zero. It is not so obvious for the symmetric tensor. The reason for this is that in forming the irreducible tensors we have "stripped out" the trace and put it into $\tilde{\mathbf{T}}_{\mathrm{Iso}}$. These traceless characteristics will be of some importance to us when we look at isotropic *versus* anisotropic Hamiltonians in Chapter 14.

As you will recall from Chapter 2, and more specifically, from eqn (2.58), a symmetric matrix can be diagonalised using orthogonal operators. Thus:

$$\hat{\mathbf{A}}^T \tilde{\mathbf{D}} \hat{\mathbf{A}} = \tilde{\Lambda}$$

$$\hat{\mathbf{A}}^T (\tilde{\mathbf{D}}_{\mathrm{Iso}} + \tilde{\mathbf{D}}_{\mathrm{Sym}}) \hat{\mathbf{A}} = \tilde{\Lambda}$$

$$= \hat{\mathbf{A}}^T \tilde{\mathbf{D}}_{\mathrm{Iso}} \hat{\mathbf{A}} + \hat{\mathbf{A}}^T \tilde{\mathbf{D}}_{\mathrm{Sym}} \hat{\mathbf{A}}$$

$$= T_{\mathrm{Iso}} \hat{\mathbf{A}}^T \tilde{\mathbf{1}} \hat{A} + \hat{\mathbf{A}}^T \tilde{\mathbf{D}}_{\mathrm{Sym}} \hat{\mathbf{A}}$$

$$= T_{\mathrm{Iso}} \tilde{\mathbf{1}} + \tilde{\Lambda}_{\mathrm{Sym}}$$

The diagonalised symmetric tensor is said to be in the principal axis system (PAS) for the tensor.

Further Reading

1. A. I. Borisenko and I. E. Tarapov, *Vector and Tensor Analysis with Applications*, Prentice-Hall, Englewood Cliffs, N.J., 1968.
2. A. M. Goodbody, *Cartesian Tensors: With Applications to Mechanics, Fluid Mechanics and Elasticity*, Halsted Press, Chichester, West Sussex, UK, 1982.
3. J. C. Kolecki, *Foundations of Tensor Analysis for Students of Physics and engineering With an Introduction to the Theory of Relativity.*, Report TP-2005-213115, NASA Science and Technical Information, 2005.
4. L. J. Mueller, *Concepts Magn. Reson., Part A*, 2011, **38A**, 221–235.

6 Probability and Statistics

Mathematicians stand on each other's shoulders.

Carl Friedrich Gauss

6.1 Simple Probability and Statistics

Let us flip a coin multiple times. In fact let us do it an infinite number of times. What is the probability of obtaining heads? We expect that this number will be:

$$p_{\text{heads}} = \lim_{n\to\infty} \frac{n_{\text{heads}}}{n_{\text{total}}} \tag{6.1}$$

Of course, after doing this experiment many times we will find that p_{heads} approaches 1/2. More generally, the probability of a given event k occurring is:

$$p_k = \lim_{n\to\infty} \frac{n_k}{n_{\text{total}}} \tag{6.2}$$

p_k can take on any value from zero to one inclusive since:

$$0 \le n_k \le n_{\text{total}}$$

Also, since:

$$\sum_k^t n_k = n_{\text{total}}$$

Essential Mathematics for NMR and MRI Spectroscopists
By Keith C. Brown

Published by the Royal Society of Chemistry, www.rsc.org

we write:

$$\sum_{k}^{t} p_k = 1$$

Let us now suppose that for an experiment with two or more possible outcomes (such as the coin-tossing experiment) that the outcome of an experiment is represented by x_k and is associated with a probability of occurring, p_k. This being the case, we can calculate a mean *expectation value* for x over many measurements:

$$\langle x \rangle = \sum_{k}^{t} x_k p_k \tag{6.3}$$

This is nothing more than the weighted average of all possible outcomes of the experiment and is encountered frequently in statistical calculations.

Another statistical entity is:

$$\langle x^2 \rangle = \sum_{k}^{t} x_k^2 p_k$$

known as the *second moment of distribution*. We can square x_k but the probability of observing it is still p_k. Thus, the mean value of x_k^2 is calculated exactly as is the mean value of x. It is not as commonly used by itself as is the mean value. However, it does appear in combination with the mean value in the *variance*:

$$\begin{aligned}
\sigma^2 = (x - \langle x \rangle)^2 &= \sum_{k}^{t} (x_k - \langle x \rangle)^2 p_k \\
&= \sum_{k}^{t} (x_k^2 - 2x_k \langle x \rangle + \langle x \rangle^2) p_k \\
&= \sum_{k}^{t} x_k^2 p_k - \sum_{k}^{t} 2x_k \langle x \rangle p_k + \sum_{k}^{t} \langle x \rangle^2 p_k \\
&= \langle x^2 \rangle - 2\langle x \rangle \langle x \rangle + \langle x \rangle^2 \\
&= \langle x^2 \rangle - 2\langle x \rangle^2 + \langle x \rangle^2 \\
&= \langle x^2 \rangle - \langle x \rangle^2
\end{aligned} \tag{6.4}$$

The square root of σ^2 is the familiar *standard deviation* from basic statistics:

$$\sigma = \sqrt{\langle x^2 \rangle - \langle x \rangle^2} \tag{6.5}$$

A common example is instructive. Consider a pair of dice. We can do multiple experiments throwing the dice and recording the sums of the two dice. The possible results are 2, 3, 4, 5, 6, 7, 8, 9, 10, 11 and 12 and the corresponding chances of these results are 1/36, 1/18, 1/12, 1/9, 5/36, 1/6, 5/36, 1/9, 1/12, 1/18, 1/36. Using these numbers and eqn (6.3), we get a mean value of 7. In other words the average result of a throw is 7. We then can calculate a variance of 5.83 using eqn (6.4) and a standard deviation of 2.42 using eqn (6.5).

$$\langle x \rangle = 7 \quad \langle x^2 \rangle = 54.83$$

$$\sigma^2 = \langle x^2 \rangle - \langle x \rangle^2$$

$$= 54.83 - 7 \times 7 = 5.83$$

$$\sigma = \sqrt{5.83} = 2.42$$

$$\sum_{x=1}^{11} p(x) = 1$$

What would be the probability of throwing seven with the dice three times in a row? From Table 6.1 we see that the probability of throwing a seven in a single throw (or experiment in probability jargon) is 1/6.

Table 6.1 Dice throw probabilities.

x	x^2	$p(x)$	$xp(x)$	$x^2p(x)$
2	4	1/36	0.06	0.11
3	9	1/18	0.17	0.5
4	16	1/12	0.33	1.33
5	25	1/9	0.56	2.78
6	36	5/36	0.83	5
7	49	1/6	1.17	8.17
8	64	5/36	1.11	8.89
9	81	1/9	1	9
10	100	1/12	0.83	8.33
11	121	1/18	0.61	6.72
12	144	1/36	0.33	4

Then, if we throw three times, the probability in each experiment is 1/6 and the probability of seeing a seven three times in a row is:

$$p(7) = \left(\frac{1}{6}\right)\left(\frac{1}{6}\right)\left(\frac{1}{6}\right) = \frac{1}{216} = 0.0046296$$

We would expect then to see three sevens in a row one in every 216 three-throw experiments over a large number of experiments.

Our variable, x, is an example of a discrete *stochastic* (or random) variable whose value is governed by a probability. It is also possible to have continuous stochastic variables such as the random orientation of a molecule in solution (which will be of interest to us later).

So far we have been computing the probabilities of one possible outcome of an experiment. We have seen what the probability of throwing a seven is singly or multiple times. However, we now ask "what would the probability of the value of the outcome being *less than or equal* to some seven?". This type of probability is of interest in many problems. For example, a statistician might want to know what the probability of a voter's age being less than or equal to 30 years of age. We can assess what the answer to the question is by looking at our dice example again. We are interested in finding the probability that we will throw a 2, 3, 4, 5, 6 or a 7. To answer this question let us temporarily consider a trivially simple and similar problem .... what is the probability that the result of the dice throw is 12 or less? Well, very obviously this probability is one! In other words, it is just the *sum* of all of the probabilities for each outcome. So, for the question of what is the probability of throwing 7 or less we simply sum the probabilities for throwing 2, 3, 4, 5, 6 and 7, which works out to 7/12. In the more general case we write:

$$p(x \leq x_i) = \sum_{x=x_1}^{x_i} p(x) \tag{6.6}$$

6.2 Permutations and Combinations

When we look at some problems involving probabilities and subsets of large numbers of objects there are various ways those subsets can be chosen and arranged. For example, suppose that we have 10 paintings but have room to hang only 4 of them. We want to know how many ways there are to arrange four of the ten paintings on the wall. For the first painting we have 10 choices, for the next we will have 9 choices and then 8 and 7 for the third and fourth choices respectively. So, to calculate how many possibilities there are we multiply these together:

$$p(10, 4) = 10 \cdot 9 \cdot 8 \cdot 7 = 5040$$

Another way to write this is:

$$p(10, 4) = \frac{10 \cdot 9 \cdot 8 \cdot 7 \cdot 6 \cdot 4 \cdot 3 \cdot 2 \cdot 1}{6 \cdot 5 \cdot 4 \cdot 3 \cdot 2 \cdot 1}$$

or more compactly, using factorial notation:

$$p(10, 4) = \frac{10!}{6!} = \frac{10!}{(10-4)!}$$

Thus, there are 5040 ways to arrange 4 pictures out of a choice of 10 on a wall. To refresh your memory, factorial means multiply the number by each of all of the preceding integers down to and including one:

$$n! = n \times (n-1) \times (n-2) \ldots 1$$

$$= \prod_{k=1}^{n} k$$

By definition, 0! is equal to one. That this must be so can be seen by considering the number of ways that we could arrange 3 pictures of a total of three. Using our formula above this would be:

$$p(3, 3) = \frac{3!}{(3-3)!} = \frac{3!}{0!} = 6$$

In order that we come up with the correct result, 6, we must *define* 0! as equal to one. Factorials quickly get out of hand for larger numbers:

$$3! = 6$$

$$10! = 3628800$$

$$50! = 3.041409 \times 10^{64}$$

$$100! = 9.33621 \times 10^{157}$$

$$1000! = 4.023872 \times 10^{2567}$$

$$10000! = 2.846259 \times 10^{35659}$$

$$1000000! = 8.263931 \times 10^{5565708}$$

The 5040 ways to arrange 4 pictures out of a choice of 10 are called *permutations* and we can generalise the concept for an arrangement of r objects chosen out of a total of n objects:

$$p(n, r) = \frac{n!}{(n-r)!} \tag{6.7}$$

Now, suppose that we are artistic philistines and do not care how the paintings are arranged on the wall but just that there are four paintings on the wall. We choose our four paintings and, within this subset, there are 4! ways to arrange them. However, we do not care about arranging them, being philistines. Therefore, we divide the number of permutations by this number to get the number of *combinations* of 4:

$$\binom{10}{4} = \frac{10!}{(10-4)!\,4!} = 210$$

or

$$C(10, 4) = \frac{10!}{(10-4!\,)4!} = 210$$

or more generally:

$$C(n, r) = \binom{n}{r} = \frac{n!}{(n-r)!\,r!} = \frac{P(n, r)}{r!} \tag{6.8}$$

Note carefully that what we have done here is to count the number of permutations $p(n, r)$ that gives us the number of distinct arrangements of r things out of a total of n. However, since we are not concerned with distinct arrangements of those r things we just divide them out of the number of permutations to get the number of combinations where arrangements do not matter. This idea will be important shortly.

When we want to know what the probability of an outcome for a series of experiments will be in which the outcomes of the individual experiments are independent of each other, we can use the *Bernoulli trial* method. Here, we define the outcome of interest as a "success" and anything else a "failure". An example of this might be "what is the probability of observing a 4 two times in 4 throws of a pair of dice?". Throwing a 4 would be a "success" and any other number would be a "failure". Of course neither is really a success or failure ... they just are what they are. From Table 6.1, the probability of throwing a four in a single throw is 1/12. The probability of not throwing a four must therefore be 11/12 since the probability of seeing something between 2 and 12 must be 1. The probability of seeing a four thrown two times is:

$$p(2) = \left(\frac{1}{12}\right)\left(\frac{1}{12}\right) = \left(\frac{1}{12}\right)^2 = \frac{1}{144} = 0.00694$$

The probability of *not* throwing a four two times is:

$$(1 - p(2)) = \left(\frac{11}{12}\right)\left(\frac{11}{12}\right) = \left(\frac{11}{12}\right)^2 = \frac{121}{144} = 0.84028$$

Now, we want to combine these into one probability statement; the probability of throwing a four two times and the probability of not throwing a four two times for a total of four trials. There is, however, one extra consideration; "how many ways can this combination of successes and failures happen?". For this simple example we can enumerate them directly:

$$ssff,\ fssf,\ fsfs$$

$$sffs,\ sfsf,\ ffss$$

There are six ways that we can potentially observe the results of our experiment ... in other words, six combinations. If we use just the probability calculations that we have so far for $p(2)$ and $(1 - p(2))$ we will get the probability for only one of these occurring. But what we desire is the probability that any of them will occur so, we finally can write our expression that tells us what the probability of observing a four rolled two times out of four tries, which includes a combination calculation:

$$\begin{aligned}
&\binom{4}{2} p^2(2)(1 - p(2))^2 \\
&= \frac{4!}{(4-2)!\,2!}\left(\frac{1}{12}\right)^2\left(\frac{11}{12}\right)^2 \\
&= 6 \times \frac{1}{144} \times \frac{121}{144} = 0.03501
\end{aligned}$$

In general we would write as the probability of observing r successes out of n individual experiments, with p_S^r equivalent to the probability of success for an individual experiment:

$$\binom{n}{r} p_S^r (1 - p_S^r)^{(n-r)} \tag{6.9}$$

Note that this collapses to our simple probability calculation if we are looking for the probability of an event occurring n times out of n experiments:

$$\begin{aligned}&\binom{n}{n} p_S^n (1 - p_S^n)^{(n-n)}\\ &= 1 \times p_S^n \times 1 = p_S^n\end{aligned}$$

6.3 Stirling's Approximation

For very large values of n we cannot conveniently calculate the value of $n!$ Our pocket calculators and computers will eventually run out of machine precision and will not be able to represent the number. One look at the list of increasing factorials above will convince you of this. We can, however, approximate the value for $n!$ when n is very large. Consider taking the logarithm of $N!$:

$$\begin{aligned}\ln(N!) &= \ln(N \times (N-1) \times (N-2) \ldots)\\ &= \ln(N) + \ln(N-1) + \ln(N-2) \ldots\\ &= \sum_{n=1}^{N} \ln(n)\end{aligned}$$

For very large N we can approximate the summation as an integral:

$$\begin{aligned}\sum_{n=1}^{N} \ln(n) &\approx \int_1^N \ln(n) \mathrm{d}n\\ &= N\ln(N) - N - (1\ln(1)) - 1)\\ &\approx N\ln(N) - N\end{aligned} \tag{6.10}$$

This is one form of *Stirling's approximation*. It is not at all useful for small values of N as the difference between $\ln(N)$ and $N \ln N - N$ is appreciable. However, as the value of N increases the difference decreases and becomes quite acceptable at very large values of N.

6.4 Probability Distributions

Our Bernoulli trial eqn (6.9) is an example of a *probability distribution* function. Let us do a series of experiments in which we throw a pair of dice 50 times and ask "what is the probability of throwing a four zero times, one time, two times . . . 50 times". To do this calculation we will use eqn (6.9) for each probability. The plot of probabilities *versus* number of dice throws is shown in Figure 6.1.

You will notice that the maximum probability on the plot is at 5 throws meaning that, for 50 throws of the dice the probability of obtaining 5 fours is the highest. Of course, the probability of throwing 50 fours is extremely low.

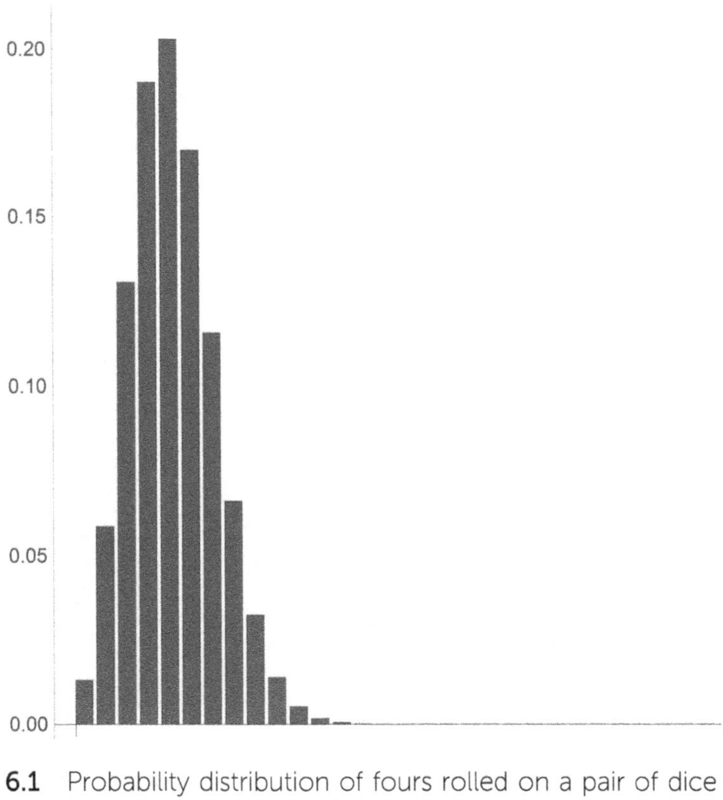

Figure 6.1 Probability distribution of fours rolled on a pair of dice for fifty experiments.

Let us now ask the same question for rolling a seven, again using 50 trials. Since the probability of throwing a seven in a single throw is double that of throwing a four we expect that in 50 throws we will have a higher probability of throwing sevens than fours. We see this reflected in the probability distribution; the maximum probability occurs at 9 throws of 7 (Figure 6.2).

If we vastly increase the number of experiments, say from 50 to 1 000 000, we will very quickly run into trouble; namely the calculation of 1 000 000!. This is a very, very large number as you can see from the list of increasing factorials in Section 6.2. Even using Stirling's approximation we would have trouble with the calculation because of the sheer size of the numbers involved. The subject of statistical mechanics, which is mostly outside the scope of this text (except, see below), deals with this situation. The interested reader is encouraged to explore the subject.

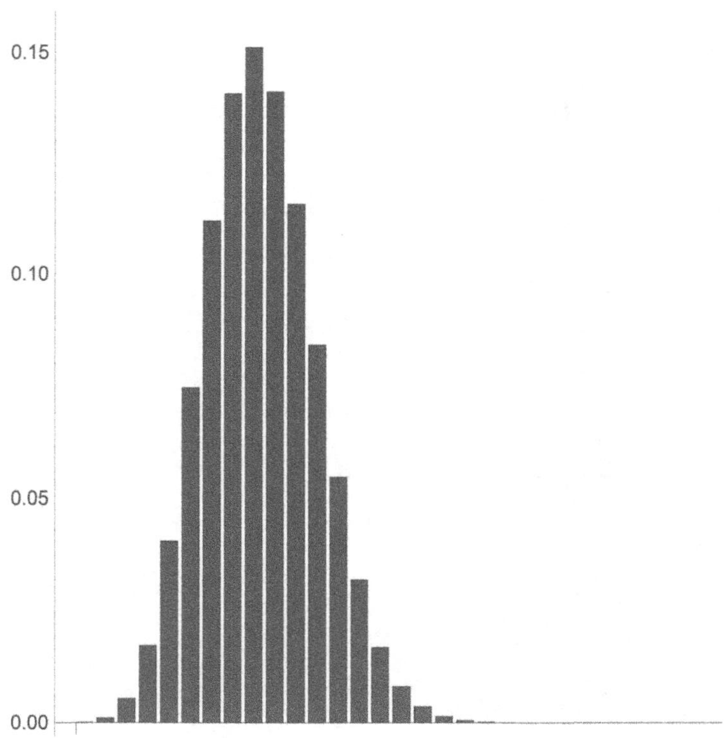

Figure 6.2 Probability distribution of sevens rolled on a pair of dice for fifty experiments.

The questions that we have asked can be expressed equivalently in the form of a function. We calculate the probability of a variable having a specific value given the probability function, *f*:

$$p(x) \propto f(x)$$

or

$$p(x) = Kf(x)$$

This is the general expression for our discrete probability distribution that we have been using. If the probability distribution is continuous we can proceed in a very analogous manner:

$$p(x)\mathrm{d}x = Kf(x)\mathrm{d}x$$

where now we are looking at an interval in the domain rather than a discrete value of x. We can apply all that we have learned about probability to this expression, including the fact that the sum of all the probabilities must be equal to one:

$$\int_{x_1}^{x_2} p(x)\mathrm{d}x = K\int_{x_1}^{x_2} f(x)\mathrm{d}x = 1$$

$$K = \frac{1}{\int_{x_1}^{x_2} f(x)\mathrm{d}x} \tag{6.11}$$

$$p(x)\mathrm{d}x = \frac{f(x)\mathrm{d}x}{\int_{x_1}^{x_2} f(x)\mathrm{d}x}$$

$p(x)$ is now referred to as the *probability density* function. This has direct consequences in the fundamental postulates of quantum mechanics as we shall see.

6.5 The Boltzmann Distribution

Suppose that we have N particles to distribute among r energy levels[†] each separated by $\Delta\epsilon$ energy units. We further suppose that this is an isolated system such that N and E, the total energy, are constant.

[†] I have not introduced quantum mechanics yet so this is formally premature but really, in today's world who but the intentionally ignorant do not know about the concept of discrete energy levels at least peripherally. Presumably, if you are reading this you will already have some small (or perhaps not-so-small) education in this area.

We will put n_i particles into energy level ϵ_i without concern for how they are arranged … in other words we will do a combination calculation. The number of ways to take n_0 particles from our pool of N particles and put them into level ϵ_0:

$$C(N, n_0) = \frac{N!}{n_0!\,(N - n_0)!}$$

Now, since we have removed n_0 particles from the pool there are N–n_0 left. Next we take n_1 particles from our depleted pool and put them into energy level ϵ_1. The number of ways to do this is:

$$C(N - n_0, n_1) = \frac{(N - n_0)!}{n_1!\,(N - n_0 - n_1)!}$$

Similarly:

$$C(N - n_1, n_2) = \frac{(N - n_0 - n_1)!}{n_2!\,(N - n_0 - n_1 - n_2)!}$$

We have done a combinations calculation for each energy level but what about the overall picture? Combining the above terms we get:

$$\begin{aligned} W &= \frac{N!}{n_0!\,(N - n_0)!} \times \frac{(N - n_0)!}{n_1!\,(N - n_0 - n_1)!} \times \frac{(N - n_0 - n_1)!}{n_2!\,(N - n_0 - n_1 - n_2)!} \cdots \\ &= \frac{N!}{n_0!\,n_1!\,n_2!\,\ldots} \\ &= \frac{N!}{\prod_{j=0}^{N-1} n_j!} \end{aligned} \tag{6.12}$$

W is called the weight or, in statistical mechanics jargon, the number of *microstates* for this particular combination of particles. A picture to illustrate this idea is appropriate. Suppose that we have three particles to assign to four different energy levels.‡ Figure 6.3 shows three possible ways to assign the three particles to the four energy levels, including assigning multiple particles to one energy

‡I am temporarily disregarding that there is a fixed energy assigned to the system. We will come back to that constraint momentarily.

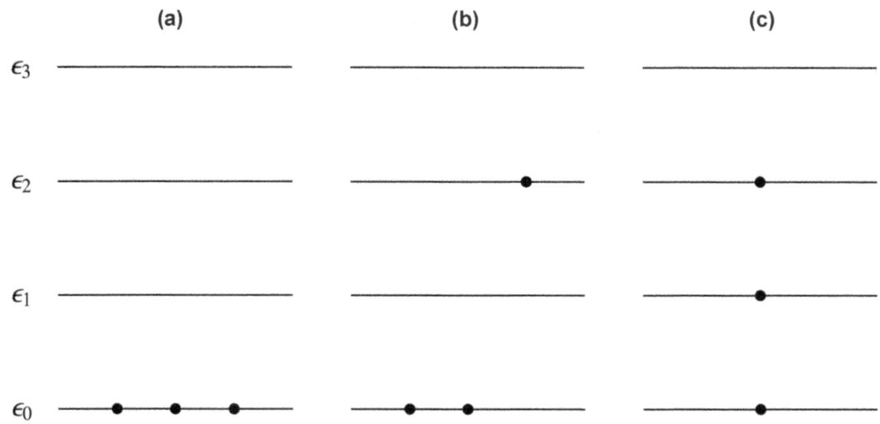

Figure 6.3 Three possible configurations of three particles distributed over four energy levels.

level. Each of these different assignments is called a *configuration* or a *macrostate*. There are, of course more possible configurations for this example but for brevity we show only three. Now, consider that for configuration *a* there is only one way to achieve this configuration … put all available particles into the lowest level. In other words, only one combination:

$$W_a = \frac{3!}{3!\,0!\,0!\,0!} = 1$$

There is only one microstate for configuration 6.3*a*. Configuration 6.3*b*, however, has three microstates corresponding to three different ways that the particles can be put into the energy levels

$$W_b = \frac{2!}{2!\,0!\,1!\,0!} = 3$$

Configuration 6.3*c* has 6 microstates:

$$W_c = \frac{3!}{1!\,1!\,1!\,0!} = 6$$

New let us make things more interesting. Consider again the three particles distributed over four energy levels with the total energy being 3ϵ where ϵ represents a quantum of energy. The possible configurations are illustrated in Figure 6.4.

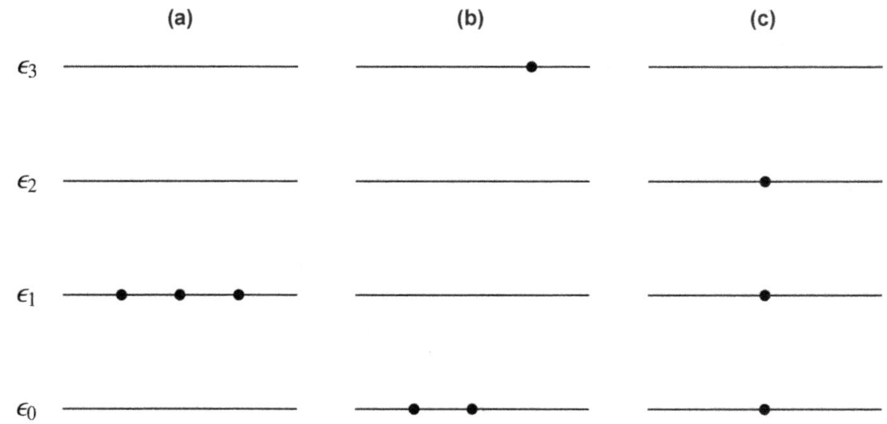

Figure 6.4 Three particles partitioned among four energy levels with total energy 3ϵ.

Note that given the energy constraint there are now only three possible configurations. We calculate the number of microstates for each configuration and the total number of configurations:

$$W_a = \frac{3!}{3!\,0!\,0!\,0!} = 1$$

$$W_b = \frac{3!}{2!\,0!\,0!\,1!} = 3$$

$$W_c = \frac{3!}{1!\,1!\,1!\,0!} = 6$$

$$W_T = W_a + W_b + W_c = 10$$

It is reasonable to expect that since most microstates are associated with configuration 6.4c that the probability of finding the particles in that configuration is the largest. That probability is written as (compare with eqn (6.11)):

$$p_i = \frac{W_i}{\sum_{i=1}^{N} W_i} = \frac{W_i}{W_T}$$

Now let us add two particles for a total of five, but keep the total energy at 3ϵ. There will still be only three possible configurations because of the energy constraint (Figure 6.5).

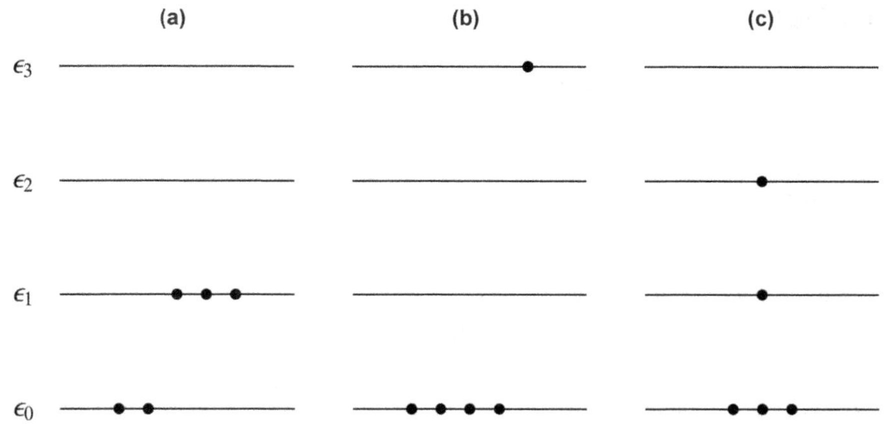

Figure 6.5 Five particles distributed among four energy levels with total energy 3ϵ.

Calculation of the total number of microstates gives:

$$W_a = \frac{5!}{2!\,3!\,0!\,0!} = 10$$

$$W_b = \frac{5!}{4!\,0!\,0!\,1!} = 5$$

$$W_c = \frac{5!}{3!\,1!\,1!\,0!} = 20$$

$$W_T = W_a + W_b + W_c = 35$$

Adding just two particles has more than tripled the number of microstates. The same calculation for $N = 10$ and all else the same gives:

$$W_a = \frac{10!}{7!\,3!\,0!\,0!} = 120$$

$$W_b = \frac{10!}{9!\,0!\,0!\,1!} = 10$$

$$W_c = \frac{10!}{8!\,1!\,1!\,0!} = 90$$

$$W_T = W_a + W_b + W_c = 220$$

and for $N = 100$:

$$W_a = \frac{100!}{97!\,3!\,0!\,0!} = 161700$$

$$W_b = \frac{100!}{99!\,0!\,0!\,1!} = 100$$

$$W_c = \frac{100!}{98!\,1!\,1!\,0!} = 9900$$

$$W_T = W_a + W_b + W_c = 171700$$

As you can see, the total number of available microstates goes up *very* rapidly with increasing numbers of particles. It is also apparent that one configuration begins to predominate over the others, making it the most probable state to be occupied.

As the numbers of energy levels and particles increase, the total number of microstates becomes literally astronomical. It is estimated that for 1000 particles with 1000 quanta of energy available that the total number of microstates is 10^{600}, more than the total number of atoms in the universe! The number of microstates associated with a mole of particles is so stupendously large as to be utterly incomprehensible. Also, as the numbers go up one configuration begins to overwhelmingly dominate the others as we saw in the simple example. The value of W will be at a maximum for that configuration and we can use this to determine where that configuration is and ultimately produce the Boltzmann distribution equation. At the maximum:

$$\frac{\partial W}{\partial n_j} = 0$$

We will find it more convenient however to use the logarithm of W:

$$\frac{\partial \ln W}{\partial n_j} = 0$$

Also, since N and E are constant:

$$N = \sum_{j=1}^{N} n_j$$

$$\frac{\partial N}{\partial n_j} = \frac{\partial}{\partial n_j} \sum_{j=1}^{N} n_j = 0$$

and

$$E=\sum_j n_j\epsilon_j$$

$$\frac{\partial E}{\partial n_j}=\frac{\partial}{\partial n_j}\sum_j n_j\epsilon_j=0$$

Why $n_j\epsilon_j$? Each particle in energy level ϵ_j has one ϵ_j quantum of energy and the total energy for that level is therefore $n_j\epsilon_j$.

Proceeding, we use eqn (6.12) in logarithmic form to get:

$$\frac{\partial W}{\partial n_j}=\frac{\partial}{\partial n_j}\ln N!-\frac{\partial}{\partial n_j}\sum \ln n_j!=0$$

and, since we are talking about very large numbers of particles, Stirling's approximation will simplify this for us:

$$\frac{\partial W}{\partial n_j}=\frac{\partial}{\partial n_j}(N\ln N-N)-\frac{\partial}{\partial n_j}\sum \ln n_j!$$

$$=\frac{\partial}{\partial n_j}N\ln N-\frac{\partial N}{\partial n_j}-\frac{\partial}{\partial n_j}\sum \ln n_j!$$

$$=0$$

We have stipulated above that N is constant and that therefore its differential is zero, as is the differential of the logarithm of N. We are then left with:

$$\frac{\partial W}{\partial n_j}=-\frac{\partial}{\partial n_j}\sum \ln n_j!=\frac{\partial}{\partial n_j}\sum \ln n_j!=0$$

Note that we have made a sign change ... whether the sign is positive or negative here, the equation is still valid. We do this for convenience later so that we will have positive values of two constants that we will be introducing. Again, using Stirling's approximation and the stipulation that the sum of n_j s is equal to N and therefore its derivative is zero:

$$\frac{\partial W}{\partial n_j} = 0 = \frac{\partial}{\partial n_j}\sum \ln(n_j \ln n_j - n_j)$$
$$= \frac{\partial}{\partial n_j}\sum \ln(n_j \ln n_j) - \frac{\partial}{\partial n_j}\sum n_j$$
$$= \left(\ln n_j + n_j \frac{1}{n_j}\right) - 1$$
$$= \ln n_j$$

We now work with our constraints and define two new functions based on them:

$$g = \sum n_j - N = 0 \Rightarrow \frac{\partial g}{\partial n_j} = 1$$
$$h = \sum n_j \epsilon_j - E = 0 \Rightarrow \frac{\partial h}{\partial n_j} = \epsilon_j$$

Adding the derivative of each of these to the derivative of W:

$$\frac{\partial W}{\partial n_j} - \alpha \frac{\partial g}{\partial n_j} + \beta \frac{\partial h}{\partial n_j} = 0$$

The constants α and β are called Lagrange multipliers[§] and the signs are such that we will end up with positive values for the constants. From our previous results we can substitute into this:

$$\ln n_j - \alpha + \beta\epsilon_j = 0$$
$$\ln n_j = \alpha - \beta\epsilon_j$$
$$n_j = e^{\alpha} e^{-\beta\epsilon_j}$$

If we make a ratio of particles in different energy levels:

$$\frac{n_j}{n_i}\frac{e^{\alpha}e^{-\beta\epsilon_j}}{e^{\alpha}e^{-\beta\epsilon_i}} = e^{-\beta(\epsilon_j - \epsilon_i)} \tag{6.13}$$

[§]For more information on Lagrange multipliers and their use see any competent calculus text. Some statistical mechanics texts also include details on the technique.

and if we take n_i to correspond to the lowest energy level that we define as zero:

$$\frac{n_j}{n_0} = e^{-\beta\epsilon_j}$$

$$\Downarrow$$

$$n_j = n_0 e^{-\beta\epsilon_j}$$

$$\Downarrow$$

$$\sum n_j = N = n_0 \sum e^{-\beta\epsilon_j}$$

$$\Downarrow$$

$$n_0 = \frac{N}{\sum e^{-\beta\epsilon_j}}$$

The term in the denominator is called the *partition function*, often symbolically represented with q:

$$q = \sum e^{-\beta\epsilon_j}$$

It is interesting to rearrange the above equations and compare the result to eqn (6.11):

$$n_j = \frac{N}{\sum e^{-\beta\epsilon_j}} e^{-\beta\epsilon_j}$$

$$\frac{n_j}{N} = \frac{e^{-\beta\epsilon_j}}{\sum e^{-\beta\epsilon_j}}$$

This is very similar in form to the general equation for probability densities and it is clear that the ratio of n_j to N is a probability.

$$p(n_j) = \frac{n_j}{N} = \frac{e^{-\beta\epsilon_j}}{\sum e^{-\beta\epsilon_j}} \tag{6.14}$$

We are left only with the problem of evaluating β. From thermodynamics,[¶] the entropy of a system is calculated:

$$S = k \ln W$$

[¶]Again, any competent text on thermodynamics will have this information in it.

and:

$$\frac{\partial S}{\partial E}=\frac{1}{T}$$

k is the proportionality constant between S and $\ln W$ and is known as Boltzmann's constant.

As above:

$$\ln W=\ln N!-\sum \ln n_j!$$

for which we apply Stirling's approximation and substitute for W in the entropy expression:

$$\begin{aligned}S&=k\left(N\ln N-N-\left(\sum n_j\ln n_j-n_j\right)\right)\\&=k\left(N\ln N-N-\sum n_j\ln n_j+\sum n_j\right)\\&=k\left(N\ln N-\sum n_j\ln n_j\right)\end{aligned}$$

From above:

$$n_j=\mathrm{e}^{\alpha}\mathrm{e}^{-\beta\epsilon_j}$$

which is substituted into the equation for S:

$$\begin{aligned}S&=k\left(N\ln N-\sum n_j\ln n_j\right)\\&=k\left(N\ln N-\sum n_j(\alpha-\beta\epsilon_j)\right)\\&=k(N\ln N-\alpha N+\beta E)\end{aligned}$$

and from this, the derivative of S with respect to E is:

$$\frac{\partial S}{\partial E}=-k\beta=\frac{1}{T}$$

$$\Downarrow$$

$$\beta=-\frac{1}{kT}$$

Finally, we have the standard form of the Boltzmann distribution function:

$$n_j=\frac{N}{q}\mathrm{e}^{-\epsilon_j/kT}\quad\text{or}\quad n_j=n_0\mathrm{e}^{-\epsilon_j/kT}\tag{6.15}$$

This will prove very useful in the development of the density operator in Chapter 17.

6.6 Random Processes and the Autocorrelation Function

Let us suppose that we have a set of variables, $X_i(t)$, each of which takes on a random value, X, that is governed by a probability. For example, each $X_i(t)$ might hold the result of a coin flip. We define this set $X_i(t)$ as a *random or stochastic process*. Let us take it a step further and specify that our set of variables is continuous and indicate this with $X(t)$. The probability density for this continuous variable is then written as $P(X, t)$; that is, the probability that variable $X(t)$ will take the value X at time t. The average value for $X(t)$ is calculated:

$$\langle X(t)\rangle = \int X(t)P(X,t)\mathrm{d}X$$

We may also define a function of variable $X(t)$, $f(X(t))$ or simply $f(t)$ that will also necessarily be random in nature and will have a similar mean calculation using the same probability density, since it is a function of $X(t)$:

$$\langle f(t)\rangle = \int f(t)P(X,t)\mathrm{d}X$$

We define a *stationary state* as one whose probability function behaves as follows:

$$P(X,t) = P(X,t + t_n)$$

In other words, the probability function at time t is the same at all other times, $t + t_n$. This would be the case with the random motions of molecules in solution that have attained a steady state. The stationary state has the interesting property that the mean value of $f(t)$ is independent of the time that it is calculated at.

$$\langle f(t)\rangle = \int f(t)P(X,t)\mathrm{d}X = \int f(t)P(X,t + t_n)\mathrm{d}X$$

Next, let us look at what will be the case for $X(t)$ and $X(t + t')$, where $t + t'$ is a very short time after t. $X(t)$ is random but since t' is very short we expect that $X(t + t')$ will not be completely independent of $X(t)$. The physical basis for thinking that this may be the case is the diffusional rotation of molecules as a result of random collisions with other molecules. We expect that for many collisions a molecule will be abruptly rotated by a small amount rather than a large jump in rotational diffusion. This means that $X(t+t')$ may be considered to be dependent on $X(t)$ for short t'.[‖] As such, we now use a *conditional probability* $P(X';t'\,|\,X;t)$ which means the probability that $X(t')=X'$ given that we know that $X(t)=X$. A stochastic process that exhibits this probability behaviour is called a Markov process.

The *autocorrelation function*, $G(t, t')$, is defined as follows:

$$\begin{aligned} G(t,t') &\equiv \langle f(t)f^*(t')\rangle \\ &= \iint f(t)f^*(t')P(X',t'\,|\,X,t)\mathrm{d}X\mathrm{d}X' \end{aligned} \tag{6.16}$$

We define it this way in order to end up with a real, positive number (see Chapter 1). It is customary to define $t'\equiv t+\tau$ and combining this with the idea of the stationary state in which the probability density does not change with time, we may "relocate" the time origin so that $t=0$ and then $t'=\tau$. In other words, G does not depend on t but rather on the *difference* between t and t' or τ as we have defined it. Our autocorrelation function then reads as:

$$\begin{aligned} G(\tau) &\equiv \langle f(0)f^*(\tau)\rangle \\ &= \iint f(0)f^*(\tau)P(X,\tau\,|\,X_0)\mathrm{d}X\mathrm{d}X_0 \end{aligned} \tag{6.17}$$

This is the form that we shall be using it in when we discuss relaxation in Chapter 20.

The autocorrelation function has several useful properties. Since we are dealing with a stationary state system we may move the time origin as we please (as we did to produce eqn (6.17)). Therefore, we can write:

$$G(\tau)=\langle f(0)f^*(\tau)\rangle=\langle f(-\tau)f^*(0)\rangle=G^*(-\tau)$$

Furthermore, if $f(t)$ is a real function then:

$$G(\tau)=G(-\tau)$$

[‖] I will leave the definition of "short" until Chapter 20 on relaxation.

This says that the autocorrelation function is symmetric and even (see Appendix A for a short discussion of even and odd functions). Also:

$$G(0) = \langle f(0) f(0) \rangle = \langle f^2(0) \rangle \geq 0$$

The function $f(t)$ correlates with itself. Finally:

$$G(\tau) \leq G(0)$$

The autocorrelation function is at a maximum at $t = 0$.

Further Reading

1. T. Engel and P. Reid, *Thermodynamics, Statistical Thermodynamics and Kinetics*, Prentice Hall, New York, 2006.
2. V. Krishnan, *Probability and Random Processes*, John Wiley & Sons, Hoboken, New Jersey, 2006.
3. M. L. Lial and C. D. Miller, *Mathematics with Applications in the Management, Natural and Social Sciences*, Scott, Foresman and Co., New York, 1983.
4. L. K. Nash, *Elements of Statistical Thermodynamics*, Addison-Wesley Publishing Company, New York, 1968.
5. M. P. Nicholas, E. Eryilmaz, F. Ferrage, D. Cowburn and R. Ghose, *Prog. Nucl. Magn. Reson. Spectrosc.*, 2010, **57**, 111–158.

7 Waves

There cannot be a language more universal and more simple, more free from errors and obscurities. . .more worthy to express the invariable relations of all natural things than mathematics.

Joseph Fourier

7.1 Characteristics of Waves

Waves or periodic oscillations have a few simple characteristics that we should be aware of before continuing. First, since waves are periodic or repeating there is a number called the *frequency* of the wave associated with it. Assuming that each wave is identical to the one before it and after it, we can define the frequency as the number of times that a certain point, say the top of a wave, passes a point in space per unit time. The SI units for this are s^{-1} and the usual symbol in the literature is the Greek ν.

The second characteristic of interest is the *wavelength*. Again, assuming identical waves, this is the distance from any arbitrary point on a wave to the corresponding point on the next (or previous) cycle. The generally used symbol for wavelength is λ and, of course, the SI units are metres (Figure 7.1).

Frequency and wavelength are related *via* the speed of the wave:

$$v = \lambda\nu \tag{7.1}$$

We often have occasion to compare two or more waves and we will say that they are *in-phase* or *out-of-phase* (Figures 7.2a and 7.2b). In-phase waves are those whose crests and troughs line up in time.

Essential Mathematics for NMR and MRI Spectroscopists
By Keith C. Brown

Published by the Royal Society of Chemistry, www.rsc.org

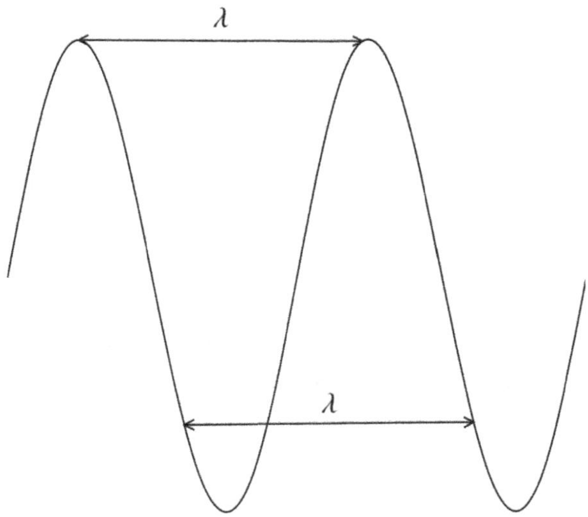

Figure 7.1 The wavelength.

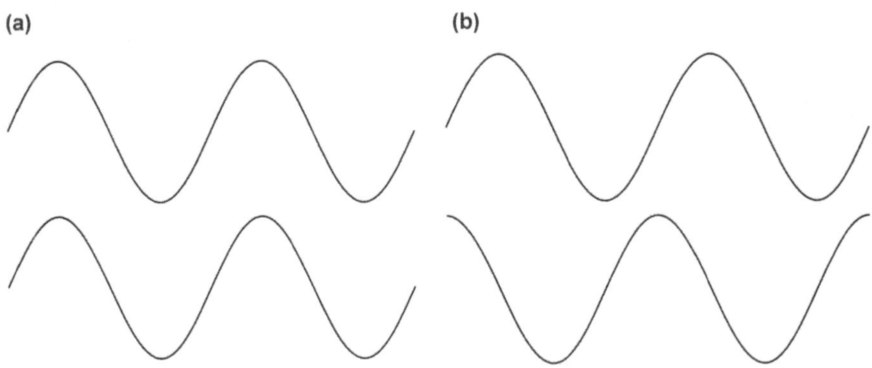

Figure 7.2 (a) In-phase waves; (b) Out-of-phase waves.

They will therefore have identical wavelengths and frequencies. Out-of-phase waves are not lined up.

In Figure 7.2b the two waves are out-of-phase by $\pi/2$ radians and correspond to a sine wave and a cosine wave. We would write the mathematical functions for these as:

$$f(t) = \sin(\omega t)$$

$$f(t) = \cos(\omega t)$$

where ω is often used to represent the frequency of the wave in radians per second and the argument to sine and cosine is therefore in the natural units of radians. The relationship between radians per second and frequencies in hertz is:[†]

$$\omega = 2\pi\nu \tag{7.2}$$

Sine and cosine waves are related by a *phase angle* of $\pi/2$ radians:

$$\cos(\omega t) = \sin\left(\omega t + \frac{\pi}{2}\right)$$

In other words, cosine waves are $\pi/2$ radians ahead of (or behind) sine waves. The phase angle is sometimes included in a general expression of the sine and cosine:

$$\begin{aligned} f(t) &= \sin(\omega t + \theta) \\ f(t) &= \cos(\omega t + \theta) \end{aligned} \tag{7.3}$$

ω is the frequency term, while θ is the phase angle.

The *period*, τ, of the wave or oscillation is the time required for one complete cycle to occur. A little thought will convince you that the frequency and the period are inversely related. Indeed, we can show this from the fundamental definition of frequency: the number of cycles per unit time. For one cycle the time will be the period and the relationship between frequency and period becomes clear:

$$\nu = \frac{1}{\tau} \tag{7.4}$$

Another way to look at periodic oscillations is to picture a rotating vector. Recall from eqn (1.7) that at any given point in time one can measure a phase angle for the rotating vector. Thus, one might write:

$$A = A_0 \exp(i\omega t) \tag{7.5}$$

where the phase angle is ωt. Projection of the vector onto the x- or y-axis will generate a sine or cosine function. Indeed, this is the picture that we use for the simple analysis of the NMR signal in Chapters 19 and 21.

[†]It would be wise to become familiar with this as we will use it very frequently in later chapters.

7.2 The Classical Wave Equation

The classical wave equation is developed[‡] from a consideration of a wave propagating (moving) in one direction along a vibrating string of length l. We can then generalise to multiple directions. We start with a function, $f(x, t)$, which describes the vertical displacement of a point in the string as the wave passes by. The velocity of the wave is v. Taking the differential of the function:

$$\mathrm{d}f = \frac{\partial f}{\partial x}\mathrm{d}x + \frac{\partial f}{\partial t}\mathrm{d}t \tag{7.6}$$

We assume that the medium is non-absorbent so that the wave retains its size and shape as it moves. In other words, the wave's maximum displacement does not decay to zero ... sometimes called an eternal wave. This being the case, eqn (7.6) can be set equal to zero since the increment in the wave displacement must vanish:

$$\frac{\partial f}{\partial x}\mathrm{d}x + \frac{\partial f}{\partial t}\mathrm{d}t = 0 \tag{7.7}$$

That this is so is a bit unclear at first. Imagine the wave moving along the string and imagine also that you are moving along with it and observing it. In other words, you are observing its evolution in both space and time. What will you see? Well, again assuming a non-absorbent medium, the shape of the wave will not change and you will see an unchanging wave and the change in wave displacement appears to be zero.

Now:

$$v = \frac{\mathrm{d}x}{\mathrm{d}t}$$

$$\mathrm{d}x = v\mathrm{d}t$$

which we substitute into eqn (7.7):

$$\left(v\frac{\partial f}{\partial x} + \frac{\partial f}{\partial t}\right)\mathrm{d}t = 0$$

[‡]See C. K. C. Lieou, *Eur. J. Phys.*, 2007, **28**, N17–N19.

Since dt can be any arbitrary increment of t the expression in the brackets must equal zero:

$$v\frac{\partial f}{\partial x}+\frac{\partial f}{\partial t}=0$$

or:

$$\frac{\partial f}{\partial x}=-\frac{1}{v}\frac{\partial f}{\partial t} \tag{7.8}$$

Taking another partial derivative with respect to x:

$$\frac{\partial^2 f}{\partial x^2}=-\frac{1}{v}\frac{\partial^2 f}{\partial x\partial t}$$

or

$$\frac{\partial^2 f}{\partial x^2}=-\frac{1}{v}\frac{\partial}{\partial x}\frac{\partial f}{\partial t}$$

We now use the equality of mixed partial derivatives (see eqn (A.42)):

$$\frac{\partial}{\partial x}\frac{\partial f}{\partial t}=\frac{\partial}{\partial t}\frac{\partial f}{\partial x}$$

and, putting this into our current equation:

$$\frac{\partial^2 f}{\partial x^2}=-\frac{1}{v}\frac{\partial}{\partial t}\frac{\partial f}{\partial x}$$

and substituting for $\partial/\partial x$ from eqn (7.8) into this:

$$\frac{\partial^2 f}{\partial x^2}=\frac{1}{v^2}\frac{\partial^2 f}{\partial t^2} \tag{7.9}$$

This is the classical wave equation in one dimension. It is easily extended to three-dimensional form:

$$\frac{\partial^2 f}{\partial x^2}+\frac{\partial^2 f}{\partial y^2}+\frac{\partial^2 f}{\partial z^2}=\frac{1}{v^2}\frac{\partial^2 f}{\partial t^2}$$

or more succinctly:

$$\nabla^2 f=\frac{1}{v^2}\frac{\partial^2 f}{\partial t^2} \tag{7.10}$$

where, as we have seen in Chapter 4, ∇^2 represents $\partial^2/\partial x^2 + \partial^2/\partial y^2 + \partial^2/\partial z^2$. This is a very general equation for all types of wave phenomena. We can, of course, solve the equation and will do so for the one-dimensional equation. The technique involves the separation of variables in which we separate the time and displacement variables into two distinct functions:

$$f(x, t) = X(x)T(t)$$

which we then substitute into eqn (7.9):

$$\frac{\partial^2}{\partial x^2}X(x)T(t) = \frac{1}{v^2}\frac{\partial^2}{\partial t^2}X(x)T(t)$$

$$T(t)\frac{\partial^2}{\partial x^2}X(x) = \frac{1}{v^2}X(x)\frac{\partial^2}{\partial t^2}T(t)$$

$$\frac{1}{X(x)}\frac{\partial^2}{\partial x^2}X(x) = \frac{1}{v^2}\frac{1}{T(t)}\frac{\partial^2}{\partial t^2}T(t) = K$$

"K" is called the *constant of separation*. How do we know this is constant? A little thought will show that, since we have separated the variables x and t so that they are on opposite sides of the equation and since they are independent of each other, a change in t will not result in a corresponding change in x and *vice versa*. Thus, in order for the expression in x to be equal to the expression in t they must each be equal to a constant.

We may now write:

$$\frac{1}{X(x)}\frac{\partial^2}{\partial x^2}X(x) = K$$

and

$$\frac{1}{v^2}\frac{1}{T(t)}\frac{\partial^2}{\partial t^2}T(t) = K$$

or, equivalently:

$$\frac{\partial^2}{\partial x^2}X(x) = KX(x)$$

and (7.11)

$$\frac{1}{v^2}\frac{\partial^2}{\partial t^2}T(t) = KT(t)$$

or, yet again, equivalently:

$$\frac{\partial^2}{\partial x^2}X(x) - KX(x) = 0$$

and

$$\frac{1}{v^2}\frac{\partial^2}{\partial t^2}T(t) - KT(t) = 0$$

The equation in $X(x)$ is sometimes referred to as the "stationary" part of the wave equation since it does not involve time. These are homogeneous, linear, 2nd-order ordinary differential equations instead of partial differential equations with standard solutions. Note that eqn (7.11) are eigenvalue equations, similar to eqn (3.45).

What are the solutions to these equations? What is the value of K? We shall proceed to answer both of these questions starting with the first equation.[§] If the equation was a first-order equation we would have something like:

$$\frac{\partial}{\partial x}X(x) = KX(x)$$

and it is well known that the solution to this equation is:

$$X(x) = e^{Kx+\phi}$$

as can shown by integration:

$$\int \frac{\mathrm{d}X(x)}{X(x)} = \int K\mathrm{d}x$$

$$\ln(X(x)) = Kx + \phi$$

$$X(x) = \mathrm{e}^{Kx+\phi}$$

Our wave equation in x is a 2nd-order equation of similar form and a little thought leads us to suspect that a solution might be something like:

$$X(x) = \mathrm{e}^{\sqrt{K}x+\phi} \tag{7.12}$$

[§]Much of this is also in the appendix; however, here we also introduce boundary conditions that lead to quantised vibrations.

as indeed it is. We can verify this by substitution into eqn (7.11):

$$\frac{\partial^2}{\partial x^2} e^{\sqrt{K}x+\phi} = Ke^{\sqrt{K}x+\phi}$$

What values of K work here? Let us put $K=0$ and our solution, eqn (7.11) into eqn (7.10):

$$\frac{\partial^2}{\partial x^2} e^{0x+\phi} = 0\ e^{0x+\phi}$$

$$\frac{\partial^2}{\partial x^2} e^{\phi} = 0$$

While this is certainly true since the derivative of a constant is zero it is not of much use to us. It is a *trivial* solution to our problem and has no physical meaning in the context of the vibrating string.

Next we try $K>0$. Since we are taking the square root of K, let us substitute α^2 for K for clarity ... a general solution will be of the form:

$$X(x) = e^{\alpha x+\phi} \quad \text{or} \quad X(x) = e^{-\alpha x+\phi}$$

since:

$$K = \alpha^2 \Rightarrow \sqrt{K} = \sqrt{\alpha} = \pm\alpha$$

If these are solutions to the equation then:

$$X(x) = e^{\alpha x+A} \quad \text{or} \quad X(x) = e^{-\alpha x+A}$$

are *also* solutions as you can easily verify by substitution. In fact we can also write:

$$\begin{aligned} X(x) &= c_1 e^{\alpha x+\phi} + c_2 e^{-\alpha x+\phi} \\ &= e^{\phi}(c_1 e^{\alpha x} + c_2 e^{-\alpha x}) \end{aligned} \tag{7.13}$$

as a solution, again verifiable by substitution:

$$\begin{aligned} \frac{\partial^2}{\partial x^2}(c_1 e^{\alpha x+\phi} + c_2 e^{-\alpha x+\phi}) &= \frac{\partial^2}{\partial x^2} c_1 e^{\alpha x+\phi} + \frac{\partial^2}{\partial x^2} c_2 e^{-\alpha x+\phi} \\ &= \alpha^2 c_1 e^{\alpha x+\phi} + \alpha^2 c_2 e^{-\alpha x+\phi} \\ &= \alpha^2 (c_1 e^{\alpha x+\phi} + c_2 e^{-\alpha x+\phi}) \end{aligned}$$

This is the most general solution to the problem. Now we introduce the boundary conditions to the problem. These follow from the fact that at either end of the vibrating string the displacement is zero. That is:

$$X(0) = 0 \; X(l) = 0$$

This being the case, the constants c_1 and c_2 in our general solution must each equal zero ... another trivial solution!

$$\begin{aligned} X(0) &= 0 = c_1 e^{\phi} + c_2 e^{\phi} = e^{\phi}(c_1 + c_2) \\ &\Downarrow \\ c_1 + c_2 &= 0 \\ X(l) &= 0 = c_1 e^{\alpha x + \phi} + c_2 e^{-\alpha x + \phi} \\ &\Downarrow \\ c_1 &= 0 = c_2 \end{aligned}$$

The *really* interesting situation for us occurs when $K < 0$. Let us see what happens in this case. If K is negative then so is α^2 and when we take the square root we get an imaginary number:

$$-K = -\alpha^2 \Rightarrow \sqrt{-K} = \sqrt{-\alpha^2} = \pm i\alpha$$

and our general solution is:

$$\begin{aligned} X(x) &= c_1 e^{i\alpha x + \phi} + c_2 e^{-i\alpha x + \phi} \\ &= e^{\phi}(c_1 e^{i\alpha x} + c_2 e^{-i\alpha x}) \\ &= e^{\phi}[(c_1 + c_2)\cos(\alpha x) + i(c_1 - c_2)\sin(\alpha x)] \end{aligned} \tag{7.14}$$

using Euler's formula (see eqn (A.1)) and gathering terms. For $X(0) = 0$:

$$\begin{aligned} X(0) &= 0 = e^{\phi}[(c_1 + c_2)\cos(0) + i(c_1 - c_2)\sin(0)] \\ &\Downarrow \\ c_1 + c_2 &= 0 \end{aligned}$$

so the cosine term vanishes. At the other end of the string, $X(l) = 0$ and:

$$X(l) = 0 = e^{\phi}[i(c_1 - c_2)\sin(\alpha l)]$$

In order for this to be true, either $i(c_1 - c_2) = 0$ or the sine term equals zero. The first would be another trivial solution (with $c_1 = c_2$). For the sine term to be zero:

$$\alpha l = \pi,\ 2\pi,\ 3\pi\ \ldots$$

or:

$$\alpha = \frac{n\pi}{l}$$

We do not include $n = 0$ because that would lead to $\alpha = 0$, yet another trivial solution.

Finally! We can now write the solution to our equation:

$$X(x) = \mathrm{e}^{\phi}\left[i(c_1 - c_2)\sin\left(\frac{n\pi x}{l}\right)\right]$$

or, since c_1 and c_2 are constants:

$$X(x) = \mathrm{e}^{\phi}\left[C\ \sin\left(\frac{n\pi x}{l}\right)\right] \tag{7.15}$$

Note that we have incorporated i into the constant C.

Now we will work with the second of our two equations arrived at after the separation of variables:

$$\frac{\partial^2}{\partial t^2}T(t) - Kv^2 T(t) = 0$$

Again, we will arrive at trivial solutions for $Kv^2 = 0$ and $Kv^2 > 0$. In this equation both v and K are constants so we might as well combine them into one constant, ω^2:

$$\frac{\partial^2}{\partial t^2}T(t) - \omega^2 T(t) = 0$$

The solution to this is similar to that above in eqn (7.14):

$$\begin{aligned} T(t) &= c_3\mathrm{e}^{i\omega t+\zeta} + c_4\mathrm{e}^{-i\omega t+\zeta} \\ &= \mathrm{e}^{\zeta}[(c_3 + c_4)\cos(\omega t) + i(c_3 - c_4)\sin(\omega t)] \\ &= \mathrm{e}^{\zeta}[A\cos(\omega t) + B\sin(\omega t)] \end{aligned} \tag{7.16}$$

where i has been included into the arbitrary constant B. There are no boundary conditions here so we can go no further with it.

Our approach to solving the wave equation was to separate the position and time functions, $X(x)$ and $T(t)$. We put them back together now using eqn (7.15) and (7.16), and recalling that:

$$\begin{aligned}
\omega^2 &= K\nu^2 = \alpha^2\nu^2 \\
\omega &= \alpha\nu = \frac{n\pi}{l}\nu \\
&\Downarrow \\
\alpha &= \frac{n\pi}{l} \\
&\text{and} \\
f(x,t) &= X(x)T(t) \\
&= e^{\phi}[C\sin(\alpha x)]e^{\zeta}[A\cos(\alpha vt) + B\sin(\alpha vt)] \\
&= e^{\phi}e^{\zeta}[C\sin(\alpha x)A\cos(\alpha vt) + C\sin(\alpha x)B\sin(\alpha vt)] \\
&= D\sin(\alpha x)\cos(\alpha vt) + E\sin(\alpha x)\sin(\alpha vt)
\end{aligned} \tag{7.17}$$

We use identities (A.11) and (A.12) to change this a bit:

$$\begin{aligned}
f(x,t) = \frac{D}{2}(\sin(\alpha x + \alpha vt) + \sin(\alpha x - \alpha vt)) \\
+ \frac{E}{2}(\cos(\alpha x - \alpha vt) - \cos(\alpha x + \alpha vt))
\end{aligned}$$

For a given x in eqn (7.17), $\sin(\alpha x)$ represents the amplitude of the function and is referred to as the *spatial amplitude* since it varies with spatial position. Since α is a function of n there are an infinite number of possible solutions that satisfy $f(x, t)$ and a more general way to write our solution is:

$$\begin{aligned}
f_n(x,t) = \sum_{n=1}^{\infty}\frac{D_n}{2}\left(\sin\left(\frac{n\pi}{l}x + \frac{n\pi}{l}vt\right) + \sin\left(\frac{n\pi}{l}x - \frac{n\pi}{l}vt\right)\right) \\
+ \sum_{n=1}^{\infty}\frac{E_n}{2}\left(\cos\left(\frac{n\pi}{l}x - \frac{n\pi}{l}vt\right) + \cos\left(\frac{n\pi}{l}x + \frac{n\pi}{l}vt\right)\right)
\end{aligned} \tag{7.18}$$

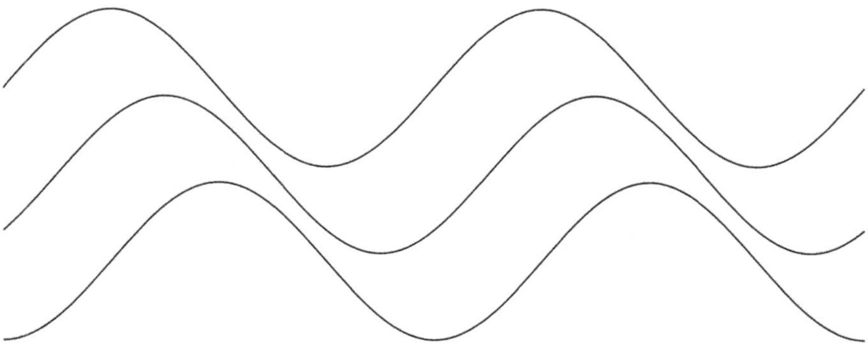

Figure 7.3 Right-travelling wave at increasing time intervals.

This solution to the wave equation is sometimes called the D'Alembert form.

Equation (7.18) represents the most general solution to the wave equation and also conveniently represents both a left-traveling wave and a right-traveling wave (see Figure 7.3). The left traveling portion is:

$$\sin(\alpha x + \alpha vt) \tag{7.19}$$

and the right-traveling portion is:

$$\sin(\alpha x - \alpha vt) \tag{7.20}$$

7.3 Superposition of Waves

That waves can be superimposed upon one another is something that we observe early in life. The ripples caused by a pair of pebbles thrown into a pond cross each other and superimpose themselves on each other as they travel through the water ... sometimes reinforcing each other and sometimes acting to cancel each other.

In this respect, eqn (7.18) is of interest. It says that the sum of the product of any or all time and position functions is a solution to the wave equation. In other words, a *superposition* of two or more waves is also a solution. This is a consequence of the linear nature of the differential eqn (7.9).

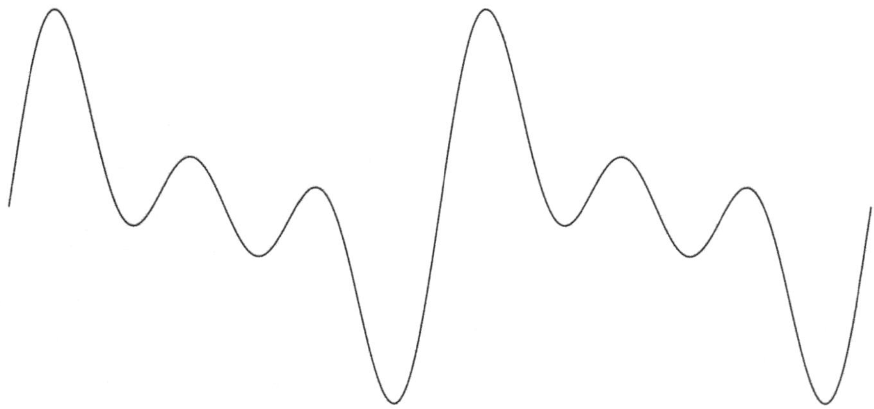

Figure 7.4 Superposition of waves at $\omega t = 0$.

Let us look at a superposition of our wave equations for $n = 1$, 2 and 3:

$$f(x,t) = F_1 \cos\left(\frac{\pi}{l}vt + \zeta\right)\sin\left(\frac{\pi x}{l}\right) + F_2 \cos\left(\frac{2\pi}{l}vt + \zeta\right)\sin\left(\frac{2\pi x}{l}\right) + F_3 \cos\left(\frac{3\pi}{l}vt + \zeta\right)\sin\left(\frac{3\pi x}{l}\right)$$

A plot of the position function portion of this is shown in Figure 7.4.

The important idea here is that the sums of individual solutions to the wave are also solutions in the same way that summing any two vectors produces a new vector in the same vector space.

7.4 Standing Waves

The derivation of the general wave equation above was shown to consist of two parts; a left-traveling wave and a right-traveling wave (eqn (7.19) and (7.20), respectively). Figure 7.3 shows the application of eqn (7.20) at increasing time intervals to illustrate the right-travelling wave.

What if we apply both eqn (7.19) and (7.20) at the *same* time? They are each solutions to the wave equation so their sum must also be a solution, as we have seen. In other words by applying both equations we construct a superposition of a left-traveling wave and a right-traveling wave.

$$\sin(\alpha x + \alpha vt) + \sin(\alpha x - \alpha vt)$$

By trigonometric identity (A.12):

$$\sin(\alpha x + \alpha vt) + \sin(\alpha x - \alpha vt) = 2\,\sin(\alpha x)\,\cos(\alpha vt) \tag{7.21}$$

This is, of course, part of our original wave equation solution, eqn (7.17) (without the constant). This form, as has been noted has the time and position functions separated, whereas the D'Alembert form has time and position integrated together in the function. We wish now to know what the time evolution of this set of waves will look like. What direction will the wave travel in?

The time evolution of this function is interesting. Consider first the sine term of eqn (7.21). Starting at $x = 0$ there will be points along the x-axis, where $\sin(\alpha x) = 0$. These are called *nodes* in wave jargon and are points along the axis that have zero amplitude . . . always. There is no up-down vibrational motion at these points. This answers the question about what direction the wave travels in . . . it does not travel in any direction but rather stays in place. If the wave were to move along the axis then these points would periodically have non-zero amplitudes. Hence the name *standing wave*.

In between the nodal points along the axis the cosine term modulates the function in time so that the amplitude here *is* periodic . . . up-down motion in other words.

Figure 7.5 shows the time evolution of our current function. The waves do not appear to move left or right but rather appear only to oscillate up and down at any given position. The superposition of the left and right-traveling waves results in the standing wave.

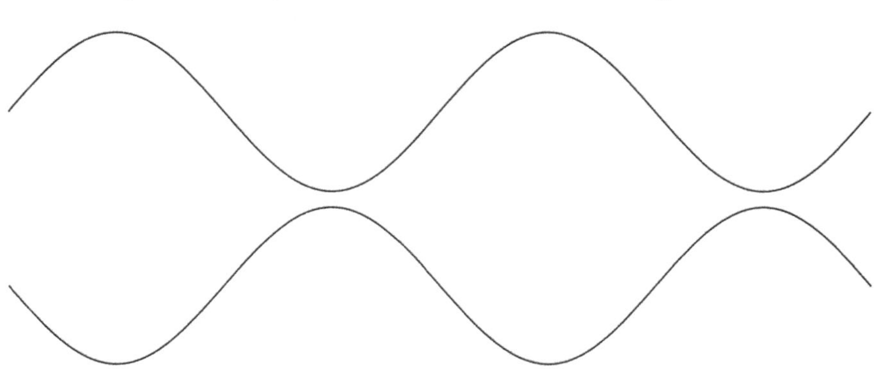

Figure 7.5 Superposition of waves at $\omega t = \pi$.

Further Reading

1. J. Billingham and A. C. King, *Wave Motion*, Cambridge University Press, Cambridge, 2000.
2. W. E. Boyce and R. C. DiPrima, *Elementary Differential Equations*, John Wiley & Sons, New York, 1992.
3. C. K. C. Lieou, *Eur. J. Phys.*, 2007, **28**, N17–N19.
4. D. A. McQuarrie, *Quantum Chemistry*, University Science Books, Sausalito, 2008.
5. A. D. Snider, *Fundamentals of Differential Equations and Boundary Value Problems*, Pearson Addison Wesley, New York, 2008.

8 The Fourier Transform

The whole is the sum of the parts.

Euclid

8.1 Fourier Series[†]

As NMR/MRI spectroscopists working with modern digital equipment we routinely acquire data in the form of the free induction decay. This is the digitised form of the analogue time domain data passing through the receiver. We wish, of course to look at these data in the frequency domain and also routinely transform the time domain data to frequency domain data for display and further analysis (Figure 8.1).

This transform of the data is accomplished using the ideas of Jean Baptiste Joseph Fourier (and others). Fourier led a very interesting life to say the least. He was a young man at the time of the French revolution[‡] and was very nearly beheaded during the reign of terror led by Robespierre. Eventually, he became acquainted with Napoleon and travelled with him and the French army to Egypt in 1798. Later, Napoleon made him prefect of Grenoble and during this time he met and mentored Jean-Francois Champollion, the person most

[†]Fourier analysis is a potential quagmire for the newcomer because of differing notations used in texts (mine included) and the sheer mass of material on the subject. The discrete Fourier transform seems especially unclear in many texts. I shall attempt here to boil it down to a readable quantity and some internal notational consistency.

[‡]"It was the best of times, it was the worst of times . . ." – Charles Dickens. Rather like the 1960s for some people.

Essential Mathematics for NMR and MRI Spectroscopists
By Keith C. Brown

Published by the Royal Society of Chemistry, www.rsc.org

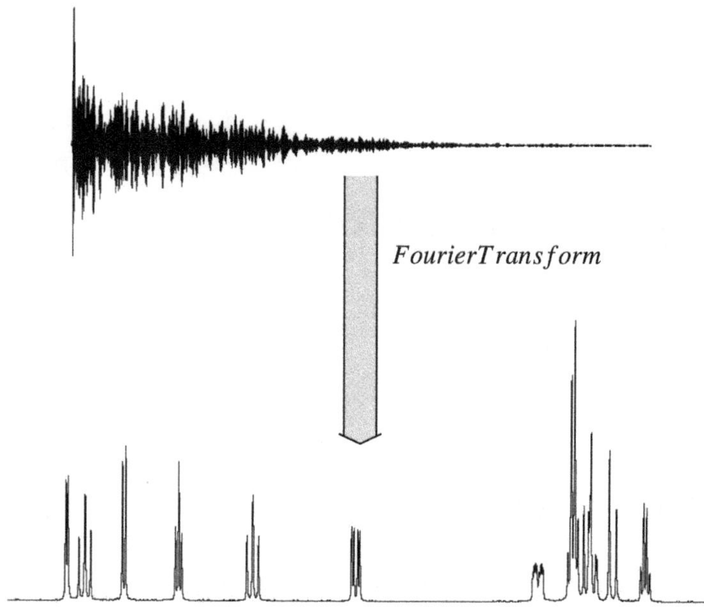

Figure 8.1 The Fourier transform of time-domain data to frequency-domain data.

responsible for the translation of the Rosetta stone,[§] which was found in Egypt by Napoleon's army. It was Fourier's studies of heat conduction that brought forward what we now call Fourier analysis and it is this work that he is most associated with.

Fourier's basic idea was that the application of infinite sums of sine and cosine functions multiplied by suitable constants can be used to represent any periodic function. To understand what is happening in our spectrometer software to accomplish this remarkable feat we must delve into the mysteries of the Fourier transform.

Our analysis of vectors showed us that we could use unit basis vectors multiplied by suitable factors to build up an overall vector that spans the vector space. This is particularly easy to visualise in two- or three-dimensional space. Mathematically, we represent this as:

$$\vec{\mathbf{A}} = \sum_i a_i \vec{\mathbf{u}}_i \tag{8.1}$$

Thus, the vector $\vec{\mathbf{A}}$ is the sum of unit basis vectors that forms the basis set from which we can build any vector. Moreover, as is obvious

[§]Currently on display at the British Museum in London, England.

in three-dimensional (and two-dimensional) space, these unit vectors, $\vec{\mathbf{u}}_i$, are orthogonal. That is, when we do an inner product of any two unit basis vectors the result will be zero:

$$\vec{\mathbf{u}}_i \cdot \vec{\mathbf{u}}_j = \|\vec{\mathbf{u}}_i\|\|\vec{\mathbf{u}}_j\|\cos\left(\frac{\pi}{2}\right) = 0 \tag{8.2}$$

This is generally true of multi-dimensional inner product spaces. The basis vectors are or can be made to be orthogonal if not already so, as was shown in Chapter 3. Of course, since these are unit vectors with length one, the inner product of a unit vector with itself is one:

$$\vec{\mathbf{u}}_i \cdot \vec{\mathbf{u}}_i = \|\vec{\mathbf{u}}_i\|\,\|\vec{\mathbf{u}}_i\|\ \cos(0) = 1 \tag{8.3}$$

Now, let us make a bit of a leap. Why not build up a function, *F*, from a set of basis functions in a way that is analogous¶ to vectors? If you look back at Chapter 3 you will see that, as long as each of the basic rules that define vectors is followed, this is reasonable.

$$\vec{\mathbf{F}} = \sum_i a_i \vec{\mathbf{f}} \tag{8.4}$$

This is what we do when using Taylor series to represent various functions:

$$\begin{aligned} e^x &= 1 + x + \frac{x^2}{2!} + \frac{x^3}{3!} + \frac{x^4}{4!} \cdots = \sum_{n=0}^{\infty} \frac{x^n}{n!} \\ \log(1+x) &= x - \frac{x^2}{2} + \frac{x^3}{3} - \frac{x^4}{4} \cdots = \sum_{n=1}^{\infty} (-1)^{n-1} \frac{x^n}{n} \end{aligned} \tag{8.5}$$

Taylor series are, however, not applicable to functions that contain discontinuities with finite magnitude. In other words, we could not model a pulse or "on-off" function using a Taylor series. We could, however, do so using a Fourier series. This type of series is excellent for modelling periodic functions that may or may not have discontinuities in them.

The Fourier series is applicable to a problem if it satisfies the Dirichlet conditions:

1. The function must be periodic. By this we mean that the function must be such that $f(t) = f(t+p)$ where p is a constant.

¶The group theory term is "isomorphous".

2. The function must be single valued and differentiable within one cycle. In other words, for each value of t there must be only one value for $f(t)$.
3. Within a cycle the function must have a finite number of maxima and minima.

As with vectors, we wish to be able to use a series of functions that are orthogonal to one another. This is the foundation upon which all of Fourier analysis is based. We can define such a set of functions using complex exponentials:

$$A = \ldots, \mathrm{e}^{-i3\omega_0 t}, \mathrm{e}^{-i2\omega_0 t}, \mathrm{e}^{-i\omega_0 t}, 1, \mathrm{e}^{-i\omega_0 t}, \mathrm{e}^{-i2\omega_0 t}, \mathrm{e}^{-i3\omega_0 t} \ldots \tag{8.6}$$

where we define ω_0 to be the fundamental radial frequency of the periodic function. To demonstrate that these functions are in fact orthogonal we do an integration over one period, $\tau = 2\pi/\omega_0$, of the function:

$$\frac{1}{\tau}\int_0^{\tau} (\mathrm{e}^{in\omega_0 t})^* \mathrm{e}^{im\omega_0 t} \mathrm{d}t \tag{8.7}$$

where we are using the complex conjugate of one of the functions. This is analogous to what is done with vectors when written in Dirac notation. It will become apparent why we multiply by $1/\tau$ momentarily.

Where does $\tau = 2\pi/\omega_0$ come from? We see from Figure 8.2 that, starting at time 0, one complete cycle or period of the function is finished at time τ. We can specify the frequency of this periodic function in two ways by using hertz (or cycles per second in the older

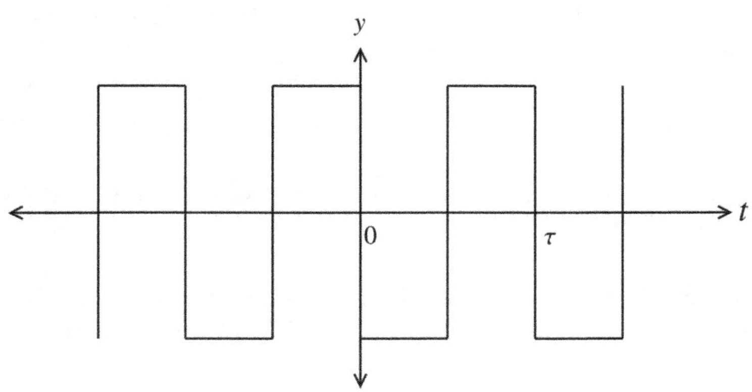

Figure 8.2 A square wave function.

nomenclature) or by using radians per second. The two are, of course, related. One cycle per second is 2π radians per second. So, to convert from hertz to radians per second, we use, from Chapter 7:

$$\omega = 2\pi\nu \tag{8.8}$$

Now that we know what the frequency is in radians per second we ask ourselves, "how long is one complete period?". It is very simply the inverse of the frequency:

$$\tau = \frac{1}{\nu} = \frac{2\pi}{\omega_0}$$

Our integral, eqn (8.7), becomes:

$$\frac{1}{\tau}\int_0^{\tau} e^{-in\omega_0 t} e^{im\omega_0 t} dt = \frac{1}{\tau}\int_0^{\tau} e^{i(m-n)\omega_0 t} dt \tag{8.9}$$

We have two possible situations now. If $m = n$ then the integral evaluates to 1:

$$\frac{1}{\tau}\int_0^{\tau} dt = \frac{t}{\tau}\bigg|_0^{\tau} = \frac{\tau}{\tau} = 1 \tag{8.10}$$

Thus, the functions are normalised.

If m is not equal to n then:

$$\frac{1}{\tau}\int_0^{\tau} e^{i(m-n)\omega_0 t} dt = \frac{e^{i\Delta\omega_0 t}}{\tau i\Delta\omega_0}\bigg|_0^{\tau} = \frac{e^{i\Delta\omega_0 t}}{2\pi i\Delta}\bigg|_0^{\tau} \tag{8.11}$$

where $\Delta = m - n$. Finishing the evaluation of the definite integral:

$$\frac{e^{i\Delta\omega_0 t}}{2\pi i\Delta}\bigg|_0^{\tau} = \frac{e^{i\Delta 2\pi} - 1}{2\pi i\Delta} = \frac{1-1}{2\pi i\Delta} = 0 \tag{8.12}$$

The exponential functions are also orthogonal. That $e^{i\Delta 2\pi}$ is equal to 1 is evident when you consider that using integral multiples of $2/\pi$ in e^{ix} must result in the cosine portion being equal to 1 and the sine portion being equal to zero:

$$e^{in2\pi} = \cos(n2\pi) + i\,\sin(n2\pi)$$

where n is any integer. You can see now why it is that the $1/\tau$ factor is needed; it is to scale the integral so that orthonormality is satisfied.

We now have a set of orthonormal functions. What do we do with them? We use them to model other functions, in a similar fashion to that of Taylor series. Since Fourier series must be periodic under one of the Dirichlet conditions, we write:

$$f(t) = \sum_{-\infty}^{\infty} C(n)\mathrm{e}^{in\omega_0 t} \tag{8.13}$$

where $f(t)$ is the function to be modelled and $C(n)$ is a constant. This is called the *synthesis* equation since it is used to synthesise or build up the function, $f(t)$.

The central problem is to assign values to all of the $C(n)$ constants. This is accomplished by multiplication of both sides of the series equation by $\mathrm{e}^{-im\omega_0 t}$ and integrating over one period:

$$\frac{1}{\tau}\int_0^{\tau} f(t)\mathrm{e}^{-im\omega_0 t}dt = \frac{1}{\tau}\int_0^{\tau} \sum_{n=-\infty}^{\infty} C(n)\mathrm{e}^{in\omega_0 t}\mathrm{e}^{-im\omega_0 t}\mathrm{d}t \tag{8.14}$$

We now interchange the integral and summation on the right side to get:

$$\frac{1}{\tau}\int_0^{\tau} f(t)\mathrm{e}^{-im\omega_0 t}\mathrm{d}t = \sum_{n=-\infty}^{\infty} C(n)\frac{1}{\tau}\int_0^{\tau} \mathrm{e}^{in\omega_0 t}\mathrm{e}^{-im\omega_0 t}\mathrm{d}t \tag{8.15}$$

There was, by the way, some considerable confusion over the validity of this step resulting in many mathematicians pondering it for many years. In fact, Fourier's original manuscript was submitted in 1806 but not published until 1822 primarily because of this!

Recalling our recent discussion of the orthonormal nature of these functions, the right side of the equation vanishes for $m \neq n$ leaving only the term with $m = n$. Of course, when $m = n$, the integral evaluates to 1, leaving one term:

$$C(m) = \frac{1}{\tau}\int_0^{\tau} f(t)\mathrm{e}^{-im\omega_0 t}\mathrm{d}t \tag{8.16}$$

This is the required evaluation of the terms in the Fourier series and is called the *analysis* equation. Note that it is necessary only to integrate over one complete period ... the actual starting and ending points do not matter as long as it is over one complete period. We could just as well write:

$$C(m) = \frac{1}{\tau}\int_{-\frac{\tau}{2}}^{\frac{\tau}{2}} f(t)\mathrm{e}^{-im\omega_0 t}\mathrm{d}t \tag{8.17}$$

An alternate way of writing the synthesis eqn (8.13) is:

$$f(t) = a_0 + \sum_{n=1}^{\infty} a_n \cos(n\omega_0 t) + \sum_{n=1}^{\infty} b_n \sin(n\omega_0 t) \tag{8.18}$$

and is sometimes called the "old Fourier representation", possibly because this was the form that Gabriel Fourier originally used in his publication. Equation (8.18) can be easily shown to be equivalent to eqn (8.13):

$$
\begin{aligned}
f(t) &= a_0 + \sum_{n=1}^{\infty} a_n \cos(n\omega_0 t) + \sum_{n=1}^{\infty} b_n \sin(n\omega_0 t) \\
&= a_0 + \sum_{n=-\infty}^{\infty} a_n \frac{(\mathrm{e}^{in\omega_0 t} + \mathrm{e}^{-in\omega_0 t})}{2} + \sum_{n=-\infty}^{\infty} a_n \frac{(\mathrm{e}^{in\omega_0 t} - \mathrm{e}^{-in\omega_0 t})}{2} \\
&= a_0 + \sum_{n=-\infty}^{\infty} \frac{(a_n - ib_n)}{2} \mathrm{e}^{in\omega_0 t} + \sum_{n=-\infty}^{\infty} \frac{(a_n + ib_n)}{2} \mathrm{e}^{-in\omega_0 t} \\
&= a_0 + \sum_{n=-\infty}^{\infty} \frac{(a_n - ib_n)}{2} \mathrm{e}^{in\omega_0 t} + \sum_{n=\infty}^{-\infty} \frac{(a_n + ib_n)}{2} \mathrm{e}^{in\omega_0 t} \\
&= \sum_{n=-\infty}^{\infty} C(n)\mathrm{e}^{in\omega_0 t} \quad \left\{ \begin{array}{ll} C(n) = \dfrac{(a_n - ib_n)}{2} & \text{if } n > 0 \\ C(n) = \dfrac{(a_n + ib_n)}{2} & \text{if } n < 0 \\ C(0) = a_0 & \text{if } n = 0 \end{array} \right\}
\end{aligned}
$$

where we have used eqn (A.22) and (A.23) and the relationship $1/i = -i$ from Chapter 1 in steps 2 and 3, respectively. Note also that in step 4 the second summation's range has changed from $n = -\infty$ to ∞ to become $n = +\infty$ to $-\infty$ and that the sign on the exponential has been changed from negative to positive. This is so that we may gather terms into the final compact equation.

Sometimes we speak of harmonics, especially with respect to the analysis of periodic functions involving sound waves. From the Fourier series point of view the harmonics are:

$$
\begin{aligned}
&h(0) = C(0) \quad \text{(average value)} \\
&h(1) = C(1)\mathrm{e}^{i\omega_0 t} + C(-1)\mathrm{e}^{-i\omega_0 t} \quad \text{(fundamental)} \\
&h(2) = C(2)\mathrm{e}^{i2\omega_0 t} + C(-2)\mathrm{e}^{-i2\omega_0 t} \quad \text{(1st harmonic)} \\
&\vdots \\
&h(n) = C(n)\mathrm{e}^{in\omega_0 t} + C(-n)\mathrm{e}^{-in\omega_0 t} \quad (n^{th} \text{ harmonic})
\end{aligned}
\tag{8.19}
$$

8.1.1 Example: The Sawtooth Function

Given the periodic function,

$$\begin{array}{cc} & \vdots \\ f(t-\tau)=f(t) & -\tau \le t \le 0 \\ f(t)=t & 0 \le t \le \tau \\ f(t+\tau)=f(t) & \tau \le t \le 2\tau \\ & \vdots \end{array}$$

we wish to find the Fourier series representation for $f(t)$. This is equivalent to finding the Fourier coefficients, $C(n)$, using eqn (8.16). Note that since we are dealing with a periodic function, its definition need only be specified over one period. Our current function represents a "sawtooth" function with a dc offset such that the amplitude of the function never dips below zero (Figure 8.3).

We begin with the analysis equation:

$$\begin{aligned} C(n) &= \frac{1}{\tau}\int_0^{\tau} f(t)\mathrm{e}^{-in\omega_0 t}\mathrm{d}t \ (n \neq 0) \\ &= \frac{\omega_0}{2\pi}\int_0^{\frac{2\pi}{\omega_0}} t\mathrm{e}^{-in\omega_0 t}\mathrm{d}t \\ &= \frac{\omega_0}{2\pi}\int_0^{\frac{2\pi}{\omega_0}} t[\cos(n\omega_0 t) + i\sin(n\omega_0 t)]\mathrm{d}t \\ &= \frac{\omega_0}{2\pi}\int_0^{\frac{2\pi}{\omega_0}} t\cos(n\omega_0 t)\mathrm{d}t + i\frac{\omega_0}{2\pi}\int_0^{\frac{2\pi}{\omega_0}} t\sin(n\omega_0 t)\mathrm{d}t \end{aligned}$$

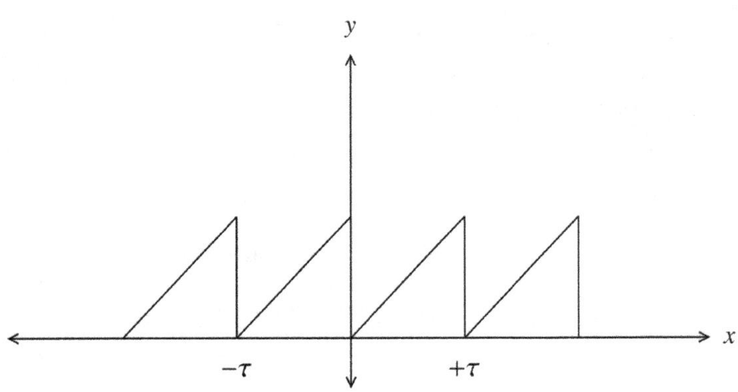

Figure 8.3 A saw-tooth wave.

The integrals are solved by integration by parts (see Appendix A) to give:

$$\frac{\omega_0}{2\pi}\left[\frac{t\sin(n\omega_0 t)}{n\omega_0}-\frac{\cos(n\omega_0 t)}{n\omega_0}\right]-i\frac{\omega_0}{2\pi}\left[\frac{t\cos(n\omega_0 t)}{n\omega_0}-\frac{\sin(n\omega_0 t)}{n\omega_0}\right]\Bigg|_0^{\frac{2\pi}{\omega_0}}$$

The first term evaluates to zero after insertion of the integration limits. The second evaluates to

$$\frac{-i}{n\omega_0} \quad \text{for} \quad n \neq 0$$

For $n=0$ the analysis equation is:

$$\begin{aligned}
C(n) &= \frac{1}{\tau}\int_0^{\tau} f(t)\mathrm{e}^{-in\omega_0 t}\mathrm{d}t \ (n=0) \\
&= \frac{\omega_0}{2\pi}\int_0^{\frac{2\pi}{\omega_0}} t\mathrm{e}^{-i0\omega_0 t}\mathrm{d}t \\
&= \frac{\omega_0}{2\pi}\int_0^{\frac{2\pi}{\omega_0}} t\mathrm{d}t \\
&= \frac{\omega_0}{2\pi}\frac{t^2}{2}\Bigg|_0^{\frac{2\pi}{\omega_0}} \\
&= \frac{\pi}{2}
\end{aligned}$$

Thus, the Fourier series that models this periodic function is:

$$\begin{aligned}
A &= \frac{\pi}{2}+\sum_{n=-\infty}^{\infty}\frac{-i}{n\omega_0}\mathrm{e}^{-in\omega_0 t} \ (n \neq 0) \\
&= \frac{\pi}{2}+\sum_{n=-\infty}^{\infty}\frac{-i}{n\omega_0}[\cos(n\omega_0 t)-i\sin(n\omega_0 t)]
\end{aligned}$$

8.2 Properties of the Time-Domain Functions

We can say some things about the Fourier coefficients, $C(n)$. First, the complex conjugate of the nth coefficient is equal to the nth coefficient (assuming $f(t)$ is real ... and for practical purposes it generally is):

$$\begin{aligned} C(n)^* &= \left[\frac{1}{\tau}\int_0^\tau f(t)\mathrm{e}^{-in\omega_0 t}\mathrm{d}t\right]^* \\ &= \frac{1}{\tau}\int_0^\tau f(t)^*\mathrm{e}^{in\omega_0 t}\mathrm{d}t \\ &= \frac{1}{\tau}\int_0^\tau f(t)\mathrm{e}^{in\omega_0 t}\mathrm{d}t \quad (f(t)=f(t)^*) \\ &= C(-n) \end{aligned} \tag{8.20}$$

We can break the coefficient, $C(n)$, into real and imaginary coefficients:

$$C(n) = a(n) + ib(n) \tag{8.21}$$

The complex conjugate of this is:

$$C(n)^* = a(n) - ib(n)$$

Using eqn (A.60) we can play with this:

$$C(n)^* = C(-n)$$

$$a(n) - ib(n) = a(-n) + ib(-n)$$

The real parts are equivalent, as are the imaginary parts:

$$a(n) = a(-n)$$

$$b(n) = -b(-n)$$

Thus, $a(n)$ is even and $b(n)$ odd (see Appendix A, eqn (A.53) and (A.54)). Let us look again at eqn (8.18):

$$f(t) = a_0 + \sum_{n=1}^{\infty} a_n\cos(n\omega_0 t) + \sum_{n=1}^{\infty} b_n\sin(n\omega_0 t)$$

and for simplicity, we will assume that $a_0 = 0$ and that all b_n s are zero as well:

$$f(t) = \sum_{n=1}^{\infty} a_n\cos(n\omega_0 t)$$

This says that the time-domain function is a composite of cosine functions in which the frequency arguments are integer multiples of a fundamental angular frequency, ω_0. This being the case, we can proceed to say:‖

$$\omega_n = n\omega_0 \Rightarrow \omega_0 = \frac{\omega_n}{n}$$

$$\tau_n = \frac{2\pi}{\omega_n}$$

where, as usual, ω represents an angular frequency and τ represents the period of oscillation. These two observations allow us to write:

$$\tau_1 = \frac{2\pi}{\omega_0} = \frac{n2\pi}{\omega_n} = n\tau_n$$

This is very interesting and not really surprising if one looks at plots of cosine waves that are integer multiples of a fundamental frequency.** Basically, it says that the period of the fundamental angular frequency is an integer multiple of each of the upper harmonic periods. It also means that:

$$f(t) = f(t+\tau_1)$$

It is easy to show that this is so:

$$f(t) = \sum_{n=1}^{\infty} a_n \cos(n\omega_0 t)$$

$$\begin{aligned} f(t+\tau_1) &= \sum_{n=1}^{\infty} a_n \cos(n\omega_0[t+\tau_1]) \\ &= \sum_{n=1}^{\infty} a_n \cos\left(n\omega_0\left[t+\frac{2\pi}{\omega_0}\right]\right) \\ &= \sum_{n=1}^{\infty} a_n \cos(n\omega_0 t + n2\pi) \\ &= \sum_{n=1}^{\infty} a_n \cos(n\omega_0 t) \\ &= f(t) \end{aligned}$$

‖Note that this says that $\omega_1 = \omega_0$. This is a consequence of the fundamental frequency being in the $n=1$ term and the fact that we have defined the fundamental frequency to be ω_0. What follows is confusing unless this is kept in mind.

**Now that very powerful computers are available to almost everyone it is easy to use symbolic algebra programs to solve equations and for visualisation of functions. There are many to choose from, both proprietary and free. I have used the freely available wxMaxima extensively for these purposes and recommend it to you.

This will also be the case for the more general time-domain function given by eqn (8.18).

8.2.1 Example: The Square Wave Function

For the function:

$$f(t)=1 \quad (0 \le t \le 1)$$
$$f(t)=-1 \quad (1 \le t \le 2)$$
$$f(t+2)=f(t)$$

which is defined over one period with $\tau=2$, $\omega_0=\pi$ (see Figure 8.4), we wish to calculate the corresponding Fourier series terms.

Visually, you can see that, as it is plotted, this is an odd function. The analysis expression is:

$$C(n)=\frac{1}{\tau}\int_0^1 f(t)\mathrm{e}^{-in\omega_0 t}\mathrm{d}t+\frac{1}{\tau}\int_1^2 f(t)\mathrm{e}^{-in\omega_0 t}\mathrm{d}t\ (n\neq 0)$$

$$=\frac{1}{2}\int_0^1 \mathrm{e}^{-in\pi t}\mathrm{d}t+\frac{1}{2}\int_1^2 -\mathrm{e}^{-in\pi t}\mathrm{d}t$$

$$=-\frac{1}{2}\frac{\mathrm{e}^{-in\pi t}}{in\pi}\bigg|_0^1+\frac{1}{2}\frac{\mathrm{e}^{-in\pi t}}{in\pi}\bigg|_1^2$$

$$=-\frac{1}{2}\frac{(\mathrm{e}^{-in\pi}-\mathrm{e}^0)}{in\pi}+\frac{1}{2}\frac{(\mathrm{e}^{-in2\pi}-\mathrm{e}^{-in\pi})}{in\pi}$$

$$=-\frac{1}{2}\frac{[\cos(n\pi)-i\sin(n\pi)-1]}{in\pi}+\frac{1}{2}\frac{[\cos(n2\pi)-i\sin(n2\pi)-\cos(n\pi)+i\sin(n\pi)]}{in\pi}$$

$$=-\frac{1}{2}\frac{\cos(n\pi)-1}{in\pi}+\frac{1}{2}\frac{[1-\cos(n\pi)]}{in\pi}$$

$$=-\frac{1}{2}\frac{\cos(n\pi)-1}{in\pi}-\frac{1}{2}\frac{\cos(n\pi)-1}{in\pi}$$

$$=-\frac{1}{2}\frac{[2\cos(n\pi)-2]}{in\pi}$$

$$=\frac{1-\cos(n\pi)}{in\pi}$$

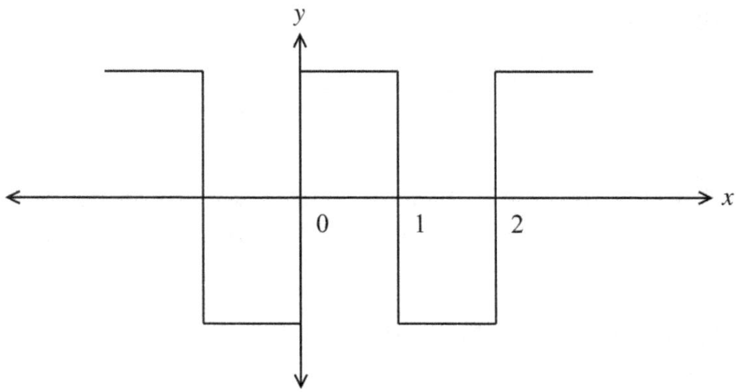

Figure 8.4 A square wave function.

Note that $\cos(n2\pi)$ will equal +1 for all values of n and that $\sin(n\pi)$ will equal zero for all values of n. For sequential values of n the cosine term will alternate between +1 and −1:

$$C(n) = \frac{1-(-1)^n}{in\pi}$$

These are the Fourier constants for n not equal to 0. For $n = 0$:

$$\begin{aligned}
C(0) &= \frac{1}{\tau}\int_0^1 f(t)\mathrm{e}^{-i0\omega_0 t}\mathrm{d}t + \frac{1}{\tau}\int_1^2 f(t)\mathrm{e}^{-i0\omega_0 t}\mathrm{d}t \\
&= \frac{1}{2}\int_0^1 1\cdot \mathrm{d}t + \frac{1}{2}\int_1^2 -1\cdot \mathrm{d}t \ (n=0) \\
&= \frac{1}{2}t\Big|_0^1 - \frac{1}{2}t\Big|_1^2 \\
&= \frac{1}{2} - 0 - 1 + \frac{1}{2} \\
&= 0
\end{aligned}$$

The Fourier series is now obtained using the synthesis equation:

$$\begin{aligned}
f(t) &= \sum_{n=-\infty}^{\infty} C(n)\mathrm{e}^{in\omega_0 t} \\
&= \sum_{n=-\infty}^{\infty} \frac{1-(-1)^n}{in\pi}\mathrm{e}^{in\omega_0 t} \\
&= \sum_{n=-\infty}^{\infty} \frac{1-(-1)^n}{in\pi}[\cos(n\pi t) + i\sin(n\pi t)]
\end{aligned}$$

For any value of n, the sine term must vanish:

$$f(t) = \sum_{n=-\infty}^{\infty} \frac{1-(-1)^n}{in\pi}\cos(n\pi t)$$

The $C(n)$ term is an odd function and cosine is an even function. Thus, the overall function, $f(t)$ is odd, as expected from eqn (A.63).

8.3 The Fourier Integral

Recall the synthesis formula, eqn (8.13):

$$f(t) = \sum_{-\infty}^{\infty} C(n)e^{in\omega_0 t}$$

This represents an infinite series of sine and cosine harmonics (see eqn (8.19)), each having some multiple of a fundamental frequency ω_0. Thus, the frequencies of the harmonics will start at ω_0 and proceed through discrete intervals ω_0, $2\omega_0$, $3\omega_0$, $4\omega_0$ ... $n\omega_0$ and the difference in frequency between each harmonic will be just $\Delta\omega = \omega_0 = 2\pi/\tau$. Thus, as τ increases $\Delta\omega$ decreases, which, to the mathematically inclined, suggests that letting t go to infinity will result in $\Delta\omega$ becoming $\delta\omega$ and rather than having discrete frequencies, all frequencies would be involved.

Next, recall the analysis eqn (8.17):

$$\begin{aligned} C(m) &= \frac{1}{\tau}\int_{-\frac{\tau}{2}}^{\frac{\tau}{2}} f(t)e^{-im\omega_0 t}dt \\ &= \frac{\omega_0}{2\pi}\int_{\frac{-\pi}{\omega_0}}^{\frac{\pi}{\omega_0}} f(t)e^{-im\omega_0 t}dt \end{aligned}$$

Let us insert this into the synthesis equation:

$$f(t) = \sum_{-\infty}^{\infty}\left(\frac{\omega_0}{2\pi}\int_{\frac{-\pi}{\omega_0}}^{\frac{\pi}{\omega_0}} f(t)e^{-in\omega_0 t}dt\right)e^{in\omega_0 t}$$

If we let t approach infinity, as discussed above, and replace the discrete frequencies, $n\omega_0$ by ω and ω_0 (or $\Delta\omega$, see above) by $d\omega$, the summation becomes an integral over ± infinity of Fourier coefficients at all frequencies:

$$f(t) = \int_{-\infty}^{\infty} \left(\frac{1}{2\pi} \int_{-\infty}^{\infty} f(t)\mathrm{e}^{-i\omega t} \right) \mathrm{e}^{i\omega t} \mathrm{d}\omega$$
$$= \frac{1}{2\pi} \int_{-\infty}^{\infty} \left(\int_{-\infty}^{\infty} f(t)\mathrm{e}^{-i\omega t} \right) \mathrm{e}^{i\omega t} \mathrm{d}\omega$$

The inner integral within the brackets, being integrated over $\mathrm{d}\omega$ by the outer integral, must be a function of ω, $F(\omega)$, and we therefore write:[††]

$$f(t) = \frac{1}{2\pi} \int_{-\infty}^{\infty} F(\omega)\mathrm{e}^{i\omega t} \mathrm{d}\omega \tag{8.22}$$

where:

$$F(\omega) = \int_{-\infty}^{\infty} f(t)\mathrm{e}^{-i\omega t} \mathrm{d}t \tag{8.23}$$

Equation (8.22) is the integral version of the synthesis equation, derived earlier, and is usually called the inverse Fourier transform integral. Equation (8.23) is the integral version of eqn (8.17), the analysis equation and is called the Fourier transform integral. Note that in the derivation we moved the $1/2\pi$ factor from the inner integral to the outer integral. If we had not done this then the $1/2\pi$ factor would have appeared in (8.23) instead. One sees these integrals written either way in the literature.[‡‡] It is even possible to see the factor split between the analysis and synthesis integrals and written as $1/\sqrt{2\pi}$ in each. Since we will be using eqn (8.23) extensively, we choose not to have to worry about this factor will use the transform integrals as written above.

[††]The conventional way of writing things is to use a lower case letter for the time-domain function and an upper case letter for the frequency-domain function. I will go with the flow and use this convention for the rest of the chapter.

[‡‡]Take care to watch for this or confusion may result. The results here may differ by a factor of $1/2\pi$ when comparing with another publication.

Since one transforms a time-domain function into a frequency-domain function and *vice versa via* these integrals they are often written in short form as *Fourier pairs*:

$$\begin{aligned} F(\omega) &\leftrightarrow 2\pi f(t) \\ f(t) &\leftrightarrow F(\omega) \end{aligned} \tag{8.24}$$

Corresponding to eqn (8.22) and (8.23), respectively.

8.4 Properties of Fourier Transforms

8.4.1 Linearity

The Fourier transform is a linear operation. So, for Fourier transform pairs:

$$\begin{aligned} x(t) &\leftrightarrow X(\omega) \\ y(t) &\leftrightarrow Y(\omega) \end{aligned}$$

we can write:

$$ax(t) + by(t) \leftrightarrow aX(\omega) + bY(\omega) \tag{8.25}$$

This is easily shown. We simply do the Fourier transform of the left side of eqn (8.25):

$$\begin{aligned} &\int_{-\infty}^{\infty} [ax(t) + by(t)]\, \mathrm{e}^{-i\omega t} \mathrm{d}t \\ &= \int_{-\infty}^{\infty} ax(t)\mathrm{e}^{-i\omega t} dt + \int_{-\infty}^{\infty} by(t)\mathrm{e}^{-i\omega t} \mathrm{d}t \\ &= a\int_{-\infty}^{\infty} x(t)\mathrm{e}^{-i\omega t} dt + b\int_{-\infty}^{\infty} y(t)\mathrm{e}^{-i\omega t} \mathrm{d}t \\ &= aX(\omega) + bY(\omega) \end{aligned}$$

8.4.2 Time Shift

Starting with the Fourier pair:

$$x(t) \leftrightarrow X(\omega)$$

we add (or subtract) a constant to t in $x(t)$ and then do the Fourier transform:

$$\int_{-\infty}^{\infty} x(t+k)\mathrm{e}^{-i\omega t}\mathrm{d}t$$

Transforming variables:

$$\bar{t} = t + k$$

$$t = \bar{t} - k$$

$$\int_{-\infty}^{\infty} x(\bar{t})\mathrm{e}^{-i\omega(\bar{t}-k)}\mathrm{d}\bar{t}$$

$$= \left[\int_{-\infty}^{\infty} x(\bar{t})\mathrm{e}^{-i\omega\bar{t}}\mathrm{d}\bar{t}\right]\mathrm{e}^{i\omega k}$$

$$= X(\omega)\mathrm{e}^{i\omega k}$$

Thus, shifting the time-domain function left or right corresponds to a phase shift in the frequency spectrum and our Fourier pair is:

$$x(t \pm k) \leftrightarrow X(\omega)\mathrm{e}^{\pm i\omega k} \tag{8.26}$$

8.4.3 Frequency Shift

What happens if we multiply $x(t)$ by a complex exponential?

$$x(t)\mathrm{e}^{i\omega_x t}$$

The Fourier transform of this is:

$$\int_{-\infty}^{\infty} x(t)\mathrm{e}^{i\omega_x t}\mathrm{e}^{-i\omega t}\mathrm{d}t$$

$$= \int_{-\infty}^{\infty} x(t)\mathrm{e}^{-i(\omega - w_x)t}\mathrm{d}t$$

$$= X(\omega - \omega_x)$$

Thus, our Fourier pair is:

$$x(t)\mathrm{e}^{\pm i\omega_x t} \leftrightarrow X\,(\omega \mp \omega_x) \tag{8.27}$$

which corresponds to a frequency shift in $X(\omega)$ when $x(t)$ is multiplied by a complex exponential. This is sometimes called the *modulation property*.

We can add to this using the sine and cosine functions. Again, we assume a Fourier pair:

$$x(t) \leftrightarrow X(\omega)$$

$$\begin{aligned}
y(t) &= x(t)\cos(\omega_x t) \\
&\Downarrow \\
Y(\omega) &= \int_{-\infty}^{\infty} y(t)\mathrm{e}^{-i\omega t}\mathrm{d}t \\
&= \int_{-\infty}^{\infty} x(t)\cos(\omega_x t)\mathrm{e}^{-i\omega t}\mathrm{d}t \\
&= \int_{-\infty}^{\infty} x(t)\frac{1}{2}(\mathrm{e}^{i\omega_x t} + \mathrm{e}^{-i\omega_x t})\mathrm{e}^{-i\omega t}\mathrm{d}t \\
&= \frac{1}{2}\int_{-\infty}^{\infty} x(t)\mathrm{e}^{-i(\omega-\omega_x)t}\mathrm{d}t + \frac{1}{2}\int_{-\infty}^{\infty} x(t)\mathrm{e}^{-i(\omega+\omega_x)t}\mathrm{d}t \\
&= \frac{1}{2}[X(\omega-\omega_x) + X(\omega+\omega_x)]
\end{aligned} \tag{8.28}$$

and for sine:

$$\begin{aligned}
y(t) &= x(t)\sin(\omega_x t) \\
&\Downarrow \\
Y(\omega) &= \int_{-\infty}^{\infty} y(t)\mathrm{e}^{-i\omega t}\mathrm{d}t \\
&= \int_{-\infty}^{\infty} x(t)\sin(\omega_x t)\mathrm{e}^{-i\omega t}\mathrm{d}t \\
&= \int_{-\infty}^{\infty} x(t)\frac{1}{2}(\mathrm{e}^{i\omega_x t} - \mathrm{e}^{-i\omega_x t})\mathrm{e}^{-i\omega t}\mathrm{d}t \\
&= \frac{1}{2}\int_{-\infty}^{\infty} x(t)\mathrm{e}^{-i(\omega-w_x)t}\mathrm{d}t - \frac{1}{2}\int_{-\infty}^{\infty} x(t)\mathrm{e}^{-i(\omega+\omega_x)t}\mathrm{d}t \\
&= \frac{1}{2}[X(\omega-\omega_x) - X(\omega+\omega_x)]
\end{aligned} \tag{8.29}$$

8.4.4 Duality

The duality property of Fourier transform integrals relies on the swapping of variables between the analysis and synthesis integrals. Starting with the inverse Fourier transform integral, eqn (8.22):

$$f(t)=\frac{1}{2\pi}\int_{-\infty}^{\infty}F(\omega)e^{i\omega t}dt$$

we swap t and ω:

$$f(\omega)=\frac{1}{2\pi}\int_{-\infty}^{\infty}F(t)e^{i\omega t}dt$$

This has the appearance of a Fourier transform integral ... almost. We must change the sign of ω in order that the sign on the exponential is correct:

$$f(-\omega)=\frac{1}{2\pi}\int_{-\infty}^{\infty}F(t)e^{-i\omega t}dt$$

or

$$2\pi f(-\omega)=\int_{-\infty}^{\infty}F(t)e^{-i\omega t}dt$$

In Fourier pair form, this is:

$$F(t)\leftrightarrow 2\pi f(-\omega)$$

8.5 Some Important Fourier Pairs

8.5.1 The Rectangular Pulse

We define a rectangular function:

$$\mathrm{Rect}\left(\frac{t}{\tau}\right)\equiv\begin{Bmatrix}1 \text{ if } |t|<\frac{\tau}{2}\\ 0 \text{ if } |t|>\frac{\tau}{2}\end{Bmatrix} \tag{8.30}$$

Then, using the Rect function:

$$F(\omega)=\int_{-\infty}^{\infty}f(t)e^{-i\omega t}dt=\int_{-\frac{\tau}{2}}^{\frac{\tau}{2}}e^{-i\omega t}dt$$

$$=\int_{-\frac{\tau}{2}}^{\frac{\tau}{2}}[\cos(\omega t)-i\sin(\omega t)]dt$$

$$=\int_{-\frac{\tau}{2}}^{\frac{\tau}{2}}\cos(\omega t)dt-i\int_{-\frac{\tau}{2}}^{\frac{\tau}{2}}\sin(\omega t)dt$$

Using eqn (A.56) and observing that cosine is an even function and sine is an odd function, this becomes:

$$F(\omega) = 2\int_0^{\frac{\tau}{2}} \cos(\omega t)\mathrm{d}t = 2\frac{\sin(\omega t)}{\omega}\Big|_0^{\frac{\tau}{2}}$$

$$= 2\frac{\sin\left(\omega\frac{\tau}{2}\right)}{\omega} = 2\frac{\sin\left(\omega\frac{\tau}{2}\right)}{\omega\times\frac{2\tau}{2\tau}}$$

$$= \tau\frac{\sin\left(\omega\frac{\tau}{2}\right)}{\omega\frac{\tau}{2}} = \tau\,\mathrm{Sa}\left(\omega\frac{\tau}{2}\right)$$

The "Sa" function (pronounced "sah") is not quite the same as the sinc function:[§§]

$$\mathrm{Sa}(x) = \frac{\sin(x)}{x}$$

$$\mathrm{sinc}(x) = \frac{\sin(\pi x)}{\pi x}$$

Sa is sometimes referred to as the unnormalised sinc function and the sinc function (as written here) as the normalised sinc function. Figure 8.5 shows a plot of the function. The oscillations to either side of the main peak are often called "sinc squiggles".

$$\mathrm{Rect}\left(\frac{t}{\tau}\right) \leftrightarrow \tau\,\mathrm{Sa}\left(\frac{\omega\tau}{2}\right) \tag{8.31}$$

or in longer form:

$$\mathrm{Rect}\left(\frac{t}{\tau}\right) = \frac{1}{2\pi}\int_{-\infty}^{\infty} \tau\,\mathrm{Sa}\left(\frac{\omega\tau}{2}\right)\mathrm{e}^{i\omega t}\mathrm{d}\omega$$

and

$$\tau\,\mathrm{Sa}\left(\frac{\omega\tau}{2}\right) = \int_{-\infty}^{\infty} \mathrm{Rect}\left(\frac{t}{\tau}\right)\mathrm{e}^{-i\omega t}\mathrm{d}t$$

from eqn (8.22) and (8.23), respectively.

What happens if we start with $F(\omega)$ and want $f(t)$? Specifically, if our frequency function is:

$$F(\omega) = \mathrm{Rect}\left(\frac{\omega}{\tau}\right)$$

[§§]Sinc is the shortened version of "sine cardinal".

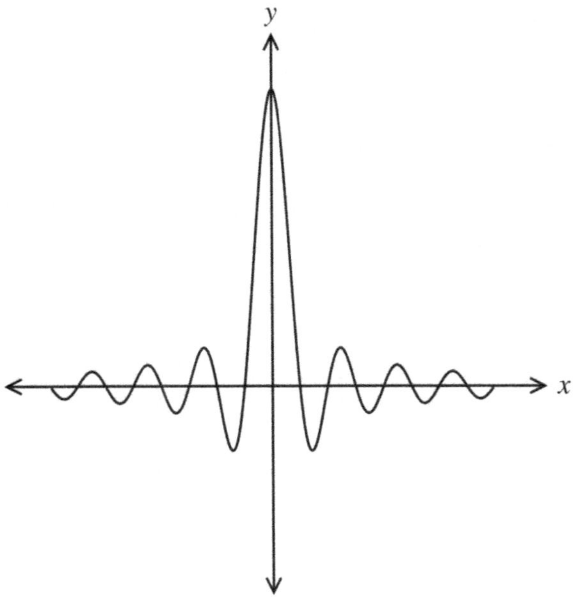

Figure 8.5 The sinc function.

what is $f(t)$? To do this we invoke the inverse Fourier transform, eqn (8.22). The procedure is very similar to the preceding and the result is the Fourier pair:

$$\frac{\tau}{2\pi}\,\mathrm{Sa}\left(\frac{\tau t}{2}\right) \leftrightarrow \mathrm{Rect}\left(\frac{\omega}{\tau}\right) \tag{8.32}$$

Or, given the linear property of transforms:

$$\tau\,\mathrm{Sa}\left(\frac{\tau t}{2}\right) \leftrightarrow 2\pi\,\mathrm{Rect}\left(\frac{\omega}{\tau}\right)$$

Contrast this with eqn (8.31).

One could also get this result using the duality property of the transform integrals:

$$f(t) \leftrightarrow F(\omega)$$

$$\mathrm{Rect}\left(\frac{t}{\tau}\right) \leftrightarrow \tau\,\mathrm{Sa}\left(\frac{\omega\tau}{2}\right)$$

$$\Downarrow$$

$$F(t) \leftrightarrow 2\pi f(-\omega)$$

$$\tau\,\mathrm{Sa}\left(\frac{\tau t}{2}\right) \leftrightarrow 2\pi\,\mathrm{Rect}(-\omega) = 2\pi\,\mathrm{Rect}(\omega)$$

and realizing that Rect($-\omega$) is equal to Rect(ω) since it is a symmetric function.

8.5.2 Single-Sided Exponential Decay

We begin with the definition of the function $U(t)$ called the *Heaviside unitary step function*:

$$U(t) = \begin{Bmatrix} 0 & \text{for } t < 0 \\ \frac{1}{2} & \text{for } t = 0 \\ 1 & \text{for } t > 0 \end{Bmatrix}$$

Now we define our single-sided exponential function using $U(t)$ (see Figure 8.6):

$$f(t) = \mathrm{e}^{-\alpha t} U(t) \quad (\alpha > 0)$$

The Fourier transform is:

$$\begin{aligned} F(\omega) &= \int_{-\infty}^{\infty} f(t) \mathrm{e}^{-i\omega t} \mathrm{d}t \\ &= \int_{-\infty}^{\infty} \mathrm{e}^{-\alpha t} \mathrm{e}^{-i\omega t} \mathrm{d}t \end{aligned}$$

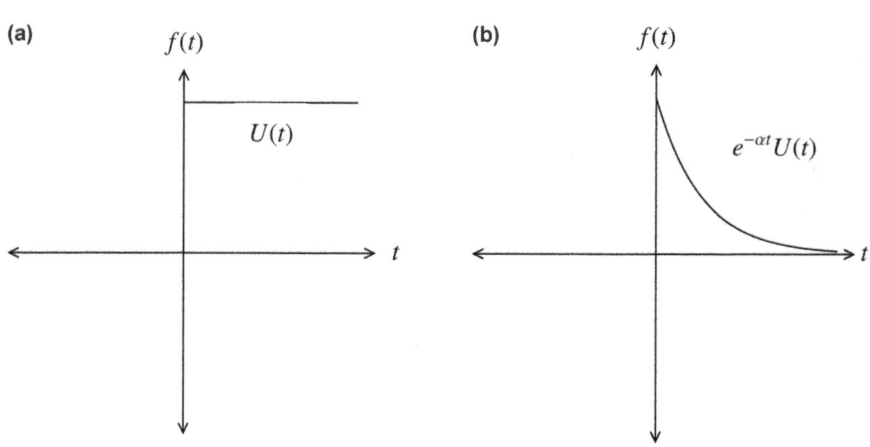

Figure 8.6 (a) the Heaviside step function; (b) the single-sided exponential decay function.

Since the integral from -infinity to zero is obviously zero we may change the integration limits to be from 0 to ∞:

$$\begin{aligned} F(\omega) &= \int_0^\infty \mathrm{e}^{-\alpha t}\mathrm{e}^{-i\omega t}\mathrm{d}t \\ &= \int_0^\infty \mathrm{e}^{-(\alpha+i\omega)t}\mathrm{d}t \\ &= \left.\frac{\mathrm{e}^{-(\alpha+i\omega)t}}{-(\alpha+i\omega)}\right|_0^\infty \\ &= \frac{1}{(\alpha+i\omega)} \end{aligned}$$

and our Fourier pair is:

$$\mathrm{e}^{-\alpha t}U(t) \leftrightarrow \frac{1}{\alpha+i\omega} \tag{8.33}$$

From the above plot you can see that $f(t)$ is neither even nor odd. We can extract the even part of the frequency function by restating it in complex format (à la eqn (8.21)):

$$\begin{aligned} F(\omega) &= \frac{1}{\alpha+i\omega} \times \frac{\alpha-i\omega}{\alpha-i\omega} \\ &= \frac{\alpha-i\omega}{\alpha^2+\omega^2} \\ &= \frac{\alpha}{\alpha^2+\omega^2} - i\frac{\omega}{\alpha^2+\omega^2} \end{aligned}$$

The real part is:

$$a(\omega) = \frac{\alpha}{\alpha^2+\omega^2} \tag{8.34}$$

and is an even function. The imaginary portion is an odd function:

$$b(\omega) = i\frac{\omega}{\alpha^2+\omega^2} \tag{8.35}$$

We shall be interested in using these functions shortly.

8.5.3 The Dirac Delta Function

This function is defined for our purpose as:

$$\delta(t) = \left\{ \begin{array}{c} 0 \quad \text{if} \quad t \neq 0 \\ \int_{-\infty}^{\infty} \delta(t)\mathrm{d}t = 1 \end{array} \right\} \tag{8.36}$$

This is a strange way to define a function since there is no assignment statement for $t=0$. This makes it hard to do the Fourier transform directly. Nonetheless, it has some very interesting and crucial properties for NMR spectroscopists.

Since the definite integral of the function is equal to one and the function is equal to zero everywhere but at $t=0$, it is reasonable to think that this represents a rectangular pulse (or impulse) of infinite height and infinitely small width at zero. We can model this using the Rect(t) function (Figure 8.7):

$$I_\tau(t) = \frac{1}{\tau}\mathrm{Rect}\left(\frac{t}{\tau}\right) \tag{8.37}$$

$I_\tau(t)$ is the *impulse function*. As τ is decreased, the width of the rectangle decreases and its height increases. Therefore, in the limit as τ approaches zero $I_\tau(t)$ approaches the Dirac delta function:

$$\delta(t) = \lim_{\tau \to 0} I_\tau(t)$$

The area under $I_\tau(t)$ is equal to one for all values of τ:

$$\int_{-\infty}^{\infty} I_\tau(t)\mathrm{d}t = 1 \tag{8.38}$$

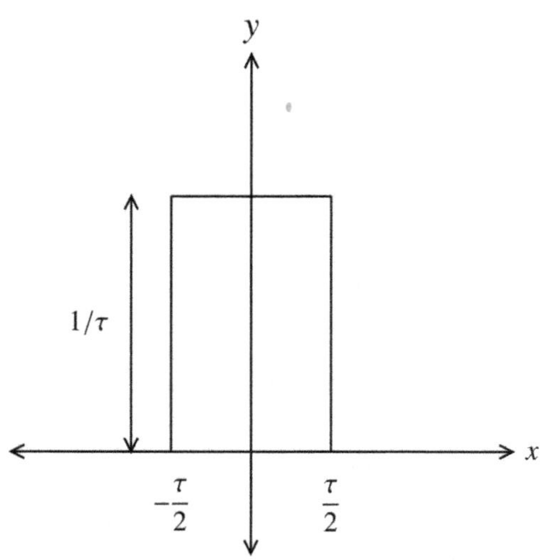

Figure 8.7 The rectangular box function.

which makes our impulse function definition consistent with the Dirac delta function's definition.

We now do the Fourier transform on the impulse function, recalling the Fourier pair for the Rect function and the fact that Fourier transforms are linear:

$$\begin{aligned} I_\tau(\omega) &= \int_{-\infty}^{\infty} I_\tau(t)e^{-i\omega t}dt \\ &= \left(\frac{1}{\tau}\right)\left[\tau\,\mathrm{Sa}\left(\frac{\omega\tau}{2}\right)\right] \\ &= \mathrm{Sa}\left(\frac{\omega\tau}{2}\right) \end{aligned}$$

and the corresponding Fourier pair is:

$$I_\tau(t) \leftrightarrow \frac{\sin\left(\frac{\omega\tau}{2}\right)}{\frac{\omega\tau}{2}}$$

Now we ask what happens as τ approaches zero:

$$F(\omega) = \lim_{\tau\to 0}\frac{\sin\left(\frac{\omega\tau}{2}\right)}{\frac{\omega\tau}{2}} = 1$$

by L'Hopital's rule (see Appendix A).

Thus, we now have the Fourier pair:

$$\delta(t) \leftrightarrow 1 \tag{8.39}$$

By reasoning similar to that leading to eqn (8.32) (or by simply using the duality property) we can also come up with:

$$1 \leftrightarrow 2\pi\delta(\omega) \tag{8.40}$$

8.5.4 The Sampling Property of the Dirac Delta Function

Let us multiply the Dirac delta function by another function of time:

$$\begin{aligned} f(t)\delta(t) &= f(t)\lim_{\tau\to 0} I_\tau(t) \\ &= \lim_{\tau\to 0} f(t)I_\tau(t) \\ &= \lim_{\tau\to 0} f(t)\frac{1}{\tau}\mathrm{Rect}\left(\frac{t}{\tau}\right) \end{aligned}$$

This is just the limit as τ approaches zero of $f(t)$ and Rect(t/τ) multiplied together. As τ gets smaller the upper and lower limits of $f(t)$ get smaller until τ reaches zero and:

$$\lim_{\tau\to 0} f(t)\frac{1}{\tau}\mathrm{Rect}\left(\frac{t}{\tau}\right) = f(0)\delta(t)$$

The delta function is said to be weighted by $f(0)$. Figure 8.8 shows this pictorially.

You can see from the figure that as τ is decreased the width of the rectangle decreases and the width of the sampled portion of the function will become smaller. In the limit when τ approaches 0 the function is non-zero at $f(0)$ only. Thus, $\delta(t)$ has "sampled" $f(t)$ at 0. The Fourier transform of this is:

$$\begin{aligned} F(\omega) &= \int_{-\infty}^{\infty} f(t)\delta(t)\mathrm{e}^{-i\omega t}\mathrm{d}t \\ &= \int_{-\infty}^{\infty} f(0)\delta(t)\mathrm{e}^{-i\omega t}\mathrm{d}t \\ &= f(0)\int_{-\infty}^{\infty} \delta(t)\mathrm{e}^{-i\omega t}\mathrm{d}t \end{aligned}$$

Using eqn (8.39) this becomes:

$$\begin{aligned} F(\omega) &= f(0)\int_{-\infty}^{\infty} \delta(t)\mathrm{e}^{-i\omega t}\mathrm{d}t \\ &= f(0) \end{aligned} \tag{8.41}$$

The sampling property can be generalised even more. First, consider our initial approach to the delta function using $I_\tau(t)$ (eqn (8.37)). The rectangular "boxes" defined by this function are centred about $t = 0$ and the delta function is centred about zero, but this need

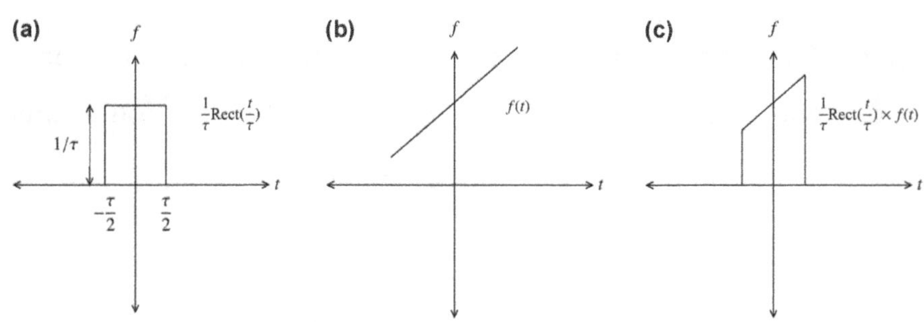

Figure 8.8 Multiplication of a function by the Rect function.

not be so. Consider the "boxes" being shifted to the right, say, and centred about ζ instead of zero (Figure 8.9).

Our impulse function would then be written:

$$I_\tau(t-\zeta)$$

Proceeding from here with the limiting process as τ approaches zero, as before, we end up now with a delta function located at $t=\zeta$. Evidently the delta function can be located at any point along the t-axis. Furthermore, we need not limit the delta function to the time domain ... we can use it in any domain including the frequency domain.

So, we generalise the sampling property in the time and frequency domains:

$$\begin{aligned} f(t)\delta(t-\zeta) &= f(\zeta)\delta(t-\zeta) \\ F(\omega)\delta(\omega-\beta) &= F(\beta)\delta(\omega-\beta) \end{aligned} \tag{8.42}$$

Now we use the sampling property (in step 2) as follows:

$$\begin{aligned} &\int_{-\infty}^{\infty} \delta(t-\zeta)\mathrm{e}^{-i\omega t}\mathrm{d}t \\ &= \int_{-\infty}^{\infty} \delta(t-\zeta)\mathrm{e}^{-i\omega\zeta}\mathrm{d}t \\ &= \mathrm{e}^{-i\omega\zeta}\int_{-\infty}^{\infty} \delta(t-\zeta)\mathrm{d}t \\ &= \mathrm{e}^{-i\omega\zeta} \end{aligned}$$

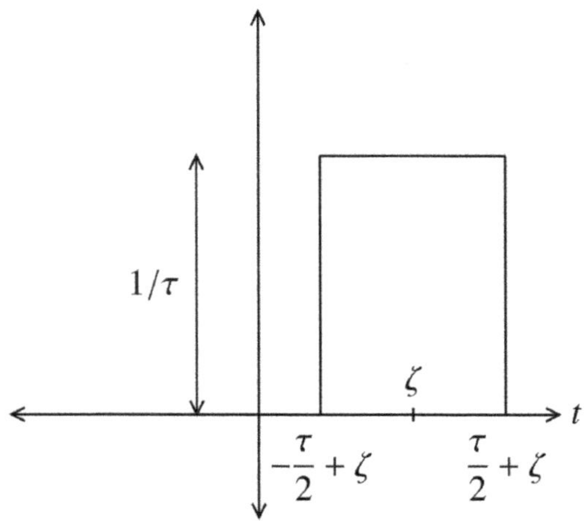

Figure 8.9 Right-shifted rectangular box function.

which gives us another Fourier pair:

$$\delta(t-\zeta)\leftrightarrow \mathrm{e}^{-\omega\zeta} \tag{8.43}$$

This is really just a generalisation of our first delta function Fourier pair. If ζ is zero in eqn (8.43) the delta function is centred about $t=0$ then the exponential evaluates to one and we have:

$$\delta(t)\leftrightarrow 1$$

as before.

Equation (8.43) is just a compact way of saying:

$$\mathrm{e}^{-i\omega\zeta}=\int_{-\infty}^{\infty}\delta(t-\zeta)\mathrm{e}^{-i\omega t}\mathrm{d}t$$

and

$$\delta(t-\zeta)=\frac{1}{2\pi}\int_{-\infty}^{\infty}\mathrm{e}^{-i\omega\zeta}\mathrm{e}^{i\omega t}\mathrm{d}\omega$$

from the definitions of the analysis and synthesis eqn (8.22) and (8.23).

Applying this to a Fourier transform:

$$\begin{aligned}F(\omega)&=\int_{-\infty}^{\infty}f(t)\delta(t-\zeta)\mathrm{e}^{-i\omega t}\mathrm{d}t\\&=f(\zeta)\int_{-\infty}^{\infty}\delta(t-\zeta)\mathrm{e}^{-i\omega t}\mathrm{d}t\\&=f(\zeta)\mathrm{e}^{-i\omega\zeta}\end{aligned}$$

8.5.5 The Complex Exponential Function

Our function is:

$$f(t)=\mathrm{e}^{i\omega_\alpha t}$$

or:

$$f(t)=1\cdot\mathrm{e}^{i\omega_\alpha t}=g(t)\ \mathrm{e}^{i\omega_\alpha t}$$

and we know that the Fourier transform of 1 is $2\pi\delta(\omega)$ from eqn (8.40) and that multiplication of $g(t)$ by $\mathrm{e}^{i\omega_\alpha t}$ gives $G(\omega-\omega_\alpha)$ from eqn (8.27). Thus, we write the Fourier pair as:

$$\mathrm{e}^{i\omega_\alpha t}\leftrightarrow 2\pi\delta(\omega-\omega_\alpha) \tag{8.44}$$

Of course, this collapses to eqn (8.40) when $\omega_\alpha=0$.

8.5.6 The Sine and Cosine Functions

The Fourier transform of these functions follows immediately from eqn (8.44), (A.22) and (A.23) in Appendix A:

$$\begin{aligned} f(t) &= \cos(\omega_\alpha t) \\ F(\omega) &= \int_{-\infty}^{\infty} f(t)\mathrm{e}^{-i\omega t}\mathrm{d}t \\ &= \int_{-\infty}^{\infty} \cos(\omega_\alpha t)\mathrm{e}^{-i\omega t}\mathrm{d}t \\ &= \int_{-\infty}^{\infty} \frac{[\mathrm{e}^{i\omega_\alpha t} + \mathrm{e}^{-i\omega_\alpha t}]}{2}\mathrm{e}^{-i\omega t}\mathrm{d}t \\ &= \frac{1}{2}\int_{-\infty}^{\infty} \mathrm{e}^{i\omega_\alpha t}\mathrm{e}^{-i\omega t}\mathrm{d}t + \frac{1}{2}\int_{-\infty}^{\infty} \mathrm{e}^{-i\omega_\alpha t}\mathrm{e}^{-i\omega t}\mathrm{d}t \\ &= \pi\delta(\omega - \omega_\alpha) + \pi\delta(\omega + \omega_\alpha) \end{aligned} \tag{8.45}$$

and:

$$\begin{aligned} f(t) &= \sin(\omega_\alpha t) \\ F(\omega) &= \int_{-\infty}^{\infty} f(t)\mathrm{e}^{-i\omega t}\mathrm{d}t \\ &= \int_{-\infty}^{\infty} \sin(\omega_\alpha t)\mathrm{e}^{-i\omega t}\mathrm{d}t \\ &= \int_{-\infty}^{\infty} \frac{[\mathrm{e}^{i\omega_\alpha t} - \mathrm{e}^{-i\omega_\alpha t}]}{2}\mathrm{e}^{-i\omega t}\mathrm{d}t \\ &= \frac{1}{2}\int_{-\infty}^{\infty} \mathrm{e}^{i\omega_\alpha t}\mathrm{e}^{-i\omega t}\mathrm{d}t - \frac{1}{2}\int_{-\infty}^{\infty} \mathrm{e}^{-i\omega_\alpha t}\mathrm{e}^{-i\omega t}\mathrm{d}t \\ &= \frac{\pi}{i}\delta(\omega - \omega_\alpha) - \frac{\pi}{i}\delta(\omega + \omega_\alpha) \end{aligned} \tag{8.46}$$

These results are of some importance to us. Recalling that $\delta(t-\zeta)$ represents a shift of t to the right by ζ units, $\delta(\omega-\omega_\alpha)$ will similarly represent a shift of the delta function to the right on the frequency scale to frequency ω_α. Therefore, eqn (8.45) is the Fourier transform of the cosine function and tells us that there are delta function responses at $+\omega_\alpha$ and $-\omega_\alpha$ (Figure 8.10).

There is a simple way to extract either of the frequencies out of these results. To get $+\omega_\alpha$ we simply add the cosine transform and sine

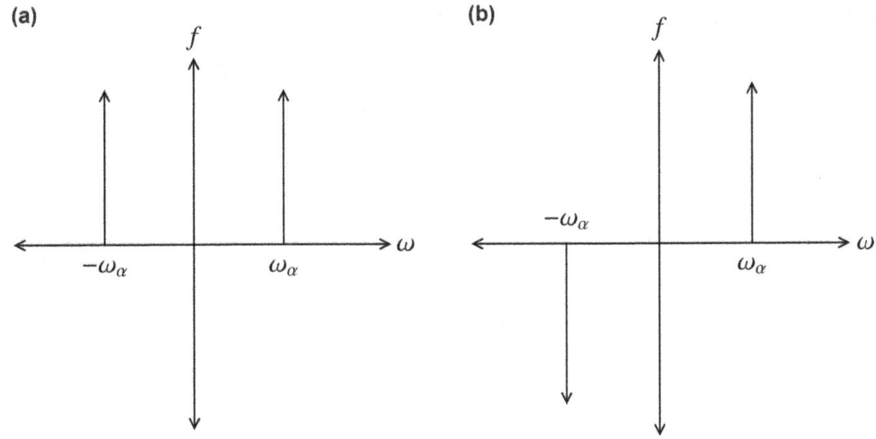

Figure 8.10 (a) delta functions for the Fourier transform of the cosine function; (b) delta functions for the Fourier transform of the sine function.

transform. Similarly, to get $-\omega_\alpha$ we subtract the sine transform from the cosine transform. This is the basis of the quadrature detection method used in NMR spectroscopy that will be discussed in more detail in Chapter 21.

8.5.7 The Free Induction Decay

This is probably the most important section of this chapter in that it deals with transforming the time-domain data from the NMR sample that is collected in the form of an exponentially decaying signal called a free induction decay or FID. Actually, two signals are usually used in a quadrature signal detection scheme.[¶¶] In the jargon of NMR spectroscopy these would be the real and imaginary or *absorption* and *dispersion* signals. These are somewhat unfortunate names for these signals because, as we shall shortly see, each channel contains both absorption and dispersion mode signal components. One channel can never give a purely absorption or dispersion signal as many texts seem to imply. In an attempt to circumvent potential confusion we shall name one signal channel the cosine channel and the other the sine channel.

The question is "how do we model these signals and what are their Fourier transforms?". Modelling the signals is easy. Referring back to

¶¶See Chapter 21 for more details on quadrature signal detection.

the discussion of the one-sided exponential decay we use the Heaviside unit step function to "turn off" everything before zero time and then the signal is just a cosine or sine function modulated (multiplied) by an exponential function such that the signal decays with time. Here, we combine the sine and cosine functions into a complex exponential for convenience:

$$f(t)_{\cos} = U(t)e^{-\alpha t}\cos(\omega_0 t)$$

or

$$f(t)_{\sin} = U(t)e^{-\alpha t}\sin(\omega_0 t)$$

To transform the cosine signal we invoke the frequency shift transform property eqn (8.28), and eqn (8.33), the Fourier transform of an exponential decay function.

$$g(t) = U(t)e^{-\alpha t}$$

$$G(\omega) = \frac{\alpha}{\alpha^2 + \omega^2} + i\frac{\omega}{\alpha^2 + \omega^2}$$

$$f(t)_{\cos} = g(t)\cos(\omega_0 t)$$

Transform property (8.28) indicates that $F(\omega)_{\cos}$ is:

$$F(\omega)_{\cos} = \int_{-\infty}^{\infty} g(t)\cos(\omega_0 t)e^{-i\omega t}dt$$
$$= \frac{1}{2}[G(\omega - \omega_0) + G(\omega + \omega_0)]$$

and substituting for $G(\omega - \omega_x)$ and $G(\omega + \omega_x)$:

$$\begin{aligned}
F(\omega)_{\cos} &= \frac{1}{2}\left[\frac{\alpha}{\alpha^2 + (\omega - \omega_0)^2} + i\frac{(\omega - \omega_0)}{\alpha^2 + (\omega - \omega_0)^2}\right] \\
&+ \frac{1}{2}\left[\frac{\alpha}{\alpha^2 + (\omega + \omega_0)^2} + i\frac{(\omega + \omega_0)}{\alpha^2 + (\omega + \omega_0)^2}\right] \\
&= \frac{1}{2}\left[\frac{\alpha}{\alpha^2 + (\omega - \omega_0)^2} + \frac{\alpha}{\alpha^2 + (\omega - \omega_0)^2}\right] \\
&+ \frac{i}{2}\left[\frac{(\omega - \omega_0)}{\alpha^2 + (\omega - \omega_0)^2} + \frac{(\omega + \omega_0)}{\alpha^2 + (\omega + \omega_0)^2}\right]
\end{aligned} \tag{8.47}$$

Similar analysis for the sine signal gives us (using eqn (8.29) this time):

$$
\begin{aligned}
F(\omega)_{\text{sin}} &= \int_{-\infty}^{\infty} g(t)\sin(\omega_0 t)\mathrm{e}^{-i\omega t}\mathrm{d}t \\
&= \frac{1}{2}[G(\omega-\omega_0) - G(\omega+\omega_0)] \\
&= \frac{1}{2}\left[\frac{\alpha}{\alpha^2+(\omega-\omega_0)^2} + i\frac{(\omega-\omega_0)}{\alpha^2+(\omega-\omega_0)^2}\right] \\
&\quad - \frac{1}{2}\left[\frac{\alpha}{\alpha^2+(\omega+\omega_0)^2} + i\frac{(\omega+\omega_x)}{\alpha^2+(\omega+\omega_0)^2}\right] \\
&= \frac{1}{2}\left[\frac{\alpha}{\alpha^2+(\omega-\omega_0)^2} - \frac{\alpha}{\alpha^2+(\omega+\omega_0)^2}\right] \\
&\quad + \frac{i}{2}\left[\frac{(\omega-\omega_0)}{\alpha^2+(\omega-\omega_0)^2} - \frac{(\omega+\omega_0)}{\alpha^2+(\omega+\omega_0)^2}\right]
\end{aligned}
\tag{8.48}
$$

Figure 8.11 plots the real part of each of eqn (8.47) and (8.48). In each of these diagrams we see a peak at $+\omega_0$ and at $-\omega_0$. In the first, both peaks are positive and in the second one is positive and one is

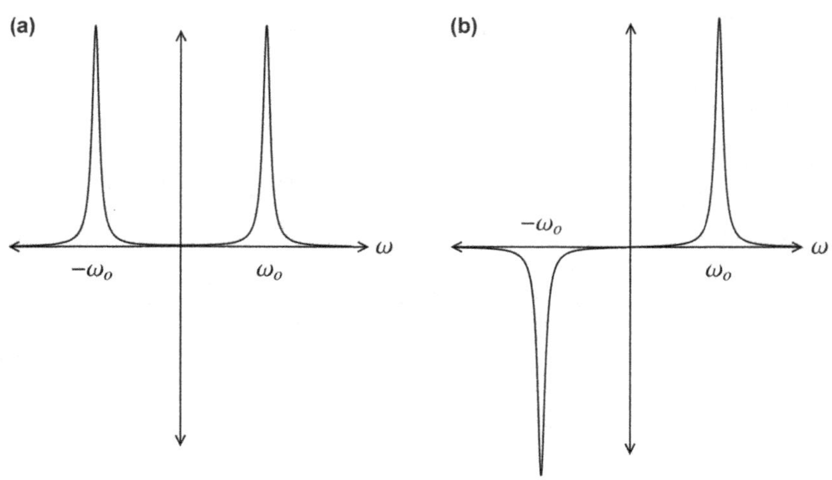

Figure 8.11 The real portions of the Fourier transform of the exponentially decaying (a) cosine and (b) sine signals.

negative. Clearly, there is a real signal associated with both the cosine and sine channels.

Summing the real parts of eqn (8.47) and (8.48) gives us our familiar NMR absorption mode signal:

$$F(\omega) = \frac{\alpha}{\alpha^2 + (\omega - \omega_0)^2}$$

whose plot is shown in Figure 8.12.

For the imaginary portions of eqn (8.47) and (8.48) we do much the same and obtain the plots shown in Figures 8.13 and 8.14.

Figures 8.12 and 8.14 correspond to what we usually think of as the absorption and dispersion signals, respectively.‖‖ We shall see in Chapter 21 how these signals are handled in the spectrometer hardware.

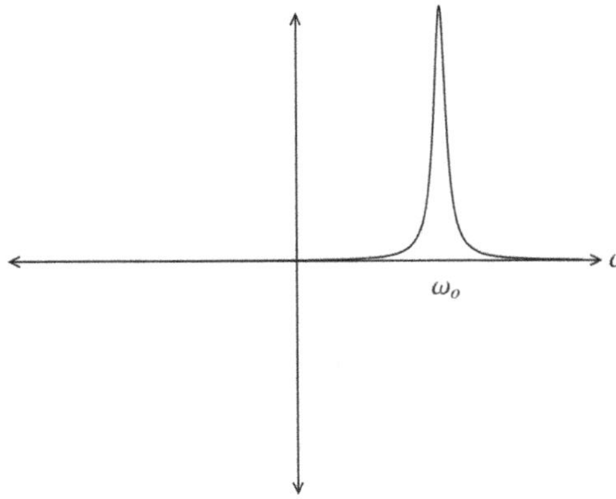

Figure 8.12 The sum of the real portions of the real portions of the Fourier transform of the sine and cosine signals.

‖‖Many texts present only these two plots and imply that they come directly from the Fourier transform of the cosine and sine channels. Indeed, if you combine the sine and cosine signals into one complex function and do the Fourier transform on it $\left(F(\omega) = \int_0^\infty e^{-\alpha t} e^{-i(\omega-\omega_x)t} dt\right)$ you will get a result that corresponds to Figures 8.11a and 8.13a. If this were true, however, it would be very difficult to explain the presence of ghost signals and the necessity of phase cycling to try to eliminate them. We shall see in Chapter 21 that our analysis of the sine and cosine signals will allow us to nicely explain ghost peaks and their suppression. As you can appreciate from the transform equations here, the cosine signal channel contains components of both the absorption and dispersion spectra, as does the sine channel. I have seen this pointed out in only one publication; Fukushima and Roeder's Experimental Pulsed NMR: A Nuts and Bolts Approach (see the Further Reading section at the end of the chapter).

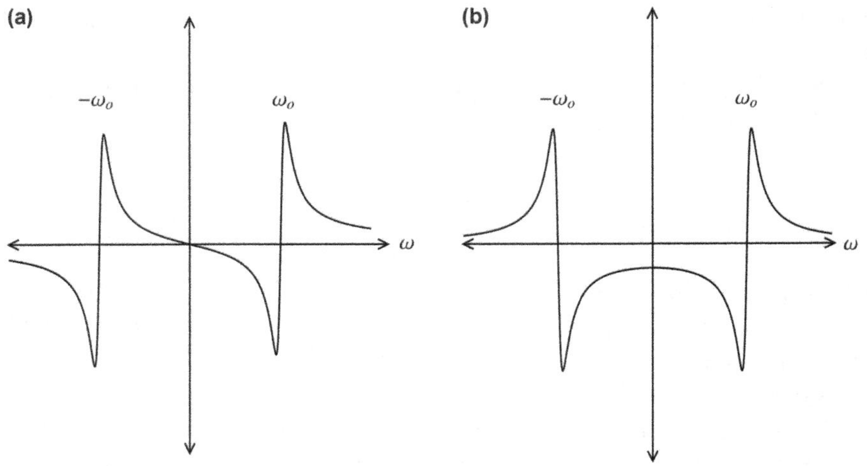

Figure 8.13 The imaginary portions of the Fourier transform of the exponentially decaying (a) cosine and (b) sine signals.

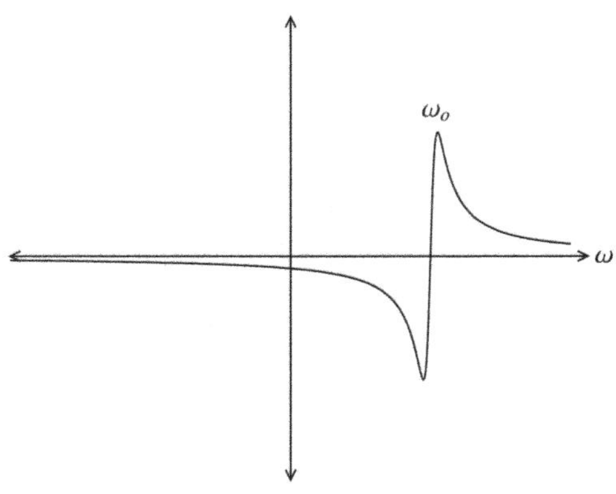

Figure 8.14 The sum of the imaginary portions of the real portions of the Fourier transform of the sine and cosine signals.

8.6 Convolution

Convolution is a difficult concept to grasp in the same sense that logarithms are difficult for most people to grasp.*** Both are very powerful and useful concepts though and so we can boil things down to a simple statement: as logarithms convert multiplication to addition so does convolution convert an integral to multiplication.

$$\ln(ab) = \ln(a) + \ln(b)$$

***It is a convoluted idea.

The convolution of frequency-domain functions is defined by:

$$F(\omega)^*G(\omega) \equiv \int_{-\infty}^{\infty} F(\omega - x)G(x)\mathrm{d}x \tag{8.49}$$

where x is used as a dummy variable. Where does this come from? Consider these two inverse Fourier transform integrals multiplied together:[†††]

$$\begin{aligned}
&\frac{1}{2\pi}\int_{-\infty}^{\infty} F(\omega)\mathrm{e}^{i\omega t}\mathrm{d}\omega \frac{1}{2\pi}\int_{-\infty}^{\infty} G(x)\mathrm{e}^{ixt}\mathrm{d}x \\
&= \frac{1}{4\pi^2}\int_{-\infty}^{\infty}\int_{-\infty}^{\infty} \mathrm{e}^{i\omega t}\mathrm{e}^{ixt}F(\omega)G(x)\mathrm{d}\omega\mathrm{d}x \\
&= \frac{1}{4\pi^2}\int_{-\infty}^{\infty}\int_{-\infty}^{\infty} \mathrm{e}^{it(\omega+x)}F(\omega)G(x)\mathrm{d}\omega\mathrm{d}x \\
&= \frac{1}{4\pi^2}\int_{-\infty}^{\infty}\left(\int_{-\infty}^{\infty} \mathrm{e}^{it(\omega+x)}F(\omega)\mathrm{d}\omega\right)G(x)\mathrm{d}x
\end{aligned}$$

Introducing a change of variables:

$$u = \omega + x, \quad \omega = u - x, \quad \mathrm{d}\omega = \mathrm{d}u$$

$$\begin{aligned}
&\frac{1}{4\pi^2}\int_{-\infty}^{\infty}\left(\int_{-\infty}^{\infty} \mathrm{e}^{itu}F(u-x)\mathrm{d}u\right)G(x)\mathrm{d}x \\
&= \frac{1}{4\pi^2}\int_{-\infty}^{\infty} \mathrm{e}^{itu}\left(\int_{-\infty}^{\infty} F(u-x)G(x)\mathrm{d}x\right)\mathrm{d}u
\end{aligned}$$

The integral in the brackets is a function of u and the outer integral is the inverse Fourier transform integral so, to make things a bit clearer, we make the substitution:

$$H(u) = \int_{-\infty}^{\infty} F(u-x)G(x)\mathrm{d}x$$

and write:

$$\begin{aligned}
&\frac{1}{4\pi^2}\int_{-\infty}^{\infty} \mathrm{e}^{itu}\left(\int_{-\infty}^{\infty} F(u-x)G(x)\mathrm{d}x\right)\mathrm{d}u \\
&= \frac{1}{4\pi^2}\int_{-\infty}^{\infty} \mathrm{e}^{itu}H(u)\mathrm{d}u \\
&= h(t)
\end{aligned}$$

[†††]You will see in a moment why I am using two different variables here.

To make us feel better, we will switch back to using ω rather than u:

$$\frac{1}{4\pi^2}\int_{-\infty}^{\infty} e^{it\omega}\left(\int_{-\infty}^{\infty} F(\omega-x)G(x)dx\right)d\omega$$

$$=\frac{1}{4\pi^2}\int_{-\infty}^{\infty} e^{it\omega}H(\omega)d\omega$$

$$=h(t)$$

From this, we define the *convolution* of $F(\omega)$ and $G(\omega)$ as in eqn (8.49):

$$F(\omega)^*G(\omega) \equiv \int_{-\infty}^{\infty} F(\omega-x)G(x)dx$$

With this definition it is possible to show that:

$$f(t)g(t) \leftrightarrow \frac{1}{2\pi}F(\omega)^*G(\omega)$$

To do so we start with the definition of the frequency-domain convolution and do an inverse Fourier transform on it:

$$\frac{1}{2\pi}\int_{-\infty}^{\infty}\left(\frac{1}{2\pi}F(\omega)^*G(\omega)\right)e^{i\omega t}d\omega$$

$$\frac{1}{2\pi}\int_{-\infty}^{\infty}\left(\frac{1}{2\pi}\int_{-\infty}^{\infty} F(\omega-x)G(x)dx\right)e^{i\omega t}d\omega$$

To proceed we reverse the order of integration and group terms in ω and x:

$$\frac{1}{2\pi}\int_{-\infty}^{\infty}\left(\frac{1}{2\pi}\int_{-\infty}^{\infty} F(\omega-x)e^{i\omega t}d\omega\right)G(x)dx$$

and let $\chi=\omega-x$ so that $\omega=\chi+x$ and $d\omega=d\chi$:

$$\frac{1}{2\pi}\int_{-\infty}^{\infty}\left(\frac{1}{2\pi}\int_{-\infty}^{\infty} F(\chi)e^{i(\chi+x)t}d\chi\right)G(x)dx$$

$$=\frac{1}{2\pi}\int_{-\infty}^{\infty}\left(\frac{1}{2\pi}\int_{-\infty}^{\infty} F(\chi)e^{i\chi t}d\chi\right)e^{ixt}G(x)dx$$

$$=\frac{1}{2\pi}\int_{-\infty}^{\infty} f(t)e^{ixt}G(x)dx$$

$$=f(t)\frac{1}{2\pi}\int_{-\infty}^{\infty} G(x)e^{ixt}dx$$

$$=f(t)g(t)$$

The term in brackets on line 2 is the inverse Fourier transform of $F(\omega)$, using the dummy variable χ, and it gives $f(t)$. The rest is the inverse Fourier transform of $G(\omega)$, using dummy variable x to give $g(t)$. Thus, we have shown:

$$f(t)g(t) \leftrightarrow \frac{1}{2\pi}F(\omega)^*G(\omega) \tag{8.50}$$

This is the most useful convolution Fourier pair for our purposes since we shall be interested in time-domain modulations (or multiplications in other words). However, one can reverse eqn (8.50):

$$f(t)^*g(t) \leftrightarrow F(\omega)\, G(\omega) \tag{8.51}$$

where now the Fourier pair involves the convolution of the time-domain functions and the product of the frequency-domain functions. The convolution definition for the time-domain functions is similar to that above for the frequency-domain functions:

$$f(t)^*g(t) \equiv \int_{-\infty}^{\infty} f(t-x)g(x)\mathrm{d}x$$

What do all of these mental gymnastics mean for us? How can we picture what the convolution is doing? One way of thinking of it is that functions f and g "smear" into each other. An easy way to picture this involves using the Rect function from Section 8.5.1 and convolving it with itself. First, however, we need to understand what $f(t-x)$ means.

The function $f(x-t)$ means the same as $f(x)$ except that $f(x-t)$ is shifted or translated (moved) to the right (or left if t is negative) by t units. $f(t-x)$ "flips" the function horizontally about t and translates it. Of course, we notice the flip only if the function is asymmetric. You can see this plotted using the function $f(x)=x^3$ in Figure 8.15.

Now we can picture what happens using the Rect function from above. Suppose that $F(\omega)$ and $G(\omega)$ both equal Rect(ω) and we are doing the convolution of these functions. The definition says that this will be:

$$F(\omega)^*G(\omega) = \int_{-\infty}^{\infty} F(\omega-x)G(x)\mathrm{d}x$$

Consider Rect $(\omega - x)$ in light of what we have just said about translated and "flipped" functions. This function will be shifted ω units left or right, depending on the sign of ω, and "flipped".

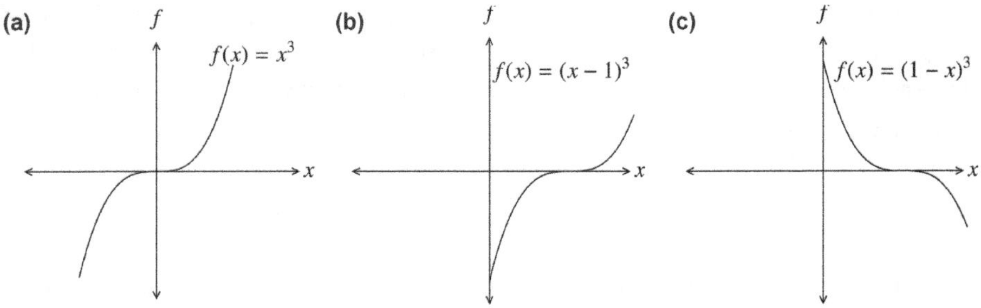

Figure 8.15 (a) function $f(x) = x^3$; (b) translation of $f(x)$; (c) $f(x)$ "flipped" and translated.

However, since the Rect function is symmetric we do not have to worry about "flipping" it visually. Rect(x) is centred at zero and extends from $-B/2$ to $B/2$. Rect($\omega - x$) is centred at $\omega - x$ and extends from $-(B+\omega)/2$ to $(B+\omega)/2$. Simply put, each is the shape of a rectangle of height 1 and length B.

The value of the convolution of $F(\omega)$ and $G(\omega)$ at fixed ω, is the integral of the product of the two functions between $-B/2$ and $B/2$ (Figure 8.16).

As you can see from Figure 8.16, the convolution of $F(\omega)$ and $G(\omega)$ is zero where the two functions do not overlap and builds linearly to a maximum at maximum overlap and then tapers off to zero again as $F(\omega - x)$ is swept from left to right. The result of this is a triangle of height B and base B (Figure 8.17).

Equation (8.50) says that the Fourier transform of the product of $f(t)$ and $g(t)$ is the convolution of $F(\omega)$ and $G(\omega)$. This idea may be applied to the free induction decay discussed in Section 8.5.7. We may think of, say, an eternal cosine signal multiplied by a decaying single-sided exponential, as in Section 8.5.7:[‡‡‡]

$$f(t)g(t) = e^{-\alpha t}\cos(\omega_x t)$$

Recall that the real portion of the Fourier transform of the single-sided decaying exponential is:

$$F(\omega) = \frac{\alpha}{\alpha^2 + \omega^2}$$

[‡‡‡]Eternal simply means going on for all of time so an eternal cosine function goes from −infinity to +infinity on the time axis.

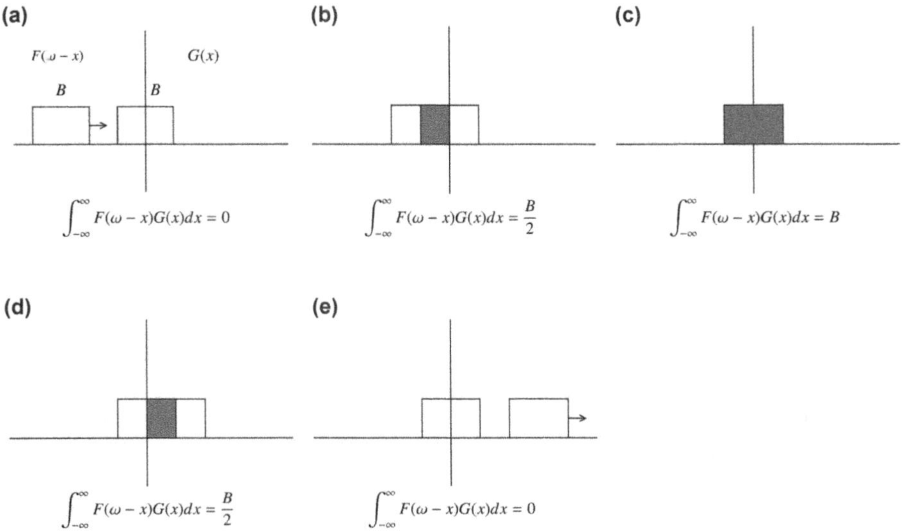

Figure 8.16 The convolution of $F(\omega)$ and $G(\omega)$.

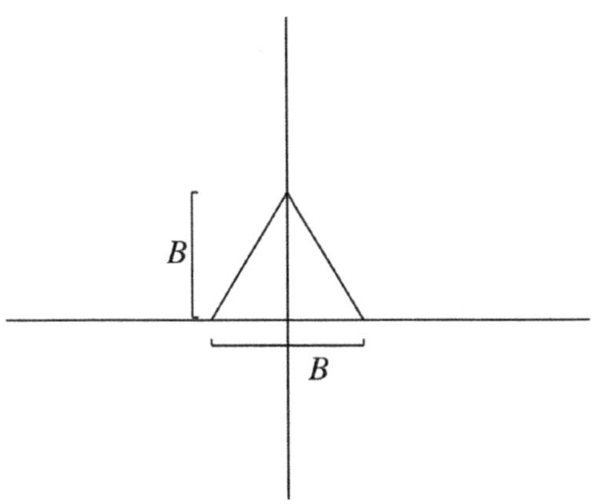

Figure 8.17 Result of the convolution of $F(\omega)$ and $G(\omega)$.

whose plot looks like that in Figure 8.18 and the Fourier transform of the eternal cosine function consists of a pair of Dirac functions at frequencies $+\omega_0$ and $-\omega_0$:

$$g(t) = \cos(\omega_x t)$$

$$G(\omega) = \pi\delta(\omega - \omega_x) + \pi\delta(\omega + \omega_x)$$

whose plot looks like Figure 8.19.

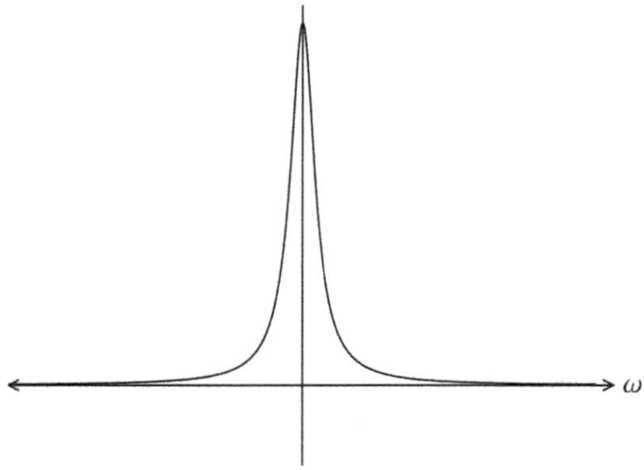

Figure 8.18 A plot of the transform of the single-sided exponential decay function.

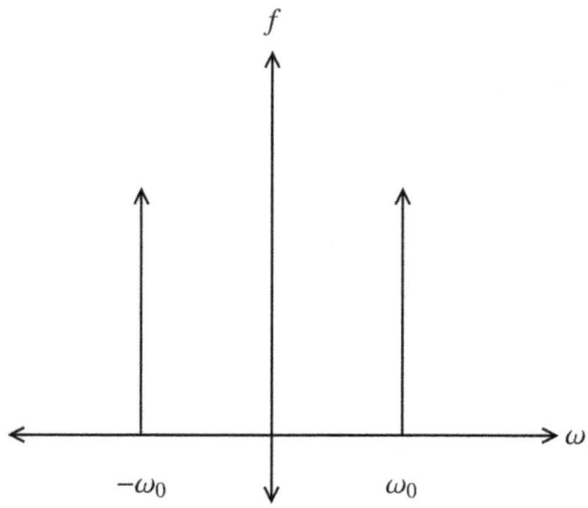

Figure 8.19 Fourier transform of the eternal cosine function.

The convolution of $F(\omega)$ and $G(\omega)$ "merges" or "smoothes" the delta functions with the frequency-domain spectrum of the exponential (Figure 8.20).

8.7 Impulse Sampling and the Nyquist Frequency

Having looked at the sampling property of the Dirac delta function (eqn (8.42)) we are in a position to begin considering discrete sampling of data and how to transform it. We begin by defining a sampling function based on our Dirac delta function.

In our modern spectrometers, a signal is sampled at regular intervals and converted to a set of digital numbers. Thus, if we

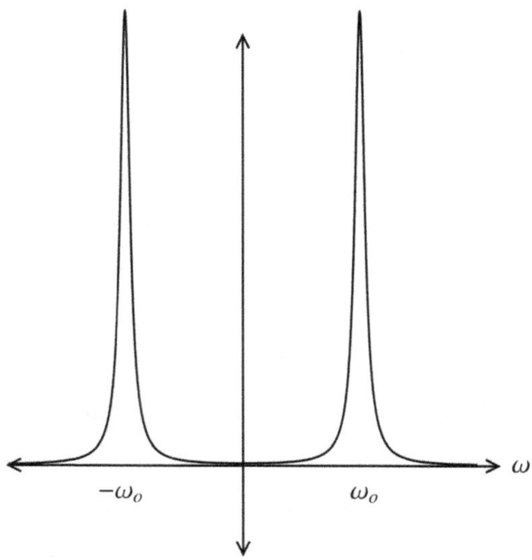

Figure 8.20 Convolution of the eternal cosine with the single-sided exponential decay function.

consider, say, a sine wave that is sampled once every T_S seconds the simplest way to model this is to consider that a periodic delta function has been multiplied into the sine wave. We may write the delta function as:

$$\Delta(t) = \sum_{k=-\infty}^{\infty} \delta(t - kT_S)$$

This is simply a sequence of delta function "pulses" at intervals of T_S and is sometimes called a "comb" function (see Figure 8.21). The period of the function being T_S the frequency is therefore $\omega_S = 2\pi/T_S$. The Fourier transform of this will be of use to us. To get this we will need to establish the Fourier series for the comb function. Using the analysis formula (8.17):

$$C(n) = \frac{1}{T_S}\int_{-\frac{T_S}{2}}^{\frac{T_S}{2}} \delta(t)\mathrm{e}^{-in\omega_S t}\mathrm{d}t$$

$$= \frac{1}{T_S}\int_{-\frac{T_S}{2}}^{\frac{T_S}{2}} \mathrm{e}^{-in\omega_S 0}\mathrm{d}t$$

$$= \frac{1}{T_S}\int_{-\frac{T_S}{2}}^{\frac{T_S}{2}} \mathrm{d}t$$

$$= \frac{1}{T_S}$$

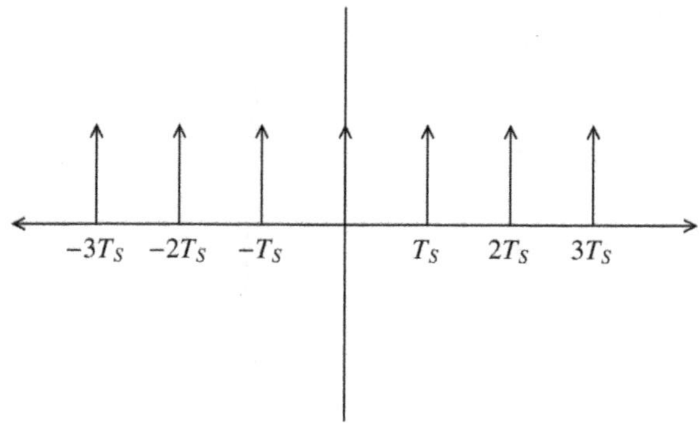

Figure 8.21 The "comb" function.

where we have used the sampling property, eqn (8.42), of the delta function in line 2. We see here that all of the coefficients in the series are $1/T_S$. Now the synthesis eqn (8.13), is used to build the Fourier series:

$$\Delta(t) = \sum_{n=-\infty}^{\infty} C(n)\mathrm{e}^{-in\omega_S t}$$

$$= \frac{1}{T_S}\sum_{n=-\infty}^{\infty} \mathrm{e}^{-in\omega_S t}$$

Finally, we use property (8.44) to write:

$$F(\omega) = \frac{2\pi}{T_S}\sum_{-\infty}^{\infty} \delta(\omega - n\omega_S)$$

$$= \omega_S \sum_{-\infty}^{\infty} \delta(\omega - n\omega_S)$$

which gives us the Fourier pair:

$$\Delta(t) \leftrightarrow \sum_{-\infty}^{\infty} \delta(\omega - n\omega_S) \qquad (8.52)$$

As we have seen, when the delta function is multiplied into the sine function it "samples" the sine function at a particular point, depending on the delta function's argument. Using the comb function extends this idea to sampling of many points from the sine function (Figure 8.22).

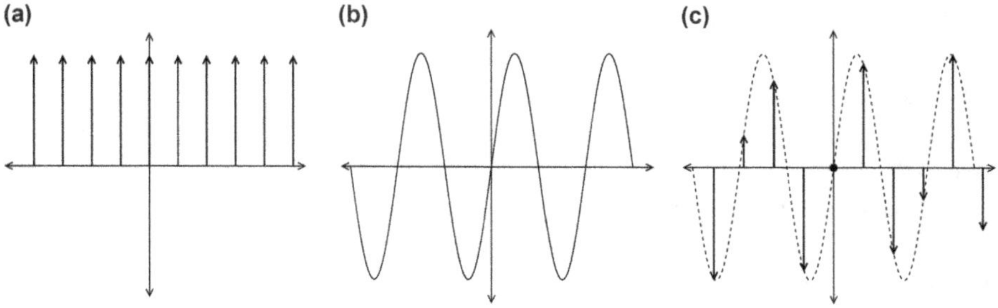

Figure 8.22 The comb function sampling a sine function at regular intervals.

Of course, we can write an equation to represent the picture in Figure 8.22c:

$$\begin{aligned} f(t)_S &= f(t)\Delta(t) \\ &= f(t)\sum_{k=-\infty}^{\infty}\delta(t-kT_S) \\ &= \sum_{k=-\infty}^{\infty} f(kT_S)\delta(t-kT_S) \end{aligned}$$

The question now is "what is the transform of $f(t)_S$?". To find this we invoke the frequency-domain convolution of $f(t)$ and $\Delta(t)$ and using the Fourier pair (8.52) to write:

$$\begin{aligned} F(\omega) &= \frac{1}{2\pi}\int_{-\infty}^{\infty} F(T)\left(\omega_S\sum_{-\infty}^{\infty}\delta(\omega-n\omega_S-T)\right)\mathrm{d}T \\ &= \frac{1}{T_S}\sum_{-\infty}^{\infty}\int_{-\infty}^{\infty} F(T)\delta(\omega-n\omega_S-T)\mathrm{d}T \quad \left(\frac{\omega_S}{2\pi}=\frac{1}{T_S}\right) \\ &= \frac{1}{T_S}\sum_{-\infty}^{\infty}\int_{-\infty}^{\infty} F(\omega-n\omega_S)\delta(\omega-n\omega_S-T)\mathrm{d}T \\ &= \frac{1}{T_S}\sum_{-\infty}^{\infty}\int_{-\infty}^{\infty} F(\omega-n\omega_S) \end{aligned} \tag{8.53}$$

This is of major importance to us. It says that the sampled frequency-domain spectrum consists of multiple copies of $F(\omega)$ each located at $n\omega_S$ and separated by ω_S. We can picture this by first considering a *band-limited* spectrum. In this situation all of the frequencies in the spectrum are contained within a band stretching from $-\omega_M$ to ω_M with the amplitude of all other frequencies being zero (Figure 8.23).

However, eqn (8.53) says that multiple copies of the spectrum are produced each located ω_S apart (Figure 8.24).

We do not want these copies to overlap or else we will observe aliasing in the spectrum in which a high frequency masquerades as a lower frequency in the spectrum. In order to avoid this we must ensure that $\omega_S \geq 2\omega_M$. Figure 8.25 shows what happens to the copies of $F(\omega)$if $\omega_S < 2\omega_M$.

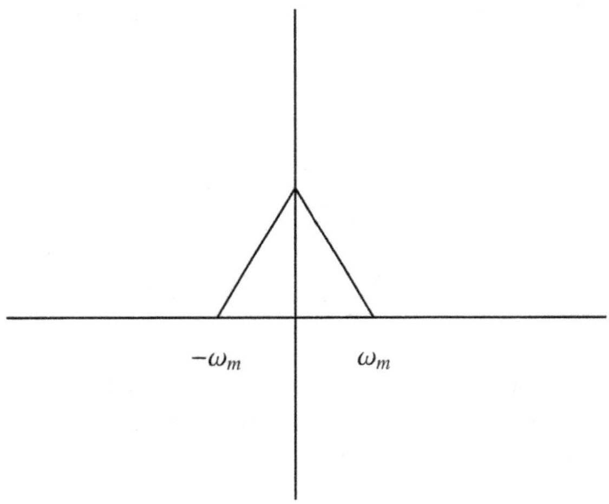

Figure 8.23 A band-limited spectrum.

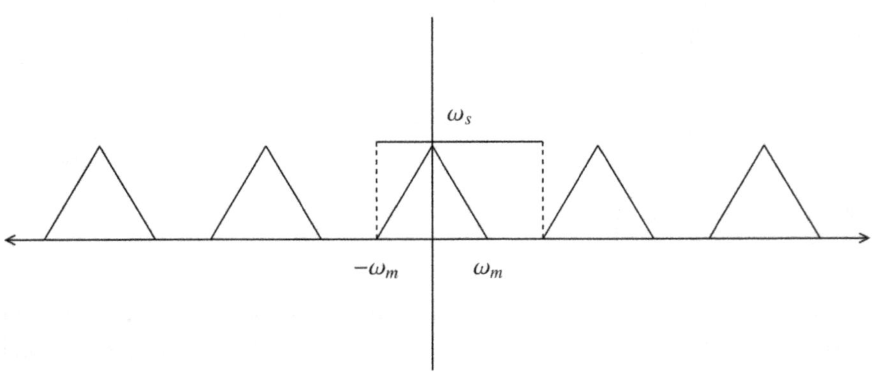

Figure 8.24 Separated copies of the spectrum.

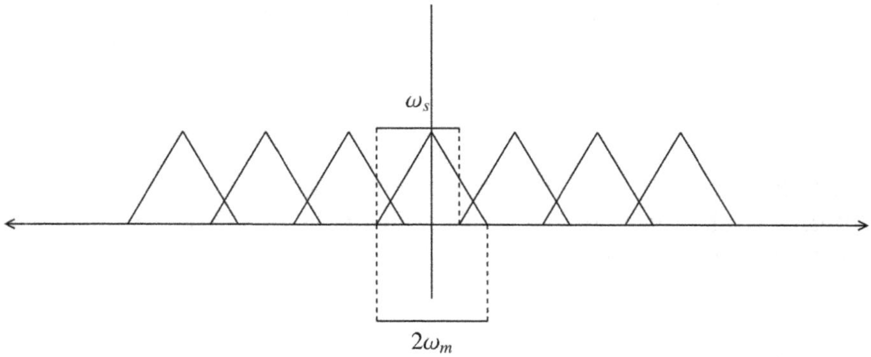

Figure 8.25 Overlapped copies of the spectrum.

The requirement that $\omega_S \geq 2\omega_M$ is called the Nyquist criterion and $2\omega_M$ is the *Nyquist frequency*. This represents the minimum frequency at which the signal can be sampled in order to avoid aliasing or overlapping of $F(\omega)$ copies.

8.8 The Discrete Fourier Transform

The modern high-resolution NMR spectrometer samples the analogue FID signal and samples and digitises it for storage in a computer. These discrete digital data points are then transformed *via* software to frequency (or spatial, see Chapter 22) domain data for further analysis by the user.

Up to this point we have been considering Fourier transforms of continuous functions integrated from $-\infty$ to $+\infty$ but digitised data are *discontinuous*. Furthermore, we do not sample the signal forever but rather do a finite number of N samples. Referring to Figure 8.22c, we see that the sine wave has been sampled at regular time intervals of Δt seconds at a sampling rate of ω_S. Generally, we sample for a finite amount of time, T, and the beginning of the sampling period is set to time zero so we can say that each point in time is given by $t_n = nT/N$ where n ranges from 0 to $N-1$, T is the total sampling time and N, of course, is the number of samples. Δt, the time interval between samples, is equal to T/N and is sometimes called the *dwell time*. One further consideration is that we stipulate that the time-domain function of interest is being sampled over exactly one period during the sampling time. Therefore, the period of $f(t)$ is T and

$\omega_0 = 2\pi/T$. Thus, we have a series of values of the time-domain function taken at regular time intervals:

$$f(0), f(t_1), f(t_2), f(t_3) \ldots f(t_{N-1})$$

Let us consider that $f(t)$ is periodic and that we have taken these samples over exactly one period. This being the case we may write, using the trapezoidal rule from Appendix A:

$$\begin{aligned} Area &= \frac{\Delta t}{2}\sum_{n=0}^{N-1}(f(n) + f(n+1)) \\ &= \frac{\Delta t}{2}(f(0) + 2f(1) + 2f(2) + 2f(3) \ldots 2f(n-1) + f(n)) \end{aligned}$$

Since this is a periodic function and we have sampled over exactly one period, $f(0) = f(n)$ and:

$$\begin{aligned} Area &= \frac{\Delta t}{2}(2f(0) + 2f(1) + 2f(2) + 2f(3) \ldots 2f(n-1)) \\ &= \Delta t(f(0) + f(1) + f(2) + f(3) \ldots f(n-1)) \\ &= \frac{T}{N}(f(0) + f(1) + f(2) + f(3) \ldots f(n-1)) \\ &= \frac{T}{N}\sum_{n=0}^{N-1} f(n) \end{aligned}$$

We use this result to estimate the Fourier coefficients of $f(t)$ starting with eqn (8.16):

$$C(k) = \frac{1}{T}\int_0^T f(t)\mathrm{e}^{-ik\omega_0 t}$$

We apply the trapezoidal rule to this to estimate the coefficients:

$$\begin{aligned} C(k) &\approx \frac{1}{T}\frac{T}{N}\sum_{n=0}^{N-1} f(n)\exp\left(\frac{-ik\omega_0 nT}{N}\right) \\ &\approx \frac{1}{N}\sum_{n=0}^{N-1} f(n)\exp\left(\frac{-i2\pi kn}{N}\right)\ \left(\omega_0 = \frac{2\pi}{T}\right) \end{aligned}$$

The summation is written as $F(k)$:

$$F(k) \equiv \sum_{n=0}^{N-1} f(n)\exp\left(\frac{-i2\pi kn}{N}\right) \quad k=0,1,2,3\ldots N-1 \tag{8.54}$$

where:

$$C(k) \approx \frac{1}{N}F(k)$$

We define eqn (8.54) as the *Discrete Fourier Transform* or DFT. Note that $f(t)$ is the continuous time-domain function and $f(n)$ is the set of N discrete samples of the time-domain function. To derive an inverse transform we first assert that the inverse transform is:

$$f(n) = \frac{1}{N}\sum_{k=0}^{N-1} F(k)\exp\left(\frac{i2\pi kn}{N}\right)$$

and show that it indeed functions as an inverse transform. We start by substituting our DFT definition for $F(k)$:

$$\frac{1}{N}\sum_{k=0}^{N-1} F(k)e^{\frac{i2\pi km}{N}} = \frac{1}{N}\sum_{k=0}^{N-1}\left(\sum_{n=0}^{N-1} f(n)\exp\left(\frac{-i2\pi kn}{N}\right)\right)\exp\left(\frac{i2\pi km}{N}\right)$$

$$= \frac{1}{N}\sum_{n=0}^{N-1}\sum_{k=0}^{N-1} f(n)\exp\left(\frac{-i2\pi kn}{N}\right)\exp\left(\frac{i2\pi km}{N}\right)$$

The two exponential terms are orthogonal ... we shall show this here. When discussing the DFT it is common to use the definition:[§§§]

$$W_N = \exp\left(\frac{i2\pi}{N}\right)$$

[§§§]It would be well for the reader to look carefully at this. I will be using it extensively as I proceed. By the way, this has the cute name "twiddle factor" in some places in the literature.

and then:

$$W_N^{-a} = \exp\left(\frac{i2\pi a}{N}\right)$$

$$W_N^{b} = \exp\left(\frac{-i2\pi b}{N}\right)$$

Note the sign convention ... a negative superscript on W implies a *positive* argument to the exponential and *vice versa*. Let us suppose that we have:

$$\sum_{k=0}^{N-1} W_N^{-ak} W_N^{-bk^*} = \sum_{k=0}^{N-1} W_N^{-(a-b)k}$$
$$= \sum_{k=0}^{N-1} (W_N^{-p})^k (p = a - b)$$

If $a \neq b$ then $p \neq 0$, $p \neq N$, $p \neq -N$ and $W_N^{-p} \neq 1$.

$$\sum_{k=0}^{N-1} (W_N^{-p})^k = \frac{W_N^{-pN} - 1}{W_N^{-p} - 1} = \frac{1 - 1}{W_N^{-p} - 1} = 0$$

where we have used the series:

$$\sum_{n=0}^{N-1} r^n = \frac{r^N - 1}{r - 1}$$

and the fact that , since p is an integer, $W_N^{-pN} = 1$:

$$W_N^{-pN} = \exp\left(\frac{i2\pi pN}{N}\right) = \exp(i2\pi p) = \cos(i2\pi p) + i\sin(i2\pi p) = 1$$

If $a = b$:

$$\sum_{k=0}^{N-1} W_N^{(a-b)k} = \sum_{k=0}^{N-1} W_N^{0k} = N$$

We see that, just as is the case with the Fourier integrals, the exponential terms are orthogonal so that the second summation term in our inverse Fourier transform vanishes for every n except $n=m$. At $n=m$ it is equal to N so:

$$\begin{aligned}
\frac{1}{N}\sum_{k=0}^{N-1}F(k)\exp\left(\frac{i2\pi km}{N}\right) &= \frac{1}{N}\sum_{n=0}^{N-1}\sum_{k=0}^{N-1}f(n)\exp\left(\frac{-i2\pi kn}{N}\right)\exp\left(\frac{i2\pi km}{N}\right)\\
&= \frac{1}{N}\sum_{n=0}^{N-1}f(n)\sum_{k=0}^{N-1}W_N^{nk}W_N^{-mk}\\
&= \frac{1}{N}\sum_{n=0}^{N-1}f(n)\sum_{k=0}^{N-1}W_N^{(n-m)k}\\
&= \frac{1}{N}f(m)N\\
&= f(m) \quad m=0,1,2,3\ldots N-1
\end{aligned} \tag{8.55}$$

In eqn (8.55), line 3, the orthogonality of $W_N^{(n-m)k}$ leads to all summation terms with $n\neq m$ equal to zero. For $n=m$, $W_N^{(n-m)k}$ is equal to N, as we saw above and, since this is the only surviving term the first summation becomes just the one at $n=m$ leaving $f(m)$.

The astute among you will be drawing parallels with the continuous Fourier transform and the inverse Fourier transform. There is indeed a parallel. Equations (8.54) and (8.55) are the discrete Fourier transform (DFT) and the inverse Discrete Fourier transforms (IDFT), respectively. We note that the Fourier transform of N discrete time-domain data points results in N discrete frequency domain points and *vice versa*. We also note that this is an *approximation* of the continuous time-domain Fourier transform and its inverse.

The discrete frequency-domain data points arrived at *via* eqn (8.54) are periodic. This is easily shown by adding N into the equation:

$$F(k) = \sum_{n=0}^{N-1}f(n)\exp\left(\frac{-i2\pi kn}{N}\right)$$

$$\Downarrow$$

$$\begin{aligned}
F(k+N) &= \sum_{n=0}^{N-1}f(n)\exp\left(\frac{-i2\pi k(n+N)}{N}\right)\\
&= \sum_{n=0}^{N-1}f(n)\left(\exp\left(\frac{-i2\pi kn}{N}\right)\right)(\exp(-i2\pi kN))\\
&= \sum_{n=0}^{N-1}f(n)\exp\left(\frac{-i2\pi kn}{N}\right)\\
&= F(k)
\end{aligned} \tag{8.56}$$

Thus, any frequency-domain data point, $F(k)$ will be replicated N data points later (or earlier).

8.8.1 Example: A Four-Data-Point Transform

Let us consider sampling a simple time-domain function $f(t)=\sin(\omega_0 t)$. We will sample over one period and take $N=4$ samples, each at $t=n\pi/2$ where n ranges from 0 to 3. To ease computations we will let $\omega_0=1$ rad/s. This being the case, the set of samples of $f(t)$ are:

$$f(n)=\begin{bmatrix}0\\1\\0\\-1\end{bmatrix}$$

We wish to transform this into a set of four $F(k)$ *via* a DFT. In order to simplify the notation it is common to replace $\exp(-i2\pi kn/N)$ with W_N^{kn}. The calculation involves doing:

$$F(0)=f(0)W_4^0+f(1)W_4^0+f(2)W_4^0+f(3)W_4^0$$

$$F(1)=f(0)W_4^0+f(1)W_4^1+f(2)W_4^2+f(3)W_4^3$$

$$F(2)=f(0)W_4^0+f(1)W_4^2+f(2)W_4^4+f(3)W_4^6$$

$$F(3)=f(0)W_4^0+f(1)W_4^3+f(2)W_4^6+f(3)W_4^9$$

This should look familiar to those who have read the previous chapters ... it is a set of equations that can be represented by a matrix calculation on a vector:

$$\begin{bmatrix}F(0)\\F(1)\\F(2)\\F(3)\end{bmatrix}=\begin{bmatrix}W_4^0 & W_4^0 & W_4^0 & W_4^0\\W_4^0 & W_4^1 & W_4^2 & W_4^3\\W_4^0 & W_4^2 & W_4^4 & W_4^6\\W_4^0 & W_4^3 & W_4^6 & W_4^9\end{bmatrix}\begin{bmatrix}f(0)\\f(1)\\f(2)\\f(3)\end{bmatrix} \tag{8.57}$$

Note that n is being incremented horizontally in the W-matrix and k is being incremented vertically. It will be important to understand this later when we discuss the FFT algorithm.

In general, we can represent this as a transformation of the time-domain f-vector to the frequency-domain F-vector *via* a W-matrix multiplication:

$$\vec{\mathbf{F}}=\hat{\mathbf{W}}\vec{\mathbf{f}}$$

Continuing with our numerical example, explicit calculation[¶¶¶] of the various W^{kn} and multiplication into the $f(n)$:

$$\begin{bmatrix} F(0) \\ F(1) \\ F(2) \\ F(3) \end{bmatrix} = \begin{bmatrix} 1 & 1 & 1 & 1 \\ 1 & -i & -1 & i \\ 1 & -1 & 1 & -1 \\ 1 & i & -1 & -i \end{bmatrix} \begin{bmatrix} 0 \\ 1 \\ 0 \\ -1 \end{bmatrix}$$

$$= \begin{bmatrix} 0 \\ -2i \\ 0 \\ 2i \end{bmatrix}$$

Note that $F(3)$ is the complex conjugate of $F(1)$; it is generally the case that $F(N-k) = F^*(k)$ (or *vice versa*):

$$F(N-k) = \sum_{n=0}^{N-1} f(n)\exp\left(\frac{-i2\pi(N-k)n}{N}\right) \quad k = 0, 1, 2, 3 \ldots N-1$$

but

$$\exp\left(\frac{-i2\pi(N-k)n}{N}\right) = \exp(-i2\pi n)\exp\left(\frac{i2\pi nk}{N}\right)$$

$$= \exp\left(\frac{i2\pi nk}{N}\right)$$

and

$$F(N-k) = \sum_{n=0}^{N-1} f(n)\exp\left(\frac{i2\pi nk}{N}\right) = F^*(k) \tag{8.58}$$

A bit of reflection will reveal that the set of $|F(nk)|$ (see eqn (1.6)) is symmetrical about $F(N/2)$. Note that this does not include $F(0)$ as it is a somewhat special component of the $F(k)$ spectrum. $F(0)$ is the amplitude of the 0 Hz or average component of the spectrum and is

[¶¶¶]An excellent task for your favourite symbolic algebra software.

simply the sum of all of the time-domain components since all of the exponential terms are 1.

$$F(0) = \sum_{n=0}^{N-1} f(n)$$

In our current case, the average (over one period) of $f(t) = \sin(\omega_0 t)$ is zero. Addition of a constant to the function would result in a non-zero value for $F(0)$. A plot of the absolute value of our four frequency-domain data points is shown in Figure 8.26a.

Of course, as we have just seen, this is symmetric about $N/2$. In this case $N/2 = 2$ and the information above $N/2$ in the spectrum is redundant. Generally only the information from 0 to $N/2$ is plotted in the frequency-domain spectrum (Figure 8.26b).

This is more like what we would expect to see in a spectrum. We know that $\sin(2\pi t)$ has a frequency of 1 Hz and we expect to see in the frequency-domain spectrum a peak at 1 hertz. What about a time-domain signal with two frequencies in it? Let us look at $f(t) = \sin(32\pi t) + 5\cos(40\pi t)$. There are two frequencies evident here, 16 Hz and 20 Hz and we expect to see these in the standard frequency-domain spectrum. The highest frequency is 20 Hz and the Nyquist theorem above says that the minimum sampling frequency must be twice this frequency, 40 Hz. We will use 64 Hz here.[‖‖] The calculated

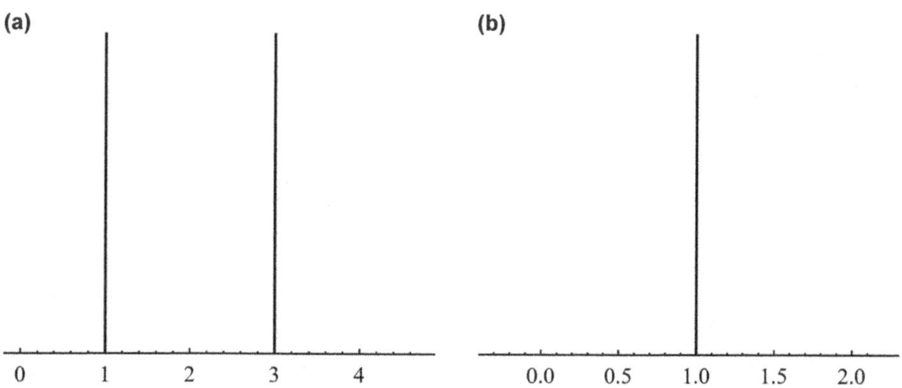

Figure 8.26 (a) Four-point DFT of $\sin(\omega_0 t)$; (b) First half of the spectrum of $\sin(\omega_0 t)$.

[‖‖]There are a lot of samples to deal with here and doing the calculations by hand is *very* tedious and error-prone. A symbolic algebra program is very useful here and elsewhere in this book ... I chose to use wxMaxima and its FFT capabilities, however, there are other excellent programs available. The FFT requires a number of samples that is a power of 2, hence the reason for using 64 samples instead of the minimum of 40.

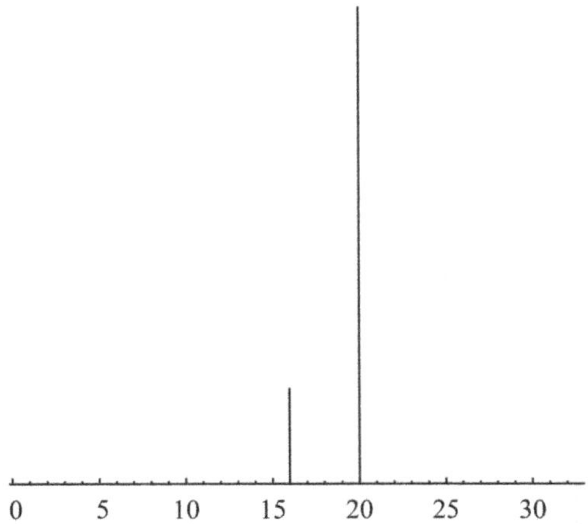

Figure 8.27 DFT of $f(t) = \sin(32t) + 5\cos(40t)$.

frequency-domain spectrum is shown in Figure 8.27. We indeed see two peaks at the expected frequencies of 16 and 20 Hz.

8.9 Real and Complex DFTs

The literature is a tangle of different ways of writing the DFT and inverse DFT equations and in discussing real and complex discrete Fourier transforms. There seems to be no standard way of doing things and one must be on one's toes when reading different books. Amidror (see the Further Reading) gives a nice overview of many of the various ways of writing things.

One of the potential sources of confusion are the terms real and imaginary. The real DFT is usually explained as a real data set being transformed to a complex frequency-domain data set, the implication being that the data set is simply a set of real numbers.**** Not so. The "real" data set is one that represents a symmetric or even function. As you can see in Appendix A, the sine and cosine functions are asymmetric and symmetric, respectively. Also, these are the basic functions used by Fourier analysis to model more complex functions.

Let us see what happens if we do a "real" DFT on a simple cosine function, $\cos(6\pi t)$. Sampling 8 times with a sampling frequency of 8 Hz (Figure 8.28).

****This is one of those "kopfschmerzen machen" topics when one is reading it for the first time.

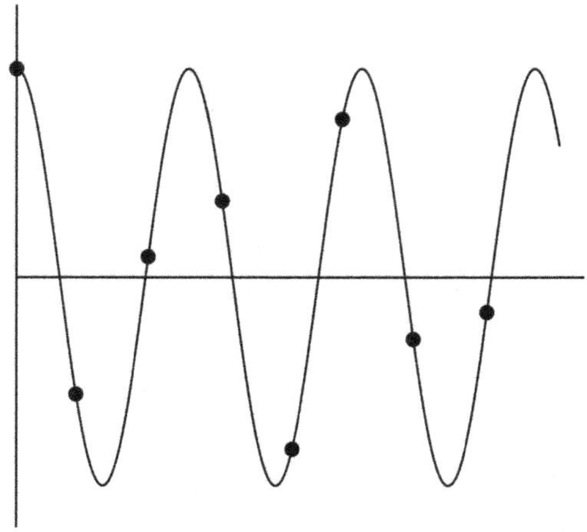

Figure 8.28 Cos($6\pi t$) sampled 8 times at 8 Hz.

We see in Figure 8.28 the continuous cosine function with the sampled data points overlaid on the function. Recall that cosine is a symmetric function. Application of a DFT to these data points produces, as before, a frequency-domain data set consisting of 8 points.

$$f(t) = [1.0 \quad -0.707 \quad 0.0 \quad 0.707 \quad -1.0 \quad 0.707 \quad 0.0 \quad -0.707]$$

$$\Downarrow DFT$$

$$F(\omega) = [0.0 \quad 0.0 \quad 0.0 \quad 4.0 \quad 0.0 \quad 4.0 \quad 0.0 \quad 0.0]$$

Note that in the spectral domain, $F(\omega)$, the numbers are real-only (meaning now, real numbers).

$F(0)$ is the 0 Hz point (sometimes called the DC point) and $F(1)$ is the 1 Hz point (see eqn (21.1) for more information on how we know this), $F(2) = 2$ Hz and $F(3) = 3$ Hz. $F(4)$ to $F(7)$ represent negative frequencies[††††] in reverse order; $F(4) = -4$ Hz, $F(5) = -3$ Hz, $F(6) = -2$ Hz and $F(7) = -1$ Hz. We do not ordinarily organise and plot frequencies in this order but rather from low frequencies to high frequencies. To do this we must rearrange the frequency-domain data points by shifting data points $F(0)$ to $F(N/2 - 1)$ to the right and wrapping

[††††]The concept of negative frequencies is a bit alien at first. However, when you consider that "0 Hz" may actually be many hundreds of megahertz as indeed it is in NMR spectroscopy, then it is much easier to understand. The DFT of a cosine function (or a sine function) alone cannot determine whether the frequency it has measured is more or less than the fundamental working frequency of the spectrometer. Therefore, it presents us with the two possibilities. To simplify things we usually call the fundamental or reference frequency used for the NMR experiment "0 Hz" and measure our frequencies with respect to this.

points $F(N/2)$ to $F(N-1)$ around to the beginning of the data set (Figure 8.29).

The amplitude/frequency plots show how this works visually in the frequency domain (Figure 8.30).

You will note that after rearrangement it is still symmetric about the centre of the plot and ... it should look a bit familiar if you have been following along in this chapter. Have a look at Figure 8.11a where you will find two frequencies plotted symmetrically about 0 Hz. Figure 8.30b is the DFT version of the cosine channel in Figure 8.11a. What will the standard frequency plot look like if we use $\sin(6\pi t)$? If you guessed something like the sine channel of Figure 8.31, you were right.

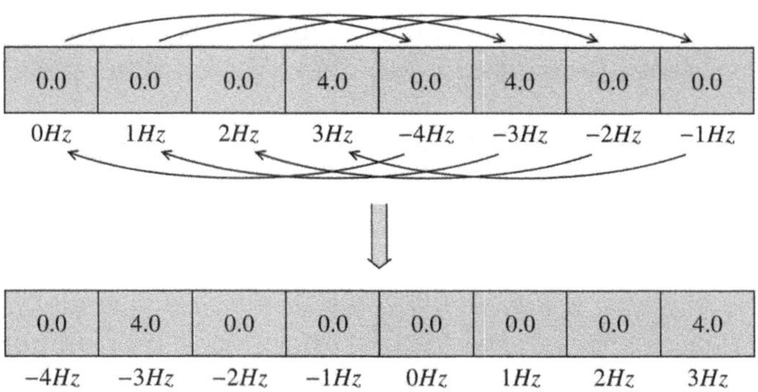

Figure 8.29 Rearranging the frequency-domain data set from low to high frequency.

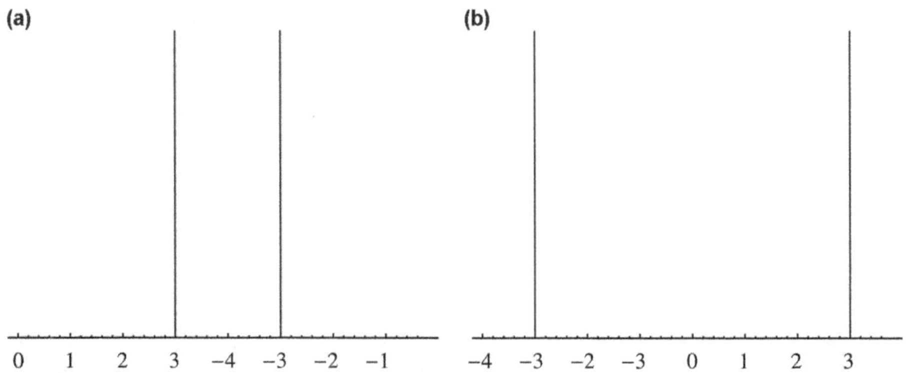

Figure 8.30 (a) DFT frequency domain plot for $\cos(6\pi t)$; (b) Rearranged standard frequency-domain plot for $\cos(6\pi t)$.

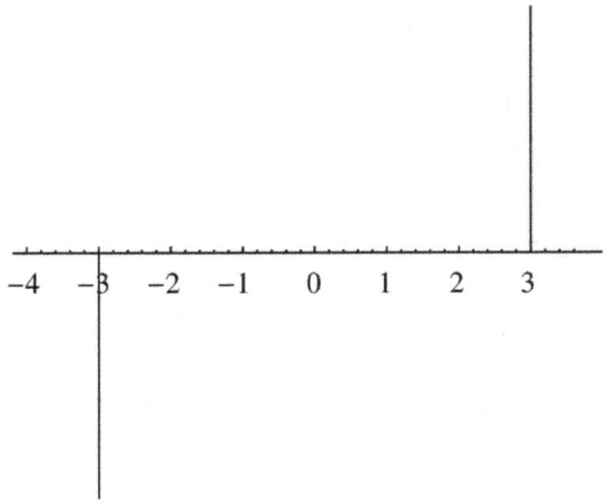

Figure 8.31 DFT standard frequency-domain plot for sin($6\pi t$).

A subtlety concerning the digitised data needs a bit of clarification. For our last example using sin($6\pi t$) the digitised data are:

$$f(t) = [0.0 \quad 0.707 \quad -1.0 \quad 0.707 \quad 0.0 \quad -0.707 \quad 1.0 \quad -0.707]$$

$$\Downarrow \text{DFT}$$

$$F(\omega) = [0.0 \quad 0.0 \quad 0.0 \quad -4.0i \quad 0.0 \quad 4.0i \quad 0.0 \quad 0.0]$$

The DFT must always produce real and imaginary numbers and we can consider that, in the case of cos($6\pi t$), all of the imaginary portions of the resulting DFT are zero, as noted above, leaving only real numbers. In the case of sin($6\pi t$) it is just the opposite. We get numbers that are either zero or imaginary. Thus, the data set resulting from the DFT consists of real and imaginary numbers combined into complex numbers. The DFT data set from sin($6\pi t$) would then be:

$$\text{Real} \quad [0.0 \quad 0.0 \quad 0.0 \quad 0.0 \quad 0.0 \quad 0.0 \quad 0.0 \quad 0.0]$$

$$\text{Imaginary} \quad [0.0i \quad 0.0i \quad 0.0i \quad -4.0i \quad 0.0i \quad 4.0i \quad 0.0i \quad 0.0i]$$

Our examples have been a bit contrived; we have shown the DFT for pure symmetric (pure real) time-domain data and pure anti-symmetric (pure imaginary) data and linked them to the real and imaginary parts of the resulting spectral data. For the symmetric case the imaginary portion of the frequency-domain data is all zero and for the anti-symmetric case the real portion of the spectral data is all zero. Although these situations are not impossible, it is most likely that any time-domain data will be a mixture of symmetric and anti-symmetric functions leading to spectral data points that are a mixture of non-zero real and imaginary parts.

What will be the sum of the sine and cosine (or imaginary and real) data? Figure 8.32 shows that, analogous to Figure 8.12, we get one peak in the spectrum.

This is what we would see if we were to transform the function $\cos(6\pi t) + i\,\sin(6\pi t)$. Thus, if we collect our time-domain data *via* two channels and arrange that these data be guaranteed to be out of phase by 90° then Fourier transform of each channel followed by addition of the resulting spectral data will give (in our simple case at least) a single peak at either a positive or negative position in the spectrum.[‡‡‡‡]

A more realistic function to model might be an exponentially decaying signal such as $[\cos(6\pi t) + i\sin(6\pi t)]\exp(-0.5t)$ (see Figure 8.33). When we do so we notice something a bit odd.

The first spectrum in Figure 8.33a shows a constant baseline offset.[§§§§] What is the origin of this effect? Consider our notation for the DFT:

$$F(k) = \sum_{n=0}^{N-1} f(n)\mathrm{e}^{-i2\pi kn/N}$$

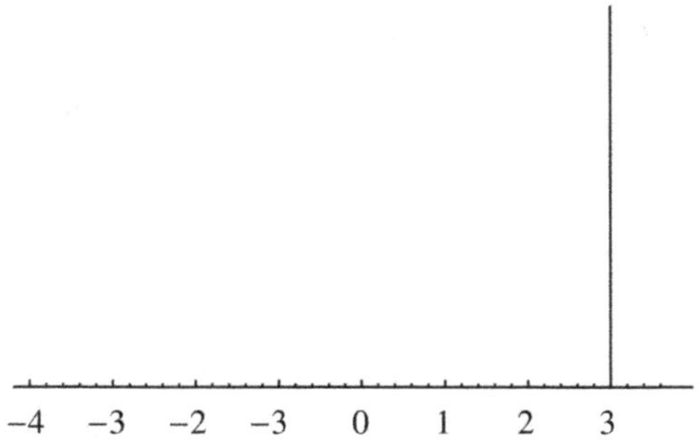

Figure 8.32 Sum of DFT standard frequency-domain plots for $\sin(6\pi t)$ and $\cos(6\pi t)$.

---

‡‡‡‡This is called quadrature detection in NMR jargon and is discussed in more detail in Chapter 21.

§§§§Baseline offsets are discussed in detail in almost no texts. I have only ever seen it in Hoch and Stern (see the Further Reading section) and my analysis is therefore necessarily heavily influenced by their discussion of the topic.

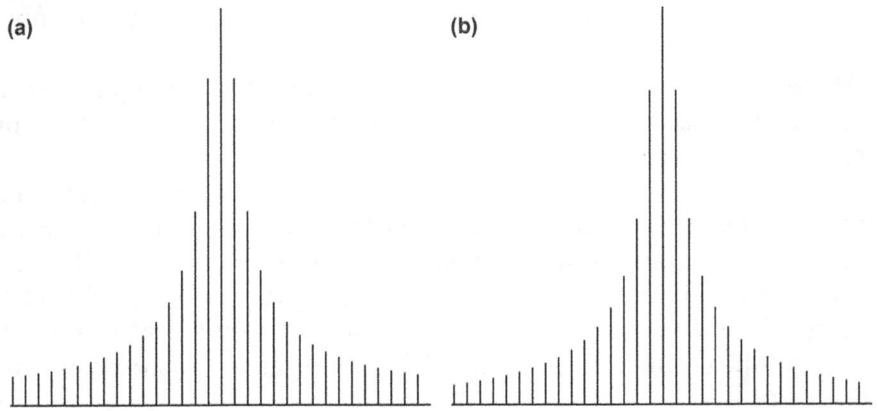

Figure 8.33 (a) DFT of $[\cos(6\pi t) + i \sin(6\pi t)]\exp(-0.5t)$; (b) first time-domain data point multiplied by 0.5.

Looking at a zero-frequency, decaying signal, an exponential function, in other words, we could write this as:

$$F(k) = \sum_{n=0}^{N-1} e^{-\pi n L \Delta t} e^{-i2\pi kn/N} = \sum_{n=0}^{N-1} z^n$$

where

$$z = e^{-\pi L \Delta t} e^{-i2\pi k/N}$$

The z-summation is a simple geometric series, as we saw above in the discussion of the orthogonal nature of the DFT exponentials:

$$\sum_{n=0}^{N-1} z^n = \frac{1 - z^N}{1 - z} = \frac{1 - e^{-\pi LN\Delta t} e^{-i2\pi k}}{1 - e^{-\pi L\Delta t} e^{-i2\pi k/N}}$$

That this is so can be seen from:

$$\sum_{n=0}^{N-1} z^n = s = 1 + z + z^2 + z^3 \ldots z^{N-1}$$

$$zs = z + z^2 + z^3 + z^4 \ldots z^N$$

$$s - zs = s(1 - z) = 1 - z^N$$

$$s = \frac{1 - z^N}{1 - z}$$

$$= \sum_{n=0}^{N-1} z^n$$

If π *LN* Δt is relatively large (a rapid exponential decay) the value of the numerator in the *z*-summation expression becomes one and we write the approximation:

$$F(k) = \sum_{n=0}^{N-1} z^n \approx \frac{1}{1 - e^{-\pi L \Delta t} e^{-i2\pi k/N}}$$

You can see clearly that this can never be zero since there is a constant term ... exp(π *L* Δt).

Another way to look at this is by examining the Fourier transform explicitly. For simplicity, we will look again at the $N = 4$ version from above:

$$\begin{bmatrix} W_4^0 & W_4^0 & W_4^0 & W_4^0 \\ W_4^0 & W_4^1 & W_4^2 & W_4^3 \\ W_4^0 & W_4^2 & W_4^4 & W_4^6 \\ W_4^0 & W_4^3 & W_4^6 & W_4^9 \end{bmatrix} \begin{bmatrix} f(0) \\ f(1) \\ f(2) \\ f(3) \end{bmatrix} = \begin{bmatrix} 1 & 1 & 1 & 1 \\ 1 & -i & -1 & i \\ 1 & -1 & 1 & -1 \\ 1 & i & -1 & -i \end{bmatrix} \begin{bmatrix} f(0) \\ f(1) \\ f(2) \\ f(3) \end{bmatrix}$$

and

$$F(0) = f(0) + f(1) + f(2) + f(3)$$

$$F(1) = f(0) - if(1) - f(2) + if(3)$$

$$F(2) = f(0) - f(1) + f(2) - f(3)$$

$$F(3) = f(0) + if(1) - f(2) - if(3)$$

Each $F(k)$ has a constant term in it ... $f(0)$, the first time-domain data point. It stands to reason that decreasing the value of $f(0)$ by some amount will decrease the constant being added into each frequency-domain data point. We do not want to simply make it zero since that would very likely result in a few negative frequency-domain data points. A good compromise is to multiply the first time-domain data point by 0.5. In spectrum (b) of Figure 8.33 the first time data point has been multiplied by 0.5 before the DFT operation and you can see that the offset is reduced.

8.10 The Fast Fourier Transform

Consider the general form of the Fourier transform matrix, W:

$$\begin{bmatrix} W_N^0 & W_N^0 & W_N^0 & \cdots & W_N^0 \\ W_N^0 & W_N^1 & W_N^2 & \cdots & W_N^{N-1} \\ W_N^0 & W_N^2 & W_N^4 & \cdots & W_N^{2(N-1)} \\ \vdots & \vdots & \vdots & \vdots & \vdots \\ W_N^0 & W_N^{N-1} & W_N^{2(N-1)} & \cdots & W_N^{(N-1)^2} \end{bmatrix}$$

and recall that we substituted W_N^{kn} for $\exp(-i2\pi kn/N)$, which is a cyclic function, as we saw earlier ... it repeats itself periodically. For example:

$$W_N^N = e^{-i2\pi N/N} = e^{-i2\pi} = 1$$

$$W_N^{\frac{N}{2}} = e^{-i2\pi N/2/N} = e^{-i\pi} = -1$$

$$W_N^{N+n} = e^{-i2\pi(N+n)/N} = e^{-i2\pi N/N}e^{-i2\pi n/N} = W_N^N W_N^n = W_N^n$$

This means that the transform matrix may be simplified somewhat in that the values of W repeat themselves periodically. However, when the matrix is used to transform the time-domain data there will be N^2 multiplications regardless of this simplification. We wish to find a more efficient (faster, in other words) way to do our DFT. The method generally used is the "divide and conquer" method. We can see a clear example of this idea in a simple sorting algorithm. Suppose we have a set of numbers that we wish to sort from lowest to highest.

5 3 4 9 1 7 6 2

We could use a brute-force approach called a selection sort and step through the list to find the lowest number and put it in its place in the list:

5	3	4	9	1	7	6	2
1	5	3	4	9	7	6	2
1	2	5	3	4	9	7	6
1	2	3	5	4	9	7	6
			⇓				
1	2	3	4	5	6	7	9

On this sample it would require a total of 28 comparisons ... 7 in the first pass, 6 in the second, 5 in the third and so on down to 1 in the last comparison. In general, for a sample of n numbers this would be:

$$(n-1)+(n-2)+(n-3)\ \ldots\ +2+1=\frac{n(n-1)}{2}$$

For large n this is approximately $n^2/2$.

Let us try a "divide and conquer" approach. We will take our list and break it up into sub-lists, which we will call 2-lists, each containing only two numbers.

$$(5\ \ 3)\ (4\ \ 9)\ (1\ \ 7)\ (6\ \ 2)$$

We now sort each 2-list:

$$(5\quad 3)\,(4\quad 9)\,(1\quad 7)\,(6\quad 2)$$

$$\Downarrow$$

$$(3\quad 5)\,(4\quad 9)\,(1\quad 7)\,(2\quad 6)$$

Next, we build higher-order 4-lists from the sorted 2-lists.

$$[(3\ \ 5)\ (4\ \ 9)]\quad \text{and}\quad [(1\ \ 7)\ (2\ \ 6)]$$

We compare the first element of each 2-list and put the smaller into the 4-list:

$$[(3\quad 5)\,(4\quad 9)]$$

$$\Downarrow$$

$$[(5)\,(4\quad 9)\,(3)]$$

Next, compare the number in the 1-list to the two in the 2-list and put the smaller of the two into the next position of the 4-list:

$$[(3\quad 5)\,(4\quad 9)]$$

$$\Downarrow$$

$$[(5)\,(4\quad 9)\,(3)]$$

$$\Downarrow$$

$$[(5)\,(9)\,(3\quad 4)]$$

And finally, in this case compare the last two and put them into the 4-list in order:

$$[(3\quad 5)\,(4\quad 9)]$$

$$\Downarrow$$

$$[(5)\,(4\quad 9)\,(3)]$$

$$\Downarrow$$

$$[(5)\,(9)\,(3\quad 4)]$$

$$\Downarrow$$

$$[3\quad 4\quad 5\quad 9]$$

The same procedure on the second two 2-lists gives us a second 4-list:

$$[3\ \ 4\ \ 5\ \ 9]\,[1\ \ 2\ \ 6\ \ 7]$$

Finally, we repeat the algorithm using now the 4-lists to get:

$$[1\ \ 2\ \ 3\ \ 4\ \ 5\ \ 6\ \ 7\ \ 9]$$

In this case we have done 17 comparisons in three steps as opposed to 28 comparisons and 8 steps for the first sort algorithm that we tried. In general, for the merge-sort the number of operations is N log N and the point here is that we have made the computation more efficient, most especially for large values of N. The same kind of thinking goes into the Fast Fourier transform.

What we are about to do is split the time-domain data set into two . . . the even and odd members of the time-domain vector. This is much as we just did in the sorting problem. Doing this, we will have two new vectors each of which will now be $N/2$ in length. It should be noted that we are considering only time-domain vectors that have $2n$ members. With this splitting of the data set in mind let us consider the W values.

$$W_{N/2}^{kn} = e^{i2\pi kn/N/2} = e^{i4\pi kn/N} = W_N^{2kn}$$

So, any powers of $W_{N/2}$ are even powers of W_N. The odd powers of W_N are handled in the following way:

$$W_N^{(2n+1)k} = W_N^{2kn} W_N^k = W_{N/2}^{kn} W_N^k$$

Our DFT is, as above:

$$F(k) = \sum_{n=0}^{N-1} f(n)\mathrm{e}^{-i2\pi kn/N}$$

$$= \sum_{n=0}^{N-1} f(n)W_N^{kn}$$

and splitting it up into even and odd components:

$$F(k) = \sum_{n=0}^{N-1} f(n)W_N^{kn} = \sum_{n=0}^{N/2-1} f(2n)W_N^{2kn} + \sum_{n=0}^{N/2-1} f(2n+1)W_N^{(2n+1)k}$$

$$= \sum_{n=0}^{N/2-1} f(2n)W_{N/2}^{nk} + \sum_{n=0}^{N/2-1} f(2n+1)W_{N/2}^{nk}W_N^k$$

$$= \sum_{n=0}^{N/2-1} f(2n)W_{N/2}^{nk} + W_N^k \sum_{n=0}^{N/2-1} f(2n+1)W_{N/2}^{nk}$$

Now, here is one of the subtle things about the FFT that is a bit tricky to pick up. For our regular DFT we have N-tuples of time-domain data going in and N-tuples of frequency-domain data coming out. It is the same for our current sums. We have $N/2$-tuples of data as input and we expect $N/2$-tuples of data coming out. So, we look at the $F(k)$ s first over the range 0 to $N/2-1$. This is just what we have currently:

$$F(k) = \sum_{n=0}^{N/2-1} f(2n)W_{N/2}^{nk} + W_N^k \sum_{n=0}^{N/2-1} f(2n+1)W_{N/2}^{nk} \qquad k = 0, 1, 2, 3 \ldots \frac{N}{2} - 1 \tag{8.59}$$

We still have $N/2$ frequency-domain data points to generate, from $N/2$ to $N-2$ ... in other words we want to write eqn (8.59) for k from $N/2$ to $N-1$. However, instead of specifying this range let us use the indices $k+N/2$ so that we can continue to sum over 0 to $N/2-1$. With this in mind let us look at the Ws for the even sum:

$$W_{N/2}^{(k+N/2)n} = W_{N/2}^{kn} W_{N/2}^{N/2} = W_{N/2}^{kn}$$

Since, as we found above:

$$W_{N/2}^{N/2} = 1$$

The first and second summation terms in eqn (8.59) do not change. What about the multiplier in front of the second sum? The only change here will be a sign change:

$$W_N^{(k+N/2)} = W_N^k W_N^{N/2} = -W_N^k$$

Since:

$$W_N^{N/2} = -1$$

Our sum formula for the second half of the frequency-domain data is therefore:

$$F\left(k+\frac{N}{2}\right) = \sum_{n=0}^{N/2-1} f(2n)W_{N/2}^{nk} - W_N^k \sum_{n=0}^{N/2-1} f(2n+1)W_{N/2}^{nk} \tag{8.60}$$

$$k = 0, 1, 2, 3 \ldots \frac{N}{2} - 1$$

Note the minus sign in front of W_N^k. This will be of importance in a moment.

Time for an example ... a four-point DFT will look like:

$$\begin{bmatrix} F(0) \\ F(1) \\ F(2) \\ F(3) \end{bmatrix} = \begin{bmatrix} W_4^0 & W_4^0 & W_4^0 & W_4^0 \\ W_4^0 & W_4^1 & W_4^2 & W_4^3 \\ W_4^0 & W_4^2 & W_4^4 & W_4^6 \\ W_4^0 & W_4^3 & W_4^6 & W_4^9 \end{bmatrix} \begin{bmatrix} f(0) \\ f(1) \\ f(2) \\ f(3) \end{bmatrix}$$

$$= \begin{bmatrix} 1 & 1 & 1 & 1 \\ 1 & W_4^1 & -1 & -W_4^1 \\ 1 & -1 & 1 & -1 \\ 1 & -W_4^1 & -1 & W_4^9 \end{bmatrix} \begin{bmatrix} f(0) \\ f(1) \\ f(2) \\ f(3) \end{bmatrix}$$

$$= \hat{\mathbf{W}}_4 \vec{\mathbf{f}}_4$$

$$\begin{bmatrix} F(0) \\ F(1) \\ F(2) \\ F(3) \end{bmatrix} = \begin{bmatrix} f(0) + f(1) + f(2) + f(3) \\ f(0) + W_4^1 f(1) - f(2) - W_4^1 f(3) \\ f(0) - f(1) + f(2) - f(3) \\ f(0) - W_4^1 f(1) - f(2) + W_4^9 f(3) \end{bmatrix}$$

Here, we have used $\hat{\mathbf{W}}_4$ to represent the 4×4 Fourier transform matrix and $\vec{\mathbf{f}}_4$ for the 4-point time-domain data vector.

Now let us try our FFT algorithm. First, we sort the even and odd members of the time-domain vector using a sorting matrix:

$$\begin{bmatrix} f(0) \\ f(2) \\ f(1) \\ f(3) \end{bmatrix} = \begin{bmatrix} 1 & 0 & 0 & 0 \\ 0 & 0 & 1 & 0 \\ 0 & 1 & 0 & 0 \\ 0 & 0 & 0 & 1 \end{bmatrix} \begin{bmatrix} f(0) \\ f(1) \\ f(2) \\ f(3) \end{bmatrix}$$

Next, we apply the $N/2$ transform matrices on the even and odd sets:

$$\begin{bmatrix} f(0) + f(2) \\ f(0) + W_2^1 f(2) \\ f(1) + f(3) \\ f(1) + W_2^1 f(3) \end{bmatrix} = \begin{bmatrix} 1 & 1 & 0 & 0 \\ 1 & W_2^1 & 0 & 0 \\ 0 & 0 & 1 & 1 \\ 0 & 0 & 1 & W_2^1 \end{bmatrix} \begin{bmatrix} f(0) \\ f(2) \\ f(1) \\ f(3) \end{bmatrix}$$

Note that the $N/2$ transform matrix is a block-matrix along the diagonal with the upper left block operating on the even data points and the lower right block operating on the odd data points. At this point we have:

$$\hat{\mathbf{W}}_2 \vec{\mathbf{f}}_{\text{even}} = \begin{bmatrix} f(0) + f(2) \\ f(0) + W_2^1 f(2) \end{bmatrix} \quad \hat{\mathbf{W}}_2 \vec{\mathbf{f}}_{\text{odd}} = \begin{bmatrix} f(1) + f(3) \\ f(1) + W_2^1 f(3) \end{bmatrix}$$

where:

$$\hat{\mathbf{W}}_2 = \begin{bmatrix} 1 & 1 \\ 1 & W_2^1 \end{bmatrix} \quad \vec{\mathbf{f}}_{\text{even}} = \begin{bmatrix} f(0) \\ f(2) \end{bmatrix} \quad \vec{\mathbf{f}}_{\text{odd}} = \begin{bmatrix} f(1) \\ f(3) \end{bmatrix}$$

More subtlety; in our $N/2$ transform step using the $\hat{\mathbf{W}}_2$ matrix (the Fourier transform 2×2 the k-indices were incremented only over 0 and 1. Ultimately, they will have to be incremented over 0 to 3 so one more step is needed. To do this, referring to eqn (8.59) and (8.60), we do not want to touch the even indices of k but for the odd indices we multiply the kth component of the first half of $\hat{\mathbf{W}}_2$ by W_4^m and the second half by $-W_4^m$:

$$\hat{\mathbf{A}}_4 = \begin{bmatrix} 1 & 0 & W_4^0 & 0 \\ 0 & 1 & 0 & W_4^1 \\ 1 & 0 & W_4^2 & 0 \\ 0 & 1 & 0 & W_4^3 \end{bmatrix} = \begin{bmatrix} 1 & 0 & 1 & 0 \\ 0 & 1 & 0 & W_4^1 \\ 1 & 0 & -1 & 0 \\ 0 & 1 & 0 & W_4^3 \end{bmatrix}$$

Note the construction of this matrix (which we shall call the *reassembly matrix* here since its application produces the final spectrum vector). It consists of two 2×2 identity matrices on the left half of the 4×4 matrix and two 2×2 diagonal matrices on the right half. The identity matrices multiply into the even data points and the right-half 2×2 matrices multiply into the odd data points, which is just what we want referring to eqn (8.60).

$$\hat{\mathbf{A}}_4 \begin{bmatrix} f(0)+f(2) \\ f(0)+W_2^1 f(2) \\ f(1)+f(3) \\ f(1)+W_2^1 f(3) \end{bmatrix} = \begin{bmatrix} 1 & 0 & 1 & 0 \\ 0 & 1 & 0 & W_4^1 \\ 1 & 0 & -1 & 0 \\ 0 & 1 & 0 & W_4^3 \end{bmatrix} \begin{bmatrix} f(0)+f(2) \\ f(0)+W_2^1 f(2) \\ f(1)+f(3) \\ f(1)+W_2^1 f(3) \end{bmatrix}$$

$$= \begin{bmatrix} f(0)+f(1)+f(2)+f(3) \\ f(0)+W_4^1 f(1)-f(2)-W_4^1 f(3) \\ f(0)-f(1)+f(2)-f(3) \\ f(0)-W_4^1 f(1)-f(2)+W_4^1 f(3) \end{bmatrix} = \begin{bmatrix} F(0) \\ F(1) \\ F(2) \\ F(3) \end{bmatrix}$$

Sure enough, exactly the same result from the regular DFT! Let us put all the FFT matrices together and examine them:

$$\begin{bmatrix} F(0) \\ F(1) \\ F(2) \\ F(3) \end{bmatrix} = \begin{bmatrix} 1 & 0 & 1 & 0 \\ 0 & 1 & 0 & W_4^1 \\ 1 & 0 & -1 & 0 \\ 0 & 1 & 0 & W_4^3 \end{bmatrix} \begin{bmatrix} 1 & 0 & 1 & 0 \\ 0 & 1 & 0 & W_4^1 \\ 1 & 0 & -1 & 0 \\ 0 & 1 & 0 & W_4^3 \end{bmatrix} \begin{bmatrix} 1 & 0 & 0 & 0 \\ 0 & 0 & 1 & 0 \\ 0 & 1 & 0 & 0 \\ 0 & 0 & 0 & 1 \end{bmatrix} \begin{bmatrix} f(0) \\ f(1) \\ f(2) \\ f(3) \end{bmatrix}$$

Lots of zeros! There are 48 entries in the three square matrices but only 20 non-zero ones. For the regular DFT there are N^2 multiplications and N^2 additions for a total of $2N^2$ operations. Splitting the time-domain data points into even and odd vectors of order $N/2$ requires $2(N/2)^2$ or $N^2/2$ multiplications and $2(N/2)^2 = N^2/2$ additions for a total of $N/2$ operations. We have halved the amount of work done!

In order to proceed we will abstract the operations that we have already discussed and write the FFT matrices in a general form:

$$\begin{bmatrix} F(0) \\ F(1) \\ \vdots \\ F(N-1) \end{bmatrix} = \hat{\mathbf{F}}_N \vec{\mathbf{f}} = \begin{bmatrix} \hat{\mathbf{1}}_{N/2} & \hat{\Delta}_{N/2} \\ \hat{\mathbf{1}}_{N/2} & \hat{\Delta}_{N/2} \end{bmatrix} \begin{bmatrix} \hat{\mathbf{F}}_{N/2} & 0 \\ 0 & \hat{\mathbf{F}}_{N/2} \end{bmatrix} \begin{bmatrix} \text{even / odd} \\ \text{sorting matrix} \end{bmatrix} \begin{bmatrix} f(0) \\ f(1) \\ \vdots \\ f(N-1) \end{bmatrix}$$

$$= \hat{\mathbf{A}}_{N/2} \begin{bmatrix} \hat{\mathbf{F}}_{N/2} & 0 \\ 0 & \hat{\mathbf{F}}_{N/2} \end{bmatrix} \begin{bmatrix} \text{even / odd} \\ \text{sorting matrix} \end{bmatrix} \vec{\mathbf{f}}$$

$$= \hat{\mathbf{A}}_{N/2} \begin{bmatrix} \hat{\mathbf{F}}_{N/2} & 0 \\ 0 & \hat{\mathbf{F}}_{N/2} \end{bmatrix} \begin{bmatrix} \vec{\mathbf{E}}_{N/2} \\ \vec{\mathbf{O}}_{N/2} \end{bmatrix}$$

where:

$$\hat{\Delta}_{N/2} = \begin{bmatrix} 1 & 0 & 0 & \cdots & 0 \\ 0 & W_N^1 & 0 & \cdots & 0 \\ 0 & 0 & W_N^2 & \cdots & 0 \\ \vdots & \vdots & \vdots & \vdots & \vdots \\ 0 & 0 & 0 & \cdots & W_N^{N/2-1} \end{bmatrix} \quad \hat{\mathbf{1}}_{N/2} = \begin{bmatrix} 1 & 0 & 0 & \cdots & 0 \\ 0 & 1 & 0 & \cdots & 0 \\ 0 & 0 & 1 & \cdots & 0 \\ \vdots & \vdots & \vdots & \vdots & \vdots \\ 0 & 0 & 0 & \cdots & 1 \end{bmatrix}$$

and $\vec{\mathbf{E}}_{N/2}$ and $\vec{\mathbf{O}}_{N/2}$ are the sorted even and odd data points.

To get the full benefit of this technique on large data sets we would break the time-domain data up into two parts, even and odd as above and then do so again and again until we are down to one data point. What is the DFT of one data point? Amazingly, just itself![¶¶¶¶] Let us expand our example to an 8-member time-domain data vector:

$$\vec{\mathbf{f}}_8 = \begin{bmatrix} f(0) \\ f(1) \\ f(2) \\ f(3) \\ f(4) \\ f(5) \\ f(6) \\ f(7) \end{bmatrix}$$

[¶¶¶¶]Verify this for yourself.

The first thing to do is reorganise into even and odd sets of data:

$$\begin{bmatrix} \vec{\mathbf{E}}_4 \\ \vec{\mathbf{O}}_4 \end{bmatrix} = \hat{\mathbf{S}}\vec{f}$$

where $\hat{\mathbf{S}}$ is the even/odd sorting matrix. Next, we multiply by the even/odd transform matrix:

$$\begin{bmatrix} \hat{\mathbf{F}}_4 & \hat{\mathbf{0}} \\ \hat{\mathbf{0}} & \hat{\mathbf{F}}_4 \end{bmatrix} \begin{bmatrix} \vec{\mathbf{E}}_4 \\ \vec{\mathbf{O}}_4 \end{bmatrix}$$

and finally by the reassembly matrix:

$$\begin{aligned} \vec{\mathbf{F}} = \hat{\mathbf{F}}_8\vec{\mathbf{f}}_8 &= \begin{bmatrix} \hat{\mathbf{1}}_4 & \hat{\Delta}_4 \\ \hat{\mathbf{1}}_4 & \hat{\Delta}_4 \end{bmatrix} \begin{bmatrix} \hat{\mathbf{F}}_4 & \hat{\mathbf{0}} \\ \hat{\mathbf{0}} & \hat{\mathbf{F}}_4 \end{bmatrix} \begin{bmatrix} \vec{\mathbf{E}}_4 \\ \vec{\mathbf{O}}_4 \end{bmatrix} \\ &= \hat{\mathbf{A}}_8 \begin{bmatrix} \vec{\mathbf{P}}_4 \\ \vec{\mathbf{Q}}_4 \end{bmatrix} \end{aligned} \tag{8.61}$$

Yet again, we find more subtlety here, this time in the form of recursion. There is a Fourier transform within a Fourier transform within a Fourier transform.[‖‖‖] To see this we look more closely at $\vec{\mathbf{P}}$ and $\vec{\mathbf{Q}}$:

$$\vec{\mathbf{P}}_4 = \hat{\mathbf{F}}_4[\vec{\mathbf{E}}_4] = \begin{bmatrix} \hat{\mathbf{1}}_2 & \hat{\Delta}_2 \\ \hat{\mathbf{1}}_2 & \hat{\Delta}_2 \end{bmatrix} \begin{bmatrix} \hat{\mathbf{F}}_2 & \hat{\mathbf{0}} \\ \hat{\mathbf{0}} & \hat{\mathbf{F}}_2 \end{bmatrix} \begin{bmatrix} \vec{\mathbf{E}}'_2 \\ \vec{\mathbf{O}}'_2 \end{bmatrix} = \begin{bmatrix} \hat{\mathbf{1}}_2 & \hat{\Delta}_2 \\ \hat{\mathbf{1}}_2 & \hat{\Delta}_2 \end{bmatrix} \begin{bmatrix} \vec{\mathbf{P}}'_2 \\ \vec{\mathbf{Q}}'_2 \end{bmatrix} = \hat{\mathbf{A}}_4 \begin{bmatrix} \vec{\mathbf{P}}'_2 \\ \vec{\mathbf{Q}}'_2 \end{bmatrix}$$

$$\mathbf{Q}_4 = \hat{\mathbf{F}}_4[\vec{\mathbf{O}}_4] = \begin{bmatrix} \hat{\mathbf{1}}_2 & \hat{\Delta}_2 \\ \hat{\mathbf{1}}_2 & \hat{\Delta}_2 \end{bmatrix} \begin{bmatrix} \hat{\mathbf{F}}_2 & \hat{\mathbf{0}} \\ \hat{\mathbf{0}} & \hat{\mathbf{F}}_2 \end{bmatrix} \begin{bmatrix} \vec{\mathbf{E}}''_2 \\ \vec{\mathbf{O}}''_2 \end{bmatrix} = \begin{bmatrix} \hat{\mathbf{1}}_2 & \hat{\Delta}_2 \\ \hat{\mathbf{1}}_2 & \hat{\Delta}_2 \end{bmatrix} \begin{bmatrix} \vec{\mathbf{P}}''_2 \\ \vec{\mathbf{Q}}''_2 \end{bmatrix} = \hat{\mathbf{A}}_4 \begin{bmatrix} \vec{\mathbf{P}}''_2 \\ \vec{\mathbf{Q}}''_2 \end{bmatrix}$$

To see why this is so look closely at eqn (8.61). We see there that:

$$\hat{\mathbf{F}}_8\vec{\mathbf{f}}_8 = \hat{\mathbf{A}}_8 \begin{bmatrix} \vec{\mathbf{P}}_4 \\ \vec{\mathbf{Q}}_4 \end{bmatrix} \tag{8.62}$$

We see also that to get to this we employ a $N/2$ Fourier transform:

$$\vec{\mathbf{P}}_4 = \hat{\mathbf{F}}_4[\vec{\mathbf{E}}_4]$$

[‖‖‖] It boggles the mind to think about this but that is the nature of recursion and the FFT algorithm.

But from eqn (8.61) we see the relationship between the size of the current $\hat{\mathbf{F}}$ matrix and the $\hat{\mathbf{A}}$ matrix so we can write:

$$\hat{\mathbf{F}}_4\left[\vec{\mathbf{E}}_4\right] = \hat{\mathbf{A}}_4\begin{bmatrix}\vec{\mathbf{P}}'_2\\ \vec{\mathbf{Q}}'_2\end{bmatrix}$$

$$\hat{\mathbf{F}}_4\left[\vec{\mathbf{O}}_4\right] = \hat{\mathbf{A}}_4\begin{bmatrix}\vec{\mathbf{P}}''_2\\ \vec{\mathbf{Q}}''_2\end{bmatrix}$$

and we can therefore substitute into eqn (8.61):

$$\vec{\mathbf{F}} = \hat{\mathbf{F}}_8\vec{\mathbf{f}}_8 = \hat{\mathbf{A}}_8\begin{bmatrix}\hat{\mathbf{A}}_2 & \hat{\mathbf{0}}\\ \hat{\mathbf{0}} & \hat{\mathbf{A}}_2\end{bmatrix}\begin{bmatrix}\vec{\mathbf{P}}'_2\\ \vec{\mathbf{Q}}'_2\\ \vec{\mathbf{P}}''_2\\ \vec{\mathbf{Q}}''_2\end{bmatrix}$$

Taking this reasoning to the limit:

$$\vec{\mathbf{F}} = \hat{\mathbf{A}}_8\begin{bmatrix}\hat{\mathbf{A}}_2 & \hat{\mathbf{0}} & \hat{\mathbf{0}} & \hat{\mathbf{0}}\\ \hat{\mathbf{0}} & \hat{\mathbf{A}}_2 & \hat{\mathbf{0}} & \hat{\mathbf{0}}\\ \hat{\mathbf{0}} & \hat{\mathbf{0}} & \hat{\mathbf{A}}_2 & \hat{\mathbf{0}}\\ \hat{\mathbf{0}} & \hat{\mathbf{0}} & \hat{\mathbf{0}} & \hat{\mathbf{A}}_2\end{bmatrix}\begin{bmatrix}\text{sorted}\\ \text{time-domain}\\ \text{data}\end{bmatrix} \tag{8.63}$$

We could take it a bit further and write an 8×8 $\hat{\mathbf{A}}_1$ matrix but this would simply multiply each sorted data point by one as noted above for the Fourier transform of a single data point.

It turns out that sorting the time-domain data by computer is spectacularly easy and fast.***** The key to the sorting algorithm is the use of binary numbers for the time-domain data indices. Let us rewrite the 8-member data set using binary indices:

$$\vec{\mathbf{f}}_8 = \begin{bmatrix}f(0)\\ f(1)\\ f(2)\\ f(3)\\ f(4)\\ f(5)\\ f(6)\\ f(7)\end{bmatrix} = \begin{bmatrix}f(000)\\ f(001)\\ f(010)\\ f(011)\\ f(100)\\ f(101)\\ f(110)\\ f(111)\end{bmatrix}$$

*****Who would do it any other way these days? Spectacularly fast and powerful computers are quite literally everywhere.

Now we swap data points that have mirror image indices.

$$\vec{\mathbf{f}}_8 = \begin{bmatrix} f(000) \\ f(100) \\ f(010) \\ f(110) \\ f(001) \\ f(101) \\ f(011) \\ f(111) \end{bmatrix} = \begin{bmatrix} f(0) \\ f(4) \\ f(2) \\ f(6) \\ f(1) \\ f(5) \\ f(3) \\ f(7) \end{bmatrix}$$

Done! Fast! The time-domain vector has been sorted into even and odd sets. No need to construct a permutation matrix as above. Also, if you look closely you can see that (in a formal sense) two swaps have been done, which is what is needed for an 8-point data set. The first swap produces two 4-member time-domain sets divided into even and odd members. The second swap does the same on each of the even and odd sets from the first swap to produce 2-member "even" and "odd" set from the 4-member even and odd sets. What is so interesting about the binary algorithm is that it does all of the required swaps in one step so that the resultant vector is immediately ready for the application of Fourier transform matrices.

Further Reading

1. I. Amidror, *Mastering the Discrete Fourier Transform in One, Two or Several Dimensions; Pitfalls and Artifacts*, Springer, London, 2013.
2. E. Chu, *Discrete and Continuous Fourier Transforms; Analysis, Applications and Fast Algorithms*, CRC Press, Boca Raton, Fla., 2008.
3. E. Fukushima and S. B. Roeder, *Experimental Pulsed NMR: A Nuts and Bolts Approach*, Addison- Wesley Publishing Co., Reading, Mass., 1981.
4. A. G. Marshall and F. R. Verdun, *Fourier Transforms in NMR, Optical and Mass Spectrometry: A User's Handbook*, Elsevier, Amsterdam, New York, 1990.
5. N. Morrison, *Introduction to Fourier Analysis*, Wiley, New York, 1994.
6. S. Smith, *The Scientist and Engineer's Guide to Digital Signal Processing*, California Technical Publishing, San Diego, 1999.

9 Introductory Classical Mechanics

Here it will be perceived from innumerable examples what is the use of mathematics for judgment in the natural sciences, and how impossible it is to philosophise correctly without it.

Galileo Galilei

9.1 Newton's Laws

Our purpose in this chapter is to lay a major part of the foundation for quantum mechanics, and more specifically, Hamiltonian mechanics from which the famous Hamiltonian of quantum mechanics is produced. All too often the Hamiltonian is simply introduced as a *fait accompli* in courses on quantum mechanics without any explanation as to where it comes from. Admittedly a full discussion of its derivation is somewhat time consuming and beyond the scope of this document but it is still important to know where it comes from and why it is used. Classical analytical mechanics has been developed over a period of about three centuries and there is, to put it mildly, much written about the subject. However, we will not be especially verbose here so the interested reader might consider looking at one or more of the Further Reading selections on the subject for more detailed information.

Mechanics (both classical and quantum) is, of course, the study of the motion of objects in space and time. One could say that much if not all of physics boils down to mechanics. We begin with the original, formalised mechanics of Isaac Newton. The three famous laws

Essential Mathematics for NMR and MRI Spectroscopists
By Keith C. Brown

Published by the Royal Society of Chemistry, www.rsc.org

of motion were postulated by Newton in his Principia Mathematica. They are (in his words followed by the modern version):

1) Newton's 1st law:
   a) Newton: Every body persists in its state of being at rest or of moving uniformly straight forward, except insofar as it is compelled to change its state by force impressed.
   b) Modern: in the absence of an external force an object will stay at rest or will move with constant velocity in a straight line.
2) Newton's 2nd law:
   a) Newton: The change of motion of a body is proportional to the impulse impressed on the body, and happens along the straight line on which that impulse is impressed.
   b) Modern: the external force on an object is equal to the mass of the object times its acceleration
3) Newton's 3rd law:
   a) Newton: To every action there is always an equal and opposite reaction: or the forces of two bodies on each other are always equal and are directed in opposite directions.
   b) Modern: when object A exerts a force F on object B, object B exerts a force $-F$ on object A.

These principles have been at the foundations of classical physics for three hundred years and have proven extremely useful in mechanics calculations for macroscopic objects.

When Newton talked about "motion" he really meant what we now call momentum and that is defined as:

$$p = mv \tag{9.1}$$

With this we can write Newton's second law:

$$F = \frac{\mathrm{d}p}{\mathrm{d}t} = m\frac{\mathrm{d}v}{\mathrm{d}t} = ma \tag{9.2}$$

In order to further dissect these equations we introduce some simplifying notation:

$$v = \frac{\mathrm{d}x}{\mathrm{d}t} = \dot{x}$$

$$a = \frac{\mathrm{d}v}{\mathrm{d}t} = \frac{\mathrm{d}^2x}{\mathrm{d}t^2} = \ddot{x}$$

Thus, we may rewrite Newton's second law:

$$F = m\frac{\mathrm{d}v}{\mathrm{d}t} = m\frac{\mathrm{d}^2x}{\mathrm{d}t^2} = m\ddot{x}$$

It is interesting to realise that in order to deal with these new ideas of motion Newton had to invent the calculus† (or fluxions and fluents as Newton called differential and integral calculus). Actually, it was simultaneously developed by Gottfried Leibniz in Hanover, Germany. However, Newton and his supporters rather unjustly attacked Leibniz saying that he had simply created another notation for the calculus rather than the principles of calculus and Leibniz was not given due credit for his contributions for many years after his death in 1716. The modern symbols used in calculus including "d" for the Latin "*diffentia*" and the elongated "S" of the integral sign after the Latin "*summa*" honour Leibniz' contributions to the calculus.

We have not included anything about directions in the above equations. If we use the idea of the position vector:

$$\vec{\mathbf{x}} = \vec{\mathbf{x}}(x_1, x_2, x_3) = x_1\vec{\mathbf{u}}_x + x_2\vec{\mathbf{u}}_y + x_3\vec{\mathbf{u}}_z$$

where x_1, x_2 and x_3 are, of course, the components of the position vector along the Cartesian axes (please see Chapter 3 for a review of vector mathematics), we can write Newton's second law in vector form:

$$\vec{\mathbf{F}} = m\frac{\mathrm{d}\vec{\mathbf{v}}}{\mathrm{d}t} = m\frac{\mathrm{d}^2\vec{\mathbf{x}}}{\mathrm{d}t^2} = m\ddot{\vec{\mathbf{x}}} = m\vec{\mathbf{a}}$$

Using Newton's laws we may calculate the future position and velocity of an object knowing only the initial position and velocity. Let us look at an example of a body moving under constant acceleration along the x-axis. We will assume an initial position at time zero of x_0 and initial velocity of v_0:

$$a = \frac{\mathrm{d}v}{\mathrm{d}t}$$

$$\int \mathrm{d}v = \int a\mathrm{d}t$$

$$v = at + C_1$$

†After reading "The Calculus Wars" listed in the Further Reading section I am inclined to credit Liebniz with the discovery of calculus. There is no doubt that he was the first to publish anything on the subject and that Newton dragged his heels for a very long time with respect to informing the world about his work. Liebniz's calculus was already firmly established on the continent by the time Newton got around to publishing and we still use Liebniz's calculus symbols today.

C_1 is, of course, the constant of integration. At time zero we can say:

$$v_0 = a \times 0 + C_1$$

$$v_0 = C_1$$

and our velocity equation now becomes:

$$v = at + v_0 \tag{9.3}$$

Similarly, we can find the position equation:

$$v = \frac{dx}{dt}$$

$$\int dx = \int v dt$$

$$x = \int (at + v_0) dt$$

$$= \int at dt + \int v_0 dt$$

$$= at^2 + v_0 t + C_2$$

Again, we find the value of C_2 from the initial position:

$$x = at^2 + v_0 t + C_2$$

$$x_0 = a \times 0 + v_0 \times 0 + C_2$$

$$x_0 = C_2$$

and the position equation is:

$$x = at^2 + v_0 t + x_0 \tag{9.4}$$

Thus, in the case of the object moving under constant acceleration we see that we can predict its position and velocity at any time in the future. Equation (9.3) and (9.4) are exceedingly powerful equations for this reason, notwithstanding their simplicity. We say that knowledge of the position and velocity of the object is knowledge of the *state* of the object.

$$(\vec{\mathbf{x}}(t), \vec{\mathbf{v}}(t))$$

Knowledge of x_0 and v_0 lets us calculate the future position and velocity (or state) of many macroscopic objects from projectiles to planets at any point in time using eqn (9.3) and (9.4). This is part of the philosophy of determinism of Descartes. Determinism essentially states that, since the beginning of time, given the initial state, the future states of the universe are, at least in principle, predictable or determinable.

Over time it was found to be more convenient to use the momentum of the object rather than the velocity to specify the state of the object. The momentum of an object is said to be *conservative* if there are no external forces acting on it. How do we know this? It is postulated in Newton's third law. What it is really saying, if somewhat obtusely, is that momentum is conserved in a physical system. Thus, when a rocket engine is activated and material moves very rapidly from the engine the rocket begins to move. However, the total momentum of the rocket and the expended fuel remains constant (in the absence of external forces). Using the momentum in place of velocity makes certain calculations easier. Thus, the state can be represented as a point in $\mathbb{R}^6$ (called phase space) since there are three components to the each of position and momentum vectors.

$$(\vec{\mathbf{x}}(t), \vec{\mathbf{p}}(t))$$

We can show the conservation of momentum with an illustrative example. Let us suppose that we have two billiard balls labelled A and B. Ball A is motionless ($v_A = 0$) and ball B is traveling with velocity $v_B = 1.5\ \mathrm{m\ s^{-1}}$ towards ball A. For simplicity we will consider them to be capable of moving in only the x-direction. We also assume them to have the same mass. What will happen when ball B strikes ball A? Newton's third law postulates that the total momentum of the two balls will remain constant. Thus, according to Newton:

$$mv_A + mv_B = k$$

$$m(v_A + v_B) = k$$

$$v_A + v_B = \frac{k}{m}$$

$$v_{Ai} + v_{Bi} = v_{Af} + v_{Bf}$$

This will be so both before and after the collision. Thus, before the collision, v_A is zero and v_B is non-zero. After the collision v_A and v_B will change but their *sum* will not. This is, of course an artificial example with the two masses being the same. More generally, using balls of different masses:

$$m_A v_A + m_B v_B = k$$

$$m_A v_{Ai} + m_B v_{Bi} = m_A v_{Af} + m_B v_{Bf}$$

In this case the sum of the momenta remains constant or is conserved. We can use this to solve problems involving the motion of objects.

9.2 Energy

What is it about moving objects that is different from motionless ones? Obviously, they are moving but beyond that, in order to change them from a motionless state to one in motion we have to give them a push. In doing so we have apparently added something to them.

In the late 17th century, Gottfried Leibniz found that for moving objects in many cases, the quantity mv^2 was constant. He called this the *vis viva* or living force of the system. Newton's third law postulated that the quantity mv is conserved and, since his influence among scientists of the time was extensive, many regarded the conservation of momentum as a fundamental principle rather than Leibniz's *vis viva*. However, later in the 19th century, long after his death, Leibniz's work was taken more seriously and the term *vis viva* was changed to energy (by Thomas Young) and the calculation of energy content was revised. Let us see how, using Newton's second law. We start by re-arranging and restating:

$$\begin{aligned} F = ma = m\frac{\mathrm{d}v}{\mathrm{d}t} &= m\frac{\mathrm{d}x}{\mathrm{d}t}\frac{\mathrm{d}v}{\mathrm{d}x} = mv\frac{\mathrm{d}v}{\mathrm{d}x} \\ F\mathrm{d}x &= mv\mathrm{d}v \\ F\mathrm{d}x &= \mathrm{d}\left(\frac{1}{2}mv^2\right) \end{aligned} \tag{9.5}$$

$F\mathrm{d}x$ is the work done on an object as a result of the application of force, F, and movement of the object through $\mathrm{d}x$. We are assuming a constant force for the moment. The other side of the equation is the change in the *kinetic energy* (or Leibniz's *vis viva*) of the object as a result of the application of the force.

$$T = \frac{1}{2}mv^2$$

$$\mathrm{d}W = F\mathrm{d}x$$

$$\mathrm{d}W = \mathrm{d}T$$

Thus, the work done on an object is equal to the change in its kinetic energy.

Now, let us assume, instead of a constant force, a position-dependent force and do some integration:

$$\int_{x_0}^{x} F(x)\mathrm{d}x = \int_{v_0}^{v} mv\mathrm{d}v$$

$$\int_{x_0}^{x} F(x)\mathrm{d}x = \frac{1}{2}mv^2 - \frac{1}{2}mv_0^2$$

$$E = \frac{1}{2}mv_0^2 = \frac{1}{2}mv^2 - \int_{x_0}^{x} F(x)\mathrm{d}x$$

We define the integral term involving $F(x)$ as the *potential energy*:

$$U(x) \equiv -\int_{x_0}^{x} F(x)\mathrm{d}x \tag{9.6}$$

and recalling the definition of kinetic energy from above:

$$E = T(v) + U(x) \tag{9.7}$$

The kinetic energy is a function of the velocity of the system and the potential energy a function of position (for moving objects). For a closed system the total energy comprised of kinetic and potential energies is conserved (or constant). Kinetic energy may be converted to potential energy and *vice versa* but the total will remain constant. As we have seen with momentum, conserved quantities are useful in simplifying calculations.

A classic example of an energy conservative system is the simple harmonic oscillator in which an object attached to a spring experiences a force $-kx$ where x is the displacement from equilibrium and k is the spring constant. Thus:

$$F = -kx = ma = mv\frac{\mathrm{d}v}{\mathrm{d}t}$$

or

$$mv\frac{\mathrm{d}v}{\mathrm{d}t} + kx = 0$$

and

$$\int mv\mathrm{d}v + \int kx\mathrm{d}x = K$$

$$\frac{1}{2}mv^2 + \frac{1}{2}kx^2 = K$$

We see the sum of the kinetic and potential energies derived here from Newton's second law and the force function, $-kx$. The sum is equal to a constant and, since its time derivative is zero, is therefore conserved.

We say that the equation:

$$ma = m\ddot{x} = -kx$$

is the equation of motion since we can use it to predict the future position and velocity of the object on the spring.

9.3 Generalised Coordinates

Assigning a position to an object involves assigning spatial coordinates to it. Most of us are very used to using Cartesian coordinates, x, y and z in two- and three-dimensional situations. We have already assigned a three-dimensional position using a vector notation:

$$\vec{\mathbf{r}} = \vec{\mathbf{r}}(x_1, x_2, x_3)$$

where, again, x_1, x_2 and x_3 are the components of the position vector along the Cartesian axes. We use this notation rather than x, y and z in order to simplify our later developments. It is also possible that this vector can depend on time as well:

$$\vec{\mathbf{r}} = \vec{\mathbf{r}}(x_1, x_2, x_3, t)$$

This being the case we can develop the velocity vector as follows:

$$\begin{aligned}\frac{\mathrm{d}}{\mathrm{d}t}\vec{\mathbf{r}} = \dot{\vec{\mathbf{r}}} &= \frac{\partial\vec{\mathbf{r}}}{\partial x_1}\frac{\partial x_1}{\partial t} + \frac{\partial\vec{\mathbf{r}}}{\partial x_2}\frac{\partial x_2}{\partial t} + \frac{\partial\vec{\mathbf{r}}}{\partial x_3}\frac{\partial x_3}{\partial t} + \frac{\partial\vec{\mathbf{r}}}{\partial t} \\ &= \sum_{i=1}^{3}\frac{\partial\vec{\mathbf{r}}}{\partial x_i}\frac{\partial x_i}{\partial t} + \frac{\partial\vec{\mathbf{r}}}{\partial t} \\ &= \sum_{i=1}^{3}\frac{\partial\vec{\mathbf{r}}}{\partial x_i}\dot{x}_i + \frac{\partial\vec{\mathbf{r}}}{\partial t}\end{aligned}$$

Of course, this can be expanded to include as many dimensions as we like:

$$\frac{\mathrm{d}}{\mathrm{d}t}\vec{\mathbf{r}} = \dot{\vec{\mathbf{r}}} = \sum_{i=1}^{n}\frac{\partial\vec{\mathbf{r}}}{\partial x_i}\dot{x}_i + \frac{\partial\vec{\mathbf{r}}}{\partial t}$$

We need *not* use Cartesian coordinates to specify the position and velocity of an object. We can just as well use, for example, spherical coordinates (see Appendix A), radial r, azimuth θ and elevation ϕ. Since the choice of coordinate system to specify position is arbitrary we can generalise our expression using q for our position coordinate:

$$\vec{\mathbf{r}} = \vec{\mathbf{r}}(q_1, q_2, \ldots t)$$

and

$$\dot{\vec{\mathbf{r}}} = \sum_{i=1}^{n} \frac{\partial \vec{\mathbf{r}}}{\partial q_i} \dot{q}_i + \frac{\partial \vec{\mathbf{r}}}{\partial t}$$

where the q_i are referred to as generalised coordinates and the $\dot{q}_i$ as generalised velocities. Generalised coordinates are discussed in much more detail in the Further Reading section and the interested reader is encouraged to investigate.

9.4 The Lagrangian

We will not go into great detail here as the Lagrangian formulation of mechanics is, for our purposes, only a stepping stone to the Hamiltonian mechanics (which will then be a stepping stone to quantum mechanics).

The Lagrange method is an alternative approach to understanding the motion of objects. At first one wonders "It seems like a lot of work for little return .. why bother?". The answer to this is twofold:

1. The *form* of the Lagrange equations remains *exactly* the same no matter what coordinate system is used. This is very convenient when applied to different coordinate systems.
2. Using the Lagrange equations one constructs a scalar (L) and deduces the equations of motion by differentiation. No vectors are involved as in Newtonian mechanics, which simplifies things.

The Lagrangian is defined as:

$$L \equiv T - U \tag{9.8}$$

where, as above, T is kinetic energy and U is potential energy. It seems at first an odd thing to consider the difference between T and U since we are used to thinking in terms of the total energy of a system, the *sum* of T and U.

We write the Euler–Lagrange equation as:

$$\frac{\mathrm{d}}{\mathrm{d}t}\left(\frac{\partial L}{\partial \dot{q}}\right) = \frac{\partial L}{\partial q} \tag{9.9}$$

Note that we can express this equation in generalised coordinates. In other words it works for any coordinate system! The interested reader might refer to Johns or Morin or Rossberg in the Further Reading list or any competent text on analytical mechanics for more information about the derivation of this equation.

Let us take our simple harmonic oscillator problem and subject it to Lagrangian treatment. First, we write the Lagrangian:

$$L = T(\dot{x}) - U(x)$$
$$= \frac{1}{2}m\dot{x}^2 + \frac{1}{2}kx^2$$

and then subject it to the Euler–Lagrange equation:

$$\frac{\mathrm{d}}{\mathrm{d}t}\left(\frac{\partial L}{\partial \dot{q}}\right) = \frac{\mathrm{d}}{\mathrm{d}t}(m\dot{x}) = m\ddot{x} = ma$$

$$\frac{\partial L}{\partial x} = kx = F$$

$$m\ddot{x} = kx$$

which is just Newton's second law, $F = ma$. This is not very instructive or useful yet. We might as well just use Newton's laws. Let us look at another problem.

Imagine a pendulum with a bob of mass m on a massless string of length l swinging through an arc of θ radians (Figure 9.1).

If we approach this using Cartesian coordinates we can come up with an equation of motion for the pendulum (also in Cartesian coordinates). We note that in this example the x and y coordinates of the pendulum bob are not independent. We therefore choose to express our equations in one of these coordinates, x in our case:

$$y = \sqrt{l^2 - x^2}$$

$$\frac{\mathrm{d}y}{\mathrm{d}t} = \dot{y} = \frac{x\dot{x}}{\sqrt{l^2 - x^2}}$$

$$l = y + h \Rightarrow h = l - y = l - \sqrt{l^2 - x^2}$$

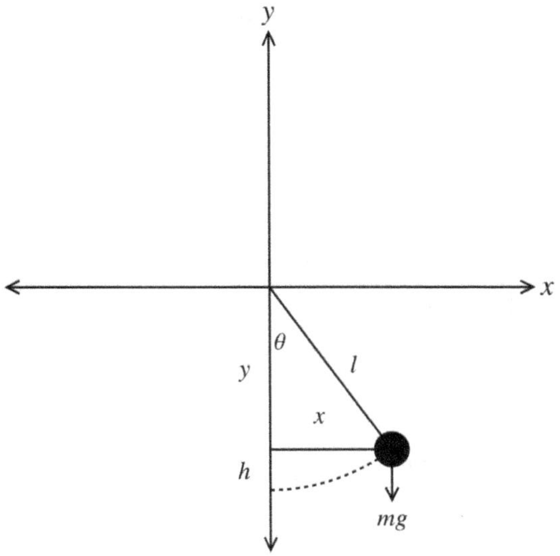

Figure 9.1 The simple pendulum.

Since we are going to build a Lagrangian we need expressions for the kinetic and potential energies:

$$T = \frac{1}{2}m(\dot{x}^2 + \dot{y}^2)$$

$$= \frac{1}{2}m\left(\dot{x}^2 + \frac{x^2\dot{x}^2}{l^2 - x^2}\right)$$

$$= \frac{1}{2}m\dot{x}^2\left(1 + \frac{x^2}{l^2 - x^2}\right)$$

$$= \frac{1}{2}m\dot{x}^2\left(\frac{l^2 - x^2 + x^2}{l^2 - x^2}\right)$$

$$= \frac{1}{2}m\dot{x}^2\left(\frac{l^2}{l^2 - x^2}\right)$$

$$U = -mgh = -mg(l - \sqrt{l^2 - x^2})$$

We now construct the Lagrangian:

$$L = T - U = \frac{1}{2}m\dot{x}^2\left(\frac{l^2}{l^2 - x^2}\right) + mg(l - \sqrt{l^2 - x^2})$$

and apply the Euler–Lagrange equation:

$$\frac{\mathrm{d}}{\mathrm{d}t}\left(\frac{\partial L}{\partial \dot{x}}\right)=\frac{\partial L}{\partial x}$$

$$\frac{\partial L}{\partial \dot{x}}=m\dot{x}\frac{\partial l^2}{\partial(l^2-x^2)}$$

$$\frac{\mathrm{d}}{\mathrm{d}t}\left(\frac{\partial L}{\partial \dot{x}}\right)=m\ddot{x}\frac{l^2}{l^2-x^2}$$

$$\frac{\partial L}{\partial x}=m\dot{x}^2x\frac{l^2}{(l^2-x^2)^2}+mg\frac{x}{\sqrt{l^2-x^2}}$$

We rearrange and come up with the equation of motion:

$$\begin{aligned} m\ddot{x}&=m\dot{x}^2\frac{l^2}{(l^2-x^2)^2}\times\frac{l^2-x^2}{l^2}+mg\frac{x}{\sqrt{l^2-x^2}}\times\frac{l^2-x^2}{l^2} \\ &=m\dot{x}^2\frac{x}{l^2-x^2}+mgx\frac{\sqrt{l^2-x^2}}{l^2} \end{aligned} \tag{9.10}$$

Now, let us do the same analysis again but this time using polar coordinates. From Figure 9.1:

$$x=l\sin(\theta)$$

$$y=-l\cos(\theta)$$

and the time derivatives (using the chain rule) of these are:

$$\dot{x}=l\theta\cos(\theta)$$

$$\dot{y}=l\theta\sin(\theta)$$

Again, we will construct a Lagrangian using kinetic and potential energies. The kinetic energy expressed in polar coordinates is:

$$\begin{aligned} T&=\frac{1}{2}m\dot{x}^2+\frac{1}{2}m\dot{y}^2 \\ &=\frac{1}{2}ml^2\dot{\theta}^2\cos^2(\theta)+\frac{1}{2}ml^2\dot{\theta}^2\sin^2(\theta) \\ &=\frac{1}{2}ml^2\dot{\theta}^2(\cos^2(\theta)+\sin^2(\theta)) \\ &=\frac{1}{2}ml^2\dot{\theta}^2 \end{aligned}$$

and the potential energy is:

$$\begin{aligned} U &= -mg(l-y) \\ &= -mg(l+\cos(\theta)) \\ &= -mgl(1+\cos(\theta)) \end{aligned}$$

The Lagrangian is:

$$L = T - U = \frac{1}{2}ml^2\dot{\theta}^2 + mgl(1+\cos(\theta))$$

Now, we write the Euler–Lagrange equation, in polar coordinates:

$$\begin{aligned} \frac{\mathrm{d}}{\mathrm{d}t}\left(\frac{\partial L}{\partial \dot{\theta}}\right) &= \frac{\partial L}{\partial \theta} \\ \frac{\partial L}{\partial \dot{\theta}} &= ml^2\dot{\theta} \\ \frac{\mathrm{d}}{\mathrm{d}t}\left(\frac{\partial L}{\partial \dot{\theta}}\right) &= ml^2\ddot{\theta} \\ \frac{\partial L}{\partial \theta} &= -mgl\sin(\theta) \\ &\text{and} \\ ml^2\ddot{\theta} &= -mgl\sin(\theta) \\ \ddot{\theta} &= \frac{-g}{l}\sin(\theta) \end{aligned} \tag{9.11}$$

This is a much simpler result than that of eqn (9.10). In this sense the use of polar coordinates is advantageous.

Note that we have used the Lagrangian with two different coordinate systems, Cartesian and polar, to produce equations of motion in those coordinate systems. This is a large part of the beauty of the Lagrangian form of mechanics … it is coordinate-independent. Note also that the differential eqn (9.11) is not one that admits solutions similar to those of the harmonic oscillator above. However, if one assumes a pendulum with small oscillations then:

$$\sin(\theta) \approx \theta$$

and eqn (9.11) becomes:

$$\ddot{\theta} = \frac{-g}{l}\theta$$

which is in the form of a differential equation with similar solutions (see Appendix A).

9.5 The Hamiltonian

The third approach to mechanics is that of the Irish mathematician Sir William Rowan Hamilton. Again, we do not wish to be rigorous or complete here, but rather to show the reader the very basics.

In the previous section we saw that the Lagrangian, L, is a function of generalised coordinates and velocities, whereas the Hamiltonian is a function of generalised coordinates and *momenta*. As with the Lagrangian, the Hamiltonian is expressed in terms of energy scalars:

$$H = T + U \tag{9.12}$$

where the kinetic energy, T, is now written in terms of momentum, instead of velocity:

$$p_i = m\dot{q}_i \Rightarrow \dot{q}_i = \frac{p_i}{m}$$

$$T = \frac{1}{2}m\dot{q}_i^2 = \frac{1}{2}m\frac{p_i^2}{m^2} = \frac{1}{2}\frac{p_i^2}{m}$$

The potential energy, U, is in the same form as in the Lagrangian. Note that H is simply the sum of the kinetic and potential energies … in other words the total energy of the system. If there are no external forces then the total energy of the system is constant or conserved. As above, the interested reader might consult the Further Reading section; in particular, both Rossberg and Goldstein give a readable treatment of the development of the Hamiltonian from the Lagrangian. We can now expand our Hamiltonian expression a bit:

$$H(q,p) = \frac{p^2}{m} + U(q) \tag{9.13}$$

using generalised coordinates and momenta. We find after suitable (and lengthy) analysis that:

$$-\dot{p}_i = \frac{\partial H}{\partial q_i} \quad \text{and} \quad \dot{q}_i = \frac{\partial H}{\partial p_i} \tag{9.14}$$

These are referred to as Hamilton's equations. Note the pleasing symmetry (except for the minus sign) in p and q.

9.6 Classical Angular Momentum

Generally, there are three broad classes of spatial motion that a body can undertake, translational, vibrational and rotational. Translational motion is "straight line" motion, vibrational motion consists of oscillations about a point and rotational motion is, well, rotational. Happily for us, many of the equations that relate to translational motion have analogues that describe rotational motion (Table 9.1).

These analogues make it easy to make the transition from thinking about translational motion to rotational motion. As NMR/MRI spectroscopists, we shall be especially interested in the angular force or torque, Γ and the angular momentum, L (not to be confused with the Lagrangian from above).

Starting very simply, a vector $\vec{\mathbf{r}}$ makes an angle θ with the x-axis and traces out a path or displacement of s units from the horizontal axis (see Figure 9.2). The angle θ is in radians, which is defined as the ratio of s to $\|\vec{\mathbf{r}}\|$; *i.e.* one radian is defined as the angle created when $s = \|\vec{\mathbf{r}}\|$.

$$\theta = \frac{s}{r} \qquad s = r\theta \tag{9.15}$$

This allows us to relate the translational velocity of the tip of $\vec{\mathbf{r}}$ to the angular velocity:

$$v = \frac{\mathrm{d}s}{\mathrm{d}t} = r\frac{\mathrm{d}\theta}{\mathrm{d}t} = r\omega \tag{9.16}$$

as well as the translational acceleration:

$$a = \frac{\mathrm{d}v}{\mathrm{d}t} = r\frac{\mathrm{d}\omega}{\mathrm{d}t} = r\alpha \tag{9.17}$$

Table 9.1 Classical linear quantities and their rotational analogues.

	Translation	Rotation
Displacement	s	θ
Velocity	$v = \frac{\mathrm{d}s}{\mathrm{d}t}$	$\omega = \frac{\mathrm{d}\theta}{\mathrm{d}t}$
Acceleration	$a = \frac{\mathrm{d}v}{\mathrm{d}t}$	$\alpha = \frac{\mathrm{d}\omega}{\mathrm{d}t}$
Force and Torque	F	Γ
Mass and Moment of Inertia	m	I
Kinetic Energy	$E_k = \frac{1}{2}mv^2$	$E_k = \frac{1}{2}I\omega^2$
Momentum	mv	$L = I\omega$

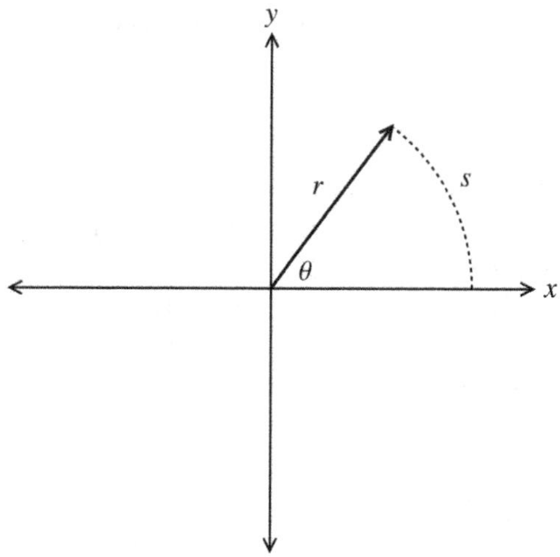

Figure 9.2 The relationship between radians and circular displacement.

The kinetic energy of motion is slightly more complicated:

$$
\begin{aligned}
v &= r\omega \\
E_k &= \frac{1}{2}mv^2 \\
&= \frac{1}{2}mr^2\omega^2
\end{aligned}
$$

This represents the kinetic energy of a single particle in the rotating body; the total energy of all the particles is:

$$E_k = \frac{1}{2}\sum mr^2\omega^2$$

The angular velocity, ω, is constant for all particles and the summation, $\sum mr^2$, is called the *moment of inertia* and is the rotational analogue of our familiar concept of mass:

$$I = \sum mr^2$$

So:

$$E_k = \frac{1}{2}I\omega^2 \qquad (9.18)$$

which looks much like the translational kinetic energy equation but is now cast in rotational variables.

When considering translational motion and work we generally use force, F, and distance (or displacement), s, to calculate the work done. Again, the rotational version of this calculation is very similar to the translational calculation. Distance, s, is replaced by an angle, θ, and force is replaced by torque, Γ:

$$W = \Gamma\theta \tag{9.19}$$

The net work done is the change in kinetic energy:

$$\mathrm{d}W = \Gamma d\theta = \mathrm{d}\left(\frac{1}{2}I\omega^2\right) = \frac{1}{2}I\mathrm{d}\omega^2$$

$$\Gamma = \frac{I\mathrm{d}\omega^2}{2\mathrm{d}\theta} = I\omega\frac{\mathrm{d}\omega}{\mathrm{d}\theta}$$

$$= I\frac{\mathrm{d}\theta}{\mathrm{d}t}\frac{\mathrm{d}\omega}{\mathrm{d}\theta} = I\frac{\mathrm{d}\omega}{\mathrm{d}t} = I\alpha$$

$$\Gamma = I\alpha$$

This leads us to angular momentum:

$$\Gamma = I\alpha = I\frac{\mathrm{d}\omega}{\mathrm{d}t} = \frac{\mathrm{d}(I\omega)}{\mathrm{d}t}$$

The product of I and ω is the rotational analogue of momentum for translational motion and is given the symbol L.

$$\Gamma = \frac{\mathrm{d}(I\omega)}{\mathrm{d}t} = \frac{\mathrm{d}L}{\mathrm{d}t}$$
$$\Gamma\mathrm{d}t = \mathrm{d}L \tag{9.20}$$

So far, there should not have been any great leaps in logic (as long as you have your calculus wits about you).

Now, let us look at some vector quantities. First, consider a tangential force vector, $\vec{\mathbf{F}}$ at a distance r from the centre of rotation and a distance vector $\vec{\mathbf{r}}$ as in Figure 9.3. It is apparent that this force will change the rotational velocity and the angular momentum. Again, this is analogous to the translational motion situation. Now consider what would happen if $\vec{\mathbf{F}}$ were to be directed not tangentially but orthogonally. In other words, if $\vec{\mathbf{F}}$ were to be directed parallel and

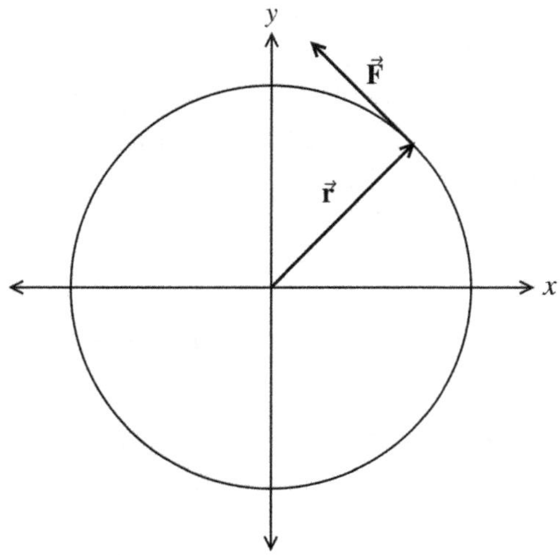

Figure 9.3 The tangential force vector, $\vec{\mathbf{F}}$.

collinear with $\vec{\mathbf{r}}$ there would be no change in the torque (however, see below). This is just what one would expect of a vector relationship between $\vec{\mathbf{r}}$ and $\vec{\mathbf{F}}$. Thus, we write the vector cross product:

$$\vec{\Gamma} = \vec{\mathbf{r}} \times \vec{\mathbf{F}} \tag{9.21}$$

Here, the magnitude of the torque is:

$$\begin{aligned}\Gamma &= \|\vec{\mathbf{r}}\|\|\vec{\mathbf{F}}\|\sin(\phi)\\ &= rF\sin(\phi)\end{aligned}$$

where ϕ is the angle between $\vec{\mathbf{r}}$ and $\vec{\mathbf{F}}$. Of course, since the torque calculation is a vector cross product, the direction of the torque is at right angles to the plane defined by $\vec{\mathbf{r}}$ and $\vec{\mathbf{F}}$. A three-dimensional picture helps us to see the relationships in Figure 9.4.

When $\vec{\mathbf{F}}$ is tangential then $\sin(\phi)$ is 1 and the torque is simply rF (Figure 9.3). When $\vec{\mathbf{F}}$ is collinear with $\vec{\mathbf{r}}$, $\sin(\phi)$ is 0 and the torque is zero. Note that here the torque is directed along the axis of rotation (however, see below).

Similar considerations for linear and angular momenta lead us to:

$$\begin{aligned}\vec{\mathbf{L}} &= \vec{\mathbf{r}} \times m\vec{\mathbf{v}}\\ &= \vec{\mathbf{r}} \times \vec{\mathbf{p}}\end{aligned} \tag{9.22}$$

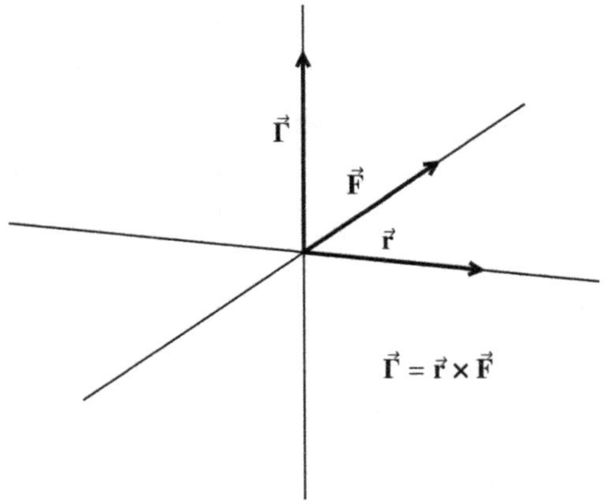

Figure 9.4 The vector relationships between radius, force and torque.

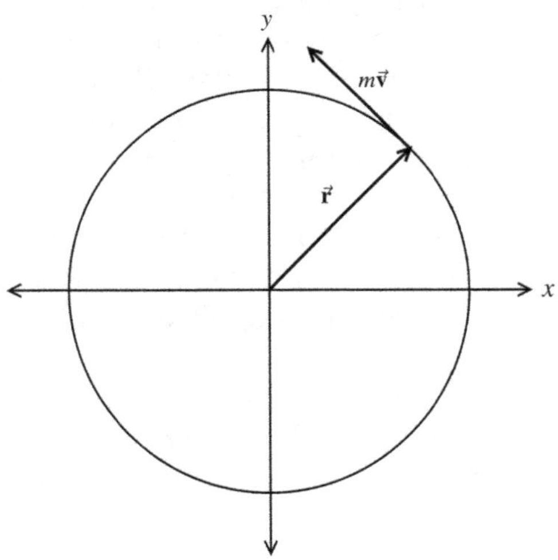

Figure 9.5 Tangential momentum.

where $\vec{\mathbf{p}}$ is linear momentum (Figure 9.5). As with the torque, $\vec{\mathbf{L}}$ is directed along the axis of rotation. We can also represent $\vec{\mathbf{L}}$ analogously to $\vec{\mathbf{p}}$ in linear mechanics:

$$\vec{\mathbf{p}} = m\vec{\mathbf{v}} \quad \text{(linear mechanics)}$$

$$\vec{\mathbf{L}} = I\vec{\omega} \quad \text{(rotational mechanics)}$$

This of course implies that $\vec{\omega}$ lies along the axis of rotation. A moment's thought will convince you that this must be so. For a rotating body with no external forces acting on it, $\vec{\omega}$ is constant. If this

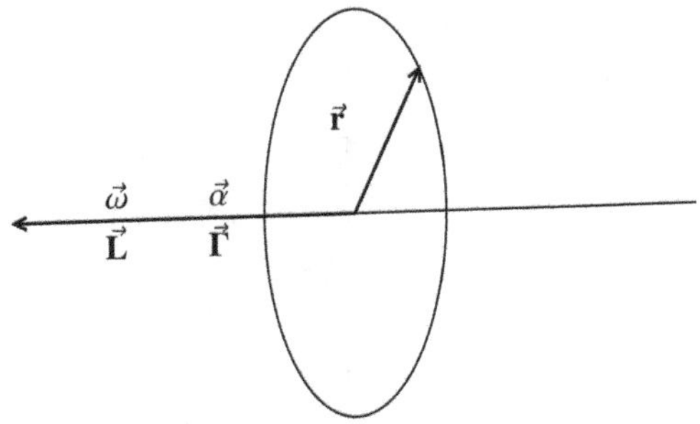

Figure 9.6 Vector angular velocity and torque.

is so then it must be directed along the rotation axis since any other direction would imply that it changes direction with the rotating body and therefore is not constant. A similar argument applies to $\vec{\alpha}$, the angular acceleration and angular momentum, $\vec{\mathbf{L}}$. It seems strange at first that these vectors are not pointed in the direction of movement but it must be so for them to be constant (Figure 9.6).

We have seen, above, what happens when there is a tangential force applied to the rotating body. A torque proportional to the force and its distance from the rotation axis is generated and is directed along the axis of rotation. In other words, the magnitude of the torque changes, but not its direction (unless the force changes direction).

It is possible to have two forces applied simultaneously to the rotating body (Figure 9.7).

In this case, the two forces are of equal magnitude and opposite direction and are called a *couple of moment*. One encounters this situation for example, when using a compass. What is the torque on the system?

$$\begin{aligned}\vec{\Gamma} &= 2\vec{\mathbf{r}} \times \vec{\mathbf{F}} \\ &= 2rF\end{aligned} \tag{9.23}$$

Since the two forces in the diagram are tangential the angle between $\vec{\mathbf{r}}$ and $\vec{\mathbf{F}}$ is 90° and the magnitude of the torque is at a maximum. Also, from the right-hand rule, the torque vector is coming out of the page towards you. If the angle that $\vec{\mathbf{F}}$ makes with $\vec{\mathbf{r}}$ changes from 90° the torque decreases as in the compass example.

What happens if the force, $\vec{\mathbf{F}}$, is directed along a line that passes through the axis of rotation? We have already seen that this must not change the magnitude of the torque. There is, however, another effect that is of crucial importance to us, called *precession*.

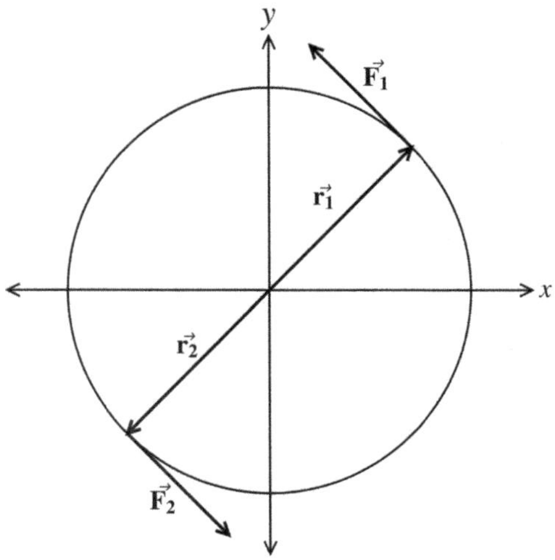

Figure 9.7 Simultaneous tangential forces.

Understanding this concept requires some pretty complex three-dimensional visualisations. The basic (somewhat crowded!) picture is shown in Figure 9.8. This is the example of the spinning top. There is a force, $m\vec{\mathbf{g}}$, on the rotating body that goes through the axis of rotation. The torque is:

$$\begin{aligned}\Gamma &= \vec{\mathbf{r}} \times \vec{\mathbf{F}} \\ &= \vec{\mathbf{r}} \times m\vec{\mathbf{g}}\end{aligned}$$

This vector cross product will be orthogonal to the plane defined by $\vec{\mathbf{r}}$ and $m\vec{\mathbf{g}}$. From eqn (9.20), the torque is related to the angular momentum or rather, the rate of change of the angular momentum. This being the case, we see that $\mathrm{d}\vec{\mathbf{L}}$ (not $\vec{\mathbf{L}}$!) is in the same direction as the torque and is at right angles to the angular momentum, $\vec{\mathbf{L}}$. As the top spins about its axis of rotation the torque causes it to precess about an axis of precession. This is the source of the wobbling motion of a top. Note that in the case of a force at right angles to the axis of rotation and not passing through the axis of rotation causes a change in the magnitude of the angular momentum but a force passing through the axis of rotation causes a change in *direction* of the angular momentum.

Now we can begin to use some of the fundamental topics discussed to this point. We look here at the classical pictures of angular momentum, magnetic moment and the classical description of NMR spectroscopy *via* the Bloch equations. You will need some familiarity

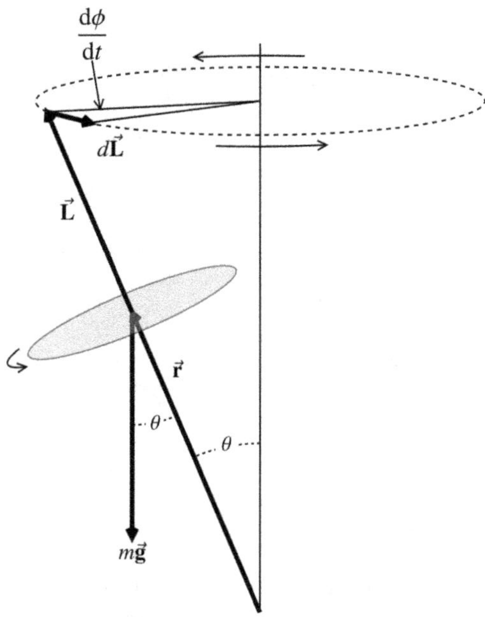

Figure 9.8 Torque, force and precession.

with vector algebra and matrix algebra to proceed with this chapter. There is no quantum mechanical interpretation used here so a full picture of the subject is not really possible and certainly, some aspects of NMR/MRI spectroscopy simply cannot be described using the following ideas. However, these topics provide a solid base from which we can begin to develop an image in our minds about how things work.

Further Reading

1. H. Goldstein, C. Poole and J. Safko, *Classical Mechanics*, Addison Wesley, San Francisco, 2002.
2. O. D. Johns, *Analytical Mechanics for Relativity and Quantum Mechanics*, Oxford University Press, London, 2005.
3. D. Morin, *Introduction to Classical Mechanics With Problems and Solutions*, Cambridge University Press, Cambridge, UK, New York, 2008.
4. R. Piziak and J. J. Mitchell, *Computers and Mathematics with Applications*, 2001, **42**, 793–805.
5. K. Rossberg, *A First Course in Analytical Mechanics*, Wiley, New York, 1983.

10 Special Relativity

The laws of nature are but the mathematical thoughts of God.

Euclid

10.1 The Lorentz Transformation Equations

Albert Einstein's Special Relativity rests on two remarkably simple postulates.[†] First, since Maxwell's equations (see Section 11.4) appear to hold in different inertial frames of reference … they hold at any time and any place on earth, Einstein postulated that the fundamental laws of physics apply in *any and all* inertial reference frames. Secondly, since Maxwell's equations clearly indicate a speed for electromagnetic waves that is independent of the reference frame, Einstein postulated that the speed of light is *constant*. From these two simple ideas some very remarkable things follow, mostly as a result of the application of the Lorentz transform equations developed by Hendrik Lorentz and re-interpreted by Einstein. These may be derived rather easily if somewhat tediously with nothing but high school algebra.

To do so we consider two inertial reference frames, F and F'. F is at "rest" and F' is moving with constant velocity v_x in the x-direction, for convenience (see Figure 10.1). An inertial frame is nothing more than a set of coordinates that are in motion at constant velocity. The requirement for constant velocity reference frames is the reason for the

[†] I know that this topic seems to be a bit out of scope with respect to NMR/MRI spectroscopy but as it turns out it is entirely relevant to the study of magnetism and therefore, as NMR/MRI spectroscopists, we should probably have at least a peripheral understanding of special relativity … if only to impress party guests.

Essential Mathematics for NMR and MRI Spectroscopists
By Keith C. Brown

Published by the Royal Society of Chemistry, www.rsc.org

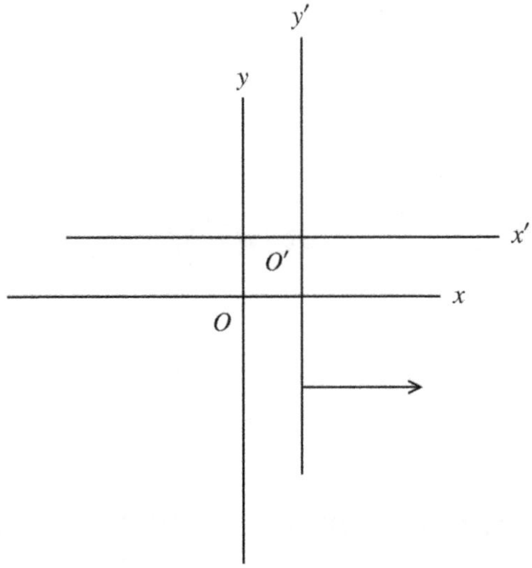

Figure 10.1 Inertial frames F and F' with F' in motion with velocity v with respect to frame F in the x-direction.

name "Special Relativity". General relativity relaxes this requirement and allows for acceleration.

We also assume, as F' moves in the positive x-direction, that at some point its origin O' coincides with the origin of F and that when this occurs, $t = t' = 0$. Let us call this the "crossing event" for later reference. We are interested in the coordinates at a point P at some time, (x, y, z, t) in F and (x', y', z', t') in F'. We shall assume that point P is at rest with respect to frame F.

From the time of Isaac Newton to the late 19th century it was assumed that for a pair of reference frames such as this the time in F and F' are the same. Thus, for this particular example to transform from frame F to F' one need only take into account the motion of F' in the x-direction. All other coordinates are equal:

$$\begin{aligned} x' &= x - v_x t \\ y' &= y \\ z' &= z \\ t' &= t \end{aligned} \tag{10.1}$$

So, assuming $v_x = 1\ \mathrm{m\,s^{-1}}$ and $x = x' = 2$ m at $t = t' = 0$ then at $t = 1$ s, $x' = 2 - 1 \times 1 = 1$ m, which shows that, from the point of view of an observer in F' point P is moving to the left. These equations are now called the Galilean transformation equations and are the "common sense" equations of everyday life. Since we can always define the direction of motion of frame F' as the x-direction we can always have this simple set of transformation equations.

What we want is a general method to transform from frame F to frame F' coordinates in view of Einstein's postulates, similar to the Galilean transformation eqn (10.1). This is accomplished by supposing that there is a set of linear functions that transforms each set of coordinates in F to a coordinate in F'. Why linear? Well, consider that a transformation between F and F' should be no different from one between F' and F. The only kind of function that accomplishes this is a linear function. Polynomials of order higher than one and transcendental functions would not work in this context. For example if $x' = ax^2$ and we wished to back-transform then we would calculate x as $x = \sqrt{x'/a}$, clearly not the same transformation as the forward one. Non-linear functions also are frequently characterised by minima, maxima and singularities, which are at odds with the concept of homogeneity of space. Linear functions do not suffer from these disadvantages.

Consider that we might write a function such that:

$$x' = f(x, y, z, t)$$

in which x' is a function of x, y, z and t. In our simplified Galilean transform it would be more like:

$$x' = f(x, t)$$

However, in the interests of being general for the moment we will develop our transformation equations with respect to all variables. With this in mind we can now write the following general linear equations:

$$x' = a_{11}x + a_{12}y + a_{13}z + a_{14}t$$

$$y' = a_{21}x + a_{22}y + a_{23}z + a_{24}t$$

$$z' = a_{31}x + a_{32}y + a_{33}z + a_{34}t$$

$$t' = a_{41}x + a_{42}y + a_{43}z + a_{44}t$$

or in matrix form:

$$\begin{bmatrix} x' \\ y' \\ z' \\ t' \end{bmatrix} = \begin{bmatrix} a_{11} & a_{12} & a_{13} & a_{14} \\ a_{21} & a_{22} & a_{23} & a_{24} \\ a_{31} & a_{32} & a_{33} & a_{34} \\ a_{41} & a_{42} & a_{43} & a_{44} \end{bmatrix} \begin{bmatrix} x \\ y \\ z \\ t \end{bmatrix}$$

In our simple example, since F' is moving in the x-direction and not the y or z directions, we may assume that y and z transform directly into y' and z' as in the Galilean transformation:

$$\begin{bmatrix} x' \\ y' \\ z' \\ t' \end{bmatrix} = \begin{bmatrix} a_{11} & a_{12} & a_{13} & a_{14} \\ 0 & 1 & 0 & 0 \\ 0 & 0 & 1 & 0 \\ a_{41} & a_{42} & a_{43} & a_{44} \end{bmatrix} \begin{bmatrix} x \\ y \\ z \\ t \end{bmatrix}$$

Also, again since F' is moving only in the x-direction, we assume that x' and t' are not affected by y or z:

$$\begin{bmatrix} x' \\ y' \\ z' \\ t' \end{bmatrix} = \begin{bmatrix} a_{11} & 0 & 0 & a_{14} \\ 0 & 1 & 0 & 0 \\ 0 & 0 & 1 & 0 \\ a_{41} & 0 & 0 & a_{44} \end{bmatrix} \begin{bmatrix} x \\ y \\ z \\ t \end{bmatrix}$$

or:

$$\begin{aligned} x' &= a_{11}x + a_{14}t \\ y' &= y \\ z' &= z \\ t' &= a_{41}x + a_{44}t \end{aligned}$$

At this point, for clarity we will disregard the transformation equations for y' and z' and change from subscripted constants to simple capitalised letters.

$$\begin{aligned} x' &= Ax + Bt \\ t' &= Cx + Dt \end{aligned} \tag{10.2}$$

or again in matrix form:

$$\begin{bmatrix} x' \\ t' \end{bmatrix} = \begin{bmatrix} A & B \\ C & D \end{bmatrix} \begin{bmatrix} x \\ t \end{bmatrix}$$

The inverse transformation from F' to F is:

$$\begin{bmatrix} x \\ t \end{bmatrix} = \begin{bmatrix} \dfrac{D}{AD - BC} & \dfrac{-B}{AD - BC} \\ \dfrac{-C}{AD - BC} & \dfrac{A}{AD - BC} \end{bmatrix} \begin{bmatrix} x' \\ t' \end{bmatrix}$$

and in equation form:

$$x = \frac{Dx' - Bt'}{AD - BC}$$
$$t = \frac{-Cx' + At'}{AD - BC} \tag{10.3}$$

See Chapter 2, Section 2.3 for more details on how the derivation of this expression is accomplished.

At the origin in F', $x' = 0$ and substituting this into the first of eqn (10.2):

$$0 = Ax + Bt$$
$$Ax = -Bt$$
$$\frac{x}{t} = v_x = -\frac{B}{A}$$
$$B = -v_x A$$

and we alter eqn (10.2) to reflect this:

$$x' = A(x - v_x t)$$
$$t' = Cx + Dt$$

Remember, v_x is the velocity of frame F' in the x-direction with respect to frame F. Similarly, we note that at the origin of F, $x = 0$, which we substitute into the first of eqn (10.3):

$$0 = \frac{Dx' - Bt'}{AD - BC}$$
$$0 = Dx' - Bt'$$
$$\frac{x'}{t'} = -v_x = \frac{B}{D} = \frac{-v_x A}{D}$$
$$A = D$$

and again, we reflect this in eqn (10.2):

$$x' = A(x - v_x t)$$
$$t' = A\left(\frac{C}{A}x + t\right)$$

Once more in the interests of clarity, we replace C/A with α:

$$\begin{aligned} x' &= A(x - v_x t) \\ t' &= A(\alpha x + t) \end{aligned} \tag{10.4}$$

We also may update eqn (10.3) to reflect our new-found equalities:

$$\begin{aligned} x &= \frac{A(x' + v_x t')}{A^2 + v_x AC} \\ t &= \frac{A(-\alpha x' + t')}{A^2 + v_x AC} \end{aligned} \tag{10.5}$$

Note the change of sign in front of v_x between eqn (10.4) and eqn (10.5). This is a reflection of the fact that from the point of view of an observer in frame F, frame F' is moving in the positive x-direction with velocity v_x and from the point of view of an observer in frame F', frame F is moving in the minus x-direction with velocity $-v_x$.

Stop for a moment to consider eqn (10.4) and (10.5). The first represents the transformation from reference frame F to frame F' that is moving in the positive x-direction with velocity v_x. The second represents the inverse transformation from F' to F. If space and time are homogeneous and isotropic then we should be able to swap primes and change signs in front of velocity terms and have exactly the same transformation equations. This means that the denominators of the fractional equations in eqn (10.5) *must* be equal to one:

$$x = \frac{A(x' + v_x t')}{A^2 + v_x AC} \Rightarrow x' = \frac{A(x - v_x t)}{A^2 + v_x AC} = A(x - v_x t)$$

It also suggests a connection between α and v_x to account for the sign change in front of α between eqn (10.4) and (10.5). Let us set the denominator equal to one:

$$A^2 + v_x AC = 1$$

or

$$A = \frac{1}{\sqrt{1 + \alpha v_x}}$$

and updating eqn (10.4):

$$\begin{aligned} x' &= \frac{x - v_x t}{\sqrt{1 + \alpha v_x}} \\ t' &= \frac{\alpha x + t}{\sqrt{1 + \alpha v_x}} \end{aligned} \tag{10.6}$$

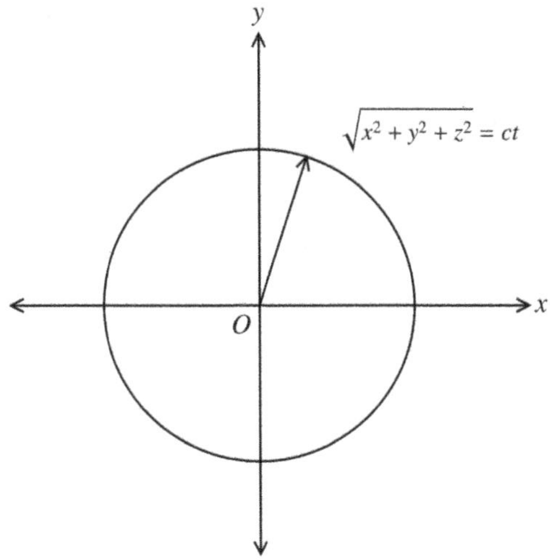

Figure 10.2 Cross section of an expanding sphere of light.

Now let us return to the "crossing event" when the origins of frames F and F' coincide. We suppose that a flash of light is emitted at this time that moves spherically outwards from the origin at speed c, the speed of light (see Figure 10.2). The square of the distance from the origin O to the sphere representing the leading edge of the flash at time t is:

$$x^2 + y^2 + z^2 = c^2t^2 \quad \text{or} \quad x^2 + y^2 + z^2 - c^2t^2 = 0 \tag{10.7}$$

Similarly, in the moving frame F' the square of the distance is:

$$x'^2 + y'^2 + z'^2 = c^2t'^2 \quad \text{or} \quad x'^2 + y'^2 + z'^2 - c^2t'^2 = 0 \tag{10.8}$$

Note that we have invoked Einstein's proposal in eqn (10.8) that the speed of light is not dependent on the reference frame but rather is constant for all observers regardless of their motion with respect to other frames. Thus, the speed of light is c in both F and F'.

We substitute for x' and t' (as well as the trivial substitutions for y' and z') from eqn (10.6) into (10.8):

$$\frac{(x - v_x t)^2}{1 + \alpha v_x} + y^2 + z^2 - \frac{c^2(\alpha x + t)^2}{1 + \alpha v_x} = 0$$

$$\frac{x^2 - 2v_x xt + v_x^2 t^2}{1 + \alpha v_x} + y^2 + z^2 - \frac{c^2(\alpha^2 x^2 + 2\alpha xt + t^2)}{1 + \alpha v_x} = 0$$

$$\frac{x^2 - 2v_x xt + v_x^2 t^2 - c^2\alpha^2 x^2 + 2c^2\alpha xt + c^2 t^2}{1 + \alpha v_x} + y^2 + z^2 = 0$$

Since we have substituted the F-frame coordinates for the F'-frame coordinates using our transformation eqn (10.6) this must now be equivalent to eqn (10.7) so we gather terms in x^2 and t^2:

$$\frac{1 - c^2\alpha^2}{1 + \alpha v_x} x^2 + \frac{v_x^2 + c^2}{1 + \alpha v_x} t^2 - \frac{v_x + c^2\alpha}{1 + \alpha v_x} 2xt + y^2 + z^2 = 0$$

The coefficient in front of each variable here must be equal to the coefficients in front of the variables in eqn (10.7) so we can write:

$$\frac{1 - c^2\alpha^2}{1 + \alpha v_x} = 1$$

$$\frac{v_x^2 + c^2}{1 + \alpha v_x} = -c^2$$

$$-\frac{v_x + c^2\alpha}{1 + \alpha v_x} 2xt = 0$$

The last of these allows us easily to solve for α. Since there is no xt term in eqn (10.7) then the numerator must be equal to zero:

$$-(v_x + c^2\alpha)2xt = 0$$

or

$$v_x + c^2\alpha = 0$$

$$\alpha = -\frac{v_x}{c^2}$$

As noted earlier, α is indeed related to the frame velocity. Substituting for α in eqn (10.6) and including the trivial transformations for y and z:

$$\begin{aligned} x' &= \frac{x - v_x t}{\sqrt{1 - \left(\frac{v_x}{c}\right)^2}} \\ y' &= y \\ z' &= z \\ t' &= \frac{-\left(\frac{v_x}{c}\right)x + t}{\sqrt{1 + \left(\frac{v_x}{c}\right)^2}} \end{aligned} \tag{10.9}$$

These are the famous Lorentz transform equations. Often they are written more compactly as:

$$\begin{aligned} x' &= \gamma(x - v_x t) \\ y' &= y \\ z' &= z \\ t' &= \gamma\left(t - \frac{v_x}{c}x\right) \end{aligned} \tag{10.10}$$

Where γ replaces $1\Big/\sqrt{1 - (v_x/c)^2}$.

Remembering the change of sign of the velocity associated with swapping primes, the inverse transformation equations are:

$$\begin{aligned} x &= \gamma(x' + v_x t') \\ y' &= y \\ z' &= z \\ t &= \gamma\left(t' + \frac{v_x}{c}x'\right) \end{aligned} \tag{10.11}$$

Hendrik Lorentz originally derived these equations in order to transform Maxwell's equations to see if they are independent of inertial reference frames. They are. Einstein, however, took things a step further and applied these transforms to everything ... not just electromagnetic waves.

10.2 Kinematic Consequences of the Lorentz Transformations

In the study of the motion of objects in the universe there are two ways to look at things. Kinematics is the study of motion without considering the causes for the motion. In other words, no forces are taken into account. Dynamics, however, includes forces in the equations. We will consider only kinematics here as it will apply directly to our discussion of the magnetic field in Chapter 11.

The consequences of the kinematic transformation equations are not intuitive. For example, consider the measurement of the lengths of objects in reference frames F and F'. In F we would measure the length of an object at rest in the x-direction as $x_2 - x_1$ at time t. Since this measurement is performed on an object at rest in its frame of reference it is known as the *proper length*. However, if a measurement of a moving object in F' is made from the point of view of F, again at time t in F, the length of the object is:

$$\begin{aligned} x_2' - x_1' &= \gamma(x_2 - v_x t) - (x_1 - v_x t) \\ &= \gamma(x_2 - x_1) \\ x_2 - x_1 &= \frac{x_2' - x_1'}{\gamma} \\ &= \sqrt{1 - \left(\frac{v_x}{c}\right)^2}(x_2' - x_1') \end{aligned} \tag{10.12}$$

or

$$\Delta x = \frac{\Delta x'}{\gamma}$$

If $0 < v_x < c$ then the measured distance in F, $x_2 - x_1$, is *less* than that measured in F', $x_2' - x_1'$. This is called the Lorentz contraction. Note that of the two sets of Lorentz transformation eqn (10.10) and (10.11), we have chosen the former since these equations contain the time t in the F frame from which the measurement is made.

A similarly non-intuitive effect occurs with time measurements. The measurement of time between two events at the same point in F' as seen from the point of view of F the measured times are:

$$
\begin{aligned}
t_2 - t_1 &= \gamma\left(t_2' - \left(\frac{v_x}{c}\right)x'\right) - \gamma\left(t_1' - \left(\frac{v_x}{c}\right)x'\right) \\
&= \gamma(t_2' - t_1') \\
&= \frac{t_2' - t_1'}{\sqrt{1 - \left(\frac{v_x}{c}\right)^2}}
\end{aligned}
\tag{10.13}
$$

or

$$\Delta t = \gamma \Delta t'$$

If $v_x < c$ then the measured time in F is greater than that measured in F'. Thus, an event that occurs in F' appears to take $t_2 - t_1$ in frame F and appears to take more time to occur than from the point of view of F'. This is called *time dilation*. Note again that we have had two choices for equations and have this time chosen eqn (10.11) since it contains the coordinate of the point in F' at which the event occurs.

How do velocities transform? How do we add velocities? If c is the maximum attainable velocity for any object with mass then we cannot simply add velocities as we might be tempted to do. For example, if there are two objects and object A is moving at $3/4c$ with respect to our laboratory frame and object B is moving at $3/4c$ with respect to object A, "common sense" tells us that with respect to the laboratory frame object B is moving at $1.5c$. Special relativity tells us that this cannot be. Let us see why.

As usual, we indicate that an object is located at coordinates x, y and z at time t in frame F. At some time $\mathrm{d}t$ later we find the object at coordinates $x + \mathrm{d}x$, $y + \mathrm{d}y$ and $z + \mathrm{d}z$; we say that it is moving with velocity $\vec{\mathbf{u}}$ having the components:[‡]

$$u_x = \frac{\partial x}{\partial t} \quad u_y = \frac{\partial y}{\partial t} \quad u_z = \frac{\partial z}{\partial t}$$

[‡]Note that I am (temporarily for this chapter) using $\vec{\mathbf{u}}$ for velocity *within* a moving frame rather than v that I am using as the velocity of the moving frame itself. Normally, I use it for basis vectors but here it seemed appropriate for this purpose. Hopefully, the context of its use throughout the text will make it apparent for what purpose it is being used.

In a frame F' that is moving at velocity v_x parallel to the x-axis in F we calculate the position at time t as:

$$x' = \gamma(x - v_x t)$$

$$y' = y$$

$$z' = z$$

$$t' = \gamma\left(t - \frac{v_x}{c}x\right)$$

and the position at time $t + \mathrm{d}t$:

$$x' + \partial x' = \gamma(x + \partial x - v_x(t + \partial t))$$

$$y' + \partial y' = y + \partial y$$

$$z' + \partial z' = z + \partial z$$

$$t' + \partial t' = \gamma\left(t + \partial t - \frac{v_x}{c}(x + \partial x)\right)$$

or:

$$\partial x' = \gamma(\partial x - v_x \partial t)$$

$$\partial y' = \partial y$$

$$\partial z' = \partial z$$

$$\partial t' = \gamma\left(\partial t - \frac{v_x}{c}\partial x\right)$$

In the frame F' the velocity components of $\vec{\mathbf{u}}$ are:

$$u'_x = \frac{\partial x'}{\partial t'} = \frac{\gamma(\partial x - v_x \partial t)}{\gamma\left(\partial t - \frac{v_x}{c}\partial x\right)} = \frac{\frac{\partial x}{\partial t} - v_x}{\left[1 - \left(\frac{v_x}{c^2}\right)\frac{\partial x}{\partial t}\right]} = \frac{u_x - v_x}{\left[1 - \left(\frac{v_x}{c^2}\right)u_x\right]}$$

$$u'_y = \frac{\partial y'}{\partial t'} = \frac{\partial y}{\gamma\left(\partial t - \frac{v_x}{c}\partial x\right)} = \frac{\frac{\partial y}{\partial t}}{\gamma\left[1 - \left(\frac{v_x}{c^2}\right)\frac{\partial x}{\partial t}\right]} = \frac{u_y}{\gamma\left[1 - \left(\frac{v_x}{c^2}\right)u_x\right]}$$

$$u'_z = \frac{u_z}{\gamma\left[1 - \left(\frac{v_x}{c^2}\right)u_x\right]}$$

and by swapping primes and changing frame velocity signs, the reverse transformation equations are:

$$u_x = \frac{u'_x + v_x}{\left[1 + \left(\frac{v_x u'_x}{c^2}\right)\right]} \quad u_y = \frac{u'_y}{\gamma\left[1 + \left(\frac{v_x u'_x}{c^2}\right)\right]} \quad u_z = \frac{u'_z}{\gamma\left[1 + \left(\frac{v_x u'_x}{c^2}\right)\right]} \tag{10.14}$$

These are the *velocity addition* formulas. Consider a train moving in the x-direction relative to the ground with velocity v_x. This is our friendly F' reference frame while the ground is our F frame. Now consider someone on the train who is walking up the aisle of a car with velocity u'_x with respect to the train. Our "common sense" says that the velocity of the person with respect to the ground (the F frame) is simply $u'_x + v_x$, however, from our equations we find that this is *not* the case. u_x, the person's velocity with respect to the ground is:

$$u_x = \frac{u'_x + v_x}{\left[1 + \left(\frac{v_x u'_x}{c^2}\right)\right]}$$

One can see that under normal circumstances v_x/c^2 is *very* small and that the calculation is well approximated by $u_x = u'_x + v_x$, the "common sense" calculation. It is interesting to note what happens if the person on the train moves at the speed of light, $u'_x = c$.:

$$u_x = \frac{c + v_x}{\left[1 + \left(\frac{v_x}{c^2}\right)c\right]} = \frac{c + v_x}{\left[1 + \frac{v_x}{c}\right]} = \frac{c + v_x}{\left[\frac{v_x + c}{c}\right]} = c$$

Thus, if the person is moving at velocity c with respect to the train (in frame F') then they are *also* moving at velocity c with respect to the ground (frame F). This very counter-intuitive result is of course what we observe for electromagnetic radiation.

10.3 Space-time

It is worth repeating the Galilean transformation eqn (10.1):

$$x' = x - v_x t$$

$$y' = y$$

$$z' = z$$

$$t' = t$$

and the Lorentz transformation eqn (10.10):

$$
\begin{aligned}
x' &= \gamma(x - v_x t) \\
y' &= y \\
z' &= z \\
t' &= \gamma\left(t - \frac{v_x}{c}x\right)
\end{aligned}
$$

for comparison. Using the Galilean transformations we consider time t' to always equal t and to be separate from the (x, y, z) coordinates that may transform according to a velocity term as does the x-coordinate in our simple example. A more general expression of the Galilean transformation is:

$$
\begin{aligned}
x' &= x - v_x t \\
y' &= y - v_y t \\
z' &= z - v_z t \\
t' &= t
\end{aligned}
$$

wherein t always transforms to t'. In fact, we would usually not consider including the last equation in the set and consider only the x, y, z transformations in three-dimensional Euclidean space. The Lorentz transformations on the other hand must include time because t' depends not only on t but on x as well in our simple example. Thus, there are four variables that transform rather than three and we consider them to correspond to a four-dimensional space. The advantage of doing this is that we can apply some of the ideas of three-dimensional vector algebra to our new four-dimensional space or *four-space*. This is indeed how many look upon things and the collection of (x, y, z, t) is called *space-time* or *Minkowski space* after the German mathematician Hermann Minkowski who worked out its geometry. Actually, in order for each dimension in space-time to have the same units we modify (x, y, z, t) to be (x, y, z, ct) and we slightly modify our Lorentz transformation equations by multiplying the time variables by c:

$$
\begin{aligned}
x' &= \gamma(x - \beta ct) \\
y' &= y \\
z' &= z \\
ct' &= \gamma(ct - \beta x)
\end{aligned}
\tag{10.15}
$$

$$\text{where } \beta \equiv \frac{v_x}{c}$$

and the back-transformation equations:

$$
\begin{aligned}
x &= \gamma(x' + \beta ct) \\
ct &= \gamma(ct' + \beta x)
\end{aligned}
\tag{10.16}
$$

β is simply the fraction of c that F' is moving at with respect to F. Note that that γ is related to β:

$$\gamma = \frac{1}{\sqrt{1 - \left(\frac{v_x}{c}\right)^2}} = \frac{1}{\sqrt{1 - \beta^2}}$$

In three-dimensional space a rotational transformation of a vector produces a new vector whose length is identical to that of the non-rotated vector. We showed this to be so in Chapter 3 and referred to it as invariance of the norm of the vector to rotation. Thus, in three-dimensional space:

$$x^2 + y^2 + z^2 = x'^2 + y'^2 + z'^2$$

where (x, y, z) and (x', y', z') are the components of the non-rotated and rotated vectors, respectively.

We might expect that this will also be the case in space-time. Let us try this and see what happens. In space-time the squares of the *four-vectors* for reference frames F and F' should be equal since, by analogy with three-dimensional Euclidean space, the sum of the squares of the vector components is the square of the norm of the vector:

$$(ct)^2 + x^2 + y^2 + z^2 = (ct')^2 + x'^2 + y'^2 + z'^2$$

However, after application of the Lorentz transformations to the primed values we get:

$$(ct)^2 + x^2 + y^2 + z^2 \neq [\gamma(ct - \beta x)]^2 + [\gamma(x - \beta ct)]^2 + y^2 + z^2$$

or

$$(ct)^2 + x^2 \neq [\gamma(ct - \beta x)]^2 + [\gamma(x - \beta ct)]^2$$

They are not, in fact, equal. The problem lies with the sign of $(ct)^2$.

Recall that when we derived the Lorentz transformation equations we used an example of a light flash moving outwards from the origin of F and said that at time t the square of the distance that the light travelled to a point (x, y, z) is:§

$$(ct)^2 = x^2 + y^2 + z^2$$

or

$$0 = x^2 + y^2 + z^2 - (ct)^2$$

To simplify things let us also consider a two-dimensional subspace in space-time, (x, ct):

$$x^2 - (ct)^2 = 0$$

We can say something similar for F':

$$x'^2 - (ct')^2 = 0$$

and application of the Lorentz transformation equations to this produces:

$$[\gamma(x - \beta ct)]^2 - [\gamma(ct - \beta x)]^2 = 0$$

$$\gamma^2(x^2 - 2c\beta xt + \beta^2(ct)^2) - \gamma^2(\beta^2 x^2 - 2c\beta xt + (ct)^2) = 0$$

$$(1 - \beta)x^2 + (\beta^2 - 1)(ct)^2 = 0$$

$$(1 - \beta)x^2 = -(\beta^2 - 1)(ct)^2 = (1 - \beta^2)(ct)^2$$

$$x^2 = (ct)^2 \Rightarrow x^2 - (ct)^2 = 0$$

Thus:

$$x'^2 - (ct')^2 = x^2 - (ct)^2$$

and we say that this quantity is invariant under Lorentz transformation. In fact, this pops up so frequently that in its difference form it is given the name *space-time interval* and is analogous to the spatial distance between two points in Euclidean space. The negative sign in front of $(ct)^2$ makes space-time fundamentally different from 3D space and it is sometimes said that space-time is *pseudo-Euclidean*. Nonetheless, calculation of the space-time interval, Δs^2, including the y and z directions, is very similar to the familiar Euclidean distance calculation:¶

$$\Delta s^2 = -(\Delta ct)^2 + (\Delta x)^2 + (\Delta y)^2 + (\Delta z)^2 \quad (10.17)$$

§Beware! Some will write this as $0 = (ct)^2 - x^2 - y^2 - z^2$, which is perfectly fine except that it will cause confusion to the uninitiated.

¶This might be viewed as the space-time version of the Pythagorean theorem.

The space-time interval is invariant to transforms between frames. That is, $\Delta s^2 = \Delta s'^2$ much as the norm of a vector is invariant to coordinate rotations, as we noted in Chapter 3.

Although we have not explicitly pointed it out, the astute reader will by now have realized that our Lorentz transformation equations can be written in vector/matrix form:

$$\begin{bmatrix} ct' \\ x' \\ y' \\ z' \end{bmatrix} = \begin{bmatrix} \gamma & -\gamma\beta & 0 & 0 \\ -\gamma\beta & \gamma & 0 & 0 \\ 0 & 0 & 1 & 0 \\ 0 & 0 & 0 & 1 \end{bmatrix} \begin{bmatrix} ct \\ x \\ y \\ z \end{bmatrix}$$

or

$$\vec{\mathbf{V}}' = \Lambda \vec{\mathbf{V}}$$

The column matrices are, as in Chapter 3, considered to be vectors and, in the context of space-time, are called *four vectors*. Just as in Chapter 3, the four vector matrix contains the components of the four-dimensional space-time vector and we sometimes see it written as:

$$\vec{\mathbf{V}} = (V_0, V_1, V_2, V_3) = (x_0, x_1, x_2, x_3)$$

and

$$V_\mu = x_\mu$$

where, by convention, $x_0 = ct$ and $(x_1, x_2, x_3) = (x, y, z)$. In terms of our Lorentz transform eqn (10.15) and (10.16), using the vector component notation we have:

$$\begin{aligned} V_0' &= \gamma(V_0 - \beta V_1) \\ V_1' &= \gamma(V_1 - \beta V_0) \\ V_2' &= V_2 \\ V_3' &= V_3 \end{aligned} \tag{10.18}$$

This gives a way to transform four vectors from one frame to another. Furthermore, the transform conserves the space-time interval as in eqn (10.17):

$$-V_0^2 + V_1^2 + V_2^2 + V_3^2 = -V_0'^2 + V_1'^2 + V_2'^2 + V_3'^2$$

in an analogous fashion to the conservation of the norm of a Euclidean vector in 3D space.

10.4 Minkowski Diagrams

The Minkowski diagram is an ingenious visual way to determine all sorts of things associated with special relativity. For example, we can use it to determine when and where events occur for observers in different inertial frames of reference. The form of the diagram is as a pair of Cartesian axes, one tilted along the 45 degree with respect to the other. The F frame consists of the x and ct axes at right angles to each other, as one would see a simple standard Cartesian axis presented. The x' and ct' axes are presented as the axes of a Cartesian coordinate system tilted on the common 45° angle. If you stare at the diagram for long enough you will see this. This is not meant to imply that there is an angle between these two frames in space. It might help to visualise at first the F' frame overlaying the F frame and then rotating the F' frame about the 45° line. The amount of rotation is dependent on the relative velocity, v_x, of frame F'.

We see from Figure 10.3 that the tangent of the angle θ, the projection angle between the x and x' axes is equal to β. Since $\beta = v_x/c$ and v_x can be no greater than c, the maximum possible rotation of the F' axes about the 45° diagonal is 90°. By the Pythagorean Theorem we can relate l in F' to γ in F:

$$l = \sqrt{\gamma^2 + \beta^2\gamma^2} = \sqrt{\left(\frac{1}{\sqrt{1-\left(\frac{v_x}{c}\right)^2}}\right)\left(\frac{\beta^2}{\sqrt{1-\left(\frac{v_x}{c}\right)^2}}\right)} = \sqrt{\left(\frac{1}{\sqrt{1-\beta^2}}\right)\left(\frac{\beta^2}{\sqrt{1-\beta^2}}\right)}$$

$$= \sqrt{\frac{1+\beta^2}{1-\beta^2}}$$

It is also possible to imagine a Minkowski diagram with an F' plane rotated about a 135°-axis (Figure 10.4).

Of what use is one of these diagrams? Let us suppose that $\beta = 0.364$.[‖] Recall that this represents the fraction of the speed of light

[‖]It seems a bit artificial, I know but I constructed the Minkowski diagrams with a projection angle of 20°, the tangent of which is 0.364. So rather than draw all new diagrams we will just use that.

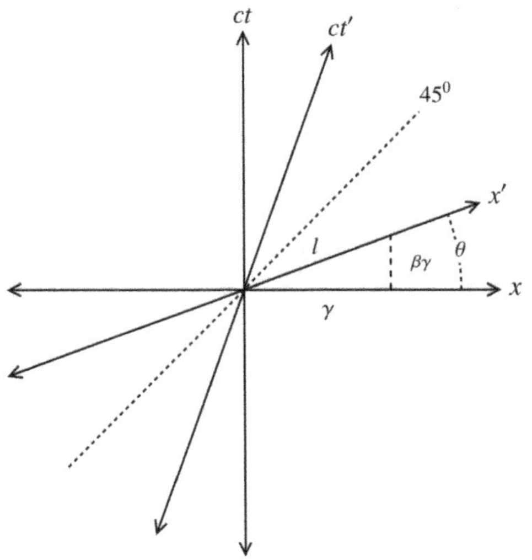

Figure 10.3 A Minkowski diagram with axes x and ct for frame F and axes x' and ct' for frame F'. The F' frame is rotated about the 45° diagonal in frame F. A length of γ in F transforms to a length of l in F'.

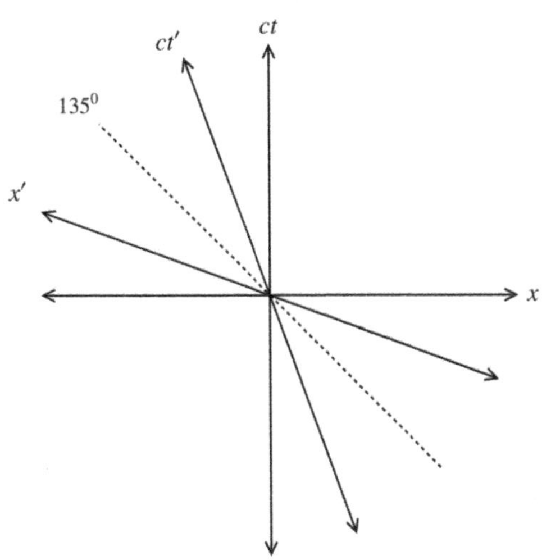

Figure 10.4 Minkowski diagram with the F' plane rotated about the 135° diagonal in plane F.

that the F' frame is moving with respect to F. This being the case we may also determine the value of γ:

$$\gamma = \sqrt{\frac{1}{1-\beta^2}}$$

Using this and our current value of β, we find that $\gamma = 1.074$. Now let us also suppose that we have an event in F that is 4 years in the future and 3 light years to the right of the origin. Using our Lorentz transformation eqn (10.15) for space-time we calculate:

$$\begin{aligned} x' &= \gamma(x - \beta ct) \\ &= 1.074(3 - 0.364 \times 4) \\ &= 1.66 \\ ct' &= \gamma(ct - \beta x) \\ &= 1.074(4 - 0.364 \times 3) \\ &= 3.12 \end{aligned}$$

To an observer in F' the event occurs at a distance of 1.66 light years and 3.12 years in the future. This is easily represented in the Minkowski diagram.

One locates the point (3,4) in the F frame using parallel lines from the x and ct axes (Figure 10.5). Then the corresponding points on the x' and ct' axes are located by drawing lines parallel to the x' and ct' axes from the point to the axes. It is also clear in the Minkowski diagram why two events that are simultaneous in F are not simultaneous in F'. Events that occur on a line that is parallel to the x-axis

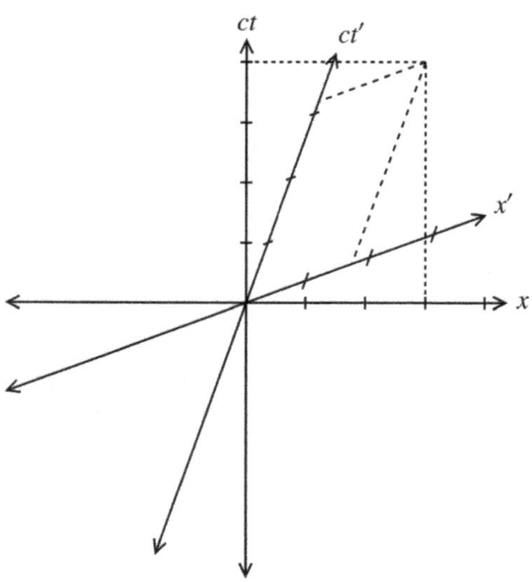

Figure 10.5 Minkowski diagram relating distance and time between two reference frames.

in F are simultaneous in F. Events that occur on a line that is parallel to the x'-axis in F' are simultaneous in F'.

In Figure 10.6 events e_1 and e_2 are simultaneous in F and events e_1 and e_3 are simultaneous in F'. However, e_1 and e_2 are not simultaneous in F'; nor are e_1 and e_3 simultaneous in F.

Time dilation is also easily portrayed (Figure 10.7). Consider point A at 3 on the ct-axis in the F frame. The corresponding point in F' on the

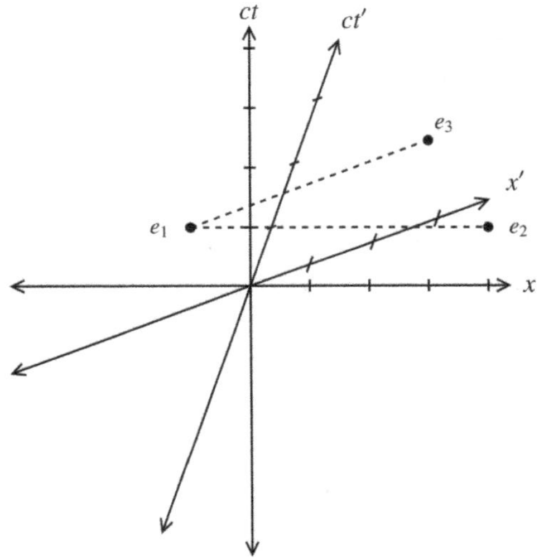

Figure 10.6 Minkowski diagram illustrating simultaneous and non-simultaneous events in space-time.

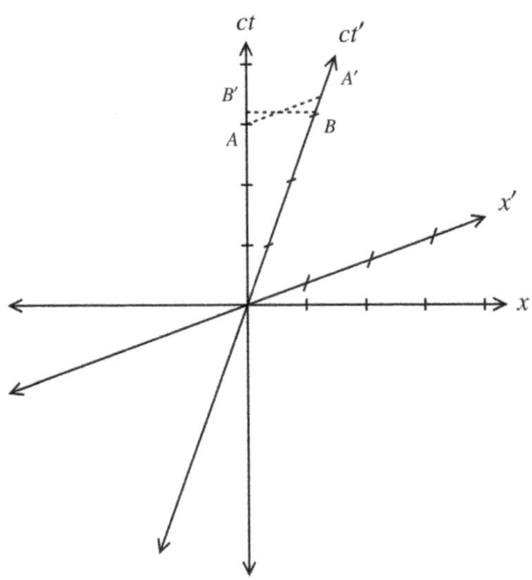

Figure 10.7 Minkowski diagram illustrating time dilation in space-time.

ct'-axis is 3.22. Thus, a process that takes 3 time units in frame F to complete appears to a viewer in F to take 3.22 time units in F'. Since there are no privileged frames it must be true that an observer in F' will see exactly the same time dilation in F! This can be seen by looking at point B at 3 on the ct' axis in F'. The corresponding point, B', on the ct axis is ... 3.22 time units. The observer in F' sees exactly the same time dilation in F.

Further Reading

1. J. Freund, *Special Relativity for Beginners: A Textbook for Undergraduates*, World Scientific, Hackensack, N.J., 2008.
2. D. J. Griffiths, *Introduction to Electrodynamics*, Pearson Prentice Hall, Upper Saddle River, New Jersey, 1999.
3. T. Helliwell, *Special Relativity*, University Science Books, Sausalito, Calif., 2010.
4. R. Skinner, *Relativity*, Blaisdel Publishing Co., Waltham, Mass., 1969.

11 Electric and Magnetic Fields

One may conceive light to spread successively, by spherical waves.

Christiaan Huygens

11.1 Coulomb's Law and the Electric Field

Charles-Augustin de Coulomb[†] discovered, experimentally, an inverse square relationship for the forces between electrically charged objects.[‡] That there is a force at all is known by all who have experienced static electricity in one form or another. The scalar form of the relationship is known as Coulomb's law:

$$F = k\frac{q_1 q_2}{r^2}$$

and says that the magnitude of the force between two charges is proportional to the size of each charge over the square of the distance between them. This force is called the *electrostatic force* between charges. Of course, as we have seen, forces can also be represented in vector form and Coulomb's law is no different:

$$\vec{\mathbf{F}} = k\frac{q_1 q_2}{r^2}\vec{\mathbf{r}}_u$$

[†]The material in this first section of this chapter may seem to be not directly related to NMR spectroscopy and what the NMR/MRI spectroscopist might need to know for her/his everyday dealings with spectrometers. However, it will be revealed in the fullness of time that there is a very deep and intimate connection between electric fields and magnetic fields that the NMR/MRI spectroscopist should probably be aware of, if only to impress colleagues.
[‡]Actually, Henry Cavendish discovered this a decade before Coulomb but neglected to publish it, as with many of his other discoveries.

Essential Mathematics for NMR and MRI Spectroscopists
By Keith C. Brown

Published by the Royal Society of Chemistry, www.rsc.org

where $\vec{\mathbf{r}}_u$ is a unit vector parallel to the straight line between the two charges constructed from the position vectors of q_1 and q_2 (see Figure 3.5). To accomplish this we have simply multiplied by a unit vector constructed *via* eqn (3.13):

$$\vec{\mathbf{F}} = k\frac{q_1 q_2}{r^2} \times \frac{\vec{\mathbf{r}}}{r} = k\frac{q_1 q_2}{r^2}\vec{\mathbf{r}}_u$$

the constant, k, is usually written (in SI units):

$$k = \frac{1}{4\pi\epsilon_0}$$

where ϵ_0 is the *permittivity* of free space. The modern SI form of Coulomb's law in vector form is:

$$\vec{\mathbf{F}} = \frac{1}{4\pi\epsilon_0}\frac{q_1 q_2}{r^2}\vec{\mathbf{r}}_u \tag{11.1}$$

What is the relationship between more than two charges separated in space? Experimentally it is found that the force on a "test charge" from a set of charges is additive:

$$\begin{aligned} F &= \frac{1}{4\pi\epsilon_0}\frac{Q_t q_1}{r^2} + \frac{1}{4\pi\epsilon_0}\frac{Q_t q_2}{r^2} + \frac{1}{4\pi\epsilon_0}\frac{Q_t q_3}{r^2}\cdots \\ &= \frac{Q_t}{4\pi\epsilon_0}\sum_i \frac{q_i}{r_i^2} \end{aligned}$$

The vector form of this is then:

$$\begin{aligned} \vec{\mathbf{F}} &= \frac{1}{4\pi\epsilon_0}\frac{Q_1 q_1}{r^2}\vec{\mathbf{r}}_{u1} + \frac{1}{4\pi\epsilon_0}\frac{Q_1 q_2}{r^2}\vec{\mathbf{r}}_{u2} + \frac{1}{4\pi\epsilon_0}\frac{Q_1 q_3}{r^2}\vec{\mathbf{r}}_{u3}\cdots \\ &= \frac{Q_t}{4\pi\epsilon_0}\sum_i \frac{q_i}{r_i^2}\vec{\mathbf{r}}_{ui} \end{aligned}$$

In our mind's eye we (artificially) separate the test charge from the other electric charge(s), which we will call the source charge(s), and this may be represented in our force equation:

$$\begin{aligned} \vec{\mathbf{F}} &= \frac{Q_t}{4\pi\epsilon_0}\sum_i \frac{q_i}{r_i^2}\vec{\mathbf{r}}_{ui} \\ &= Q_t\vec{\mathbf{E}} \end{aligned} \tag{11.2}$$

where

$$\vec{\mathbf{E}} \equiv \frac{1}{4\pi\epsilon_0}\sum_i \frac{q_i}{r_i^2}\vec{\mathbf{r}}_{ui}$$

$\vec{\mathbf{E}}$ is here defined in terms of the Coulomb force and a test charge, Q_t, and is called the *electric field*. Note that, like the Coulomb force, the electric field is additive; that is, the field at any point in space is the sum of the field contributions from all of the non-test charges. We can easily calculate the magnitude of the electric field at a point in space since eqn (11.2) is in terms of a unit vector:

$$E = \frac{1}{4\pi\epsilon_0}\frac{Q_s}{r^2}$$

where Q_s represents the source charge.

So far, our equations have presumed a collection of discrete point charges. For a continuous distribution of charge our summation sign simply becomes an integral:

$$\vec{\mathbf{E}} = \frac{1}{4\pi\epsilon_0}\int \frac{1}{r^2}\vec{\mathbf{r}}_u \mathrm{d}q$$

There are three general types of charge distributions, charge distributed along a line, charge distributed over an area and charge distributed through a volume. We then add the distribution expression into the integral. For example, for volume distribution we can write in differential terms:

$$\mathrm{d}q = \rho(x, y, z)\mathrm{d}V \tag{11.3}$$

where we use $\rho(x, y, z)$ to represent the (continuous) distribution of charge throughout the volume as a function of position. The integral becomes:

$$\vec{\mathbf{E}} = \frac{1}{4\pi\epsilon_0}\int \frac{\rho(x, y, z)}{r^2}\vec{\mathbf{r}}_u \mathrm{d}V \tag{11.4}$$

How do we picture the electric field? There are two ways. The first is to use essentially the same view as the force fields pictured in Chapter 4 using vectors to indicate the magnitude and direction of the field at any point in space. This is a rather clumsy way to show the field as it is difficult to show the field at many points in space; the vectors take up so much room in the picture. The second, more elegant way is to use *field lines*. These are lines whose tangent is in the direction of the field vector at that point in space. So, for a

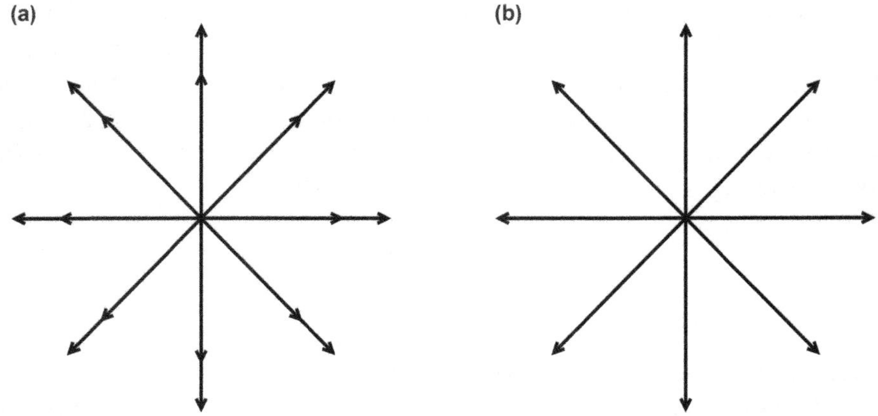

Figure 11.1 (a) Vector representation of the electric field of a positive charge; (b) Field-line representation of the electric field of a positive charge.

simple single positive charge the two pictures are shown in Figures 11.1a and 11.1b.

For a single negative charge the direction of the vectors is reversed.

11.2 Field Flux and Gauss's Law

Let us visualise the electric field line idea in 3D for a single charge and surround the charge with a sphere (Figure 11.2).

We see that the field lines are directed radially outward from the charge as we would expect from Coulomb's law. Recalling the discussion in Chapter 4 concerning flux and specifically, eqn (4.26), we can write an equation for the flux of the electric field lines through the surface of the sphere (or any other shape as we shall see). Our flux equation is:

$$\Phi = \iint \vec{\mathbf{E}} \cdot \mathrm{d}\vec{\sigma}$$

or

$$\Phi = \iint \vec{\mathbf{E}} \cdot \vec{\mathbf{n}} \mathrm{d}\sigma$$

where $\mathrm{d}\vec{\sigma}$ is a surface area infinitesimal vector that is orthogonal to the surface and $\vec{\mathbf{n}}$ is a unit vector orthogonal to the surface. What is

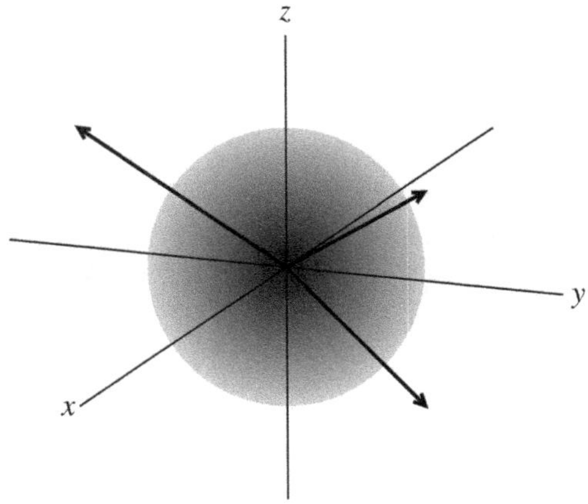

Figure 11.2 Field lines from charge q.

the flux for our current example? We set up the integral from the flux equation using our expression for $\vec{\mathbf{E}}$, eqn (11.2) and, since we are considering a sphere, spherical coordinates (see Appendix A):

$$\begin{aligned}\Phi &= \iint \vec{\mathbf{E}} \cdot \mathrm{d}\vec{\sigma} \\ &= \int_0^{2\pi}\int_0^{\pi}\left(\frac{1}{4\pi\epsilon_0}\frac{q}{r^2}\vec{\mathbf{r}}_u\right)\cdot(\vec{\mathbf{r}}_u r^2 \sin(\theta)\mathrm{d}\theta\mathrm{d}\phi) \\ &= \int_0^{2\pi}\int_0^{\pi}\frac{q}{4\pi\epsilon_0}\sin(\theta)\mathrm{d}\theta\mathrm{d}\phi \\ &= \frac{q}{\epsilon_0}\end{aligned}$$

It is interesting to note that the result of this calculation does not depend on r which will be of some small importance later. Another interesting thing here is that it does not matter what shape the enclosing object is. If we were to enclose the sphere that itself encloses the charge, q, with say, a cube, the same number of field lines would pass through the cube as pass through the sphere (Figure 11.3).

Thus, for a single charge enclosed by a shape the total flux through the shape is q/ϵ_0, *regardless* of the shape.

What is the case for more than one charge enclosed by a surface? The electric field is a vector quantity and is therefore additive (eqn (11.2)) so we have every right to expect that the flux for a collection of charges is:

$$\Phi = \sum_i \frac{q_i}{\epsilon_0}$$

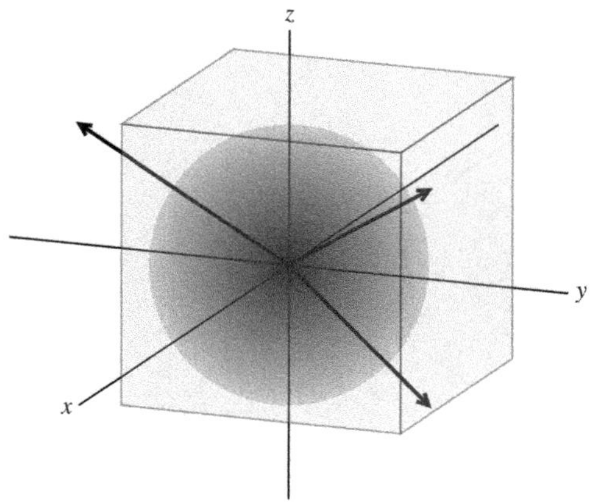

Figure 11.3 Field flux through a cube.

and for a charge distribution $\rho(x, y, z)$ throughout a volume:

$$\Phi = \frac{1}{\epsilon_0}\iiint_V \rho(x, y, z)\mathrm{d}V$$
$$= \frac{Q_s}{\epsilon_0}$$
$$= \frac{1}{\epsilon_0}\sum_i \frac{q_i}{\epsilon_0}$$

where here, Q_s represents the total charge enclosed by a surface:

$$Q_s = \iiint_V \rho(x, y, z)\mathrm{d}V$$

and $\iiint_V$ is the integral over the entire volume.

What we have just calculated is from Gauss's law (see Chapter 4) and relates the total flux through a surface completely enclosing a charge to the total charge. In other words the flux and the charge are proportional to each other. A statement of Gauss's law would be:

$$\iint \vec{\mathbf{E}} \cdot \mathrm{d}\vec{\sigma} = \frac{Q_s}{\epsilon_0}$$

or (11.5)

$$\oiint_S \vec{\mathbf{E}} \cdot \mathrm{d}\vec{\sigma} = \frac{Q_s}{\epsilon_0}$$

Recall that $\oiint_S$ is the integral over an entire surface.

Now, here is where all of the work of Chapter 4 begins to pay off. Recalling Gauss's divergence theorem (eqn (4.43)) and the definition of Q_s from above, we can write:

$$\oiint_S \vec{\mathbf{E}} \cdot d\vec{\sigma} = \iiint_V \vec{\nabla} \cdot \vec{\mathbf{E}} dV = \frac{Q_s}{\epsilon_0} = \frac{1}{\epsilon_0} \iiint_V \rho(x, y, z) dV$$

From this we write:

$$\iiint_V \vec{\nabla} \cdot \vec{\mathbf{E}} dV = \frac{1}{\epsilon_0} \iiint_V \rho(x, y, z) dV$$

Since these two integrals are equal their arguments must also be equal:

$$\vec{\nabla} \cdot \vec{\mathbf{E}} = \frac{\rho(x, y, z)}{\epsilon_0} \tag{11.6}$$

This is an alternate, differential expression of Gauss's law and as we will recall from Chapter 4, is the divergence of $\vec{\mathbf{E}}$.

A curious result occurs when we do the divergence calculation for $\vec{\mathbf{E}}$ directly, using a simple one-charge system. Our general expression for the electric field around this single charge is:

$$\vec{\mathbf{E}} = \frac{q}{4\pi\epsilon_0} \frac{\vec{\mathbf{r}}_u}{r^2}$$

From Figure 11.2 and our understanding of the meaning of the divergence from Chapter 4 we expect to find a non-zero value for the divergence of $\vec{\mathbf{E}}$. Indeed, from eqn (11.6) (using a single charge rather than a charge distribution) we expect to get a value of q/ϵ_0 for a single charge. The divergence calculation (in spherical coordinates for simplicity) would appear to be a simple calculation (using eqn (4.36)):

$$\vec{\nabla} \cdot \vec{\mathbf{E}} = \left(\frac{q}{4\pi\epsilon_0}\right) \vec{\nabla} \cdot \left(\frac{\vec{\mathbf{r}}_u}{r^2}\right)$$

$$= \left(\frac{q}{4\pi\epsilon_0}\right) \frac{1}{r^2} \frac{\partial\left(r^2 \frac{1}{r^2}\right)}{\partial r} = 0$$

This is a surprise! We were expecting a non-zero value for this calculation. What is wrong here? The problem is that at $r = 0$ there is a discontinuity in our electric field equation. We are dividing by zero to get an infinite electric field at $r = 0$. Gauss's law gave us a non-zero result since it does not depend on r as noted above and tells us that the value of the flux is non-zero. The technique for dealing with this problem is to use the Dirac delta function of Chapter 8:

$$\delta(x) \equiv \begin{Bmatrix} 0 & \text{if } x \neq 0 \\ \int_{-\infty}^{\infty} \delta(x) & \mathrm{d}x = 1 \end{Bmatrix}$$

Furthermore, we saw (eqn (8.42)) that the Dirac delta function can be used to "sample" a function at a specific point:

$$\int F(x)\delta(x-a)\mathrm{d}x = F(a)\int \delta(x-a)\mathrm{d}x = F(a)$$

We use this to model the fact that the divergence is zero everywhere except at the origin and we use the delta function to "pick out" the value of the divergence at the origin.[§]

$$\left(\frac{q}{4\pi\epsilon_0}\right)\vec{\nabla}\cdot\left(\frac{\vec{\mathbf{r}}_u}{r^2}\right) = \left(\frac{q}{4\pi\epsilon_0}\right)4\pi\delta(\vec{\mathbf{r}}) = \frac{q}{\epsilon_0}$$

where we have defined:

$$4\pi\delta(\vec{\mathbf{r}}) \equiv \vec{\nabla}\cdot\left(\frac{\vec{\mathbf{r}}_u}{r^2}\right)$$

How about the curl of $\vec{\mathbf{E}}$? We expect, based on the visual picture (Figure 11.2) that the value of the curl of $\vec{\mathbf{E}}$ should be zero. Again, let us test this using spherical coordinates and eqn (4.37)}:

$$\begin{aligned}\vec{\nabla}\times\vec{\mathbf{E}} = {} & \frac{\vec{\mathbf{u}}_r}{r\sin(\theta)}\left(\frac{\partial\sin(\theta)E_\phi}{\partial\theta} - \frac{\partial E_\theta}{\partial\varphi}\right) \\ & + \frac{\vec{\mathbf{u}}_\theta}{r}\left(\frac{1}{\sin(\theta)}\frac{\partial E_r}{\partial\varphi} - \frac{\partial rE_\phi}{\partial r}\right) \\ & + \frac{\vec{\mathbf{u}}_\phi}{r}\left(\frac{\partial rE_\theta}{\partial r} - \frac{\partial E_r}{\partial\theta}\right) \\ = {} & 0\end{aligned}$$

To see that this is so, one must realise that all of the spherical components of $\vec{\mathbf{E}}$ are zero except E_r and that E_r depends only on r.

[§]This is one of the major uses of the Dirac delta function in physics. It is frequently used to handle otherwise intractable problems in which there is discontinuity of some kind.

Then, each derivative in our spherical coordinate curl expression is zero. We could also arrive at this result using Stokes' theorem from Chapter 4. The line integral around a closed path in an electric field is:

$$\begin{aligned}\oint \vec{\mathbf{E}} \cdot \mathrm{d}\vec{\mathbf{l}} &= \left(\frac{q}{4\pi\epsilon_0}\right)\int_a^b \frac{1}{r^2}\mathrm{d}r \\ &= \left(\frac{q}{4\pi\epsilon_0}\right)\left(\frac{1}{r_a}-\frac{1}{r_b}\right) \\ &= 0\end{aligned} \tag{11.7}$$

indicating that the electric field is conservative. Also from Stokes' theorem we find therefore that:

$$\nabla \times \vec{\mathbf{E}} = 0 \tag{11.8}$$

since for

$$\iint_S (\vec{\nabla} \times \vec{\mathbf{E}}) \cdot \vec{\mathbf{n}}\mathrm{d}\sigma = \oint_C \vec{\mathbf{E}} \cdot \mathrm{d}\vec{\mathbf{l}} = 0$$

to be true, $\vec{\nabla} \times \vec{\mathbf{E}}$ must be zero.

Now, recalling from Chapter 4 that $\vec{\nabla}\phi$ (where ϕ is a scalar function) is a vector and from eqn (4.22) that $\vec{\nabla} \times (\vec{\nabla}\phi) = 0$, our result for the curl of $\vec{\mathbf{E}}$ indicates that we can generate $\vec{\mathbf{E}}$ from a scalar function. For the electric field this function is called the *electric potential* and is defined by:

$$\begin{aligned}\phi(\vec{\mathbf{r}}) &\equiv -\int_R^r \vec{\mathbf{E}} \cdot \mathrm{d}\vec{\mathbf{l}} \\ &= \int_R^r (\vec{\nabla}\phi) \cdot \mathrm{d}\vec{\mathbf{l}} \\ \vec{\mathbf{E}} &\equiv -\vec{\nabla}\phi\end{aligned} \tag{11.9}$$

(see also eqn (4.18)). The minus sign is by convention and the reason for it will shortly become evident. R is the reference point … we are free to choose any point we like. Let us calculate the electric potential at a point r in a field generated by a single charge q using infinity as our reference point:

$$
\begin{aligned}
\phi(r) &= -\int_{\infty}^{r} \vec{\mathbf{E}} \cdot \mathrm{d}\vec{\mathbf{l}} \\
&= -\frac{q}{4\pi\epsilon_0}\int_{\infty}^{r} \frac{\vec{\mathbf{r}}_u}{r^2} \cdot \mathrm{d}\vec{\mathbf{l}} \\
&= \frac{q}{4\pi\epsilon_0}\frac{1}{r}\bigg|_{\infty}^{r} = \frac{q}{4\pi\epsilon_0}\left(\frac{1}{r} - \frac{1}{\infty}\right) \\
&= \frac{q}{4\pi\epsilon_0}\frac{1}{r}
\end{aligned}
\tag{11.10}
$$

We see two things in this result. The first is the reason for using infinity as the lower integrating limit . . . it removes one of the definite integral terms. What we are really calculating here is the potential difference between two points in the electric field:

$$
\begin{aligned}
-\int_{r1}^{r2} \vec{\mathbf{E}} \cdot \mathrm{d}\vec{\mathbf{l}} &= -\frac{q}{4\pi\epsilon_0}\int_{r1}^{r2} \frac{\vec{\mathbf{r}}_u}{r^2} \cdot \mathrm{d}\vec{\mathbf{l}} \\
&= -\frac{q}{4\pi\epsilon_0}\frac{1}{r}\bigg|_{r1}^{r2} = \frac{q}{4\pi\epsilon_0}\left(\frac{1}{r_2} - \frac{1}{r_1}\right) \\
&= \phi(r_2) - \phi(r_1)
\end{aligned}
$$

The second thing we note is the reason for the negative sign in the definition of the electric potential . . . it results in a positive value for the potential.

We can generalise this for a charge distribution rather than a point charge:

$$
\phi(r) = \frac{1}{4\pi\epsilon_0}\int_V \frac{\rho}{r}\mathrm{d}V \tag{11.11}
$$

From eqn (11.2) we recall that the force on a test charge is:

$$
\vec{\mathbf{F}} = Q_t\vec{\mathbf{E}}
$$

and indeed, serves to define the electric field. Let us alter eqn (11.7) using the test charge:

$$
\oint Q_t\vec{\mathbf{E}} \cdot \mathrm{d}\vec{\mathbf{l}} = \oint \vec{\mathbf{F}} \cdot \mathrm{d}\vec{\mathbf{l}} = 0
$$

Since the electric field is conservative and the line integral along a closed curve is zero so also must the line integral for the force on a test charge along the same closed curve. The units for these integrals are energy units and calculate the work done in moving a test charge along the closed curve. Since the total work done is zero there must be no movement of charge along the closed curve. Translating this to the real world, it means that a static electric field will not cause a current to flow in an electric circuit.

The line integral of the electric field defines the electromotive force (EMF) and is the work done on a test charge divided by the magnitude of the test charge.

$$EMF \equiv \oint \vec{\mathbf{E}} \cdot d\vec{\mathbf{l}} \tag{11.12}$$

This is of course, zero for a static electric field. We shall encounter this later in the context of time-varying fields and discover it to be non-zero.

11.3 The Magnetic Field

Magnetism has been known for a very long time; at least as long ago as the time of Aristotle.[¶] For the last thousand years or so the magnetic compass has been a staple of the navigator's tool chest. It was not until the early 1800s, however, that a serious scientific investigation of the phenomenon was undertaken. Oersted, Ampere, Gauss, Biot, Savart, Faraday, Maxwell and Einstein all contributed to our understanding of magnetism. Today, we know that electrical currents and magnetic fields are inextricably linked and any study of one should also include the other; they are really two facets of the same phenomenon as we shall see.

It was Oersted who discovered that a compass needle can be deflected by an electric current. This inspired others to investigate the phenomenon and it was Ampere who then found that there is a force between two wires that are conducting an electric current (Figure 11.4).

If the current is flowing in the same direction in the two wires they are attracted towards each other and if the current flow is opposite in

[¶]What we want to be able to accomplish in this section is to be able to calculate the magnetic field strength at all points along the centre line of a solenoid. Our modern superconducting spectrometer magnets all use solenoids to generate the field and it is definitely of interest to us as NMR/MRI spectroscopists to know how to do this. The path to this goal may seem a bit torturous but stay with it and we will arrive at our destination.

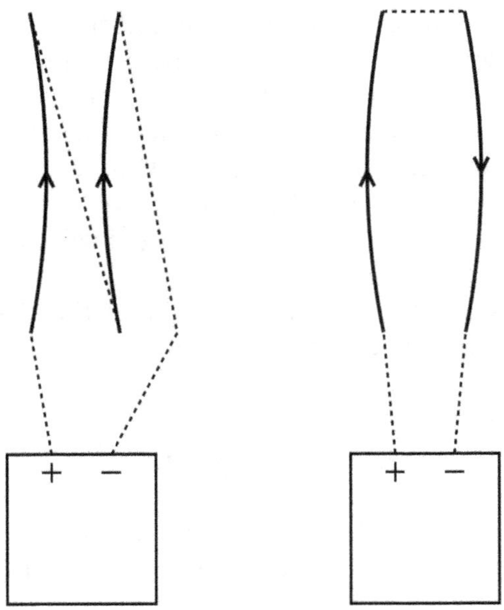

Figure 11.4 Current flow in pairs of wires and magnetic forces between them.

each wire they are repelled from each other. That it is not a Coulomb force of the type we have seen is evident because there is no static charge on the wires. Also, reversing the direction of current flow in one of the wires changes the direction of the force by 180°, not what one would expect of the static electrical charges that we have seen so far.

Whatever is happening here it is clearly not electrostatic in nature and we therefore cannot attribute the force of attraction and repulsion to an electric field. That a compass needle is deflected by the current in a wire leads us to believe that what we observe is a magnetic phenomenon due to the flow of electricity through the wires. Similar to the case with a static electric charge, we define a magnetic field, $\vec{\mathbf{B}}$. We also note that using the compass we find that the field points around the wire perpendicular to the flow of current and not in the direction of the force between the two wires as is the case with Coulomb forces (Figure 11.5).

This perpendicular relationship leads one to think that there might be a cross product association between the current, the field and the force and indeed we find that this is so:

$$\vec{\mathbf{F}} = q(\vec{\mathbf{v}} \times \vec{\mathbf{B}}) \tag{11.13}$$

where q is a test charge, $\vec{\mathbf{v}}$ is the velocity of the test charge and $\vec{\mathbf{B}}$ is the magnetic field vector at the test charge. This force is referred to as the

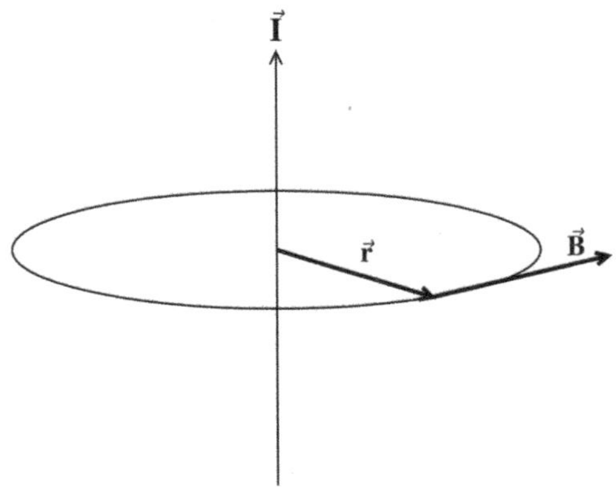

Figure 11.5 The magnetic field around a current-conducting wire.

magnetostatic force and is a bit of a misnomer since as we have seen the charges involved must be in motion in order to observe it. One can test this relationship easily using a cathode ray tube and a permanent magnet.‖ The beam of electrons will be deflected in a direction that is orthogonal to both the velocity of the electrons and the magnetic field.

A more detailed picture of what is happening can be obtained by considering separated charges in motion (Figure 11.6).

The force due to the magnetic field generated by the motion of charge q_2 on charge q_1 is obtained by an expansion of eqn (11.13):

$$\vec{\mathbf{F}}_1 = K\frac{q_1 q_2}{r_{12}^2}\vec{\mathbf{v}}_1 \times (\vec{\mathbf{v}}_2 \times \vec{\mathbf{r}}_u)$$

where $\vec{\mathbf{F}}_1$ is the force on q_1 due to the magnetic field at q_1. As usual, $\vec{\mathbf{r}}_u$ is a unit vector between q_1 and q_2. The form of this expression is similar to Coulomb's law (see eqn (11.1)) but rather than having to consider only one vector there are now three vectors to worry about! We do much the same with this equation as with Coulomb's law and define a field called the magnetic field:

$$\begin{aligned}\vec{\mathbf{F}}_1 &= q_1\vec{\mathbf{v}}_1 \times K\frac{q_2}{r_{12}^2}(\vec{\mathbf{v}}_2 \times \vec{\mathbf{r}}_u)\\ &= q_1\vec{\mathbf{v}}_1 \times \vec{\mathbf{B}}_{q1}\end{aligned}$$

‖Those of you who may recall the grand days of NMR equipment that could be repaired on-site will know that using a crt-oscilloscope in the proximity of the NMR magnet was a bit of a challenge, due largely to the effect on the oscilloscope beam by the field described by eqn (11.13).

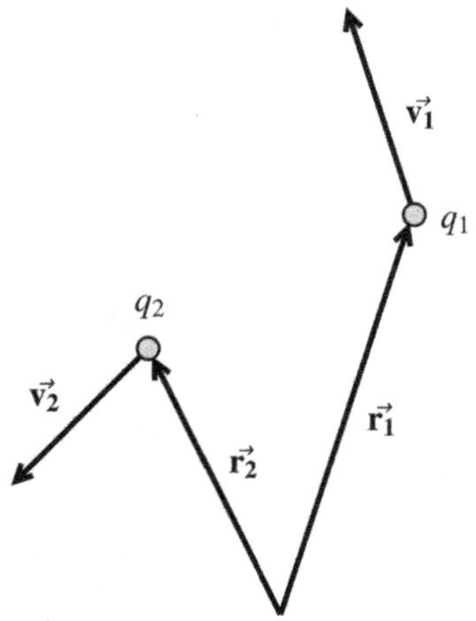

Figure 11.6 Charges q_1 and q_2 and their positions and velocities.

which is of the same form as eqn (11.13). $\vec{\mathbf{B}}_{q_1}$ refers to the magnetic field at q_1 and in terms of q_2, $\vec{\mathbf{v}}_2$ and $\vec{\mathbf{r}}_u$ is:

$$\vec{\mathbf{B}}_{q1} = K\frac{q_2}{r_{12}^2}(\vec{\mathbf{v}}_2 \times \vec{\mathbf{r}}_u)$$

As it is written here, our test charge is q_1 and our field generating charge is q_2. Let us take things a step further and instead of using a single charge to generate a magnetic field at q_1 use a set of charges:

$$\vec{\mathbf{B}} = \sum_{i=1} K\frac{q_i}{r_i^2}(\vec{\mathbf{v}}_i \times \vec{\mathbf{r}}_{ui})$$

In this equation we have dropped the subscript q_1 on $\vec{\mathbf{B}}$ and now assume $\vec{\mathbf{B}}$ to be the field at the test charge (formerly q_1). The q_i are all charges in motion, each with its own velocity $\vec{\mathbf{v}}_i$ and distance from the test charge, $\vec{\mathbf{r}}_{ui}$. Taking things to the limit we can write in integral form:

$$\vec{\mathbf{B}} = K\int\frac{(\vec{\mathbf{v}} \times \vec{\mathbf{r}}_u)}{r^2}\mathrm{d}q = K\int\frac{\rho(\vec{\mathbf{v}} \times \vec{\mathbf{r}}_u)}{r^2}\mathrm{d}V$$

recalling that the charge density ρ is $\mathrm{d}q/\mathrm{d}V$. As for Coulomb's law we state that the value of K as written here is:

$$K = \frac{\mu_0}{4\pi}$$

where μ_0 is the *permeability* of free space. So our formula for the magnetic field becomes:

$$\vec{\mathbf{B}} = \frac{\mu_0}{4\pi}\int \frac{\rho(\vec{\mathbf{v}} \times \vec{\mathbf{r}}_u)}{r^2} \mathrm{d}V \tag{11.14}$$

A full consideration of the forces on a test charge would also include any electric fields from the other charges as well and the full expression of these forces is encapsulated in the Lorentz force law:

$$\vec{\mathbf{F}} = q\vec{\mathbf{E}} + q\vec{\mathbf{v}} \times \vec{\mathbf{B}} \tag{11.15}$$

Since we are dealing with electrically charged particles that are in motion when considering magnetic fields and since often this motion is through a conducting wire we will consider the concept of electric current as it applies here. Electric current is loosely defined as the number of coulombs of electrons flowing past a point on wire per unit time:

$$I = \frac{\mathrm{d}q}{\mathrm{d}t} \tag{11.16}$$

The units are coulombs per second or amperes. Oddly, the direction of current flow is taken as the direction of flow of *positive* charge, even though electrons are negatively charged.**

This definition of current implicitly assumes that the charge density in the conducting wire is constant, however, it is possible to envision situations where this is not the case. We therefore make a more general definition of current using the concept of *current density*. Start by considering the charge density, ρ, whose units are coulombs per m^3. If the charge is moving through the wire then so is the charge density. We use this to define the current density:

$$\vec{\mathbf{J}} \equiv \rho\vec{\mathbf{v}} \tag{11.17}$$

$\vec{\mathbf{J}}$ is the volume current density and $\vec{\mathbf{v}}$ is the average charge velocity through the wire (more on this later). Evidently, the units for the volume current density are coulombs per m^2 s. Proceeding, we picture a volume element in our current-conductor of area, da (not necessarily perpendicular to the flow of charge) through which charge flows into and of length dl (Figure 11.7).

**This is a result of the guess that Benjamin Franklin made as to the sign of the two possible charges ... he had a 50% chance of guessing correctly such that things worked out and missed!

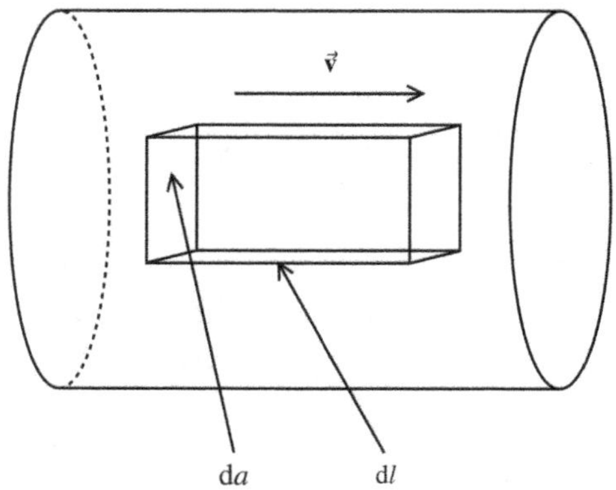

Figure 11.7 Volume element in a current-conducting wire.

The total amount of charge in the volume element is the charge density times the volume:

$$\mathrm{d}q = \rho \mathrm{d}a\ \mathrm{d}l\ \cos(\theta)$$

Why the cosine? The angle θ is the angle between the area vector and the velocity vector ... the current flowing through da will depend on this angle. Replacing dl with vdt:

$$\mathrm{d}q = \rho\ v\mathrm{d}a\ \mathrm{d}t\ \cos(\theta)$$

or:

$$\frac{\mathrm{d}q}{\mathrm{d}t} = \rho v \mathrm{d}a \cos(\theta)$$

and in vector notation:

$$\frac{\mathrm{d}q}{\mathrm{d}t} = \vec{\mathbf{J}} \cdot \mathrm{d}\vec{\mathbf{a}}$$

This is the current in our volume element contained within the wire. The total current passing through the wire is:

$$I = \int_S \vec{\mathbf{J}} \cdot \mathrm{d}\vec{\mathbf{a}} \tag{11.18}$$

"S" refers to the cross-sectional area of the wire. Of course, if the current density is constant throughout the conductor then eqn (11.18) collapses to:

$$I = \vec{\mathbf{J}} \cdot \vec{\mathbf{a}}$$

Invoking eqn (4.43), the Divergence theorem, with respect to eqn (11.18) we can write:

$$I = \int_S \vec{\mathbf{J}} \cdot \mathrm{d}\vec{\mathbf{a}} = \int_V \vec{\nabla} \cdot \vec{\mathbf{J}} \mathrm{d}V$$

This is to be taken as the current flowing *out* of a volume since the area vector $\mathrm{d}\vec{\mathbf{a}}$ points out of the surface in question. The flow occurs at the expense of charge within the volume; an outward flow will decrease the charge within the volume:

$$\int_V \vec{\nabla} \cdot \vec{\mathbf{J}} \mathrm{d}V = -\frac{\mathrm{d}q}{\mathrm{d}t} = -\frac{\mathrm{d}}{\mathrm{d}t} \int_V \rho \mathrm{d}V$$

Since this is true for any volume we may equate the integrands:

$$\vec{\nabla} \cdot \vec{\mathbf{J}} = -\frac{\mathrm{d}\rho}{\mathrm{d}t}$$

In the case of a steady flow of current into and out of a volume in a conductor this becomes:

$$\vec{\nabla} \cdot \vec{\mathbf{J}} = 0 \tag{11.19}$$

Since $\mathrm{d}\rho/\mathrm{d}t$ is zero for a constant flow of charge through a conductor.

Recall from Chapter 4 the Helmholtz theorem that stated that every vector can be decomposed into a part that may be generated from the gradient of a scalar function and a part generated from the curl of a vector function (see eqn (4.22)). We know from above that the electric field can be generated from the gradient of a scalar function and we now consider that the magnetic field may be obtained from the curl of a vector potential, $\vec{\mathbf{A}}$:

$$\vec{\mathbf{B}} = \vec{\nabla} \times \vec{\mathbf{A}} \tag{11.20}$$

Immediately we see (using eqn (4.21)) that the divergence of $\vec{\mathbf{B}}$ must be zero if this is the case:

$$\vec{\nabla} \cdot \vec{\mathbf{B}} = \vec{\nabla} \cdot (\vec{\nabla} \times \vec{\mathbf{A}}) = 0 \tag{11.21}$$

Let us make a guess as to what a reasonable vector potential will look like. We know that the curl of $\vec{\mathbf{A}}$ is $\vec{\mathbf{B}}$ so a reasonable guess would be a function whose curl gives us $\vec{\mathbf{B}}$. Recall that the form of the potential function for the electric field was (see eqn (11.11)):

$$\phi(\vec{\mathbf{r}}) = \frac{1}{4\pi\epsilon_0}\int_V \frac{\rho}{r}\mathrm{d}V$$

Given what we have already seen regarding the magnetic field, we might expect something similar for the magnetic vector potential. Let us take the magnetic field formula, eqn (11.14) and make it into something like a vector potential function:

$$\begin{aligned}\vec{\mathbf{A}} &= \frac{\mu_0}{4\pi}\int \frac{\rho\vec{\mathbf{v}}}{r}\mathrm{d}V \\ &= \frac{\mu_0}{4\pi}\int \frac{\vec{\mathbf{J}}}{r}\mathrm{d}V\end{aligned}$$

The curl of this function will generate $\vec{\mathbf{B}}$:

$$\begin{aligned}\vec{\nabla}\times\vec{\mathbf{A}} &= \frac{\mu_0}{4\pi}\int\left(\vec{\nabla}\times\frac{\vec{\mathbf{J}}}{r}\right)\mathrm{d}V \\ &= \frac{\mu_0}{4\pi}\int\left(\frac{1}{r}\vec{\nabla}\times\vec{\mathbf{J}} - \vec{\mathbf{J}}\times\vec{\nabla}\frac{1}{r}\right)\mathrm{d}V \\ &\quad \text{(by identity 5, chapter 4)} \\ &= -\frac{\mu_0}{4\pi}\int\left(\vec{\mathbf{J}}\times\vec{\nabla}\frac{1}{r}\right)\mathrm{d}V \\ &= -\frac{\mu_0}{4\pi}\int\frac{\vec{\mathbf{J}}\times\vec{\mathbf{r}}_u}{r^2}\mathrm{d}V\end{aligned}$$

Of course, this is not a unique vector potential function as was mentioned in Chapter 4. There is a subtlety here that bears close examination. You will note that the first cross product in the second line vanishes. Why? This may best be explained with a picture (see Figure 11.8).

We are calculating the vector potential at the point where the test charge Q_t is located and the $\vec{\nabla}$ operator differentiates with respect to the position vector $\vec{\mathbf{r}}$. However, $\vec{\mathbf{J}}$ refers to the other charges, q_1 to q_4 in Figure 11.8, that are in motion and has nothing whatsoever to do with

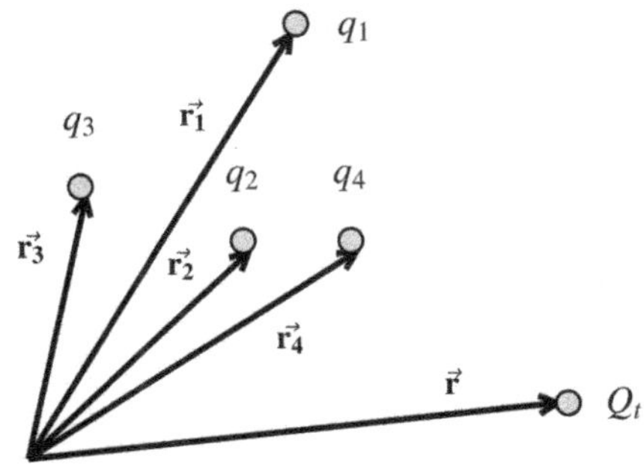

Figure 11.8 Position vectors of a collection of charges.

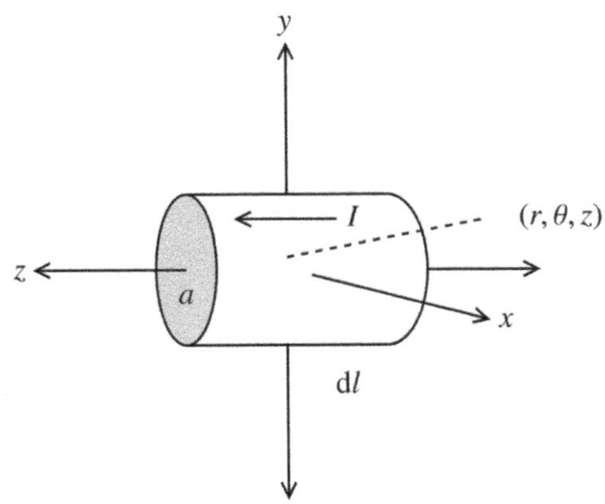

Figure 11.9 Charge-carrying conductor.

Q_t. Therefore, as far as $\vec{\nabla}$ is concerned, $\vec{\mathbf{J}}$ is a constant that differentiates to zero. Hence the disappearance of the first cross product.

Now consider the diagram of an element of a current-conducting wire of cross-sectional area a and length $\mathrm{d}l$ (Figure 11.9).

We wish to calculate the vector potential function $\vec{\mathbf{A}}(x, y, z)$ at any point in space around the current-conducting element. From our previous considerations, the current is $I = \vec{\mathbf{J}} \cdot \vec{\mathbf{a}}$ and the volume of the element is $\mathrm{d}V = \vec{\mathbf{a}} \cdot \mathrm{d}\vec{\mathbf{l}}$ so that $\vec{\mathbf{J}}\,\mathrm{d}V = I\,\mathrm{d}\vec{\mathbf{l}}$, which we substitute into the expression for the vector potential:

$$\begin{aligned}\vec{\mathbf{A}} &= \frac{\mu_0}{4\pi}\int \frac{\vec{\mathbf{J}}}{r}\mathrm{d}V \\ &= \frac{\mu_0}{4\pi} I \int \frac{1}{r}\mathrm{d}\vec{\mathbf{l}}\end{aligned} \tag{11.22}$$

The magnetic field is the curl of this:

$$\vec{\mathbf{B}} = \vec{\nabla} \times \vec{\mathbf{A}} = \frac{\mu_0}{4\pi} I \int \vec{\nabla} \times \left(\frac{1}{r} \mathrm{d}\vec{\mathbf{l}}\right)$$

$$= \frac{\mu_0}{4\pi} I \int \left(\frac{1}{r} \vec{\nabla} \times \mathrm{d}\vec{\mathbf{l}} - \mathrm{d}\vec{\mathbf{l}} \times \vec{\nabla} \frac{1}{r}\right)$$

(from identity 5, chapter 4)

$$= \frac{\mu_0}{4\pi} I \int -\mathrm{d}\vec{\mathbf{l}} \times \vec{\nabla} \frac{1}{r} \tag{11.23}$$

$$= \frac{\mu_0}{4\pi} I \int -\mathrm{d}\vec{\mathbf{l}} \times \frac{-\vec{\mathbf{r}}_u}{r^2}$$

$$= \frac{\mu_0}{4\pi} I \int \frac{\mathrm{d}\vec{\mathbf{l}} \times \vec{\mathbf{r}}_u}{r^2}$$

This is (finally!) the Biot–Savart law describing the magnetic field due to a current flowing through a wire. The equivalent form of the Biot–Savart law using the current density is:

$$\vec{\mathbf{B}} = \frac{\mu_0}{4\pi} I \int \frac{\vec{\mathbf{J}} \times \vec{\mathbf{r}}_u}{r^2} \mathrm{d}V \tag{11.24}$$

We will use this to calculate the field inside a solenoid but first, we look at the field due to a long current-conducting wire. According to eqn (11.23) the magnetic field around a wire is at right angles to the current flow and the position vector. Figure 11.5 shows the relationships between current flow, position vector and magnetic field. An alternate view would be to look along the current-conducting wire with the current flowing towards you (Figure 11.10).

Thus, the field "curls" around the wire and its magnitude decreases as the distance from the wire increases. We also note a circular symmetry, indicating that the field strength (but not the direction) is the same in any direction from the wire. Given the current, what is the value of $\vec{\mathbf{B}}$ at a distance R from the wire? This calculation amounts to adding up all of the contributions from the current elements due to the various $\mathrm{d}\vec{\mathbf{l}}$ elements along the wire. For simplicity we shall assume a wire of infinite length (Figure 11.11).

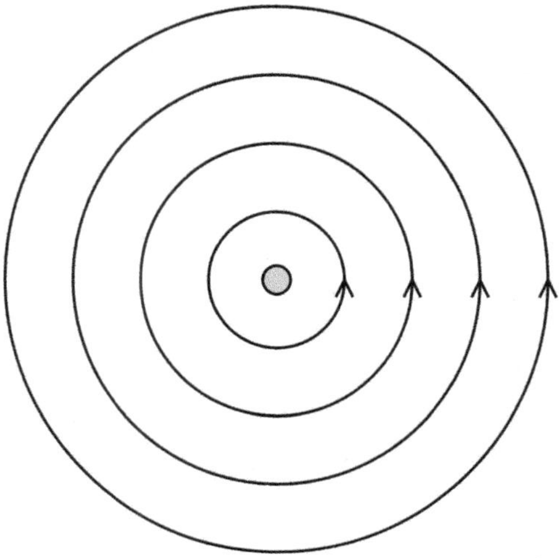

Figure 11.10 Magnetic field lines around a conductor.

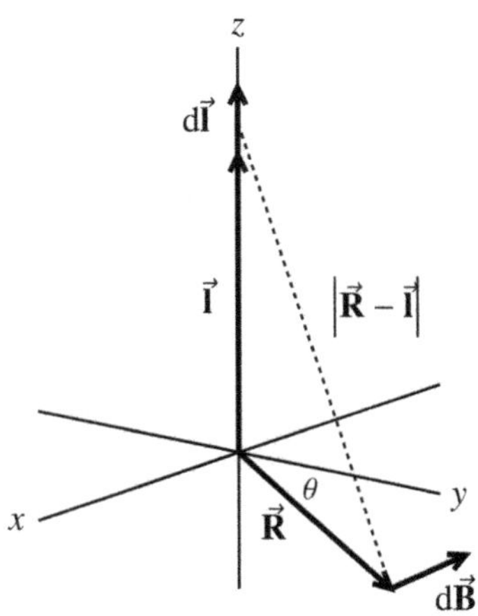

Figure 11.11 Field strength at distance R from a conductor.

It will be convenient to use cylindrical coordinates (see Appendix A) here with the z-axis coincident with the wire. We start with the Biot–Savart law:

$$\vec{\mathbf{B}} = \frac{\mu_0}{4\pi} I \int \frac{\mathrm{d}\vec{\mathbf{l}} \times \vec{\mathbf{r}}_u}{r^2}$$

$$= \frac{\mu_0}{4\pi} I \int \frac{\mathrm{d}\vec{\mathbf{l}} \times \vec{\mathbf{r}}}{r^3}$$

$$\left(\text{where } \vec{\mathbf{r}}_u = \frac{\vec{\mathbf{r}}}{\|\vec{\mathbf{r}}\|} = \frac{\vec{\mathbf{r}}}{r}\right)$$

In terms of cylindrical coordinates:

$$\vec{\mathbf{l}} = a\vec{\mathbf{u}}_z$$

$$\mathrm{d}\mathbf{l} = \vec{\mathbf{u}}_z \mathrm{d}a$$

$$\vec{\mathbf{R}} = b\vec{\mathbf{u}}_\rho$$

$$r = \|\vec{\mathbf{R}} - \vec{\mathbf{l}}\| = \|b\vec{\mathbf{u}}_\rho - a\vec{\mathbf{u}}_z\| = (b^2 - a^2)^{1/2}$$

and the Biot–Savart expression is, after substitution and integrating between the points P_1 and P_2 on the wire (and recalling the relationships between the unit vectors, Appendix A):

$$\vec{\mathbf{B}} = \frac{\mu_0}{4\pi} I \int \frac{\mathrm{d}\vec{\mathbf{l}} \times \vec{\mathbf{r}}}{r^3} = \frac{\mu_0}{4\pi} I \int_{P_1}^{P_2} \frac{\vec{\mathbf{u}}_z \times (b\vec{\mathbf{u}}_\rho - a\vec{\mathbf{u}}_z)}{\|b\vec{\mathbf{u}}_\rho - a\vec{\mathbf{u}}_z\|^{3/2}} \mathrm{d}a$$

$$= \frac{\mu_0}{4\pi} I \int_{P_1}^{P_2} \frac{\vec{\mathbf{u}}_z \times (b\vec{\mathbf{u}}_\rho - a\vec{\mathbf{u}}_z)}{(b^2 + a^2)^{3/2}} \mathrm{d}a = \frac{\mu_0 \vec{\mathbf{u}}_\phi}{4\pi} I \int_{P_1}^{P_2} \frac{1}{(b^2 + a^2)^{3/2}} \mathrm{d}a$$

Essentially, we have set the length of vector $\vec{\mathbf{l}}$ to be a, and the vector $\vec{\mathbf{R}}$ to be b. We continue, using trigonometric substitution. From Figure 11.11:

$$a = b\tan(\theta)$$

$$\mathrm{d}a = \frac{b}{\cos^2(\theta)} \mathrm{d}\theta$$

$$\sqrt{a^2 + b^2} = \frac{b}{\cos(\theta)}$$

and substituting:

$$\vec{\mathbf{B}} = \frac{\mu_0 \vec{\mathbf{u}}_\phi}{4\pi} I \int_{P1}^{P2} \frac{1}{(b^2 + a^2)^{3/2}} \mathrm{d}a$$

$$= \frac{\mu_0 \vec{\mathbf{u}}_\phi}{4\pi} I \int_{\theta 1}^{\theta 2} \frac{\cos^3(\theta)}{b^3} \frac{b}{\cos^2(\theta)} \mathrm{d}\theta$$

$$= \frac{\mu_0 \vec{\mathbf{u}}_\phi}{4\pi b^2} I \int_{\theta 1}^{\theta 2} \cos(\theta) \mathrm{d}\theta \quad (b \text{ is constant})$$

$$= \frac{\mu_0 \vec{\mathbf{u}}_\phi}{4\pi b^2} I[\sin(\theta_2) - \sin(\theta_1)]$$

The current-conducting wire being of infinite length, θ_1 and θ_2 are set to minus and plus $\pi/2$, respectively, and the expression becomes:

$$\vec{\mathbf{B}} = \frac{\mu_0 I}{2\pi b^2}\vec{\mathbf{u}}_\phi \qquad (11.25)$$

and we see that, indeed, the $\vec{\mathbf{B}}$-field is directed in a circle around the current-conducting wire since $\vec{\mathbf{u}}_\phi$ is the unit vector orthogonal to both the z-axis and the radial. We see also that the strength of the field varies inversely with the distance from the wire.

Now, instead of a long straight wire let us look at a circular current-conducting wire (Figure 11.12).

In this case we want to know the $\vec{\mathbf{B}}$-field on the central axis of a circular conducting wire of radius a and at a point b units along the axis from the plane of the wire. Using cylindrical coordinates we find:

$$\vec{\mathbf{R}} = a\vec{\mathbf{u}}_\rho$$

$$\vec{\mathbf{k}} = b\vec{\mathbf{u}}_z$$

$$\mathrm{d}\vec{\mathbf{l}} = a\vec{\mathbf{u}}_\phi \mathrm{d}\phi$$

$$r = \|\vec{\mathbf{k}} - \vec{\mathbf{R}}\| = \|b\vec{\mathbf{u}}_z - a\vec{\mathbf{u}}_\rho\| = (b^2 + a^2)^{1/2}$$

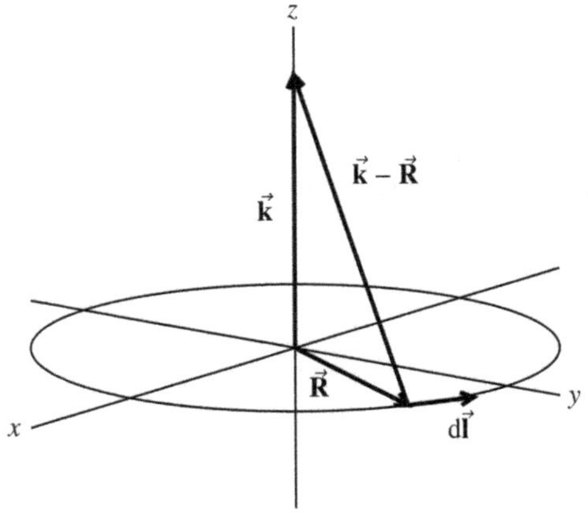

Figure 11.12 Field strength along the axis of a circular conductor.

$$\begin{aligned}\vec{\mathbf{B}} &= \frac{\mu_0}{4\pi} I \int \frac{\mathbf{d}\vec{\mathbf{l}} \times \vec{\mathbf{r}}}{r^3} \\ &= \frac{\mu_0}{4\pi} I \int_0^{2\pi} \frac{a\vec{\mathbf{u}}_\phi \times (b\vec{\mathbf{u}}_z - a\vec{\mathbf{u}}_\rho)}{(a^2+b^2)^{3/2}} \mathrm{d}\phi \\ &= \frac{\mu_0}{4\pi} I \int_0^{2\pi} \frac{ab\vec{\mathbf{u}}_\rho + a^2\vec{\mathbf{u}}_z}{(a^2+b^2)^{3/2}} \mathrm{d}\phi\end{aligned}$$

Why is $\mathbf{d}\vec{\mathbf{l}} = a\vec{\mathbf{u}}_\phi\, \mathrm{d}\phi$? Although it is not detailed in Appendix A, for cylindrical coordinates the reasoning is the same as that for $rd\theta$ in spherical coordinates, which *is* explained in Appendix A.

We have made use of the orthogonality of the cylindrical coordinate basis vectors and the fact that $\vec{\mathbf{u}}_\phi \times \vec{\mathbf{u}}_\rho = -\vec{\mathbf{u}}_z$ (see Appendix A). Continuing, we make use of the transformation properties of the basis sets (yet again using Appendix A):

$$\begin{aligned}\vec{\mathbf{B}} &= \frac{\mu_0}{4\pi} I \int_0^{2\pi} \frac{ab[\cos(\phi)\vec{\mathbf{u}}_x + \sin(\phi)\vec{\mathbf{u}}_y] + a^2\vec{\mathbf{u}}_z}{(a^2+b^2)^{3/2}} \mathrm{d}\phi \\ &= \frac{\mu_0 I}{4\pi} \frac{a^2}{(a^2+b^2)^{3/2}} \vec{\mathbf{u}}_z \int_0^{2\pi} \mathrm{d}\phi + \frac{\mu_0 I}{4\pi} \frac{ab}{(a^2+b^2)^{3/2}} \int_0^{2\pi} [\cos(\phi)\vec{\mathbf{u}}_x + \sin(\phi)\vec{\mathbf{u}}_y] \mathrm{d}\phi \\ &= \frac{\mu_0 I}{4\pi} \frac{a^2}{(a^2+b^2)^{3/2}} \vec{\mathbf{u}}_z [2\pi - 0] \\ &\quad + \frac{\mu_0 I}{4\pi} \frac{ab}{(a^2+b^2)^{3/2}} [\sin(2\pi) - \sin(0) + \cos(0) - \cos(2\pi)] \\ &= \frac{\mu_0 I}{2} \frac{a^2}{(a^2+b^2)^{3/2}} \vec{\mathbf{u}}_z\end{aligned}$$

Thus, the field along the centre line of the circular current-conducting wire is:

$$\begin{aligned}\vec{\mathbf{B}} &= \frac{\mu_0 I}{2} \frac{a^2}{(a^2+b^2)^{3/2}} \vec{\mathbf{u}}_z \\ &= \frac{\mu_0 I}{2} \frac{a^2}{r^3} \vec{\mathbf{u}}_z\end{aligned} \tag{11.26}$$

We see that the field is directed along the z-axis as we might expect, based on the symmetry of the problem, and that if $b = 0$ the field strength is at a maximum in the centre of the circular ring of wire ($r = a$):

$$\vec{\mathbf{B}} = \frac{\mu_0 I}{2a}\vec{\mathbf{u}}_z$$

The *solenoid calculation* is just an extension of the circular wire calculation. To do so we consider that the helical arrangement of wire in the solenoid may be approximated by multiple circular elements and then add up the contributions at a point from all of these circular elements. We will suppose that we have a solenoid of N turns of wire, is L units in length and, as in the case of the circular current-conducting wire, radius a. Our task is to calculate the field at a point p along the central axis of the solenoid due to the contribution of all of the current-rings that make up the solenoid. Again, we will use the cylindrical coordinate system and set our origin at the point P for convenience (Figure 11.13).[††]

The number of turns of wire in a circular element of dz' width is:

$$\frac{N}{L}\mathrm{d}z'$$

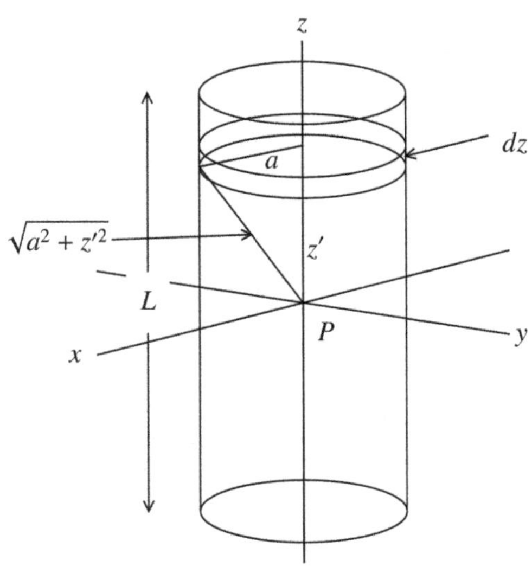

Figure 11.13 Field strength along the axis of a solenoid.

[††]This is the solenoid coil in your NMR/MRI magnet!

Using this in our expression for the field due to a single circular current-conducting wire, eqn (11.26):

$$\mathrm{d}\vec{\mathbf{B}} = \frac{\mu_0 NI}{2L}\frac{a^2}{\left(a^2 + z'^2\right)^{3/2}}\vec{\mathbf{u}}_z \mathrm{d}z'$$

or

$$\vec{\mathbf{B}} = \frac{\mu_0 NIa^2}{2L}\int\frac{1}{\left(a^2 + z'^2\right)^{3/2}}\vec{\mathbf{u}}_z \mathrm{d}z'$$

This integral is solvable by the standard trigonometric substitution technique. From Figure 11.13 we see that we may write:

$$z' = a\tan(\theta)$$

$$\mathrm{d}z' = \frac{a}{\cos^2(\theta)}\mathrm{d}\theta$$

$$\cos(\theta) = \frac{a}{\sqrt{a^2 + z'^2}}$$

and substitute into our current expression:

$$\begin{aligned}
\vec{\mathbf{B}} &= \frac{\mu_0 NI}{2L}\int\frac{a^2}{\left(a^2 + z'^2\right)^{3/2}}\vec{\mathbf{u}}_z \mathrm{d}z' \\
&= \frac{\mu_0 NI}{2L}\vec{\mathbf{u}}_z\int\frac{a^3}{\left(a^2 + z'^2\right)^{3/2}\cos^2(\theta)}\mathrm{d}\theta \\
&= \frac{\mu_0 NI}{2L}\vec{\mathbf{u}}_z\int_{\theta_1}^{\theta_2}\frac{\cos^3(\theta)}{\cos^2(\theta)}\mathrm{d}\theta \\
&= \frac{\mu_0 NI}{2L}\vec{\mathbf{u}}_z\int_{\theta_1}^{\theta_2}\cos(\theta)\mathrm{d}\theta \\
&= \frac{\mu_0 NI}{2L}\vec{\mathbf{u}}_z\left[\sin(\theta_2) - \sin(\theta_1)\right]
\end{aligned}$$

Because of where we have set the origin in our example, θ_1 is below the x, y-plane and θ_2 is above the plane (Figure 11.14).

The calculation of θ_1 is:

$$\theta_1 = \frac{-z'}{\sqrt{a^2 + z'^2}}$$

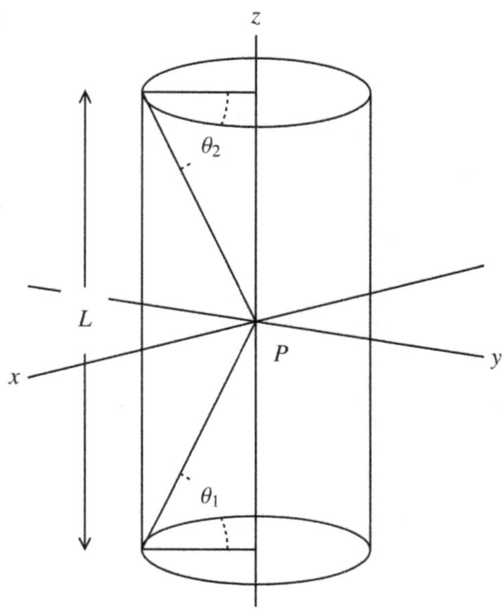

Figure 11.14 Field strength along the axis of a solenoid.

and is therefore negative. From the trigonometric identities in Appendix A:

$$\sin(\theta_1) = -\sin(-\theta_1)$$

which we substitute into our current expression:

$$\begin{aligned}\vec{\mathbf{B}} &= \frac{\mu_0 NI}{2L}\vec{\mathbf{u}}_z[\sin(\theta_2) - \sin(\theta_1)] \\ &= \frac{\mu_0 NI}{2L}\vec{\mathbf{u}}_z[\sin(\theta_2) + \sin(-\theta_1)]\end{aligned}$$

If L is large then $-\theta_1$ and θ_2 each approach $\pi/2$ (as long as the point p is not too close to either end) and our final expression for the magnetic field along the centre line of the centre of a solenoid is:

$$\vec{\mathbf{B}} = \frac{\mu_0 NI}{L}\vec{\mathbf{u}}_z \tag{11.27}$$

In the section on the electric field we determined from eqn (11.6) and (11.8) that the electric field is irrotational (see Chapter 4 for a discussion of this term). In order to do the same thing for the magnetic field we must determine the values of both the divergence and the curl of $\vec{\mathbf{B}}$, as was done for the electric field. We have seen (eqn

(11.21)) that the divergence of the magnetic field is zero and it is left only to determine the curl of $\vec{\mathbf{B}}$. We begin by asking "what is the line integral of $\vec{\mathbf{B}}$ on a line surrounding a current-conducting wire?". Let us start for simplicity with a circular line of radius a around the infinite-length wire.

$$\vec{\mathbf{B}} = \frac{\mu_0 I \vec{\mathbf{u}}_\phi}{2\pi a}$$

$$\mathrm{d}\vec{\mathbf{l}} = \mathrm{d}a\vec{\mathbf{u}}_\rho + a\mathrm{d}\phi\vec{\mathbf{u}}_\phi + \mathrm{d}z\vec{\mathbf{u}}_z$$

$$\oint \vec{\mathbf{B}} \cdot \mathrm{d}\vec{\mathbf{l}} = \frac{\mu_0 I}{2\pi a}\oint \vec{\mathbf{u}}_\phi \cdot \mathrm{d}\vec{\mathbf{l}} = \frac{\mu_0 I}{2\pi a}\int_0^{2\pi} a\mathrm{d}\phi(\mathrm{d}l = b\mathrm{d}\phi)$$

$$= \frac{\mu_0 I}{2\pi a}[a2\pi - a \cdot 0]$$

$$= \mu_0 I$$

As it turns out, we need not have used a circular line around the wire ... any shape would have sufficed as long as it surrounded the wire. For a line going straight out from the wire or going parallel to the wire the line integral is zero since the inner product of $\vec{\mathbf{u}}_\phi$ and the component in those directions is zero. Only the components in the direction of $\vec{\mathbf{u}}_\phi$ contribute and even though the field decreases as the radius increases the circumference of any associated circular component increases. Thus, the shape of the line is irrelevant. Now, recalling eqn (11.18), the general definition of current, we write:

$$\oint \vec{\mathbf{B}} \cdot \mathrm{d}\vec{\mathbf{l}} = \mu_0 I = \mu_0 \int \vec{\mathbf{J}} \cdot \mathrm{d}\vec{\mathbf{a}}$$

Using Stokes' theorem once again we write:

$$\oint \vec{\mathbf{B}} \cdot \mathrm{d}\vec{\mathbf{l}} = \int (\vec{\nabla} \times \vec{\mathbf{B}}) \cdot \mathrm{d}\vec{\mathbf{a}} = \mu_0 \int \vec{\mathbf{J}} \cdot \mathrm{d}\vec{\mathbf{a}}$$

Since the area integrals are the equal and apply to any area da their arguments must also be equal and therefore:

$$\vec{\nabla} \times \vec{\mathbf{B}} = \mu_0 \vec{\mathbf{J}} \tag{11.28}$$

This is Ampere's Law. Since the curl of $\vec{\mathbf{B}}$ is non-zero and the divergence of $\vec{\mathbf{B}}$ is zero the magnetic field must therefore be a solenoidal vector field. At this point we are almost literally a hair's breadth away from Maxwell's laws.

11.4 Maxwell's Laws

This is a brief section for interest's sake only and will not have direct bearing on any of our NMR considerations.‡‡

We have, currently, four equations dealing with electrostatics and magnetostatics:

$$\begin{aligned}
&\text{(i)}\ \vec{\nabla}\times\vec{\mathbf{E}}=0 \quad \text{(no name)}\\
&\text{(ii)}\ \vec{\nabla}\cdot\vec{\mathbf{E}}=\rho/\epsilon_0 \quad \text{(Gauss's Law)}\\
&\text{(iii)}\ \vec{\nabla}\times\vec{\mathbf{B}}=\mu_0\vec{\mathbf{J}} \quad \text{(Ampere's Law)}\\
&\text{(iv)}\ \vec{\nabla}\cdot\vec{\mathbf{B}}=0 \quad \text{(no name)}
\end{aligned} \tag{11.29}$$

These are Maxwell's equations for a static situation. In other words, a situation in which neither the electric or magnetic field is changing with time. What do they look like when this is not the case? To answer this we have to start by looking at a critical set of observations made by Michael Faraday in the 1830s.

There were essentially three important experiments that he performed. First he moved a magnet near a wire that was connected to a galvanometer for measuring current and noted that a current flowed in the circuit (Figure 11.15). Moving the magnet in the opposite direction resulted in a current flowing again but in the opposite direction to the first time.

In the second experiment the magnet was left motionless and instead the wire was moved through the magnet's field. Essentially the same observations were made … when the wire moves a current flows. The third experiment involved keeping both the magnet and wire motionless and changing the strength of the field. Again a current was observed to flow. The direction of flow depended on whether the magnetic field was being increased or decreased. Thus, Faraday saw that the current in the wire was induced by a *changing* magnetic field.

To put these observations in a modern, mathematical context, recall the flux calculation over a surface from Chapter 4

‡‡I could not resist including this. To the physicist and the mathematician alike it is breathtaking.

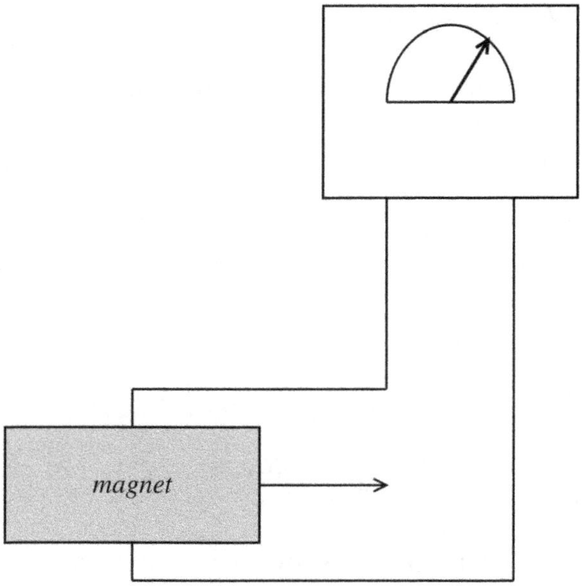

Figure 11.15 Faraday's electric current induction experiment.

(see eqn (4.23)). We can write for the flux of the magnetic field $\vec{\mathbf{B}}$ over a surface S:

$$\Phi = \int_S \vec{\mathbf{B}} \cdot \mathrm{d}\vec{\sigma}$$

Also recall that the flux is the component of the field that is normal to the surface.

Moving the magnet past the wire or moving the wire past the magnet or changing the magnetic field creates an *electromotive force* that drives the current through the wire for as long as the field is changing. We express this as:

$$EMF = -\frac{\mathrm{d}\Phi}{\mathrm{d}t} \tag{11.30}$$

and call it Faraday's Law. This is an empirical law derived from experiment and cannot be obtained by purely mathematical techniques.

Recalling eqn (11.12):

$$EMF = \oint \vec{\mathbf{E}} \cdot \mathrm{d}\vec{\mathbf{l}} = -\frac{\mathrm{d}\Phi}{\mathrm{d}t} = -\frac{\mathrm{d}}{\mathrm{d}t}\int_S \vec{\mathbf{B}} \cdot \mathrm{d}\vec{\sigma}$$

We have discovered something very interesting here. The line integral around a closed curve for the (static) electric field as we have seen it so far is equal to zero. Here we have, for a time-varying magnetic field, a non-zero value for the line integral. What this means in the real world is that there will be a non-zero electromotive force around the circuit and that electrical current will flow around the curve. It also shows that there is an intimate connection between the electric field and the magnetic field. Just how intimate will be revealed in the next section of this chapter.

Why the negative sign in Faraday's Law? This is a consequence of another experimental observation that is formalised as Lenz's Law and helps us determine the direction of flow of current. Basically, it says that when a time-varying magnetic field induces a current in a wire the current will flow in such a direction so as to create a magnetic field that *opposes* the change in external magnetic field. You can picture what this means in a diagram of a circular wire (Figure 11.16).

In Figure 11.16a the field is decreasing, which induces an electric current in the direction such that the induced current produces a field that opposes the decrease in $\vec{\mathbf{B}}$. You can verify this for yourself *via* the right-hand rule for currents and magnetic fields. In Figure 11.16b the situation is just the opposite. Increasing $\vec{\mathbf{B}}$ induces a current that itself produces a field that opposes the original increase in $\vec{\mathbf{B}}$. The negative sign in Faraday's Law accounts for this.

If we consider a conductor that is not moving with respect to a magnetic field and that the field that is changing with respect to time we may write, *via* Stokes' theorem:

$$\oint \vec{\mathbf{E}} \cdot \mathrm{d}\vec{\mathbf{l}} = \int_S (\vec{\nabla} \times \vec{\mathbf{E}}) \cdot \mathrm{d}\vec{\sigma} = -\int_S \frac{\mathrm{d}\vec{\mathbf{B}}}{\mathrm{d}t} \cdot \mathrm{d}\vec{\sigma}$$

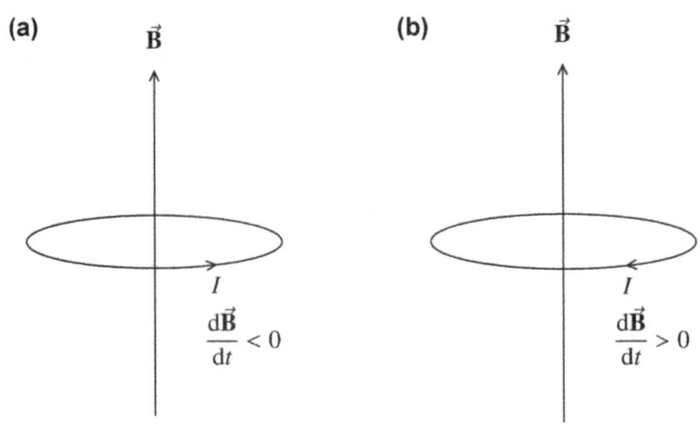

Figure 11.16 Current flow directions for time-varying magnetic fields.

The argument of the last two integrals must be equal since they are valid for any surface S:

$$\vec{\nabla} \times \vec{\mathbf{E}} = -\frac{\mathrm{d}\vec{\mathbf{B}}}{\mathrm{d}t}$$

This is a modification of the first of eqn (11.29) and Maxwell's form of Faraday's Law. We now have:

$$\begin{aligned} &\text{(i) } \vec{\nabla} \times \vec{\mathbf{E}} = -\frac{\mathrm{d}\vec{\mathbf{B}}}{\mathrm{d}t} \quad \text{(Faraday's Law)} \\ &\text{(ii) } \vec{\nabla} \cdot \vec{\mathbf{E}} = \rho/\epsilon_0 \quad \text{(Gauss's Law)} \\ &\text{(iii) } \vec{\nabla} \times \vec{\mathbf{B}} = \mu_0 \vec{\mathbf{J}} \quad \text{(Ampere's Law)} \\ &\text{(iv) } \vec{\nabla} \cdot \vec{\mathbf{B}} = 0 \quad \text{(no name)} \end{aligned} \tag{11.31}$$

There remains but one more modification to make and we will have Maxwell's equations for non-static fields. There exists a major problem with the third of eqn (11.31), Ampere's Law. Our discussions on vector calculus in Chapter 4 indicated that the divergence of a curl must be equal to zero. This works nicely for eqn (11.31)(i):

$$\vec{\nabla} \cdot (\vec{\nabla} \times \vec{\mathbf{E}}) = -\vec{\nabla} \cdot \frac{\mathrm{d}\vec{\mathbf{B}}}{\mathrm{d}t} = -\frac{\mathrm{d}}{\mathrm{d}t}(\vec{\nabla} \cdot \vec{\mathbf{B}}) = 0$$

The left side is zero by virtue of eqn (4.22) as is the right by virtue of eqn (11.31)(iv). We run into problems, however, when we try this with Ampere's Law:

$$\vec{\nabla} \cdot (\vec{\nabla} \times \vec{\mathbf{B}}) = \mu_0 \, (\vec{\nabla} \cdot \vec{\mathbf{J}})$$

Again, by virtue of eqn (4.22) the left side must be zero but what about the right side of the expression? Let us evaluate it:

$$\begin{aligned} \mu_0(\vec{\nabla} \cdot \vec{\mathbf{J}}) &= -\mu_0\left(\frac{\partial \rho}{\partial t}\right) = -\,\mu_0 \frac{\partial}{\partial t}(\epsilon_0 \vec{\nabla} \cdot \vec{\mathbf{E}}) \\ &= -\vec{\nabla} \cdot \left(\mu_0 \epsilon_0 \frac{\partial \vec{\mathbf{E}}}{\partial t}\right) \end{aligned}$$

In the static situation we saw that in general we can expect that $\mathrm{d}\rho/\mathrm{d}t = 0$ since otherwise there would be a build-up of charge in a conductor. In the time-varying situation we cannot expect that this will be so; the current density will be changing. Maxwell's solution was childishly simple[§§] ... just add a term in to compensate:

$$\vec{\nabla} \times \vec{\mathbf{B}} = \mu_0 \vec{\mathbf{J}} + \mu_0 \epsilon_0 \frac{\partial \vec{\mathbf{E}}}{\partial t}$$

This works! For both the static and time-varying situations the divergence of this expression evaluates to zero. So, we finally have the modern form of Maxwell's four famous equations for time-varying fields:

$$\begin{aligned}
&\text{(i) } \vec{\nabla} \times \vec{\mathbf{E}} = -\frac{d\vec{\mathbf{B}}}{dt} \quad \text{(Faraday's Law)} \\
&\text{(ii) } \vec{\nabla} \cdot \vec{\mathbf{E}} = \rho/\epsilon_0 \quad \text{(Gauss's Law)} \\
&\text{(iii) } \vec{\nabla} \times \vec{\mathbf{B}} = \mu_0 \vec{\mathbf{J}} + \mu_0 \epsilon_0 \frac{\partial \vec{\mathbf{E}}}{\partial t} \quad \text{(Ampere's Law - altered)} \\
&\text{(iv) } \vec{\nabla} \cdot \vec{\mathbf{B}} = 0 \quad \text{(no name)}
\end{aligned} \tag{11.32}$$

Now for the sublime part ... the most beautiful result. Consider a region of space without any charge and no flow of charge. $\rho = 0$ and $\vec{\mathbf{J}} = \vec{\mathbf{0}}$ in this region of space and eqn (11.32) become:

$$\begin{aligned}
&\text{(i) } \vec{\nabla} \times \vec{\mathbf{E}} = -\frac{d\vec{\mathbf{B}}}{dt} \\
&\text{(ii) } \vec{\nabla} \cdot \vec{\mathbf{E}} = \rho/\epsilon_0 \\
&\text{(iii) } \vec{\nabla} \times \vec{\mathbf{B}} = \mu_0 \epsilon_0 \frac{\partial \vec{\mathbf{E}}}{\partial t} \\
&\text{(iv) } \vec{\nabla} \cdot \vec{\mathbf{B}} = 0
\end{aligned}$$

Let us take the curl of the first and third equations:

$$\vec{\nabla} \times (\vec{\nabla} \times \vec{\mathbf{E}}) = -\frac{\partial}{\partial t}(\vec{\nabla} \times \vec{\mathbf{B}}) = -\mu_0 \epsilon_0 \frac{\partial^2 \vec{\mathbf{E}}}{\partial t^2}$$

$$\vec{\nabla} \times (\vec{\nabla} \times \vec{\mathbf{B}}) = \mu_0 \epsilon_0 \frac{\partial}{\partial t}(\vec{\nabla} \times \vec{\mathbf{E}}) = -\mu_0 \epsilon_0 \frac{\partial^2 \vec{\mathbf{B}}}{\partial t^2}$$

[§§] A lot of the great ideas in physics seem to be simple.

and using identity 7 from Chapter 4, Section 4.1:

$$\vec{\nabla} \times (\vec{\nabla} \times \vec{\mathbf{a}}) = \vec{\nabla}\,(\vec{\nabla} \cdot \vec{\mathbf{a}}) - \nabla^2 \vec{\mathbf{a}}$$

we have:

$$\vec{\nabla} \times (\vec{\nabla} \times \vec{\mathbf{E}}) = \vec{\nabla}(\vec{\nabla} \times \vec{\mathbf{E}}) - \nabla^2 \vec{\mathbf{E}} = -\nabla^2 \vec{\mathbf{E}} = -\mu_0 \epsilon_0 \frac{\partial^2 \vec{\mathbf{E}}}{\partial t^2}$$

$$\vec{\nabla} \times (\vec{\nabla} \times \vec{\mathbf{B}}) = \vec{\nabla}(\vec{\nabla} \times \vec{\mathbf{B}}) - \nabla^2 \vec{\mathbf{B}} = -\nabla^2 \vec{\mathbf{B}} = -\mu_0 \epsilon_0 \frac{\partial^2 \vec{\mathbf{B}}}{\partial t^2}$$

since $\vec{\nabla} \cdot \vec{\mathbf{E}}$ and $\vec{\nabla} \cdot \vec{\mathbf{B}}$ are zero. Rearranging we get:

$$\frac{\partial^2 \vec{\mathbf{E}}}{\partial t^2} - \frac{1}{\mu_0 \epsilon_0} \nabla^2 \vec{\mathbf{E}} = 0$$

$$\frac{\partial^2 \vec{\mathbf{B}}}{\partial t^2} - \frac{1}{\mu_0 \epsilon_0} \nabla^2 \vec{\mathbf{B}} = 0$$

Do these look familiar? If you have looked at Chapter 7 you will realise that these are wave equations … electromagnetic waves! They are of the same form as eqns (7.11). The term $1/\mu_0\epsilon_0$ corresponds to v^2 in eqn (7.11), which is the square of the velocity of the wave. Electromagnetic waves as we know travel at the speed of light because they are light so we can substitute c for $1/\mu_0\epsilon_0$ and write:

$$\begin{aligned} \frac{\partial^2 \vec{\mathbf{E}}}{\partial t^2} - c^2 \nabla^2 \vec{\mathbf{E}} = 0 \\ \frac{\partial^2 \vec{\mathbf{B}}}{\partial t^2} - c^2 \nabla^2 \vec{\mathbf{B}} = 0 \end{aligned} \qquad (11.33)$$

where, of course, c is the speed of light in a vacuum and is equal to $1/\sqrt{\mu_0\epsilon_0}$. Magnificent! One of the greatest triumphs of theoretic physics.

11.5 Special Relativity and Magnetism

Surely as magnetic resonance spectroscopists it behoves us to know what a magnetic field is and where it comes from. We use our magnets every day and most do not appreciate the intimacy of the relationship between magnetism and electricity.[¶¶] Einstein's Special Relativity seems to most to be an esoteric collection of phenomena that occur only at very high and usually unobtainable velocities, however, anyone who has ever experienced an electric motor in operation has experienced the phenomenon of special relativity; that is, almost everyone on earth. We shall attempt to display that relationship here in as simple a fashion as possible.

The basic principles of special relativity that we need are given in Chapter 10. There are also very detailed descriptions given in Griffiths and in Rosser, to which the interested reader is directed. Also, the following owes much to Purcell's exposition on the subject which makes this portion of the text no different from many. His is a lucid and simple picture. See the references for further details.

As we saw in Figure 11.4, when the current in two wires is flowing in opposite directions there is a magnetic force of attraction between them and that when the current flow is in the same direction there is a repulsive force between the wires. There is, therefore, a requirement that charges be in motion for there to be a magnetic force between them. However, there is a subtle problem with our explanation of magnetism. We have said that for a charge to experience the force due to a magnetic field it must be moving. However, if we change our reference frame such that the frame is moving along with the charge then the charge cannot experience a magnetic force, as it is motionless in its new frame. However, the force cannot simply vanish because we have changed our viewpoint because an observer in either frame of reference must see the charge move. So, if the force is not magnetic what could it be?

Let us consider a wire that can conduct a current. The wire consists of atoms that have mobile, negatively charged electrons and stationary, positively charged ions. Let us also consider a test charge that is a short distance, R, away from the wire (Figure 11.17).

As we have drawn it, the test charge will not experience a force of either electrostatic or magnetic origin since the wire is electrically neutral and there is no current flow and the test charge is motionless

[¶¶]This material is not really necessary for an understanding of the workings of the NMR/MRI spectrometer but consider that, as background information, it will expand your knowledge and appreciation of the wonders of NMR/MRI spectroscopy and physics in general. The mathematics of special relativity itself is not difficult to understand contrary to popular myth. It is rather, the counter-intuitive results of the theory that are difficult to embrace.

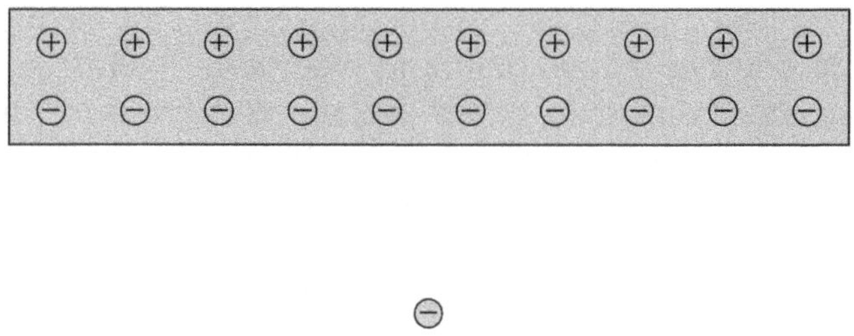

Figure 11.17 Wire with a test charge, *q*, located nearby. All charges are motionless.

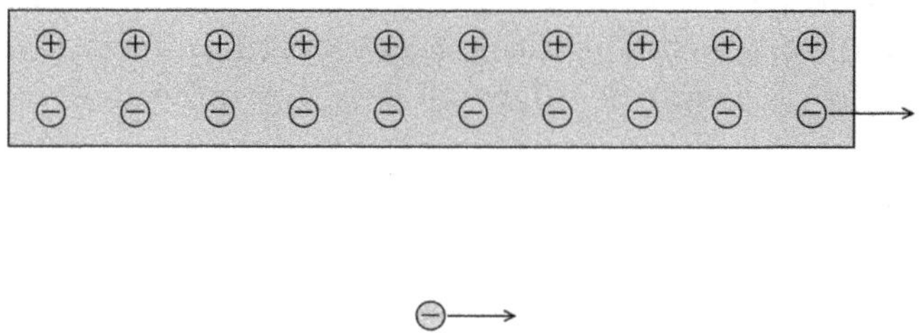

Figure 11.18 Wire conducting a current with a test charge, *q*, moving in the same direction as the flow of current. The reference frame is the laboratory frame.

with respect to the wire. We will call this reference frame the laboratory frame in which the wire itself is motionless.

Now, let us suppose that there is a current-flow to the right and that the charge in the wire is moving, both with velocity v. Furthermore, for simplicity let the test charge also move to the right with velocity v (Figure 11.18). From what we have already seen, the test charge should experience a magnetic force directed towards the wire, however, let us pretend that we have never before experienced magnetism and are just now working out the consequences of our "newly" discovered principles of special relativity.

Note that we are still in the laboratory frame in this diagram.

One of the fundamental results of special relativity is that when an object is in motion with respect to another reference frame, its length is contracted (see Chapter 10). The amount of contraction is calculated using:

$$l' = l\sqrt{1 - \left(\frac{v}{c}\right)^2}$$

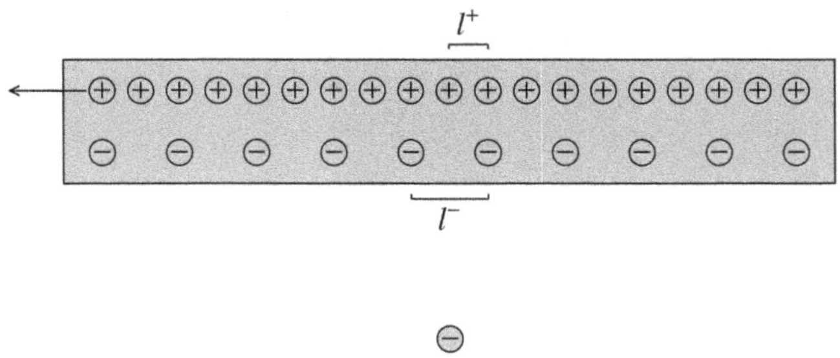

Figure 11.19 The current and charge motion in the test charge reference frame. Both the test charge and the negative charges in the wire are motionless in this frame and the positive charges appear to move in the opposite direction to the laboratory frame current flow direction.

where l' is the length of the moving object as measured in the laboratory frame and l is the length of the object at rest in the laboratory frame. v is the speed of the moving object and c is the speed of light. We will use l and l' to represent the average distance between charges in the wire.

Now we switch our frame of reference to be that of the moving test charge (and the current in the wire) and call it the test charge frame (Figure 11.19). The test charge is now "motionless" and from its point of view the negative charges in the wire are also motionless but the positive charges are moving to the left. Furthermore, the average distance between positive charges will be less than in the laboratory frame and their linear density will therefore increase. On the other hand, the linear density of the negative charges will decrease in the test charge frame since on going from the laboratory frame to the test charge frame they become motionless.

$$l^- = l\sqrt{1 - \left(\frac{v}{c}\right)^2}$$

$$l^+ = \frac{l}{\sqrt{1 - \left(\frac{v}{c}\right)^2}}$$

We see from the equations and the diagram that the linear density of the positive charges increases and the linear density of the negative charges decreases from the test charge's point of view and it will therefore see a net positive charge on the wire. This will result in the

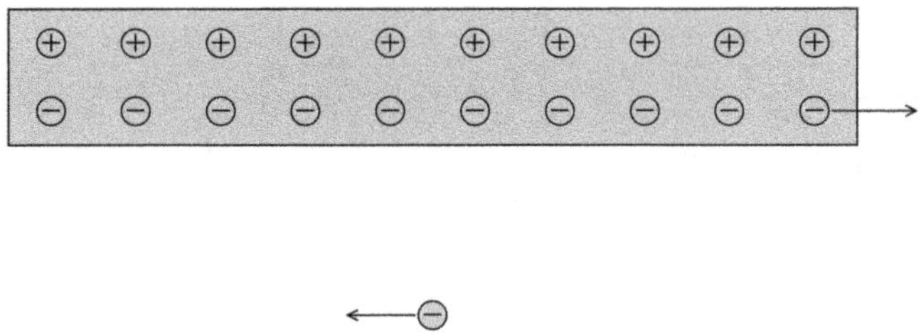

Figure 11.20 Wire conducting a current with a test charge, q, moving in the opposite direction to the flow of current. The reference frame is the laboratory frame.

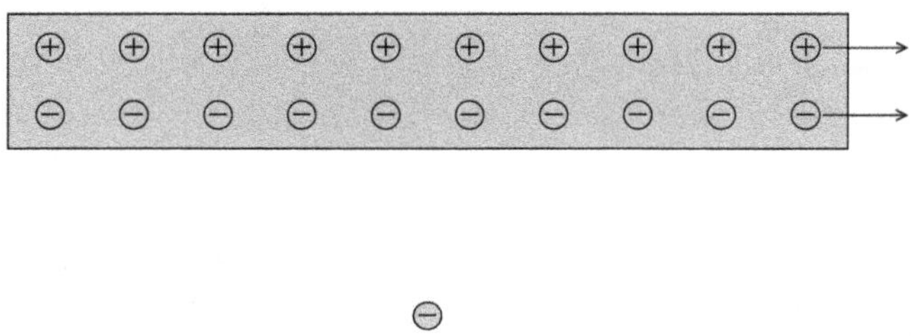

Figure 11.21 The test charge frame. The charge is motionless and the charges in the wire appear to move in the same direction with approximately the same velocity.

test charge experiencing an *electrostatic* force directed towards the wire.

Now, let us consider the laboratory frame picture in which the test charge is moving to the left with velocity $-v$ and the current is moving to the right with velocity v (Figure 11.20).

If we move to the test charge frame (see Figure 11.21) then the test charge is motionless and the negative-charge current is moving to the right with a larger-than-v velocity. The simple "common sense" view would be that the current velocity is $2v$ with respect to the test charge by simple velocity addition. Within the framework of special relativity we have seen that this is not true, however, it turns out that current velocities in wires are really very, very small compared to the speed of light (see below) so, assuming that the test charge is not moving at high velocity we may simply approximate the current velocity as $2v$. Also, the positive charges will be moving to the right with velocity v.

Note that in this situation both the positive and negative charges in the wire are in motion with respect to the test charge. The diagram does not show any length contractions of the charges. Let us calculate what they will be. For the negative charges the test charge frame velocity is $2v$ and the length contraction is (recalling the laboratory frame expansion above):

$$l^{+} = \frac{l}{\sqrt{1-\left(\frac{v}{c}\right)^2}}\sqrt{1-\left(\frac{2v}{c}\right)^2}$$

$$= l\left[1-\left(\frac{v^2}{c^2}\right)\right]^{-1/2}\left[1-\left(\frac{4v^2}{c^2}\right)\right]^{1/2}$$

$$\approx l\left(1+\frac{v^2}{2c^2}\right)\left(1-\frac{2v^2}{c^2}\right) \quad \text{(binomial expansion approximation)}$$

$$= l\left(1-\frac{3v^2}{2c^2}-\frac{v^4}{c^4}\right)$$

The length contraction of the negative charges is:

$$l^{-} = l\sqrt{1-\left(\frac{v}{c}\right)^2}$$

We see that the length contraction of the negative charges is larger than that of the positive charges. The negative charge density is therefore *higher* than the positive charge density and there is a net negative charge on the wire *from the test charge's point of view* resulting in an electrostatic force on the test charge pointing away from the wire.

In our (laboratory frame) picture of the electric field of a motionless point charge we have viewed a spherically symmetric arrangement of field vectors about the charge involved (Figure 11.1). Another way of looking at it is that at any point on a sphere surrounding the point charge the magnitude of the field is the same and is directed normal to the surface. If we now consider what will happen as the point charge begins to move relative to the rest frame we will see that the electric field magnitude will decrease in the direction parallel to the charge's motion. The field lines undergo expansion perpendicular to the direction of motion (see Figure 11.22). Thus, on the surface of the oblate sphere the field magnitudes are not the same everywhere as they are in the rest frame.

This is a picture that we can show mathematically (of course!). But first we will retreat a bit and do some electric field flux calculations. Consider a large flat sheet of evenly distributed charge with charge density ρ. Here, the charge density is q/A rather than q/V. A moment's reflection will convince you that the electric field vectors can only be

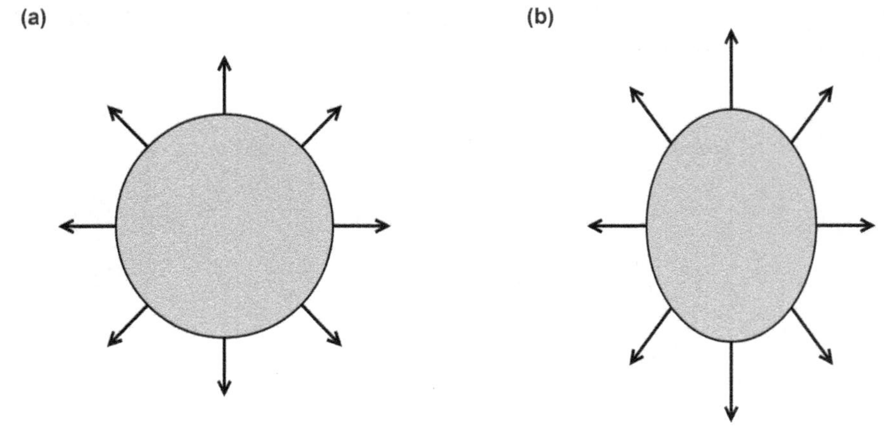

Figure 11.22 (a) Electric field around a motionless electric charge; (b) Electric field around a charge in motion as viewed in the laboratory frame.

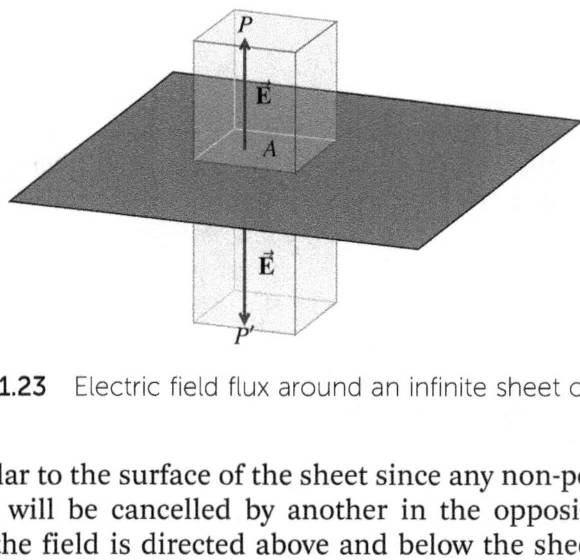

Figure 11.23 Electric field flux around an infinite sheet of charge.

perpendicular to the surface of the sheet since any non-perpendicular component will be cancelled by another in the opposite direction. Of course, the field is directed above and below the sheet. Now consider a square-faced rectangular volume intersecting the sheet perpendicularly (Figure 11.23).

Since the field is oriented perpendicularly at all points on the sheet only the faces of the volume with area A will have any electric field flux. Also, since the field and the area vector are collinear we may say that the magnitude of the field flux is simply the magnitude of the field at P and P' times the area, A:

$$\Phi = E_P A$$

The total flux through the volume is the sum of the flux above and below the sheet ... $2EA$. Now recall that we calculated that for a collection of charges at a point have a flux of Q_S/ϵ_0 (see eqn (11.5)). Therefore:

$$2E_P A = \frac{Q_S}{\epsilon_0} = \frac{\rho A}{\epsilon_0}$$

and:

$$E_P = \frac{\rho}{2\epsilon_0}$$

We see that the magnitude of the electric field has *no dependence on the distance from the sheet*. Fascinating! Now we extend this picture to two flat sheets of charge each parallel to the x, y-plane and each having opposite charges (Figure 11.24).

In this case the electric field flux between the two sheets is the sum of the individual fluxes for each sheet.

$$E = E_+ + E_- = \frac{\rho}{\epsilon_0}$$

Now we suppose that, from our point of view, these parallel sheets of charge are moving to our right in the x-direction. The amount of charge on the sheet does not change but the length of the sheet does change ... it foreshortens, as we have seen in Chapter 10.

This relativistic contraction in the direction of motion has the effect of increasing the charge density by the same factor, γ, as the contraction:

$$\rho' = \gamma\rho$$

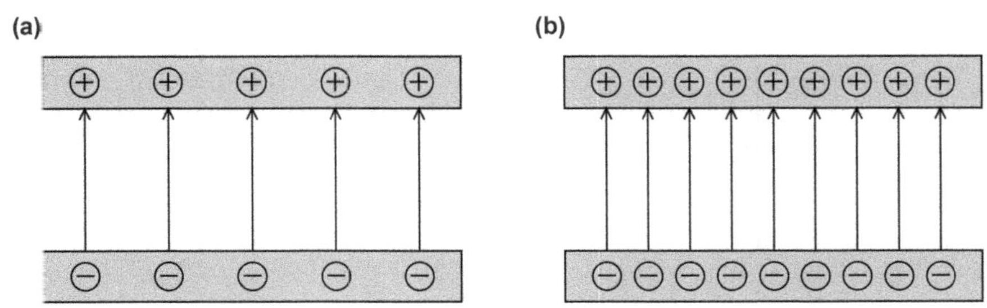

Figure 11.24 (a) Electric field flux between two oppositely charged plane sheets, motionless in the laboratory frame; (b) Electric field flux between two oppositely charged plane sheets in motion from the laboratory frame point of view.

and the electric field in the z-direction changes as:

$$E' = E'_+ + E'_- = \frac{\rho'}{\epsilon_0} = \frac{\gamma\rho}{\epsilon_0}$$

The field component perpendicular to the direction of motion increases, $E'_\perp = \gamma E_\perp$ but the field component parallel to the direction of motion does not change, $\mathbf{E}'_\parallel = E_\parallel$.

Now, consider an electric field vector whose x-component is in the direction of motion of the F' frame. The vector is drawn in F and F' as in Figure 11.25a.

To make the analysis easier we shall assume that the origins of the motionless and moving frames coincide at $t=0$ and $t'=0$, and that at these times $x=y=z=0$ and $x'=y'=z'=0$. In the motionless frame, F, with a charge Q located at the origin, the electric field magnitude components are:

$$E_x = \frac{Q}{r^2}\cos(\phi) \quad E_z = \frac{Q}{r^2}\sin(\phi)$$

or

$$E_x = \frac{Qx}{(x^2+z^2)^{3/2}}\cos(\phi) \quad E_z = \frac{Qz}{(x^2+z^2)^{3/2}}\sin(\phi)$$

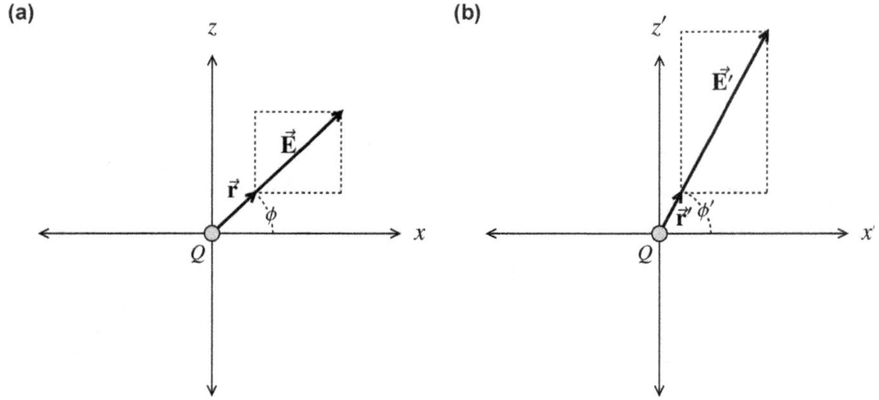

Figure 11.25 (a) Electric field vector about a motionless charge in the laboratory frame of reference; (b) Electric field vector about a charge in motion, in the laboratory frame.

The Lorentz coordinate transformations (eqns (10.12)) are:

$$x' = \gamma x - \gamma\beta ct$$
$$y' = y$$
$$z' = z$$
$$t' = \gamma t - \frac{\gamma\beta x}{c}$$

where, as in Chapter 10, we have substituted β for v_x/c. We wish to find E'_x and E'_z and will need to find the expressions for them in F'. Recall from Chapter 10 that we do this by swapping primes and changing velocity signs:

$$x = \gamma x' + \gamma\beta ct'$$
$$y = y'$$
$$z = z'$$
$$t = \gamma t' + \frac{\gamma\beta x'}{c}$$

With this (and remembering that we are considering the frames F and F' to be "crossing" with their origins overlapping so that $t = t' = 0$ we can write:

$$E'_{x'} = \frac{\gamma Q x'}{((\gamma x')^2 + z'^2)^{3/2}}\cos(\phi) \quad E'_{z'} = \frac{Q z'}{((\gamma x')^2 + z'^2)^{3/2}}\sin(\phi)$$

The total electric field strength is (remembering that $\gamma = 1/(1-\beta^2)$:

$$\begin{aligned} E'^2 &= \left(\frac{\gamma Q x'}{((\gamma x')^2 + z'^2)^{3/2}}\right)^2 + \left(\frac{Q z'}{((\gamma x')^2 + z'^2)^{3/2}}\right)^2 \\ &= \gamma^2 Q^2 \frac{(x'^2 + z'^2)}{(\gamma^2 x'^2 + z'^2)^3} \\ &= \gamma^2 Q^2 \frac{(x'^2 + z'^2)}{\gamma^6\left(x'^2 + \dfrac{z'^2}{\gamma^2}\right)^3} \\ &= Q^2 \frac{(x'^2 + z'^2)(1-\beta^2)^2}{(\gamma x'^2 + (1-\beta^2)z'^2)^3} \end{aligned}$$

$$= Q^2 \frac{(x'^2 + z'^2)(1 - \beta^2)^2}{(\gamma x'^2 + z'^2 - \beta z'^2)^3}$$

$$= Q^2 \frac{(x'^2 + z'^2)(1 - \beta^2)^2}{(x'^2 + z'^2)^3 \left(1 - \frac{\beta z'^2}{x'^2 + z'^2}\right)^3}$$

$$= \frac{Q^2(1 - \beta^2)^2}{(x'^2 + z'^2)^2 \left(1 - \frac{\beta z'^2}{x'^2 + z'^2}\right)^3} \tag{11.34}$$

$$= \frac{Q^2(1 - \beta^2)^2}{r^4(1 - \beta^2 \sin(\phi'))^3}$$

$$E' = \frac{Q(1 - \beta^2)}{r^2(1 - \beta^2 \sin(\phi'))^{3/2}}$$

All of these considerations about the relationship between electric and magnetic fields is in itself fascinating but there is another consideration that makes it even more so ... the drift velocity of electrons in a current-carrying wire. We are naturally tempted to think that the velocity of the electrons in the wire is extremely high, perhaps approaching the speed of light. However, this is demonstrably not so. Their average speed is actually quite small! Consider a section of current-conducting wire of length l and cross-sectional area A (Figure 11.26).

The number of mobile electrons contained within this volume is:

$$N_e = nAl$$

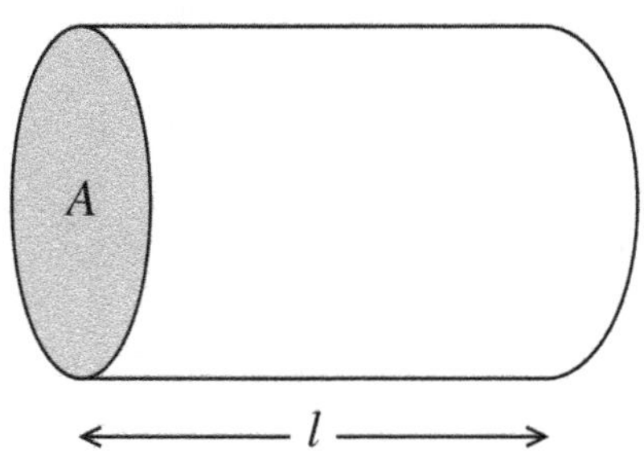

Figure 11.26 A portion of a conductor of length l.

where n is the density of mobile electrons in Coulombs m^{-3} and the total charge in Coulombs on these electrons is:

$$\text{total charge} = nAlQ$$

We can substitute for the length term by recognizing that the length is equal to vt, where v is the drift velocity of the electrons in m s^{-1} that are being accelerated by an EMF.

$$\text{charge} = nAvtQ$$

Current is defined as charge per unit time so:

$$I = \frac{nAvtQ}{t}$$
$$= nAvQ$$

and solving for drift velocity:

$$v = \frac{I}{nAQ}$$

If we do a calculation with actual numbers we encounter an amazing result. Assuming a copper wire of 1 mm diameter and a current of 10 A flowing through it we calculate:

$$v = \frac{10\,\text{C}\,\text{s}^{-1}}{8.49 \times 10^{28}\text{m}^{-3} \cdot 1.96 \times 10^{-5}\text{m}^2 \cdot 1.6 \times 10^{-19}\,\text{C}}$$
$$= 0.000038\frac{\text{m}}{\text{s}}$$

That's pretty slow!! If you think about it for a bit it means that the β term is very small . . . the drift velocity is *nowhere near* the speed of light. Even at this slow speed, however, there is a measurable magnetic field. Imagine the magnitude of the magnetic field for a current flow at a relativistic speed. It would be *stupendously* large. Also, this calculation points out that special relativity is not just a phenomenon that is relegated to very esoteric physics laboratories but rather directly affects each of us in our daily lives.

11.6 Electric Dipoles

Let us consider a pair of equal and opposite charges located at a distance of d units from each other.

We wish to know the electric field, E, at any point P around these two charges. To do this we will first calculate the potential at point P and then from this, the electric field by calculating the gradient of the potential.

In Figure 11.27, $\vec{\mathbf{r}}$ is the position vector of point P and runs from the midpoint between the two charges to the point.[|||] We begin by writing that the electric potential (eqn (11.10)) at point P as the sum of the potentials from each of the two charges:

$$\begin{aligned}\phi(\vec{\mathbf{r}}) &= \frac{q}{4\pi\epsilon_0}\left(\frac{1}{r_+} - \frac{1}{r_-}\right)\\ &= \frac{q}{4\pi r\epsilon_0}\left(\frac{r}{r_+} - \frac{r}{r_-}\right)\end{aligned}$$

Using the law of cosines (eqn (A.21)) we find from Figure 11.27:

$$\begin{aligned}r_+^2 &= r^2 + \left(\frac{d}{2}\right)^2 - 2\frac{d}{2} r\cos(\theta)\\ &= r^2\left[1 + \left(\frac{d}{2r}\right)^2 - \frac{d}{r}\cos(\theta)\right]\end{aligned}$$

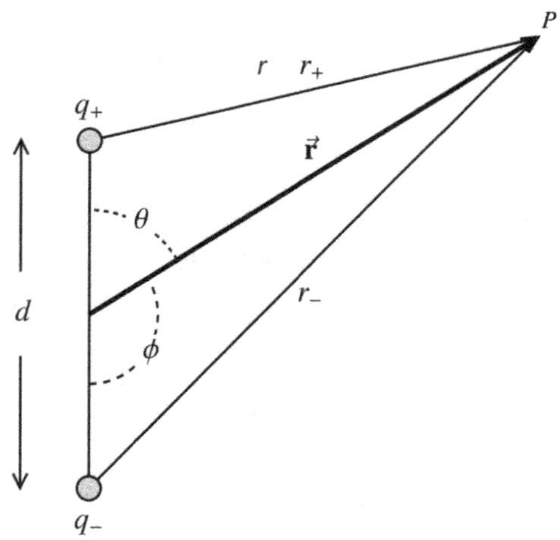

Figure 11.27 Equal and opposite charges separated by distance d.

[|||] Note that this is a subtle way of saying that the origin is at the mid-point between the two charges. We shall also assume the z-axis is coincident with the line connecting the two charges and that the positive direction is towards the positive charge.

A major simplification is to assume that the point P is at a large distance compared to the separation of the charges ... in other words, $r \gg d$. Using this assumption:

$$r_+^2 \approx r^2\left(1 - \frac{d}{r}\cos(\theta)\right)$$

$$\frac{r}{r_+} = \left(1 - \frac{d}{r}\cos(\theta)\right)^{1/2}$$

The binomial expansion (eqn (A.6)) allows us to write:

$$\begin{aligned}\frac{r}{r_+} &= \left(1 - \frac{d}{r}\cos(\theta)\right)^{1/2}\\ &= 1 + \frac{d}{2r}\cos(\theta) + \frac{3d^2}{2r^2}\cos^2(\theta)\ \ldots\end{aligned}$$

and, since we are assuming $r \gg d$:

$$\frac{r}{r_+} \approx 1 + \frac{d}{2r}\cos(\theta)$$

For r_-, after realizing that $\phi = \pi - \theta$ and that $\cos(\phi) = \cos(\pi - \theta) = -\cos(\theta)$, we find that:

$$\frac{r}{r_-} \approx 1 - \frac{d}{2r}\cos(\theta)$$

Substituting into the electric potential expression from above:

$$\begin{aligned}f(\vec{\mathbf{r}}) &= \frac{q}{4\pi r\epsilon_0}\left(\frac{r}{r_+} - \frac{r}{r_-}\right)\\ &= \frac{q}{4\pi r\epsilon_0}\left[1 + \frac{d}{2r}\cos(\theta) - \left(1 - \frac{d}{2r}\cos(\theta)\right)\right]\\ &= \frac{qd}{4\pi r^2\epsilon_0}\cos(\theta)\end{aligned}$$

The term qd is called the *dipole moment*, p, and is characterised by the distance and the magnitude of one of the charges. We may write this equation using the inner product of the vectors $\vec{\mathbf{p}}$ and $\vec{\mathbf{r}}$:

$$f(\vec{\mathbf{r}}) = \frac{qd}{4\pi r^2 \epsilon_0}\cos(\theta)$$
$$= \frac{\vec{\mathbf{p}}\cdot\vec{\mathbf{r}}}{4\pi r^3 \epsilon_0}$$

The vector $\vec{\mathbf{p}}$ is coincident with the line connecting the two charges and points towards the positive charge ... along the $+z$-axis.***

The electric field may now be calculated from the gradient of the electric potential expression that we have arrived at. From eqn (11.9):†††

$$\vec{\mathbf{E}} = -\vec{\nabla} f(\vec{\mathbf{r}})$$

Calculating using the spherical coordinate version of $\vec{\nabla}$ (eqn (4.32)):

$$\begin{aligned}\vec{\mathbf{E}} &= -\vec{\nabla} f(\vec{\mathbf{r}}) = -\vec{\mathbf{u}}_r\frac{\partial f(\vec{\mathbf{r}})}{\partial r} - \frac{\vec{\mathbf{u}}_\theta}{r}\frac{\partial f(\vec{\mathbf{r}})}{\partial \theta} \quad \text{(no } \phi \text{ variable)} \\ &= -(-2)\frac{p}{4\pi r^3\epsilon_0}\cos(\theta)\vec{\mathbf{u}}_r - \left(\frac{1}{r}\right)\frac{-p}{4\pi r^2\epsilon_0}\sin(\theta)\vec{\mathbf{u}}_\theta \\ &= \frac{p}{4\pi r^3\epsilon_0}[2\cos(\theta)\vec{\mathbf{u}}_r + \sin(\theta)\vec{\mathbf{u}}_\theta]\end{aligned} \tag{11.35}$$

This version of the electric field equation is in spherical coordinates. One also sees it written in coordinate-free form. To write this we start by noting that the dipole moment vector is $\vec{\mathbf{p}} = p\vec{\mathbf{u}}_z$. Converting to $\vec{\mathbf{u}}_z$ from polar coordinates (eqn (A.31)):

$$\hat{\mathbf{u}}_z = \cos(\theta)\ \vec{\mathbf{u}}_r - \sin(\theta)\vec{\mathbf{u}}_\theta$$

and substituting for $\vec{\mathbf{u}}_z$ in $\vec{\mathbf{p}} = p\vec{\mathbf{u}}_z$:

$$\begin{aligned}\vec{\mathbf{p}} &= p\vec{\mathbf{u}}_z \\ &= p\cos(\theta)\mathbf{u}_r - p\sin(\theta)\mathbf{u}_\theta\end{aligned}$$

***A word of caution here. In the last section we used p for momentum but here, in the current context we shall use it to indicate the electric dipole moment. Hopefully, the text of the discussion should make it clear what we are using the symbol for.

†††Do not forget the minus sign!

and:

$$
\begin{aligned}
\vec{\mathbf{p}} \cdot \vec{\mathbf{u}}_r &= p\vec{\mathbf{u}}_z \cdot \vec{\mathbf{u}}_r \\
&= p[\cos(\theta)\vec{\mathbf{u}}_r - \sin(\theta)\vec{\mathbf{u}}_\theta] \cdot \vec{\mathbf{u}}_r \\
&= p\cos(\theta)
\end{aligned}
$$

From our expression for the electric field we now generate the coordinate-free form of the electric field equation:

$$
\begin{aligned}
\vec{\mathbf{E}} &= \frac{p}{4\pi r^3 \epsilon_0}[2\cos(\theta)\vec{\mathbf{u}}_r + \sin(\theta)\vec{\mathbf{u}}_\theta] \\
&= \frac{p}{4\pi r^3 \epsilon_0}[3p\cos(\theta)\vec{\mathbf{u}}_r - p\cos(\theta)\vec{\mathbf{u}}_r + p\sin(\theta)\vec{\mathbf{u}}_\theta] \\
&= \left(\frac{p}{4\pi r^3 \epsilon_0}\right) 3(\vec{\mathbf{p}} \cdot \vec{\mathbf{u}}_r)\vec{\mathbf{u}}_r - \vec{\mathbf{p}}
\end{aligned}
$$

or, without the unit vector, $\vec{\mathbf{u}}_r$:

$$
\begin{aligned}
\vec{\mathbf{E}} &= \left(\frac{p}{4\pi r^3 \epsilon_0}\right) 3(\vec{\mathbf{p}} \cdot \vec{\mathbf{u}}_r)\vec{\mathbf{u}}_r - \vec{\mathbf{p}} \\
&= \left(\frac{p}{4\pi \epsilon_0}\right)\left(\frac{3(\vec{\mathbf{p}} \cdot \vec{\mathbf{r}})\vec{\mathbf{r}}}{r^5} - \frac{\vec{\mathbf{p}}}{r^3}\right)
\end{aligned}
\tag{11.36}
$$

Let us now consider an electric dipole located in a uniform electric field (Figure 11.28).

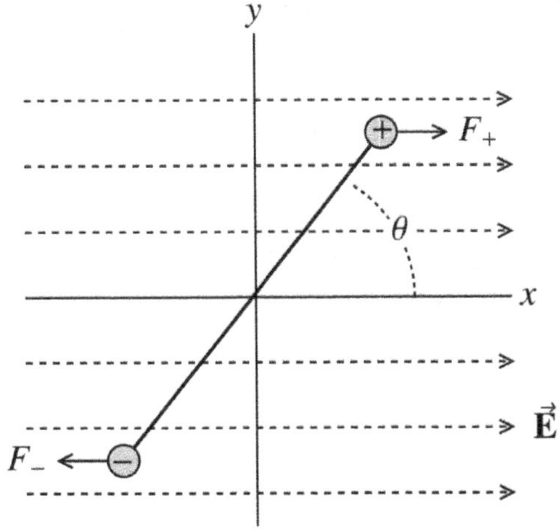

Figure 11.28 An electric dipole in an external electric field.

Each end of the dipole will experience an equal and opposite force directed towards its appropriate field direction ... this is a *couple of moment* as mentioned in Chapter 10. In Figure 11.28 we shall again assume the origin of the electric dipole to be at the centre point of the line between the two opposite charges and the x and y axes as shown. The electric dipole moment vector, $\vec{\mathbf{p}}$ is directed along the inter-charge line and the distance between the charges is d. We can write this in terms of unit Cartesian vectors:

$$\begin{aligned}\vec{\mathbf{p}} &= p\vec{\mathbf{u}}_r \\ &= qd[\cos(\theta)\vec{\mathbf{u}}_x + \sin(\theta)\vec{\mathbf{u}}_y]\end{aligned}$$

We can now write the torque on the electric dipole:

$$\begin{aligned}\vec{\Gamma} &= \vec{\mathbf{r}}_+ \times \vec{\mathbf{F}}_+ + \vec{\mathbf{r}}_- \times \vec{\mathbf{F}}_- \\ &= \left(\frac{d}{2}\cos(\theta)\vec{\mathbf{u}}_x + \frac{d}{2}\sin(\theta)\vec{\mathbf{u}}_y\right) \\ &\quad \times F_+\vec{\mathbf{u}}_x + \left(-\frac{d}{2}\cos(\theta)\vec{\mathbf{u}}_x - \frac{d}{2}\sin(\theta)\vec{\mathbf{u}}_y\right) \times (-F_-\vec{\mathbf{u}}_x) \\ &= \frac{dF_+}{2}\sin(\theta)(-\vec{\mathbf{u}}_z) + \frac{dF_-}{2}\sin(\theta)(-\vec{\mathbf{u}}_z) \\ &= dF_+\sin(\theta)(-\vec{\mathbf{u}}_z)\end{aligned}$$

Why the minus signs in front of the unit vectors? Look at eqn (3.20) for the answer to this question. Here, we have equated F_+ and F_- to F since their magnitudes are the same. From eqn (11.2) the force F is equal to qE that we substitute into our current equation (for magnitude only):

$$\begin{aligned}\Gamma &= dF\sin(\theta) \\ &= dqE\sin(\theta) \\ &= pE\sin(\theta)\end{aligned}$$

This points us towards a general expression for the torque experienced by an electric dipole in an external electric field:

$$\vec{\Gamma} = \vec{\mathbf{p}} \times \vec{\mathbf{E}}$$

From eqn (9.19) the work done by a torque on a rotating object is:

$$\begin{aligned}dW &= -\Gamma d\theta \\ &= -pE\sin(\theta)d\theta\end{aligned}$$

The minus sign is there because the torque is such that it opposes an increase in θ. In other words, the forces will rotate the dipole in Figure 11.28 clockwise such that the angle θ decreases. The minus sign corresponds to this sense of rotation.

Now we calculate the work done in rotating the dipole:

$$W = \int_{\theta_1}^{\theta_2} -pE \sin t(\theta) d\theta$$
$$= -pE \cos(\theta)\Big|_{\theta_1}^{\theta_2}$$

The change in potential energy is the negative of the work done:

$$\Delta U = U_2 - U_1 = -W = pE[\cos(\theta_2) - \cos(\theta_1)]$$

and if we define the potential energy at $\theta_1 = \pi$ to be zero (*i.e.* $U_1 = 0$) then the energy of the electric dipole in an external electric field is:

$$U = -pE\cos(\theta) = -\vec{\mathbf{p}} \cdot \vec{\mathbf{E}} \tag{11.37}$$

We are now in a position to say what the interaction energy of one electric dipole is in the field of another electric dipole. Using eqn (11.36) as the field of dipole moment 2 we use eqn (11.37) for the calculation:

$$\begin{aligned} U &= -\vec{\mathbf{p}}_1 \cdot \vec{\mathbf{E}}_2 \\ &= -\vec{\mathbf{p}}_1 \cdot \left(\frac{1}{4\pi\epsilon_0}\right)\left(\frac{3(\vec{\mathbf{p}}_2 \cdot \vec{\mathbf{r}}_2)\vec{\mathbf{r}}_2}{r^5} - \frac{\vec{\mathbf{p}}_2}{r^3}\right) \\ &= \left(\frac{1}{4\pi\epsilon_0}\right)\left(\frac{\vec{\mathbf{p}}_1 \cdot \vec{\mathbf{p}}_2}{r^3} - \frac{3(\vec{\mathbf{p}}_2 \cdot \vec{\mathbf{r}}_2)(\vec{\mathbf{p}}_1 \cdot \vec{\mathbf{r}}_2)}{r^5}\right) \end{aligned} \tag{11.38}$$

11.7 Magnetic Dipoles

The idea of the magnetic moment is similar to that of the electric dipole moment and comes from some basic observations of the physical nature of electric currents and magnetic fields. We wish here to describe the interaction between a magnetic moment and a static magnetic field and also with other magnetic moments, parallel to the last section on electric dipole moments. The most basic observation is that a charge, q, moving through a magnetic field, $\vec{\mathbf{B}}$, with velocity, $\vec{\mathbf{v}}$ experiences a force, $\vec{\mathbf{F}}$, which is directed at right angles to both $\vec{\mathbf{v}}$ and $\vec{\mathbf{B}}$ (Figure 11.29).

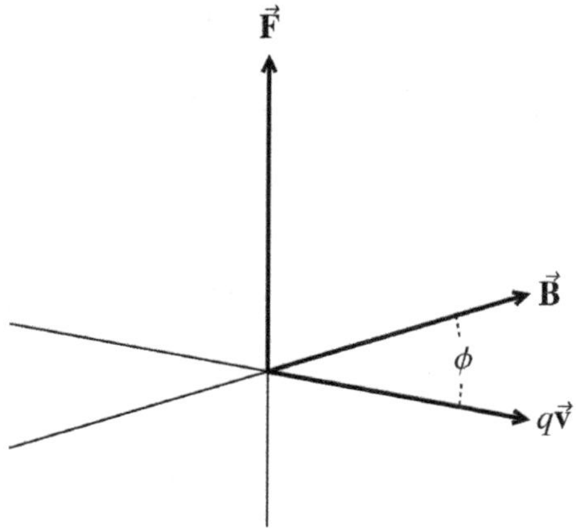

Figure 11.29 Force on a moving charged particle due to magnetic field, $\vec{\mathbf{B}}$.

The vector equation for this is (see eqn (3.18)):

$$\vec{\mathbf{F}} = q\vec{\mathbf{v}} \times \vec{\mathbf{B}} \tag{11.39}$$

This is, of course, a vector cross product from which we can get the magnitude of $\vec{\mathbf{F}}$ quite easily (see eqn (3.19)):

$$F = qvB\,\sin(\phi)$$

where ϕ is the angle between $\vec{\mathbf{v}}$ and $\vec{\mathbf{B}}$ as in Figure 11.29. We can write the vector equation in terms of infinitesimals:

$$\mathrm{d}\vec{\mathbf{F}} = \mathrm{d}q(\vec{\mathbf{v}} \times \vec{\mathbf{B}})$$

and since $\vec{\mathbf{v}}$ represents the linear velocity of charge per unit time, $\mathrm{d}l/\mathrm{d}t$, we can write:

$$\mathrm{d}\vec{\mathbf{F}} = \frac{\mathrm{d}q}{\mathrm{d}t}(\mathrm{d}\vec{\mathbf{l}} \times \vec{\mathbf{B}})$$

If we integrate over the length of the wire to get the total force of the current we get:

$$\vec{\mathbf{F}} = I\,\vec{\mathbf{l}} \times \vec{\mathbf{B}} \tag{11.40}$$

where $I = \mathrm{d}q/\mathrm{d}t$.

Now, if $\vec{\mathbf{l}}$ and $\vec{\mathbf{B}}$ are at right angles to each other then we can calculate the magnitude of $\vec{\mathbf{F}}$ directly:

$$F = IlB\sin\left(\frac{\pi}{2}\right)$$

where I is the current, l is the length of the wire and B is the magnitude of the magnetic field. We assume that B is constant over the length of the wire.

Now, let us take things a step further and imagine a current conducting coil in an external magnetic field and that the coil makes an angle ϕ with the field, $\vec{\mathbf{B}}$ (Figure 11.30).

Figure 11.30 is a side view of the coil that is at an angle to the field, $\vec{\mathbf{B}}$, and is able to rotate about an axis coming out towards the viewer. Figure 11.31 is the top view. Here, the field, $\vec{\mathbf{B}}$, in the second diagram is directed out of the page towards the viewer. The current carrying coil is rectangular in shape with sides of length a and b. From eqn (11.40) the magnitude of the horizontal force, F, is then given by:

$$F = IaB$$

assuming a current of I amperes moving through the coil and that the field, B, is at right angles to the movement of current. The force on sides b is:

$$F' = IbB\sin(\phi)$$

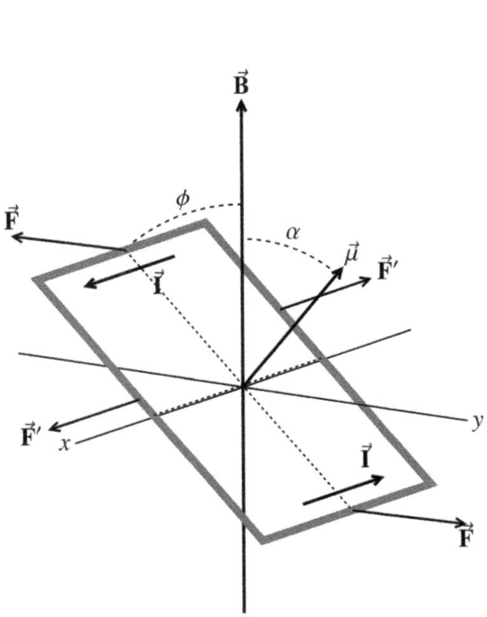

Figure 11.30 A current loop in an external magnetic field. Forces F' are equal and opposite to each other. Forces F constitute a coupled pair and lead to rotation of the coil about the x-axis.

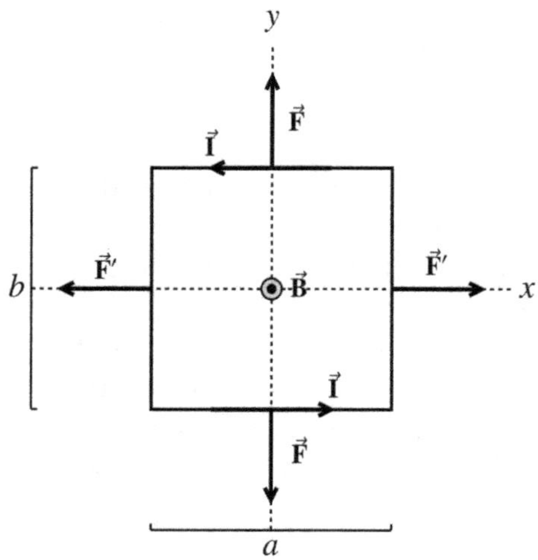

Figure 11.31 The current loop of Figure 11.30 viewed from above. The external magnetic field is coming out of the diagram towards the viewer.

Forces F' on sides b cancel each other. Forces F are equal in magnitude but are *not* collinear (Figure 11.30) so that they constitute a couple of moment. From eqn (9.23) the torque that this produces is:

$$\begin{aligned}\vec{\Gamma} &= 2\vec{\mathbf{r}} \times \vec{\mathbf{F}} \\ &= 2rF\sin(\alpha) \\ &= 2\frac{1}{2}bF\sin(\alpha) \\ &= bF\sin(\alpha) \\ &= IaBb\sin(\alpha)\end{aligned}$$

which will cause the coil to begin to rotate. The torque vector will be along the axis of rotation that, in the diagram, is collinear with F'. a times b is the area inside the coil so:

$$\Gamma = IAB\sin(\alpha)$$

where A represents the area. $I\vec{\mathbf{A}}$ is defined as the *magnetic moment*, $\vec{\mu}$ where $\vec{\mathbf{A}}$ is the vector area. The vector area is the usual area of a surface with direction perpendicular to the surface. Thus, the magnetic moment of our coil lies along an axis perpendicular to the coil.

$$\Gamma = \mu B\sin(\alpha)$$

where sin(α) is the angle between the magnetic moment and the applied field. In vector notation:

$$\begin{aligned}\vec{\Gamma} &= \vec{\mu} \times \vec{\mathbf{B}} \\ &= \frac{\mathrm{d}\vec{\mathbf{L}}}{\mathrm{d}t}\end{aligned} \tag{11.41}$$

A commonly used term is the *gyromagnetic ratio* (or the magnetogyric ratio depending on to whom you are talking). For a classically spinning (periodic) system we can calculate what this is:

$$\mu = IA = \underbrace{\frac{q}{t}\pi r^2}_{(3)} = \underbrace{\frac{qv}{2\pi r}\pi r^2}_{(4)} = \frac{q}{2m}mvr = \frac{q}{2m}L$$

where we have used eqn (9.15) with $\theta = 2\pi$ and $s = vt$ to convert from term three to term four and the magnitude version of eqn (9.22) to reach the last term. $q/2m$ is the gyromagnetic ratio.

$$\begin{aligned}\mu &= \frac{q}{2m}L \\ &= \gamma L\end{aligned} \tag{11.42}$$

or in vector and differential form:

$$\begin{aligned}\vec{\mu} &= \gamma\vec{\mathbf{L}} \\ \frac{\mathrm{d}\vec{\mu}}{\mathrm{d}t} &= \gamma\frac{\mathrm{d}\vec{\mathbf{L}}}{\mathrm{d}t} = \gamma(\vec{\mu} \times \vec{\mathbf{B}})\end{aligned} \tag{11.43}$$

We see from this that the magnetic moment is proportional to the angular momentum with the constant of proportionality being γ.

Finally, we are in a position to discuss the magnetic moment of a spin which, as NMR/MRI spectroscopists, we are naturally very interested in. The really fascinating feature of spins is that they contain intrinsic angular momentum as a fundamental property in the same way that mass and charge are fundamental physical properties. Hence, the use of the word "spin". This is a property of matter that was wholly unknown less than a hundred years ago.

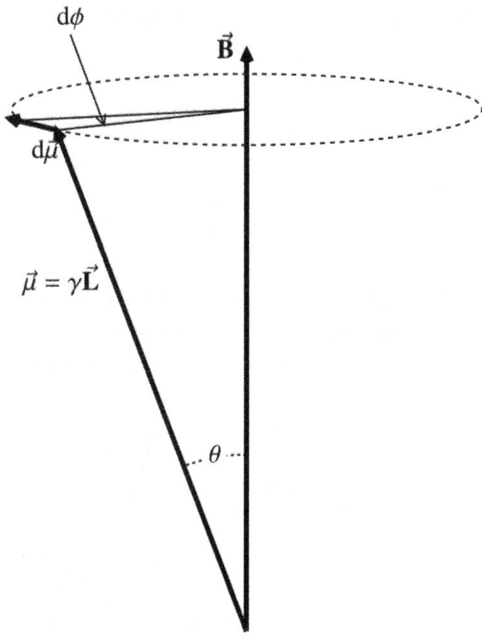

Figure 11.32 The magnetic moment of a "spin" in an external magnetic field.

Our spin is labelled with its magnetic moment, $\vec{\mu}$, which depends on its particular gyromagnetic ratio (Figure 11.32). The torque that $\vec{\mu}$ experiences is (eqn (11.41)):

$$\vec{\Gamma} = \frac{\mathrm{d}\vec{\mathbf{L}}}{\mathrm{d}t} = \vec{\mu} \times \vec{\mathbf{B}}$$

or:

$$\begin{aligned} \mathrm{d}\vec{\mathbf{L}} &= \vec{\Gamma} dt \\ &= \vec{\mu} \times \vec{\mathbf{B}}\mathrm{d}t \end{aligned}$$

We can use the gyromagnetic ratio:

$$\mathrm{d}\vec{\mathbf{L}} = \vec{\mu} \times \vec{\mathbf{B}}\mathrm{d}t$$

$$\mathrm{d}\vec{\mu} = \gamma(\vec{\mu} \times \vec{\mathbf{B}})\mathrm{d}t$$

and

$$\mathrm{d}\mu = \gamma\mu B \sin(\theta)\mathrm{d}t$$

Also from Figure 11.32:

$$\frac{r}{\|\vec{\mu}\|} = \frac{r}{\mu} = \sin(\theta) \Rightarrow r = \mu\sin(\theta)$$

$$\begin{aligned} \mathrm{d}\mu &= r\mathrm{d}\phi \\ &= \mu\sin(\theta)\mathrm{d}\phi \\ &= \gamma\mu B\sin(\theta)\mathrm{d}t \quad (\mathrm{d}\phi = \gamma B\mathrm{d}t) \end{aligned}$$

and therefore:

$$\begin{aligned} \frac{\mathrm{d}\phi}{\mathrm{d}t} &= \gamma B \\ \omega &= -\frac{\mathrm{d}\phi}{\mathrm{d}t} = -\gamma B \end{aligned} \tag{11.44}$$

where ω is the angular precession frequency. Why the negative sign? From Figure 11.32 the direction of precession is left-handed compared to the direction of $\vec{\mathbf{B}}$. Using the right-hand rule and aligning the thumb with $\vec{\mathbf{B}}$, we see that the "positive" direction is opposite to that of $\mathrm{d}\phi/\mathrm{d}t$. In the context of NMR spectroscopy, ω is the *Larmor frequency* of precession of a spin and eqn (11.44) is the Larmor equation. Thus, if we know the values of the gyromagnetic ratio and the applied field we can calculate the precession frequency. For a proton in a 11.7 Tesla field this turns out to be 500 MHz. Just to confuse things, the preceding applies to nuclei with a positive γ … things are opposite for nuclei with a negative γ.

The energy of a magnetic moment in an external field may be calculated using the differential version of eqn (9.19):

$$\mathrm{d}W = \Gamma\mathrm{d}\theta$$

and doing the indefinite integral, we get:

$$\begin{aligned} W &= \int \Gamma d\theta \\ &= \int \mu B\sin(\theta)\mathrm{d}\theta \\ &= -\mu B\cos(\theta) + C \end{aligned} \tag{11.45}$$

If we arbitrarily take the zero point of energy to be at $\theta = \pi/2$, then we can write for the energy:

$$E = -\mu B\cos(\theta)$$

or, in dot product vector notation:

$$E = -\vec{\mu}\cdot\vec{\mathbf{B}} \tag{11.46}$$

This is a fundamental equation in NMR spectroscopy, needed for building the quantum mechanical Hamiltonian operator. The definite integral of the above energy eqn (11.45) done between $\theta = 0$ and $\theta = \pi$ gives the energy difference between the lowest and highest energy of the spinning magnetic moment:

$$\begin{aligned} W &= \int_0^{\pi} \mu B\sin(\theta)\mathrm{d}\theta \\ &= -\mu B\cos(\theta)|_0^{\pi} \\ &= 2\mu B \end{aligned}$$

Equation (11.46) expresses the classical energy associated with the interaction of a magnetic moment and a static magnetic field. Note that it is very similar in form to that of the energy of an electric dipole moment in an external electric field eqn (11.37):

$$U = -\vec{\mathbf{p}}\cdot\vec{\mathbf{E}}$$

Now we ask "what is the field around a magnetic moment?" and "what is the energy of interaction between two magnetic moments?". This is much the same as we did in the previous section on electric dipole moments. Recall that in Section 11.6 we started by finding the scalar potential function around the electric dipole and taking its gradient to find the corresponding electric field followed by use of the energy of an electric dipole moment in an external field to find the interaction energy. Here, we will do much the same for the magnetic field around a magnetic dipole. The major difference is that we cannot start with a scalar potential function but rather must find the vector potential function around the magnetic moment. We do so by considering a diagram of a circular current-carrying wire and the magnetic field generated by it (Figure 11.33).

The distance of point P from the dipole origin is r while the distance from the current-carrying wire is R. For simplicity we use a

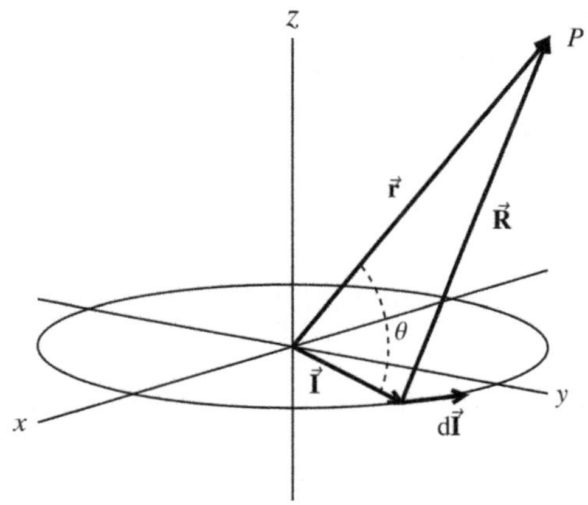

Figure 11.33 The off-axis field of a magnetic dipole generated by a circular current loop.

circular current of radius l. Using our magnetic vector potential (see eqn (11.22)):

$$\vec{\mathbf{A}} = \frac{\mu_0}{4\pi} I \oint \frac{1}{R} \mathrm{d}\vec{\mathbf{l}}$$

For the electric dipole we used a geometric argument to develop our equations. However, there is an alternate approach called the multipole expansion that we shall explore. We wish, ultimately, to find an expression in terms of r for the magnetic field at P but the vector potential equation is in terms of R. The multipole expansion will take care of this for us.

Using the law of cosines (Appendix A) we write, from Figure 11.33:

$$R^2 = r^2 + b^2 - 2rl\cos(\theta) = r^2\left[1 + \left(\frac{b}{r}\right)^2 - 2\left(\frac{b}{r}\right)\cos(\theta)\right]$$

$$R = r\sqrt{1 + \left(\frac{b}{r}\right)^2 - 2\left(\frac{b}{r}\right)\cos(\theta)}$$

$$= r\sqrt{1 + x}$$

where

$$x = \left(\frac{b}{r}\right)^2 - 2\left(\frac{b}{r}\right)\cos(\theta) = a(a - 2\cos(\theta))$$

We have temporarily substituted a for l/r for clarity. If $l < r$, which is usually the case, then we may use the binomial expansion (see Appendix A) on the square root portion of this equation:

$$\frac{1}{R} = \frac{1}{r(1+x)^{1/2}}$$
$$= \frac{1}{r}\left(1 - \frac{1}{2}x + \frac{3}{8}x^2 - \frac{15}{16}x^3 \ldots\right)$$

Inserting the expression for x back into this we get:

$$\frac{1}{R} = \frac{1}{r}\left[1 - \frac{1}{2}a(a - 2\cos(\theta)) + \frac{3}{8}a^2(a - 2\cos(\theta))^2 - \frac{15}{16}a^3(a - 2\cos(\theta))^3 \ldots\right]$$

and expanding:‡‡‡

$$\frac{1}{R} = \frac{1}{r}\left[1 + a\cos(\theta) + a^2\frac{3\cos^2(\theta) - 1}{2} + a^3\frac{5\cos 3(\theta) - 3\cos(\theta)}{2} \ldots\right]$$

We now substitute this for $1/R$ in the vector potential equation:

$$\vec{\mathbf{A}} = \frac{\mu_0}{4\pi}I\frac{1}{r}\oint\left[1 + a\cos(\theta) + a^2\frac{3\cos^2(\theta) - 1}{2} + a^3\frac{5\cos 3(\theta) - 3\cos(\theta)}{2} \ldots\right]\mathrm{d}\vec{\mathbf{l}}$$

Note that $1/r$ has been taken out of the integral since it is constant. Expanding this into several integrals:

$$\vec{\mathbf{A}} = \frac{\mu_0}{4\pi}I\frac{1}{r}\oint\mathrm{d}\vec{\mathbf{l}} + \frac{\mu_0}{4\pi}I\frac{1}{r}\oint a\cos(\theta)\mathrm{d}\vec{\mathbf{l}}$$
$$+ \frac{\mu_0}{4\pi}I\frac{1}{r}\oint a^2\frac{3\cos^2(\theta) - 1}{2}\mathrm{d}\vec{\mathbf{l}} + \frac{\mu_0}{4\pi}I\frac{1}{r}\oint a^3\frac{5\cos 3(\theta) - 3\cos(\theta)}{2}\mathrm{d}\vec{\mathbf{l}} \ldots$$

The first integral is the *monopole* term, the second the *dipole* term, the third the *quadrupole* term and so on. In this case the first term evaluates to zero since the integral:

$$\oint\mathrm{d}\vec{\mathbf{l}} = 0$$

‡‡‡This is a *very* tedious (and therefore error-prone) expansion, although straightforward. I recommend using symbolic algebra software such as wxMaxima or Mathematica for this.

Why is this? If you simply add all of the dls around the current loop they will cancel each other out, adding up to zero. This is a reflection of the fact that there are (probably) no magnetic monopoles; there is always a north and south pole to a magnet. The dominant term therefore is the second one, the dipole term:

$$\begin{aligned}
\vec{\mathbf{A}} &= \frac{\mu_0}{4\pi} I \frac{1}{r} \oint a \cos(\theta) \mathrm{d}\vec{\mathbf{l}} \\
&= \frac{\mu_0}{4\pi} I \frac{1}{r^2} \oint b \cos(\theta) \mathrm{d}\vec{\mathbf{l}} \quad \left(a = \frac{b}{r}\right) \\
&= \frac{\mu_0}{4\pi} I \frac{1}{r^2} \oint \vec{\mathbf{u}}_r \cdot \vec{\mathbf{b}} \mathrm{d}\vec{\mathbf{l}}
\end{aligned}$$

The last line is arrived at by realising that the angle θ is between vectors $\vec{\mathbf{l}}$ and $\vec{\mathbf{r}}$ (Figure 11.33) and that it must therefore also be the angle between $\vec{\mathbf{l}}$ and a unit vector parallel to $\vec{\mathbf{r}}$. Recalling eqn (4.52), we may now write:

$$\begin{aligned}
\vec{\mathbf{A}} &= \frac{\mu_0}{4\pi} I \frac{1}{r^2} \oint (\vec{\mathbf{u}}_r \cdot \vec{\mathbf{b}}) \mathrm{d}\vec{\mathbf{l}} \\
&= \frac{\mu_0}{4\pi} I \frac{1}{r^2} \vec{\mathbf{a}} \times \vec{\mathbf{u}}_r
\end{aligned}$$

Vector $\vec{\mathbf{a}}$ is the area vector of the circular current loop.

Multiplication of $\vec{\mathbf{a}}$ by the current, I, gives us the magnetic moment, μ:

$$\vec{\mathbf{A}} = \frac{\mu_0}{4\pi} \frac{\vec{\mu} \times \vec{\mathbf{u}}_r}{r^2} \tag{11.47}$$

Calculation of the magnetic field of the dipole and the energy of the dipole now proceeds in much the same way as for the electric dipole. As we have see above (eqn (11.20)), the magnetic field is the curl of $\vec{\mathbf{A}}$:

$$\begin{aligned}
\vec{\mathbf{B}} &= \vec{\nabla} \times \vec{\mathbf{A}} \\
&= \frac{\mu_0}{4\pi r^2} \vec{\nabla} \times (\vec{\mu} \times \vec{\mathbf{u}}_r)
\end{aligned}$$

Recall that the direction of the magnetic moment vector, μ, is along the z-axis of the coordinate system in Figure 11.33 and that of $\vec{\mathbf{u}}_r$ is along the direction of $\vec{\mathbf{r}}$, the direction of the vector resulting from their cross product must be orthogonal to them. In spherical coordinates this would be $\vec{\mathbf{u}}_\phi$ and we rewrite $\vec{\mathbf{A}}$ to reflect this:

$$\begin{aligned}\vec{\mathbf{B}} &= \frac{\mu_0}{4\pi r^2}\mu(\vec{\mathbf{u}}_z \times \vec{\mathbf{u}}_r)(\vec{\mu} = \mu\vec{\mathbf{u}}_z) \\ &= \frac{\mu_0}{4\pi r^2}\mu\sin(\theta)\vec{\mathbf{u}}_\phi\end{aligned}$$

Using the spherical coordinates version of $\vec{\nabla}$ (eqn (4.31)) we can now calculate $\vec{\mathbf{B}}$:

$$\begin{aligned}\vec{\mathbf{B}} &= \frac{\mu_0}{4\pi}\mu\vec{\nabla} \times \left(\frac{\sin(\theta)\vec{\mathbf{u}}_\phi}{r^2}\right) \\ &= \frac{\mu_0\mu}{4\pi}\left[\frac{\mathbf{u}_r}{r\sin(\theta)}\left(\frac{\partial \sin(\theta)A_\phi}{\partial\theta}\right) - \frac{\mathbf{u}_\theta}{r}\left(\frac{\mathrm{d}rA_r}{\mathrm{d}r}\right)\right] \\ &= \frac{\mu_0\mu}{4\pi}\left[\frac{\mathbf{u}_r}{r^3\sin(\theta)}\left(\frac{\partial \sin^2(\theta)}{\partial\theta}\right) - \frac{\mathbf{u}_\theta\sin(\theta)}{r}\left(\frac{\mathrm{d}}{\mathrm{d}r}\frac{1}{r}\right)\right] \\ &= \frac{\mu_0\mu}{4\pi}\left(\frac{\mathbf{u}_r}{r^3\sin(\theta)}(2\sin(\theta)\cos(\theta)) + \frac{\mathbf{u}_\theta\sin(\theta)}{r}\right) \\ &= \frac{\mu_0\mu}{4\pi r^3}(2\cos(\theta)\mathbf{u}_r + \sin(\theta)\mathbf{u}_\theta) = \vec{\mathbf{B}}\end{aligned}$$

This is strikingly similar to the field expression for an electric dipole (see eqn (11.35)). As with the electric dipole we may write the equation for $\vec{\mathbf{B}}$ in coordinate-free form. The technique for doing so is identical to that for the electric dipole (see eqn (11.36) and its development), the only major difference being that we are now using the value of the magnetic dipole moment, μ, instead of the value for the electric dipole moment, ϵ. We will not explicitly write out the development here but rather just state it.

$$\begin{aligned}\vec{\mathbf{B}} &= \left(\frac{\mu_0}{4\pi}\right)3(\vec{\mu}\cdot\vec{\mathbf{u}}_r)\vec{\mathbf{u}}_r - \vec{\mu} \\ &= \left(\frac{\mu_0}{4\pi}\right)\left(\frac{3(\vec{\mu}\cdot\vec{\mathbf{r}})\vec{\mathbf{r}}}{r^5} - \frac{\vec{\mu}}{r^3}\right)\end{aligned} \tag{11.48}$$

Our ultimate goal here is to find an expression for the energy of a magnetic dipole in the field of another magnetic dipole, much as we did for the electric dipole. From eqn (11.46) the energy of a magnetic dipole in an external magnetic field is:

$$E = \vec{\mu}\cdot\vec{\mathbf{B}}$$

and as with an electric dipole in the field of another electric dipole:

$$\begin{aligned} E &= -\vec{\mu}_1 \cdot \vec{\mathbf{B}}_2 \\ &= -\vec{\mu}_1 \cdot \left(\frac{\mu_0}{4\pi}\right)\left(\frac{3(\vec{\mu}_2 \cdot \vec{\mathbf{r}}_2)\vec{\mathbf{r}}_2}{r^5} - \frac{\vec{\mu}_2}{r^3}\right) \\ &= \left(\frac{\mu_0}{4\pi}\right)\left(\frac{\vec{\mu}_1 \cdot \vec{\mu}_2}{r^3} - \frac{3(\vec{\mu}_1 \cdot \vec{\mathbf{r}}_2)(\vec{\mu}_2 \cdot \vec{\mathbf{r}}_2)}{r^5}\right) \end{aligned} \tag{11.49}$$

Further Reading

1. S. W. Anwane, *Fundamentals of Electromagnetic Fields*, Infinity Science Press, Hingham, Mass., 2007.
2. W. B. Cheston, *Elementary Theory of Electric and Magnetic Fields*, John Wiley & Sons, London, New York, Sydney, 1964.
3. D. J. Griffiths, *Introduction to Electrodynamics*, Pearson Prentice Hall, Upper Saddle River, New Jersey, 1999.
4. B. Guru and H. Hiziroglu, *Electromagnetic Field Theory*, Cambridge University Press, Cambridge, UK, New York, 2004.
5. E. Purcell, *Electricity and Magnetism*, McGraw-Hill, New York, 1985.
6. W. G. V. Rosser, *Classical Electromagnetism via Relativity: an alternative approach to Maxwell's equations*, Plenum Press, New York, 1968.

12 The Magnetic Moment and the Bloch Equations

What has philosophy got to do with measuring anything? It's the mathematicians you have to trust, and they measure the skies like we measure a field.

Galileo Galilei

12.1 The Bloch Equations

Using the ideas developed in Chapter 11 concerning magnetic moments and angular momentum, we can develop a description of the time evolution of nuclear magnetic moments or just the nuclear magnetisation.[†] These equations are named after Felix Bloch, one of the giants of early NMR spectroscopy along with Edward Purcell and Isador Rabi.

So far, we have considered only a single spin and its magnetic moment. What about a large collection of spins that have been placed in a magnetic field? We now define a quantity called the *bulk magnetic moment*. This is the macroscopic version of the microscopic magnetic moments of the nuclei in the sample and is the vector sum of all magnetic moments in the bulk of the material of interest:

$$\vec{\mathbf{M}} = \sum_{i}^{N} \vec{\mu}_i \tag{12.1}$$

We might picture this as in Figure 12.1.

[†]This discussion of the Bloch equations proceeds with some major simplifications. However, the results that are obtained are entirely satisfactory for our purposes. For a more rigorous exploration of the Bloch equations, see Bain in the Further Reading section.

Essential Mathematics for NMR and MRI Spectroscopists
By Keith C. Brown

Published by the Royal Society of Chemistry, www.rsc.org

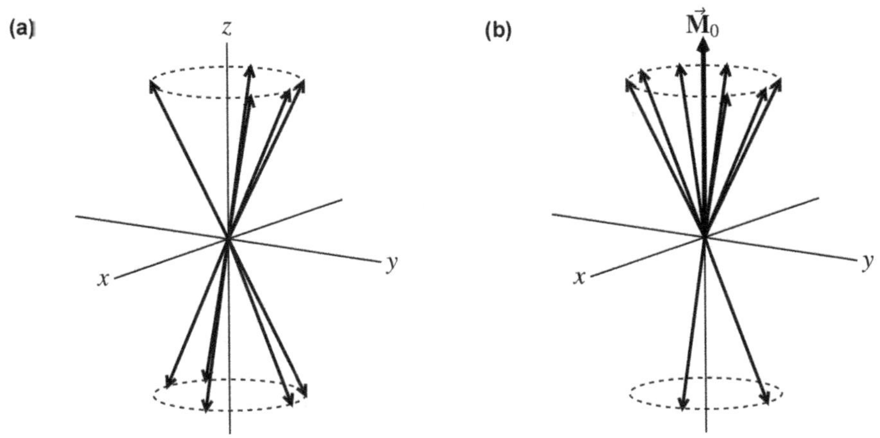

Figure 12.1 (a) Randomly oriented magnetic moments; (b) Excess spins oriented with the external field whose resultant is the net magnetic moment.

Of course, at the instant that they are put into the field there will be a completely random distribution of magnetic moments in all directions. The bulk magnetic moment (or net magnetisation or net polarisation in NMR jargon) is therefore zero. After some time, however, this changes and a net magnetic moment, $\vec{\mathbf{M}}$, appears, aligned with the external magnetic field, $\vec{\mathbf{B}}_0$, and the spin system is said to be at equilibrium (Figure 12.1b). Chapter 20 deals with the mechanism of this process in more detail.

Now we look closely at the net magnetisation. At equilibrium, this magnetisation vector is aligned parallel to the *z*-axis as is the external magnetic field, $\vec{\mathbf{B}}_0$. Note that we are now subscripting $\vec{\mathbf{B}}$ since we will shortly be dealing with two different magnetic fields and must keep track of them. We will label the magnetisation vector in the equilibrium state as $\vec{\mathbf{M}}_0$. Imagine that we now move this vector away from its equilibrium position in some as yet unspecified way and that the vector, now labelled $\vec{\mathbf{M}}$ since it is not in the equilibrium position, and is not necessarily collinear with the *x*, *y* or *z* axes. We can break $\vec{\mathbf{M}}$ down into its component vectors along the axes (Figure 12.2).

Vector $\vec{\mathbf{M}}$, being at an angle with respect to the *z*-axis and therefore the applied field $\vec{\mathbf{B}}_0$ as well, will precess about the *z*-axis, just as the individual magnetic moments will. Assuming for the moment that there will be no relaxation of $\vec{\mathbf{M}}$ back to the equilibrium position along the *z*-axis and that it will precess indefinitely we can write:

$$\frac{d\vec{\mathbf{M}}}{dt} = \gamma\vec{\mathbf{M}} \times \vec{\mathbf{B}}_0 \tag{12.2}$$

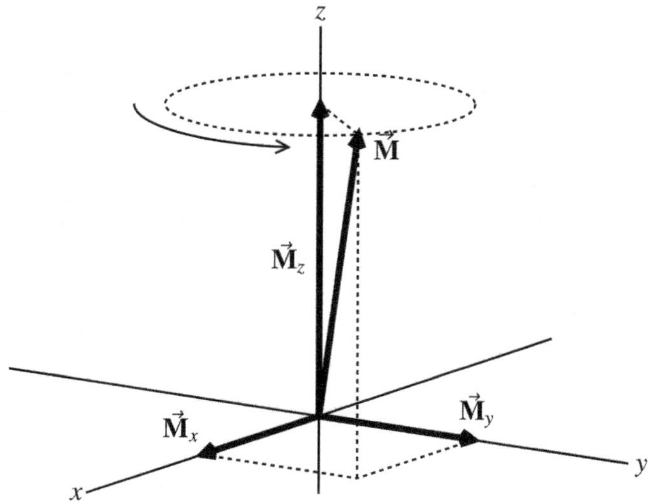

Figure 12.2 The component vectors of $\vec{\mathbf{M}}$.

which is the macroscopic version of eqn (11.43). Using our determinant method (see eqn (3.18)) to expand this we get:

$$\begin{aligned}\frac{\mathrm{d}\vec{\mathbf{M}}}{\mathrm{d}t} &= \gamma \det\begin{bmatrix}\vec{\mathbf{u}}_x & \vec{\mathbf{u}}_y & \vec{\mathbf{u}}_z \\ M_x & M_y & M_z \\ B_{0x} & B_{0y} & B_{0z}\end{bmatrix} \\ &= \gamma(M_yB_{0z} - M_zB_{0y})\vec{\mathbf{u}}_x - \gamma(M_xB_{0z} - M_zB_{0x})\vec{\mathbf{u}}_y + \gamma(M_xB_{0y} - M_yB_{0x})\vec{\mathbf{u}}_z \\ &= \gamma[(M_yB_{0z})\vec{\mathbf{u}}_x - (M_xB_{0z})\vec{\mathbf{u}}_y + 0\vec{\mathbf{u}}_z] \\ &= \gamma[(M_yB_{0z})\vec{\mathbf{u}}_x - (M_xB_{0z})\vec{\mathbf{u}}_y]\end{aligned} \tag{12.3}$$

remembering that the field $\vec{\mathbf{B}}_0$ is parallel to the z-axis and has no components in the x, y-plane. $\vec{\mathbf{u}}_x$, $\vec{\mathbf{u}}_y$ and $\vec{\mathbf{u}}_z$ are of course unit Cartesian basis vectors. This result tells us that the magnetisation changes in the x, y- plane but not along the z-axis. In other words the magnetisation vector, $\vec{\mathbf{M}}$, precesses about the z-axis at the Larmor frequency, as shown in Figure 12.2. Since the projection of $\vec{\mathbf{M}}$ onto the z-axis will (obviously) not precess about the z-axis we can write:

$$\frac{\mathrm{d}M_z}{\mathrm{d}t} = 0 \tag{12.4}$$

This should also be evident from the fact that $\mathrm{d}\vec{\mathbf{M}}/\mathrm{d}t$ is at right angles to the plane defined by $\vec{\mathbf{M}}$ and $\vec{\mathbf{B}}_0$ and will therefore have no component on the z-axis. $\vec{\mathbf{M}}_z$ is called the *longitudinal magnetisation.*

What about the magnetisation in the transverse (the x, y) plane? eqn (12.3) indicates that all of the change in $\vec{\mathbf{M}}$ is in the transverse plane. From eqn (12.3) it is reasonable to write:

$$\begin{aligned}\frac{\mathrm{d}M_x}{\mathrm{d}t} &= \gamma M_y B_{0z}\\ &= M_y \omega_0\end{aligned} \tag{12.5}$$

using the magnitude of the x-component of $\mathrm{d}\vec{\mathbf{M}}/\mathrm{d}t$ and eqn (11.44). Also:

$$\begin{aligned}\frac{\mathrm{d}M_y}{\mathrm{d}t} &= \gamma M_x B_{0z}\\ &= M_x \omega_0\end{aligned} \tag{12.6}$$

Recombining eqn (12.5) and (12.6) back into a vector equation:

$$\frac{\mathrm{d}\vec{\mathbf{M}}}{\mathrm{d}t} = \vec{\omega}_0 \times \vec{\mathbf{M}} \tag{12.7}$$

where we have again used eqn (11.44):

$$\begin{gathered}\vec{\omega}_0 = -\gamma \vec{\mathbf{B}}_0\\ \text{or}\\ \vec{\mathbf{B}}_0 = -\frac{\vec{\omega}_0}{\gamma}\end{gathered} \tag{12.8}$$

You might be forgiven for asking "where did the negative sign come from?". This was briefly addressed in the discussion of eqn (11.44), however, there is now another way to approach the question:

$$\begin{aligned}\frac{\mathrm{d}\vec{\mathbf{M}}}{\mathrm{d}t} &= \vec{\omega}_0 \times \vec{\mathbf{M}}\\ &= -\vec{\mathbf{M}} \times \vec{\omega}_0\\ &= \vec{\mathbf{M}} \times (-\vec{\omega}_0)\\ &= \vec{\mathbf{M}} \times \gamma \vec{\mathbf{B}}_0\\ &= \gamma \vec{\mathbf{M}} \times \vec{\mathbf{B}}_0\end{aligned}$$

Eqn (12.2) and (12.7) refer to the special (and artificial) case in which there is a magnetisation $\vec{\mathbf{M}}$ in an external field $\vec{\mathbf{B}}_0$ and there is no relaxation taking place. They do, however, serve as a starting point for our further considerations.

Before we take into account relaxation let us look at the situation when a linearly oscillating electromagnetic field, $\vec{\mathbf{B}}_x$, moving back and forth along the x-axis is applied at right angles to the external field, $\vec{\mathbf{B}}_0$. Conceptually, the $\vec{\mathbf{B}}_x$ field may then be broken up into two counter-rotating vectors the sum of which always is equal to $\vec{\mathbf{B}}_x$ (see Section 19.3 for more information). We will (temporarily) call these new vectors $\vec{\mathbf{B}}_{1L}$ and $\vec{\mathbf{B}}_{1R}$ for the moment and assign a rotational frequency of $+\omega_1$ and $-\omega_1$ to them (Figure 12.3). The x and y components of these two counter-rotating vectors are contained in:

$$\vec{\mathbf{B}}_{1L} = B_1[\cos(\omega_1 t)\vec{\mathbf{u}}_x + \sin(\omega_1 t)\vec{\mathbf{u}}_y]$$
$$\vec{\mathbf{B}}_{1R} = B_1[\cos(\omega_1 t)\vec{\mathbf{u}}_x - \sin(\omega_1 t)\vec{\mathbf{u}}_y] \qquad (12.9)$$
$$\vec{\mathbf{B}}_{1L} + \vec{\mathbf{B}}_{1R} = 2B_1\cos(\omega_1 t)\vec{\mathbf{u}}_x = \vec{\mathbf{B}}_x$$

where $\vec{\mathbf{B}}_{1L}$ (arbitrarily) uses $+\omega_1$ and $\vec{\mathbf{B}}_{1R}$ uses $-\omega_1$.

Only one of these two $\vec{\mathbf{B}}_1$ fields need concern us since the other is rotating with a negative frequency relatively speaking and does not

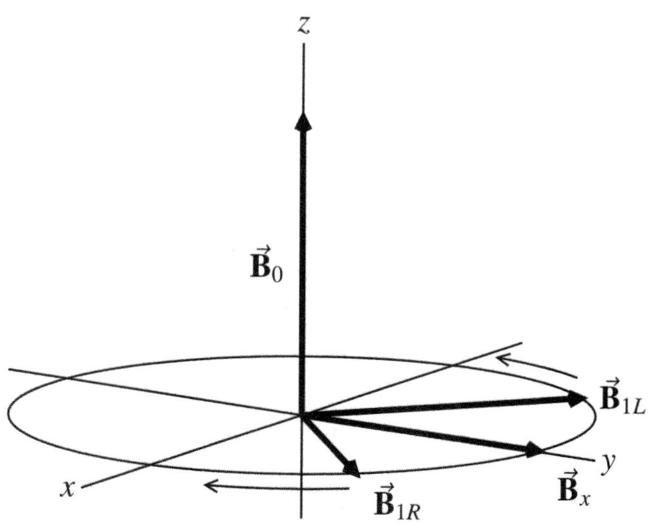

Figure 12.3 Counter-rotating applied magnetic fields.

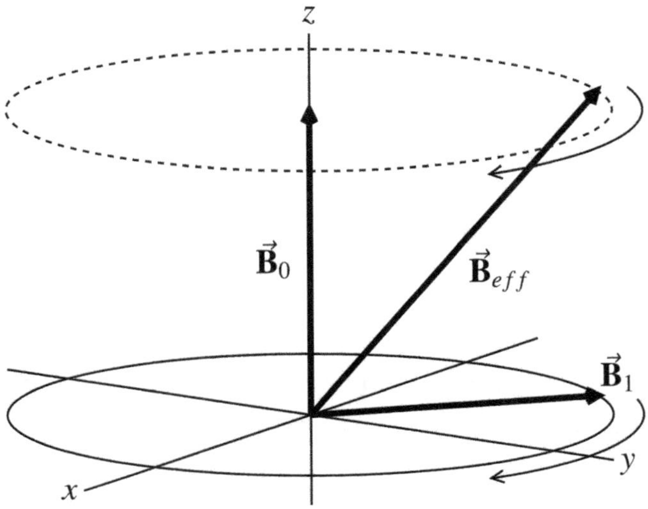

Figure 12.4 Effective magnetic field, $\vec{\mathbf{B}}_{\text{eff}}$.

affect any spins. We choose one and ignore the other. Let us choose $\vec{\mathbf{B}}_{1R}$ and simply call it $\vec{\mathbf{B}}_1$ for the rest of the discussion.

$$\vec{\mathbf{B}}_1 = B_1[\cos(\omega_1 t)\vec{\mathbf{u}}_x - \sin(\omega_1 t)\vec{\mathbf{u}}_y\,] \tag{12.10}$$

So, now we have two fields to consider, $\vec{\mathbf{B}}_0$ and $\vec{\mathbf{B}}_1$. $\vec{\mathbf{B}}_0$ is, of course, static and aligned along the z-axis (or perhaps more accurately, the z-axis is aligned with $\vec{\mathbf{B}}_0$) and $\vec{\mathbf{B}}_1$ is at 90° to the z-axis and rotating about the z-axis at ω_1 radians per second. We can call the sum of these two vectors the *effective field*, $\vec{\mathbf{B}}_{\text{eff}}$ (Figure 12.4).

Obviously, since $\vec{\mathbf{B}}_1$ is rotating about the z-axis, $\vec{\mathbf{B}}_{\text{eff}}$ is rotating about the z-axis as well. The size of $\vec{\mathbf{B}}_1$ is much exaggerated in Figure 12.4. Generally, $\|\vec{\mathbf{B}}_1\| \ll \|\vec{\mathbf{B}}_0\|$ so that $\vec{\mathbf{B}}_{\text{eff}}$ is only very slightly off of the z-axis. Our new simple Bloch equations (still without relaxation taken into account) are:

$$\begin{aligned}
\frac{\mathrm{d}\vec{\mathbf{M}}}{\mathrm{d}t} &= \gamma\vec{\mathbf{M}} \times \vec{\mathbf{B}}_{\text{eff}} \\
&= \gamma\vec{\mathbf{M}} \times (\vec{\mathbf{B}}_0 + \vec{\mathbf{B}}_1) \\
&= \gamma\vec{\mathbf{M}} \times \vec{\mathbf{B}}_0 + \gamma\vec{\mathbf{M}} \times \vec{\mathbf{B}}_1 \\
&= \gamma \det\begin{bmatrix} \vec{\mathbf{u}}_x & \vec{\mathbf{u}}_y & \vec{\mathbf{u}}_z \\ M_x & M_y & M_z \\ B_{0x} & B_{0y} & B_{0z} \end{bmatrix} + \gamma \det\begin{bmatrix} \vec{\mathbf{u}}_x & \vec{\mathbf{u}}_y & \vec{\mathbf{u}}_z \\ M_x & M_y & M_z \\ B_{1x} & B_{1y} & B_{1z} \end{bmatrix}
\end{aligned}$$

$$= \gamma(M_y B_{0z} - M_z B_{0y})\vec{\mathbf{u}}_x + \gamma(M_y B_{1z} - M_z B_{1y})\vec{\mathbf{u}}_x$$

$$- \gamma(M_x B_{0z} - M_z B_{0x})\vec{\mathbf{u}}_y - \gamma(M_x B_{1z} - M_z B_{1x})\vec{\mathbf{u}}_y$$

$$+ \gamma(M_x B_{0y} - M_y B_{0x})\vec{\mathbf{u}}_z + \gamma(M_x B_{1y} - M_y B_{1x})\vec{\mathbf{u}}_z$$

$$= \gamma(M_y B_{0z} - M_z B_{1y})\vec{\mathbf{u}}_x$$

$$- \gamma(M_x B_{0z} - M_z B_{1x})\vec{\mathbf{u}}_y$$

$$+ \gamma(M_x B_{1y} - M_y B_{1x})\vec{\mathbf{u}}_z$$

remembering that, as in eqn (12.3), $\vec{\mathbf{B}}_0$ lies along the z-axis and $\vec{\mathbf{B}}_1$ is in the transverse plane and therefore $\vec{\mathbf{B}}_{0x}$, $\vec{\mathbf{B}}_{0y}$ and $\vec{\mathbf{B}}_{1z}$ are zero. The rates of change of each component of $\vec{\mathbf{M}}$ are:

$$\begin{aligned} \frac{dM_x}{dt} &= \gamma(M_y B_0 - M_z B_{1y}) \\ \frac{dM_y}{dt} &= -\gamma(M_x B_0 - M_z B_{1x}) \\ \frac{dM_z}{dt} &= \gamma(M_x B_{1y} - M_y B_{1x}) \end{aligned} \tag{12.11}$$

Also as in eqn (12.3), $\vec{\mathbf{B}}_0$ and $\vec{\mathbf{B}}_{0z}$ are equal since $\vec{\mathbf{B}}_0$ is collinear with the z-axis ... hence the change from $\vec{\mathbf{B}}_{0z}$ to $\vec{\mathbf{B}}_0$ in eqn (12.11). Using eqn (12.9) we can modify eqn (12.11):

$$\begin{aligned} \frac{dM_x}{dt} &= \gamma(M_y B_0 - M_z B_1 \sin(\omega_1 t)) \\ \frac{dM_y}{dt} &= -\gamma(M_x B_0 - M_z B_1 \cos(\omega_1 t)) \\ \frac{dM_z}{dt} &= \gamma(M_x B_1 \sin(\omega_1 t) - M_y B_1 \cos(\omega_1 t)) \end{aligned} \tag{12.12}$$

Again, note that these equations do not yet include relaxation and assume the presence of $\vec{\mathbf{B}}_0$ and $\vec{\mathbf{B}}_1$.

The perceptive reader will recognise that we have introduced the oscillating $\vec{\mathbf{B}}_1$ field without saying why. Something completely astounding and counter-intuitive happens when we do this! First, let us invoke the rotating frame of reference from

Chapter 4. Specifically, let us use eqn (4.54) and our magnetisation vector $\vec{\mathbf{M}}$:

$$\left(\frac{\mathrm{d}\vec{\mathbf{M}}}{\mathrm{d}t}\right)_{\mathrm{LAB}} = \left(\frac{\mathrm{d}\vec{\mathbf{M}}}{\mathrm{d}t}\right)_{\mathrm{ROT}} + (\vec{\omega}_1 \times \vec{\mathbf{M}})$$

or

$$\left(\frac{\mathrm{d}\vec{\mathbf{M}}}{\mathrm{d}t}\right)_{\mathrm{ROT}} = \left(\frac{\mathrm{d}\vec{\mathbf{M}}}{\mathrm{d}t}\right)_{\mathrm{LAB}} - (\vec{\omega}_1 \times \vec{\mathbf{M}})$$

Recall from eqn (4.54) that $\vec{\omega}_1 \times \vec{\mathbf{M}}$ refers to the rotation of the x, y and z axes at vector frequency $\vec{\omega}_1$. We are completely free to choose what that frequency is so we choose the frequency of the rotating $\vec{\mathbf{B}}_1$ field, $\vec{\omega}_1$ (hence the subscript "1").

We also realise that the *LAB* portion of this equation can be replaced by eqn (12.7):

$$\begin{aligned}
\left(\frac{\mathrm{d}\vec{\mathbf{M}}}{\mathrm{d}t}\right)_{\mathrm{ROT}} &= (\vec{\omega}_0 \times \vec{\mathbf{M}}) - (\vec{\omega}_1 \times \vec{\mathbf{M}}) \\
&= \gamma(\vec{\mathbf{M}} \times \vec{\mathbf{B}}_0) + (\vec{\mathbf{M}} \times \vec{\omega}_1) \\
&= \vec{\mathbf{M}} \times (\gamma\vec{\mathbf{B}}_0 + \vec{\omega}_1) \\
&= \gamma\vec{\mathbf{M}} \times \left(\vec{\mathbf{B}}_0 + \frac{\vec{\omega}_1}{\gamma}\right)
\end{aligned}$$

Note that, since the order of $\vec{\mathbf{M}}$ and $\vec{\omega}_1$ were reversed, the sign of their cross product changes (see Chapter 3).

At this point we have the rate of change of $\vec{\mathbf{M}}$ due to $\vec{\mathbf{B}}_0$ in the rotating frame of reference. Let us complete things by putting $\vec{\mathbf{B}}_1$ into the equation now:

$$\left(\frac{\mathrm{d}\vec{\mathbf{M}}}{\mathrm{d}t}\right)_{\mathrm{ROT}} = \gamma\vec{\mathbf{M}} \times \left(\vec{\mathbf{B}}_0 + \frac{\vec{\omega}_1}{\gamma} + \vec{\mathbf{B}}_1\right) \tag{12.13}$$

and define a "reduced field" (using eqn (12.8)):

$$\begin{aligned}
\vec{\mathbf{B}}_{\mathrm{red}} &\equiv \vec{\mathbf{B}}_0 + \frac{\vec{\omega}_1}{\gamma} \\
&= -\frac{\vec{\omega}_0}{\gamma} + \frac{\vec{\omega}_1}{\gamma} \qquad (12.14) \\
&= \frac{\vec{\omega}_1 - \vec{\omega}_0}{\gamma}
\end{aligned}$$

and put it into eqn (12.13):

$$\begin{aligned}\left(\frac{\mathrm{d}\vec{\mathbf{M}}}{\mathrm{d}t}\right)_{\mathrm{ROT}} &= \gamma\vec{\mathbf{M}} \times \left(\frac{\vec{\omega}_1 - \vec{\omega}_0}{\gamma} + \vec{\mathbf{B}}_1\right) \\ &= \gamma\vec{\mathbf{M}} \times (\vec{\mathbf{B}}_{\mathrm{red}} + \vec{\mathbf{B}}_1) \\ &= \gamma\vec{\mathbf{M}} \times \vec{\mathbf{B}}_r \quad (\vec{\mathbf{B}}_r \equiv \vec{\mathbf{B}}_{\mathrm{red}} + \vec{\mathbf{B}}_1)\end{aligned} \tag{12.15}$$

where $\vec{\mathbf{B}}_r$ is the *resultant* field (Figure 12.5). This is the equation of motion of the magnetisation in the rotating frame that you will immediately recognise and has *exactly* the same form as our previously developed equation for the laboratory frame, eqn (12.2).

So, we have $\vec{\omega}_0$, $\vec{\omega}_1$, $\vec{\mathbf{M}}$ and $\vec{\mathbf{B}}_1$. ω_0 is the Larmor frequency of a spin in the static field, $\vec{\mathbf{B}}_0$, and is itself constant since the static field is ... static.‡ $\vec{\mathbf{M}}$ is the magnetisation due to the sum of the magnetic moments in the collection of (many, many) spins in the sample. $\vec{\mathbf{B}}_1$ is the rotating magnetic component of the applied electromagnetic field. ω_1 is the frequency of the rotating frame of reference that is also the frequency of the $\vec{\mathbf{B}}_1$ rf-field.

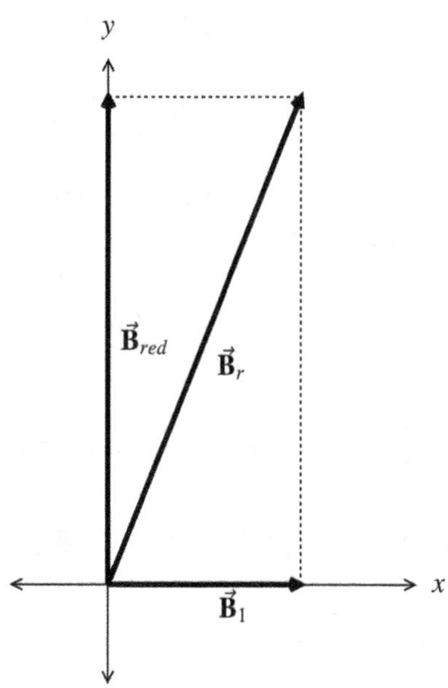

Figure 12.5 Resultant magnetic field.

‡I am being a bit sloppy here since I am moving back and forth between the frequency vector, $\vec{\omega}$, and the magnitude of the frequency vector, ω. Hopefully you will be able to move along with me.

What will happen if we vary ω_1 from zero up to ω_0?[§] The term $(\vec{\omega}_1 - \vec{\omega}_0)/\gamma$ in eqn (12.15) corresponds to our reduced field, $\vec{\mathbf{B}}_{\text{red}}$, and will at first be equal to $\vec{\mathbf{B}}_0$ when ω_1 is zero. As ω_1 increases, the value of $\vec{\mathbf{B}}_{\text{red}}$ decreases and the vector sum of $\vec{\mathbf{B}}_{\text{red}}$ and $\vec{\mathbf{B}}_1$, which we have denoted by $\vec{\mathbf{B}}_r$ begins to tilt slowly towards the *xy*-plane. As ω_1 approaches ω_0 the rate of tilt increases until at $\omega_1 - \omega_0 = 0$, $\vec{\mathbf{B}}_r$ is entirely in the *xy*-plane and is collinear with $\vec{\mathbf{B}}_1$. In other words, $\vec{\mathbf{B}}_{\text{red}}$ *vanishes*! Thus, when the frequency of the $\vec{\mathbf{B}}_1$ rf-field irradiation (and equivalently, the frequency of the rotating frame), ω_1, is equivalent to the Larmor frequency of the spin, ω_0, the external field vanishes from the rotating frame of reference. A very amazing and unexpected result!

Furthermore, if we vary ω_1 slowly the magnetisation, $\vec{\mathbf{M}}$, will precess about $\vec{\mathbf{B}}_r$ and end up precessing about $\vec{\mathbf{B}}_1$, which is located in the transverse plane. When $\omega_1 - \omega_0 = 0$ we say that we are at (or on) resonance $\cdots$ hence the name for our discipline, nuclear magnetic resonance. Also, since $\vec{\mathbf{M}}$ now sees no external field in the rotating frame its internal precession frequency about the *z*-axis is zero relative to the rotating frame. You can see this in eqn (12.14) when $\omega_1 - \omega_0 = 0$.

Our equations so far have assumed the projections of $\vec{\mathbf{M}}$ onto the laboratory frame axes, M_x, M_y, and M_z (see Figure 12.2). We ask now, what are the projections of $\vec{\mathbf{M}}$ and $\vec{\mathbf{B}}_r$ onto the rotating frame? We assume that the rotating frame frequency is that of $\vec{\mathbf{B}}_1$ and that $\vec{\mathbf{B}}_1$ is aligned along the rotating x'-axis. We will use x' and y' to denote the *rotating* axes and *x* and *y* for the *static* axes. Of course, the *z*-axis is the same in both frames.

Figure 12.6 shows that $\vec{\mathbf{B}}_1$ is aligned along the x'-axis and $\vec{\mathbf{B}}_{\text{red}}$ is aligned along the *z*-axis. Figure 12.7 shows the vector components of $\vec{\mathbf{M}}$ in the rotating frame. With reference to Figure 12.6 we recognise that the magnitude of $\vec{\mathbf{B}}_r$ along the rotating frame axes is that of $\vec{\mathbf{B}}_1$ along the x'-axis, zero along the y'-axis and $\vec{\mathbf{B}}_{\text{red}}$ or $(\vec{\omega}_1 - \vec{\omega}_0)/\gamma$ along the *z*-axis. eqn (12.15) can be expanded using these components:

$$\left(\frac{d\vec{\mathbf{M}}}{dt}\right)_{\text{ROT}} = \gamma\vec{\mathbf{M}} \times \vec{\mathbf{B}}_r = \gamma \det \begin{bmatrix} \vec{\mathbf{u}}'_x & \vec{\mathbf{u}}'_y & \vec{\mathbf{u}}'_z \\ M'_x & M'_y & M_z \\ B_1 & 0 & \dfrac{\omega_1 - \omega_0}{\gamma} \end{bmatrix}$$

$$= \gamma\left(M'_y \frac{\omega_1 - \omega_0}{\gamma}\right)\vec{\mathbf{u}}'_x - \gamma\left(M'_x \frac{\omega_1 - \omega_0}{\gamma} - M_z B_1\right)\vec{\mathbf{u}}'_y - \gamma M'_y B_1 \vec{\mathbf{u}}'_z \tag{12.16}$$

[§]We do not actually do this experimentally although I suppose it could be done. The (by now) old CW NMR spectrometers used to do something like this in that they would scan through a range of frequencies looking for "resonances". Here, however, this is just meant to be one of those thought-experiments that Albert Einstein was so fond of doing.

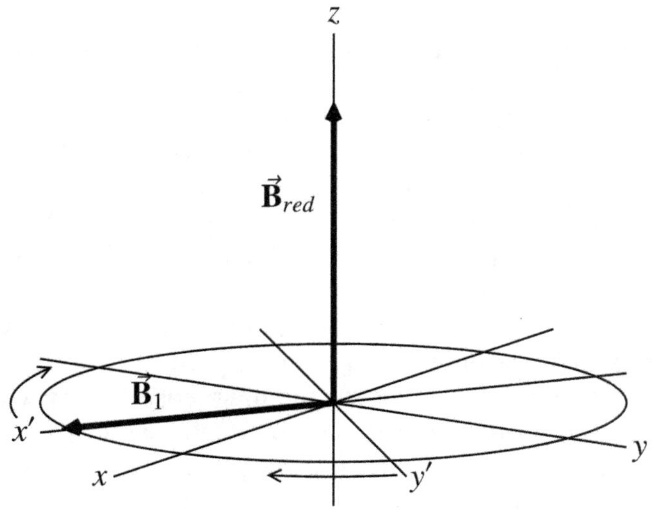

Figure 12.6 Static and rotating frames of reference.

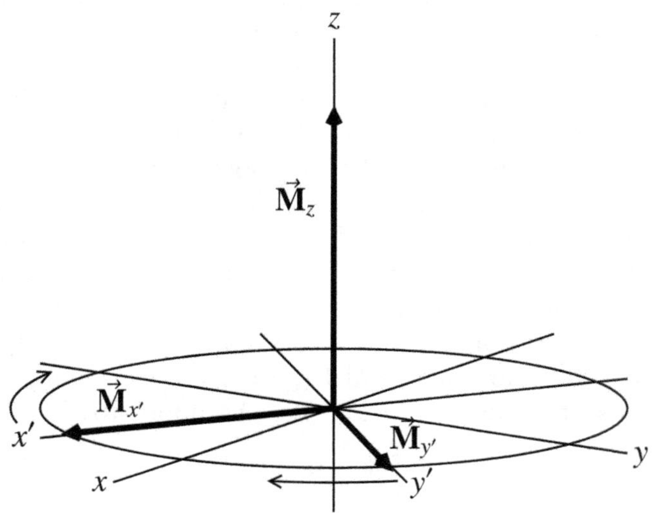

Figure 12.7 Magnetisation component vectors in the static and rotating frames.

The components of d**M**/dt along the (rotating) coordinate axes are:

$$\frac{\mathrm{d}M_x'}{\mathrm{d}t} = M_y'(\omega_1 - \omega_0) \tag{12.17}$$

$$\frac{\mathrm{d}M_y'}{\mathrm{d}t} = -M_x'(\omega_1 - \omega_0) + \gamma M_z B_1 \tag{12.18}$$

$$\frac{\mathrm{d}M_z}{\mathrm{d}t} = -\gamma M_y' B_1 \tag{12.19}$$

These are the Bloch equations "anchored" to the rotating frame, rather than the laboratory or static frame (eqn (12.12)). Note that the sine and cosine terms are now gone. This is reasonable since $\vec{\mathbf{B}}_1$ is now "motionless" in the rotating frame and therefore should naturally be expected to be time-invariant in this frame of reference ($\vec{\mathbf{B}}_1$ that is, not the magnetisation, $\vec{\mathbf{M}}$).

Now, finally, we will include relaxation.[¶] What is relaxation? As it turns out, there are two types that will be of interest to us here. First, if you do an inversion recovery experiment you can watch the change of M_z from a non-equilibrium state back to equilibrium where it is equal to M_0. The description of this experiment is detailed in Chapter 21. The change from non-equilibrium to equilibrium is easily modelled, as it turns out, by an exponential function and we can write it very simply as:

$$M_0 - M_z = A\mathrm{e}^{-t/T_1}$$

or

$$\ln(M_0 - M_z) = A' - \frac{t}{T_1}$$

where M_0 is the equilibrium magnetisation (and is constant) and M_z is the magnetisation at time t (and is therefore variable). We model it this way so that we have a positive quantity that decays towards zero in a way that is entirely analogous to radioactive decay. A and A' are constants whose values we need not be concerned about here since they will disappear. Now, differentiating with respect to time:

$$-\frac{1}{M_0 - M_z}\frac{\mathrm{d}M_z}{\mathrm{d}t} = -\frac{1}{T_1}$$

or (12.20)

$$\frac{\mathrm{d}M_z}{\mathrm{d}t} = \frac{M_0 - M_z}{T_1}$$

Do not confuse this with eqn (12.12)! eqn (12.20) refers to the rate of change of M_z due to relaxation processes *only*. What about M_x and M_y?

[¶]See Chapter 20 for a more detailed description of relaxation in NMR spectroscopy.

In the xy-plane there is a similar (mathematically at least) decay. Magnetisation here decays according to:

$$M_x = Ae^{-t/T_2}$$

or

$$\ln(M_x) = A' - \frac{t}{T_2}$$
$$\frac{1}{M_x}\frac{dM_x}{dt} = -\frac{1}{T_2} \tag{12.21}$$
$$\frac{dM_x}{dt} = -\frac{M_x}{T_2}$$

and likewise for M_y:

$$\frac{dM_y}{dt} = -\frac{M_y}{T_2} \tag{12.22}$$

Thus, T_1 and T_2 are the relaxation constants in two *different* exponential, first-order decay processes. T_1 models the decay process from non-equilibrium back to equilibrium and T_2 models the decoherence decay process in the transverse plane. A bit of thought will show that T_2 cannot be larger than T_1 since, when all spins are back in the equilibrium state, there can be no magnetisation left in the transverse plane.

If we add these in to our current Bloch eqn ((12.17), (12.18) and (12.19)) we get:

$$\frac{dM'_x}{dt} = M'_y(\omega_1 - \omega_0) - \frac{M'_x}{T_2} \tag{12.23}$$

$$\frac{dM'_y}{dt} = -M'_x(\omega_1 - \omega_0) + \gamma M_z B_1 - \frac{M'_y}{T_2} \tag{12.24}$$

$$\frac{dM_z}{dt} = -\gamma M'_y B_1 - \frac{M_z - M_0}{T_1} \tag{12.25}$$

In the old CW (continuous-wave) spectrometers, one swept either the frequency ω_1 or the field $\vec{\mathbf{B}}_1$ until the resonance condition was reached, as was discussed above. This was done slowly so as to maintain *steady-state* conditions. Under these circumstances, M_x', M_y' and M_z' remained constant in the rotating frame. Since the derivative of a constant is zero, we can write:

$$0 = M_y'(\omega_1 - \omega_0) - \frac{M_x'}{T_2}$$

$$0 = -M_x'(\omega_1 - \omega_0) + \gamma M_z B_1 - \frac{M_y'}{T_2}$$

$$0 = -\gamma M_y' B_1 - \frac{M_z - M_0}{T_1}$$

These being three equations in three variables, M_x', M_y' and M_z', we can solve for each of the magnetisation components using our matrix algebra from Chapter 2. First, to simplify things a bit we will rewrite our Bloch equations:

$$-M_x' + T_2(\omega_1 - \omega_0)M_y' + 0M_z = 0$$

$$-T_2(\omega_1 - \omega_0)M_x' - M_y' + \gamma T_2 B_1 M_z = 0$$

$$0M_x' - \gamma T_1 B_1 M_y' - M_z = -M_0$$

Here, we have multiplied by T_1 or T_2 as needed to get rid of the fractions and added $0M_x'$ and $0M_z$ for clarity. It is still cluttered so let us simplify a bit further by temporarily letting:

$$a = T_2(\omega_1 - \omega_0)$$

$$b = \gamma T_2 B_1$$

$$c = \gamma T_1 B_1$$

The set of equations now looks like:

$$-M_x' + aM_y' + 0M_z = 0$$

$$-aM_x' - M_y' + bM_z = 0$$

$$0M_x' - cM_y' - M_z = -M_0$$

and when we put them into matrix form:

$$\begin{bmatrix} -1 & a & 0 \\ -a & -1 & b \\ 0 & -c & -1 \end{bmatrix} \begin{bmatrix} M_x' \\ M_y' \\ M_z \end{bmatrix} = \begin{bmatrix} 0 \\ 0 \\ -M_0 \end{bmatrix}$$

Referring to the information on solving equations *via* matrices from Chapter 2 we construct the inverse matrix of the coefficients of $\vec{\mathbf{M}}$, being very careful about the signs of cofactors and transposing to get the adjoint matrix:‖

$$\frac{\operatorname{adj}\begin{bmatrix}-1 & a & 0\\ -a & -1 & b\\ 0 & -c & -1\end{bmatrix}}{\det\begin{bmatrix}-1 & a & 0\\ -a & -1 & b\\ 0 & -c & -1\end{bmatrix}}\times\begin{bmatrix}-1 & a & 0\\ -a & -1 & b\\ 0 & -c & -1\end{bmatrix}\times\begin{bmatrix}M'_x\\ M'_y\\ M_z\end{bmatrix}=\frac{\operatorname{adj}\begin{bmatrix}-1 & a & 0\\ -a & -1 & b\\ 0 & -c & -1\end{bmatrix}}{\det\begin{bmatrix}-1 & a & 0\\ -a & -1 & b\\ 0 & -c & -1\end{bmatrix}}\times\begin{bmatrix}0\\ 0\\ -M_0\end{bmatrix}$$

which becomes:

$$\begin{bmatrix}M'_x\\ M'_y\\ M_z\end{bmatrix}=\frac{\operatorname{adj}\begin{bmatrix}-1 & a & 0\\ -a & -1 & b\\ 0 & -c & -1\end{bmatrix}}{\det\begin{bmatrix}-1 & a & 0\\ -a & -1 & b\\ 0 & -c & -1\end{bmatrix}}\times\begin{bmatrix}0\\ 0\\ -M_0\end{bmatrix}$$

$$\det\begin{bmatrix}-1 & a & 0\\ -a & -1 & b\\ 0 & -c & -1\end{bmatrix}=-1(1+bc)-a(a-0)$$

$$=-(bc+1+a^2)$$

$$\operatorname{adj}\begin{bmatrix}-1 & a & 0\\ -a & -1 & b\\ 0 & -c & -1\end{bmatrix}=\begin{bmatrix}1+bc & a & ab\\ -a & 1 & b\\ ac & -c & 1+a^2\end{bmatrix}$$

So we have:

$$\begin{bmatrix}M'_x\\ M'_y\\ M_z\end{bmatrix}=\frac{\begin{bmatrix}1+bc & a & ab\\ -a & 1 & b\\ ac & -c & 1+a^2\end{bmatrix}}{-(bc+1+a^2)}\times\begin{bmatrix}0\\ 0\\ -M_0\end{bmatrix}$$

$$=\frac{\begin{bmatrix}-M_0ab\\ -M_0b\\ -M_0(1+a^2)\end{bmatrix}}{-(bc+1+a^2)}$$

‖You could do this by hand or use a symbolic algebra program such as Mathematica or wxMaxima for these calculations.

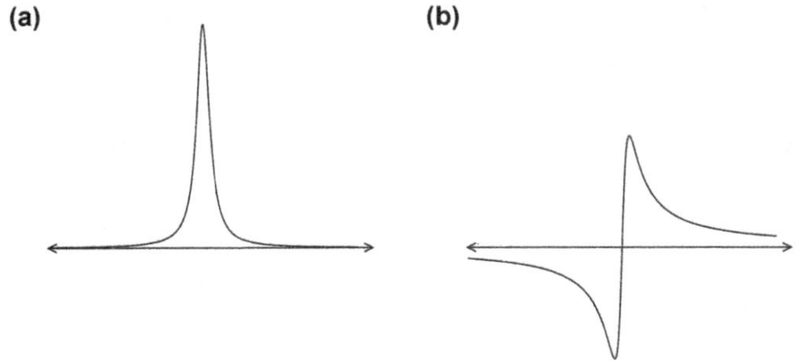

Figure 12.8 (a) M_y' versus ω_1; (b) M_x' versus ω_1.

and we finally arrive at our solutions (after substituting for a, b and c from our definitions and factoring -1 from the numerator and denominator):

$$M_x' = \frac{\gamma M_0 T_2^2 B_1(\omega_1 - \omega_0)}{\gamma^2 T_1 T_2 B_1^2 + 1 + T_2^2(\omega_1 - \omega_0)^2}$$

$$M_y' = \frac{\gamma M_0 T_2 B_1}{\gamma^2 T_1 T_2 B_1^2 + 1 + T_2^2(\omega_1 - \omega_0)^2}$$

$$M_z = \frac{M_0\left[1 + T_2^2(\omega_1 - \omega_0)^2\right]}{\gamma^2 T_1 T_2 B_1^2 + 1 + T_2^2(\omega_1 - \omega_0)^2}$$

It is generally the case that the magnitude of $\vec{\mathbf{B}}_1$ is only a few milligauss and if T_1 and T_2 are only a few seconds each:

$$\gamma^2 T_1 T_2 B_1^2 \ll 1$$

and our equations in the x, y-plane can be written approximately as:

$$M_x' = \frac{\gamma M_0 T_2^2 B_1(\omega_1 - \omega_0)}{1 + T_2^2(\omega_1 - \omega_0)^2}$$

$$M_y' = \frac{\gamma M_0 T_2 B_1}{1 + T_2^2(\omega_1 - \omega_0)^2}$$

A plot of M_y' *vs.* ω_1 is shown in Figure 12.8a and is called the *absorption line shape*. M_x' *vs.* ω_1 is shown in Figure 12.8b and is called the *dispersion line shape*.

12.2 The Vector Model of NMR Spectroscopy

Strictly speaking this topic is not so much mathematical as it is pictorial, however, it follows directly from the development of the Bloch equations and is very useful in introductory NMR spectroscopy. The pictures that it provides will help us greatly in understanding the more obtuse theoretical discourses to follow.

We have spoken of the equilibrium magnetisation and pictured it as a vector in Cartesian space in Figure 12.1a. We have also spoken of how it is that this vector can move through this Cartesian space *via* the effect of an applied linearly oscillating magnetic field, $\vec{\mathbf{B}}_1$.

Let us suppose that we have a spin system at equilibrium in an external magnetic field and we apply a short $\vec{\mathbf{B}}_1$ field pulse (that is, the rf-field is turned on for a very short time and then off again). How can we picture what will happen given what we have discussed so far? We have seen that, in the rotating frame of reference, if the $\vec{\mathbf{B}}_1$ frequency is at or very near ω_1, the intrinsic Larmor frequency of the spin, the external field $\vec{\mathbf{B}}_0$ effectively collapses to zero (eqn (12.15) with $\omega_0 = \omega_1$). Thus, what the equilibrium magnetisation "sees" in the rotating frame of reference is a *static* $\vec{\mathbf{B}}_1$ field oriented along an axis in the x, y-plane (Figure 12.9).

What will happen? The equilibrium magnetisation will precess about $\vec{\mathbf{B}}_1$ of course, just as it precesses about the $\vec{\mathbf{B}}_0$ field in the laboratory frame.

In Fourier transform NMR spectroscopy, turning on the $\vec{\mathbf{B}}_1$ field for enough time that the magnetisation rotates by 90° into the x, y-plane is commonly referred to as issuing a 90° pulse. Similarly, a 180° pulse effects a rotation of the magnetisation through 180°.

When we turn off the $\vec{\mathbf{B}}_1$ field what happens? If the magnetisation has been rotated into the x, y-plane it will now precess about $\vec{\mathbf{B}}_{red}$, which will be zero if the pulse is "on resonance" and non-zero otherwise. On-resonance generally means $\omega_0 = \omega_1$ in eqn (12.15). For

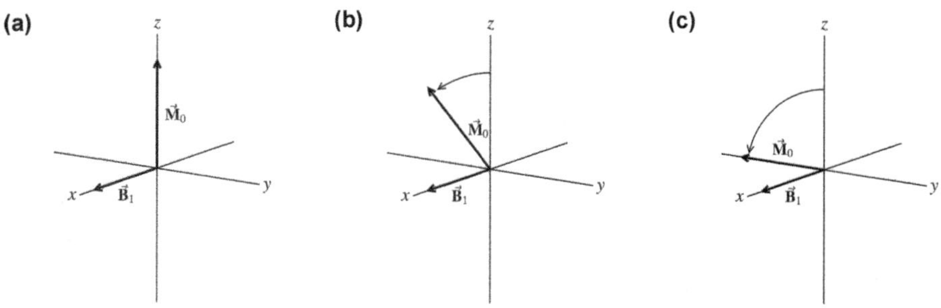

Figure 12.9 (a) Application of a $\vec{\mathbf{B}}_1$ field to the equilibrium magnetisation vector, $\vec{\mathbf{M}}_0$; (b) Rotation of $\vec{\mathbf{M}}_0$ towards the transverse plane under the influence of the $\vec{\mathbf{B}}_1$ field; (c) Rotation of the equilibrium magnetisation vector by 90°.

a set of spins with differing Larmor frequencies only one spin can experience an on-resonance pulse while the others will experience varying degrees of $\vec{\mathbf{B}}_{\text{red}}$. It is easiest to begin describing pulse sequences with on-resonance pulses. So, we have a spin whose magnetisation is in the x, y-plane as a result of an on-resonance 90° pulse. This being the case, in the rotating frame there will be no apparent *chemical shift evolution* since the Larmor frequency of the spin is the same as the rotating frame frequency. The magnetisation $\vec{\mathbf{M}}$ will appear to be motionless.

Scalar coupling evolution can be represented vectorially as a splitting of the vector into two or more vectors with differing rotational frequencies in the rotating frame. If we imagine for convenience an on-resonance 90° pulse issued to a spin that is split into a doublet in the spectrum, our vector model will have the vector in the x,y-plane split into two "sub vectors", one rotating clockwise in the rotating frame and one rotating counter-clockwise. Why is this? An on-resonance pulse is issued at the chemical shift frequency of the spin. Thus, the "zero" frequency is the chemical shift frequency; that is, in the rotating frame the chemical shift of the spin appears to be zero. However, for a doublet in an NMR spectrum the chemical shift is half-way between the two peaks observed. Thus, one peak is slightly behind the rotating frame frequency and one is slightly ahead by the same amount. Therefore, in the rotating frame we picture two counter-rotating vectors (Figure 12.10).

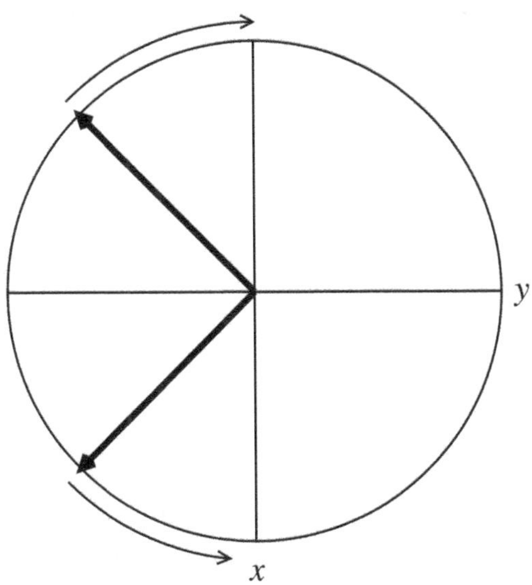

Figure 12.10 Vector representation of scalar coupling.

Note that we have changed our perspective here for convenience. We are now looking down from "above" the x, y-plane with the z-axis coming out of the page towards us. The coupling constant, usually measured in hertz, is the frequency difference between the peaks of the multiplet in the spectrum. For a doublet this is J Hz and in our picture, one vector rotates with relative frequency $+J/2$ Hz and one with relative frequency $-J/2$ Hz. Why $J/2$? Remember that we are thinking in the rotating frame of reference and that the chemical shift frequency is zero in that frame of reference (for an on-resonance pulse). Since the frequency difference between the doublet peaks is J Hz, one of the two doublet peaks will rotate at $+J/2$ Hz and the other at $-J/2$ Hz *relative to the chemical shift frequency.*

Now, suppose we wait for a period of time after the $\vec{\mathbf{B}}_1$ field is turned off. Furthermore, let us again assume an on-resonance pulse for simplicity and clarity and a doublet with coupling constant J Hz. If the time period is $1/(2J)$ seconds where will the doublet vectors be? We need to calculate how far each vector will have rotated:

$$\begin{aligned}\theta &= \omega \times t \\ &= 2\pi J \times \frac{1}{2J} \\ &= \pi\end{aligned}$$

This gives the angle between the two counter-rotating vectors. Half of this then will be the angle that one of the vectors will have rotated through:

$$\begin{aligned}\theta &= \omega \times t \\ &= \pi J \times \frac{1}{2J} \\ &= \frac{\pi}{2}\end{aligned}$$

Thus, one vector that is rotating clockwise will rotate $+90°$ and the other $-90°$ (Figure 12.11).

We can apply these simple principles to several pulse sequences. The J-modulation experiment consists of pulses on two channels, usually ^{1}H and ^{13}C (Figure 12.12).

The effect of the ^{1}H decoupler is to collapse multiplets into singlets. In other words, multiplet vectors collapse to a single vector located at the chemical shift frequency. So, in this pulse sequence there is coupling evolution during the first delay period of $1/J$ seconds but not

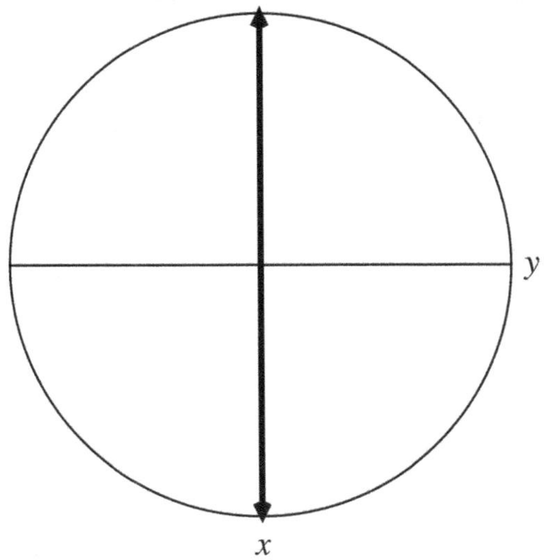

Figure 12.11 Counter-rotating scalar coupling vectors, rotated through 90° each, after initial excitation by an rf pulse.

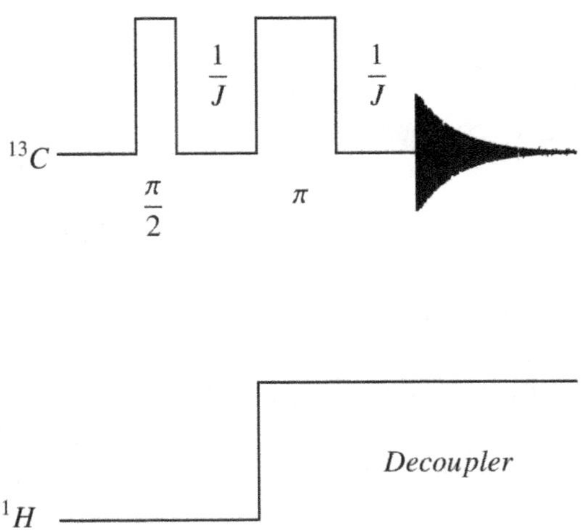

Figure 12.12 The *J*-modulation pulse sequence.

during the second delay period (Figure 12.13). How is this useful? During the first delay period the two counter-rotating vectors of the doublet will each rotate by 180° and will be collinear along (say) the *y*-axis. The 180° pulse on ^{13}C rotates them back to where they started. The decoupler collapses them to a single vector that will now exhibit only chemical shift evolution. Assuming on-resonance pulses, this will be zero so that the vector is motionless in the rotating frame

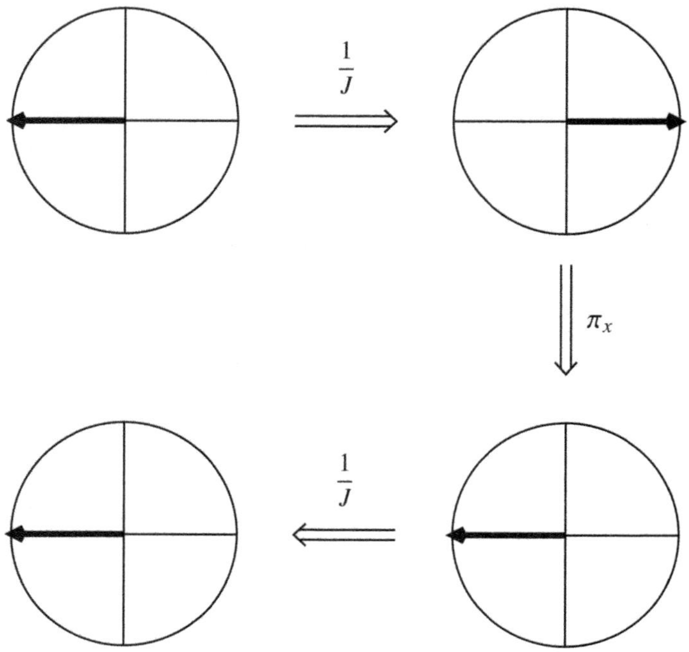

Figure 12.13 Vector picture of the spin behaviour as a result of the application of the J-modulation pulse sequence for an on-resonance doublet.

during the second delay period. Acquisition then gives a spectrum with a (decoupled) ^{13}C peak in it.

The interesting part of all of this is what happens when we consider now a triplet, again with coupling constant J Hz. Still assuming an on-resonance pulse for convenience, the vector representing the centre peak will be motionless in the rotating frame since it is also at the chemical shift of the multiplet and the other two vectors will rotate at $+J$ Hz and $-J$ Hz with respect to the chemical shift frequency. This is just twice that of the doublet so each of the "outer" vectors will rotate twice as far in $1/J$ seconds. Using our APT pulse sequence analysis from above the vector diagrams will look as shown in Figure 12.14.

The vector at acquisition time is now pointed 180° away from where the double vector was. This will result in a peak in the ^{13}C spectrum that is 180° out of phase with the doublet peak. The APT experiment is a very useful way to sort out the multiplicity of ^{13}C peaks.

The simple vector model is very useful for introducing the analysis of pulse sequences but there comes a point at which it breaks down.

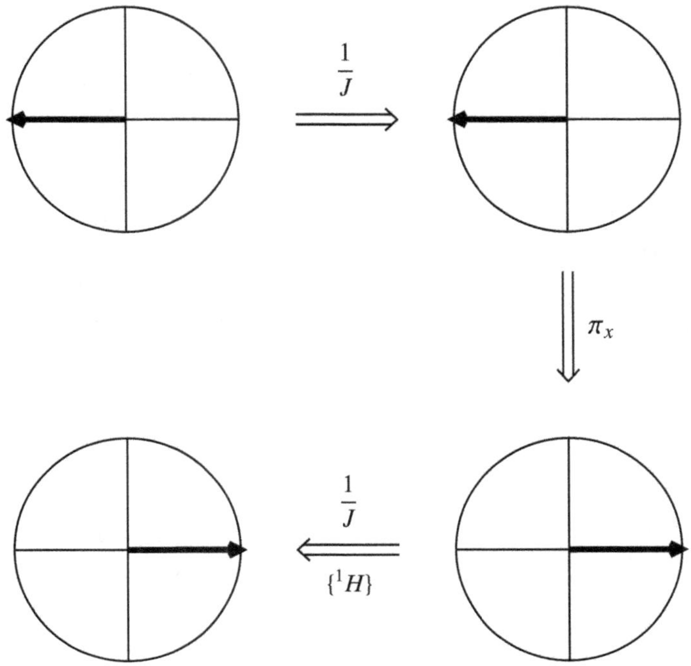

Figure 12.14 Vector picture of the spin behaviour as a result of the application of the *J*-modulation pulse sequence for an on-resonance triplet.

Specifically, when multiple quantum coherence** is generated it is not possible to follow the evolution of the magnetisation using the simple vector model.†† Fear not, however. There are other ways to do so and the vector-picture will be very useful in introducing them.

Further Reading

1. A. D. Bain, *J. Magn. Res.*, 2010, **206**, 227–240.
2. H. J. Bernstein, *High Resolution Nuclear Magnetic Resonance*, McGraw-Hill Book Company, New York, 1959.
3. F. Bloch, *Phys. Rev.*, 1949, **70**, 460–473.

**More on this topic in Chapters 18 and 19.

††This is not strictly true. The vector-model has been extended to include pictorial representations of multiple quantum coherence. See for example, D.P. Goldenberg, *Concepts in Magnetic Resonance Part A*, 2010, Vol. **36A**(2) 49. I find these extended pictorial systems confusing and would rather use such full-blown quantum mechanical methods as the product operator model.

4. J. R. Bolton and J. A. Weil, *Electron Paramagnetic Resonance*, John Wiley & Sons Ltd., Hoboken, NJ, 2nd edn, 2007.
5. F. A. Bovey, L. Jelinski and P. A. Mirau, *Nuclear Magnetic Resonance Spectroscopy*, Academic Press, San Diego, 1988.
6. P. K. Madhu and A. Kumar, *J. Magn. Res. Ser. A*, 1994, **114**, 201–212.
7. P. K. Madhu and A. Kumar, *Concepts Magn. Reson.*, 1997, **9**, 1–12.
8. M. L. Martin, G. J. Martin and J. J. Jean-Jacques Delpuech, *Practical NMR Spectroscopy*, Heyden and Son Ltd., London, 1980.
9. G. A. Morris and P. B. Chilvers, *J. Magn. Res. Ser. A*, 1994, **107**, 236–238.
10. J. H. Nelson, *Nuclear Magnetic Resonance Spectroscopy*, Prentice Hall, Upper Saddle River, NJ, 2003.
11. D. Shaw, *Fourier Transform NMR Spectroscopy*, Elsevier, Amsterdam, New York, 1984.
12. C. P. Slichter, *Principles of Magnetic Resonance*, Springer-Verlag, New York, Berlin, Heidelberg, 3rd edn, 1992.
13. H. C. Torrey, *Phys. Rev.*, 1949, **76**, 1059–1068.

13 Introductory Quantum Mechanics

In these matters the only certainty is that there is nothing certain.

Pliny the Elder

13.1 Historical

The full story of the early development of quantum mechanics is fascinating and full of opinionated characters, triumphs, tragedy and irony. There was much back-and-forth discussion and several avenues of thought as to how to develop the new physics. It was by no means a straightforward progression to modern quantum physics. In fact, the arguments about the meaning of the equations involved is still ongoing.

We begin with J.J. Thomson. He, of course, discovered the electron in 1897 at Cambridge University. This was quite a disturbing thing to find out since it was thought at the time that atoms were indivisible (as indeed, the very word "atom" means) and yet here was a small particle coming out of an atom. Evidently, atoms are divisible. Since the electron was determined to have a negative charge and since atoms are neutral it was thought that the electrons were embedded in a sphere of positive charge to make up the atom. This was the short-lived "plum pudding" model of the atom. Ironically, J.J. Thomson's son George Thomson later showed that the electron can behave as a wave rather than a particle. We return to the electron shortly.

At the end of the nineteenth century, physics had achieved astounding success with the application of Newton's laws (see Chapter 9) to the movement of bodies in space and the elucidation

Essential Mathematics for NMR and MRI Spectroscopists
By Keith C. Brown

Published by the Royal Society of Chemistry, www.rsc.org

of the characteristics of electromagnetic radiation *via* Maxwell's equations (Chapter 11). In the sciences and most especially, in physics we like to model physical phenomena mathematically as it allows us to predict what will happen in the future. For two centuries Newton's laws were the fundamental starting point for modelling all sorts of physical phenomena and were very successful and are still fully adequate for macroscopic phenomena. It was then, perhaps natural for physicists to think that there was not much of a fundamental nature left to discover but rather small things that would fit into the big picture as described by the physics of the time. There were of course some minor problems but they would ultimately be solved and all would be well. One of these was the problem of blackbody radiation.

All bodies in the universe radiate electromagnetic energy of a spectrum of frequencies. Measurement of this spectrum of frequencies shows two main things. First, for very low and very high frequencies the intensity of the radiation is low and at intermediate frequencies there is a frequency at which the intensity is at a maximum, ν_m. Secondly, the maximum ν_m changes as the temperature changes. When T is increased ν_m increases. Many of us are familiar with this phenomenon in connection with electric heaters that glow as they warm up. A plot of typical measurements of this spectrum is shown in Figure 13.1.

Attempts to model this resulted in the Rayleigh–Jeans law that worked for large values of wavelength but increasingly deviated from

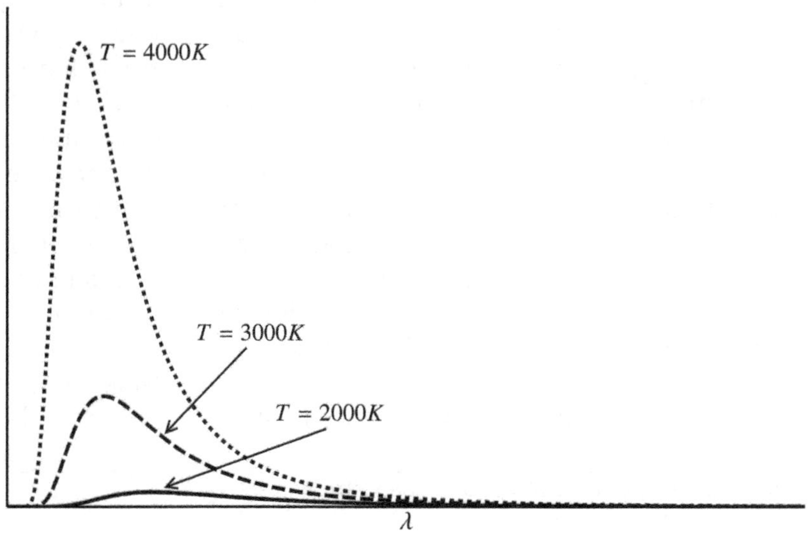

Figure 13.1 Blackbody radiation spectrum.

the observed spectrum as wavelength was decreased (Figure 13.2). As the Rayleigh–Jeans law was based on the physics of the day this was very much a puzzle to be solved.

The idealised absorber/emitter of radiation is the blackbody and has spectral intensity plots as shown in the previous figures. It was thought at the time of Max Planck, in 1900, that the radiation of electromagnetic energy was due to the oscillation of charges in the surface of the blackbody and that there was no restriction on the energies of the oscillations. The Rayleigh–Jeans law was built on this assumption. Max Planck, however, took the very bold step of saying "what if the energies are not continuous but rather restricted to certain values?". Using this idea he put together a new equation to predict radiation intensities as a function of wavelength and it was successful at predicting radiation intensities over a wide spectrum of frequencies. Using Planck's new equation it was possible to replicate exactly the observed data of Figure 13.2. Central to his derivation was the idea that the energy of the oscillators is proportional to their frequency:

$$E = h\nu$$

$$\text{or} \tag{13.1}$$

$$E = h\frac{\omega}{2\pi} = \hbar\omega \quad \left(\hbar = \frac{h}{2\pi}\right)$$

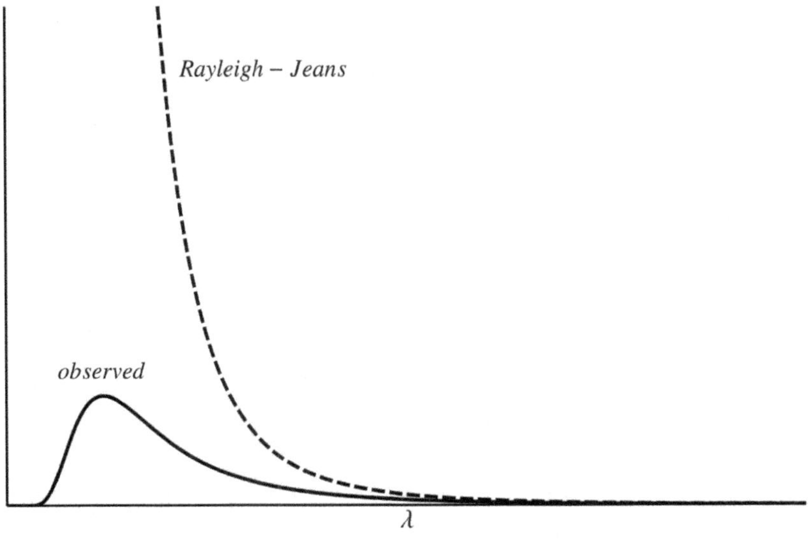

Figure 13.2 Rayleigh–Jeans law (dashed line) and experimental data (solid line).

where h is the constant of proportionality now known as Planck's constant. Planck wrote that he thought of this as merely a "trick" in order to derive a workable equation and did not put much further thought into it.

In 1905 Albert Einstein wrote a paper discussing another curious phenomenon observed earlier by Phillip Lenard in 1902, the photoelectric effect. When a metallic surface under vacuum is exposed to electromagnetic radiation above a certain threshold frequency, the light is absorbed and electrons are emitted. The salient observations were:

1. For a given metal and frequency of incident radiation, the rate at which photoelectrons are ejected is directly proportional to the intensity of the incident light.
2. For a given metal, there exists a certain minimum frequency of incident radiation below which no photoelectrons can be emitted. This frequency is called the *threshold* frequency.
3. Above the threshold frequency, the maximum kinetic energy of the emitted electron is *independent* of the intensity of the incident light but depends on the frequency of the incident light.

These were very curious observations with respect to the understanding of electromagnetic waves (light) at the time. Maxwell's equations indicate that the intensity of the electromagnetic wave is directly related to the energy of the electromagnetic wave. Thus, it was reasonable to expect that increasing the intensity of the light would eventually result in the removal of electrons from the metal regardless of the frequency. Observation 2 counters this idea. There is a minimum frequency required for ejection of electrons. If the frequency of light is below this threshold frequency it does not matter what the intensity is; no electrons will be emitted. Also flying in the face of conventional understanding was observation 3. Again, one would expect that the higher the intensity of the incident radiation the larger the kinetic energy of the emitted electrons would be. Not so. Instead, increasing the intensity resulted in an increase in the number of emitted electrons not their kinetic energy.

In order to explain this Einstein postulated that the incident light was not doing so in a wave fashion but rather as particles, which we now call photons. Furthermore, he took Planck's idea of the oscillators with energies proportional to their frequencies and applied it to these particles. Thus, each photon is deemed to have an energy of $E = h\nu$ and the intensity of the incident light on the metal is related to the number of photons, not the energy of the light. Each photon then interacts with the electrons in the metal and, if the photon has enough energy, possibly ejects an electron from the metal. If the frequency of the photon is below the threshold frequency this is equivalent to saying that the photon does not carry enough energy to eject the electron. The basic idea is that a photon with energy $h\nu$ interacts with an electron and transfers its energy to the electron. If the energy is sufficient the electron will be removed from the surface of the metal and have a kinetic energy that is directly related to the energy of the photon:

$$h\nu = E_k + W$$

$$E_k = h\nu - W$$

where W is the work function or the energy required to remove an electron from the metal atom in question. The particle idea combined with Planck's equation explains the photoelectric effect exceptionally well, whereas the old electromagnetic wave ideas of Maxwell do not.

Equation (13.1), commonly known as Planck's equation, leads us to a new way of thinking about things. We can, when we hear the word "frequency", we think "energy" and *vice versa*.

Further evidence for the particle nature of light came in 1923 from Arthur Compton. His observation was that when X-rays are beamed at electrons they interact in such a way as to impart momentum to the electrons and scatter the X-rays. Furthermore, the scattered X-rays are at longer wavelengths or lower frequencies. The wave picture of electromagnetic radiation cannot explain this but the particle picture of light does so very nicely. An X-ray photon has momentum, some of which it can impart to the electron. In doing so it loses energy and will, according to Planck's equation, decrease in frequency.

In 1908, Ernest Rutherford's lab staff, recent PhD Hans Gieger and undergraduate student Ernest Marsden performed an experiment in which alpha rays were directed at a very thin sheet of gold foil. The expected result based on Thomson's plum pudding model of the atom was that most of the alpha rays would pass directly through the foil with some minimal scattering a few degrees off the path of the beam. To their surprise a very small amount of alpha rays were deflected almost backwards towards the source. Since alpha rays are positively charged particles this means that they encountered a small region of concentrated positive charge in the gold atoms in the foil. In other words the nucleus! This led to the Rutherford model of the atom with negatively charged electrons orbiting a positively charged nucleus and is the visual picture still used in many situations today such as company logos.

Another perplexing problem at the time was that of the hydrogen emission spectrum. When subjecting an evacuated tube containing a bit of hydrogen to a high voltage light is emitted. When this light is passed through a spectrometer prism it is split into discrete lines. In the visible region of the light spectrum these lines are known as the Balmer series (Figure 13.3).

Balmer was able to empirically fit the positions of these lines into an equation:

$$\nu = \frac{1}{\lambda} = 109680\left(\frac{1}{2^2} - \frac{1}{n^2}\right)$$

There are of course other lines in the hydrogen spectrum in the ultraviolet and infrared regions. The Swiss spectroscopist Johannes Rydberg produced a general equation for all lines in the hydrogen spectrum:

$$\nu = \frac{1}{\lambda} = 109680\left(\frac{1}{n_f^2} - \frac{1}{n_i^2}\right)$$

This is a beautifully simple equation but it does not really tell us anything about *why* we can do this calculation.

Enter Neils Bohr. He had been working in the Rutherford laboratory at about the same time as the scattering experiment and was very familiar with the Rutherford model of the nuclear atom. He was

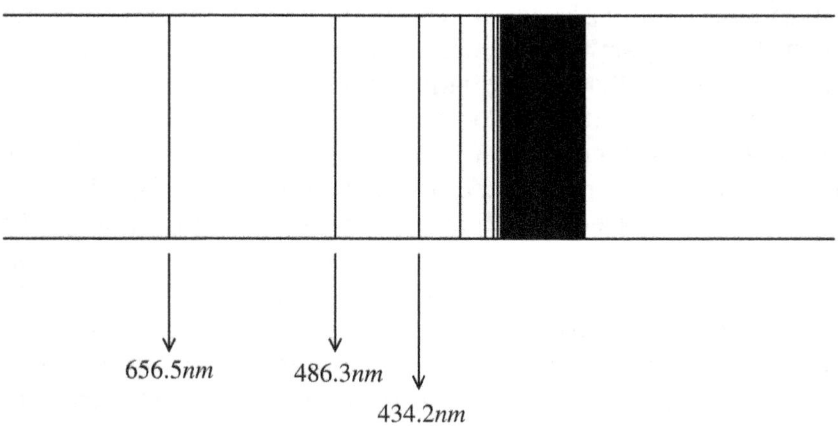

Figure 13.3 The Balmer series of the hydrogen emission spectrum.

puzzling over the hydrogen atom emission spectrum in 1913 and came up with the idea that the orbits of the electrons might be fixed. The essence of his idea is:

1. The electrons can only travel in certain orbits at a certain discrete set of distances from the nucleus with specific energies.
2. The electrons do not continuously radiate energy as they orbit the nucleus. Normally, an accelerating charged particle radiates electromagnetic energy. They can only gain and lose energy by switching from one allowed orbit to another, absorbing or emitting light with a frequency ν determined by the energy difference of the levels according to the Planck equation.

This was a very crude model of the atom and left many questions to be answered (Figure 13.4). However, it was successful in explaining and reproducing the Rydberg equation for the emission spectrum of hydrogen and perhaps more importantly pointed the way for those who followed Bohr.

Prince Louis de Broglie of the French nobility and a trained physicist had the very brilliant idea† in 1924 that if light could exhibit the characteristics of waves under some circumstances and particles

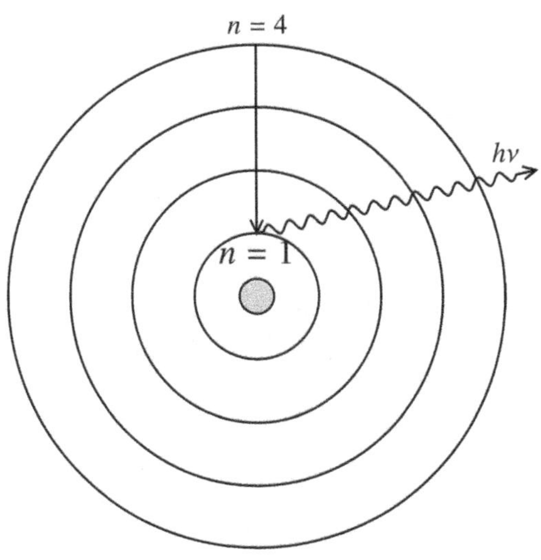

Figure 13.4 The Bohr model of the atom.

†A brilliantly simple (or simply brilliant) idea as most of the really good ones seem to be.

under others then might not the same be possible for particles such as electrons. In a dazzling synthesis of Planck's equation and Einstein's matter/energy equation from special relativity he derived what the wavelength of the particle must be. Starting with Einstein's (now) famous equation:

$$E = mc^2$$

$$\frac{E}{c} = mc$$

and Planck's equation:

$$E = h\nu = \frac{hc}{\lambda}$$

$$\frac{E}{c} = \frac{h}{\lambda}$$

and:

$$\frac{h}{\lambda} = mc$$

$$\lambda = \frac{h}{mc}$$

This equation applies to light but might it not carry over into the world of matter? If so then we simply write:

$$\lambda = \frac{h}{mv} \tag{13.2}$$

This was, of course, a very big leap and was not shown to be true conclusively for some years (by George Thomson[‡] in Britain and Davidson and Germer in the USA). Nonetheless, it paved the way for further developments, specifically, Erwin Schrodinger's quantum mechanics.

In early 1926, while thinking about the Bohr atomic model and transitions of electrons between states, Werner Heisenberg came up with a method of doing quantum calculations based on matrix methods. One of the most unusual things to come out of this was the

[‡]This is one of the classic ironies in physics. J.J. Thomson discovered the first subatomic particle and his son, George, showed that it also behaves as a wave.

fact that the multiplication of matrices is generally non-commutative (as we saw in Chapter 2). Thus, with $\hat{\mathbf{p}}$ being the matrix representation of momentum and $\hat{\mathbf{q}}$ the matrix form of position coordinates:

$$\hat{\mathbf{p}}\hat{\mathbf{q}} - \hat{\mathbf{q}}\hat{\mathbf{p}} \neq \hat{\mathbf{0}}$$

and was to become the genesis of the Heisenberg uncertainty principle.

Also in 1926, Erwin Schrodinger produced a method of doing quantum calculations based on differential calculus founded on the de Broglie idea of constrained "matter waves". His quantum mechanics is the more intuitive of the two main methods (if that can be said of either), the other of course being Heisenberg's matrix mechanics. Erwin Schrodinger viewed his functions (see below) as referring to "matter waves", however, an alternate interpretation suggested by Max Born was that they represented the *probability* of finding the electron in a region of space. Born's view has prevailed to this day and is part of the Copenhagen interpretation of quantum mechanics.

The development of quantum mechanics took place during the first thirty years of the 20th century with the findings of those up to Heisenberg's time referred to as "old quantum mechanics" and those of Heisenberg and afterwards "new quantum mechanics". In the old quantum mechanics physicists were trying to hold on to the classical pictures of microscopic matter developed over the previous centuries. The orbiting electron of Rutherford and Bohr is an example. With Heisenberg and those afterwards these pictures were abandoned in favour of a more abstract vision founded in the mathematical description of microscopic events. There was, of course, much discussion and argument about what the new equations mean. Einstein famously quipped "God does not play dice" in connection with the probabilistic interpretation of the wave equation. The birth of modern quantum mechanics can be dated from the 5th Solvay conference in Brussels in 1927 where most of the major figures in European physics gathered to hammer out the details of the new physics[§] (see Figure 13.5).

We have not discussed the property of matter called spin", which was discovered in conjunction with the development of quantum mechanics. This we will leave until a later chapter as it is to us as NMR/MRI spectroscopists, a rather special property.

[§]Oh, to be a fly on the wall at that meeting! The meanings of the equations are *still* being argued about.

Figure 13.5 The participants at the 1927 Solvay conference.

13.2 The Basics of Quantum Mechanics

This was always a very confusing topic in school ... and probably rightly so. For the beginner, there is so much in the way of unfamiliar mathematics that is difficult to relate to "reality" that it sometimes seems hopeless to even begin to comprehend the topic. You will *not* end up understanding quantum mechanics but you will be in good company such as the physicist Richard Feynman who said "nobody understands it". Nonetheless, to the reader we say "stay the course" and keep reading. Hopefully we shall be able to answer some or even all of your questions.

We usually begin talking about modern quantum mechanics by referring to the stationary Schrodinger equation:

$$\hat{\mathbf{H}}\Psi = E\Psi \tag{13.3}$$

which is not derived but rather postulated (see below). It is "stationary" in the same sense as the "stationary" part of the classical wave equation of Chapter 7. $\hat{\mathbf{H}}$ is the Hamiltonian operator, E is the total energy (a scalar quantity) and Ψ is the "wave function" that is postulated to contain all of the knowable information about the system of interest. It is therefore, as you can see, an eigenvalue equation. We cannot derive it but we can rationalise it by thinking about the three-dimensional wave equation starting with a three-dimensional version of a stationary wave equation (7.9):¶

$$\nabla^2\Psi = -\alpha^2\Psi$$

where:

$$\nabla^2 = \frac{\partial^2}{\partial x^2} + \frac{\partial^2}{\partial y^2} + \frac{\partial^2}{\partial z^2}$$

and

$$\alpha = \frac{2\pi}{\lambda}$$

Now, using de Broglie's postulate concerning the wavelength of a particle:

$$\lambda = \frac{h}{p} = \frac{h}{mv}$$

$$\alpha = \frac{2\pi p}{h} = \frac{2\pi mv}{h}$$

our wave equation now becomes:

$$\nabla^2\Psi = \frac{-4\pi^2 m^2 v^2}{h^2}\Psi$$

Using the Hamiltonian expression from classical mechanics (see Chapter 9):

$$E = T + U = \frac{p^2}{2m} + U = \frac{1}{2}mv^2 + U$$

we rearrange to get:

$$mv^2 = 2(E - U)$$

¶Adapted from J. Baggot's exposition. See the Further Reading section.

and then insert into our current wave equation:

$$\nabla^2\Psi = \frac{-8\pi^2 m}{h^2}(E - U)\Psi \Rightarrow \left(\frac{-h^2}{8\pi^2 m}\nabla^2 + U\right)\Psi = E\Psi$$

or

$$\hat{\mathbf{H}}\Psi = E\Psi$$

where

$$\hat{\mathbf{H}} \equiv \frac{-h^2}{8\pi^2 m}\nabla^2 + U$$

to get the Schrodinger wave equation. $\hat{\mathbf{H}}$ is an operator that operates on Ψ. One will frequently see a modification of this:

$$\left(\frac{-\hbar^2}{2m}\nabla^2 + U\right)\Psi = E\Psi$$
$$\hbar \equiv \frac{h}{2\pi} \quad \text{(reduced Planck constant)} \tag{13.4}$$

This version is helpful for seeing that the first term is a kinetic energy term:

$$\frac{-\hbar^2}{2m}\nabla^2 = \frac{-\hbar^2}{2m}\left(\frac{\partial^2}{\partial x^2} + \frac{\partial^2}{\partial y^2} + \frac{\partial^2}{\partial z^2}\right)$$

where:

$$\frac{-\hbar^2}{2m}\frac{\partial^2}{\partial x^2} \Rightarrow \frac{p_x^2}{2m} \tag{13.5}$$

corresponds to the kinetic energy in the x-coordinate. Accordingly, when we construct a quantum mechanics Hamiltonian for a particular problem we generally replace the classical p_x with $-i\hbar\partial/\partial x$ (see below). Why the i, you ask? To generate the minus sign in eqn (13.5). We will use this expression for the momentum operator when we (finally) begin to talk about NMR spectroscopy.

13.3 The Postulates of Quantum Mechanics

We proceed now with the postulates of quantum mechanics upon which everything is based. Postulates are either "self-evident" in the same sense that it is self-evident that heat flows from hot objects to cold objects, or they are taken on trust and must survive experimental test if they are to be kept. So we can only state them here, not prove them since they are "unprovable" except by repeated experimental verification. We can, however, try to show why they were postulated in the first place.

The usual introductory exposition of the postulates uses the wave functions of Erwin Schrodinger from above, however, for the purposes of this document, it is useful to cast them in the more general form of Dirac. That is, we shall consider our wave functions as state vectors (which they are by definition, see Chapter 3) and use the algebra of vectors and Dirac's notation to develop our equations. This has the advantage that, even though the concept of the wave function and working with it is somewhat foreign to the uninitiated, working with vectors is usually not. One can always refer to the familiar concepts of Euclidean vector algebra and apply them to state vectors; the fundamental algebra is identical. Thus, we can rewrite the Schrodinger eqn (13.3) in its Dirac form:

$$\hat{\mathbf{H}}|\Psi\rangle = E|\Psi\rangle$$

13.3.1 Postulate I

The state of a system at a given time is fully described by a state vector, $|\Psi\rangle$, in a Hilbert space. $|\Psi\rangle$ is the "ket" form of the vector. The corresponding "bra" form is $\langle\Psi|$ and in combination with the ket is used to form the inner product:

$$\langle\Psi|\Psi\rangle$$

which evaluates to a scalar. Also, since this is a Hilbert space (see Chapter 3) we have that:

$$\langle\Psi|\Psi\rangle = (\langle\Psi|\Psi\rangle)^*$$

(see eqn (3.46)) and the scalar is therefore real.

Since we are dealing with a vector we must be able to express it as a linear combination of any other vectors in the Hilbert space.

In particular, we can represent the vector as a combination of orthonormal basis vectors:

$$|\Psi\rangle = \sum_i c_i|\phi_i\rangle = \sum_i |\phi_i\rangle\langle\phi_i|\Psi\rangle$$

where (13.6)

$$\langle\phi_i|\Psi\rangle = c_i$$

much as in eqn (3.4). This is the well-known *superposition principle* of quantum mechanics and since (obviously) $|\Psi\rangle = |\Psi\rangle$ we can say that:

$$\sum_i |\phi_i\rangle\langle\phi_i| = \hat{\mathbf{1}} \tag{13.7}$$

which is known as the *closure relation*. This is very helpful in developing further equations.

13.3.2 Postulate II

To every physical observable there corresponds a linear, Hermitian operator, $\hat{\mathbf{A}}$, acting on $|\Psi\rangle$. To write the operator explicitly we first write the classical version of the observable in Cartesian coordinates and linear momentum terms, if applicable. This means that to go from a classical expression to quantum mechanical operator we do (as in eqn (13.5)):

$$x \Rightarrow x$$

and (13.8)

$$\frac{p_x}{2m} \Rightarrow -i\hbar\frac{\partial}{\partial x}$$

In our discussion of vectors in Chapter 3 we saw that operators can be represented by matrices. This is also so here and we will develop this extensively in our explorations of NMR spectroscopy.

13.3.3 Postulate III

The only possible values that can result from measurements of a physical observable A are the eigenvalues a_i of the equation:

$$\hat{\mathbf{A}}|\phi_i\rangle = a_i|\phi_i\rangle$$

where $\hat{\mathbf{A}}$ is the operator corresponding to the observable and $|\phi_i\rangle$ is the state eigenvector. The operator $\hat{\mathbf{A}}$ is, from postulate II, a Hermitian operator, which means that the eigenvalues, a_i, are real, as we saw must be the case for these types of mathematical objects in Chapter 3.

13.3.4 Postulate IV

If $|\Psi\rangle$ is the normalised state vector of a system, then the average value (or expectation value) of a physical observable, A is:

$$\langle A\rangle = \langle\Psi|\hat{\mathbf{A}}|\Psi\rangle$$

We can rationalise this using the Born postulate: the square of the modulus of $|\Psi\rangle$, $|\Psi\rangle^2$, is the probability (probability density actually) of observing a particle in a small volume of a Hilbert space.

This originally was proposed by Max Born to explain the diffraction of electrons *via* Schrodinger's wave-function formulation of quantum mechanics. Of course, the probability of finding the particle anywhere from $-\infty$ to $+\infty$ must be 1 so the normalised wave function is written:

$$\int \||\Psi\rangle\|^2 \mathrm{d}\tau = \langle\Psi|\Psi\rangle = 1$$

The Born postulate is useful for explaining where the average value postulate comes from. We use $\||\Psi\rangle\|^2$ in order to guarantee a positive number, which of course we must have if it is to be interpreted as a probability.

We expand $|\Psi\rangle$ in its (orthonormal) basis set:

$$|\Psi\rangle = \sum_i c_i|\phi_i\rangle$$

$$\langle\Psi|\Psi\rangle = 1 = \sum_i c_i^*\langle\phi_i|\sum_j c_j|\phi_j\rangle$$

$$= \sum_i\sum_j c_i^* c_j\langle\phi_i|\phi_j\rangle$$

$$= \sum_i\sum_j c_i^* c_j\delta_{ij}$$

$$= \sum_i |c_i|^2$$

where δ_{ij} is the Kronecker delta from Chapter 3. We see that the sum of the squares of the moduli of the basis vector coefficients is one. Similarly:

$$
\begin{aligned}
\langle A\rangle &= \langle \hat{\mathbf{A}} \mid \Psi\rangle\rangle \\
&= \sum_i c_i^* \langle \phi_i | \hat{\mathbf{A}} \sum_j c_j |\phi_j\rangle \\
&= \sum_i c_i^* \langle \phi_i | \sum_j a_j c_j |\phi_j\rangle \\
&= \sum_i \sum_j c_i^* a_j c_j \langle \phi_i \mid \phi_j\rangle \\
&= \sum_i \sum_j c_i^* a_j c_j \delta_{ij} \\
&= \sum_i |c_i|^2 a_i
\end{aligned}
$$

This is very reminiscent of the weighted average calculation of Chapter 6, eqn (6.3). From simple statistics the weighted average, $\langle A\rangle$, over a number of measurements is:

$$\langle A\rangle = \sum_i P_i a_i$$

where P_i is the probability of obtaining measurement a_i. Also, recall that:

$$\sum_i P_i = 1$$

Evidently, the square of the moduli of the basis coefficients correspond to probabilities.

13.3.5 Postulate V

The time dependence of the state vector is expressed in the time-dependent Schrodinger equation:

$$\hat{\mathbf{H}} \mid \Psi\rangle(t) = i\hbar \frac{\partial}{\partial t} \mid \Psi\rangle(t)$$

or

$$\frac{\partial}{\partial t} \mid \Psi\rangle(t) = \frac{-i}{\hbar} \hat{\mathbf{H}} \mid \Psi\rangle(t)$$

We will use this later in our discussion of the time evolution of a set of spins.

The time-dependent Schrodinger equation can be easily solved (see Appendix A) since it is a simple first-order differential equation:

$$\frac{\partial}{\partial t}|\Psi\rangle(t) = \frac{-i}{\hbar}\hat{\mathbf{H}}|\Psi\rangle(t)$$

$$\Downarrow$$

$$|\Psi\rangle(t) = \exp\left(-\frac{i}{\hbar}\hat{\mathbf{H}}\right)|\Psi\rangle(t)$$

and, using eqn (3.62) and the superposition of basis functions:

$$|\Psi\rangle(t) = \sum \exp\left(-\frac{i}{\hbar}E_n t\right)|\phi_n\rangle \tag{13.9}$$

13.4 Vector Notation

Recalling the vector information from above, we note that both our wave functions and vectors can be referred to as being orthogonal (or can be made so). This leads to the idea that we can think of and represent our wave functions as vectors *via* Dirac notation. Indeed, much of the literature refers to wave functions as vectors. So, for a superposition of basis states:

$$|\Psi\rangle = \sum c_i|\phi_i\rangle \tag{13.10}$$

we recall from Chapter 3 that we can represent this ket as a column matrix vector:

$$|\Psi\rangle = \begin{bmatrix} c_1 \\ c_2 \\ c_3 \\ \vdots \end{bmatrix}$$

and the corresponding bra is represented as a row matrix:

$$\langle\Psi| = [c_1^* \quad c_2^* \quad c_3^* \quad \ldots]$$

Note that now, since we are using complex numbers, we use the complex conjugate in the transpose, as was done in Chapter 3. We can calculate our inner product by matrix multiplication:

$$\langle\Psi|\Psi\rangle = [c_1^* \quad c_2^* \quad c_3^* \quad \ldots] \begin{bmatrix} c_1 \\ c_2 \\ c_3 \\ \vdots \end{bmatrix} = \sum_i c_i^* c_i$$

From Chapter 3, we can represent our operator with a matrix:

$$\hat{\mathbf{A}} = \begin{bmatrix} A_{11} & A_{12} & \cdots \\ A_{21} & A_{22} & \cdots \\ \vdots & \vdots & \ddots \end{bmatrix} \tag{13.11}$$

Projection operators (also discussed in Chapter 3) are constructed from the outer products of members of the basis vector set:

$$\mathbf{P}_i = |\phi_i\rangle\langle\phi_i| \tag{13.12}$$

We will defer the discussion of the time evolution of the states of a system until we have developed the specific form of our NMR Hamiltonian operator in the next chapter.

13.5 The Uncertainty Principle

When we are dealing with a statistical sample of things we generally calculate an average or expectation value and perhaps a standard deviation or something similar. From Chapter 6 we have:

$$\begin{aligned} \langle x\rangle &= \sum_{k=1}^{t} x_k p_k \\ \langle x\rangle^2 &= \sum_{k=1}^{t} x_k^2 p_k \\ \sigma = \Delta x &= \sqrt{x - \langle x\rangle} \end{aligned} \tag{13.13}$$

where p_k is the probability of measurement x_k being observed.

Let us consider a similar calculation for a pair of non-commuting Hermitian operators, $\hat{\mathbf{A}}$ and $\hat{\mathbf{B}}$. First, we define the standard deviation for each operator in a fashion analogous to that of eqn (6.4):

$$\Delta\hat{\mathbf{A}} = \hat{\mathbf{A}} - \langle A\rangle$$

$$\Delta\hat{\mathbf{B}} = \hat{\mathbf{B}} - \langle B\rangle$$

Since $\hat{\mathbf{A}}$ is Hermitian and $\langle A\rangle$ is a real scalar, we can say that $\Delta\hat{\mathbf{A}}$ is also Hermitian. Assume that the action of these operators on the state vector ket, $|\Psi\rangle$, is:

$$\begin{aligned}\Delta\hat{\mathrm{A}}\,|\Psi\rangle &= |\Phi\rangle\\ \Delta\hat{\mathbf{B}}\,|\Psi\rangle &= |\Theta\rangle\end{aligned} \qquad (13.14)$$

Similarly, the bra operations will be:

$$\begin{aligned}\langle\Psi|\Delta\hat{\mathbf{A}} &= \langle\Psi\Delta\hat{\mathbf{A}}^{T}| = \langle\Phi|\\ \langle\Psi|\Delta\hat{\mathbf{B}} &= \langle\Psi\varDelta\hat{\mathbf{B}}^{T}| = \langle\Theta|\end{aligned} \qquad (13.15)$$

since $\Delta\vec{\mathbf{A}}$ and $\Delta\hat{\mathbf{B}}$ are both Hermitian. With reference to Chapter 3, we invoke the Cauchy–Schwartz inequality (eqn (3.10)):

$$\begin{gathered}\langle\Psi|\Delta\hat{\mathbf{A}}^{2}|\Psi\rangle\langle\Psi|\Delta\hat{\mathbf{B}}^{2}|\Psi\rangle \geq |\langle\Psi|\Delta\hat{\mathbf{A}}\Delta\hat{\mathbf{B}}|\Psi\rangle|^{2}\\ \text{or}\\ |\langle\Phi|\Phi\rangle|\,|\langle\Theta|\Theta\rangle| \geq |\langle\Phi|\Theta\rangle|^{2}\end{gathered} \qquad (13.16)$$

This can be seen to be so when we consult our bra and ket operations using $\Delta\vec{\mathbf{A}}$ and $\Delta\hat{\mathbf{B}}$ from eqn (13.14) and (13.15):

$$\langle(\Delta A)^{2}\rangle = \langle\Psi|(\Delta\hat{\mathbf{A}})^{2}|\Psi\rangle = \langle\Psi|\Delta\hat{\mathbf{A}}\Delta\hat{\mathbf{A}}|\Psi\rangle = \langle\Phi|\Phi\rangle$$

$$\langle(\Delta B)^{2}\rangle = \langle\Psi|(\Delta\hat{\mathbf{B}})^{2}|\Psi\rangle = \langle\Psi|\Delta\hat{\mathbf{B}}\Delta\hat{\mathbf{B}}|\Psi\rangle = \langle\Theta|\Theta\rangle$$

Now, since the left side of eqn (13.16) is real and positive (see eqn (3.50)):

$$\langle\Phi|\Phi\rangle \geq 0$$

and

$$\langle\Theta|\Theta\rangle \geq 0$$

the right side must *also* be real and positive:

$$|\langle\Phi|\Theta\rangle|^2 \geq 0$$

We cannot say the same about $\langle\Phi|\Theta\rangle$ alone ... it will generally be a complex number. However, since $\langle\Phi|\Theta\rangle$ and $\langle\Theta|\Phi\rangle$ are conjugates of each other, we can write:

$$|\langle\Phi|\Theta\rangle|^2 = \langle\Phi|\Theta\rangle\langle\Theta|\Phi\rangle$$

which guarantees that the product of the two inner products, $\langle\Phi|\Theta\rangle$ and $\langle\Theta|\Phi\rangle$, is real and we can then extend the inequality a bit:

$$\langle(\Delta A)^2\rangle\langle(\Delta B)^2\rangle = |\langle\Phi|\Phi\rangle|\,|\langle\Theta|\Theta\rangle| \geq |\langle\Phi|\Theta\rangle|^2 = \langle\Phi|\Theta\rangle\langle\Theta|\Phi\rangle$$

or

$$\langle(\Delta A)^2\rangle\langle(\Delta B)^2\rangle \geq \langle\Phi|\Theta\rangle\langle\Theta|\Phi\rangle$$

Now from our definitions of Φ and Θ we write:

$$\begin{aligned}
\langle\Phi|\Theta\rangle &= \langle\Psi|\Delta\hat{\mathbf{A}}\Delta\hat{\mathbf{B}}|\Psi\rangle = \langle\Delta A\Delta B\rangle \\
&= \langle\Psi|(\hat{\mathbf{A}} - \langle A\rangle)(\hat{\mathbf{B}} - \langle B\rangle)|\Psi\rangle \\
&= \langle\Psi|\hat{\mathbf{A}}\hat{\mathbf{B}} - \langle A\rangle\hat{\mathbf{B}} - \hat{\mathbf{A}}\langle B\rangle + \langle A\rangle\langle B\rangle|\Psi\rangle \\
&= \langle\Psi|\hat{\mathbf{A}}\hat{\mathbf{B}}|\Psi\rangle - \langle\Psi|\langle A\rangle\hat{\mathbf{B}}|\Psi\rangle - \langle\Psi|\hat{\mathbf{A}}\langle B\rangle|\Psi\rangle + \langle\Psi|\langle A\rangle\langle B\rangle|\Psi\rangle \\
&= \langle AB\rangle - \langle A\rangle\langle B\rangle - \langle A\rangle\langle B\rangle + \langle A\rangle\langle B\rangle \\
&= \langle AB\rangle - \langle A\rangle\langle B\rangle
\end{aligned}$$

and similarly:

$$\langle\Theta|\Phi\rangle = \langle\Delta B\Delta A\rangle = \langle BA\rangle - \langle A\rangle\langle B\rangle$$

When looking at all of these symbols we can sometimes be overwhelmed and forget that something such as $\langle\Phi|\Theta\rangle$ is simply a number, in our case a complex number, and must follow all of the algebraic rules for complex numbers. For clarity, we will temporarily indicate this by substituting z for $\langle\Phi|\Theta\rangle$ and z^* for $\langle\Theta|\Phi\rangle$ where the asterisk means, of course, the complex conjugate. Thus, we can write:

$$zz^* = |z|^2$$

using eqn (1.6). From eqn (1.4) we have:

$$z - z^{\star} = 2i\,\mathrm{Im}\,z$$

or

$$\mathrm{Im}\,z = \frac{z - z^{\star}}{2}i$$

and we can rewrite our inequality:

$$\langle(\Delta A)^2\rangle\langle(\Delta B)^2\rangle \geq \mathrm{Re}\,z^2 + \mathrm{Im}\,z^2$$

This being the case, since the left-hand part of the inequality is greater than or equal to the right-hand part, we can write that the left-hand side is greater than or equal to the real or imaginary parts *individually* and the inequality will still be valid. We choose the imaginary part because we wish to develop an expression with a commutator in it. Choosing the real part would give us an anti-commutator.

$$\langle(\Delta A)^2\rangle\langle(\Delta B)^2\rangle \geq \mathrm{Im}\,z^2$$

$$\langle(\Delta A)^2\rangle\langle(\Delta B)^2\rangle \geq \left(\frac{z - z^{*}}{2i}\right)^2$$

$$\langle(\Delta A)^2\rangle\langle(\Delta B)^2\rangle \geq \left(\frac{\langle\Phi\,|\,\Theta\rangle - \langle\Theta\,|\,\Phi\rangle}{2i}\right)^2$$

Now, using our calculations from above we write:

$$\langle(\Delta A)^2\rangle\langle(\Delta B)^2\rangle \geq \left(\frac{\langle AB\rangle - \langle A\rangle\langle B\rangle - \langle BA\rangle + \langle A\rangle\langle B\rangle}{2i}\right)^2$$

$$\langle(\Delta A)^2\rangle\langle(\Delta B)^2\rangle \geq \left(\frac{\langle AB\rangle - \langle BA\rangle}{2i}\right)^2$$

$$\langle(\Delta A)^2\rangle\langle(\Delta B)^2\rangle \geq \left(\frac{\langle AB - BA\rangle}{2i}\right)^2 \tag{13.17}$$

$$\langle(\Delta A)^2\rangle\langle(\Delta B)^2\rangle \geq \left(\frac{[A, B]}{2i}\right)^2$$

or

$$\langle\Delta A\rangle\langle\Delta B\rangle \geq \frac{[A, B]}{2i}$$

This final result (after much work!) is the generalised version of Heisenberg's famous uncertainty principle. For the operators $\hat{\mathbf{x}}$ and $\hat{\mathbf{p}}$ the commutator operation on $|\Psi\rangle$ is:

$$[\hat{\mathbf{x}},\hat{\mathbf{p}}]\,|\Psi\rangle=(\hat{\mathbf{x}}\hat{p}-\hat{\mathbf{p}}\hat{x})|\Psi\rangle=(x)\left(i\hbar\frac{\partial}{\partial x}\right)|\Psi\rangle-\left(i\hbar\frac{\partial}{\partial x}\right)(x)|\Psi\rangle$$

$$=i\hbar\left(x\frac{\partial|\Psi\rangle}{\partial x}-\frac{\partial}{\partial x}(x)\,|\Psi\rangle\right)$$

$$=i\hbar\left(x\frac{\partial|\Psi\rangle}{\partial x}-|\Psi\rangle-x\frac{\partial}{\partial x}\,|\Psi\rangle\right)\quad\text{(using the product rule)}$$

$$=i\hbar\,|\Psi\rangle$$

and:

$$[\hat{\mathbf{x}},\hat{\mathbf{p}}]\,|\Psi\rangle=i\hbar\,|\Psi\rangle$$

From eqn (13.17):

$$\Delta x\Delta p\geq\frac{[\hat{\mathbf{x}},\hat{\mathbf{p}}]}{2i}$$

$$\text{or}\tag{13.18}$$

$$\Delta x\Delta p\geq\frac{\hbar}{2}$$

which is the common expression of the Heisenberg uncertainty principle with respect to position and momentum.

The commutator is useful for telling us whether or not two operators have common eigenvectors (see eqn (3.58)). This being the case the commutator of the operators will be zero and from eqn (13.17) there can potentially be no uncertainty in measurements corresponding to these operators:

$$\langle\Delta A\rangle\langle\Delta B\rangle\geq 0$$

However, if the commutator is non-zero, as in the case of operators without simultaneous eigenvectors, then eqn (13.17) applies.

Most commonly, we are introduced to the uncertainty principle *via* eqn (13.18), but other operator uncertainty relationships can be written as well. The commutation relationship is the key. However, there is one uncertainty relationship that does not involve operators, at least one term is not an operator. This is the energy–time uncertainty relationship:

$$\Delta E \Delta t \geq \frac{\hbar}{2}$$

Time is not an operator in the quantum mechanics formalism but rather a parameter (at least, in non-relativistic quantum mechanics). Therefore we cannot think of a commutation relationship between energy and time. Nonetheless, the relationship can be rationalised and is of use to us in looking at free-induction decay times and linewidths. We start by assuming an observable, $\hat{\mathbf{O}}$, and computing its expectation value's time-derivative:

$$\begin{aligned}\frac{\mathrm{d}\langle O\rangle}{\mathrm{d}t} &= \frac{\mathrm{d}\langle\Psi|\hat{\mathbf{O}}|\Psi\rangle}{\mathrm{d}t} \\ &= \left\langle\frac{\mathrm{d}\Psi}{\mathrm{d}t}\middle|\hat{\mathbf{O}}\,|\Psi\rangle + \langle\Psi|\,\frac{\mathrm{d}\hat{\mathbf{O}}}{\mathrm{d}t}\,|\Psi\rangle + \langle\Psi|\,\hat{\mathbf{O}}\middle|\frac{\mathrm{d}\Psi}{\mathrm{d}t}\right\rangle\end{aligned}$$

using the product rule. Recalling postulate V:

$$\frac{\partial}{\partial t}\,|\Psi\rangle = -\frac{i}{\hbar}\hat{\mathbf{H}}\,|\Psi\rangle$$

$$\frac{\mathrm{d}\langle O\rangle}{\mathrm{d}t} = \frac{i}{\hbar}\langle\hat{\mathbf{H}}\Psi\,|\,\hat{\mathbf{O}}|\,\Psi\rangle + \left\langle\frac{\mathrm{d}\hat{\mathbf{O}}}{\mathrm{d}t}\right\rangle - \frac{i}{\hbar}\langle\Psi\,|\hat{\mathbf{O}}|\,\hat{\mathbf{H}}\Psi\rangle$$

and since $\hat{\mathbf{H}}$ is Hermitian (see eqn (2.57) and (3.49)):

$$\begin{aligned}\frac{\mathrm{d}\langle O\rangle}{\mathrm{d}t} &= \frac{i}{\hbar}\langle\Psi\,|\,\hat{\mathbf{H}}\hat{\mathbf{O}}\,|\,\Psi\rangle + \left\langle\frac{\mathrm{d}\hat{\mathbf{O}}}{\mathrm{d}t}\right\rangle - \frac{i}{\hbar}\langle\Psi\,|\,\hat{\mathbf{O}}\hat{\mathbf{H}}\,|\,\Psi\rangle \\ &= \frac{i}{\hbar}\langle[\hat{\mathbf{H}},\hat{\mathbf{O}}]\rangle + \left\langle\frac{\mathrm{d}\hat{\mathbf{O}}}{\mathrm{d}t}\right\rangle\end{aligned}$$

Assuming there to be no time-dependence of the operator $\hat{\mathbf{O}}$:

$$\frac{\mathrm{d}\langle O\rangle}{\mathrm{d}t} = \frac{i}{\hbar}\langle[\hat{\mathbf{H}},\hat{\mathbf{O}}]\rangle$$

$$\langle[\hat{\mathbf{H}},\hat{\mathbf{O}}]\rangle = \frac{\hbar}{i}\frac{\mathrm{d}\langle O\rangle}{\mathrm{d}t}$$

Now, using eqn (13.17) and substituting H for A and O for B:

$$\langle(\Delta H)^2\rangle\langle(\Delta O)^2\rangle \geq \left(\frac{\langle[H,O]\rangle}{2i}\right)^2 = \left(\frac{1}{2i}\frac{\hbar}{i}\right)^2\left(\frac{\mathrm{d}\langle O\rangle}{\mathrm{d}t}\right)^2$$

$$= \left(\frac{\hbar}{2}\right)^2\left(\frac{\mathrm{d}\langle O\rangle}{\mathrm{d}t}\right)^2$$

or

$$\langle\Delta H\rangle\langle\Delta O\rangle \geq \left|\frac{-\hbar}{2}\right|\left|\frac{\mathrm{d}\langle O\rangle}{\mathrm{d}t}\right| = \frac{-\hbar}{2}\left|\frac{\mathrm{d}\langle O\rangle}{\mathrm{d}t}\right|$$

We end our development of the energy–time uncertainty relation by making a couple of definitions:

$$\Delta E \equiv \Delta H$$

$$\Delta t \equiv \frac{\Delta O}{\mathrm{d}\langle O\rangle/\mathrm{d}t}$$

and our equation now reads:

$$\langle\Delta E\rangle\langle\Delta t\rangle \geq \frac{\hbar}{2} \tag{13.19}$$

This is the form of the energy–time uncertainty principle that we introduced above. It is, however, as noted, not really the same as the "standard" uncertainty principle because of the time term. The interpretation to be applied to it is that Δt is the amount of time for observable O to change significantly.

How this affects us as NMR/MRI spectroscopists is in how the time-domain free-induction decay (FID) and the frequency-domain spectrum are related; in particular, how the decay of the FID and the linewidths are related. Anyone with a bit of experience in NMR spectroscopy will have noted that a rapidly decaying FID results in wider linewidths. We can rationalise this using the energy–time uncertainty relationship; a rapidly decaying FID corresponds to a small Δt that would mean that ΔE would be larger. Frequency and energy are related by Planck's equation, so an uncertainty in energy translates to an uncertainty or spread in frequencies, which is just what is observed; larger ΔE, larger $\Delta\nu$.

Further Reading

1. J. Baggot, *The Meaning of Quantum Theory*, Oxford University Press, London, 1992.
2. P. A. M. Dirac, *Quantum Mechanics*, Oxford University Press, London, Third edn., 1947.
3. D. J. Griffiths, *Introduction to Electrodynamics*, Pearson Prentice Hall, Upper Saddle River, New Jersey, 1999.
4. R. I. G. Hughes, *The Structure and Interpretation of Quantum Mechanics*, Harvard University Press, Cambridge, Mass., 1989.
5. S. Jones, *The Quantum Ten*, Thomas Allen Publishers, New York, 2008.
6. I. N. Levine, *Quantum Chemistry*, Allyn and Bacon Inc., Boston, Second edn., 1974.
7. D. A. McQuarrie, *Quantum Chemistry*, University Science Books, Sausalito, 2008.
8. R. Shankar, *Principles of Quantum Mechanics*, Kluwer Academic Publishers, New York, 1994.
9. E. G. Steward and S. M. McMurray, *Quantum Mechanics: Its Early Development and the Road to Entanglement*, Imperial College Press, London, 2008.

14 The Hamiltonians of NMR

Newton advanced, with one gigantic stride, from the regions of twilight into the noon day of science.

Thomas Young

14.1 The NMR Hamiltonian

As we have seen in Chapter 9, the Hamiltonian of classical mechanics is the energy of a conservative system and is the sum of the kinetic and potential energies of the particle in question. Thus, a system may lose kinetic energy as it gains potential energy and *vice versa* but the total remains constant:

With respect to the topic of this book, nuclear magnetic resonance spectroscopy, there is only a kinetic energy term due to "spin" and no potential energy term to worry about which simplifies things appreciably.

We saw in the last chapter that the quantum mechanical Hamiltonian can be easily rationalised using the standard wave equation and the de Broglie postulate concerning the wavelike properties of matter. In our case, since there is no potential energy to worry about, the Hamiltonian of Chapter 13 can be simplified:

$$-\frac{\hbar^2}{2m}\nabla^2\Psi = E\Psi$$

It should be obvious from this expression that the Hamiltonian is a scalar quantity and thus, any Hamiltonians that we write must evaluate to a scalar.

Essential Mathematics for NMR and MRI Spectroscopists
By Keith C. Brown

Published by the Royal Society of Chemistry, www.rsc.org

Our NMR Hamiltonians can be expressed in a very general form:†

$$H = C \cdot \vec{\mathbf{I}}\tilde{\mathbf{A}}\vec{\mathbf{K}} \tag{14.1}$$

where C is a constant, $\vec{\mathbf{I}}$ is a spin angular momentum vector operator, $\tilde{\mathbf{A}}$ is a rank-2 Cartesian tensor that is related to a molecular property and may be direction dependent. $\vec{\mathbf{K}}$ is possibly another angular momentum vector operator similar to $\vec{\mathbf{I}}$ or may be the static magnetic field vector. Our discussion of tensor algebra leads us to conclude that the result of the dot product operations produces a scalar, as it should. Each of the following Hamiltonians can be cast in this form.

14.2 The Zeeman Field Hamiltonian

The Zeeman field Hamiltonian is the simplest of our set of energy operators and leads to a picture of the basic energy levels in the spin system. This operator is built from the fundamental classical energy equation developed in Chapter 11, using eqn (11.43) and (11.46):

$$\begin{aligned} E &= -\vec{\mu} \cdot \vec{\mathbf{B}} \\ &= -\gamma \vec{\mathbf{L}} \cdot \vec{\mathbf{B}} \end{aligned} \tag{14.2}$$

The classical equation as we have written it uses L as the angular momentum variable where generally in NMR, I is used as the angular momentum variable. We will make this switch here. The z-axis is normally chosen to be parallel to the static field axis so that there are no components of the static field along the x or y axes. This being the case, the energy simplifies:

$$\vec{\mathbf{I}} \cdot \vec{\mathbf{B}} = \begin{bmatrix} I_x & I_y & I_y \end{bmatrix} \begin{bmatrix} 0 \\ 0 \\ B_z \end{bmatrix} = I_z B_z$$

$$\Downarrow$$

$$E = -\gamma I_z B_z$$

†See the Further Reading list for more information about this type of format for the Hamiltonian.

Using this as our guide we postulate a Hamiltonian by replacing E with $\hat{\mathbf{H}}$ and I_z with the corresponding operator $\hat{\mathbf{I}}_z$ (see Section 15.5 for a more detailed discussion of this Hamiltonian):

$$\hat{\mathbf{H}} = -\gamma B_0 \hat{\mathbf{I}}_z \tag{14.3}$$

such that the Schrodinger equation is satisfied:‡

$$\hat{\mathbf{H}}|\Psi\rangle = E|\Psi\rangle = -\gamma B_0 \hat{\mathbf{I}}_z|\Psi\rangle$$

In this case the Hamiltonian follows the format of eqn (14.1) in that $C = \gamma$, $K = B_0$ and $\tilde{\mathbf{A}}$ is the unit tensor:

$$H = -\gamma \vec{\mathbf{I}} \cdot \tilde{\mathbf{1}} \cdot \vec{\mathbf{B}}_0$$

but simplifies to eqn (14.3).

For a spin-1/2 we show formally in Chapter 15 that there are two spin states, frequently labelled α and β, and that the eigenvalues generated by the operator $\hat{\mathbf{I}}_z$ on these states are, respectively, $+1/2\hbar$ and $-1/2\hbar$ in energy notation.§ Thus, there are two energy states that the spin-1/2 nucleus can be in. Actually, quantum mechanics does not indicate that it is in either one or the other but rather is in a superposition of the two states (see eqn (13.6)) until we make a measurement. We can calculate what the energy difference between the states is:

$$\begin{aligned}\hat{\mathbf{H}}|\alpha\rangle &= -\gamma B_0 \hat{\mathbf{I}}_z|\alpha\rangle \\ &= -\frac{\hbar\gamma B_0}{2} = E_\alpha \\ \hat{\mathbf{H}}|\beta\rangle &= -\gamma B_0 \hat{\mathbf{I}}_z|\beta\rangle \\ &= \frac{\hbar\gamma B_0}{2} = E_\beta \\ \Delta E &= E_\beta - E_\alpha \\ &= \frac{\hbar\gamma B_0}{2} - \left(-\frac{\hbar\gamma B_0}{2}\right) \\ &= \hbar\gamma B_0\end{aligned}$$

We see that the energy difference between the two states is dependent on the magnitude of the static field, B_0. The sign of the energy difference

‡Note that we have inserted the subscript "0" into the B term as is generally the practice, to differentiate it from other possible fields.

§Energy notation and frequency notation differ merely by the inclusion or exclusion, respectively, of $\hbar$ in the expression.

also depends on the sign of γ ... here we have assumed a positive sign, as is the case for the commonly observed nuclei, ^{1}H and ^{13}C.

Recalling that $\omega_0 = -\gamma B_0$ from eqn (11.44) it is possible to recast the Hamiltonian as:

$$\hat{\mathbf{H}} = \omega_0 \hat{\mathbf{I}}_z \tag{14.4}$$

in which case a similar analysis of energy states will give us:

$$\Delta E = -\,\omega_0 B_0$$

14.3 The Chemical Shift Hamiltonian

When the first NMR experiments were performed a curious thing was noted. Rather than observing resonances at a single frequency, as implied by eqn (14.4), different resonance frequencies were observed for different compounds. How could this be? Would not hydrogen resonances, for example, appear at one frequency only, and not several, as was observed?

The answer to this turns out to be that the electronic environment around the nucleus affects the "perceived" static field at the nucleus. Thus, the external field applied to the molecule is not what its nuclei experience but rather a modified field called the *local field*. The local field is very sensitive to the electronic environment around the nucleus and so can easily be different at different nuclei. Therefore, if the local magnetic field that different nuclei experience is different then their resonant frequencies will be different.

The chemical-shift Hamiltonian is a modified version of the Zeeman Hamiltonian:

$$H_i = -\gamma_i \vec{\mathbf{I}}_i \cdot \tilde{\sigma}_i \cdot \vec{\mathbf{B}}_0$$

where the unit tensor has been replaced with the *shielding tensor*, $\tilde{\sigma}$, which is generally different for each unique nucleus in the molecule (as is indicated here with the subscript i). The canny reader will note that the z-subscript in the angular momentum operator has been dropped here. The reason for this is that the interaction of $\tilde{\sigma}_i$ with $\vec{\mathbf{B}}_0$ produces new field vectors, $\vec{\mathbf{B}}_i$, for each spin, which are not necessarily of the same magnitude and direction as $\vec{\mathbf{B}}_0$:¶

$$\tilde{\sigma}_i \cdot \vec{\mathbf{B}}_0 = \vec{\mathbf{B}}_i$$

and the Hamiltonian for spin i becomes:

$$H_i = -\gamma_i \hat{\mathbf{I}}_i \cdot \vec{\mathbf{B}}_i$$

¶Recall from Chapter 5 that the right (or left) inner product of a rank 2 tensor with a vector produces a new vector.

where here $\vec{\mathbf{B}}_i$ is the effective field at nucleus i. Since the $\vec{\mathbf{B}}_i$ fields are not necessarily oriented along the z-axis we cannot collapse the angular momentum operators down to just the $\hat{\mathbf{I}}_z$ operator, as we did for the Zeeman Hamiltonian. The effect of this is that a nucleus may resonate at differing frequencies depending on the orientation of a solid crystal of the material containing the nucleus in the external applied field, $\vec{\mathbf{B}}_0$.‖ In a powder sample consisting of many minute crystals oriented at random angles the spectrum will consist of a mixture of all of the possible single-crystal spectra over a range of frequencies. This is the *anisotropic* situation.

This is not what most of those newly initiated to NMR spectroscopy would predict. Most of us are introduced to solution NMR spectroscopy in which the chemical-shift of a nucleus does not change. Indeed, we use the chemical-shift as an indicator of the environment of the nucleus we are looking at; a CH_3 hydrogen usually shows up at approximately 1 ppm in the spectrum, whereas a $C{=}CH_2$ hydrogen would show up at around 4 to 5 ppm. The reason for this is that in solution there is chaotic molecular tumbling occurring that averages out the anisotropic portion of the shielding tensor. Recalling the discussion of tensors and, specifically, the breaking up of a tensor into isotropic, asymmetric and modified symmetric portions (see eqn (5.9)), motional averaging collapses the asymmetric and modified symmetric irreducible tensors to zero leaving only the isotropic tensor:

$$\tilde{\sigma} = \tilde{\sigma}_{\text{iso}} + \tilde{\sigma}_{\text{asym}} + \tilde{\sigma}_{\text{mod}}$$

$$\Downarrow$$

$$\tilde{\sigma} = \tilde{\sigma}_{\text{iso}} = \sigma_{\text{iso}} \begin{bmatrix} 1 & 0 & 0 \\ 0 & 1 & 0 \\ 0 & 0 & 1 \end{bmatrix}$$

σ_{iso} is equal to $1/3\text{Tr}(\tilde{\sigma})$ (see Chapter 5) The Hamiltonian for the isotropic (solution) situation is then:

$$\begin{aligned} H_i &= -\gamma\sigma_i \hat{\mathbf{I}}_i \cdot \vec{\mathbf{B}}_0 \\ \hat{\mathbf{H}}_i &= -\gamma\sigma_i B_0 \hat{\mathbf{I}}_{iz} \\ &= -\gamma_i B_i \hat{\mathbf{I}}_{iz} \\ &= \omega_i \hat{\mathbf{I}}_{iz} \end{aligned} \tag{14.5}$$

‖See Levitt for a simple explanation of this along with illuminating Figures.

where σ_i is the isotropic shielding constant. Thus, in solution each distinct nucleus has its own resonant frequency that is not dependent on the orientation of the molecule with respect to the applied field.

14.4 The Dipolar Coupling Hamiltonian

The classical energy of interaction between a pair of magnetic dipoles was seen in Chapter 11 to be:

$$E = \left(\frac{\mu_0}{4\pi}\right)\left(\frac{\vec{\mu}_1 \cdot \vec{\mu}_2}{r^3} - \frac{3(\vec{\mu}_1 \cdot \vec{\mathbf{r}}_2)(\vec{\mu}_2 \cdot \vec{\mathbf{r}}_2)}{r^5}\right)$$

As with the Zeeman Hamiltonian, in order to construct the Hamiltonian we substitute $\gamma\hat{\mathbf{I}}$ for μ and H for E to produce the dipolar coupling Hamiltonian (In frequency notation!):

$$\begin{aligned} H &= \left(\frac{\mu_0}{4\pi}\right)(\gamma_1\gamma_2)\left(\frac{\hat{\mathbf{I}}_1 \cdot \hat{\mathbf{I}}_2}{r^3} - \frac{3(\hat{\mathbf{I}}_1 \cdot \vec{\mathbf{r}}_2)(\hat{\mathbf{I}}_2 \cdot \vec{\mathbf{r}}_2)}{r^5}\right) \\ &= \left(\frac{\mu_0}{4\pi}\right)\frac{(\gamma_1\gamma_2)}{r^3}(\hat{\mathbf{I}}_1 \cdot \hat{\mathbf{I}}_2 - 3(\hat{\mathbf{I}}_1 \cdot \vec{\mathbf{u}}_r)(\hat{\mathbf{I}}_2 \cdot \vec{\mathbf{u}}_r)) \end{aligned} \tag{14.6}$$

The compact tensor notation version of this Hamiltonian is:

$$H = \left(\frac{\mu_0}{4\pi}\right)(\gamma_1\gamma_2)\vec{\mathbf{I}}_1 \cdot \tilde{\mathbf{D}} \cdot \vec{\mathbf{I}}_2 \tag{14.7}$$

where $\tilde{\mathbf{D}}$ is the dipolar coupling tensor. $\tilde{\mathbf{D}}$ is symmetric and traceless (unlike the shielding tensor), the latter property meaning that in isotropic situations such as in solutions dipolar coupling becomes zero, which makes it possible to do high-resolution NMR spectroscopy. $(\mu_0/4\pi)(\gamma_1\gamma_2)/r^3$ is called the dipolar coupling constant.

The dipolar coupling tensor is of some direct importance in solid-state NMR spectroscopy (to put it mildly) but, since the tensor collapses for molecules in solution because of rapid molecular tumbling, it has a relatively minor impact on the appearance of high-resolution solution spectra. It is of importance, however, when discussing relaxation and NOE effects for these spectra.

14.5 The Indirect (Scalar) Coupling Hamiltonian

As well as the observation that nuclei in differing electronic environments are observed at different frequencies or chemical shifts in the spectrum, there is the observation that some peaks in the spectrum are split into two, three and more peaks ... even in solution where dipolar couplings average to zero. This splitting was explained** by invoking an indirect interaction between two (or more) nuclei *via* the bonding electron cloud. Furthermore, it was observed that the magnitude of this "splitting" of peaks in the spectrum is independent of applied field strength. This effect is called by various names; indirect coupling, J-coupling and scalar coupling are common examples.

As with our other Hamiltonians we may express the indirect coupling version in its most general form using tensor notation:

$$H = 2\pi \vec{\mathbf{I}}_1 \cdot \tilde{\mathbf{J}} \cdot \vec{\mathbf{I}}_2$$

We include the 2π factor in the Hamiltonian since the units of J are generally given in hertz and we wish to convert to "natural units" of radians per second. In this form, the coupling tensor $\tilde{\mathbf{J}}$ is dependent on the angle that the molecule makes with the external magnetic field and would be applicable to solids spectroscopy. However, as with the chemical-shift, in solution, rapid motional averaging occurs and the asymmetric and symmetric portions of the tensor collapse to zero leaving only the isotropic tensor. The Hamiltonian then is written (for two interacting spins):

$$H = 2\pi J_{12} \vec{\mathbf{I}}_1 \vec{\mathbf{I}}_2 \tag{14.8}$$

where J_{12} is equal to $1/3\mathrm{Tr}(\tilde{\mathbf{J}})$ (see Chapter 5).

14.6 The RF-Field Hamiltonian

Recalling the discussion of the Bloch equations from Section 12.1, the laboratory frame $\vec{\mathbf{B}}_1$ rf-field is time-dependent (eqn (12.10)):

$$\vec{\mathbf{B}}_1 = B_1[\cos(\omega_1 t)\,\vec{\mathbf{u}}_x - \sin(\omega_1 t)\vec{\mathbf{u}}_y]$$

Also, recall that we selected one of two counter-rotating fields as our $\vec{\mathbf{B}}_1$ rf-field and that this equation represents that choice.

Ingeniously I think. See N.F. Ramsey and E.M. Purcell, *Phys. Rev.*, **85, 1952, 143.

Since the field is time-dependent we expect the energy to be time-dependent as well:

$$
\begin{aligned}
E &= -\vec{\mu} \cdot \vec{\mathbf{B}}_1 \\
&= -\gamma \vec{\mathbf{I}} \cdot \vec{\mathbf{B}}_1 \\
&= \gamma \begin{bmatrix} I_x & I_y & I_z \end{bmatrix} \begin{bmatrix} B_1 \cos(\omega_1 t) \\ B_1 \sin(\omega_1 t) \\ 0 \end{bmatrix} \\
&= -\gamma B_1 [\cos(\omega_1 t) I_x + \sin(\omega_1 t) I_y]
\end{aligned}
$$

As with eqn (14.3), we replace E, I_x and I_y with their operator counterparts to generate the Hamiltonian:

$$\hat{\mathbf{H}} - \gamma B_1 [\cos(\omega_1 t) \hat{\mathbf{I}}_x + \sin(\omega_1 t) \hat{\mathbf{I}}_y]$$

As it stands, this equation is not yet completely general . . . there is one last thing we can do to make it so. In trigonometric functions we can have two angles as arguments to the function, the second being called a phase angle:

$$\cos(\omega_1 + \phi)$$

ω_1 sets the frequency and ϕ sets the phase of the oscillation. We can insert a phase angle into our Hamiltonian as well:

$$\hat{\mathbf{H}} - \gamma B_1 [\cos(\omega_1 t + \phi) \hat{\mathbf{I}}_x + \sin(\omega_1 t + \phi) \hat{\mathbf{I}}_y]$$

which gives us our general laboratory frame rf-field Hamiltonian. What does the rotating frame $\vec{\mathbf{B}}_1$ Hamiltonian look like? ω_1 here refers to the rotational frequency of the $\vec{\mathbf{B}}_1$ field with respect to the static laboratory frame axes. In the rotating frame the axes are rotating about the *z*-axis and we generally choose the frequency of rotation to be ω_1. This being the case the relative rotational frequency of the $\vec{\mathbf{B}}_1$ field is zero and the rf Hamiltonian becomes:

$$\hat{\mathbf{H}} - \gamma B_1 [\cos(\phi) \hat{\mathbf{I}}_x + \sin(\phi) \hat{\mathbf{I}}_y] \tag{14.9}$$

Modern NMR spectrometers are able to manipulate the phase angle ϕ such that only one of the two angular momentum operators is applied. Thus, if $\phi = 0$ then the Hamiltonian is:

$$\hat{\mathbf{H}} = -\gamma B_1 \hat{\mathbf{I}}_x$$

Generally, one sees phase angles of 0, $\pi/2$, π and $3\pi/2$ used when a $\vec{\mathbf{B}}_1$ field is applied. We will have more to say about this in Chapter 19.

Further Reading

1. D. M. Grant, *Encyclopedia of Nuclear Magnetic Resonance*, John Wiley & Sons, New York, vol. 2, 1996.
2. M. H. Levitt, *Spin Dynamics*, John Wiley & Sons, Chichester, 2001.
3. S. A. Smith, W. E. Palke and J. T. Gerig, *Concepts Magn. Reson.*, 1912, **5**, 151–177.
4. S. A. Smith, W. E. Palke and J. T. Gerig, *Concepts Magn. Reson.*, 1992, **4**, 107–144.
5. S. A. Smith, W. E. Palke and J. T. Gerig, *Concepts Magn. Reson.*, 1992, **4**, 181–204.
6. R. E. Wasylishen, in *Encyclopedia of Nuclear Magnetic Resonance*, John Wiley & Sons, New York, vol. 3, 1996.

15 The Quantum Mechanics of Angular Momentum

There are only two kinds of certain knowledge: Awareness of our own existence and the truths of mathematics.

Jean le Rond d'Alembert

15.1 A Brief History

The Stern–Gerlach experiment, performed in 1922, showed that a beam of silver atoms passed through an inhomogeneous magnetic field was split into two and only two beams. Why silver atoms? Silver atoms have a single unpaired electron and, of course, are neutral so that they can be passed through a magnetic field without tracing a curved path (see Figure 11.29). Why an inhomogeneous field? An inhomogeneous magnetic field will have a field gradient that will result in a force in the direction of the gradient. This would not be so in a homogeneous field. The Bohr–Sommerfield theory of the atom at the time proposed quantised orbital angular momenta and the experiment was designed to test this hypothesis (*not* spin angular momentum as many texts seem to imply ... spin was unknown at the time). The reasoning was that if the angular momentum behaved classically then as a silver atom passed through the inhomogeneous field there would be a larger deflection at one end of the magnetic dipole than the other and the atom would be deflected by an angle depending on the orientation of the dipole as it entered the field.

Essential Mathematics for NMR and MRI Spectroscopists
By Keith C. Brown

Published by the Royal Society of Chemistry, www.rsc.org

Thus, one would expect to see an even spread of deflections. Quantised orbital angular momenta on the other hand would result in only specific deflections. Furthermore, orbital angular momentum quantum calculations of the time indicated that if the electron behaved "quantum mechanically" it would have in integral angular momentum quantum number and would be in one of an odd number of states; one of three, five, seven, *etc.* Thus, the expectation was that if orbital angular momentum is quantised one should see the beam of atoms deflected such that there were an odd number of patches on the detector. This was not observed. There were *two* patches that could not be explained by the understanding of the time. We now know, however, that it was actually the first observation of the effects of *intrinsic* angular momentum or spin.

The idea that microscopic particles have a property corresponding to angular momentum was first proposed by two young Dutch-American graduate students George Uhlenbeck and Samuel Goudsmid in 1925. The background to this hypothesis was the observation that, when in a magnetic field, the hydrogen emission spectrum (see Figure 13.3) lines were each split into a fine structure of lines around the positions where the (non-magnetic) lines were observed. Quantum mechanics was successful in explaining the non-magnetic hydrogen emission spectrum but was unable to account for the observed splittings. The suggestion that the hydrogen electron behaves as though it has another type of angular momentum and, since it is charged, a magnetic moment led to the injection of the concept into the then still developing field of quantum mechanics. Wolfgang Pauli later named it "spin", which is still used today.

An interesting side story is that Uhlenbeck and Goudsmit sent their hypothesis for publication and then had a discussion about it with Hendrik Lorentz (of Lorentz transform fame), one of the most eminent physicists of the time who argued that electron-spin would be impossible. They then tried to stop publication but were too late. Of course, they were right and Lorentz was wrong. Even experts can make mistakes. Also interesting is that spin angular momentum does not make itself evident in the Schrodinger formalism of quantum mechanics but that it *does* appear Dirac's relativistic formulation of quantum mechanics.

15.2 General Angular Momentum

The following discussion is very involved and uses much of the vector algebra and complex number algebra that we have learned so far.[†] Note that we switch from using the operator $\hat{\mathbf{J}}$ for general angular momentum to using $\hat{\mathbf{I}}$ for spin angular momentum part way through the chapter. The form of the equations for both are identical, only the symbol changes.

In classical physics, the angular momentum of a particle is given by the vector cross product, $\vec{\mathbf{L}} = \vec{\mathbf{r}} \times \vec{\mathbf{p}}$, where $\vec{\mathbf{r}}$ is the position vector of the particle and $\vec{\mathbf{p}}$ is the tangential linear momentum, $m\vec{\mathbf{v}}$ (see Figure 9.6).

The components of $\vec{\mathbf{L}}$ are (by the determinant method, Chapter 2):

$$\vec{\mathbf{L}} = \vec{\mathbf{r}} \times \vec{\mathbf{p}} = \begin{vmatrix} \mathbf{u}_x & \mathbf{u}_y & \mathbf{u}_z \\ x & y & z \\ p_x & p_y & p_z \end{vmatrix} = \mathbf{u}_x(yp_z - zp_y) - \mathbf{u}_y(zp_x - xp_z) + \mathbf{u}_z(xp_y - yp_x)$$

and the magnitude of each component is:

$$L_x = yp_z - zp_y$$

$$L_y = zp_x - xp_z$$

$$L_z = xp_y - yp_x$$

To get the corresponding Hamiltonian operators we replace the momentum term, p, with

$$-i\hbar\frac{\partial}{\partial x}\left(\text{or } -i\hbar\frac{\partial}{\partial y} \text{ or } -i\hbar\frac{\partial}{\partial z}\right)$$

(see Chapter 13, eqn (13.8)). Of course, i is the root of minus one and $\hbar$ is the reduced Planck constant or Dirac constant and is equal to $h/2\pi$. Why do this? Recall from the chapter on waves that natural frequency units in radians per second and ordinary frequency units in cycles per second or hertz are related by eqn (7.2):

$$2\pi\omega = \nu$$

$$\omega = \frac{\nu}{2\pi}$$

[†]The substance of this section owes much to discussions with Professor J. A. Weil and to his book, (see the Further Reading list at the chapter's end). There is much here and you will certainly be excused for being intimidated by the quantity. However, there is nothing that has not been explained in detail in previous chapters and as long as you have a reasonably firm grasp of complex numbers and vectors there will be no surprises (well, maybe one or two).

In dividing h by 2π we are specifying that our frequency results will be in natural frequency units or radians per second. The coordinates remain the same:

$$\hat{\mathbf{J}}_x = -ih\left(y\frac{\partial}{\partial z} - z\frac{\partial}{\partial y}\right)$$

$$\hat{\mathbf{J}}_y = -ih\left(z\frac{\partial}{\partial x} - x\frac{\partial}{\partial z}\right) \tag{15.1}$$

$$\hat{\mathbf{J}}_z = -ih\left(x\frac{\partial}{\partial y} - y\frac{\partial}{\partial x}\right)$$

We have switched to the operator symbol $\hat{\mathbf{J}}$ from the vector symbol $\vec{\mathbf{L}}$ here, as $\hat{\mathbf{J}}$ is the general angular momentum operator symbol. One often sees $\vec{\mathbf{L}}$ used for orbital angular momentum, $\hat{\mathbf{S}}$ for electron spin angular momentum and $\hat{\mathbf{I}}$ for nuclear spin angular momentum, as well as the more general $\hat{\mathbf{J}}$.[‡]

We can also define $\hat{\mathbf{J}}^2$ by summing the squares of each of the angular momentum component operators:

$$\hat{\mathbf{J}}^2 = \hat{\mathbf{J}}_x^2 + \hat{\mathbf{J}}_y^2 + \hat{\mathbf{J}}_z^2 \tag{15.2}$$

Do these operators commute? The commutator of $\hat{\mathbf{J}}_x$ and $\hat{\mathbf{J}}_y$ is (see Appendix A for a refresher on commutator algebra):

$$[\hat{\mathbf{J}}_x, \hat{\mathbf{J}}_y] = \hat{\mathbf{J}}_x\hat{\mathbf{J}}_y - \hat{\mathbf{J}}_y\hat{\mathbf{J}}_x$$

Using the above definitions eqn (15.1):

$$\begin{aligned}
\hat{\mathbf{J}}_x\hat{\mathbf{J}}_y &= -\hbar^2\left(y\frac{\partial}{\partial z} - z\frac{\partial}{\partial y}\right)\left(z\frac{\partial}{\partial x} - x\frac{\partial}{\partial z}\right) \\
&= -\hbar^2\left[\left(y\frac{\partial}{\partial z}\right)\left(z\frac{\partial}{\partial x}\right) - \left(z\frac{\partial}{\partial y}\right)\left(z\frac{\partial}{\partial x}\right) - \left(y\frac{\partial}{\partial z}\right)\left(x\frac{\partial}{\partial z}\right) + \left(z\frac{\partial}{\partial y}\right)\left(x\frac{\partial}{\partial z}\right)\right] \\
&= -\hbar^2\left[yz\frac{\partial^2}{\partial z\partial x} - z^2\frac{\partial^2}{\partial y\partial x} - yx\frac{\partial^2}{\partial z^2} + zx\frac{\partial^2}{\partial y\partial z}\right]
\end{aligned}$$

What happened to the "i"? It is there, hiding in the minus sign. Do not forget that we have multiplied $\hat{\mathbf{J}}_x$ times $\hat{\mathbf{J}}_y$, each of which contains i and the product of i times i is equal to -1. Using the

[‡]To sow further confusion, it is common in the NMR literature to use $\hat{\mathbf{I}}$ and $\hat{\mathbf{S}}$ to represent spins of different γ. Hopefully, the context will make it clear what I am talking about.

product rule of the differential calculus (see Appendix A), the first term is:

$$\frac{\partial^2 yz}{\partial z \partial x} = y\frac{\partial^2 z}{\partial z \partial x} + z\frac{\partial^2 y}{\partial z \partial x} = y\frac{\partial}{\partial x} + z\frac{\partial^2 y}{\partial z \partial x}$$

so that the last line becomes:

$$\hat{\mathbf{J}}_x\hat{\mathbf{J}}_y = -\hbar^2\left[y\frac{\partial}{\partial x} + yz\frac{\partial^2}{\partial z \partial x} - z^2\frac{\partial^2}{\partial y \partial x} - yx\frac{\partial^2}{\partial z^2} + zx\frac{\partial^2}{\partial y \partial z}\right]$$

similarly:

$$\hat{\mathbf{J}}_y\hat{\mathbf{J}}_x = -\hbar^2\left[zy\frac{\partial^2}{\partial x \partial z} - xy\frac{\partial^2}{\partial z^2} - z^2\frac{\partial^2}{\partial x \partial y} + x\frac{\partial}{\partial y} + zx\frac{\partial^2}{\partial z \partial y}\right]$$

and so:

$$\begin{aligned}
\hat{\mathbf{J}}_x\hat{\mathbf{J}}_y - \hat{\mathbf{J}}_y\hat{\mathbf{J}}_x &= -\hbar^2\left(y\frac{\partial}{\partial x} - x\frac{\partial}{\partial y}\right) \\
&= \hbar^2\left(x\frac{\partial}{\partial y} - y\frac{\partial}{\partial x}\right) \\
&= -i^2\hbar^2\left(x\frac{\partial}{\partial y} - y\frac{\partial}{\partial x}\right) \\
&= i\hbar\hat{\mathbf{J}}_z
\end{aligned}$$

using the definition of $\hat{\mathbf{J}}_z$ above and the fact that $i^2 = -1$. Similarly, we can write:

$$\begin{aligned}
\left[\hat{\mathbf{J}}_x, \hat{\mathbf{J}}_y\right] &= i\hbar\hat{\mathbf{J}}_z \\
\left[\hat{\mathbf{J}}_z, \hat{\mathbf{J}}_x\right] &= i\hbar\hat{\mathbf{J}}_y \\
\left[\hat{\mathbf{J}}_y, \hat{\mathbf{J}}_z\right] &= i\hbar\hat{\mathbf{J}}_x
\end{aligned} \tag{15.3}$$

Note the cyclic permutation in these commutation relationships and that the components of the angular momenta do not commute with each other. This means that we cannot assign definite values to any two of these at once as we saw in our discussion of the uncertainty principle in Chapter 13. Put another way, there are no functions that are simultaneously eigenfunctions of all three angular momentum operators. One will often see these commutation relationships written in "frequency" notation. Recalling from Chapter 13 that Planck's energy equation relates the energy of a photon to its frequency $E = h\nu$, where ν is in hertz. Alternatively:

$$E = \frac{h\omega}{2\pi} = \hbar\omega$$

where ω is in radians per second. Dividing by $\hbar$ gives the frequency notation, which is more convenient for NMR spectroscopists.[§] We shall, however, have occasion to temporarily switch back to the energy notation in a later chapter. If we divide our commutation relationships, eqn (15.3), by $\hbar$ we get the frequency notation form of the angular momentum commutators:

$$\begin{aligned} \frac{E}{\hbar} &= \omega \\ \left[\hat{\mathbf{J}}_x, \hat{\mathbf{J}}_y\right] &= i\hat{\mathbf{J}}_z \\ \left[\hat{\mathbf{J}}_z, \hat{\mathbf{J}}_x\right] &= i\hat{\mathbf{J}}_y \\ \left[\hat{\mathbf{J}}_y, \hat{\mathbf{J}}_z\right] &= i\hat{\mathbf{J}}_x \end{aligned} \tag{15.4}$$

We can do the kind of calculations of the commutation relationships between $\hat{\mathbf{J}}^2$ and $\hat{\mathbf{J}}_x$, $\hat{\mathbf{J}}_y$ and $\hat{\mathbf{J}}_z$. When we do so, using the above definitions for these operators, we get:

$$\begin{aligned} \left[\hat{\mathbf{J}}^2, \hat{\mathbf{J}}_x\right] &= 0 \\ \left[\hat{\mathbf{J}}^2, \hat{\mathbf{J}}_y\right] &= 0 \\ \left[\hat{\mathbf{J}}^2, \hat{\mathbf{J}}_z\right] &= 0 \end{aligned}$$

In these cases we see that $\hat{\mathbf{J}}^2$ commutes with all three angular momentum operators. Thus, it is possible to measure simultaneous values using $\hat{\mathbf{J}}^2$ and one of the angular momentum operators. Equivalently, there exists an eigenfunction that can be used simultaneously with $\hat{\mathbf{J}}^2$ and one of the angular momentum operators (see eqn (3.58)).

[§]This is perhaps, not so important in the MRI world since there, the spectroscopist is not dealing directly with the frequency domain but rather k-space and position. See Chapter 22 for more information.

Now, this discussion of eigenfunctions and angular momentum operators naturally leads one to ask "what are the eigenvalues of these eigenfunctions?". Since there is a set of eigenfunctions that satisfies *both* $\hat{\mathbf{J}}^2$ and one of $\hat{\mathbf{J}}_x$, $\hat{\mathbf{J}}_y$ or $\hat{\mathbf{J}}_z$, we can try to find out what it might be or to find the corresponding eigenvalues. The choice that is generally made is to use $\hat{\mathbf{J}}^2$ and $\hat{\mathbf{J}}_z$.

In the following discussion it is convenient to introduce a couple of new operator definitions. These are the so-called *ladder operators* or *raising operator* and *lowering operator*:

$$\hat{\mathbf{J}}_+ = \hat{\mathbf{J}}_x + i\hat{\mathbf{J}}_y \quad \text{and} \quad \hat{\mathbf{J}}_- = \hat{\mathbf{J}}_x - i\hat{\mathbf{J}}_y \tag{15.5}$$

Their commutation relationships are:

$$\begin{aligned}
&[\hat{\mathbf{J}}^2, \hat{\mathbf{J}}_+] = [\hat{\mathbf{J}}^2, \hat{\mathbf{J}}_-] = 0 \\
&[\hat{\mathbf{J}}_z, \hat{\mathbf{J}}_+] = \hat{\mathbf{J}}_+ \\
&[\hat{\mathbf{J}}_z, \hat{\mathbf{J}}_-] = -\hat{\mathbf{J}}_- \\
&[\hat{\mathbf{J}}_+, \hat{\mathbf{J}}_-] = 2\hat{\mathbf{J}}_z
\end{aligned} \tag{15.6}$$

Inspection of their definitions above will show that:

$$\hat{\mathbf{J}}_+ = \hat{\mathbf{J}}_-^* \quad \text{and} \quad \hat{\mathbf{J}}_- = \hat{\mathbf{J}}_+^*$$

Now let us try for some eigenvalues of $\hat{\mathbf{J}}^2$ and $\hat{\mathbf{J}}_z$. First, remembering that these two operators have common eigenfunctions, we will (temporarily) call the eigenvalues of $\hat{\mathbf{J}}^2$ and $\hat{\mathbf{J}}_z$, λ_j and λ_m, respectively, and we can write:

$$\begin{aligned}
&\hat{\mathbf{J}}^2 |j, m\rangle = \lambda_j |j, m\rangle \\
&\hat{\mathbf{J}}_z |j, m\rangle = \lambda_m |j, m\rangle
\end{aligned} \tag{15.7}$$

This is our by now familiar Dirac notation that has been used here to represent the common eigenfunction of $\hat{\mathbf{J}}^2$ and $\hat{\mathbf{J}}_z$. $|j, m\rangle$ represents the spin wave function with the emphasis placed on j and m, rather that writing $|\Psi\rangle$. It will help us to keep our focus on the eigenvalues of $\hat{\mathbf{J}}^2$ and $\hat{\mathbf{J}}_z$. Another way that one sees this written in the literature is:

$$\begin{aligned}
&\hat{\mathbf{J}}^2 \Psi_{j,m} = \lambda_j \Psi_{j,m} \\
&\hat{\mathbf{J}}_z \Psi_{j,m} = \lambda_m \Psi_{j,m}
\end{aligned}$$

The Dirac notation is rather more compact and is useful in developing the matrix representation of operators. Note that as usual, these are or can be made to be orthonormal functions where:

$$\langle j', m' | j, m\rangle = 1 \quad \text{if} \quad j' = j \quad \text{and} \quad m' = m$$

or

$$\langle j', m' | j, m\rangle = 0 \quad \text{if} \quad j' \neq j \quad \text{and/or} \quad m' \neq m$$

You will need to remember these when we talk about matrix representations of operators.

Now, from the definition of $\hat{\mathbf{J}}^2$ (eqn (15.2)):

$$\hat{\mathbf{J}}_x^2 + \hat{\mathbf{J}}_y^2 = \hat{\mathbf{J}}^2 - \hat{\mathbf{J}}_z^2$$

and, using this in our eigenvalue eqn (15.7):

$$\left(\hat{\mathbf{J}}_x^2 + \hat{\mathbf{J}}_y^2\right)|j, m\rangle = \left(\hat{\mathbf{J}}^2 - \hat{\mathbf{J}}_z^2\right)|j, m\rangle$$

$$= (\lambda_j - \lambda_m^2)|j, m\rangle$$

Since $\hat{\mathbf{J}}_x$ and $\hat{\mathbf{J}}_y$ correspond to experimental observables (and are therefore Hermitian) they must give real, non-negative numbers so that $(\hat{\mathbf{J}}_x^2 + \hat{\mathbf{J}}_y^2)$ will also give real, non-negative numbers:

$$(\lambda_j - \lambda_m^2) \geq 0 \tag{15.8}$$

Recalling the commutator $[\hat{\mathbf{J}}_z, \hat{\mathbf{J}}_+] = \hat{\mathbf{J}}_+$ or $\hat{\mathbf{J}}_z\hat{\mathbf{J}}_+ - \hat{\mathbf{J}}_+\hat{\mathbf{J}}_z = \hat{\mathbf{J}}_+$ (from eqn (15.6)) we can write:

$$\langle j, m' | \hat{\mathbf{J}}_z\hat{\mathbf{J}}_+ - \hat{\mathbf{J}}_+\hat{\mathbf{J}}_z | j, m\rangle = \langle j, m' | \hat{\mathbf{J}}_+ | j, m\rangle$$

where j has the same value in the bra and ket but m' may (or may not) be different from m. That this is so is because:

$$\hat{\mathbf{J}}_z\hat{\mathbf{J}}_+|j, m\rangle = (\hat{\mathbf{J}}_+ + \hat{\mathbf{J}}_+\hat{\mathbf{J}}_z)|j, m\rangle$$

(using $\hat{\mathbf{J}}_z\hat{\mathbf{J}}_+ - \hat{\mathbf{J}}_+\hat{\mathbf{J}}_z = \hat{\mathbf{J}}_+$ from above)

$$= (\hat{\mathbf{J}}_+ + \hat{\mathbf{J}}_+\lambda_m)|j, m\rangle$$

$$= (\lambda_m + 1)\hat{\mathbf{J}}_+|j, m\rangle$$

This shows us that $\hat{\mathbf{J}}_z\hat{\mathbf{J}}_+$, operating on the eigenfunction, $|j, m\rangle$, produces an eigenvalue of λ_m *plus* one. This is a very important result for us! Let us think about it for a moment. The result of $\hat{\mathbf{J}}_z|j, m\rangle$

is $\lambda_m|\,j,\,m\rangle$ ... we said this earlier in eqn (15.7). Contrast this with $\hat{\mathbf{J}}_z\,\hat{\mathbf{J}}_+|j,\,m\rangle$ which gives $(\lambda_m+1)\,\hat{\mathbf{J}}_+|j,\,m\rangle$. If you consider $\hat{\mathbf{J}}_+|j,\,m\rangle$ as a whole to be just another eigenfunction of $\hat{\mathbf{J}}_z$, then this result indicates that $\hat{\mathbf{J}}_+$ *must* operate on $|\,j,\,m\rangle$ to give $|j,\,m+1\rangle$:

$$\hat{\mathbf{J}}_+|\,j,\,m\rangle=\zeta_m|\,j,\,m+1\rangle \tag{15.9}$$

Only m is affected by this operation ... j, related to the total angular momentum, is unaffected. Furthermore, there must be a whole set of eigenfunctions of $\hat{\mathbf{J}}_z$ whose eigenvalues are separated from each other by one. We shall look at the value of ζ_m momentarily.

By a similar argument:

$$\hat{\mathbf{J}}_-|\,j,\,m\rangle=\xi_m|\,j,\,m-1\rangle \tag{15.10}$$

Now, since we have shown that $(\lambda_j-\lambda_m^2)\geq 0$ (see eqn (15.8)), this means that there must be a maximum and a minimum value for λ_m. In other words:

$$-\sqrt{\lambda_j}\leq\lambda_m\leq\sqrt{\lambda_j}$$

If we try to use the raising operator on an eigenfunction that already has a maximum value of m we cannot generate one with $m_{\text{max}}+1$ and we are forced to write:

$$\begin{aligned}\hat{\mathbf{J}}_+|j,m_{\text{max}}\rangle&=0\\ \hat{\mathbf{J}}_-|j,m_{\text{min}}\rangle&=0\end{aligned} \tag{15.11}$$

Now:

$$\hat{\mathbf{J}}_-\hat{\mathbf{J}}_+|j,m_{\text{max}}\rangle=(\hat{\mathbf{J}}^2-\hat{\mathbf{J}}_z^2-\hat{\mathbf{J}}_z)|j,m_{\text{max}}\rangle$$

$$\text{(from the definitions of }\hat{\mathbf{J}}_-,\hat{\mathbf{J}}_+\text{ and }\hat{\mathbf{J}}^2)$$

$$=(\lambda_j-\lambda_{m(\text{max})}^2-\lambda_{m(\text{max})})=0$$

Why is it equal to zero?

$$\hat{\mathbf{J}}_-\hat{\mathbf{J}}_+|j,m_{\text{max}}\rangle=\hat{\mathbf{J}}_-(\hat{\mathbf{J}}_+|j,m_{\text{max}}\rangle)=\hat{\mathbf{J}}_-(0)=0$$

so, assuming that $|j,\,m_{\text{max}}\rangle$ has a non-zero value:

$$(\lambda_j-\lambda_{m(\text{max})}^2-\lambda_{m(\text{max})})=0$$

or:

$$\lambda_j=\lambda_{m(\text{max})}^2+\lambda_{m(\text{max})}$$

$$=\lambda_{m(\text{max})}(\lambda_{m(\text{max})}+1)$$

Similarly, for:

$$\hat{\mathbf{J}}_{+}\,\hat{\mathbf{J}}_{-}|\,j,\,m_{\text{min}}\rangle$$

we obtain:

$$\lambda_j = \lambda_{m(\text{min})}\,(\lambda_{m(\text{min})} - 1)$$

and we can then say:

$$\lambda_{m(\text{max})}^2 + \lambda_{m(\text{max})} = \lambda_{m(\text{min})}^2 + \lambda_{m(\text{min})}$$

This can *only* be true if $\lambda_{m(\text{max})} = -\,\lambda_{m(\text{min})}$.

Combined with our knowledge that the eigenvalues of $\hat{\mathbf{J}}_z|\,j,\,m\rangle$ differ from each other by *exactly* one, the fact that there is a maximum and a minimum eigenvalue for the set of eigenfunctions and the fact that the maximum value equals the negative of the minimum value means that the eigenvalues of $\hat{\mathbf{J}}_z|\,j,\,m\rangle$ can take only the values $\ldots -3/2,\ -1,\ -1/2,\ 0,\ +1/2,\ +1,\ +3/2$. No other values are possible. Thus, there are $2\lambda_{m(\text{max})}$ possible values of m. It is customary to define $j = \lambda_{m(\text{max})}$ and $-j = \lambda_{m(\text{min})}$. This gives us:

$$\lambda_j = j(j+1)$$

Now, since j is defined in terms of $\lambda_{m(\text{max})}$ and λ_m ranges from $\lambda_{m(\text{min})}$ to $\lambda_{m(\text{max})}$ in steps of one, the values that j can take are 0, 1/2, 1, 3/2, 2 . . .

The possible values of λ_m depend on j and the fact that they differ by one from each other. So, if $j = 1$, then λ_m or m for short, can be -1, 0, $+1$. More generally, for a given value of j, m ranges from $-j$ to $+j$ in steps of one.

So now we have the eigenvalues of $\hat{\mathbf{J}}^2$ and $\hat{\mathbf{J}}_z$, $j(j+1)$ and m respectively. We have so far said nothing, essentially, about ζ_m and ζ_m except that they are the eigenvalues of $\hat{\mathbf{J}}_{+}$ and $\hat{\mathbf{J}}_{-}$, respectively, as you may recall:

$$\langle j,m|\hat{\mathbf{J}}_{-}\hat{\mathbf{J}}_{+}|j,m\rangle = \zeta_m\langle j,m|\hat{\mathbf{J}}_{-}|j,m+1\rangle = \zeta_m\xi_m\langle j,m|j,m\rangle = \zeta_m\xi_{m+1}$$

Also, you will remember that:

$$\langle j,m|\hat{\mathbf{J}}_{-}\hat{\mathbf{J}}_{+}|j,m\rangle = \langle j,m|\hat{\mathbf{J}}^2 - \hat{\mathbf{J}}_z^2 - \hat{\mathbf{J}}_z|j,m\rangle$$

and since we now know what the eigenvalues of $\hat{\mathbf{J}}^2$ and $\hat{\mathbf{J}}_z$ are:

$$\begin{aligned}\langle j,m|\hat{\mathbf{J}}^2 - \hat{\mathbf{J}}_z^2 - \hat{\mathbf{J}}_z|j,m\rangle &= j(j+1) - m^2 - m\langle j,m\,|\,j,m\rangle\\ &= j(j+1) - m(m+1)\\ &= \zeta_m\xi_{m+1}\end{aligned}$$

We can also call up the Hermitian operator "trick" to evaluate $\langle j, m|\hat{\mathbf{J}}_{-}\hat{\mathbf{J}}_{+}|j, m\rangle$ a third way:

$$\begin{aligned}\langle j,m|\hat{\mathbf{J}}_{-}\hat{\mathbf{J}}_{+}|j,m\rangle &= \zeta_m\langle j,m|\hat{\mathbf{J}}_{-}|j,m+1\rangle \\ &= \zeta_m\left(\langle j,m+1|\hat{\mathbf{J}}_{+}|j,m\rangle\right)^* \\ &= \zeta_m\zeta_m^*\langle j,m+1\,|\,j,m+1\rangle\end{aligned}$$

$$(\text{since}\,\hat{\mathbf{J}}_{+} = \hat{\mathbf{J}}_{-}^*)$$

So, we now have:

$$\zeta_m\zeta^*{}_m = \zeta_m\ \xi_{m+1}$$

And therefore:

$$\zeta_m^* = \xi_{m+1}$$

$$\zeta_m\zeta_m^* = |\zeta_m|^2 = j(j+1) - m(m+1)$$

$$\zeta_m = \sqrt{j(j+1) - m(m+1)}$$

Similarly, for $\langle j, m|\hat{\mathbf{J}}_{+}\hat{\mathbf{J}}_{-}|j, m\rangle$:

$$\xi_m = \sqrt{j(j+1) - m(m-1)}$$

In summary we now have four eigenvalue equations:

1. $\hat{\mathbf{J}}_z|j,\ m\rangle = m|j,\ m\rangle$ and there are $2j+1$ possible eigenvalues of $\hat{\mathbf{J}}_z$.
2. $\hat{\mathbf{J}}^2|j,\ m\rangle = j(j+1)|j,\ m\rangle$
3. $\hat{\mathbf{J}}_{-}|j,m\rangle = \sqrt{j(j+1) - m(m-1)}|j,m-1\rangle$
4. $\hat{\mathbf{J}}_{+}|j,m\rangle = \sqrt{j(j+1) - m(m+1)}|j,m+1\rangle$

Also, we can derive expressions using $\hat{\mathbf{J}}_x$ and $\hat{\mathbf{J}}_y$ from the definitions of the ladder operators:

$$\begin{aligned}\hat{\mathbf{J}}_x &= \frac{\hat{\mathbf{J}}_{+} + \hat{\mathbf{J}}_{-}}{2} \\ \hat{\mathbf{J}}_y &= \frac{\hat{\mathbf{J}}_{+} - \hat{\mathbf{J}}_{-}}{2i}\end{aligned} \tag{15.12}$$

So:

$$\begin{aligned}\hat{\mathbf{J}}_x|j,m\rangle &= \frac{\hat{\mathbf{J}}_+ + \hat{\mathbf{J}}_-}{2}|j,m\rangle \\ &= \frac{\hat{\mathbf{J}}_+}{2}|j,m\rangle + \frac{\hat{\mathbf{J}}_-}{2}|j,m\rangle \\ &= \frac{1}{2}\sqrt{j(j+1)-m(m+1)}|j,m+1\rangle + \frac{1}{2}\sqrt{j(j+1)-m(m-1)}|j,m-1\rangle\end{aligned} \tag{15.13}$$

Similarly:

$$\begin{aligned}\hat{\mathbf{J}}_y|j,m\rangle &= \frac{\hat{\mathbf{J}}_+ - \hat{\mathbf{J}}_-}{2i}|j,m\rangle \\ &= \frac{\hat{\mathbf{J}}_+}{2i}|j,m\rangle - \frac{\hat{\mathbf{J}}_-}{2i}|j,m\rangle \\ &= \frac{1}{2i}\sqrt{j(j+1)-m(m+1)}|j,m+1\rangle - \frac{1}{2i}\sqrt{j(j+1)-m(m-1)}|j,m-1\rangle\end{aligned}$$

or (remembering that $1/i = -i$):

$$\begin{aligned}\hat{\mathbf{J}}_y|j,m\rangle &= \frac{\hat{\mathbf{J}}_+ - \hat{\mathbf{J}}_-}{2i}|j,m\rangle \\ &= \frac{\hat{\mathbf{J}}_+}{2i}|j,m\rangle - \frac{\hat{\mathbf{J}}_-}{2i}|j,m\rangle \\ &= -\frac{i}{2}\sqrt{j(j+1)-m(m+1)}|j,m+1\rangle + \frac{i}{2}\sqrt{j(j+1)-m(m-1)}|j,m-1\rangle\end{aligned} \tag{15.14}$$

The orthonormality of the eigenfunctions will considerably simplify these expressions when looking at operator matrices.

The symbol, $\hat{\mathbf{J}}$, is a generalised symbol for an angular momentum operator. When one wishes to talk about a specific type of angular momentum one generally changes the symbol for the operator. For example, when the orbital angular momentum of an electron is being discussed one would use $\hat{\mathbf{L}}$ or when the nuclear spin is of interest, $\hat{\mathbf{I}}$ or $\hat{\mathbf{S}}$ would be used.

15.3 Angular Momentum Operators in Spherical Coordinates

The classical expression for angular momentum is:

$$\vec{\mathbf{L}} = \vec{\mathbf{r}} \times \vec{\mathbf{p}}$$

And we saw in eqn (15.1) that the general angular momentum operators are, in Cartesian coordinates:

$$\begin{aligned}
\hat{\mathbf{J}}_x &= -i\hbar\left(y\frac{\partial}{\partial z} - z\frac{\partial}{\partial y}\right) \\
\hat{\mathbf{J}}_y &= -i\hbar\left(z\frac{\partial}{\partial x} - x\frac{\partial}{\partial z}\right) \\
\hat{\mathbf{J}}_z &= -i\hbar\left(x\frac{\partial}{\partial y} - y\frac{\partial}{\partial x}\right)
\end{aligned} \tag{15.15}$$

It is occasionally useful to convert these operators to spherical coordinates. We wish to convert the operators $\partial/\partial x$, $\partial/\partial y$ and $\partial/\partial z$ to $\partial/\partial r$, $\partial/\partial \theta$ and $\partial/\partial \phi$. To do so we write the partial differential operator equations:

$$\begin{aligned}
\frac{\partial}{\partial x} &= \frac{\partial r}{\partial x}\frac{\partial}{\partial r} + \frac{\partial \theta}{\partial x}\frac{\partial}{\partial \theta} + \frac{\partial \phi}{\partial x}\frac{\partial}{\partial \phi} \\
\frac{\partial}{\partial y} &= \frac{\partial r}{\partial y}\frac{\partial}{\partial r} + \frac{\partial \theta}{\partial y}\frac{\partial}{\partial \theta} + \frac{\partial \phi}{\partial y}\frac{\partial}{\partial \phi} \\
\frac{\partial}{\partial z} &= \frac{\partial r}{\partial z}\frac{\partial}{\partial r} + \frac{\partial \theta}{\partial z}\frac{\partial}{\partial \theta} + \frac{\partial \phi}{\partial z}\frac{\partial}{\partial \phi}
\end{aligned} \tag{15.16}$$

much as was done in eqn (4.27). Appendix A, eqn (A.32), (A.33) and (A.34) show the partial differential equations for $\partial r/\partial x$, $\partial \theta/\partial x$, *etc.* and we use them here in eqn (15.16)

$$\begin{aligned}
\frac{\partial}{\partial x} &= \sin(\theta)\cos(\phi)\frac{\partial}{\partial r} + \frac{\cos(\theta)\cos(\phi)}{r}\frac{\partial}{\partial \theta} - \frac{\sin(\phi)}{r\sin(\theta)}\frac{\partial}{\partial \phi} \\
\frac{\partial}{\partial y} &= \sin(\theta)\sin(\phi)\frac{\partial}{\partial r} + \frac{\cos(\theta)\sin(\phi)}{r}\frac{\partial}{\partial \theta} + \frac{\cos(\phi)}{r\sin(\theta)}\frac{\partial}{\partial \phi} \\
\frac{\partial}{\partial z} &= \cos(\theta)\frac{\partial}{\partial r} - \frac{\sin(\theta)}{r}\frac{\partial}{\partial \theta} + 0
\end{aligned}$$

Now, we substitute these (and the appropriate terms in eqn (A.29) in Appendix A) into eqn (15.15) to obtain the spherical coordinate version of the angular momentum operators:

$$\begin{aligned}
\hat{\mathbf{J}}_x &= -i\hbar\left(y\frac{\partial}{\partial z} - z\frac{\partial}{\partial y}\right) \\
&= -i\hbar\left[r\sin(\theta)\sin(\phi)\left(\cos(\theta)\frac{\partial}{\partial r} - \frac{\sin(\theta)}{r}\frac{\partial}{\partial \theta}\right)\right. \\
&\quad \left. -r\cos(\theta)\left(\sin(\theta)\sin(\phi)\frac{\partial}{\partial r} + \frac{\cos(\theta)\sin(\phi)}{r}\frac{\partial}{\partial \theta} + \frac{\cos(\phi)}{r\sin(\theta)}\frac{\partial}{\partial \phi}\right)\right] \\
&= -i\hbar\left[-\cos^2(\theta)\sin(\phi)\frac{\partial}{\partial \theta} - \sin^2(\theta)\sin(\phi)\frac{\partial}{\partial \theta} - \frac{\cos(\theta)\cos(\phi)}{\sin(\theta)}\frac{\partial}{\partial \phi}\right] \\
&= -i\hbar\left[-\sin(\phi)\frac{\partial}{\partial \theta} - \frac{\cos(\theta)\cos(\phi)}{\sin(\theta)}\frac{\partial}{\partial \phi}\right]
\end{aligned} \tag{15.17}$$

Similarly, $\hat{\mathbf{J}}_y$ and $\hat{\mathbf{J}}_z$ are:

$$\begin{aligned}
\hat{\mathbf{J}}_y &= -i\hbar\left[\cos(\phi)\frac{\partial}{\partial \theta} - \frac{\cos(\theta)\cos(\phi)}{\sin(\theta)}\frac{\partial}{\partial \phi}\right] \\
\hat{\mathbf{J}}_z &= -i\hbar\frac{\partial}{\partial \phi}
\end{aligned} \tag{15.18}$$

The ladder operators are:

$$\begin{aligned}
\hat{\mathbf{J}}_+ &= \hat{\mathbf{J}}_x + i\hat{\mathbf{J}}_y \\
&= -i\hbar\left[-\sin(\phi)\frac{\partial}{\partial \theta} - \frac{\cos(\theta)\cos(\phi)}{\sin(\theta)}\frac{\partial}{\partial \phi}\right] - i^2\hbar\left[\cos(\phi)\frac{\partial}{\partial \theta} - \frac{\cos(\theta)\sin(\phi)}{\sin(\theta)}\frac{\partial}{\partial \phi}\right] \\
&= \left[i\hbar\sin(\phi)\frac{\partial}{\partial \theta} - i^2\hbar\cos(\phi)\frac{\partial}{\partial \theta}\right] + \left[i\hbar\frac{\cos(\theta)\cos(\phi)}{\sin(\theta)}\frac{\partial}{\partial \phi} + i^2\hbar\frac{\cos(\theta)\sin(\phi)}{\sin(\theta)}\frac{\partial}{\partial \phi}\right] \\
&= \hbar(\cos(\phi) + i\sin(\phi))\frac{\partial}{\partial \theta} + \hbar(\cos(\phi) + i\sin(\phi))\left(\frac{i\cos(\theta)}{\sin(\theta)}\frac{\partial}{\partial \phi}\right) \\
&= \hbar e^{i\phi}\left(\frac{\partial}{\partial \theta} + \frac{i\cos(\theta)}{\sin(\theta)}\frac{\partial}{\partial \phi}\right) \\
\hat{\mathbf{J}}_- &= \hat{\mathbf{J}}_x - i\hat{\mathbf{J}}_y = \hbar e^{-i\phi}\left(\frac{\partial}{\partial \theta} - \frac{i\cos(\theta)}{\sin(\theta)}\frac{\partial}{\partial \phi}\right)
\end{aligned} \tag{15.19}$$

15.4 Angular Momentum Operator Matrices

We can do some expectation value calculations using the angular momentum operators that we previously developed:

$$\begin{aligned}\langle j,m|\hat{\mathbf{J}}_z|j,m\rangle &= \langle j,m|m|j,m\rangle\\ &= m\langle j,m|j,m\rangle\\ &= m\end{aligned}$$

(assuming $|j, m\rangle$ is normalised).

As we have seen, generally, for a spin with spin quantum number j, there are $2j+1$ values for m, ranging from $-j$ to $+j$ in increments of one. For example, for a spin of 1, m can be −1, 0 or +1. We can use this to represent all of the possible expectation values for an operator in a matrix. So, for a single spin with $j=1$ we would use a 3×3 matrix (3 because of the three values of m). The columns in the matrix would be labelled $|j, +j\rangle$ to $|j, -j\rangle$ in steps of one or just $+j$ to $-j$, which are the possible values of m. For our spin with $j=1$ these would be 1, 0, −1. The rows are labelled similarly except that bras are used …$\langle j, +j|$ to $\langle j, -j|$.

Each matrix element is then the expectation value of the operator using the corresponding bra and ket. So, for row 1 and column 1 the expectation value of $\hat{\mathbf{J}}_z$ would be:

$$\langle 1,1|\hat{\mathbf{J}}_z|1,1\rangle = \langle 1,1|1|1,1\rangle = 1\langle 1,1|1,1\rangle = 1$$

The whole $\hat{\mathbf{J}}_z$ matrix for our $j=1$ spin is:

$$\hat{\mathbf{J}}_z = \begin{array}{c} \\ \langle 1,1| \\ \langle 1,0| \\ \langle 1,-1| \end{array}\begin{array}{c} \langle 1,1| \;\; \langle 1,0| \;\; \langle 1,-1| \\ \begin{bmatrix} 1 & 0 & 0 \\ 0 & 0 & 0 \\ 0 & 0 & -1 \end{bmatrix}\end{array}$$

The basis functions in bra and ket form are shown beside the appropriate rows and above the appropriate columns. Columns 2 and 3, row 1 are zero since the eigenfunctions are not orthogonal and collapse to zero.

We can now write out the generalised expectation values for all of our angular momentum operators for all non-zero matrix elements:

$$\begin{aligned}\langle j,m|\hat{\mathbf{J}}_z|j,m\rangle &= \langle j,m|m|j,m\rangle \\ &= m\langle j,m|j,m\rangle \\ &= m\end{aligned}$$

$$\begin{aligned}\langle j,m|\hat{\mathbf{J}}^2|j,m\rangle &= \langle j,m|j(j+1)|j,m\rangle \\ &= j(j+1)\langle j,m|j,m\rangle \\ &= j(j+1)\end{aligned}$$

$$\begin{aligned}\langle j,m+1|\hat{\mathbf{J}}_+|j,m\rangle &= \langle j,m+1|\sqrt{j(j+1)-m(m+1)}|j,m+1\rangle \\ &= \sqrt{j(j+1)-m(m+1)}\langle j,m+1|j,m+1\rangle \\ &= \sqrt{j(j+1)-m(m+1)}\end{aligned}$$

and similarly:

$$\langle j,m-1|\hat{\mathbf{J}}_-|j,m\rangle = \sqrt{j(j+1)-m(m-1)}$$

There are two ways that the element of $\hat{\mathbf{J}}_x$ can be constructed, depending on the value of m in the bra:

$$\begin{aligned}\langle j,m+1|\hat{\mathbf{J}}_x|j,m\rangle &= \langle j,m+1|\frac{\sqrt{j(j+1)-m(m+1)}}{2}|j,m+1\rangle \\ &= \frac{\sqrt{j(j+1)-m(m+1)}}{2}\langle j,m+1|j,m+1\rangle \\ &= \frac{\sqrt{j(j+1)-m(m+1)}}{2}\end{aligned}$$

or

$$\begin{aligned}\langle j,m-1|\hat{\mathbf{J}}_x|j,m\rangle &= \langle j,m-1|\frac{\sqrt{j(j+1)-m(m-1)}}{2}|j,m-1\rangle \\ &= \frac{\sqrt{j(j+1)-m(m-1)}}{2}\langle j,m-1|j,m-1\rangle \\ &= \frac{\sqrt{j(j+1)-m(m-1)}}{2}\end{aligned}$$

Similarly for $\hat{\mathbf{J}}_y$:

$$\langle j, m+1|\hat{\mathbf{J}}_y|j, m\rangle = \frac{i}{2}\sqrt{j(j+1) - m(m+1)}$$

or

$$\langle j, m-1|\hat{\mathbf{J}}_y|j, m\rangle = \frac{i}{2}\sqrt{j(j+1) - m(m-1)}$$

This is a *lot* of calculation for what turns out to be some really rather simple matrices. In the case of spin-1/2 nuclei (spin-1/2 refers of course to $j = 1/2$) we can calculate what the expectation values of the nuclear spin angular momentum operators, $\hat{\mathbf{I}}_z$, $\hat{\mathbf{I}}_x$ and $\hat{\mathbf{I}}_y$ are. For $\hat{\mathbf{I}}_z$ the calculation is particularly simple. Since:

$$\left\langle \frac{1}{2}, \frac{1}{2}\middle|\hat{\mathbf{I}}_z\middle|\frac{1}{2}, \frac{1}{2}\right\rangle = \frac{1}{2}$$

$$\left\langle \frac{1}{2}, \frac{1}{2}\middle|\hat{\mathbf{I}}_z\middle|\frac{1}{2}, -\frac{1}{2}\right\rangle = 0$$

$$\left\langle \frac{1}{2}, -\frac{1}{2}\middle|\hat{\mathbf{I}}_z\middle|\frac{1}{2}, -\frac{1}{2}\right\rangle = -\frac{1}{2}$$

and

$$\left\langle \frac{1}{2}, -\frac{1}{2}\middle|\hat{\mathbf{I}}_z\middle|\frac{1}{2}, \frac{1}{2}\right\rangle = 0$$

the matrix is:

$$\hat{\mathbf{I}}_z = \begin{bmatrix} \frac{1}{2} & 0 \\ 0 & -\frac{1}{2} \end{bmatrix} = \frac{1}{2}\begin{bmatrix} 1 & 0 \\ 0 & -1 \end{bmatrix}$$

For $\hat{\mathbf{I}}_x$ the calculation is a bit more complicated (but not much so!). First, we visualise the basis functions alongside the initially empty matrix:

$$\begin{matrix} & \langle\frac{1}{2},\frac{1}{2}| & \langle\frac{1}{2},-\frac{1}{2}| \\ \langle\frac{1}{2},\frac{1}{2}| & 0 & 0 \\ \langle\frac{1}{2},-\frac{1}{2}| & 0 & 0 \end{matrix}$$

As you will recall from a few lines back, there are two expressions for the expectation value of $\hat{\mathbf{I}}_x$ ($\hat{\mathbf{J}}_x$ actually, since it was a general angular momentum operator then) depending on whether m increases by one or decreases by one. If you look at row 1, column 1, this is

$$\left\langle \frac{1}{2}, \frac{1}{2} \right| \hat{\mathbf{I}}_x \left| \frac{1}{2}, \frac{1}{2} \right\rangle$$

You cannot use the first of the two equations since this would raise 1/2 to 3/2 but m cannot be larger than j (remember the maximum value of m is $+j$ and the minimum value is $-j$). If you use the second of the two expressions you will get

$$\left\langle \frac{1}{2}, \frac{1}{2} \right| \sqrt{j(j+1) - m(m-1)} \left| \frac{1}{2}, -\frac{1}{2} \right\rangle$$
$$= \sqrt{j(j+1) - m(m-1)} \left\langle \frac{1}{2}, \frac{1}{2} \middle| \frac{1}{2}, -\frac{1}{2} \right\rangle$$

and since the basis functions are assumed orthonormal to one another the expression

$$\left\langle \frac{1}{2}, \frac{1}{2} \middle| \frac{1}{2}, -\frac{1}{2} \right\rangle = 0$$

Similar calculations for the other matrix elements lead to the $\hat{\mathbf{I}}_x$ matrix:

$$\hat{\mathbf{I}}_x = \begin{bmatrix} 0 & \frac{1}{2} \\ \frac{1}{2} & 0 \end{bmatrix} = \frac{1}{2} \begin{bmatrix} 0 & 1 \\ 1 & 0 \end{bmatrix}$$

and for $\hat{\mathbf{I}}_y$:

$$\hat{\mathbf{I}}_y = \begin{bmatrix} 0 & -\frac{i}{2} \\ \frac{i}{2} & 0 \end{bmatrix} = \frac{1}{2} \begin{bmatrix} 0 & -i \\ i & 0 \end{bmatrix}$$

These, along with the matrix representing $\hat{\mathbf{I}}_z$ are the well-known Pauli matrices. For spin-1/2 nuclei we often abbreviate $\left| \frac{1}{2}, \frac{1}{2} \right\rangle$ to $|\alpha\rangle$ and $\left| \frac{1}{2}, -\frac{1}{2} \right\rangle$ to $|\beta\rangle$ and the nuclear spin operators are $\hat{\mathbf{I}}_x$ or $\hat{\mathbf{I}}_y$ or $\hat{\mathbf{I}}_z$. There are, of course, corresponding ladder operators, $\hat{\mathbf{I}}_+$ and $\hat{\mathbf{I}}_-$.

Hopefully, from the above discussion and the discussion of angular momentum operators you can see that:

$$\begin{aligned}
&\hat{\mathbf{I}}_z|\alpha\rangle = \frac{1}{2}|\alpha\rangle \quad \hat{\mathbf{I}}_z|\beta\rangle = -\frac{1}{2}|\beta\rangle \\
&\hat{\mathbf{I}}_+|\alpha\rangle = 0 \quad \hat{\mathbf{I}}_+|\beta\rangle = |\alpha\rangle \\
&\hat{\mathbf{I}}_-|\alpha\rangle = |\beta\rangle \quad \hat{\mathbf{I}}_-|\beta\rangle = 0
\end{aligned} \tag{15.20}$$

One of those frustrating "it can be shown that" statements that one runs across is "it can be shown that" $\hat{\mathbf{I}}_x|\alpha\rangle = \frac{1}{2}|\beta\rangle$ but then it is not shown or stated where it is shown. We show it here using the ladder operator definitions:

$$\begin{aligned}
&\hat{\mathbf{I}}_+ = \hat{\mathbf{I}}_x + i\hat{\mathbf{I}}_y \\
&\hat{\mathbf{I}}_- = \hat{\mathbf{I}}_x - i\hat{\mathbf{I}}_y
\end{aligned} \tag{15.21}$$

Adding these together and solving for $\hat{\mathbf{I}}_x$:

$$\hat{\mathbf{I}}_x = \frac{\hat{\mathbf{I}}_+ + \hat{\mathbf{I}}_-}{2} \tag{15.22}$$

So:

$$\hat{\mathbf{I}}_x|\alpha\rangle = \frac{\hat{\mathbf{I}}_+ + \hat{\mathbf{I}}_-}{2}|\alpha\rangle$$

and since these are linear operators:

$$\hat{\mathbf{I}}_x|\alpha\rangle = \frac{1}{2}\hat{\mathbf{I}}_+|\alpha\rangle + \frac{1}{2}\hat{\mathbf{I}}_-|\alpha\rangle$$

Remembering our definitions of α and β and the effect of the ladder operators from above:

$$\hat{\mathbf{I}}_x|\alpha\rangle = 0 + \frac{1}{2}|\beta\rangle$$

The "0" comes about because the eigenvalue of $\hat{\mathbf{I}}_+|\alpha\rangle$ is zero for spin-1/2 nuclei. This is, of course, because the ladder operator, $\hat{\mathbf{I}}_+$ increases m by one but for the $|\alpha\rangle$ spin state m is already as high as it can be. Thus, operating on it with a "raising" operator returns nothing. Thus:

$$\hat{\mathbf{I}}_x|\alpha\rangle = \frac{1}{2}|\beta\rangle \tag{15.23}$$

and similarly:

$$\begin{aligned}\hat{\mathbf{I}}_x|\beta\rangle &= \frac{1}{2}|\alpha\rangle\\ \hat{\mathbf{I}}_y|\alpha\rangle &= \frac{i}{2}|\beta\rangle\\ \hat{\mathbf{I}}_y|\beta\rangle &= -\frac{i}{2}|\alpha\rangle\end{aligned} \tag{15.24}$$

As you can see, $|\alpha\rangle$ and $|\beta\rangle$ are *not* eigenfunctions of $\hat{\mathbf{I}}_x$ and $\hat{\mathbf{I}}_y$ as we should expect to be the case since they do not commute with $\hat{\mathbf{I}}_z$.

15.5 Expectation Values

We can now write the state of our single spin-1/2 as a superposition of basis states $|\alpha\rangle$ and $|\beta\rangle$:

$$|\Psi\rangle = c_\alpha|\alpha\rangle + c_\beta|\beta\rangle$$

followed by an expectation value calculation:

$$\begin{aligned}\langle I_z\rangle &= \left[\langle\alpha|c_\alpha^* + \langle\beta|c_\beta^*\right]\hat{\mathbf{I}}_z\left[c_\alpha|\alpha\rangle + c_\beta|\beta\rangle\right]\\ &= \left[\langle\alpha|c_\alpha^* + \langle\beta|c_\beta^*\right]\left[\frac{1}{2}c_\alpha|\alpha\rangle - \frac{1}{2}c_\beta|\beta\rangle\right]\\ &= \frac{1}{2}c_\alpha^* c_\alpha\langle\alpha\,|\,\alpha\rangle - \frac{1}{2}c_\alpha^* c_\beta\langle\alpha\,|\,\beta\rangle + \frac{1}{2}c_\beta^* c_\alpha\langle\beta\,|\,\alpha\rangle - \frac{1}{2}c_\beta^* c_\beta\langle\beta\,|\,\beta\rangle\\ &= \frac{1}{2}c_\alpha^* c_\alpha - 0 + 0 - \frac{1}{2}c_\beta^* c_\beta\\ &= \frac{1}{2}(c_\alpha^* c_\alpha - c_\beta^* c_\beta)\\ &= \frac{1}{2}(c_\alpha^2 - c_\beta^2)\end{aligned} \tag{15.25}$$

where c_α and c_β represent the probabilities of finding the spin in state $|\alpha\rangle$ or $|\beta\rangle$, respectively. Thus, the expectation value is half the difference between these two probabilities. Similarly for $\hat{\mathbf{I}}_x$ and $\hat{\mathbf{I}}_y$:

$$\begin{aligned}\langle I_x\rangle &= \left[\langle\alpha|c_\alpha^* + \langle\beta|c_\beta^*\right]\hat{\mathbf{I}}_x\left[c_\alpha|\alpha\rangle + c_\beta|\beta\rangle\right] \\ &= \left[\langle\alpha|c_\alpha^* + \langle\beta|c_\beta^*\right]\left[\frac{1}{2}c_\alpha|\beta\rangle + \frac{1}{2}c_\beta|\alpha\rangle\right] \\ &= \frac{1}{2}c_\alpha^*c_\alpha\langle\alpha\,|\,\beta\rangle + \frac{1}{2}c_\alpha^*c_\beta\langle\alpha\,|\,\alpha\rangle + \frac{1}{2}c_\beta^*c_\alpha\langle\beta\,|\,\beta\rangle + \frac{1}{2}c_\beta^*c_\beta\langle\beta\,|\,\alpha\rangle \qquad (15.26) \\ &= 0 + \frac{1}{2}c_\alpha^*c_\beta + \frac{1}{2}c_\beta^*c_\alpha + 0 \\ &= \frac{1}{2}(c_\alpha^*c_\beta + c_\beta^*c_\alpha)\end{aligned}$$

$$\begin{aligned}\langle I_y\rangle &= \left[\langle\alpha|c_\alpha^* + \langle\beta|c_\beta^*\right]\hat{\mathbf{I}}_y\left[c_\alpha|\alpha\rangle + c_\beta|\beta\rangle\right] \\ &= \left[\langle\alpha|c_\alpha^* + \langle\beta|c_\beta^*\right]\left[\frac{i}{2}c_\alpha|\beta\rangle - \frac{i}{2}c_\beta|\alpha\rangle\right] \\ &= \frac{i}{2}c_\alpha^*c_\alpha\langle\alpha\,|\,\beta\rangle - \frac{i}{2}c_\alpha^*c_\beta\langle\alpha\,|\,\alpha\rangle + \frac{i}{2}c_\beta^*c_\alpha\langle\beta\,|\,\beta\rangle - \frac{i}{2}c_\beta^*c_\beta\langle\beta\,|\,\alpha\rangle \qquad (15.27) \\ &= 0 - \frac{i}{2}c_\alpha^*c_\beta + \frac{i}{2}c_\beta^*c_\alpha + 0 \\ &= \frac{i}{2}(c_\beta^*c_\alpha - c_\alpha^*c_\beta)\end{aligned}$$

From the discussion of complex numbers in Chapter 1, we recall that complex numbers may be represented in Argand notation or polar notation. If we choose polar notation we can write our complex coefficients as:

$$c_\alpha = r_\alpha e^{i\phi_\alpha} \quad c_\alpha^* = r_\alpha e^{-i\phi_\alpha}$$

$$c_\beta = r_\beta e^{i\phi_\beta} \quad c_\beta^* = r_\beta e^{-i\phi_\beta}$$

and then rewrite our expectation values in polar notation and using trigonometric identities (A.22) and (A.23) from Appendix A:

$$\begin{aligned}\langle I_z\rangle &= \frac{1}{2}(c_\alpha^2 - c_\beta^2) \\ &= \frac{1}{2}(r_\alpha^2 e^{i2\phi_\alpha} - r_\beta^2 e^{i2\phi_\beta})\end{aligned} \qquad (15.28)$$

$$\begin{aligned}
\langle I_x \rangle &= \frac{1}{2}(c_\alpha^* c_\beta + c_\beta^* c_\alpha) \\
&= \frac{1}{2}\left(r_\alpha \mathrm{e}^{-i\phi_\alpha} r_\beta \mathrm{e}^{i\phi_\beta} + r_\beta \mathrm{e}^{-i\phi_\beta} r_\alpha \mathrm{e}^{i\phi_\alpha}\right) \\
&= \frac{1}{2}\left[r_\alpha r_\beta (\mathrm{e}^{i(\phi_\beta - \phi_\alpha)} + \mathrm{e}^{-i(\phi_\beta - \phi_\alpha)})\right] \\
&= r_\alpha r_\beta \cos(\phi_\beta - \phi_\alpha)
\end{aligned} \tag{15.29}$$

$$\begin{aligned}
\langle I_y \rangle &= \frac{1}{2}(c_\beta^* c_\alpha - c_\alpha^* c_\beta) \\
&= \frac{i}{2}\left(r_\beta \mathrm{e}^{-i\phi_\beta} r_\alpha \mathrm{e}^{i\phi_\alpha} - r_\alpha \mathrm{e}^{-i\phi_\alpha} r_\beta \mathrm{e}^{i\phi_\beta}\right) \\
&= \frac{i}{2}\left[r_\alpha r_\beta (\mathrm{e}^{-i(\phi_\beta - \phi_\alpha)} - \mathrm{e}^{i(\phi_\beta - \phi_\alpha)})\right] \\
&= r_\alpha r_\beta \sin(\phi_\beta - \phi_\alpha)
\end{aligned} \tag{15.30}$$

We shall have occasion to look more closely at these expressions in a later chapter.

15.6 Time Evolution

In Chapter 13 we developed the stationary Schrodinger equations in terms of the superposition principle. We did not, however, discuss the time evolution of states but rather deferred it until we have a specific form for the Hamiltonian operator. We will now develop the Hamiltonians for $\hat{\mathbf{I}}_z$, $\hat{\mathbf{I}}_x$, and $\hat{\mathbf{I}}_y$ and, coupling them with the effect of these operators on the states of a single spin-1/2 (eqn (15.20), (15.23) and (15.24)); we shall be able to obtain some expressions for the behaviour of a spin system with respect to time. We will not include relaxation in our calculations and so will produce equations that are artificial models of the physics of a spin system. Nonetheless, they will prove very useful to us.

Recall from eqn (11.46) that the classical expression for the energy of a magnetic moment in an external magnetic field is:

$$E = -\vec{\mu} \cdot \vec{\mathbf{B}}$$

where $\vec{\mu}$ is the magnetic moment vector and $\vec{\mathbf{B}}$ is the external field vector. Recall also from eqn (11.42) that the magnetic moment is proportional to the angular momentum:

$$\begin{aligned}\vec{\mu} &= \frac{q}{2m}\vec{\mathbf{L}} \\ &= \gamma\vec{\mathbf{L}} \\ &= \gamma L_x + \gamma L_y + \gamma L_z \\ &= \mu_x + \mu_y + \mu_z\end{aligned}$$

where γ is the gyromagnetic ratio. What we want to do, of course, is convert this to the corresponding Hamiltonian expression. Substituting the proportionality equation into the energy equation:

$$E = -\gamma L_x B_x - \gamma L_y B_u - \gamma L_z B_z$$

and from the beginning of this chapter:

$$\begin{aligned}L_x &= y p_z - z p_y \\ L_y &= z p_x - x p_z \\ L_z &= x p_y - y p_x\end{aligned}$$

Our energy equation now can be written:

$$E = -\gamma B_x\,(y p_z - z p_y) - \gamma B_u\,(z p_x - x p_z) - \gamma B_z\,(x p_y - y p_x) \tag{15.31}$$

Remember that this is still the classical energy expression. If we want the analogous quantum mechanical operators we replace the momenta, p_x, p_y and p_z with $-i\hbar(\partial/\partial x)$, $-i\hbar(\partial/\partial y)$ and $-i\hbar(\partial/\partial z)$, respectively. So, the quantum mechanical energy Hamiltonian for a magnetic moment in an external field is:

$$\begin{aligned}H = &-i\hbar\gamma B_x\left(y\frac{\partial}{\partial z} - z\frac{\partial}{\partial y}\right) \\ &-i\hbar\gamma B_y\left(z\frac{\partial}{\partial x} - x\frac{\partial}{\partial z}\right) \\ &-i\hbar\gamma B_z\left(x\frac{\partial}{\partial y} - y\frac{\partial}{\partial x}\right)\end{aligned}$$

or, using eqn (15.1):

$$\hat{\mathbf{H}} = -\gamma\hbar B_x\hat{\mathbf{I}}_x - \gamma\hbar B_y\hat{\mathbf{I}}_y - \gamma\hbar B_z\hat{\mathbf{I}}_z$$

where we have used $\hat{\mathbf{I}}$ as the symbol for angular momentum operator instead of $\hat{\mathbf{J}}$. In practice, the direction of the external field is defined as the z-direction and there will therefore be no x or y component to the external field. The B_z component is often written B_0, as we have seen in the discussion of the Bloch equations. Thus, the Hamiltonian simplifies to:

$$H = -\gamma\hbar B_0\hat{\mathbf{I}}_z \tag{15.32}$$

Also, referring to eqn (11.44), we can write the simplified Hamiltonian in terms of the Larmor precession frequency:

$$H = \omega_0\hbar\hat{\mathbf{I}}_z \tag{15.33}$$

The fifth postulate of quantum mechanics listed in Chapter 13 gives us the time-dependent Schrodinger equation:

$$H|\Psi(t)\rangle = i\hbar\frac{\mathrm{d}}{\mathrm{d}t}|\Psi(t)\rangle$$

or in units of $\hbar$:

$$H|\Psi(t)\rangle = i\frac{\mathrm{d}}{\mathrm{d}t}|\Psi(t)\rangle$$

$$\frac{\mathrm{d}}{\mathrm{d}t}|\Psi(t)\rangle = -iH|\Psi(t)\rangle$$

Since we now have an explicit Hamiltonian we can insert it into the equation, we write:

$$\frac{\mathrm{d}}{\mathrm{d}t}|\Psi(t)\rangle = -i\omega_0\hat{\mathbf{I}}_z|\Psi(t)\rangle$$

Substituting the superposition state terms gives us:

$$\frac{\mathrm{d}}{\mathrm{d}t}\left(c_\alpha(t)|\alpha\rangle + c_\beta(t)|\beta\rangle\right) = -i\omega_0\hat{\mathbf{I}}_z\left(c_\alpha(t)|\alpha\rangle + c_\beta(t)|\beta\rangle\right)$$

Note here that we have assumed the time dependence to be wholly in the coefficients c_α and c_β. This is similar to the situation with classical coordinate systems. There, the basis vectors do not change with time but the coordinates of position vectors do change with time when describing motion. In other words, the coefficients of the position vectors change but the basis vectors are time invariant.¶

¶This need not be the case. We can reverse things and have the basis vectors change with time and the coordinate coefficients be static, however, this is not the usual approach, at least for introductory magnetic resonance spectroscopy.

We assume the basis states $|\alpha\rangle$ and $|\beta\rangle$ to be time-invariant as well, which means the time-dependence is in the coefficients.

We now cast the problem as an inner product calculation using $\langle\alpha|$ as the bra:

$$\langle\alpha|\frac{\mathrm{d}}{\mathrm{d}t}\left[c_\alpha(t)|\alpha\rangle + c_\beta(t)|\beta\rangle\right] = \langle\alpha| - i\omega_0\hat{\mathbf{I}}_z\left[c_\alpha(t)|\alpha\rangle + c_\beta(t)|\beta\rangle\right]$$

$$\frac{\mathrm{d}}{\mathrm{d}t}c_\alpha(t)\langle\alpha\,|\,\alpha\rangle + \frac{\mathrm{d}}{\mathrm{d}t}c_\beta(t)\langle\alpha\,|\,\beta\rangle = \langle\alpha| - i\omega_0\hat{\mathbf{I}}_z c_\alpha(t)|\alpha\rangle + \langle\alpha| - i\omega_0\hat{\mathbf{I}}_z c_\beta(t)|\beta\rangle$$

$$\frac{\mathrm{d}c_\alpha(t)}{\mathrm{d}t} = \frac{-i\omega_0 c_a(t)}{2}\langle\alpha\,|\,\alpha\rangle + \frac{i\omega_0 c_\beta(t)}{2}\langle\alpha\,|\,\beta\rangle$$

$$\frac{\mathrm{d}c_\alpha(t)}{\mathrm{d}t} = \frac{-i\omega_0 c_a(t)}{2}$$

This is a simple first-order differential equation that is easily solved (see Appendix A):

$$c_\alpha(t) = c_\alpha(0)\exp\left(\frac{-i\omega_0 t}{2}\right) \tag{15.34}$$

A similar analysis (using $\langle\beta|$ for c_β gives us:

$$c_\beta(t) = c_\beta(0)\exp\left(\frac{i\omega_0 t}{2}\right)$$

So we now have:

$$|\Psi\rangle = c_\alpha(0)\exp\left(\frac{-i\omega_0 t}{2}\right)|\alpha\rangle + c_\beta(0)\exp\left(\frac{i\omega_0 t}{2}\right)|\beta\rangle \tag{15.35}$$

We ask now "what is the time-dependence, if any, of the expectation values?". For $\langle I_z\rangle$ (and recalling eqn (15.25)):

$$\begin{aligned}
\langle I_z\rangle(t) &= \frac{1}{2}\left(c_\alpha^*(t)c_\alpha(t) - c_\beta^*(t)c_\beta(t)\right)\\
&= \frac{1}{2}\left(\left[c_\alpha(0)e^{\frac{-i\omega_0 t}{2}}\right]^*\left[c_\alpha(0)e^{\frac{-i\omega_0 t}{2}}\right] - \left[c_\beta(0)e^{\frac{i\omega_0 t}{2}}\right]^*\left[c_\beta(0)e^{\frac{i\omega_0 t}{2}}\right]\right)\\
&= \frac{1}{2}\left[c_\alpha(0)^*e^{\frac{i\omega_0 t}{2}}c_\alpha(0)e^{\frac{-i\omega_0 t}{2}} - c_\beta(0)^*e^{\frac{-i\omega_0 t}{2}}c_\beta(0)e^{\frac{i\omega_0 t}{2}}\right]\\
&= \frac{1}{2}\left[c_\alpha(0)^*c_\alpha(0) - c_\beta(0)^*c_\beta(0)\right]
\end{aligned}$$

The $c(0)$ s are constants, which means that under the influence of $\vec{\mathbf{B}}_0$ the expectation value of $\hat{\mathbf{I}}_z$ is constant. Thinking in terms of the NMR vector model of Chapter 12, this is exactly what we would anticipate. For $\langle I_x \rangle$ and $\langle I_y \rangle$ (recalling eqn (15.26) and (15.27)):

$$\langle I_x \rangle(t) = \frac{1}{2}\left(c_\alpha^\star(t)c_\beta(t) + c_\beta^\star(t)c_\alpha(t)\right)$$

$$= \frac{1}{2}\left(\left[c_\alpha(0)\mathrm{e}^{\frac{-i\omega_0 t}{2}}\right]^\star \left[c_\beta(0)\mathrm{e}^{\frac{i\omega_0 t}{2}}\right] + \left[c_\beta(0)\mathrm{e}^{\frac{i\omega_0 t}{2}}\right]^\star \left[c_\alpha(0)\mathrm{e}^{\frac{-i\omega_0 t}{2}}\right]\right)$$

$$= \frac{1}{2}\left[c_\alpha(0)^\star c_\beta(0)\mathrm{e}^{i\omega_0 t} + c_\beta(0)^\star c_\alpha(0)\mathrm{e}^{-i\omega_0 t}\right]$$

$$= \frac{1}{2}\left[c_\alpha(0)^\star c_\beta(0)(\cos(\omega_0 t) + i\sin(\omega_0 t)) + c_\beta(0)^\star c_\alpha(0)(\cos(\omega_0 t) - i\sin(\omega_0 t))\right]$$

$$= \cos(\omega_0 t)\frac{1}{2}\left(c_\alpha(0)^\star c_\beta(0) + c_\beta(0)^\star c_\alpha(0)\right) - \sin(\omega_0 t)\frac{i}{2}\left(c_\beta(0)^\star c_\alpha(0) - c_\alpha(0)^\star c_\beta(0)\right)$$

$$= \langle I_x \rangle(0)\cos(\omega_0 t) - \langle I_y \rangle(0)\sin(\omega_0 t)$$

$$\langle I_y \rangle(t) = \frac{i}{2}\left(c_\beta^\star(t)c_\alpha(t) - c_\alpha^\star(t)c_\beta(t)\right)$$

$$= \frac{i}{2}\left(\left[c_\beta(0)\mathrm{e}^{\frac{i\omega_0 t}{2}}\right]^\star \left[c_\alpha(0)\mathrm{e}^{\frac{-i\omega_0 t}{2}}\right] - \left[c_\alpha(0)\mathrm{e}^{\frac{-i\omega_0 t}{2}}\right]^\star \left[c_\beta(0)\mathrm{e}^{\frac{i\omega_0 t}{2}}\right]\right)$$

$$= \frac{i}{2}\left[c_\beta(0)^\star c_\alpha(0)\mathrm{e}^{-i\omega_0 t} - c_\alpha(0)^\star c_\beta(0)\mathrm{e}^{i\omega_0 t}\right]$$

$$= \frac{i}{2}\left[c_\beta(0)^\star c_\alpha(0)(\cos(\omega_0 t) - i\sin(\omega_0 t)) - c_\alpha(0)^\star c_\beta(0)(\cos(\omega_0 t) + i\sin(\omega_0 t))\right]$$

$$= \cos(\omega_0 t)\frac{i}{2}\left(c_\beta(0)^\star c_\alpha(0) - c_\alpha(0)^\star c_\beta(0)\right) + \sin(\omega_0 t)\frac{1}{2}\left(c_\alpha(0)^\star c_\beta(0) + c_\beta(0)^\star c_\alpha(0)\right)$$

$$= \langle I_y \rangle(0)\cos(\omega_0 t) + \langle I_x \rangle(0)\sin(\omega_0 t)$$

Recalling the precession of a magnetic moment from Chapter 12 and specifically, Figure 11.32, these are exactly what we would expect under the influence of an external magnetic field, $\vec{\mathbf{B}}_0$. The first equation says that $\langle I_x \rangle(t)$ is a combination of $\langle I_x \rangle(0)$ and $\langle I_y \rangle(0)$, and that as time progresses the mixture of the two changes in an oscillatory fashion. In terms of the vector model of Chapter 12, we can picture a vector rotating in a right-hand sense about the z-axis.

This sense of rotation is a bit difficult to visualise. At first glance it appears that it should be a left-hand rotation but keep in mind that the equation is describing the time-evolution of $\langle I_x \rangle$ that is a scalar value (or a magnitude if you like) not a vector, and that the value of $\langle I_x \rangle(t)$ is a magnitude calculation that depends on $\langle I_x \rangle(0)$, $\langle I_y \rangle(0)$ and the amount of time that has passed since $t=0$. The result of the calculation tells us the magnitude of the projection of the vector onto the $\langle I_x \rangle(t)$ -axis in Figure 15.1.

So, at $t=0$:

$$\begin{aligned}\langle I_x \rangle(0) &= \langle I_x \rangle(0)\cos(\omega_0 0) - \langle I_y \rangle(0)\sin(\omega_0 0)\\ &= \langle I_x \rangle(0)\end{aligned}$$

At $t=\pi/(2\omega_0)$:

$$\begin{aligned}\langle I_x \rangle\left(\frac{\pi}{2\omega_0}\right) &= \langle I_x \rangle(0)\cos\left(\omega_0\frac{\pi}{2\omega_0}\right) - \langle I_y \rangle(0)\sin\left(\omega_0\frac{\pi}{2\omega_0}\right)\\ &= -\langle I_y \rangle(0)\end{aligned}$$

and (the scalar) $\langle I_x \rangle$ is equal to the negative of the initial value of $\langle I_y \rangle$. This can only be the case if, in a vector-picture the vector describes a rotation in the right-handed sense. You can use Figure 15.1 to help in seeing this. The second equation says much the same for $\langle I_y \rangle$ and implies the same handedness in the associated vector rotation.

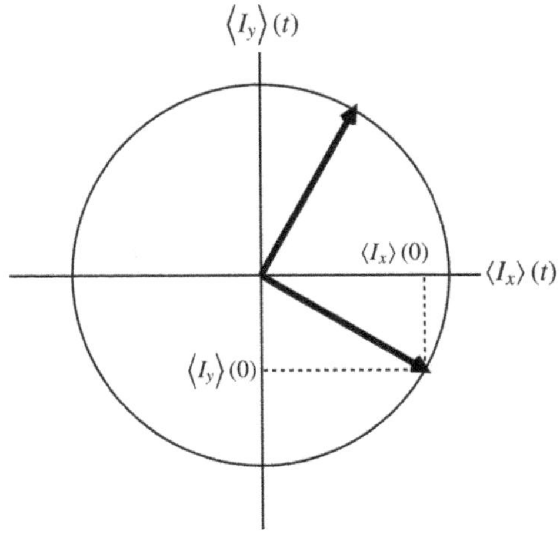

Figure 15.1 Vector picture of the time evolution of $\langle I_x \rangle$.

We have just described the situation with respect to the time evolution of the system under the influence of an external magnetic field. There is another field to consider ... the applied $\vec{\mathbf{B}}_1$ field. In other words we now wish to know about the time evolution of the spin system under the influence of a pulse of rf energy.

The Hamiltonian that we use is very similar to the one we used to determine the time dependence of the spin system on the external field. Recall that the $\vec{\mathbf{B}}_1$ field is at right angles to the $\vec{\mathbf{B}}_0$ field and, as is usual, the coordinate system is such that the z-axis is parallel to the $\vec{\mathbf{B}}_0$ field. Also, recall that in the rotating frame of reference, the $\vec{\mathbf{B}}_0$ field disappears from the equations (at the rotating frame frequency, that is). So, using the rotating frame of reference we can write a Hamiltonian expression:

$$\hat{\mathbf{H}} = \omega_1 \hat{\mathbf{I}}_x$$

Of course, we have arbitrarily assumed that the $\vec{\mathbf{B}}_1$ field is aligned with the x-axis but then we are free to choose where the x and y axes lie in the transverse plane. We could just as well have chosen the $\vec{\mathbf{B}}_1$ field to lie along the y-axis in which case we would write:

$$\hat{\mathbf{H}} = \omega_1 \hat{\mathbf{I}}_y$$

Note that the frequency here is ω_1, not ω_0, because the frequency of rotation is determined by the rf-field strength and the $\vec{\mathbf{B}}_1$-field is generally appreciably different from that of the $\vec{\mathbf{B}}_0$ external field. Our analysis proceeds much as above:

$$\langle\alpha|\frac{\mathrm{d}}{\mathrm{d}t}\left[c_\alpha(t)|\alpha\rangle + c_\beta(t)|\beta\rangle\right] = \langle\alpha| - i\omega_1\hat{\mathbf{I}}_x\left[c_\alpha(t)|\alpha\rangle + c_\beta(t)|\beta\rangle\right]$$

$$\frac{\mathrm{d}}{\mathrm{d}t}c_\alpha(t)\langle\alpha|\alpha\rangle + \frac{\mathrm{d}}{\mathrm{d}t}c_\beta(t)\langle\alpha|\beta\rangle = \langle\alpha| - i\omega_1\hat{\mathbf{I}}_x c_\alpha(t)|\alpha\rangle + \langle\alpha| - i\omega_1\hat{\mathbf{I}}_x c_\beta(t)|\beta\rangle$$

$$\frac{\mathrm{d}c_\alpha(t)}{\mathrm{d}t} = \frac{-i\omega_1 c_a(t)}{2}\langle\alpha|\beta\rangle - \frac{i\omega_1 c_\beta(t)}{2}\langle\alpha|\alpha\rangle$$

$$\frac{\mathrm{d}c_\alpha(t)}{\mathrm{d}t} = \frac{-i\omega_1 c_\beta(t)}{2}$$

and similarly, using $\langle\beta|$ instead of $\langle\alpha|$:

$$\frac{\mathrm{d}c_\beta(t)}{\mathrm{d}t} = \frac{-i\omega_1 c_\alpha(t)}{2}$$

These are *coupled differential equations* ... their solutions are:‖

$$c_\alpha(t) = c_\alpha(0)\cos\left(\frac{\omega_1 t}{2}\right) - ic_\beta(0)\sin\left(\frac{\omega_1 t}{2}\right)$$
$$c_\beta(t) = c_\beta(0)\cos\left(\frac{\omega_1 t}{2}\right) - ic_\alpha(0)\sin\left(\frac{\omega_1 t}{2}\right) \quad (15.36)$$

Now, what do the expectation values look like? Starting with eqn (15.25) and using eqn (15.36) and trigonometric identities (A.14) and (A.15) from Appendix A:

$$\begin{aligned}
\langle I_z\rangle(t) &= \frac{1}{2}(c_\alpha^*(t)c_\alpha(t) - c_\beta^*(t)c_\beta(t))\\
&= \frac{1}{2}\left[\left(c_\alpha^*(0)\cos\left(\frac{\omega_1 t}{2}\right) + ic_\beta^*(0)\sin\left(\frac{\omega_1 t}{2}\right)\right)\left(c_\alpha(0)\cos\left(\frac{\omega_1 t}{2}\right) - ic_\beta(0)\sin\left(\frac{\omega_1 t}{2}\right)\right)\right]\\
&\quad -\frac{1}{2}\left[\left(c_\beta^*(0)\cos\left(\frac{\omega_1 t}{2}\right) + ic_\alpha^*(0)\sin\left(\frac{\omega_1 t}{2}\right)\right)\left(c_\beta(0)\cos\left(\frac{\omega_1 t}{2}\right) - ic_\alpha(0)\sin\left(\frac{\omega_1 t}{2}\right)\right)\right]\\
&= \frac{1}{2}[c_\alpha^*(0)c_\alpha(0) - c_\beta^*(0)c_\beta(0)]\left(\cos^2\left(\frac{\omega_1 t}{2}\right) - \sin^2\left(\frac{\omega_1 t}{2}\right)\right)\\
&\quad +\frac{1}{2}[c_\beta^*(0)c_\alpha(0) - c_\alpha^*(0)c_\beta(0)]\left(i\sin\left(\frac{\omega_1 t}{2}\right) - \cos\left(\frac{\omega_1 t}{2}\right)\right)\\
&= \frac{1}{2}[c_\alpha^*(0)c_\alpha(0) - c_\beta^*(0)c_\beta(0)]\cos(\omega_1 t)\\
&\quad +\frac{1}{2}[c_\beta^*(0)c_\alpha(0) - c_\alpha^*(0)c_\beta(0)]\sin(\omega_1 t)
\end{aligned}$$

Using eqn (15.26) and (15.27):

$$\langle I_z\rangle(t) = \langle I_z\rangle(0)\cos(\omega_1 t) + \langle I_y\rangle(0)\sin(\omega_1 t)$$

Similar calculations for $\langle I_x\rangle(t)$ and $\langle I_y\rangle(t)$ give:

$$\langle I_x\rangle(t) = \langle I_x\rangle(0)$$
$$\langle I_y\rangle(t) = \langle I_x\rangle(0)\cos(\omega_1 t) - \langle I_z\rangle(0)\sin(\omega_1 t)$$

We can now see that using the Hamiltonian $H = \omega_1 I_x$ on our single spin system, the expectation values correspond to a rotation of a

‖You are encouraged to work this out for yourself using the example in Appendix A (being very, very careful with signs!).

vector about the x-axis at a frequency of ω_1 radians per second. As might be anticipated, a vector lying along the x-axis will not rotate about the x-axis and this is reflected in the expectation value for $\hat{\mathbf{I}}_x$. A vector lying along the y-axis will be rotated towards the $+z$-axis and a vector lying along the $+z$-axis will be rotated towards the $-y$-axis.

Given the results of the above analyses for I_z and I_x, and our vector picture, we can predict what the results will be for I_y:

$$\langle I_z \rangle(t) = \langle I_z \rangle(0)\cos(\omega_1 t) - \langle I_x \rangle(0)\sin(\omega_1 t)$$

$$\langle I_y \rangle(t) = \langle I_y \rangle(0)$$

$$\langle I_x \rangle(t) = \langle I_x \rangle(0)\cos(\omega_1 t) + \langle I_z \rangle(0)\sin(\omega_1 t)$$

Further Reading

1. P. J. Hore, J. A. Jones and S. Wimperis, *NMR: The Toolkit*, Oxford Science Publications, London, 2009.
2. J. Keeler, *Understanding NMR Spectroscopy*, John Wiley & Sons, Chichester, 2nd edn, 2010.
3. I. N. Levine, *Quantum Chemistry*, Allyn and Bacon Inc., Boston, 2nd edn, 1974.
4. N. Zetilli, *Quantum Mechanics*, John Wiley & Sons, Chichester, 2nd edn, 2009.
5. J. R. Bolton and J. A. Weil, *Electron Paramagnetic Resonance*, John Wiley & Sons Ltd., Hoboken, NJ, 2nd edn, 2007.

16 NMR Rotation Operators

Each problem that I solved became a rule, which served afterwards to solve other problems.

Rene Descartes

16.1 Rotation in Euclidean Space

As NMR/MRI spectroscopists we are very much interested in rotations. After all, we are dealing with "spins" and the mathematics that deals with angular momentum is therefore of primary importance to understanding them. We have already seen the application of rotations to vectors in Chapter 3 but the topic is of such importance to us that we will delve into it a bit more deeply here.

Rotations are very easily visualised for an object in Cartesian coordinate space. For simplicity, let us look at the rotation of a 2D vector (Figure 16.1).

This picture is actually of a counter-clockwise rotation of coordinate axes with the vector remaining "motionless" rather than a vector rotation but it is equivalent to a rotation of the vector in the opposite direction. We speak of *active rotations* and *passive rotations* to distinguish between the two. The active rotation refers to the rotation of the vector within the (static) coordinate system such that its new vector components are paired up with the same unit vectors as the starting vector. Figures 3.19 and 3.20 illustrate an active rotation. In other words, there is no change of coordinate system or basis. A passive rotation,† on the other hand, rotates the coordinate system (as in Figure 16.1) leaving

†We shall be mainly concerned with passive rotations in our studies of NMR mathematics.

Essential Mathematics for NMR and MRI Spectroscopists
By Keith C. Brown

Published by the Royal Society of Chemistry, www.rsc.org

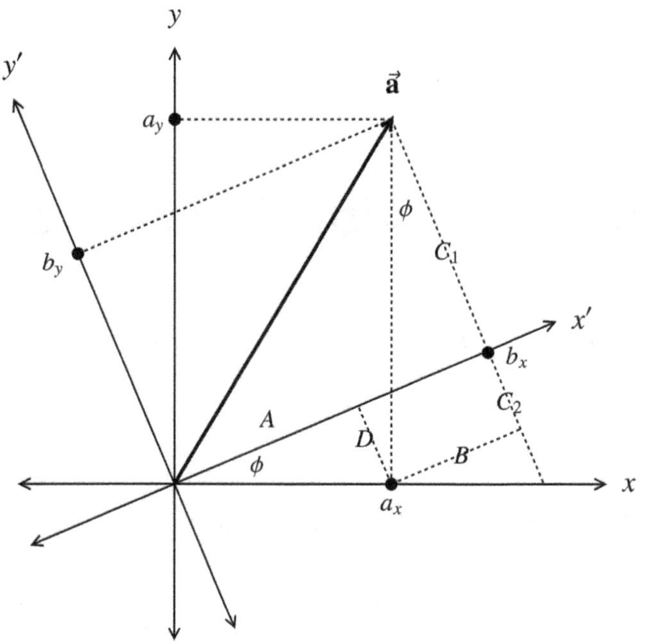

Figure 16.1 Analysis of a vector rotation.

the vector unchanged; in this case we say that the vector is *invariant* to rotations. By this we mean that rotating the orthogonal coordinates does not change the vector; its norm remains constant (see eqn (3.39)).

The fundamental problem in rotation mathematics is, given starting values of the components of vector $\vec{\mathbf{a}}$, a_x and a_y in the coordinate axes x and y, what will the new values b_x and b_y be in the new coordinate axes x' and y'. We have already seen in eqn (3.38) that the answer to this question for a 2D Euclidean vector is:

$$\begin{bmatrix} b_x \\ b_y \end{bmatrix} = \begin{bmatrix} \cos(\varphi) & \sin(\varphi) \\ -\sin(\varphi) & \cos(\varphi) \end{bmatrix} \begin{bmatrix} a_x \\ a_y \end{bmatrix}$$

or even more compactly:

$$\vec{\mathbf{b}} = \hat{\mathbf{R}}(\phi)\,\vec{\mathbf{a}}$$

Another way to look at this is to characterise the vector with a complex number (Figure 16.2).

Recalling from the discussion of complex numbers in Chapter 1 we can algebraically write:

$$|\vec{\mathbf{r}}| = x + iy$$

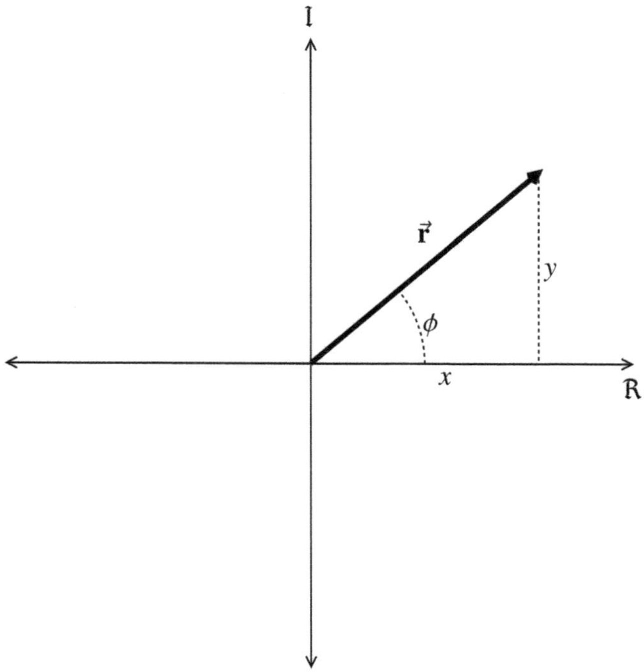

Figure 16.2 Vector $\vec{\mathbf{r}}$ depicted as being composed of complex coordinates.

and simple trigonometry gives us:

$$x = |\vec{\mathbf{r}}|\cos(\theta)$$

$$y = |\vec{\mathbf{r}}|\sin(\theta)$$

Substituting into the first equation:

$$\begin{aligned} r &= |\vec{\mathbf{r}}|\cos(\theta) + i|\vec{\mathbf{r}}|\sin(\theta) \\ &= |\vec{\mathbf{r}}|[\cos(\theta) + i\sin(\theta)] \\ &= |\vec{\mathbf{r}}|\mathrm{e}^{i\theta} \end{aligned}$$

where we have used the Euler identity from Appendix A.

We now have an exponential representation of our vector. What about rotation? Let us multiply $|\vec{\mathbf{r}}|\ \mathrm{e}^{i\theta}$ by e^{ω}:

$$r' = |\vec{\mathbf{r}}|\ \mathrm{e}^{i\theta}\,\mathrm{e}^{\omega} = |\vec{\mathbf{r}}|\ \mathrm{e}^{i(\theta+\omega)}$$

This must represent a new vector that has phase angle $\theta + \omega$ (see Figure 16.3).

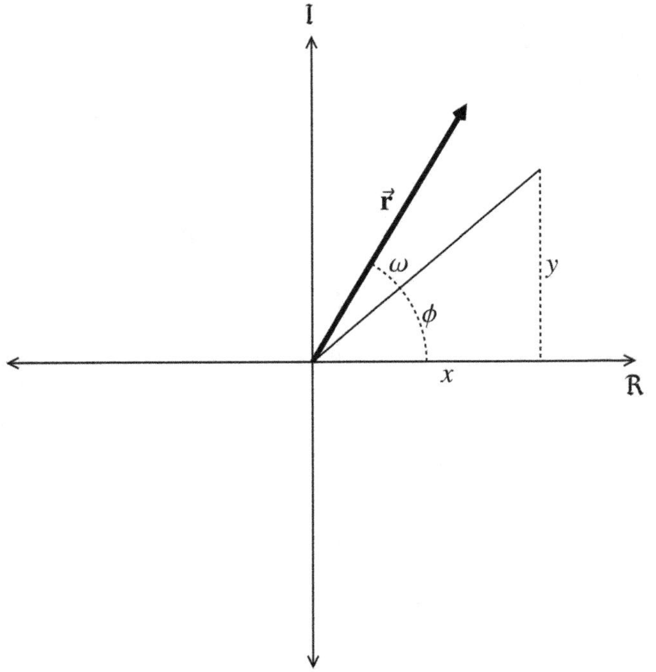

Figure 16.3 Active rotation of vector $\vec{\mathbf{r}}$ by ω radians.

So, our original vector, at phase angle θ has been rotated by ω radians counter-clockwise. e^{ω} can be considered to have operated on the original vector, $|\vec{\mathbf{r}}|\ e^{i\theta}$, to produce a rotation of that vector. Obviously, to produce a clockwise rotation we would use $-\omega$ instead. Note that as shown in Figure 16.3 this is an active rotation. For a passive rotation in which the coordinate axes rotate counter-clockwise we would write:

$$r' = |\vec{\mathbf{r}}|\ e^{i\theta}\ e^{\omega} = |\vec{\mathbf{r}}|\ e^{i(\theta-\omega)}$$

16.2 The Euler Rotation Matrix

The Euler angles were first introduced by Leonhard Euler to describe the orientation of a body in space.[‡] They are specified using rotation angles α the angle of rotation of the x and y axes about the z-axis, β the rotation angle of the z' and x' axes about the y'-axis and γ, the angle of

[‡]It must be pointed out that there it appears that there is no standard way of specifying Euler angles and that one will see what amounts to a confusing mess in the literature regarding this topic. I am using the right-handed coordinate system method used by Rose here and shall continue to do so everywhere in this text.

rotation of the x'' and y'' axes about the z''-axis. This is much easier to see pictorially. The first rotation produces new axes x' and y' and leaves the z-axis unchanged (however, we will still label it z' coincident with z) (Figure 16.4).

The next rotation is about the newly created y'-axis, which creates a new z'' and another new x''-axis (and y'' coincident with y'). Note that we have changed our point of view a bit so that we can clearly see the rotation (Figure 16.5).

The final rotation through angle γ takes place about the z''-axis and produces new x''' and y''' axes (and z''' coincident with z'') (see Figure 16.6).

Since these are successive rotations we can represent them with matrices as in Chapter 2:

$$\hat{\mathbf{R}}_E = \hat{\mathbf{R}}_{z''}(\gamma)\hat{\mathbf{R}}_{y'}(\beta)\hat{\mathbf{R}}_z(\alpha)$$

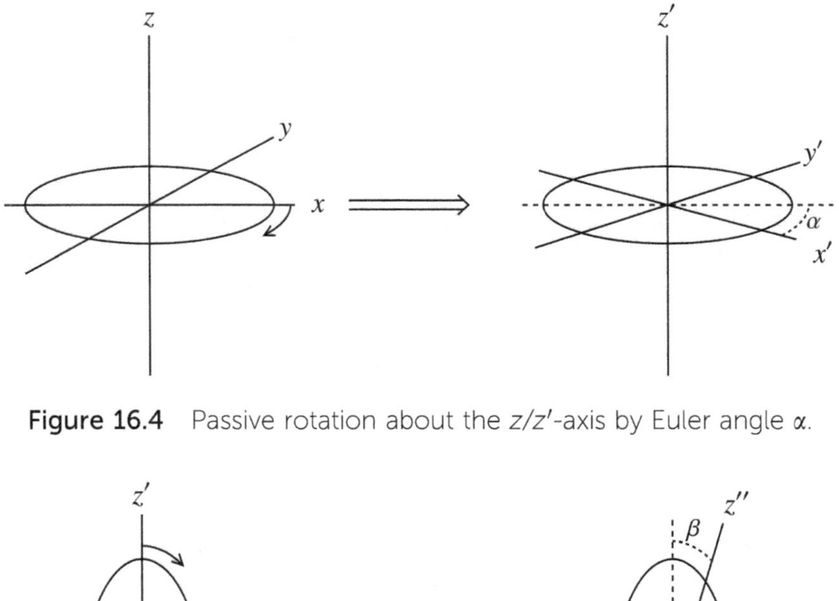

Figure 16.4 Passive rotation about the z/z'-axis by Euler angle α.

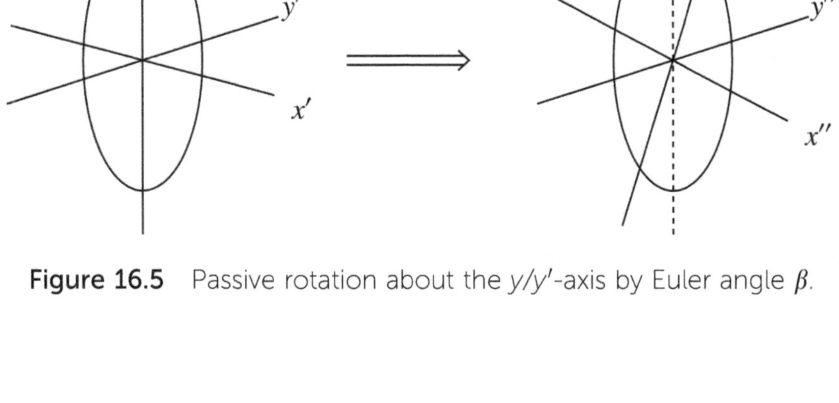

Figure 16.5 Passive rotation about the y/y'-axis by Euler angle β.

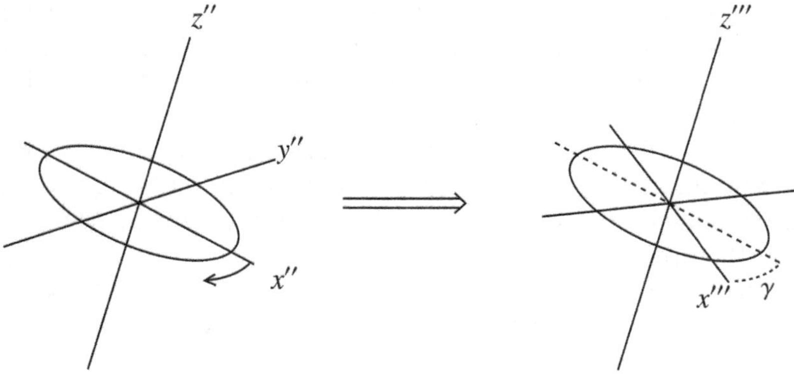

Figure 16.6 Passive rotation about the z''/z'''-axis by Euler angle γ.

Of course, the reverse rotations to regenerate the original co-ordinates use the inverse matrices:

$$\hat{\mathbf{R}}_E^{-1} = \hat{\mathbf{R}}_z^{-1}(\alpha)\hat{\mathbf{R}}_{y'}^{-1}(\beta)\hat{\mathbf{R}}_{z''}^{-1}(\gamma)$$

and

$$\hat{\mathbf{R}}_E^{-1}\hat{\mathbf{R}}_E = \hat{\mathbf{1}}$$

We have already seen some of this in the form of eqn (3.37). This was for a two-dimensional rotation, however, we could view it as an Euler rotation in which $\phi = \alpha$, $\beta = 0$ and $\gamma = 0$. That is, a rotation about the z-axis only. In this case the rotation matrix is:[§]

$$\hat{\mathbf{R}}_z(\alpha) = \begin{bmatrix} \cos(\alpha) & \sin(\alpha) & 0 \\ -\sin(\alpha) & \cos(\alpha) & 0 \\ 0 & 0 & 1 \end{bmatrix} \tag{16.1}$$

This is a unitary matrix, as you will recall from Chapter 2, which means that its transpose and inverse are equivalent; a characteristic of all rotation matrices. Therefore, we expect that rotation matrices $\hat{\mathbf{R}}_{y'}(\beta)$, $\hat{\mathbf{R}}_{z'}(\gamma)$ and the overall Euler rotation matrix $\hat{\mathbf{R}}_E$ should each be unitary. Taking our cue from eqn (16.1) we can take a try at writing $\hat{\mathbf{R}}_{y'}(\beta)$, $\hat{\mathbf{R}}_{z'}(\gamma)$ and $\hat{\mathbf{R}}_E$.

$$\hat{\mathbf{R}}_{y'}(\beta) = \begin{bmatrix} \cos(\beta) & 0 & \sin(\beta) \\ 0 & 1 & 0 \\ -\sin(\beta) & 0 & \cos(\beta) \end{bmatrix} \quad \hat{\mathbf{R}}_{z''}(\gamma) = \begin{bmatrix} \cos(\gamma) & \sin(\gamma) & 0 \\ -\sin(\gamma) & \cos(\gamma) & 0 \\ 0 & 0 & 1 \end{bmatrix}$$

[§]You can easily verify this using the direction cosine idea and the fact that the final z-axis is the same as the initial z-axis. Give it a try.

$$\hat{\mathbf{R}}_E = \hat{\mathbf{R}}_{z''}(\gamma)\hat{\mathbf{R}}_{y'}(\beta)\hat{\mathbf{R}}_z(\alpha)$$

$$= \begin{bmatrix} \cos(\gamma) & \sin(\gamma) & 0 \\ -\sin(\gamma) & \cos(\gamma) & 0 \\ 0 & 0 & 1 \end{bmatrix} \begin{bmatrix} \cos(\beta) & 0 & \sin(\beta) \\ 0 & 1 & 0 \\ -\sin(\beta) & 0 & \cos(\beta) \end{bmatrix} \begin{bmatrix} \cos(\alpha) & \sin(\alpha) & 0 \\ -\sin(\alpha) & \cos(\alpha) & 0 \\ 0 & 0 & 1 \end{bmatrix}$$

$$= \begin{bmatrix} \cos(\gamma) & \sin(\gamma) & 0 \\ -\sin(\gamma) & \cos(\gamma) & 0 \\ 0 & 0 & 1 \end{bmatrix} \begin{bmatrix} \cos(\alpha)\cos(\beta) & \sin(\alpha)\cos(\beta) & \sin(\beta) \\ -\sin(\alpha) & \cos(\alpha) & 0 \\ -\cos(\alpha)\sin(\beta) & -\sin(\alpha)\sin(\beta) & \cos(\beta) \end{bmatrix}$$

$$= \begin{bmatrix} c(\alpha)c(\beta)c(\gamma) - s(\alpha)s(\gamma) & s(\alpha)c(\beta)c(\gamma) - c(\alpha)s(\gamma) & s(\beta)c(\gamma) \\ -c(\alpha)c(\beta)s(\gamma) - s(\alpha)c(\gamma) & -s(\alpha)c(\beta)s(\gamma) - c(\alpha)c(\gamma) & -s(\beta)s(\gamma) \\ -c(\alpha)s(\beta) & -s(\alpha)c(\beta) & c(\beta) \end{bmatrix}$$

Here, we have shortened "cos" to "*c*" and "sin" to "*s*" for brevity. The inverse of $\hat{\mathbf{R}}_E$ times $\hat{\mathbf{R}}_E$ is equal to the unit matrix.¶

Two types of rotation matrices exist, one with determinant equal to +1 and one with determinant equal to −1. The former is called a *proper rotation* matrix and the latter an *improper rotation* matrix. The proper rotation is a rotation pure and simple, whereas the improper rotation can be thought of as a combination of a rotation *and* a reflection. The determinant of $\hat{\mathbf{R}}_z(\alpha)$ is easily calculated to be 1 (see Chapter 2). Similarly, the determinants of $\hat{\mathbf{R}}_{z''}(\gamma)$ and $\hat{\mathbf{R}}_{y'}(\beta)$ are equal to 1 and from eqn (2.38) it is seen that the determinant of the product of these matrices is also 1. In other words the determinant of $\hat{\mathbf{R}}_E$ is equal to 1 and is therefore a proper rotation matrix.

The matrix $\hat{\mathbf{R}}_E$ that we have developed here is not to be considered as "the" Euler rotation matrix but rather as one of many possible matrices that accomplish the same thing. Our Euler rotation matrix above has been developed from a rotation about the *z*-axis followed by a rotation about the y'-axis and lastly by a rotation about the z''-axis, which we will call a $zy'z''$ rotation. We could have accomplished the same thing by starting with a rotation about the *x*-axis followed by the y'-axis and then the x''-axis or $xy'x''$. There are a total of 24 possible sets of rotation matrices; six choices for the first axis (both positive and negative), 4 choices for the next axis (positive and negative again)

¶The calculation is much too tedious and space-filling to include here. You can either take my word for it (I *have* actually checked it) or you can verify it for yourself. If you choose the latter course I would suggest investing time in learning the use of a computer algebra program.

and one for the third axis (the same one as the first). Note that swapping operators does not result in the same rotation, which means the rotation matrices do not commute.

$$\hat{\mathbf{R}}_{y'}(\beta)\hat{\mathbf{R}}_z(\alpha) \neq \hat{\mathbf{R}}_z(\alpha)\hat{\mathbf{R}}_{y'}(\beta)$$

16.3 Angular Momentum Rotation Operators

Rotation operators for angular momentum are slightly more complicated. We start with the time-dependent Schrodinger equation (postulate V, Chapter 13):

$$\frac{\mathrm{d}}{\mathrm{d}t}|\Psi\rangle = -i\hat{\mathbf{H}}|\Psi\rangle$$

in Dirac notation where $\hat{\mathbf{H}}$ is the Hamiltonian operator. Assuming no time-dependence in the Hamiltonian, this is just a simple 1st-order differential equation with a simple solution (see Appendix A):

$$|\Psi\rangle(t) = \exp(-i\hat{\mathbf{H}}t)|\Psi\rangle(0)$$

where $|\Psi\rangle(0)$ is the state of the system at time 0 and $\exp(-i\hat{\mathbf{H}}t)$ is the exponential form of the Hamiltonian operator.

Now let us see what happens if we calculate the expectation value for the operator, $\hat{\mathbf{I}}_x$ under the influence of a magnetic field. Our Hamiltonian operator becomes $-\gamma\hbar B_0\hat{\mathbf{I}}_z$ or, in frequency units, $-\omega_0\hat{\mathbf{I}}_z$ and the expectation value is:

$$\begin{aligned}
\langle I_x\rangle(t) &= \langle\Psi(t)|\hat{\mathbf{I}}_x|\Psi(t)\rangle \\
&= \left\langle \exp(-i\hat{\mathbf{H}}t)\Psi(0)\middle|\hat{\mathbf{I}}_x\middle|\exp(-i\hat{\mathbf{H}}t)\Psi(0)\right\rangle \\
&= \langle\Psi(0)|\exp(i\hat{\mathbf{H}}t)\hat{\mathbf{I}}_x\exp(-i\hat{\mathbf{H}}t)|\Psi(0)\rangle \\
&= \langle\Psi(0)|\exp(-i\omega_0\hat{\mathbf{I}}_z t)\hat{\mathbf{I}}_x\exp(i\omega_0\hat{\mathbf{I}}_z t)|\Psi(0)\rangle \\
&= \langle\Psi(0)|\hat{\mathbf{I}}'_x|\Psi(0)\rangle
\end{aligned}$$

using eqn (3.48) and where we have defined:

$$\hat{\mathbf{I}}'_x \equiv \exp(-i\omega_0\hat{\mathbf{I}}_z t)\ \hat{\mathbf{I}}_x\ \exp\ (i\omega_0\hat{\mathbf{I}}_z t)$$

$\hat{\mathbf{I}}'_x$ is related to $\hat{\mathbf{I}}_x$ by a rotation about the z-axis by $\omega_0 t = \theta$. To show this we start with the following definition:

$$g(\theta) \equiv \exp(-i\theta\hat{\mathbf{I}}_z)\hat{\mathbf{I}}_x\ \exp(i\theta\hat{\mathbf{I}}_z)$$

The first derivative of $g(\theta)$ (using the Product Rule, eqn (A.37) in Appendix A) is:

$$\begin{aligned} g'(\theta) &= \frac{\mathrm{d}}{\mathrm{d}\theta}[\exp(-i\theta\hat{\mathbf{I}}_z)]\hat{\mathbf{I}}_x\exp(i\theta\hat{\mathbf{I}}_z) + \exp(-i\theta\hat{\mathbf{I}}_z)\frac{\mathrm{d}}{\mathrm{d}\theta}[\hat{\mathbf{I}}_x]\exp(i\theta\hat{\mathbf{I}}_z) \\ &\quad + \exp(-i\theta\hat{\mathbf{I}}_z)\hat{\mathbf{I}}_x\frac{\mathrm{d}}{\mathrm{d}\theta}[\exp(i\theta\hat{\mathbf{I}}_z)] \\ &= -i\hat{\mathbf{I}}_z\exp(-i\theta\hat{\mathbf{I}}_z)\hat{\mathbf{I}}_x\exp(i\theta\hat{\mathbf{I}}_z) + 0 + \exp(-i\theta\hat{\mathbf{I}}_z)\hat{\mathbf{I}}_x i\hat{\mathbf{I}}_z\exp(i\theta\hat{\mathbf{I}}_z) \\ &= \exp(-i\theta\hat{\mathbf{I}}_z)(-i\hat{\mathbf{I}}_z\hat{\mathbf{I}}_x + i\hat{\mathbf{I}}_x\hat{\mathbf{I}}_z)\exp(i\theta\hat{\mathbf{I}}_z) \end{aligned}$$

Remembering our angular momentum commutators in frequency notation (eqn (15.4)) and being careful with minus signs:

$$-i[\hat{\mathbf{I}}_z, \hat{\mathbf{I}}_x] = -i\hat{\mathbf{I}}_z\hat{\mathbf{I}}_x + i\hat{\mathbf{I}}_x\hat{\mathbf{I}}_z = \hat{\mathbf{I}}_y$$

we substitute this into our current equation for $g'(\theta)$:

$$g'(\theta) = \exp(-i\theta\hat{\mathbf{I}}_z)\ \hat{\mathbf{I}}_y\ \exp(i\theta\hat{\mathbf{I}}_z)$$

Similarly, the second derivative is (again, being very careful with minus signs):

$$\begin{aligned} g''(\theta) &= \exp(-i\theta\hat{\mathbf{I}}_z)(-i\hat{\mathbf{I}}_z\hat{\mathbf{I}}_y + i\hat{\mathbf{I}}_y\hat{\mathbf{I}}_z)\exp(i\theta\hat{\mathbf{I}}_z) \\ &= \exp(-i\theta\hat{\mathbf{I}}_z)i(-\hat{\mathbf{I}}_z\hat{\mathbf{I}}_y + \hat{\mathbf{I}}_y\hat{\mathbf{I}}_z)\exp(i\theta\hat{\mathbf{I}}_z) \\ &= \exp(-i\theta\hat{\mathbf{I}}_z)i[\hat{\mathbf{I}}_y, \hat{\mathbf{I}}_z]\exp(i\theta\hat{\mathbf{I}}_z) \\ &= \exp(-i\theta\hat{\mathbf{I}}_z)i(i\hat{\mathbf{I}}_x)\exp(i\theta\hat{\mathbf{I}}_z) \\ &= -\exp(-i\theta\hat{\mathbf{I}}_z)\hat{\mathbf{I}}_x\exp(i\theta\hat{\mathbf{I}}_z) \\ &= -g(\theta) \end{aligned}$$

or, in standard differential equation form:

$$g''(\theta) + g(\theta) = 0$$

The most general solution to this equation is (see Appendix A):

$$\begin{aligned} g(\theta) &= A\cos(\theta) + B\sin(\theta) \\ &= \exp(-i\theta\hat{\mathbf{I}}_z)\hat{\mathbf{I}}_x\exp(i\theta\hat{\mathbf{I}}_z) \end{aligned}$$

and:

$$g'(\theta) = -A\sin(\theta) + B\cos(\theta)$$

$$= \exp(-i\theta\hat{\mathbf{I}}_z)\hat{\mathbf{I}}_y\exp(i\theta\hat{\mathbf{I}}_z)$$

From these we see that A equals $g(0)$, which is just $\hat{\mathbf{I}}_x$, and B equals $g'(0)$, which is $\hat{\mathbf{I}}_y$:

$$g(0) = A\cos(0) + B\sin(0) = \exp(0)\hat{\mathbf{I}}_x\exp(0)$$

$$A\cdot 1 + B\cdot 0 = 1\cdot\hat{\mathbf{I}}_x\cdot 1$$

$$A = \hat{\mathbf{I}}_x$$

$$g'(0) = -A\sin(0) + B\cos(0) = \exp(0)\hat{\mathbf{I}}_y\exp(0)$$

$$A\cdot 0 + B\cdot 1 = 1\cdot\hat{\mathbf{I}}_y\cdot 1$$

$$B = \hat{\mathbf{I}}_y$$

so:

$$g(\theta) = \hat{\mathbf{I}}_x\cos(\theta) + \hat{\mathbf{I}}_y\sin(\theta)$$

and:

$$\hat{\mathbf{I}}_x' = \exp(i\theta\,\hat{\mathbf{I}}_z)\,\hat{\mathbf{I}}_x\exp(-i\theta\hat{\mathbf{I}}_z) = \hat{\mathbf{I}}_x\cos(\theta) + \hat{\mathbf{I}}_y\sin(\theta\,)$$

Similarly:

$$\hat{\mathbf{I}}_y' = \exp(i\theta\hat{\mathbf{I}}_z)\,\hat{\mathbf{I}}_y\exp(-i\theta\hat{\mathbf{I}}_z) = \hat{\mathbf{I}}_y\cos(\theta) - \hat{\mathbf{I}}_x\sin(\theta)$$

This should look familiar to those who have read Chapter 2. Look above to the very beginning where vectors in Cartesian space were being discussed. This equation has exactly the same form as the simple rotation equations. Evidently, an expression such as:

$$\exp(i\theta\hat{\mathbf{I}}_z)\,\hat{\mathbf{I}}_x\exp(-i\theta\hat{\mathbf{I}}_z)$$

corresponds to a rotation about the z-axis by θ radians. Sometimes you will see this expression shortened to:

$$\hat{\mathbf{R}}(\theta)\,\hat{\mathbf{I}}_x\hat{\mathbf{R}}(-\theta)$$

The norm assigns to each vector a length. Because of this, the norm is often known as the magnitude. After a rotation in a space, Hilbert or otherwise, the norm necessarily remains the same or invariant ... as one would expect. Also, the vector products between any two vectors must be invariant to rotation operations.

Now, since we can say:

$$\hat{\mathbf{I}}'_x = \hat{\mathbf{R}}(\theta)\hat{\mathbf{I}}_x\hat{\mathbf{R}}(-\theta)$$

this may be jogging your memory. Referring to Chapter 2 in general and eqn (2.56) in particular, we see that this is actually a unitary transformation. $\hat{\mathbf{R}}_z(\theta)$ is the complex conjugate of $\hat{\mathbf{R}}_z(-\theta)$ as well as being the inverse. We may, of course, represent $\hat{\mathbf{R}}_z(\theta)$ (and $\hat{\mathbf{R}}_z(-\theta)$) with a matrix:

$$\hat{\mathbf{R}}(\theta) = \exp(-i\theta\hat{\mathbf{I}}_z) = \hat{\mathbf{1}} + i\theta\hat{\mathbf{I}}_z + \frac{(i\theta\hat{\mathbf{I}}_z)^2}{2} \dots$$

Further Reading

1. R. M. Brannon, Rotation (draft manuscript), http://www.mech.utah.edu/~brannon/public/rotation.pdf).
2. M. E. Rose, *Elementary Theory of Angular Momentum*, John Wiley & Sons, New York, 1957.
3. R. N. Zare, *Angular Momentum: Understanding Spatial Aspects in Chemistry and Physics*, Wiley Interscience, New York, 1988.

17 Density Operators and Matrices

As long as algebra and geometry have been separated, their progress have been slow and their uses limited; but when these two sciences have been united, they have lent each mutual forces, and have marched together towards perfection.

Joseph-Louis Lagrange

17.1 The Density Matrix

One of the fundamental postulates of quantum mechanics indicates that the wavefunction for the spin system, contains all of the information that can be known about the system it describes.† In other words, it is as complete a description of the state of the system as we can ever hope to have. For a spin-1/2 system (which is most often encountered in NMR spectroscopy) there are two spin states corresponding to the two possible eigenvalues of $\hat{\mathbf{I}}_z$ with Ψ as the eigenfunction. See Chapter 15 for an explanation of this. These states do not change with time and are said to be time invariant. The convention is to name the state with eigenvalue $= +1/2$ "α" and the state with eigenvalue $= -1/2$ "β".

Sometimes, it will be erroneously noted that a spin must be in either of these two states. However, there is no justification for this

†There are many complex-looking matrices and convoluted trigonometric expressions in this chapter. The point of this is, ultimately, to show you that there is an easier way to approach things in a later chapter. The expression "No pain, no gain" comes to mind when considering the material here. For the mathematically timid, all will be well. No Fear!

Essential Mathematics for NMR and MRI Spectroscopists
By Keith C. Brown

Published by the Royal Society of Chemistry, www.rsc.org

assumption ... nowhere in quantum mechanics is this required. Rather, we may think of a spin-1/2 as being in a superposition state that is a mixture of the two states, α and β:

$$\Psi = c_\alpha|\alpha\rangle + c_\beta|\beta\rangle \tag{17.1}$$

with c_α and c_β being interpreted as the probabilities of measuring the corresponding state. This is one of the most mysterious aspects of quantum mechanics. It says that a spin can be both clockwise and counter-clockwise at the same time![‡] Of course, the sum of c_α and c_β is one. It is only when we do an experiment to see which state it is in will it tell us definitely which one it is in. Note the similarity between eqn (17.1) and (3.4). We may treat the wavefunction as a vector and apply all of the vector operations that we have discussed. $|\alpha\rangle$ and $|\beta\rangle$ may be thought of as basis vectors and represented very simply as column matrices:

$$|\alpha\rangle = \begin{bmatrix} 1 \\ 0 \end{bmatrix} \quad |\beta\rangle = \begin{bmatrix} 0 \\ 1 \end{bmatrix}$$

As we have seen in Chapter 13, states α and β can normally be considered to be time-invariant so all of the time-related changes are contained in the coefficients c_α and c_β. From Chapter 3 we know that we may represent the bra and ket state functions as row and column matrices (or vectors):

$$\langle\Psi| = \begin{bmatrix} c_\alpha^* & c_\beta^* \end{bmatrix} \quad |\Psi\rangle = \begin{bmatrix} c_\alpha \\ c_\beta \end{bmatrix}$$

Here we have emphasised the complex conjugate nature of the row vector. Also, as we have seen, the expectation value for an operator operating on Ψ is:

$$\langle A\rangle = \langle\Psi|\hat{\mathbf{A}}|\Psi\rangle$$

or, in terms of the superposition state in matrix-vector notation:

$$\langle A\rangle = \begin{bmatrix} c_\alpha^* & c_\beta^* \end{bmatrix} \begin{bmatrix} \langle\alpha|A|\alpha\rangle & \langle\alpha|A|\beta\rangle \\ \langle\beta|A|\alpha\rangle & \langle\beta|A|\beta\rangle \end{bmatrix} \begin{bmatrix} c_\alpha \\ c_\beta \end{bmatrix}$$

[‡]Actually, there is a relatively new movement afoot in physics that is promoting the idea of multiple universes in which each state exists separately from the others in separate universes! Sure, why not. It is just as "reasonable" as saying that a system can exist in multiple states simultaneously. Quantum mechanics is kind of strange that way.

Or more compactly (see eqn (13.11):

$$\langle A\rangle = \begin{bmatrix} c_\alpha^* & c_\beta^* \end{bmatrix} \begin{bmatrix} A_{\alpha\alpha} & A_{\alpha\beta} \\ A_{\beta\alpha} & A_{\beta\beta} \end{bmatrix} \begin{bmatrix} c_\alpha \\ c_\beta \end{bmatrix}$$

Multiplication of these matrices gives:

$$\langle \mathrm{A}\rangle = c_\alpha c_\alpha^* A_{\alpha\alpha} + c_\beta c_\alpha^* A_{\beta\alpha} + c_\alpha c_\beta^* A_{\alpha\beta} + c_\beta c_\beta^* A_{\beta\beta} \tag{17.2}$$

Now, the reader should bear with us because this is one of those areas of NMR mathematics that has a certain beauty to it. If you look at the preceding equation you see several terms involving "c" that are generated from the row and column matrices. You can generate each of these terms by multiplying the column and row matrices together in "reverse" order:

$$\begin{bmatrix} c_\alpha^* & c_\beta^* \end{bmatrix} \begin{bmatrix} c_\alpha \\ c_\beta \end{bmatrix} = \begin{bmatrix} c_\alpha c_\alpha^* & c_\alpha c_\beta^* \\ c_\beta c_\alpha^* & c_\beta c_\beta^* \end{bmatrix}$$

In terms of the wavefunction, Ψ this could be written:

$$|\Psi\rangle\langle\Psi|$$

Recalling the discussion of vectors in Chapter 3, this is an outer product with all of the characteristics discussed there. Now, if we multiply this square matrix of coefficients into the operator matrix we get:

$$\begin{bmatrix} c_\alpha c_\alpha^* & c_\alpha c_\beta^* \\ c_\beta c_\alpha^* & c_\beta c_\beta^* \end{bmatrix} \begin{bmatrix} A_{\alpha\alpha} & A_{\alpha\beta} \\ A_{\beta\alpha} & A_{\beta\beta} \end{bmatrix}$$

$$= \begin{bmatrix} c_\alpha c_\alpha^* A_{\alpha\alpha} + c_\alpha c_\beta^* A_{\beta\alpha} & c_\alpha c_\alpha^* A_{\alpha\beta} + c_\alpha c_\beta^* A_{\beta\beta} \\ c_\beta c_\alpha^* A_{\alpha\alpha} + c_\beta c_\beta^* A_{\beta\alpha} & c_\beta c_\alpha^* A_{\alpha\beta} + c_\beta c_\beta^* A_{\beta\beta} \end{bmatrix}$$

The trace of this matrix is:

$$c_\alpha c_\alpha^* A_{\alpha\alpha} + c_\beta c_\alpha^* A_{\beta\alpha} + c_\alpha c_\beta^* A_{\alpha\beta} + c_\beta c_\beta^* A_{\beta\beta}$$

the *same* result as above from the superposition of states analysis.

Since $|\Psi\rangle\langle\Psi|$ is in the form of a square matrix and considering our previous discussion of matrices and operators, we suspect that it too can be called an operator. This is indeed the case … it is called the density operator[§] and given the symbol ρ in general quantum

[§]This is not a particularly descriptive name for this entity. Possibly more pertinent would be statistical operator but … it is what it is. The word "density" refers to probability densities like those discussed in Chapter 6.

mechanics and, frequently, σ in spin quantum mechanics. σ is referred to as the "reduced" density operator ... it operates only on spin coordinates rather than spatial and spin coordinates for ρ.

$$\hat{\sigma} = |\Psi\rangle\langle\Psi|$$

Using this definition and what has just been shown we can say:

$$\langle A\rangle = \langle\Psi|\hat{\text{A}}|\Psi\rangle = \text{Tr}(\hat{\sigma}\hat{\mathbf{A}}) \tag{17.3}$$

The diagonal terms, diagonal being from upper left to lower right, are called *populations* and in this case correspond to populations in the α and β states. The off-diagonals are referred to as *coherences* or perhaps a bit more descriptively, coherent superpositions of eigenstates.

It is easy to verify (see eqn (2.23)) by matrix multiplication that:

$$\text{Tr}(\hat{\sigma}\hat{\mathbf{A}}) = \text{Tr}(\hat{\mathbf{A}}\hat{\sigma})$$

The matrix form of $\hat{\sigma}$ is called the density matrix.

$$\hat{\sigma} = \begin{bmatrix} c_\alpha c_\alpha^* & c_\alpha c_\beta^* \\ c_\beta c_\alpha^* & c_\beta c_\beta^* \end{bmatrix}$$

or

$$\hat{\sigma} = \begin{bmatrix} \sigma_{\alpha\alpha} & \sigma_{\alpha\beta} \\ \sigma_{\beta\alpha} & \sigma_{\beta\beta} \end{bmatrix} \tag{17.4}$$

or

$$\hat{\sigma} = \begin{bmatrix} \langle\alpha|\sigma_{\alpha\alpha}|\alpha\rangle & \langle\alpha|\sigma_{\alpha\beta}|\beta\rangle \\ \langle\beta|\sigma_{\beta\alpha}|\alpha\rangle & \langle\beta|\sigma_{\beta\beta}|\beta\rangle \end{bmatrix}$$

Since the coefficients are complex we may write this in polar notation as well (see eqn (1.7)):

$$\hat{\sigma} = \begin{bmatrix} r_\alpha r_\alpha & r_\alpha r_\beta e^{i(\phi_\alpha-\phi_\beta)} \\ r_\alpha r_\alpha e^{i(\phi_\beta-\phi_\alpha)} & r_\beta r_\beta \end{bmatrix} \tag{17.5}$$

So ... we have an expression that contains time-variable and time-invariant terms, $\hat{\sigma}$ and $\hat{\mathbf{A}}$, respectively. Knowledge of the density operator and the requisite operator for the property of interest will allow one to calculate an expectation value for that property.

Since the sum of c_α^2 and c_β^2 is equal to one, as mentioned above, we can say that generally, the trace of the density matrix is equal to one:

$$\mathrm{Tr}(\hat{\sigma}) = \sum_i c_i c_i^* = \sum_i c_i^2 = 1$$

for a single spin.

17.2 Pure and Mixed States

The last section discussed the density operator in the context of a single spin-1/2 only, with spin-states α and β. Indeed, when we introduce quantum mechanical ideas into NMR/MRI spectroscopy we generally start with a single spin and progress to multiple spins but still in a single-spin system. However, we do not usually deal with single one-spin systems in the laboratory except under very exotic conditions. Most often we deal with a vast multitude of spins (an ensemble in statistical mechanics terms) in a typical sample and we have no right to expect that each and every one of them will be in exactly the same state as we begin our experiment. On the other hand, if somehow they *were* all in the exactly the same state we would refer to that as the *pure state*.[¶] By this definition a single spin is in a pure state.

What is most likely, however, is that our spins will not all be in the same state this is referred to as a *mixed state*. From eqn (15.29) and (15.30), the polar notations for the expectation values of $\hat{\mathbf{I}}_x$ and $\hat{\mathbf{I}}_y$ are:

$$\langle I_x \rangle = r_\alpha r_\beta \cos(\phi_\beta - \phi_\alpha)$$

$$\langle I_y \rangle = r_\alpha r_\beta \sin(\phi_\beta - \phi_\alpha)$$

For a mixed state at equilibrium we expect that there will be a random distribution of phase angles ϕ that will lead to average values of zero for the sine and cosine terms and zero therefore for the expectation values. In terms of the vector-picture of NMR spectroscopy in Chapter 12 this makes sense since, at equilibrium there is only z-magnetisation and none in the transverse x,y-plane.

Of special interest is the mixed state equilibrium density matrix since this is the starting point for analysis of real laboratory

¶This is so stupendously unlikely that it would probably qualify as a miracle, as any student of statistical mechanics will attest to.

experiments. Looking at eqn (17.5) we see that, for the same reasons as for the expectation values of $\hat{\mathbf{I}}_x$ and $\hat{\mathbf{I}}_y$, the off-diagonal elements of the density matrix are expected to be zero. There should be a random distribution of phase angles ϕ_α and ϕ_β such that the average value of the coefficients is zero. This is the major difference between the pure state density matrix (eqn (17.4) and (17.5)) and the mixed state density matrix. Obviously, we could not say the same thing about a single isolated spin. Our single-spin ensemble equilibrium density matrix now looks like this:

$$\hat{\sigma} = \begin{bmatrix} \bar{c}_\alpha \bar{c}_\alpha^* & 0 \\ 0 & \bar{c}_\alpha \bar{c}_\alpha^* \end{bmatrix}$$

and it remains to find out what is the value of the diagonal elements at equilibrium after which we will be in a position to be able to begin to analyse our NMR pulse-sequences.‖

As mentioned in the previous section the diagonal elements of the density matrix are interpreted as populations … population fractions actually, but the term population is generally used in the literature. Population fractions are very conveniently computed using the Boltzmann distribution function (see Chapter 6):

$$\frac{n_i}{N_T} = \frac{1}{2}\exp\left(\frac{-E_i}{kT}\right)$$

So, for an ensemble of spins-1/2 the population fractions are:

$$\bar{c}_\alpha \bar{c}_\alpha^* = \frac{n_\alpha}{N_T} = \frac{1}{2}\exp\left(\frac{-E_\alpha}{kT}\right)$$

$$\bar{c}_\beta \bar{c}_\beta^* = \frac{n_\beta}{N_T} = \frac{1}{2}\exp\left(\frac{-E_\beta}{kT}\right)$$

and our density matrix now looks like:

$$\hat{\sigma} = \begin{bmatrix} \frac{n_\alpha}{N_T} & 0 \\ 0 & \frac{n_\beta}{N_T} \end{bmatrix}$$

‖Note that we are now using overbars to indicate ensemble averages in the density matrix components.

Referring to eqn (A.28) in Appendix A, let us rewrite the Boltzmann expression:

$$n_\alpha = \frac{N_T}{2}\exp\left(\frac{-E_\alpha}{kT}\right) \approx N_T\left(1 - \frac{E_\alpha}{kT}\right)$$

$$n_\beta = \frac{N_T}{2}\exp\left(\frac{-E_\beta}{kT}\right) \approx N_T\left(1 - \frac{E_\beta}{kT}\right)$$

The magnetic field Hamiltonian from Chapter 15 is, in energy notation rather than frequency notation (see eqn (15.33)):

$$\hat{\mathbf{H}} = -\hbar\gamma B_0 \hat{\mathbf{I}}_z$$

and from eqn (15.20):

$$\hat{\mathbf{I}}_z|\alpha\rangle = \frac{1}{2}|\alpha\rangle \quad \hat{\mathbf{I}}_z|\beta\rangle = -\frac{1}{2}|\beta\rangle$$

Combining these we have:

$$\hat{\mathbf{H}}|\alpha\rangle = -\hbar\gamma B_0 \hat{\mathbf{I}}_z|\alpha\rangle = -\frac{1}{2}\hbar\gamma B_0|\alpha\rangle$$

$$\hat{\mathbf{H}}|\beta\rangle = -\hbar\gamma B_0 \hat{\mathbf{I}}_z|\beta\rangle = \frac{1}{2}\hbar\gamma B_0|\beta\rangle$$

and

$$E_\alpha = -\frac{1}{2}\hbar\gamma B_0 \quad E_\beta = \frac{1}{2}\hbar\gamma B_0$$

Using these in our approximate expression for the population of each state:

$$n_\alpha = \frac{N_T}{2}\left(1 - \frac{E_\alpha}{kT}\right) = \frac{N_T}{2}\left(1 + \frac{\hbar\gamma B_0}{2kT}\right)$$

$$n_\beta = \frac{N_T}{2}\left(1 - \frac{E_\beta}{kT}\right) = \frac{N_T}{2}\left(1 - \frac{\hbar\gamma B_0}{2kT}\right)$$

The average population is then simply:

$$n_{av} = \frac{n_\alpha + n_\beta}{2} = \frac{N_T/2 + N_T/2}{2} = \frac{N_T}{2}$$

and the population difference is:

$$n_{\text{diff}} = n_\alpha - n_\beta = \frac{\hbar\gamma B_0}{kT}$$

Note that the population difference is proportional to the gyromagnetic ratio and the external field and inversely proportional to the absolute temperature. Since the population difference directly affects the intensity of the NMR signal, each of these variables has an effect on the signal strength. The last two expressions allow us to write for n_α and n_β:

$$n_\alpha = n_{av} + \frac{n_{\text{diff}}}{2}$$

$$n_\beta = n_{av} - \frac{n_{\text{diff}}}{2}$$

and inserting these into the density matrix:

$$\begin{aligned}
\hat{\sigma} &= \frac{1}{N_T}\begin{bmatrix} n_{av} + \frac{n_{\text{diff}}}{2} & 0 \\ 0 & n_{av} - \frac{n_{\text{diff}}}{2} \end{bmatrix} \\
&= \frac{1}{N_T}\begin{bmatrix} n_{av} & 0 \\ 0 & n_{av} \end{bmatrix} + \frac{1}{N_T}\begin{bmatrix} \frac{n_{\text{diff}}}{2} & 0 \\ 0 & -\frac{n_{\text{diff}}}{2} \end{bmatrix} \\
&= \frac{n_{av}}{N_T}\begin{bmatrix} 1 & 0 \\ 0 & 1 \end{bmatrix} + \frac{n_{\text{diff}}}{N_T}\begin{bmatrix} \frac{1}{2} & 0 \\ 0 & -\frac{1}{2} \end{bmatrix} \\
&= \frac{n_{av}}{N_T}\hat{\mathbf{1}} + \frac{n_{\text{diff}}}{N_T}\hat{\mathbf{I}}_z
\end{aligned}$$

We see here the equilibrium density matrix expressed as a linear combination of matrices, much as vectors may be constructed from basis vectors. This particular expression pertains to the equilibrium state and so contains only $\hat{\mathbf{I}}_z$ and $\hat{\mathbf{1}}$. A more general expression of the density matrix is as a linear combination of our three angular momentum operator matrices from Chapter 15 and the unit matrix:

$$\hat{\sigma} = a_E\hat{\mathbf{E}} + a_x\hat{\mathbf{I}}_x + a_y\hat{\mathbf{I}}_y + a_z\hat{\mathbf{I}}_z \qquad (17.6)$$

In this context the unit matrix is usually written as $\hat{\mathbf{E}}$. Each of these matrices is trace orthogonal to the others (see eqn (2.25)). Thus, eqn

(17.6) is very similar to the familiar vectors of Chapter 3. One may write the matrix form of $\hat{\sigma}$ using eqn (17.6):

$$\hat{\sigma} = \begin{bmatrix} a_E + a_z & a_x - ia_y \\ a_x + ia_y & a_E - a_z \end{bmatrix} = \begin{bmatrix} \sigma_{\alpha\alpha} & \sigma_{\alpha\beta} \\ \sigma_{\beta\alpha} & \sigma_{\beta\beta} \end{bmatrix}$$

Of course, at equilibrium, a_x and a_y are zero:

$$\hat{\sigma} = \begin{bmatrix} a_E + a_z & 0 \\ 0 & a_E - a_z \end{bmatrix} = \begin{bmatrix} \dfrac{n_{av} + n_{\text{diff}}}{N_T} & 0 \\ 0 & \dfrac{n_{av} - n_{\text{diff}}}{N_T} \end{bmatrix}$$

The unit matrix, when operated on by another matrix simply regenerates the other matrix and is of little interest. We are more concerned with $\hat{\mathbf{I}}_z$ since it will deliver the most interesting information when acted upon by an appropriate operator. Therefore, we usually just leave the unit matrix out of the equilibrium density operator (and the non-equilibrium ones for that matter) for simplicity:

$$\hat{\sigma} = \frac{n_{\text{diff}}}{N_T}\hat{\mathbf{I}}_z$$

Furthermore, since the coefficient n_{diff}/N_T does not have any bearing on the time behaviour of the density matrix it is usually the case that it too is left out. So, the final expression for the equilibrium density matrix is very simple:

$$\hat{\sigma} = \hat{\mathbf{I}}_z \tag{17.7}$$

17.3 The Time Evolution of the Density Matrix

What we *really* want to know in NMR spectroscopy is what happens as time proceeds. How do we calculate how a system will evolve in time? We have seen that in the Schrodinger representation, the time-dependence is carried in the coefficients of the basis vectors. In turn, these coefficients make up the density matrix and we can therefore develop an expression involving the density matrix for the time-evolution of our spin system. Using the fifth postulate of quantum mechanics from Chapter 13 (and remembering that $|\Psi\rangle$ is the complex conjugate of $|\Psi\rangle$):

$$\frac{\partial}{\partial t}|\Psi\rangle(t) = -\frac{i}{\hbar}\hat{\mathbf{H}}|\Psi\rangle(t)$$

$$\frac{\partial}{\partial t}\langle\Psi(t)| = \frac{i}{\hbar}\hat{\mathbf{H}}\langle\Psi(t)|$$

and using the above definition of the density matrix:

$$\begin{aligned}
\frac{\partial}{\partial t}|\Psi\rangle(t)\langle\Psi|(t) &= \left(\frac{\partial}{\partial t}|\Psi\rangle(t)\right)\langle\Psi|(t) + |\Psi\rangle(t)\left(\frac{\partial}{\partial t}\langle\Psi|(t)\right) \\
&= \frac{i}{\hbar}[\hat{\mathbf{H}}|\Psi\rangle(t)\langle\Psi|(t) - |\Psi\rangle(t)\langle\Psi|(t)\hat{\mathbf{H}}] \\
&= -\frac{i}{\hbar}[\hat{\mathbf{H}}\hat{\sigma}(t) - \hat{\sigma}(t)\hat{\mathbf{H}}] \\
&= -\frac{i}{\hbar}[\hat{\mathbf{H}}, \hat{\sigma}(t)]
\end{aligned} \tag{17.8}$$

This is known as the Liouville–Von Neumann equation. What we want from this is the time-dependence of $\hat{\sigma}$. With this and the appropriate operator we can easily determine the state of a spin system at any time … in other words, we can follow the effects of rf-pulses and time-evolution periods in NMR pulse sequences. The solution of the Liouville–Von Neumann equation is:

$$\hat{\sigma}(t) = \exp(-i\hat{\mathbf{H}}t)\ \hat{\sigma}(t_0)\ \exp\ (i\hat{\mathbf{H}}t) \tag{17.9}$$

Note that it is assumed that the Hamiltonian is time-invariant in arriving at this solution. It is easily verified that this is indeed a solution to the Liouville–Von Neumann equation by differentiation of eqn (17.9):

$$\begin{aligned}
\frac{d\hat{\sigma}(t)}{dt} &= \left[\frac{d\exp(-i\hat{\mathbf{H}}t)}{dt}\right]\hat{\sigma}(t_0)\exp(i\hat{\mathbf{H}}t) + \exp(-i\hat{\mathbf{H}}t)\hat{\sigma}(t_0)\left[\frac{d\exp(-i\hat{\mathbf{H}}t)}{dt}\right] \\
&= -i\hat{\mathbf{H}}\exp(-i\hat{\mathbf{H}}t)\hat{\sigma}(t_0)\exp(i\hat{\mathbf{H}}t) + \exp(-i\hat{\mathbf{H}}t)\hat{\sigma}(t_0)\exp(i\hat{\mathbf{H}}t)i\hat{\mathbf{H}} \\
&= -i[\hat{\mathbf{H}}\hat{\sigma}(t) - \hat{\sigma}(t)\hat{\mathbf{H}}] \\
&= -i[\hat{\mathbf{H}}, \hat{\sigma}(t)]
\end{aligned}$$

See Section 16.3 for an explanation of how we arrived at line 3 from line 2. For a series of times during which the Hamiltonian is time-invariant we may apply eqn (17.9) sequentially. For time-period 1:

$$\hat{\sigma}(t_1) = \exp(-i\hat{\mathbf{H}}_1 t_1)\ \hat{\sigma}(t_0)\ \exp\ (i\hat{\mathbf{H}}_1 t_1)$$

and then for time-period 2:

$$\begin{aligned}\hat{\sigma}(t_2) &= \exp(-i\hat{\mathbf{H}}_2 t_2)\hat{\sigma}(t_1)\exp(i\hat{\mathbf{H}}_2 t_2)\\ &= \exp(-i\hat{\mathbf{H}}_2 t_2)\exp(-i\hat{\mathbf{H}}_1 t_1)\hat{\sigma}(t_0)\exp(i\hat{\mathbf{H}}_1 t_1)\exp(i\hat{\mathbf{H}}_2 t_2)\end{aligned}$$

etc.

Now, the observant reader will notice that eqn (17.9) bears more than a passing resemblance to a matrix-transformation such as in eqn (2.55). Indeed, this is so and furthermore, it turns out that this is an Hermitian transformation as shown in eqn (2.57). We may show that eqn (17.9) can be expressed in matrix form. We do this for the angular momentum Hamiltonian operator $\hat{\mathbf{I}}_z$ from Chapter 15:

$$\exp(i\omega_0\hat{\mathbf{I}}_z t) = \hat{\mathbf{1}} + i\omega_0 t\hat{\mathbf{I}}_z + \frac{(i\omega_0 t)^2\hat{\mathbf{I}}_z^2}{2!} + \frac{(i\omega_0 t)^3\hat{\mathbf{I}}_z^3}{3!}\ \ldots$$

Also from Chapter 15:

$$\hat{\mathbf{I}}_z = \frac{1}{2}\begin{bmatrix}1 & 0\\ 0 & -1\end{bmatrix}$$

and substituting:

$$\exp(i\omega_0\hat{\mathbf{I}}_z t) = \hat{\mathbf{1}} + i\omega_0 t\begin{bmatrix}1 & 0\\ 0 & -1\end{bmatrix} + \frac{(i\omega_0 t)^2}{4\cdot 2!}\begin{bmatrix}1 & 0\\ 0 & -1\end{bmatrix}^2 + \frac{(i\omega_0 t)^3}{8\cdot 3!}\begin{bmatrix}1 & 0\\ 0 & -1\end{bmatrix}^3\ \ldots$$

$$\hat{\mathbf{I}}_z^2 = \frac{1}{4}\begin{bmatrix}1 & 0\\ 0 & -1\end{bmatrix}^2 = \frac{1}{4}\begin{bmatrix}1 & 0\\ 0 & -1\end{bmatrix} = \frac{1}{4}\hat{\mathbf{1}}$$

$$\hat{\mathbf{I}}_z^3 = \frac{1}{8}\begin{bmatrix}1 & 0\\ 0 & -1\end{bmatrix}^3 = \frac{1}{8}\begin{bmatrix}1 & 0\\ 0 & -1\end{bmatrix} = \frac{1}{8}\hat{\mathbf{I}}_z$$

Generally, for $\hat{\mathbf{I}}_z$:

$$\hat{\mathbf{I}}_z^n = \frac{1}{2^n}\hat{\mathbf{1}}\quad \text{(for } n \text{ even)}$$

$$\hat{\mathbf{I}}_z^n = \frac{1}{2^n}\hat{\mathbf{I}}_z\quad \text{(for } n \text{ odd)}$$

Substituting into the expansion and separating even and odd terms:

$$\exp(i\omega_0\hat{\mathbf{I}}_z t) = \left(1 - \frac{\left(\frac{\omega_0 t}{2}\right)^2}{2!} + \frac{\left(\frac{\omega_0 t}{2}\right)^4}{4!} \cdots\right)\begin{bmatrix} 1 & 0 \\ 0 & 1 \end{bmatrix}$$

$$+ i\left(\frac{\omega_0 t}{2} - \frac{\left(\frac{\omega_0 t}{2}\right)^3}{3!} + \frac{\left(\frac{\omega_0 t}{2}\right)^5}{5!} \cdots\right)\begin{bmatrix} 1 & 0 \\ 0 & -1 \end{bmatrix}$$

$$= \cos\left(\frac{\omega_0 t}{2}\right)\begin{bmatrix} 1 & 0 \\ 0 & 1 \end{bmatrix} + i\sin\left(\frac{\omega_0 t}{2}\right)\begin{bmatrix} 1 & 0 \\ 0 & -1 \end{bmatrix}$$

$$= \begin{bmatrix} \cos\left(\frac{\omega_0 t}{2}\right) + i\sin\left(\frac{\omega_0 t}{2}\right) & 0 \\ 0 & \cos\left(\frac{\omega_0 t}{2}\right) - i\sin\left(\frac{\omega_0 t}{2}\right) \end{bmatrix}$$

Similarly:

$$\exp(-i\omega_0\hat{\mathbf{I}}_z t) = \begin{bmatrix} \cos\left(\frac{\omega_0 t}{2}\right) - i\sin\left(\frac{\omega_0 t}{2}\right) & 0 \\ 0 & \cos\left(\frac{\omega_0 t}{2}\right) + i\sin\left(\frac{\omega_0 t}{2}\right) \end{bmatrix}$$

A faster way to do this for diagonal matrices is to use the identity developed in Chapter 2:

$$\text{if } \mathbf{A} = \begin{bmatrix} a & 0 \\ 0 & b \end{bmatrix} \text{ then } \mathrm{e}^{\alpha\hat{\mathbf{A}}} = \begin{bmatrix} e^{\alpha a} & 0 \\ 0 & e^{\alpha b} \end{bmatrix}$$

$$\mathrm{e}^{i\omega_0 t\hat{\mathbf{I}}_z} = \begin{bmatrix} \mathrm{e}^{i\omega_0 t/2} & 0 \\ 0 & \mathrm{e}^{-i\omega_0 t/2} \end{bmatrix} = \begin{bmatrix} \cos\left(\frac{i\omega_0 t}{2}\right) + i\sin\left(\frac{i\omega_0 t}{2}\right) & 0 \\ 0 & \cos\left(\frac{i\omega_0 t}{2}\right) - i\sin\left(\frac{i\omega_0 t}{2}\right) \end{bmatrix}$$

Noting that the terms in these matrices are conjugates, we can simplify things somewhat:

$$\text{let } \alpha = \cos\left(\frac{i\omega_0 t}{2}\right) + \sin\left(\frac{i\omega_0 t}{2}\right)$$

$$\text{and } \alpha^* = \cos\left(\frac{i\omega_0 t}{2}\right) - \sin\left(\frac{i\omega_0 t}{2}\right)$$

then

$$\exp(i\omega_0 t\hat{\mathbf{I}}_z) = \begin{bmatrix} \alpha & 0 \\ 0 & \alpha^* \end{bmatrix} \quad \exp(-i\omega_0 t\hat{\mathbf{I}}_z) = \begin{bmatrix} \alpha^* & 0 \\ 0 & \alpha \end{bmatrix}$$

We now have the matrix versions of the transformation operators that will tell us what the time behaviour of the density matrix is. Using the density matrix from eqn (17.7) and our transformation matrices:

$$\hat{\sigma}(t) = \frac{1}{2}\begin{bmatrix} \alpha^* & 0 \\ 0 & \alpha \end{bmatrix}\begin{bmatrix} 1 & 0 \\ 0 & -1 \end{bmatrix}\begin{bmatrix} \alpha & 0 \\ 0 & \alpha^* \end{bmatrix}$$

$$= \frac{1}{2}\begin{bmatrix} \alpha\alpha^* & 0 \\ 0 & -\alpha\alpha^* \end{bmatrix}$$

$$= \frac{1}{2}\begin{bmatrix} 1 & 0 \\ 0 & -1 \end{bmatrix}$$

This is a trivial example showing that the time evolution of the equilibrium density matrix under the influence of $\hat{\mathbf{I}}_z$ is zero. Our next task will be to show a non-trivial example of the density matrix under the influence of the $\hat{\mathbf{I}}_x$ operator. To do this we need the matrix form of $\exp(i\omega_1 t\hat{\mathbf{I}}_x)$. This can be accomplished in much the same way as for $\exp(i\omega_0 t\hat{\mathbf{I}}_z)$:

$$\hat{\mathbf{H}}_x = \omega_1 \hat{\mathbf{I}}_x \quad \text{and} \quad \hat{\mathbf{I}}_x = \frac{1}{2}\begin{bmatrix} 0 & 1 \\ 1 & 0 \end{bmatrix}$$

$$\exp(i\omega_1 t\hat{\mathbf{I}}_x) = \hat{\mathbf{1}} + \frac{i\omega_1 t}{2}\begin{bmatrix} 0 & 1 \\ 1 & 0 \end{bmatrix} + \frac{(i\omega_1 t)^2}{4 \cdot 2!}\begin{bmatrix} 0 & 1 \\ 1 & 0 \end{bmatrix}^2 + \frac{(i\omega_1 t)^3}{8 \cdot 2!}\begin{bmatrix} 0 & 1 \\ 1 & 0 \end{bmatrix}^3 \cdots$$

$$\hat{\mathbf{I}}_x^2 = \frac{1}{4}\begin{bmatrix} 0 & 1 \\ 1 & 0 \end{bmatrix}^2 = \frac{1}{4}\begin{bmatrix} 1 & 0 \\ 0 & 1 \end{bmatrix} = \frac{1}{4}\hat{\mathbf{1}}$$

$$\hat{\mathbf{I}}_x^3 = \frac{1}{8}\begin{bmatrix} 0 & 1 \\ 1 & 0 \end{bmatrix}^3 = \frac{1}{8}\begin{bmatrix} 0 & 1 \\ 1 & 0 \end{bmatrix} = \frac{1}{8}\hat{\mathbf{I}}_x$$

$$\exp(i\omega_1 t\hat{\mathbf{I}}_x) = \left(1 - \frac{\left(\frac{\omega_1 t}{2}\right)^2}{2!} + \frac{\left(\frac{\omega_1 t}{2}\right)^4}{4!} \cdots\right)\begin{bmatrix} 1 & 0 \\ 0 & 1 \end{bmatrix}$$

$$+ i\left(\frac{\omega_1 t}{2} - \frac{\left(\frac{\omega_1 t}{2}\right)^3}{3!} + \frac{\left(\frac{\omega_1 t}{2}\right)^5}{5!} \cdots\right)\begin{bmatrix} 0 & 1 \\ 1 & 0 \end{bmatrix}$$

$$= \cos\left(\frac{\omega_1 t}{2}\right)\begin{bmatrix} 1 & 0 \\ 0 & 1 \end{bmatrix} + i\sin\left(\frac{\omega_1 t}{2}\right)\begin{bmatrix} 0 & 1 \\ 1 & 0 \end{bmatrix}$$

$$= \begin{bmatrix} \cos\left(\frac{\omega_1 t}{2}\right) & i\sin\left(\frac{\omega_1 t}{2}\right) \\ i\sin\left(\frac{\omega_1 t}{2}\right) & \cos\left(\frac{\omega_1 t}{2}\right) \end{bmatrix}$$

and, similarly

$$\exp(-i\omega_1 t\hat{\mathbf{I}}_x) = \begin{bmatrix} \cos\left(\frac{\omega_1 t}{2}\right) & -i\sin\left(\frac{\omega_1 t}{2}\right) \\ -i\sin\left(\frac{\omega_1 t}{2}\right) & \cos\left(\frac{\omega_1 t}{2}\right) \end{bmatrix}$$

We can now write the effect of $\hat{\mathbf{I}}_x$ on the simplified equilibrium density matrix, $\hat{\mathbf{I}}_z$:

$$\text{let } \alpha = \cos\left(\frac{\omega_1 t}{2}\right),\ \beta = i\sin\left(\frac{\omega_1 t}{2}\right),\ \beta^* = -i\sin\left(\frac{\omega_1 t}{2}\right)$$

then

$$\alpha^2 = \cos^2\left(\frac{\omega_1 t}{2}\right),\ \alpha\beta = i\cos\left(\frac{\omega_1 t}{2}\right)\sin\left(\frac{\omega_1 t}{2}\right)$$

$$\beta\beta^* = \sin^2\left(\frac{\omega_1 t}{2}\right),\ \alpha\beta^* = -i\cos\left(\frac{\omega_1 t}{2}\right)\sin\left(\frac{\omega_1 t}{2}\right)$$

$$\exp(i\omega_1 t\hat{\mathbf{I}}_x) = \begin{bmatrix} \alpha & \beta \\ \beta & \alpha \end{bmatrix} \quad \exp(-i\omega_1 t\hat{\mathbf{I}}_x) = \begin{bmatrix} \alpha & \beta^* \\ \beta^* & \alpha \end{bmatrix}$$

and

$$\begin{aligned} \hat{\sigma}(t) &= \exp(-i\omega_1\hat{\mathbf{I}}_x t)\hat{\mathbf{I}}_z\exp(i\omega_1\hat{\mathbf{I}}_x t) \\ &= \frac{1}{2}\begin{bmatrix} \alpha & \beta^* \\ \beta^* & \alpha \end{bmatrix}\begin{bmatrix} 1 & 0 \\ 0 & -1 \end{bmatrix}\begin{bmatrix} \alpha & \beta \\ \beta & \alpha \end{bmatrix} \\ &= \frac{1}{2}\begin{bmatrix} \alpha^2 - \beta\beta^* & \alpha\beta - \alpha\beta^* \\ \alpha\beta^* - \alpha\beta & \beta\beta^* - \alpha^2 \end{bmatrix} \\ &= \frac{1}{2}\begin{bmatrix} \cos(\omega_1 t) & i\sin(\omega_1 t) \\ -i\sin(\omega_1 t) & -\cos(\omega_1 t) \end{bmatrix} \end{aligned}$$

The last line is reached using trigonometric identities (A.14) and (A.15). ω_1 is an angular frequency in radians per second and t is time so the arguments of the sine and cosine functions are an angle. It would be very natural to ask "what if the angle were $\pi/2$ radians?" or "what if the angle were π radians?" The answers to these questions are easily obtained:

$$\text{for } \omega_1 t = \frac{\pi}{2}$$

$$\hat{\sigma}(t) = \frac{1}{2}\begin{bmatrix} 0 & i \\ -i & 0 \end{bmatrix} = -\hat{\mathbf{I}}_y$$

$$\text{for } \omega_1 t = \pi$$

$$\hat{\sigma}(t) = \frac{1}{2}\begin{bmatrix} -1 & 0 \\ 0 & 1 \end{bmatrix} = -\hat{\mathbf{I}}_z$$

These results are identical to those that we saw in the vector-picture description of NMR spectroscopy in Chapter 12. When t is such that $\omega_1 t$ equals $\pi/2$ radians the equilibrium density matrix $\hat{\mathbf{I}}_z$ is transformed under the influence of $\hat{\mathbf{I}}_x$ to become $-\hat{\mathbf{I}}_y$. The corresponding vector-picture is that the equilibrium magnetisation rotates about the x-axis under the influence of a $\vec{\mathbf{B}}_1$-field aligned along the x-axis for $\pi/2$ radians and comes to rest on the $-y$-axis. As we would expect, if t is long enough we can rotate the equilibrium magnetisation π radians to end up pointing along the $-z$-axis and this is indeed reflected in our calculation.

What is the time-behaviour of magnetisation in the transverse plane? In other words, after $\hat{\mathbf{I}}_z$ evolves under the influence of, say, $\hat{\mathbf{I}}_x$ what then? The vector model says that it will rotate about the z-axis under the influence of $\vec{\mathbf{B}}_0$, the external field. Let us see if our density matrix model says the same. We start with $-\hat{\mathbf{I}}_y$ as our density matrix and apply a rotation under the influence of $\hat{\mathbf{I}}_z$:

$$\text{let } \alpha = \cos\left(\frac{\omega_0 t}{2}\right) + i\sin\left(\frac{\omega_0 t}{2}\right)$$

$$\text{and } \alpha^* = \cos\left(\frac{\omega_0 t}{2}\right) - i\sin\left(\frac{\omega_0 t}{2}\right)$$

then

$$\exp(i\omega_0 \hat{\mathbf{I}}_z t) = \begin{bmatrix} \alpha & 0 \\ 0 & \alpha^* \end{bmatrix} \quad \exp(-i\omega_0 \hat{\mathbf{I}}_z t) = \begin{bmatrix} \alpha^* & 0 \\ 0 & \alpha \end{bmatrix}$$

$$\hat{\sigma}(0) = -\hat{\mathbf{I}}_y$$

$$\begin{aligned}\hat{\sigma}(t) &= \exp(-i\omega_0\hat{\mathbf{I}}_z t)(-\hat{\mathbf{I}}_y)\exp(i\omega_0\hat{\mathbf{I}}_z t)\\ &= \frac{1}{2}\begin{bmatrix}\alpha^* & 0\\ 0 & \alpha\end{bmatrix}\begin{bmatrix}0 & i\\ -i & 0\end{bmatrix}\begin{bmatrix}\alpha & 0\\ 0 & \alpha^*\end{bmatrix}\\ &= \frac{1}{2}\begin{bmatrix}0 & i\alpha^{*2}\\ -i\alpha^2 & 0\end{bmatrix}\\ &= \frac{1}{2}\begin{bmatrix}0 & i\left(\cos\left(\frac{\omega_0 t}{2}\right) - i\sin\left(\frac{\omega_0 t}{2}\right)\right)^2\\ i\left(\cos\left(\frac{\omega_0 t}{2}\right) + i\sin\left(\frac{\omega_0 t}{2}\right)\right)^2 & 0\end{bmatrix}\\ &= \frac{1}{2}\begin{bmatrix}0 & i(\cos(\omega_0 t) - i\sin(\omega_0 t))\\ i(\cos(\omega_0 t) + i\sin(\omega_0 t)) & 0\end{bmatrix}\end{aligned}$$

where again, we have used trigonometric identities (A.14) and (A.15) to reach the last line. If t is such that ω_0 is $\pi/2$ radians then this becomes:

$$\hat{\sigma}(t) = \frac{1}{2}\begin{bmatrix}0 & 1\\ 1 & 0\end{bmatrix}$$

From the point of view of the vector model of NMR spectroscopy this is exactly what is expected to happen. In the vector model, magnetisation along the $-y$-axis will rotate in a right-hand sense towards the x-axis.

With this picture in mind we refer to the matrix expression of the exponentials as rotation matrices. Most often we use $\pi/2$ and π rotations and it is convenient to specify the rotation matrices for these operations and to introduce a rotation operator notation. For rotation about the x-axis:

$$\text{if } \omega_1 t = \frac{\pi}{2}$$

$$\exp\left(\frac{i\pi\hat{\mathbf{I}}_x}{2}\right) = \hat{\mathbf{R}}_{90x} = \frac{1}{\sqrt{2}}\begin{bmatrix}1 & i\\ i & 1\end{bmatrix}$$

$$\text{if } \omega_1 t = \pi$$

$$\exp(i\pi\hat{\mathbf{I}}_x) = \hat{\mathbf{R}}_{180x} = \begin{bmatrix}0 & i\\ i & 0\end{bmatrix}$$

Similarly, for rotation about the y-axis:

$$\text{if } \omega_1 t = \frac{\pi}{2}$$

$$\exp\left(\frac{i\pi\hat{\mathbf{I}}_y}{2}\right) = \hat{\mathbf{R}}_{90y} = \frac{1}{\sqrt{2}}\begin{bmatrix} 1 & 1 \\ -1 & 1 \end{bmatrix}$$

$$\text{if } \omega_1 t = \pi$$

$$\exp(i\pi\hat{\mathbf{I}}_y) = \hat{\mathbf{R}}_{180y} = \begin{bmatrix} 0 & -1 \\ 1 & 0 \end{bmatrix}$$

Now, let us look more closely at our previous result for an x-phase rotation:

$$\exp(-i\omega_1\hat{\mathbf{I}}_x t)\hat{\mathbf{I}}_z \exp(i\omega_1\hat{\mathbf{I}}_x t) = \frac{1}{2}\begin{bmatrix} \cos(\omega_1 t) & i\sin(\omega_1 t) \\ -i\sin(\omega_1 t) & -\cos(\omega_1 t) \end{bmatrix}$$

Manipulation of the right-hand side matrix gives:

$$\frac{1}{2}\begin{bmatrix} \cos(\omega_1 t) & i\sin(\omega_1 t) \\ -i\sin(\omega_1 t) & -\cos(\omega_1 t) \end{bmatrix} = \frac{1}{2}\begin{bmatrix} 1 & 0 \\ 0 & -1 \end{bmatrix}\cos(\omega_1 t) + \frac{1}{2}\begin{bmatrix} 0 & i \\ -i & 0 \end{bmatrix}\sin(\omega_1 t)$$

$$= \hat{\mathbf{I}}_z \cos(\omega_1 t) - \hat{\mathbf{I}}_y \sin(\omega_1 t)$$

and we can write:

$$\exp(-i\omega_1\hat{\mathbf{I}}_x t)\hat{\mathbf{I}}_z \exp(i\omega_1\hat{\mathbf{I}}_x t) = \hat{\mathbf{I}}_z\cos(\omega_1 t) - \hat{\mathbf{I}}_y \sin(\omega_1 t) \tag{17.10}$$

This identity will be of considerable use to us along with others that we will detail in the next chapter.

This is as far as we can go for a single-spin system. In the next section we will tackle multiple spin systems.

17.4 Multiple-Spin Density Matrices

So far, we have considered only a single spin. In an NMR experiment a single spin will show chemical-shift behaviour but never coupling evolution. To see scalar coupling we must have at least two spins in our spin system and later we will encounter "passive" and "active" coupling that involves a minimum of three spins in the spin system.

We could use the technique of Chapter 13 to build our density matrices starting with the wavefunction. Assuming two spins-1/2 labelled "I" and "S", that an "I" operator does not affect an "S" spin and the usual orthogonal relationships, we can write:

$$\Psi = c_{\alpha_I\alpha_S}|\alpha_I\alpha_S\rangle + c_{\alpha_I\beta_S}|\alpha_I\beta_S\rangle + c_{\beta_I\alpha_S}|\beta_I\alpha_S\rangle + c_{\beta_I\beta_S}|\beta_I\beta_S\rangle$$

$$\hat{\mathbf{I}}_z|\alpha_I\alpha_S\rangle = \frac{1}{2}|\alpha_I\alpha_S\rangle$$

$$\hat{\mathbf{I}}_z|\beta_I\alpha_S\rangle = -\frac{1}{2}|\beta_I\alpha_S\rangle$$

Using this we write out the expectation values. With two spins there will be a total of sixteen possible calculations. Because of orthogonality many of the results will be zero, so we will show only the non-zero results here:

$$\langle\alpha_I, \alpha_S|\hat{\mathbf{I}}_z|\alpha_I, \alpha_S\rangle = \frac{1}{2}$$

$$\langle\alpha_I, \beta_S|\hat{\mathbf{I}}_z|\alpha_I, \beta_S\rangle = \frac{1}{2}$$

$$\langle\beta_I, \alpha_S|\hat{\mathbf{I}}_z|\beta_I, \alpha_S\rangle = -\frac{1}{2}$$

$$\langle\beta_I, \beta_S|\hat{\mathbf{I}}_z|\beta_I, \beta_S\rangle = -\frac{1}{2}$$

The corresponding two-spin $\hat{\mathbf{I}}_z$ matrix is:

$$\hat{\mathbf{I}}_z = \frac{1}{2}\begin{bmatrix} 1 & 0 & 0 & 0 \\ 0 & 1 & 0 & 0 \\ 0 & 0 & -1 & 0 \\ 0 & 0 & 0 & -1 \end{bmatrix}$$

This method will work for all of the one-spin operators, $\hat{\mathbf{I}}_z$, $\hat{\mathbf{I}}_x$, $\hat{\mathbf{I}}_y$, $\hat{\mathbf{S}}_z$, $\hat{\mathbf{S}}_x$ and $\hat{\mathbf{S}}_y$. However, it is somewhat tedious and there is an easier way to obtain our one-spin (and higher) density matrices. In Chapter 3 we saw that vector spaces may be merged by performing an outer product calculation between the two spaces and the same thing can be done with our one-spin matrices. In order to generate the appropriate matrices we need only do direct products between the one-spin systems that are coupled (see eqn (2.8)).

Our one-spin density matrices consist of $\hat{\mathbf{I}}_z$, the equilibrium density matrix and $\hat{\mathbf{I}}_x$, $\hat{\mathbf{I}}_y$ and formally, $\hat{\mathbf{E}}$ the unit matrix. Of course, if you try to do a rotation of $\hat{\mathbf{E}}$ you will end up with $\hat{\mathbf{E}}$ no matter which rotation operator you use. We saw that the density matrix can be thought of as a linear combination of matrices (eqn (17.6)):

$$\hat{\sigma} = a_E\hat{\mathbf{E}} + a_x\hat{\mathbf{I}}_x + a_y\hat{\mathbf{I}}_y + a_z\hat{\mathbf{I}}_z$$

To arrive at a set of two-spin operators we do a series of outer products of each of the one-spin matrices with each other. Let us label the spins "*I*" and "*S*". There are four matrices for each spin and therefore the total number of possible product combinations is 16. For example, to get the two-spin matrix $\hat{\mathbf{I}}_{2z}$ we do the outer product of $\hat{\mathbf{I}}_{1z}$ and $\hat{\mathbf{E}}_1$:**

$$\hat{\mathbf{I}}_{1z} \otimes \hat{\mathbf{E}}_1 = \frac{1}{2}\begin{bmatrix} 1 & 0 \\ 0 & -1 \end{bmatrix} \otimes \begin{bmatrix} 1 & 0 \\ 0 & 1 \end{bmatrix} = \frac{1}{2}\begin{bmatrix} 1 & 0 & 0 & 0 \\ 0 & 1 & 0 & 0 \\ 0 & 0 & -1 & 0 \\ 0 & 0 & 0 & -1 \end{bmatrix} = \hat{\mathbf{I}}_{2z}$$

and similarly:

$$\hat{\mathbf{I}}_{1x} \otimes \hat{\mathbf{E}}_1 = \frac{1}{2}\begin{bmatrix} 0 & 1 \\ 1 & 0 \end{bmatrix} \otimes \begin{bmatrix} 1 & 0 \\ 0 & 1 \end{bmatrix} = \frac{1}{2}\begin{bmatrix} 0 & 0 & 1 & 0 \\ 0 & 0 & 0 & 1 \\ 1 & 0 & 0 & 0 \\ 0 & 1 & 0 & 0 \end{bmatrix} = \hat{\mathbf{I}}_{2x}$$

$$\hat{\mathbf{I}}_{1y} \otimes \hat{\mathbf{E}}_1 = \frac{1}{2}\begin{bmatrix} 0 & -i \\ i & 0 \end{bmatrix} \otimes \begin{bmatrix} 1 & 0 \\ 0 & 1 \end{bmatrix} = \frac{1}{2}\begin{bmatrix} 0 & 0 & -i & 0 \\ 0 & 0 & 0 & -i \\ i & 0 & 0 & 0 \\ 0 & i & 0 & 0 \end{bmatrix} = \hat{\mathbf{I}}_{2y}$$

To obtain the two-spin $\hat{\mathbf{S}}$ matrices we do the same type of outer product calculation except that we reverse the order and multiply by $\hat{\mathbf{E}}_1$ first:

$$\hat{\mathbf{E}}_1 \otimes \hat{\mathbf{S}}_{1z} = \begin{bmatrix} 1 & 0 \\ 0 & 1 \end{bmatrix} \otimes \frac{1}{2}\begin{bmatrix} 1 & 0 \\ 0 & -1 \end{bmatrix} = \frac{1}{2}\begin{bmatrix} 1 & 0 & 0 & 0 \\ 0 & -1 & 0 & 0 \\ 0 & 0 & 1 & 0 \\ 0 & 0 & 0 & -1 \end{bmatrix} = \hat{\mathbf{S}}_{2z}$$

**I am temporarily using 1 and 2 in the subscript to differentiate between the one-spin and two-spin operators.

$$\hat{\mathbf{E}}_1 \otimes \hat{\mathbf{S}}_{1x} = \begin{bmatrix} 1 & 0 \\ 0 & 1 \end{bmatrix} \otimes \frac{1}{2}\begin{bmatrix} 0 & 1 \\ 1 & 0 \end{bmatrix} = \frac{1}{2}\begin{bmatrix} 0 & 1 & 0 & 0 \\ 1 & 0 & 0 & 0 \\ 0 & 0 & 0 & 1 \\ 0 & 0 & 1 & 0 \end{bmatrix} = \hat{\mathbf{S}}_{2x}$$

$$\hat{\mathbf{E}}_1 \otimes \hat{\mathbf{S}}_{1y} = \begin{bmatrix} 1 & 0 \\ 0 & 1 \end{bmatrix} \otimes \frac{1}{2}\begin{bmatrix} 0 & -i \\ i & 0 \end{bmatrix} = \frac{1}{2}\begin{bmatrix} 0 & -i & 0 & 0 \\ i & 0 & 0 & 0 \\ 0 & 0 & 0 & -i \\ 0 & 0 & i & 0 \end{bmatrix} = \hat{\mathbf{S}}_{2y}$$

That accounts for six one-spin matrices and as mentioned there are a total of sixteen, including the unit matrix. The next ten matrices are the result of outer products of $\hat{\mathbf{I}}_z$, $\hat{\mathbf{I}}_x$, $\hat{\mathbf{I}}_y$, $\hat{\mathbf{S}}_z$, $\hat{\mathbf{S}}_x$, $\hat{\mathbf{S}}_y$ and $\hat{\mathbf{E}}$ with itself. These are all two-spin matrices except $\hat{\mathbf{E}}$, which is a zero-spin matrix. We will now drop the use of numeric subscripts, it being understood that the arguments of the outer product operator "⊗" refer to one-spin operators and their resulting matrices refer to two-spin operators.[††]

$$\hat{\mathbf{I}}_z \otimes \hat{\mathbf{S}}_z = \frac{1}{2}\begin{bmatrix} 1 & 0 \\ 0 & -1 \end{bmatrix} \otimes \frac{1}{2}\begin{bmatrix} 1 & 0 \\ 0 & -1 \end{bmatrix} = \frac{1}{4}\begin{bmatrix} 1 & 0 & 0 & 0 \\ 0 & -1 & 0 & 0 \\ 0 & 0 & -1 & 0 \\ 0 & 0 & 0 & 1 \end{bmatrix} = \hat{\mathbf{I}}_z\hat{\mathbf{S}}_z$$

$$\hat{\mathbf{I}}_x \otimes \hat{\mathbf{S}}_z = \frac{1}{2}\begin{bmatrix} 0 & 1 \\ 1 & 0 \end{bmatrix} \otimes \frac{1}{2}\begin{bmatrix} 1 & 0 \\ 0 & -1 \end{bmatrix} = \frac{1}{4}\begin{bmatrix} 0 & 0 & 1 & 0 \\ 0 & 0 & 0 & -1 \\ 1 & 0 & 0 & 0 \\ 0 & -1 & 0 & 0 \end{bmatrix} = \hat{\mathbf{I}}_x\hat{\mathbf{S}}_z$$

$$\hat{\mathbf{I}}_y \otimes \hat{\mathbf{S}}_z = \frac{1}{2}\begin{bmatrix} 0 & -i \\ i & 0 \end{bmatrix} \otimes \frac{1}{2}\begin{bmatrix} 1 & 0 \\ 0 & -1 \end{bmatrix} = \frac{1}{4}\begin{bmatrix} 0 & 0 & -i & 0 \\ 0 & 0 & 0 & i \\ -i & 0 & 0 & 0 \\ 0 & i & 0 & 0 \end{bmatrix} = \hat{\mathbf{I}}_y\hat{\mathbf{S}}_z$$

[††]I will only be discussing one- and two-spin operators in this section of the text so this is reasonable. Were we to be considering three-spin operators then this would not work. You could not assume that the arguments of the operator are both one-spin operators.

$$\hat{\mathbf{I}}_x \otimes \hat{\mathbf{S}}_x = \frac{1}{2}\begin{bmatrix} 0 & 1 \\ 1 & 0 \end{bmatrix} \otimes \frac{1}{2}\begin{bmatrix} 0 & 1 \\ 1 & 0 \end{bmatrix} = \frac{1}{4}\begin{bmatrix} 0 & 0 & 0 & 1 \\ 0 & 0 & 1 & 0 \\ 0 & 1 & 0 & 0 \\ 1 & 0 & 0 & 0 \end{bmatrix} = \hat{\mathbf{I}}_x\hat{\mathbf{S}}_x$$

$$\hat{\mathbf{I}}_y \otimes \hat{\mathbf{S}}_x = \frac{1}{2}\begin{bmatrix} 0 & -i \\ i & 0 \end{bmatrix} \otimes \frac{1}{2}\begin{bmatrix} 0 & 1 \\ 1 & 0 \end{bmatrix} = \frac{1}{4}\begin{bmatrix} 0 & 0 & 0 & i \\ 0 & 0 & i & 0 \\ 0 & -i & 0 & 0 \\ -i & 0 & 0 & 0 \end{bmatrix} = \hat{\mathbf{I}}_y\hat{\mathbf{S}}_x$$

$$\hat{\mathbf{I}}_y \otimes \hat{\mathbf{S}}_y = \frac{1}{2}\begin{bmatrix} 0 & -i \\ i & 0 \end{bmatrix} \otimes \frac{1}{2}\begin{bmatrix} 0 & -i \\ i & 0 \end{bmatrix} = \frac{1}{4}\begin{bmatrix} 0 & 0 & 0 & -1 \\ 0 & 0 & 1 & 0 \\ 0 & 1 & 0 & 0 \\ -1 & 0 & 0 & 0 \end{bmatrix} = \hat{\mathbf{I}}_y\hat{\mathbf{S}}_y$$

$$\hat{\mathbf{I}}_z \otimes \hat{\mathbf{S}}_x = \frac{1}{2}\begin{bmatrix} 1 & 0 \\ 0 & -1 \end{bmatrix} \otimes \frac{1}{2}\begin{bmatrix} 0 & 1 \\ 1 & 0 \end{bmatrix} = \frac{1}{4}\begin{bmatrix} 0 & 1 & 0 & 0 \\ 1 & 0 & 0 & 0 \\ 0 & 0 & 0 & -1 \\ 0 & 0 & -1 & 0 \end{bmatrix} = \hat{\mathbf{I}}_z\hat{\mathbf{S}}_x$$

$$\hat{\mathbf{I}}_z \otimes \hat{\mathbf{S}}_y = \frac{1}{2}\begin{bmatrix} 1 & 0 \\ 0 & -1 \end{bmatrix} \otimes \frac{1}{2}\begin{bmatrix} 0 & -i \\ i & 0 \end{bmatrix} = \frac{1}{4}\begin{bmatrix} 0 & -i & 0 & 0 \\ i & 0 & 0 & 0 \\ 0 & 0 & 0 & i \\ 0 & 0 & -i & 0 \end{bmatrix} = \hat{\mathbf{I}}_z\hat{\mathbf{S}}_y$$

$$\hat{\mathbf{E}} \otimes \hat{\mathbf{E}} = \begin{bmatrix} 1 & 0 \\ 0 & 1 \end{bmatrix} \otimes \begin{bmatrix} 1 & 0 \\ 0 & 1 \end{bmatrix} = \begin{bmatrix} 1 & 0 & 0 & 0 \\ 0 & 1 & 0 & 0 \\ 0 & 0 & 1 & 0 \\ 0 & 0 & 0 & 1 \end{bmatrix} = \hat{\mathbf{E}}$$

Note the ordering of $\hat{\mathbf{I}}$ and $\hat{\mathbf{S}}$ in these calculations. The order does not matter but consistency *does* matter.[‡‡] Exchanging label order would also result in an exchange of matrices.

[‡‡]You will notice, if you compare my two-spin matrices with those of Mateescu and Valeriu (see the Further Reading section at the end of the chapter) that they are not the same. This is because they have used the opposite order in their book.

There is one other matrix that will be handy a bit later. It is the combination of matrices $\hat{\mathbf{I}}_x\hat{\mathbf{S}}_x$, $\hat{\mathbf{I}}_y\hat{\mathbf{S}}_y$ and $\hat{\mathbf{I}}_z\hat{\mathbf{S}}_z$:

$$\hat{\mathbf{I}}\hat{\mathbf{S}} = \hat{\mathbf{I}}_x\hat{\mathbf{S}}_x + \hat{\mathbf{I}}_y\hat{\mathbf{S}}_y + \hat{\mathbf{I}}_z\hat{\mathbf{S}}_z$$

$$= \frac{1}{4}\begin{bmatrix} 0 & 0 & 0 & 1 \\ 0 & 0 & 1 & 0 \\ 0 & 1 & 0 & 0 \\ 1 & 0 & 0 & 0 \end{bmatrix} + \frac{1}{4}\begin{bmatrix} 0 & 0 & 0 & -1 \\ 0 & 0 & 1 & 0 \\ 0 & 1 & 0 & 0 \\ -1 & 0 & 0 & 0 \end{bmatrix} + \frac{1}{4}\begin{bmatrix} 1 & 0 & 0 & 0 \\ 0 & -1 & 0 & 0 \\ 0 & 0 & -1 & 0 \\ 0 & 0 & 0 & 1 \end{bmatrix} \tag{17.11}$$

$$= \frac{1}{4}\begin{bmatrix} 1 & 0 & 0 & 0 \\ 0 & -1 & 2 & 0 \\ 0 & 2 & -1 & 0 \\ 0 & 0 & 0 & 1 \end{bmatrix}$$

The reader may verify that these matrices are all Hermitian, normalised and trace orthogonal. This being the case, from the discussion in Chapter 2 of matrix transforms we expect that the trace of our density matrices is invariant to transformation because the transformation matrices are also Hermitian and this is indeed so. We have seen that we can transform from $\hat{\mathbf{I}}_z$ to, say, $\hat{\mathbf{I}}_x$ and a quick perusal of these matrices will show their traces to be equivalent (except for $\hat{\mathbf{E}}$). Their determinants are also equal.

Also of interest to us are the commutation properties of these operators. Do $\hat{\mathbf{I}}_x$ and $\hat{\mathbf{S}}_x$ commute? How does this affect our calculations? Let us tackle the first question. The commutator of $\hat{\mathbf{I}}_x$ and $\hat{\mathbf{S}}_x$ we can get easily by matrix additions:

$$\left[\hat{\mathbf{I}}_x, \hat{\mathbf{S}}_x\right] = \frac{1}{2}\left(\begin{bmatrix} 0 & 0 & 1 & 0 \\ 0 & 0 & 0 & 1 \\ 1 & 0 & 0 & 0 \\ 0 & 1 & 0 & 0 \end{bmatrix}\begin{bmatrix} 0 & 1 & 0 & 0 \\ 1 & 0 & 0 & 0 \\ 0 & 0 & 0 & 1 \\ 0 & 0 & 1 & 0 \end{bmatrix} - \begin{bmatrix} 0 & 1 & 0 & 0 \\ 1 & 0 & 0 & 0 \\ 0 & 0 & 0 & 1 \\ 0 & 0 & 1 & 0 \end{bmatrix}\begin{bmatrix} 0 & 0 & 1 & 0 \\ 0 & 0 & 0 & 1 \\ 1 & 0 & 0 & 0 \\ 0 & 1 & 0 & 0 \end{bmatrix}\right)$$

$$= \begin{bmatrix} 0 & 0 & 0 & 1 \\ 0 & 0 & 1 & 0 \\ 0 & 1 & 0 & 0 \\ 1 & 0 & 0 & 0 \end{bmatrix} - \begin{bmatrix} 0 & 0 & 0 & 1 \\ 0 & 0 & 1 & 0 \\ 0 & 1 & 0 & 0 \\ 1 & 0 & 0 & 0 \end{bmatrix} = \hat{\mathbf{0}}$$

So, $\hat{\mathbf{I}}_x$ and $\hat{\mathbf{S}}_x$ commute. This means that the order in which we apply a pulse to the spins does not matter. We could write either:

$$\hat{\sigma}(t) = \exp(-i\omega_I\hat{\mathbf{I}}_x t)\ \exp(-i\omega_S\hat{\mathbf{S}}_x t)\ (\hat{\mathbf{I}}_x + \hat{\mathbf{S}}_x)\ \exp\ (i\omega_S\hat{\mathbf{S}}_x t)\ \exp(i\omega_I\hat{\mathbf{I}}_x t)$$

wherein we apply a $\hat{\mathbf{S}}_x$ operator rotation followed by a $\hat{\mathbf{I}}_x$ operator rotation or we could write:

$$\hat{\sigma}(t) = \exp(-i\omega_S\hat{\mathbf{S}}_x t)\ \exp(-i\omega_I\hat{\mathbf{I}}_x t)(\hat{\mathbf{I}}_x + \hat{\mathbf{S}}_x)\ \exp\ (i\omega_I\hat{\mathbf{I}}_x t)\ \exp(i\omega_S\hat{\mathbf{S}}_x t)$$

in which the order of application of the $\hat{\mathbf{I}}_x$ and $\hat{\mathbf{S}}_x$ operators has been reversed. Each will give the same result. This is a general result with commuting operators ... the order of application is not relevant to the outcome of a set of rotations. How do we know this? Looking back at the Liouville–von Neumann equation we see that it is very compactly expressed (in frequency notation) in the form of a commutation calculation:

$$\frac{\mathrm{d}\hat{\sigma}(t)}{\mathrm{d}t} = i[\hat{\mathbf{H}}, \hat{\sigma}(t)]$$

This says that if the operator $\hat{\mathbf{H}}$ and the density operator $\hat{\sigma}(t)$ commute then the time evolution of $\hat{\sigma}(t)$ is zero. In other words, $\hat{\mathbf{H}}$ has no effect on $\hat{\sigma}(t)$ and *vice versa*. You can verify that this is so for yourself with the matrices that we developed. We shall see more of this later.

What about the rotation operators for a two-spin system? We can choose to construct them as we did for a single-spin using the expansion of the exponential function, however, there is simpler way to arrive at them. For the two-spin rotation operators we simply write the direct product, just as we did when constructing the two-spin operators above:

$$\exp(\pm i\omega_I\hat{\mathbf{I}}_x t) \otimes \exp(\pm i\omega_I\hat{\mathbf{I}}_x t)$$

$$= \begin{bmatrix} \cos\left(\frac{\omega_I t}{2}\right) & \pm i\sin\left(\frac{\omega_I t}{2}\right) \\ \pm i\sin\left(\frac{\omega_I t}{2}\right) & \cos\left(\frac{\omega_I t}{2}\right) \end{bmatrix} \otimes \begin{bmatrix} \cos\left(\frac{\omega_I t}{2}\right) & \pm i\sin\left(\frac{\omega_I t}{2}\right) \\ \pm i\sin\left(\frac{\omega_I t}{2}\right) & \cos\left(\frac{\omega_I t}{2}\right) \end{bmatrix}$$

$$= \begin{bmatrix} c^2 & \pm ics & \pm ics & -s^2 \\ \pm ics & c^2 & -s^2 & \pm ics \\ \pm ics & -s^2 & c^2 & \pm ics \\ -s^2 & \pm ics & \pm ics & c^2 \end{bmatrix}$$

where for brevity we have replaced $\cos(\omega_I t/2)$ and $\sin(\omega_I t/2)$ with c and s, respectively. Since this is a rotation operator corresponding to a pulse and since most pulses rotate the magnetisation either $\pi/2$ radians or π radians we may simplify this particular rotation matrix:

$$\overset{\text{for } \pi/2}{\begin{bmatrix} 0 & 0 & 0 & -1 \\ 0 & 0 & -1 & 0 \\ 0 & -1 & 0 & 0 \\ -1 & 0 & 0 & 0 \end{bmatrix}} \quad \overset{\text{for } \pi}{\begin{bmatrix} 1 & 0 & 0 & 0 \\ 0 & 1 & 0 & 0 \\ 0 & 0 & 1 & 0 \\ 0 & 0 & 0 & 1 \end{bmatrix}}$$

The reader may verify that these rotation matrices work as advertised.

Now that we have a two-spin system we must enlarge our free-precession Hamiltonian to include both Zeeman terms and a scalar coupling term:

$$\hat{\mathbf{H}}_{IS} = \omega_I \hat{\mathbf{I}}_z + \omega_S \hat{\mathbf{S}}_z + 2\pi J \hat{\mathbf{I}}\hat{\mathbf{S}}$$

As it stands, the Zeeman terms do *not* commute with the scalar coupling term:

$$[\hat{\mathbf{I}}_z, \hat{\mathbf{I}}\hat{\mathbf{S}}] = \frac{1}{8}\left(\begin{bmatrix} 1 & 0 & 0 & 0 \\ 0 & 1 & 0 & 0 \\ 0 & 0 & -1 & 0 \\ 0 & 0 & 0 & -1 \end{bmatrix} \begin{bmatrix} 1 & 0 & 0 & 0 \\ 0 & -1 & 2 & 0 \\ 0 & 2 & -1 & 0 \\ 0 & 0 & 0 & 1 \end{bmatrix} - \begin{bmatrix} 1 & 0 & 0 & 0 \\ 0 & -1 & 2 & 0 \\ 0 & 2 & -1 & 0 \\ 0 & 0 & 0 & 1 \end{bmatrix} \begin{bmatrix} 1 & 0 & 0 & 0 \\ 0 & 1 & 0 & 0 \\ 0 & 0 & -1 & 0 \\ 0 & 0 & 0 & -1 \end{bmatrix} \right)$$

$$= \frac{1}{8} \begin{bmatrix} 0 & 0 & 0 & 0 \\ 0 & 0 & 4 & 0 \\ 0 & -4 & 0 & 0 \\ 0 & 0 & 0 & 0 \end{bmatrix} \neq \hat{\mathbf{0}}$$

however, we shall modify things so that they do commute. Consider the two-spin Hamiltonian matrix (see eqn (13.11)):

$$\begin{bmatrix} \langle \alpha,\alpha|\hat{\mathbf{H}}|\alpha,\alpha\rangle & \langle \alpha,\alpha|\hat{\mathbf{H}}|\alpha,\beta\rangle & \langle \alpha,\alpha|\hat{\mathbf{H}}|\beta,\alpha\rangle & \langle \alpha,\alpha|\hat{\mathbf{H}}|\beta,\beta\rangle \\ \langle \alpha,\beta|\hat{\mathbf{H}}|\alpha,\alpha\rangle & \langle \alpha,\beta|\hat{\mathbf{H}}|\alpha,\beta\rangle & \langle \alpha,\beta|\hat{\mathbf{H}}|\beta,\alpha\rangle & \langle \alpha,\beta|\hat{\mathbf{H}}|\beta,\beta\rangle \\ \langle \beta,\alpha|\hat{\mathbf{H}}|\alpha,\alpha\rangle & \langle \beta,\alpha|\hat{\mathbf{H}}|\alpha,\beta\rangle & \langle \beta,\alpha|\hat{\mathbf{H}}|\beta,\alpha\rangle & \langle \beta,\alpha|\hat{\mathbf{H}}|\beta,\beta\rangle \\ \langle \beta,\beta|\hat{\mathbf{H}}|\alpha,\alpha\rangle & \langle \beta,\beta|\hat{\mathbf{H}}|\alpha,\beta\rangle & \langle \beta,\beta|\hat{\mathbf{H}}|\beta,\alpha\rangle & \langle \beta,\beta|\hat{\mathbf{H}}|\beta,\beta\rangle \end{bmatrix}$$

Since the Hamiltonian is a linear operator we can break up this matrix into three parts:

$$\begin{bmatrix} \langle\alpha,\alpha|\omega_I\hat{\mathbf{I}}_z|\alpha,\alpha\rangle & \langle\alpha,\alpha|\omega_I\hat{\mathbf{I}}_z|\alpha,\beta\rangle & \langle\alpha,\alpha|\omega_I\hat{\mathbf{I}}_z|\beta,\alpha\rangle & \langle\alpha,\alpha|\omega_I\hat{\mathbf{I}}_z|\beta,\beta\rangle \\ \langle\alpha,\beta|\omega_I\hat{\mathbf{I}}_z|\alpha,\alpha\rangle & \langle\alpha,\beta|\omega_I\hat{\mathbf{I}}_z|\alpha,\beta\rangle & \langle\alpha,\beta|\omega_I\hat{\mathbf{I}}_z|\beta,\alpha\rangle & \langle\alpha,\beta|\omega_I\hat{\mathbf{I}}_z|\beta,\beta\rangle \\ \langle\beta,\alpha|\omega_I\hat{\mathbf{I}}_z|\alpha,\alpha\rangle & \langle\beta,\alpha|\omega_I\hat{\mathbf{I}}_z|\alpha,\beta\rangle & \langle\beta,\alpha|\omega_I\hat{\mathbf{I}}_z|\beta,\alpha\rangle & \langle\beta,\alpha|\omega_I\hat{\mathbf{I}}_z|\beta,\beta\rangle \\ \langle\beta,\beta|\omega_I\hat{\mathbf{I}}_z|\alpha,\alpha\rangle & \langle\beta,\beta|\omega_I\hat{\mathbf{I}}_z|\alpha,\beta\rangle & \langle\beta,\beta|\omega_I\hat{\mathbf{I}}_z|\beta,\alpha\rangle & \langle\beta,\beta|\omega_I\hat{\mathbf{I}}_z|\beta,\beta\rangle \end{bmatrix}$$

$$+\begin{bmatrix} \langle\alpha,\alpha|\omega_S\hat{\mathbf{S}}_z|\alpha,\alpha\rangle & \langle\alpha,\alpha|\omega_S\hat{\mathbf{S}}_z|\alpha,\beta\rangle & \langle\alpha,\alpha|\omega_S\hat{\mathbf{S}}_z|\beta,\alpha\rangle & \langle\alpha,\alpha|\omega_S\hat{\mathbf{S}}_z|\beta,\beta\rangle \\ \langle\alpha,\beta|\omega_S\hat{\mathbf{S}}_z|\alpha,\alpha\rangle & \langle\alpha,\beta|\omega_S\hat{\mathbf{S}}_z|\alpha,\beta\rangle & \langle\alpha,\beta|\omega_S\hat{\mathbf{S}}_z|\beta,\alpha\rangle & \langle\alpha,\beta|\omega_S\hat{\mathbf{S}}_z|\beta,\beta\rangle \\ \langle\beta,\alpha|\omega_S\hat{\mathbf{S}}_z|\alpha,\alpha\rangle & \langle\beta,\alpha|\omega_S\hat{\mathbf{S}}_z|\alpha,\beta\rangle & \langle\beta,\alpha|\omega_S\hat{\mathbf{S}}_z|\beta,\alpha\rangle & \langle\beta,\alpha|\omega_S\hat{\mathbf{S}}_z|\beta,\beta\rangle \\ \langle\beta,\beta|\omega_S\hat{\mathbf{S}}_z|\alpha,\alpha\rangle & \langle\beta,\beta|\omega_S\hat{\mathbf{S}}_z|\alpha,\beta\rangle & \langle\beta,\beta|\omega_S\hat{\mathbf{S}}_z|\beta,\alpha\rangle & \langle\beta,\beta|\omega_S\hat{\mathbf{S}}_z|\beta,\beta\rangle \end{bmatrix}$$

$$+\begin{bmatrix} \langle\alpha,\alpha|2\pi J\hat{\mathbf{I}}\hat{\mathbf{S}}|\alpha,\alpha\rangle & \langle\alpha,\alpha|2\pi J\hat{\mathbf{I}}\hat{\mathbf{S}}|\alpha,\beta\rangle & \langle\alpha,\alpha|2\pi J\hat{\mathbf{I}}\hat{\mathbf{S}}|\beta,\alpha\rangle & \langle\alpha,\alpha|2\pi J\hat{\mathbf{I}}\hat{\mathbf{S}}|\beta,\beta\rangle \\ \langle\alpha,\beta|2\pi J\hat{\mathbf{I}}\hat{\mathbf{S}}|\alpha,\alpha\rangle & \langle\alpha,\beta|2\pi J\hat{\mathbf{I}}\hat{\mathbf{S}}|\alpha,\beta\rangle & \langle\alpha,\beta|2\pi J\hat{\mathbf{I}}\hat{\mathbf{S}}|\beta,\alpha\rangle & \langle\alpha,\beta|2\pi J\hat{\mathbf{I}}\hat{\mathbf{S}}|\beta,\beta\rangle \\ \langle\beta,\alpha|2\pi J\hat{\mathbf{I}}\hat{\mathbf{S}}|\alpha,\alpha\rangle & \langle\beta,\alpha|2\pi J\hat{\mathbf{I}}\hat{\mathbf{S}}|\alpha,\beta\rangle & \langle\beta,\alpha|2\pi J\hat{\mathbf{I}}\hat{\mathbf{S}}|\beta,\alpha\rangle & \langle\beta,\alpha|2\pi J\hat{\mathbf{I}}\hat{\mathbf{S}}|\beta,\beta\rangle \\ \langle\beta,\beta|2\pi J\hat{\mathbf{I}}\hat{\mathbf{S}}|\alpha,\alpha\rangle & \langle\beta,\beta|2\pi J\hat{\mathbf{I}}\hat{\mathbf{S}}|\alpha,\beta\rangle & \langle\beta,\beta|2\pi J\hat{\mathbf{I}}\hat{\mathbf{S}}|\beta,\alpha\rangle & \langle\beta,\beta|2\pi J\hat{\mathbf{I}}\hat{\mathbf{S}}|\beta,\beta\rangle \end{bmatrix}$$

The first two may be simplified into familiar forms:

$$\begin{bmatrix} \langle\alpha,\alpha|\omega_I\hat{\mathbf{I}}_z|\alpha,\alpha\rangle & \langle\alpha,\alpha|\omega_I\hat{\mathbf{I}}_z|\alpha,\beta\rangle & \langle\alpha,\alpha|\omega_I\hat{\mathbf{I}}_z|\beta,\alpha\rangle & \langle\alpha,\alpha|\omega_I\hat{\mathbf{I}}_z|\beta,\beta\rangle \\ \langle\alpha,\beta|\omega_I\hat{\mathbf{I}}_z|\alpha,\alpha\rangle & \langle\alpha,\beta|\omega_I\hat{\mathbf{I}}_z|\alpha,\beta\rangle & \langle\alpha,\beta|\omega_I\hat{\mathbf{I}}_z|\beta,\alpha\rangle & \langle\alpha,\beta|\omega_I\hat{\mathbf{I}}_z|\beta,\beta\rangle \\ \langle\beta,\alpha|\omega_I\hat{\mathbf{I}}_z|\alpha,\alpha\rangle & \langle\beta,\alpha|\omega_I\hat{\mathbf{I}}_z|\alpha,\beta\rangle & \langle\beta,\alpha|\omega_I\hat{\mathbf{I}}_z|\beta,\alpha\rangle & \langle\beta,\alpha|\omega_I\hat{\mathbf{I}}_z|\beta,\beta\rangle \\ \langle\beta,\beta|\omega_I\hat{\mathbf{I}}_z|\alpha,\alpha\rangle & \langle\beta,\beta|\omega_I\hat{\mathbf{I}}_z|\alpha,\beta\rangle & \langle\beta,\beta|\omega_I\hat{\mathbf{I}}_z|\beta,\alpha\rangle & \langle\beta,\beta|\omega_I\hat{\mathbf{I}}_z|\beta,\beta\rangle \end{bmatrix}$$

$$=\frac{\omega_I}{2}\begin{bmatrix} 1 & 0 & 0 & 0 \\ 0 & 1 & 0 & 0 \\ 0 & 0 & -1 & 0 \\ 0 & 0 & 0 & -1 \end{bmatrix} = \omega_I\hat{\mathbf{I}}_z$$

$$\begin{bmatrix} \langle\alpha,\alpha|\omega_S\hat{\mathbf{S}}_z|\alpha,\alpha\rangle & \langle\alpha,\alpha|\omega_S\hat{\mathbf{S}}_z|\alpha,\beta\rangle & \langle\alpha,\alpha|\omega_S\hat{\mathbf{S}}_z|\beta,\alpha\rangle & \langle\alpha,\alpha|\omega_S\hat{\mathbf{S}}_z|\beta,\beta\rangle \\ \langle\alpha,\beta|\omega_S\hat{\mathbf{S}}_z|\alpha,\alpha\rangle & \langle\alpha,\beta|\omega_S\hat{\mathbf{S}}_z|\alpha,\beta\rangle & \langle\alpha,\beta|\omega_S\hat{\mathbf{S}}_z|\beta,\alpha\rangle & \langle\alpha,\beta|\omega_S\hat{\mathbf{S}}_z|\beta,\beta\rangle \\ \langle\beta,\alpha|\omega_S\hat{\mathbf{S}}_z|\alpha,\alpha\rangle & \langle\beta,\alpha|\omega_S\hat{\mathbf{S}}_z|\alpha,\beta\rangle & \langle\beta,\alpha|\omega_S\hat{\mathbf{S}}_z|\beta,\alpha\rangle & \langle\beta,\alpha|\omega_S\hat{\mathbf{S}}_z|\beta,\beta\rangle \\ \langle\beta,\beta|\omega_S\hat{\mathbf{S}}_z|\alpha,\alpha\rangle & \langle\beta,\beta|\omega_S\hat{\mathbf{S}}_z|\alpha,\beta\rangle & \langle\beta,\beta|\omega_S\hat{\mathbf{S}}_z|\beta,\alpha\rangle & \langle\beta,\beta|\omega_S\hat{\mathbf{S}}_z|\beta,\beta\rangle \end{bmatrix}$$

$$= \frac{\omega_S}{2}\begin{bmatrix} 1 & 0 & 0 & 0 \\ 0 & -1 & 0 & 0 \\ 0 & 0 & 1 & 0 \\ 0 & 0 & 0 & -1 \end{bmatrix} = \omega_S\hat{\mathbf{S}}_z$$

and building on this we write for the third matrix using eqn (17.11):

$$\begin{bmatrix} \langle\alpha,\alpha|2\pi J\hat{\mathbf{I}}\hat{\mathbf{S}}|\alpha,\alpha\rangle & \langle\alpha,\alpha|2\pi J\hat{\mathbf{I}}\hat{\mathbf{S}}|\alpha,\beta\rangle & \langle\alpha,\alpha|2\pi J\hat{\mathbf{I}}\hat{\mathbf{S}}|\beta,\alpha\rangle & \langle\alpha,\alpha|2\pi J\hat{\mathbf{I}}\hat{\mathbf{S}}|\beta,\beta\rangle \\ \langle\alpha,\beta|2\pi J\hat{\mathbf{I}}\hat{\mathbf{S}}|\alpha,\alpha\rangle & \langle\alpha,\beta|2\pi J\hat{\mathbf{I}}\hat{\mathbf{S}}|\alpha,\beta\rangle & \langle\alpha,\beta|2\pi J\hat{\mathbf{I}}\hat{\mathbf{S}}|\beta,\alpha\rangle & \langle\alpha,\beta|2\pi J\hat{\mathbf{I}}\hat{\mathbf{S}}|\beta,\beta\rangle \\ \langle\beta,\alpha|2\pi J\hat{\mathbf{I}}\hat{\mathbf{S}}|\alpha,\alpha\rangle & \langle\beta,\alpha|2\pi J\hat{\mathbf{I}}\hat{\mathbf{S}}|\alpha,\beta\rangle & \langle\beta,\alpha|2\pi J\hat{\mathbf{I}}\hat{\mathbf{S}}|\beta,\alpha\rangle & \langle\beta,\alpha|2\pi J\hat{\mathbf{I}}\hat{\mathbf{S}}|\beta,\beta\rangle \\ \langle\beta,\beta|2\pi J\hat{\mathbf{I}}\hat{\mathbf{S}}|\alpha,\alpha\rangle & \langle\beta,\beta|2\pi J\hat{\mathbf{I}}\hat{\mathbf{S}}|\alpha,\beta\rangle & \langle\beta,\beta|2\pi J\hat{\mathbf{I}}\hat{\mathbf{S}}|\beta,\alpha\rangle & \langle\beta,\beta|2\pi J\hat{\mathbf{I}}\hat{\mathbf{S}}|\beta,\beta\rangle \end{bmatrix}$$

$$= \frac{\pi J}{2}\begin{bmatrix} 1 & 0 & 0 & 0 \\ 0 & -1 & 2 & 0 \\ 0 & 2 & -1 & 0 \\ 0 & 0 & 0 & 1 \end{bmatrix} = \frac{\pi J}{2}\hat{\mathbf{I}}\hat{\mathbf{S}}$$

Recombining the three matrices:

$$\frac{\omega_I}{2}\begin{bmatrix} 1 & 0 & 0 & 0 \\ 0 & 1 & 0 & 0 \\ 0 & 0 & -1 & 0 \\ 0 & 0 & 0 & -1 \end{bmatrix} + \frac{\omega_S}{2}\begin{bmatrix} 1 & 0 & 0 & 0 \\ 0 & -1 & 0 & 0 \\ 0 & 0 & 1 & 0 \\ 0 & 0 & 0 & -1 \end{bmatrix} + \frac{\pi J}{2}\begin{bmatrix} 1 & 0 & 0 & 0 \\ 0 & -1 & 2 & 0 \\ 0 & 2 & -1 & 0 \\ 0 & 0 & 0 & 1 \end{bmatrix}$$

$$= \begin{bmatrix} \frac{1}{2}(\omega_I+\omega_S)+\frac{\pi J}{2} & 0 & 0 & 0 \\ 0 & \frac{1}{2}(\omega_I+\omega_S)-\frac{\pi J}{2} & \pi J & 0 \\ 0 & \pi J & -\frac{1}{2}(\omega_I+\omega_S)-\frac{\pi J}{2} & 0 \\ 0 & 0 & 0 & -\frac{1}{2}(\omega_I+\omega_S)+\frac{\pi J}{2} \end{bmatrix}$$

We see that in addition to diagonal eigenvalues there are off-diagonal ones. We can greatly simplify things by assuming that $\pi J \leq \omega_I - \omega_S$. This is known as the *weak coupling* assumption. Leaving out the off-diagonal elements we have:

$$\hat{\mathbf{H}}_{IS} = \begin{bmatrix} \frac{1}{2}(\omega_I + \omega_S) + \frac{\pi J}{2} & 0 & 0 & 0 \\ 0 & \frac{1}{2}(\omega_I - \omega_S) - \frac{\pi J}{2} & 0 & 0 \\ 0 & 0 & -\frac{1}{2}(\omega_I - \omega_S) - \frac{\pi J}{2} & 0 \\ 0 & 0 & 0 & -\frac{1}{2}(\omega_I + \omega_S) + \frac{\pi J}{2} \end{bmatrix}$$

$$= \frac{\omega_I}{2}\begin{bmatrix} 1 & 0 & 0 & 0 \\ 0 & 1 & 0 & 0 \\ 0 & 0 & -1 & 0 \\ 0 & 0 & 0 & -1 \end{bmatrix} + \frac{\omega_S}{2}\begin{bmatrix} 1 & 0 & 0 & 0 \\ 0 & -1 & 0 & 0 \\ 0 & 0 & 1 & 0 \\ 0 & 0 & 0 & -1 \end{bmatrix} + \frac{\pi J}{2}\begin{bmatrix} 1 & 0 & 0 & 0 \\ 0 & -1 & 0 & 0 \\ 0 & 0 & -1 & 0 \\ 0 & 0 & 0 & 1 \end{bmatrix}$$

$$= \frac{\omega_I}{2}\hat{\mathbf{I}}_z + \frac{\omega_S}{2}\hat{\mathbf{S}}_z + \frac{\pi J}{2}\hat{\mathbf{I}}_z\hat{\mathbf{S}}_z$$

Now, happily, we find that $\hat{\mathbf{I}}_z$ and $\hat{\mathbf{S}}_z$ commute with $\hat{\mathbf{I}}_z\hat{\mathbf{S}}_z$, which means that when applying them as rotation operators the order of application is immaterial. The weak coupling approximation applies frequently in homonuclear spin systems unless the chemical shifts of the I and S spins are not much different and it always applies in heteronuclear spin systems since the Larmor frequencies of spins I and S are generally very far apart and hence their chemical shift differences are much, much larger than their coupling constants.

We can derive the pulse and chemical shift rotation operators by direct products of one-spin operators as we did above. However, since for a single-spin there is no coupling rotation operator we cannot do this for two-spin coupling and must resort to the expansion of the exponential function:

$$\exp(\pm i\pi Jt\hat{\mathbf{I}}_z\hat{\mathbf{S}}_z) = \hat{\mathbf{1}} \pm i\pi Jt\hat{\mathbf{I}}_z\hat{\mathbf{S}}_z + \frac{(\pm i\pi Jt)^2}{2\,!}(\hat{\mathbf{I}}_z\hat{\mathbf{S}}_z)^2 + \frac{(\pm i\pi Jt)^3}{3\,!}(\hat{\mathbf{I}}_z\hat{\mathbf{S}}_z)^3 \ldots$$

$$(\hat{\mathbf{I}}_z\hat{\mathbf{S}}_z)^2 = \frac{1}{4}\begin{bmatrix} 1 & 0 & 0 & 0 \\ 0 & 1 & 0 & 0 \\ 0 & 0 & 1 & 0 \\ 0 & 0 & 0 & 1 \end{bmatrix} = \frac{1}{4}\hat{\mathbf{1}} \quad (\hat{\mathbf{I}}_z\hat{\mathbf{S}}_z)^3 = \frac{1}{8}\begin{bmatrix} 1 & 0 & 0 & 0 \\ 0 & -1 & 0 & 0 \\ 0 & 0 & -1 & 0 \\ 0 & 0 & 0 & 1 \end{bmatrix} = \frac{1}{8}\hat{\mathbf{I}}_z\hat{\mathbf{S}}_z$$

and in general:

$$(\hat{\mathbf{I}}_z\hat{\mathbf{S}}_z)^n = \frac{1}{2^n}\begin{bmatrix} 1 & 0 & 0 & 0 \\ 0 & 1 & 0 & 0 \\ 0 & 0 & 1 & 0 \\ 0 & 0 & 0 & 1 \end{bmatrix} = \frac{1}{2^n}\hat{\mathbf{1}} \quad (n\,\text{even})$$

$$(\hat{\mathbf{I}}_z\hat{\mathbf{S}}_z)^n = \frac{1}{2^n}\begin{bmatrix} 1 & 0 & 0 & 0 \\ 0 & -1 & 0 & 0 \\ 0 & 0 & -1 & 0 \\ 0 & 0 & 0 & 1 \end{bmatrix} = \frac{1}{2^n}\hat{\mathbf{I}}_z\hat{\mathbf{S}}_z \quad (n\,\text{odd})$$

$$\exp(\pm i\pi Jt\hat{\mathbf{I}}_z\hat{\mathbf{S}}_z) = \hat{\mathbf{1}} + \frac{\left(\frac{\pi Jt}{2}\right)^2}{2!}\hat{\mathbf{1}} + \frac{\left(\frac{\pi Jt}{2}\right)^4}{4!}\hat{\mathbf{1}} \ldots$$

$$\pm i\left(\frac{\pi Jt}{2}\right)\hat{\mathbf{I}}_z\hat{\mathbf{S}}_z \mp i\frac{\left(\frac{\pi Jt}{2}\right)^3}{3!}\hat{\mathbf{I}}_z\hat{\mathbf{S}}_z \pm i\frac{\left(\frac{\pi Jt}{2}\right)^5}{5!}\hat{\mathbf{I}}_z\hat{\mathbf{S}}_z \ldots$$

$$= \begin{bmatrix} \cos\left(\frac{\pi Jt}{2}\right) \pm i\sin\left(\frac{\pi Jt}{2}\right) & 0 & 0 & 0 \\ 0 & \cos\left(\frac{\pi Jt}{2}\right) \mp i\sin\left(\frac{\pi Jt}{2}\right) & 0 & 0 \\ 0 & 0 & \cos\left(\frac{\pi Jt}{2}\right) \mp i\sin\left(\frac{\pi Jt}{2}\right) & 0 \\ 0 & 0 & 0 & \cos\left(\frac{\pi Jt}{2}\right) \pm i\sin\left(\frac{\pi Jt}{2}\right) \end{bmatrix}$$

What is the effect of these rotation operators on, say, $\hat{\mathbf{I}}_y$? Let us do the multiplication with:

$$a = \cos\left(\frac{\pi J t}{2}\right) + i\sin\left(\frac{\pi J t}{2}\right)$$

$$a^{*} = \cos\left(\frac{\pi J t}{2}\right) - i\sin\left(\frac{\pi J t}{2}\right)$$

$$\hat{\sigma}(t) = \begin{bmatrix} a^{*} & 0 & 0 & 0 \\ 0 & a & 0 & 0 \\ 0 & 0 & a & 0 \\ 0 & 0 & 0 & a^{*} \end{bmatrix} \frac{1}{4} \begin{bmatrix} 0 & 0 & -i & 0 \\ 0 & 0 & 0 & -i \\ i & 0 & 0 & 0 \\ 0 & i & 0 & 0 \end{bmatrix} \begin{bmatrix} a & 0 & 0 & 0 \\ 0 & a^{*} & 0 & 0 \\ 0 & 0 & a^{*} & 0 \\ 0 & 0 & 0 & a \end{bmatrix}$$

$$= \frac{1}{4} \begin{bmatrix} 0 & 0 & -ia^{*2} & 0 \\ 0 & 0 & 0 & -ia^{*2} \\ -ia^{2} & 0 & 0 & 0 \\ 0 & -ia^{2} & 0 & 0 \end{bmatrix}$$

$$= \frac{1}{4} \begin{bmatrix} 0 & 0 & i\cos(\pi J t) - \sin(\pi J t) & 0 \\ 0 & 0 & 0 & i\cos(\pi J t) + \sin(\pi J t) \\ i\cos(\pi J t) - \sin(\pi J t) & 0 & 0 & 0 \\ 0 & i\cos(\pi J t) + \sin(\pi J t) & 0 & 0 \end{bmatrix}$$

where we have used trigonometric identities (A.14) and (A.15) to arrive at the last line. The last matrix may be split into two matrices:

$$\frac{1}{4} \begin{bmatrix} 0 & 0 & -i & 0 \\ 0 & 0 & 0 & -i \\ i & 0 & 0 & 0 \\ 0 & i & 0 & 0 \end{bmatrix} \cos(\pi J t) + \frac{1}{4} \begin{bmatrix} 0 & 0 & -1 & 0 \\ 0 & 0 & 0 & 1 \\ -1 & 0 & 0 & 0 \\ 0 & 1 & 0 & 0 \end{bmatrix} \sin(\pi J t) \qquad (17.12)$$

$$= \hat{\mathbf{I}}_y \cos(\pi J t) - \hat{\mathbf{I}}_x \hat{\mathbf{S}}_z \sin(\pi J t)$$

$\hat{\mathbf{I}}_y$ evolves into $-\hat{\mathbf{I}}_x\hat{\mathbf{S}}_z$ under the influence of the scalar coupling operator, $\hat{\mathbf{I}}_z\hat{\mathbf{S}}_z$.

Now that we have the tools needed let us see what happens if we pulse a homonuclear[§§] two-spin system and let it evolve after the pulse. First, we apply a $\pi/2$ pulse to the equilibrium state of the spin system using the $\hat{\mathbf{I}}_x$ and $\hat{\mathbf{S}}_x$ exponential operators:

$$\begin{aligned}\hat{\sigma}(1) &= \exp\left(-i\frac{\pi}{2}\hat{\mathbf{S}}_x\right)\exp\left(-i\frac{\pi}{2}\hat{\mathbf{I}}_x\right)(\hat{\mathbf{I}}_z+\hat{\mathbf{S}}_z)\exp\left(i\frac{\pi}{2}\hat{\mathbf{I}}_x\right)\exp\left(i\frac{\pi}{2}\hat{\mathbf{S}}_x\right)\\ &= -(\hat{\mathbf{I}}_y+\hat{\mathbf{S}}_y)\\ &= -\hat{\mathbf{F}}_y\end{aligned}$$

Terms such as $\hat{\mathbf{I}}_y$ that evolve under the direction of exponential operators are frequently referred to as "magnetisation" in the NMR literature. We will follow the same convention here.

Next, we let the system evolve under chemical shift evolution (the Zeeman Hamiltonian) and scalar coupling. As we have just seen, using the weak coupling assumption we may order the exponential rotations as we please. First, using an identity similar to the one in eqn (17.10), the chemical shifts:[¶¶]

$$\begin{aligned}\hat{\sigma}(2) &= \exp(-i\omega_S\hat{\mathbf{S}}_z t_2)\exp(-i\omega_I\hat{\mathbf{I}}_z t_2)(-\hat{\mathbf{I}}_y-\hat{\mathbf{S}}_y)\exp(i\omega_I\hat{\mathbf{I}}_z t_2)\exp(i\omega_S\hat{\mathbf{S}}_z t_2)\\ &= -\hat{\mathbf{I}}_y\cos(\omega_I t_2)-\hat{\mathbf{S}}_y\cos(\omega_S t_2)+\hat{\mathbf{I}}_x\sin(\omega_I t_2)+\hat{\mathbf{S}}_x\sin(\omega_S t_2)\end{aligned}$$

You can use the vector model here to picture what is happening. $\hat{\mathbf{I}}_y$ magnetisation corresponding to a vector aligned with the $-y$-axis rotates (or evolves in NMR language) in a right-hand sense towards the $+x$-axis with chemical shift frequency ω_I. The same applies to the $\hat{\mathbf{S}}_y$ magnetisation. We follow this up with the coupling evolution calculation:

$$\begin{aligned}\hat{\sigma}(3) &= \exp(-i\pi J\hat{\mathbf{I}}_z\hat{\mathbf{S}}_z t_2)\hat{\sigma}(2)\exp(+i\pi J\hat{\mathbf{I}}_z\hat{\mathbf{S}}_z t_2)\\ &= -\hat{\mathbf{I}}_y\cos(\omega_I t_2)\cos(\pi J t_2)+\hat{\mathbf{I}}_x\hat{\mathbf{S}}_z\cos(\omega_I t_2)\sin(\pi J t_2)\\ &\quad-\hat{\mathbf{S}}_y\cos(\omega_S t_2)\cos(\pi J t_2)+\hat{\mathbf{I}}_z\hat{\mathbf{S}}_x\cos(\omega_S t_2)\sin(\pi J t_2)\\ &\quad+\hat{\mathbf{I}}_x\sin(\omega_I t_2)\cos(\pi J t_2)+\hat{\mathbf{I}}_y\hat{\mathbf{S}}_z\sin(\omega_I t_2)\sin(\pi J t_2)\\ &\quad+\hat{\mathbf{S}}_x\sin(\omega_S t_2)\cos(\pi J t_2)+\hat{\mathbf{I}}_z\hat{\mathbf{S}}_y\sin(\omega_S t_2)\sin(\pi J t_2)\end{aligned}$$

[§§]Generally, in the literature, I and S refer to heteronuclear species and one sees, for homonuclear systems, numbered subscripts such as I_1 and I_2. I find it easier, however, to keep track of I and S rather than subscripted I_s.

[¶¶]We will reveal this identity and others in the next chapter.

In going from $\hat{\sigma}(2)$ to $\hat{\sigma}(3)$ we note that $-\hat{\mathbf{I}}_y$ "splits" into $-\hat{\mathbf{I}}_y$ and $-\hat{\mathbf{I}}_x\hat{\mathbf{S}}_z$ magnetisations as does each of the other terms in $\hat{\sigma}(2)$. See Chapter 18 for a nice visualisation of this. We may verify that this happens using the full-blown matrix method. Also, we note that the trigonometric terms of $\hat{\sigma}(2)$ are carried through into $\hat{\sigma}(3)$.

What is the expectation value of, say, the y-magnetisation? We can use eqn (17.3) for this:

$$\langle F_y \rangle = \langle I_y + S_y \rangle = \mathrm{Tr}(\hat{\sigma}(3)(\hat{\mathbf{I}}_y + \hat{\mathbf{S}}_y))$$

However, we would need to do some *very* tedious matrix multiplications to arrive at the $\hat{\sigma}(3)$ matrix so we will take a shorter route. We need the matrices for $\hat{\sigma}(3)$ and $\hat{\mathbf{I}}_y + \hat{\mathbf{S}}_y$ for this calculation. The second is easy ... we simply add together our matrices for $\hat{\mathbf{I}}_y$ and $\hat{\mathbf{S}}_y$ from above:

$$(\hat{\mathbf{I}}_y + \hat{\mathbf{S}}_y) = \frac{1}{2}\begin{bmatrix} 0 & 0 & -i & 0 \\ 0 & 0 & 0 & -i \\ i & 0 & 0 & 0 \\ 0 & i & 0 & 0 \end{bmatrix} + \frac{1}{2}\begin{bmatrix} 0 & -i & 0 & 0 \\ i & 0 & 0 & 0 \\ 0 & 0 & 0 & -i \\ 0 & 0 & i & 0 \end{bmatrix} = \frac{1}{2}\begin{bmatrix} 0 & -i & -i & 0 \\ i & 0 & 0 & -i \\ i & 0 & 0 & -i \\ 0 & i & i & 0 \end{bmatrix}$$

$\hat{\sigma}(3)$ is almost as easy if a bit more time-consuming. The density matrix after the first $\pi/2$ pulse is $\hat{\sigma}(1)$:

$$\hat{\sigma}(1) = -(\hat{\mathbf{I}}_y + \hat{\mathbf{S}}_y)$$

To convert this to $\hat{\sigma}(3)$ we need only apply the rotation operator using the weak-coupling, free-precession Hamiltonian $\hat{\mathbf{H}}_{IS}$:[||]

$$\text{let} \quad \gamma_1 = \frac{1}{2}(\omega_I + \omega_S) + \frac{\pi J}{2} \quad \gamma_2 = \frac{1}{2}(\omega_I - \omega_S) - \frac{\pi J}{2}$$

$$\gamma_3 = -\frac{1}{2}(\omega_I - \omega_S) - \frac{\pi J}{2} \quad \gamma_4 = -\frac{1}{2}(\omega_I + \omega_S) + \frac{\pi J}{2}$$

$$\hat{\mathbf{H}}_{IS} = \begin{bmatrix} \gamma_1 & 0 & 0 & 0 \\ 0 & \gamma_2 & 0 & 0 \\ 0 & 0 & \gamma_3 & 0 \\ & 0 & 0 & \gamma_4 \end{bmatrix}$$

[||]Recall that this operator accomplishes *both* chemical-shift and coupling evolutions for a two-spin system. There is no need therefore to generate the $\hat{\sigma}(2)$ matrix. Using this we can convert directly from $\hat{\sigma}(1)$ to $\hat{\sigma}(3)$.

Since $\hat{\mathbf{H}}_{IS}$ is diagonal, the form of the rotation operator is simple (hence the reason for wanting a diagonal matrix):

$$\exp(\pm i\hat{\mathbf{H}}_{IS}t) = \begin{bmatrix} e^{(\pm i\gamma_1)} & 0 & 0 & 0 \\ 0 & e^{(\pm i\gamma_2)} & 0 & 0 \\ 0 & 0 & e^{(\pm i\gamma_3)} & 0 \\ 0 & 0 & 0 & e^{(\pm i\gamma_4)} \end{bmatrix}$$

The final state, $\hat{\sigma}(3)$, is then arrived at by application of these rotation operators to $\hat{\sigma}(1)$:

$$\hat{\sigma}(3) = \begin{bmatrix} e^{(-i\gamma_1 t)} & 0 & 0 & 0 \\ 0 & e^{(-i\gamma_2 t)} & 0 & 0 \\ 0 & 0 & e^{(-i\gamma_3 t)} & 0 \\ 0 & 0 & 0 & e^{(-i\gamma_4 t)} \end{bmatrix} \frac{1}{2} \begin{bmatrix} 0 & -i & -i & 0 \\ i & 0 & 0 & -i \\ i & 0 & 0 & -i \\ 0 & i & i & 0 \end{bmatrix} \begin{bmatrix} e^{(i\gamma_1 t)} & 0 & 0 & 0 \\ 0 & e^{(i\gamma_2 t)} & 0 & 0 \\ 0 & 0 & e^{(i\gamma_3 t)} & 0 \\ 0 & 0 & 0 & e^{(i\gamma_4 t)} \end{bmatrix}$$

$$= \begin{bmatrix} 0 & ie^{i(\gamma_2-\gamma_1)t} & ie^{i(\gamma_3-\gamma_1)}t & 0 \\ -ie^{i(\gamma_2-\gamma_1)t} & 0 & 0 & ie^{i(\gamma_4-\gamma_2)t} \\ -ie^{i(\gamma_3-\gamma_1)t} & 0 & 0 & ie^{i(\gamma_4-\gamma_3)t} \\ 0 & -ie^{i(\gamma_4-\gamma_2)}t & -ie^{i(\gamma_4-\gamma_3)t} & 0 \end{bmatrix}$$

$$= \begin{bmatrix} 0 & ie^{-i(\omega_S+\pi J)t} & ie^{-i(\omega_I+\pi J)t} & 0 \\ -ie^{-i(\omega_S+\pi J)t} & 0 & 0 & ie^{-i(\omega_I-\pi J)t} \\ -ie^{-i(\omega_I+\pi J)t} & 0 & 0 & ie^{-i(\omega_S-\pi J)t} \\ 0 & -ie^{-i(\omega_I-\pi J)t} & -ie^{-i(\omega_S-\pi J)t} & 0 \end{bmatrix}$$

The perceptive reader will most likely see what is coming. $\hat{\sigma}(3)$ contains terms that modulate at four frequencies, $\omega_1 \pm \pi J$ and $\omega_S \pm \pi J$ as we might expect to see for a pair of coupled nuclei that each split into a doublet in the NMR spectrum. Nonetheless, we press on to the final calculation using a few substitutions for further simplification:

$$\text{let } \alpha_I^+ = \omega_I + \pi J \quad \alpha_I^- = \omega_I - \pi J$$

$$\alpha_S^+ = \omega_S + \pi J \quad \alpha_S^- = \omega_S - \pi J$$

$$\langle F_y \rangle = \mathrm{Tr}(\hat{\sigma}(3)(\hat{\mathbf{I}}_y + \hat{\mathbf{S}}_y))$$

$$= \mathrm{Tr}\left(\frac{1}{2}\begin{bmatrix} 0 & ie^{-i\alpha_S^+ t} & ie^{-i\alpha_I^+ t} & 0 \\ -ie^{-i\alpha_S^+ t} & 0 & 0 & ie^{-i\alpha_I^- t} \\ -ie^{-i\alpha_I^+ t} & 0 & 0 & ie^{-i\alpha_S^- t} \\ 0 & -ie^{-i\alpha_I^- t} & -ie^{-i\alpha_S^- t} & 0 \end{bmatrix} \frac{1}{2}\begin{bmatrix} 0 & -i & -i & 0 \\ i & 0 & 0 & -i \\ i & 0 & 0 & -i \\ 0 & i & i & 0 \end{bmatrix}\right)$$

We do not really need to calculate the entire matrix . . . just the diagonal elements since the trace picks out only these elements anyway.***

$$\langle F_y \rangle = \frac{1}{4}(-e^{-i\alpha_S^+ t} - e^{-i\alpha_I^+ t} - e^{-i\alpha_S^+ t} - e^{-i\alpha_I^- t} - e^{-i\alpha_I^+ t} - e^{-i\alpha_S^- t} - e^{-i\alpha_I^- t} - e^{-i\alpha_S^- t})$$

$$= \frac{1}{2}(-e^{-i\alpha_S^+ t} - e^{-i\alpha_I^+ t} - e^{-i\alpha_I^- t} - e^{-i\alpha_S^- t})$$

$$= -\frac{1}{2}(e^{-i(\omega_I + \pi J)t} + e^{-i(\omega_I - \pi J)t} + e^{-i(\omega_S + \pi J)t} + e^{-i(\omega_S - \pi J)t})$$

This represents oscillations at four different frequencies, as mentioned above. The expectation values for $\langle I_x + S_x \rangle$, $\langle 2I_xS_z + 2I_yS_z \rangle$ and $\langle 2I_zS_x + 2I_zS_y \rangle$ using the appropriate Hamiltonian rotation operator and the appropriate density operators may be calculated in a very similar fashion.

The NMR spectrum of two coupled spins under the weak-coupling assumption consists of four peaks of equal intensity, two at $\omega_I \pm \pi J$ and two at $\omega_S \pm \pi J$ (Figure 17.1).

The reader will by now (or more likely, some time ago) appreciate that working through these calculations using matrices is very time consuming and will be wondering if there is an easier way to do things. There is, in fact, a much simpler method that we have already seen in bits and pieces and which we will encounter fully in the next chapter.

17.5 Spin-One

One occasionally encounters nuclei of other than spin-1/2 in one's NMR spectroscopy career.††† We may construct their density matrices in much the same manner as our spin-1/2 matrices. For a spin-one

***Signs are of importance in this calculation. Be careful!
†††Actually, more than occasionally if you use deuterated solvents.

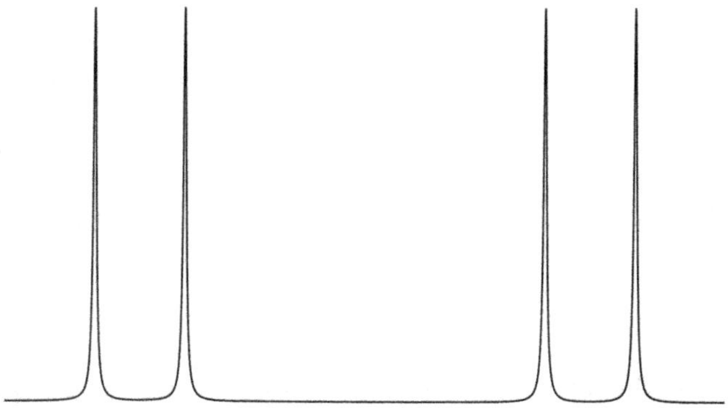

Figure 17.1 Spectrum of coupled spins *I* and *S* under the weak-coupling approximation.

nucleus there are three possible spin-states that we can find it in. We will label them α, β and γ, and as usual, assume them to be orthogonal and normalised. For a single spin-one nucleus the $\hat{\mathbf{I}}_z$ matrix will be a 3×3 matrix which we can construct after recalling the discussion in Chapter 15:

$$\langle 1,1|\hat{\mathbf{I}}_z|1,1\rangle = 1 = \langle\alpha|\hat{\mathbf{I}}_z|\alpha\rangle$$

$$\langle 1,0|\hat{\mathbf{I}}_z|1,0\rangle = 0 = \langle\beta|\hat{\mathbf{I}}_z|\beta\rangle$$

$$\langle 1,-1|\hat{\mathbf{I}}_z|1,-1\rangle = -1 = \langle\gamma|\hat{\mathbf{I}}_z|\gamma\rangle$$

$$\langle 1,1|\hat{\mathbf{I}}_z|1,0\rangle = 0 = \langle\alpha|\hat{\mathbf{I}}_z|\beta\rangle$$

$$\text{etc.}$$

The single-spin spin-1 $\hat{\mathbf{I}}_z$ matrix is then:

$$\hat{\mathbf{I}}_z = \begin{bmatrix} 1 & 0 & 0 \\ 0 & 0 & 0 \\ 0 & 0 & -1 \end{bmatrix}$$

Also, following the recipes of Chapter 15, we find:

$$\hat{\mathbf{I}}_x = \frac{1}{\sqrt{2}}\begin{bmatrix} 0 & 1 & 0 \\ 1 & 0 & 1 \\ 0 & 1 & 0 \end{bmatrix} \quad \hat{\mathbf{I}}_y = \frac{1}{\sqrt{2}}\begin{bmatrix} 0 & -i & 0 \\ i & 0 & -i \\ 0 & i & 0 \end{bmatrix}$$

What are the rotation operators associated with these matrices.[‡‡‡] The $\hat{\mathbf{I}}_z$ rotation operator is trivial to determine since the $\hat{\mathbf{I}}_z$ density matrix is diagonal:

$$\exp(\pm i\omega\hat{\mathbf{I}}_z t) = \begin{bmatrix} e^{\pm i\omega t} & 0 & 0 \\ 0 & 0 & 0 \\ 0 & 0 & e^{\mp i\omega t} \end{bmatrix}$$

The rotation operator matrix for $\hat{\mathbf{I}}_x$ is obtained in our usual way with one small twist:

$$\exp(i\omega_I\hat{\mathbf{I}}_x t) = \hat{\mathbf{1}} + i\omega_I t\hat{\mathbf{I}}_x + \frac{(i\omega_I t)^2}{2!}\hat{\mathbf{I}}_x^2 + \frac{(i\omega_I t)^3}{3!}\hat{\mathbf{I}}_x^3 + \frac{(i\omega_I t)^4}{4!}\hat{\mathbf{I}}_x^4 \ldots$$

$$= \begin{bmatrix} 1 & 0 & 0 \\ 0 & 1 & 0 \\ 0 & 0 & 1 \end{bmatrix} + i\omega_I t\frac{1}{\sqrt{2}}\begin{bmatrix} 0 & 1 & 0 \\ 1 & 0 & 1 \\ 0 & 1 & 0 \end{bmatrix} - \frac{(\omega_I t)^2}{2!}\left(\frac{1}{\sqrt{2}}\right)^2\begin{bmatrix} 0 & 1 & 0 \\ 1 & 0 & 1 \\ 0 & 1 & 0 \end{bmatrix}^2 \ldots$$

A bit of thought about it (and perhaps some matrix multiplication) will verify that:

$$\hat{\mathbf{I}}_x^{2n} = \hat{\mathbf{I}}_x^2 \text{ and } \hat{\mathbf{I}}_x^{2n+1} = \hat{\mathbf{I}}_x$$

Note that here, unlike the case with spins-1/2, $\hat{\mathbf{I}}_x^{2n} = \hat{\mathbf{I}}_x^2$ rather than $\hat{\mathbf{I}}_x^{2n} = \hat{\mathbf{1}}$. We will not be able to use $\hat{\mathbf{1}}$. in our spin-one expansion in the same way as with our spin-1/2 expansions. We may rewrite our expansion as follows:

$$\exp(i\omega_I\hat{\mathbf{I}}_x t) = \hat{\mathbf{1}} + \left(1 - 1 - \frac{(\omega_I t)^2}{2!} + \frac{(\omega_I t)^4}{4!} \ldots\right)\hat{\mathbf{I}}_x^2$$
$$+ i\left(\omega_I t - \frac{(\omega_I t)^3}{3!} + \frac{(\omega_I t)^5}{5!} \ldots\right)\hat{\mathbf{I}}_x$$
$$= \hat{\mathbf{1}} + [\cos(\omega_I t) - 1]\hat{\mathbf{I}}_x^2 + i\sin(\omega_I t)\hat{\mathbf{I}}_x$$

[‡‡‡]This is very sparsely treated if at all in most texts.

Note that here we have "played a mathematical trick" by injecting $1-1$ into the first part of the expansion in order to be able to write it in cosine form. In matrix form this is:

$$\exp(i\omega_I\hat{\mathbf{I}}_x t) = \begin{bmatrix} 1 & 0 & 0 \\ 0 & 1 & 0 \\ 0 & 0 & 1 \end{bmatrix} + \frac{\cos(\omega_I t)-1}{2}\begin{bmatrix} 1 & 0 & 1 \\ 0 & 2 & 0 \\ 1 & 0 & 1 \end{bmatrix} + \frac{i\sin(\omega_I t)}{\sqrt{2}}\begin{bmatrix} 0 & 1 & 0 \\ 1 & 0 & 1 \\ 0 & 1 & 0 \end{bmatrix}$$

$$= \begin{bmatrix} \frac{1}{2}[\cos(\omega_I t)-1]+1 & \frac{1}{\sqrt{2}}i\sin(\omega_I t) & \frac{1}{2}[\cos(\omega_I t)-1] \\ \frac{1}{\sqrt{2}}i\sin(\omega_I t) & [\cos(\omega_I t)-1]+1 & \frac{1}{\sqrt{2}}i\sin(\omega_I t) \\ \frac{1}{2}[\cos(\omega_I t)-1] & \frac{1}{\sqrt{2}}i\sin(\omega_I t) & \frac{1}{2}[\cos(\omega_I t)-1]+1 \end{bmatrix}$$

$$= \begin{bmatrix} \frac{1}{2}[\cos(\omega_I t)+1] & \frac{1}{\sqrt{2}}i\sin(\omega_I t) & \frac{1}{2}[\cos(\omega_I t)-1] \\ \frac{1}{\sqrt{2}}i\sin(\omega_I t) & \cos(\omega_I t) & \frac{1}{\sqrt{2}}i\sin(\omega_I t) \\ \frac{1}{2}[\cos(\omega_I t)-1] & \frac{1}{\sqrt{2}}i\sin(\omega_I t) & \frac{1}{2}[\cos(\omega_I t)+1] \end{bmatrix}$$

Similarly:

$$\exp(-i\omega_I\hat{\mathbf{I}}_x t) = \begin{bmatrix} \frac{1}{2}[\cos(\omega_I t)+1] & -\frac{1}{\sqrt{2}}i\sin(\omega_I t) & \frac{1}{2}[\cos(\omega_I t)-1] \\ -\frac{1}{\sqrt{2}}i\sin(\omega_I t) & \cos(\omega_I t) & -\frac{1}{\sqrt{2}}i\sin(\omega_I t) \\ \frac{1}{2}[\cos(\omega_I t)-1] & -\frac{1}{\sqrt{2}}i\sin(\omega_I t) & \frac{1}{2}[\cos(\omega_I t)+1] \end{bmatrix}$$

For a $\pi/2$ rotation the matrix is:

$$\exp\left(\pm i\frac{\pi}{2}t\right) = \begin{bmatrix} \frac{1}{2} & \pm\frac{i}{\sqrt{2}} & -\frac{1}{2} \\ \pm\frac{i}{\sqrt{2}} & 0 & \pm\frac{i}{\sqrt{2}} \\ -\frac{1}{2} & \pm\frac{i}{\sqrt{2}} & \frac{1}{2} \end{bmatrix}$$

and the rotation of $\hat{\mathbf{I}}_z$ under the influence of $\hat{\mathbf{I}}_x$ is given by:

$$\exp\left(-i\frac{\pi}{2}\hat{\mathbf{I}}_x t\right)\hat{\mathbf{I}}_z \exp\left(i\frac{\pi}{2}\hat{\mathbf{I}}_x t\right)$$

$$= \begin{bmatrix} \frac{1}{2} & -\frac{i}{\sqrt{2}} & -\frac{1}{2} \\ -\frac{i}{\sqrt{2}} & 0 & -\frac{i}{\sqrt{2}} \\ -\frac{1}{2} & -\frac{i}{\sqrt{2}} & \frac{1}{2} \end{bmatrix} \begin{bmatrix} 1 & 0 & 0 \\ 0 & 0 & 0 \\ 0 & 0 & -1 \end{bmatrix} \begin{bmatrix} \frac{1}{2} & \frac{i}{\sqrt{2}} & -\frac{1}{2} \\ \frac{i}{\sqrt{2}} & 0 & \frac{i}{\sqrt{2}} \\ -\frac{1}{2} & \frac{i}{\sqrt{2}} & \frac{1}{2} \end{bmatrix}$$

$$= \frac{1}{\sqrt{2}} \begin{bmatrix} 0 & i & 0 \\ -i & 0 & i \\ 0 & -i & 0 \end{bmatrix} = -\hat{\mathbf{I}}_y$$

The reader is encouraged to try the same type of operation with a $\hat{\mathbf{I}}_y$ rotation matrix and verify that its application to the spin one $\hat{\mathbf{I}}_z$ density matrix will produce the $\hat{\mathbf{I}}_x$ matrix. The reader may also verify that, just as for spin-1/2, one may write:

$$\exp(-i\omega_I\hat{\mathbf{I}}_x t)\hat{\mathbf{I}}_z \exp(i\omega_I\hat{\mathbf{I}}_x t) = \cos(\omega_I t)\,\hat{\mathbf{I}}_z - \sin(\omega_I t)\,\hat{\mathbf{I}}_y$$

We now wish to model a spin system that has *mixed* spins in it as this is the most commonly encountered situation. So, let us look at spins I and S, specifying that spin I is a spin-1/2 nucleus and spin S is a spin-one nucleus.[§§§] We will assume them to be scalar coupled. What are the corresponding density matrices? We find them easily, as

[§§§]This is a bit artificial in that the Hamiltonian for most non-spin-1/2 nuclei includes a quadrupolar component that has a large influence on the outcome of the calculation. However, in the case of deuterium that is a spin-one nucleus, this is not the case. It acts much like a spin-1/2 nucleus.

above, by direct product multiplication of the appropriate I and S operator matrices:¶¶¶

$$\hat{\mathbf{I}}_z \otimes \hat{\mathbf{E}} = \frac{1}{2}\begin{bmatrix} 1 & 0 \\ 0 & -1 \end{bmatrix} \otimes \begin{bmatrix} 1 & 0 & 0 \\ 0 & 1 & 0 \\ 0 & 0 & 1 \end{bmatrix} = \frac{1}{2}\begin{bmatrix} 1 & 0 & 0 & 0 & 0 & 0 \\ 0 & 1 & 0 & 0 & 0 & 0 \\ 0 & 0 & 1 & 0 & 0 & 0 \\ 0 & 0 & 0 & -1 & 0 & 0 \\ 0 & 0 & 0 & 0 & -1 & 0 \\ 0 & 0 & 0 & 0 & 0 & -1 \end{bmatrix} = \hat{\mathbf{I}}_z$$

$$\hat{\mathbf{E}} \otimes \hat{\mathbf{S}}_z = \begin{bmatrix} 1 & 0 \\ 0 & 1 \end{bmatrix} \otimes \begin{bmatrix} 1 & 0 & 0 \\ 0 & 0 & 0 \\ 0 & 0 & -1 \end{bmatrix} = \begin{bmatrix} 1 & 0 & 0 & 0 & 0 & 0 \\ 0 & 0 & 0 & 0 & 0 & 0 \\ 0 & 0 & -1 & 0 & 0 & 0 \\ 0 & 0 & 0 & 1 & 0 & 0 \\ 0 & 0 & 0 & 0 & 0 & 0 \\ 0 & 0 & 0 & 0 & 0 & -1 \end{bmatrix} = \hat{\mathbf{S}}_z$$

Note that, as above, it is understood that the arguments to the direct product operator $\otimes$ on the left side are in the original spin-spaces and the operators on the right are in the expanded spin-spaces. So, $\hat{\mathbf{I}}_x \otimes \hat{\mathbf{E}}$ contains the spin-1/2 $\hat{\mathbf{I}}_z$ and the spin-one $\hat{\mathbf{E}}$, whereas on the right side of the expression, $\hat{\mathbf{I}}_z$ is in the expanded two-spin space. The two $\hat{\mathbf{I}}_z$s are *not* the same. For the two-spin operators the appropriate normalisation constant is added:

$$\hat{\mathbf{I}}_x \otimes \hat{\mathbf{E}} = \frac{1}{2}\begin{bmatrix} 0 & 1 \\ 1 & 0 \end{bmatrix} \otimes \begin{bmatrix} 1 & 0 & 0 \\ 0 & 1 & 0 \\ 0 & 0 & 1 \end{bmatrix} = \frac{1}{2}\begin{bmatrix} 0 & 0 & 0 & 1 & 0 & 0 \\ 0 & 0 & 0 & 0 & 1 & 0 \\ 0 & 0 & 0 & 0 & 0 & 1 \\ 1 & 0 & 0 & 0 & 0 & 0 \\ 0 & 1 & 0 & 0 & 0 & 0 \\ 0 & 0 & 1 & 0 & 0 & 0 \end{bmatrix} = \hat{\mathbf{I}}_x$$

$$\hat{\mathbf{E}} \otimes \hat{\mathbf{S}}_x = \begin{bmatrix} 1 & 0 \\ 0 & 1 \end{bmatrix} \otimes \begin{bmatrix} 0 & 1 & 0 \\ 1 & 0 & 1 \\ 0 & 1 & 0 \end{bmatrix} = \begin{bmatrix} 0 & 1 & 0 & 0 & 0 & 0 \\ 1 & 0 & 1 & 0 & 0 & 0 \\ 0 & 1 & 0 & 0 & 0 & 0 \\ 0 & 0 & 0 & 0 & 1 & 0 \\ 0 & 0 & 0 & 1 & 0 & 1 \\ 0 & 0 & 0 & 0 & 1 & 0 \end{bmatrix} = \hat{\mathbf{S}}_x$$

¶¶¶A word of warning here. Those of you who perhaps have read or are reading Mateescu's otherwise excellent book mentioned in the references will note that the matrices obtained here are not the same as his. His explanation of the origin of these mixed-spin matrices is a bit sparse but close inspection will show that, using our current symbols, he is putting S (the spin-one nucleus) first followed by I (the spin-one half nucleus), whereas we do the opposite here. As noted earlier, the order is not of importance but consistency in the order is important.

The twelve remaining matrices are obtained in exactly the same way (except the unit matrix $\hat{\mathbf{E}}$ of course).

The rotation operators may be obtained in the same way as for spin-1/2 and one-spin spin-one operators. We will leave this an exercise for the reader.

17.6 Density Matrices and Perturbation Calculations

After a spin system has come to equilibrium anything that changes this equilibrium is called a *perturbation* ... an rf pulse is a characteristic example of this. The way that this is typically handled is to write a Hamiltonian consisting of a time-independent part and a time-dependent component. See Atkins and Friedman in the Further Reading section for an example of how this is done.

An alternate approach to the problem uses the density matrix. As with the first approach it involves a composite Hamiltonian.‖‖‖ We write a suitably altered equation of motion for the density matrix using eqn (17.8) as our model:

$$\frac{\mathrm{d}\hat{\sigma}(t)}{\mathrm{d}t} = i[\hat{\mathbf{H}}_0 + \hat{\mathbf{H}}(t), \hat{\sigma}(t)] \tag{17.13}$$

$\hat{\mathbf{H}}(t)$ is assumed to represent a much smaller interaction than $\hat{\mathbf{H}}_0$ ... we have already seen something like this with respect to the $\vec{\mathbf{B}}_0$ and $\vec{\mathbf{B}}_1$ fields. In our analysis of the effects of the $\vec{\mathbf{B}}_1$ field we assumed it to be much smaller than the external $\vec{\mathbf{B}}_0$ field. Also, generally $\hat{\mathbf{H}}(t)$ is assumed to be due to an external interaction or perturbation ... again, such as a $\vec{\mathbf{B}}_1$ field. Recall that the solution to this differential equation, if $\hat{\mathbf{H}}(t)$ is zero, is:

$$\hat{\sigma}(t) = \exp(-i\hat{\mathbf{H}}_0(t)\ \hat{\sigma}(0)\exp(i\hat{\mathbf{H}}_0(t)$$

We now define a new quantity:

$$\hat{\sigma}(t) \equiv \exp(-i\hat{\mathbf{H}}_0 t)\hat{\sigma}^{\#}(t)\exp(i\hat{\mathbf{H}}_0 t)$$

$$\text{or} \tag{17.14}$$

$$\hat{\sigma}^{\#}(t) \equiv \exp(i\hat{\mathbf{H}}_0 t)\hat{\sigma}(t)\exp(-i\hat{\mathbf{H}}_0 t)$$

‖‖‖This is loosely adapted from Slichter's exposition (see the references) with expanded explanations. I do not use $\hat{\sigma}^*$ as he does since it might possibly be confused with the complex conjugate. I will use $\hat{\sigma}^{\#}$ instead. Also, Slichter appears to be a bit fast and loose with variable notation ... I will try to be a bit more careful here. This material may seem somewhat peripheral to NMR spectroscopy but it will be very useful in the discussion of relaxation in Chapter 20. Beautifully horrible (or horribly beautiful, depending on your point of view)!

Obviously, at $t = 0$, $\sigma(t)$ and $\hat{\sigma}^{\#}(t)$ are identical. We wish to produce an equation that will be of use to us when we begin to analyse relaxation processes in Chapter 20. To accomplish this we will do a second-order perturbation calculation using $\hat{\sigma}^{\#}(t)$. We continue by substituting eqn (17.14) into the left side of eqn (17.13) and differentiating:

$$\frac{\mathrm{d}}{\mathrm{d}t}[\exp(-i\hat{\mathbf{H}}_0 t)\hat{\sigma}^{\#}(t)\exp(i\hat{\mathbf{H}}_0 t)]$$

$$= -i\hat{\mathbf{H}}_0\exp(-i\hat{\mathbf{H}}_0 t)\hat{\sigma}^{\#}(t)\exp(i\hat{\mathbf{H}}_0 t) + \exp(-i\hat{\mathbf{H}}_0 t)\frac{\mathrm{d}\hat{\sigma}^{\#}(t)}{\mathrm{d}t}exp(i\hat{\mathbf{H}}_0 t)$$

$$+ \exp(-i\hat{\mathbf{H}}_0 t)\hat{\sigma}^{\#}(t)\exp(i\hat{\mathbf{H}}_0 t)i\hat{\mathbf{H}}_0$$

$$= -i\hat{\mathbf{H}}_0\hat{\sigma}(t) + \exp(-i\hat{\mathbf{H}}_0 t)\frac{\mathrm{d}\hat{\sigma}^{\#}(t)}{\mathrm{d}t}\exp(i\hat{\mathbf{H}}_0 t) + i\hat{\sigma}(t)\hat{\mathbf{H}}_0$$

$$= i[\hat{\mathbf{H}}_0, \hat{\sigma}(t)] + \exp(-i\hat{\mathbf{H}}_0 t)\frac{\mathrm{d}\hat{\sigma}^{\#}(t)}{\mathrm{d}t}\exp(i\hat{\mathbf{H}}_0 t)$$

This is then equivalent to the right side of eqn (17.13), which we expand with the aid of the commutator algebra from Appendix A:

$$i[\hat{\mathbf{H}}_0, \hat{\sigma}(t)] + \exp(-i\hat{\mathbf{H}}_0 t)\frac{\mathrm{d}\hat{\sigma}^{\#}(t)}{\mathrm{d}t}\exp(i\hat{\mathbf{H}}_0 t) = i[\hat{\mathbf{H}}_0, \hat{\sigma}(t)] + i[\hat{\mathbf{H}}(t), \hat{\sigma}(t)]$$

$$\Downarrow \tag{17.15}$$

$$\exp(-i\hat{\mathbf{H}}_0 t)\frac{\mathrm{d}\hat{\sigma}^{\#}(t)}{\mathrm{d}t}\exp(i\hat{\mathbf{H}}_0 t) = i[\hat{\mathbf{H}}(t), \hat{\sigma}(t)]$$

The commutator $i[\hat{\mathbf{H}}_0, \hat{\sigma}(t)]$ vanishes from both sides. For convenience, we similarly define:

$$\hat{\mathbf{H}}^{\#}(t) \equiv \exp(i\hat{\mathbf{H}}_0 t)\ \hat{\mathbf{H}}(t)\exp(-i\hat{\mathbf{H}}_0 t)$$

and we then multiply each side of eqn (17.15) from the left by $\exp(i\hat{\mathbf{H}}_0 t)$ and the right by $\exp(-i\hat{\mathbf{H}}_0 t)$:

$$\exp(i\hat{\mathbf{H}}_0 t)\exp(-i\hat{\mathbf{H}}_0 t)\frac{\mathrm{d}\hat{\sigma}^{\#}(t)}{\mathrm{d}t}\exp(i\hat{\mathbf{H}}_0 t)\exp(-i\hat{\mathbf{H}}_0 t) = \exp(i\hat{\mathbf{H}}_0 t)i[\hat{\mathbf{H}}(t), \hat{\sigma}(t)]\exp(-i\hat{\mathbf{H}}_0 t)$$

$$\Downarrow$$

$$\frac{\mathrm{d}\hat{\sigma}^{\#}(t)}{\mathrm{d}t} = i\exp(i\hat{\mathbf{H}}_0 t)\hat{\mathbf{H}}(t)\hat{\sigma}(t)\exp(-i\hat{\mathbf{H}}_0 t) - i\exp(i\hat{\mathbf{H}}_0 t)\hat{\sigma}(t)\hat{\mathbf{H}}(t)\exp(-i\hat{\mathbf{H}}_0 t)$$

and substitute our above definition of $\hat{\sigma}(t)$ in terms of $\hat{\sigma}^{\#}(t)$ and our definition of $\hat{\mathbf{H}}^{\#}(t)$ for $\exp(i\hat{\mathbf{H}}_0 t)\,\hat{\mathbf{H}}(t)\exp(-i\hat{\mathbf{H}}_0 t)$:

$$\begin{aligned}\frac{\mathrm{d}\hat{\sigma}^{\#}(t)}{\mathrm{d}t} &= i\exp(i\hat{\mathbf{H}}_0 t)\hat{\mathbf{H}}(t)\hat{\sigma}(t)\exp(-i\hat{\mathbf{H}}_0 t) - i\exp(i\hat{\mathbf{H}}_0 t)\hat{\sigma}(t)\hat{\mathbf{H}}(t)\exp(-i\hat{\mathbf{H}}_0 t)\\ &= i\exp(i\hat{\mathbf{H}}_0 t)\hat{\mathbf{H}}(t)\exp(-i\hat{\mathbf{H}}_0 t)\hat{\sigma}^{\#}(t)\exp(i\hat{\mathbf{H}}_0 t)\exp(-i\hat{\mathbf{H}}_0 t)\\ &\quad - i\exp(i\hat{\mathbf{H}}_0 t)\exp(-i\hat{\mathbf{H}}_0 t)\hat{\sigma}^{\#}(t)\exp(i\hat{\mathbf{H}}_0 t)\hat{\mathbf{H}}(t)\exp(-i\hat{\mathbf{H}}_0 t)\\ &= i\exp(i\hat{\mathbf{H}}_0 t)\hat{\mathbf{H}}(t)\exp(-i\hat{\mathbf{H}}_0 t)\hat{\sigma}^{\#}(t) - i\hat{\sigma}^{\#}(t)\exp(i\hat{\mathbf{H}}_0 t)\hat{\mathbf{H}}(t)\exp(-i\hat{\mathbf{H}}_0 t)\\ &= i\hat{\mathbf{H}}^{\#}(t)\hat{\sigma}^{\#}(t) - i\hat{\sigma}^{\#}(t)\hat{\mathbf{H}}^{\#}(t)\\ &= i[\hat{\mathbf{H}}^{\#}(t), \hat{\sigma}^{\#}(t)]\end{aligned} \tag{17.16}$$

This is known as the *interaction representation* that is a method for separating the static operator $\hat{\mathbf{H}}_0$ from the time-dependent operator, $\hat{\mathbf{H}}(t)$, similar to the rotating-frame discussion of Chapter 12 in which $\vec{\mathbf{B}}_0$ (and therefore its associated Hamiltonian) vanished. The difference here is that, unlike the rotating frame situation, $\hat{\mathbf{H}}(t)$ retains its time dependence. Our definitions of $\hat{\sigma}^{\#}(t)$ and $\hat{\mathbf{H}}^{\#}(t)$ are specific examples of a more general definition of the interaction representation of operators:

$$\hat{\mathbf{O}}^{\#} = \exp(i\hat{\mathbf{H}}t)\,\hat{\mathbf{O}}\,\exp(-i\hat{\mathbf{H}}t) \tag{17.17}$$

Solving for $\hat{\sigma}^{\#}(t)$:

$$\begin{gathered}\frac{\mathrm{d}\hat{\sigma}^{\#}(t)}{\mathrm{d}t} = i[\hat{\mathbf{H}}^{\#}(t), \hat{\sigma}^{\#}(t)]\\ \Downarrow\\ \int_0^t \mathrm{d}\hat{\sigma}^{\#}(t') = i\int_0^t [\hat{\mathbf{H}}^{\#}(t'), \hat{\sigma}^{\#}(t')]\mathrm{d}t'\\ \hat{\sigma}^{\#}(t')\Big|_0^t = i\int_0^t [\hat{\mathbf{H}}^{\#}(t'), \hat{\sigma}^{\#}(t')]\mathrm{d}t'\\ \hat{\sigma}^{\#}(t) - \hat{\sigma}^{\#}(0) = i\int_0^t [\hat{\mathbf{H}}^{\#}(t'), \hat{\sigma}^{\#}(t')]\mathrm{d}t'\\ \hat{\sigma}^{\#}(t) = \hat{\sigma}^{\#}(0) + i\int_0^t [\hat{\mathbf{H}}^{\#}(t'), \hat{\sigma}^{\#}(t')]\mathrm{d}t'\end{gathered} \tag{17.18}$$

and we solve approximately by replacing $\hat{\sigma}^{\#}(t')$ with $\hat{\sigma}^{\#}(0)$:

$$\hat{\sigma}^{\#}(t) = \hat{\sigma}^{\#}(0) + i\int_0^t [\hat{\mathbf{H}}^{\#}(t'), \hat{\sigma}^{\#}(0)]\mathrm{d}t'$$

and then, for a better second-order approximation, iteratively substitute this back into eqn (17.18) for $\hat{\sigma}^{\#}(t')$:

$$\hat{\sigma}^{\#}(t') = \hat{\sigma}^{\#}(0) + i\int_0^t \left[\hat{\mathbf{H}}^{\#}(t'), \hat{\sigma}^{\#}(0) + i\int_0^t [\hat{\mathbf{H}}^{\#}(t''), \hat{\sigma}^{\#}(0)]\right]\mathrm{d}t''\mathrm{d}t'$$

$$= \hat{\sigma}^{\#}(0) + i\int_0^t \left[\hat{\mathbf{H}}^{\#}(t'), \hat{\sigma}^{\#}(0)\right]\mathrm{d}t' + i^2\int_0^t\int_0^{t'} [\hat{\mathbf{H}}^{\#}(t'), [\hat{\mathbf{H}}^{\#}(t''), \hat{\sigma}^{\#}(0)]]\mathrm{d}t''\mathrm{d}t'$$

Differentiating:

$$\frac{\mathrm{d}\hat{\sigma}^{\#}(t)}{\mathrm{d}t} = i[\hat{\mathbf{H}}^{\#}(t'), \hat{\sigma}^{\#}(0)] + i^2\int_0^t [\hat{\mathbf{H}}^{\#}(t), [\hat{\mathbf{H}}^{\#}(t'), \hat{\sigma}^{\#}(0)]]\mathrm{d}t' \qquad (17.19)$$

Some further manipulations … using the matrix elements of the density matrix in the form delimited by eqn (3.51) and the time evolution results in eqn (17.9) for the time-invariant Hamiltonian, we may say:

$$\begin{aligned}\langle m|\hat{\sigma}(t)|n\rangle &= \langle\phi_m|\hat{\sigma}(t)|\phi_n\rangle \\ &= \langle\phi_m|\exp(-i\hat{\mathbf{H}}t)\hat{\sigma}(0)\exp(i\hat{\mathbf{H}}t)|\phi_n\rangle \\ &= \langle\exp(i\hat{\mathbf{H}}t)\phi_m|\hat{\sigma}(0)|\exp(i\hat{\mathbf{H}}t)\phi_n\rangle \\ &= \exp(-iE_m t)\langle\phi_m|\hat{\sigma}(0)|\phi_n\rangle\exp(iE_n t) \\ &= \exp(i(E_n - E_m)t)\langle\phi_m|\hat{\sigma}(0)|\phi_n\rangle\end{aligned} \qquad (17.20)$$

where ϕ represents a basis wavefunction. Replacing $\hat{\sigma}(0)$ in eqn (17.20) with $\hat{\sigma}^{\#}(t)$ we can write:

$$\langle m|\hat{\sigma}(t)|n\rangle = \exp(i[E_n - E_m]t)\ \langle\phi_m|\hat{\sigma}^{\#}(t)|\phi_n\rangle \qquad (17.21)$$

and similarly for $\hat{\mathbf{H}}^{\#}$:

$$\langle m|\hat{\mathbf{H}}(t)|n\rangle = \exp(i[E_n - E_m]t)\ \langle\phi_m|\hat{\mathbf{H}}^{\#}(t)|\phi_n\rangle \qquad (17.22)$$

We shall return to these equations in Chapter 20.

17.7 Strong Coupling

We would now like to examine a situation in which we might profitably use the density matrix/operator idea. Recall that when we consider two scalar-coupled spins the full Hamiltonian matrix is:

$$\hat{\mathbf{H}}_{IS} = \begin{bmatrix} \frac{1}{2}(\omega_I + \omega_S) + \frac{\pi J}{2} & 0 & 0 & 0 \\ 0 & \frac{1}{2}(\omega_I + \omega_S) - \frac{\pi J}{2} & \pi J & 0 \\ 0 & \pi J & -\frac{1}{2}(\omega_I + \omega_S) - \frac{\pi J}{2} & 0 \\ 0 & 0 & 0 & -\frac{1}{2}(\omega_I + \omega_S) + \frac{\pi J}{2} \end{bmatrix}$$

with diagonal and off-diagonal elements and the corresponding wavefunction is:

$$\Psi = c_{\alpha_I\alpha_S}|\alpha_I, \alpha_S\rangle + c_{\alpha_I\beta_S}|\alpha_I, \beta_S\rangle + c_{\beta_I\alpha_S}|\beta_I, \alpha_S\rangle + c_{\beta_I\beta_S}|\beta_I, \beta_S\rangle$$

As for the single-spin system, we may represent the basis vectors for a two-spin system very simply:

$$|\alpha_I, \alpha_S\rangle = \begin{bmatrix} 1 \\ 0 \\ 0 \\ 0 \end{bmatrix} \quad |\alpha_I, \beta_S\rangle = \begin{bmatrix} 0 \\ 1 \\ 0 \\ 0 \end{bmatrix} \quad |\beta_I, \alpha_S\rangle = \begin{bmatrix} 0 \\ 0 \\ 1 \\ 0 \end{bmatrix} \quad |\beta_I, \beta_S\rangle = \begin{bmatrix} 0 \\ 0 \\ 0 \\ 1 \end{bmatrix}$$

The weak-coupling assumption simply removes the off-diagonals that eases calculations and is a reasonable assumption in most cases, especially for heteronuclear spin-systems. Occasionally, however, weak coupling is not an appropriate assumption. In homonuclear spin systems in which two coupled nuclei have nearly identical chemical shifts we must include the off-diagonals in the Hamiltonian.

We may diagonalise this matrix using the technique developed in Chapter 2 (specifically, see eqn (2.45)). First, to simplify the calculation we recognise that the first and last terms of the Hamiltonian will not change during a diagonalisation procedure as there are no off-diagonals associated with them.**** Only the second and third

****If you do not believe this then try the diagonalisation on the full matrix ... the first and last terms will be unchanged. It is much easier to work only on the sub-matrix. This can be easily done using symbolic algebra software.

terms will change so we look only at the sub-matrix containing those terms and their associated off-diagonals:

$$\mathbf{H}_{sub} = \begin{bmatrix} \frac{\delta}{2} - \frac{\pi J}{2} & \pi J \\ \pi J & -\frac{\delta}{2} - \frac{\pi J}{2} \end{bmatrix}$$

where for brevity we have temporarily substituted δ for $\omega_I - \omega_S$. Using the general equation for eigenvalues of a 2×2 matrix developed in Chapter 2 (eqn (2.61)):

$$\lambda = \frac{-\pi J \pm \sqrt{(\pi J)^2 - (\pi J)^2 + \delta^2 + 2(\pi J)^2}}{2}$$

$$= \pm\frac{1}{2}\sqrt{4(\pi J)^2 + \delta^2} - \frac{\pi J}{2}$$

and the full diagonalised Hamiltonian matrix is:

$$\hat{\mathbf{H}}_{diag} =$$

$$\begin{bmatrix} \frac{1}{2}(\omega_I + \omega_S) + \frac{\pi J}{2} & 0 & 0 & 0 \\ 0 & \frac{1}{2}\sqrt{4(\pi J)^2 + \delta^2} - \frac{\pi J}{2} & 0 & 0 \\ 0 & 0 & -\frac{1}{2}\sqrt{4(\pi J)^2 + \delta^2} - \frac{\pi J}{2} & 0 \\ 0 & 0 & 0 & -\frac{1}{2}(\omega_I + \omega_S) + \frac{\pi J}{2} \end{bmatrix}$$

A close look at this shows us that under the weak-coupling assumption this matrix collapses to the weak-coupling version from above. Also, note that according to eqn (2.52) the trace of $\hat{\mathbf{H}}_{diag}$ should be the same as that of $\hat{\mathbf{H}}_{IS}$ and so it is (check for yourself).

The matrices that diagonalise $\hat{\mathbf{H}}_{IS}$ are of interest:

$$\hat{\mathbf{H}}_{IS}\hat{\mathbf{X}} = \hat{\mathbf{X}}\hat{H}_{diag}$$

$$\hat{\mathbf{X}}^{-1}\hat{\mathbf{H}}_{IS}\hat{\mathbf{X}} = \hat{\mathbf{H}}_{diag}$$

in direct analogy with eqn (2.48). $\hat{\mathbf{H}}_{diag}$, of course, contains the eigenvalues of the eigenvectors contained in $\hat{\mathbf{X}}$. Matrices $\hat{\mathbf{X}}$ and $\hat{\mathbf{X}}^{-1}$ perform a unitary transformation on $\hat{\mathbf{H}}_{IS}$ to produce the diagonalised matrix, $\hat{\mathbf{H}}_{diag}$. As we saw in Chapter 2, the diagonalisation procedure may be likened to a rotation operation.

If we diagonalise the Hamiltonian matrix then the associated wavefunctions must also be altered. To see that this is so, consider the fundamental Schrodinger eigenvalue equation:

$$\hat{\mathbf{H}}|\Psi\rangle = E|\Psi\rangle$$

Let us apply $\hat{\mathbf{X}}^{-1}$ to each side of the equation:

$$\hat{\mathbf{X}}^{-1}\hat{\mathbf{H}}_{\mathrm{IS}}|\Psi\rangle = \hat{\mathbf{X}}^{-1}E|\Psi\rangle = E\hat{\mathbf{X}}^{-1}|\Psi\rangle$$

Recalling that $\hat{\mathbf{X}}\hat{\mathbf{X}}^{-1} = \hat{\mathbf{1}}$, we insert into this equation:

$$\hat{\mathbf{X}}^{-1}\hat{\mathbf{H}}_{IS}\hat{\mathbf{X}}\hat{X}^{-1}|\Psi\rangle = E\hat{\mathbf{X}}^{-1}|\Psi\rangle$$

$$\Downarrow$$

$$\hat{\mathbf{H}}_{diag}\hat{\mathbf{X}}^{-1}|\Psi\rangle = E\hat{\mathbf{X}}^{-1}|\Psi\rangle$$

$$\Downarrow$$

$$\hat{\mathbf{H}}_{diag}|\Psi'\rangle = E|\Psi'\rangle$$

You can see that in order to maintain the validity of the Schrodinger equation we must alter the original wavefunctions. In our example above we diagonalised the sub-matrix, $\hat{\mathbf{H}}_{sub}$ for simplicity since the first and last terms along the diagonal of $\hat{\mathbf{H}}_{IS}$ remain unaffected by the diagonalisation procedure. Thus, the full matrices $\hat{\mathbf{X}}$ and $\hat{\mathbf{X}}^{-1}$ are:

$$\hat{\mathbf{X}} = \begin{bmatrix} 1 & 0 & 0 & 0 \\ 0 & \cos(\phi) & \sin(\phi) & 0 \\ 0 & -\sin(\phi) & \cos(\phi) & 0 \\ 0 & 0 & 0 & 1 \end{bmatrix} \quad \hat{\mathbf{X}}^{-1} = \begin{bmatrix} 1 & 0 & 0 & 0 \\ 0 & \cos(\phi) & -\sin(\phi) & 0 \\ 0 & \sin(\phi) & \cos(\phi) & 0 \\ 0 & 0 & 0 & 1 \end{bmatrix}$$

where the value of ϕ is given by eqn (2.59):

$$\phi = \frac{1}{2}\arctan\left(\frac{4\pi J}{2}\right)$$

Application of $\hat{\mathbf{X}}^{-1}$ to the original basis vectors above gives:

$$\hat{\mathbf{X}}^{-1}\begin{bmatrix} 1 \\ 0 \\ 0 \\ 0 \end{bmatrix} = \begin{bmatrix} 1 \\ 0 \\ 0 \\ 0 \end{bmatrix} \quad \hat{\mathbf{X}}^{-1}\begin{bmatrix} 0 \\ 1 \\ 0 \\ 0 \end{bmatrix} = \begin{bmatrix} 0 \\ \cos(\phi) \\ \sin(\phi) \\ 0 \end{bmatrix} \quad \hat{\mathbf{X}}^{-1}\begin{bmatrix} 0 \\ 0 \\ 1 \\ 0 \end{bmatrix} = \begin{bmatrix} 0 \\ -\sin(\phi) \\ \cos(\phi) \\ 0 \end{bmatrix}$$

$$\hat{\mathbf{X}}^{-1}\begin{bmatrix} 0 \\ 0 \\ 0 \\ 1 \end{bmatrix} = \begin{bmatrix} 0 \\ 0 \\ 0 \\ 1 \end{bmatrix}$$

Let us calculate the expectation value, $\langle F_y \rangle$, for the strong-coupling case and see how it might differ from the weak-coupling case. We do the calculation in the normal way:

$$\langle F_y \rangle = \langle I_y + S_y \rangle = \mathrm{Tr}(\hat{\sigma}\hat{\mathbf{F}}_y)$$

As with the weak-coupling example above we start with the equilibrium density matrix, $\hat{\sigma}(0)$:

$$\hat{\sigma}(0) = \hat{\mathbf{F}}_z = (\hat{\mathbf{I}}_z + \hat{\mathbf{S}}_z)$$

Application of a $\pi/2$ pulse to both spins will rotate them such that the new density matrix, $\hat{\sigma}(1)$, is:

$$\hat{\sigma}(1) = -\hat{\mathbf{F}}_y = -(\hat{\mathbf{I}}_y + \hat{\mathbf{S}}_y)$$

Next, to get to $\hat{\sigma}(2)$ we apply the strong-coupling Hamiltonian, $\hat{\mathbf{H}}_{IS}$. However, since it is not diagonal it will be difficult to produce an exponential matrix operator. As we have seen, diagonal matrices lend themselves to easy construction of exponential operator matrices. Therefore, we wish to use the diagonalised version of $\hat{\mathbf{H}}_{IS}$, $\hat{\mathbf{H}}_{diag}$:

$$\hat{\sigma}(2) = \exp(-i\hat{\mathbf{H}}_{\mathrm{IS}}t)\ \hat{\sigma}(1)\exp(i\hat{\mathbf{H}}_{\mathrm{IS}}t)$$

We make use of eqn (2.61) here to modify the Hamiltonian exponential because it is not necessarily diagonal and we would like to do that since the exponential of a diagonal matrix is easy to find as we discovered in Chapter 2.

$$\exp(i\hat{\mathbf{H}}_{IS}) = \hat{\mathbf{X}}\ \exp(i\hat{\mathbf{H}}_{diag})\ \hat{\mathbf{X}}^{-1}$$

Using this we modify the previous expression:

$$\hat{\sigma}(2) = \hat{\mathbf{X}}\ \exp(-i\hat{\mathbf{H}}_{diag}t)\ \hat{\mathbf{X}}^{-1}\ \hat{\sigma}(1)\hat{\mathbf{X}}\ \exp(i\hat{\mathbf{H}}_{diag}t)\ \hat{\mathbf{X}}^{-1}$$

We know the form of each of these (many) matrices from above, however, their calculation in this expression is quite tedious. Fortunately, we can simplify things somewhat. The expectation value of $\langle F_y \rangle$ is:

$$\begin{aligned}
\langle F_y \rangle &= \mathrm{Tr}(\hat{\sigma}(2)\hat{\mathbf{F}}_y) \\
&= \mathrm{Tr}(\hat{\mathbf{X}}\exp(-i\hat{\mathbf{H}}_{diag}t)\hat{\mathbf{X}}^{-1}\hat{\sigma}(1)\hat{\mathbf{X}}\exp(i\hat{\mathbf{H}}_{diag}t)\hat{\mathbf{X}}^{-1}\hat{\mathbf{F}}_y) \\
&= -\mathrm{Tr}(\hat{\mathbf{X}}\exp(-i\hat{\mathbf{H}}_{diag}t)\hat{\mathbf{X}}^{-1}\hat{\mathbf{F}}_y\hat{\mathbf{X}}\exp(i\hat{\mathbf{H}}_{diag}t)\hat{\mathbf{X}}^{-1}\hat{\mathbf{F}}_y)
\end{aligned}$$

Cyclic permutation (see eqn (2.23)) of $\hat{\mathbf{X}}$ gives us:

$$\langle F_y \rangle = -\,\mathrm{Tr}\,(\exp(-i\hat{\mathbf{H}}_{diag}t)\,[\hat{\mathbf{X}}^{-1}\,\hat{\mathbf{F}}_y\,\hat{\mathbf{X}}]\exp(i\hat{\mathbf{H}}_{diag}t)\,[\hat{\mathbf{X}}^{-1}\,\hat{\mathbf{F}}_y\,\hat{\mathbf{X}}])$$

The matrix grouping $[\hat{\mathbf{X}}^{-1}\,\hat{\mathbf{F}}_y\,\hat{\mathbf{X}}]$ is:††††

$$\begin{bmatrix} 1 & 0 & 0 & 0 \\ 0 & \cos(\phi) & \sin(\phi) & 0 \\ 0 & -\sin(\phi) & \cos(\phi) & 0 \\ 0 & 0 & 0 & 1 \end{bmatrix} \frac{1}{4} \begin{bmatrix} 0 & -i & -i & 0 \\ i & 0 & 0 & -i \\ i & 0 & 0 & -i \\ 0 & i & i & 0 \end{bmatrix} \begin{bmatrix} 1 & 0 & 0 & 0 \\ 0 & \cos(\phi) & -\sin(\phi) & 0 \\ 0 & \sin(\phi) & \cos(\phi) & 0 \\ 0 & 0 & 0 & 1 \end{bmatrix}$$

$$= \frac{1}{4} \begin{bmatrix} 0 & -(c+s) & -(c-s) & 0 \\ c+s & 0 & 0 & -(c+s) \\ c-s & 0 & 0 & -(c-s) \\ 0 & c+s & c-s & 0 \end{bmatrix} = \hat{\mathbf{D}}$$

where as usual we have shortened $\cos(\phi)$ and $\sin(\phi)$ to c and s, respectively. The last part of the calculation is to obtain the matrix from which the trace is extracted using $\hat{\mathbf{H}}_{\mathrm{diag}}$ from above:

$$\langle F_y \rangle = \mathrm{Tr}(\exp(-i\hat{\mathbf{H}}_{diag}t)\hat{\mathbf{D}}\exp(i\hat{\mathbf{H}}_{diag}t)\hat{\mathbf{D}})$$

$$\exp(i\hat{\mathbf{H}}_{diag}t) = \begin{bmatrix} e^{\frac{i}{2}[(\omega_i-\omega_S)+\pi J]t} & 0 & 0 & 0 \\ 0 & e^{\frac{i}{2}\left[\sqrt{4(\pi J)^2+(\omega_i-\omega_S)^2}-\pi J\right]t} & 0 & 0 \\ 0 & 0 & e^{\frac{i}{2}\left[-\sqrt{4(\pi J)^2+(\omega_i-\omega_S)^2}-\pi J\right]t} & 0 \\ 0 & 0 & 0 & e^{\frac{i}{2}[-(\omega_i-\omega_S)+\pi J]t} \end{bmatrix}$$

††††To avoid yet more tedious calculations I again invite you to use wxMaxima or your favourite computer algebra program to do these calculations.

After very much tedious calculation:[‡‡‡‡]

$$\begin{aligned}\langle F_y\rangle &= \mathrm{Tr}(\exp(-i\hat{\mathbf{H}}_{diag}t)\hat{\mathbf{D}}\exp(i\hat{\mathbf{H}}_{diag}t)\hat{\mathbf{D}})\\
&= \frac{1}{4}(1+\sin(2\phi))\left(\mathrm{e}^{\frac{i}{2}\left[(\omega_I+\omega_S)-\sqrt{4(\pi J)^2+(\omega_I-\omega_S)^2}+2\pi J\right]t}+\mathrm{e}^{\frac{-i}{2}\left[(\omega_I+\omega_S)-\sqrt{4(\pi J)^2+(\omega_I-\omega_S)^2}+2\pi J\right]t}\right)\\
&+\frac{1}{4}(1+\sin(2\phi))\left(\mathrm{e}^{\frac{i}{2}\left[(\omega_I+\omega_S)+\sqrt{4(\pi J)^2+(\omega_I-\omega_S)^2}-2\pi J\right]t}+\mathrm{e}^{\frac{-i}{2}\left[(\omega_I+\omega_S)+\sqrt{4(\pi J)^2+(\omega_I-\omega_S)^2}-2\pi J\right]t}\right)\\
&+\frac{1}{4}(1-\sin(2\phi))\left(\mathrm{e}^{\frac{i}{2}\left[(\omega_I+\omega_S)+\sqrt{4(\pi J)^2+(\omega_I-\omega_S)^2}+2\pi J\right]t}+\mathrm{e}^{\frac{-i}{2}\left[(\omega_I+\omega_S)+\sqrt{4(\pi J)^2+(\omega_I-\omega_S)^2}+2\pi J\right]t}\right)\\
&+\frac{1}{4}(1-\sin(2\phi))\left(\mathrm{e}^{\frac{i}{2}\left[(\omega_I+\omega_S)-\sqrt{4(\pi J)^2+(\omega_I-\omega_S)^2}-2\pi J\right]t}+\mathrm{e}^{\frac{-i}{2}\left[(\omega_I+\omega_S)-\sqrt{4(\pi J)^2+(\omega_I-\omega_S)^2}-2\pi J\right]t}\right)\end{aligned}$$

This can be simplified somewhat by the application of eqn (A.22) from Appendix A to our current result:

$$\begin{aligned}\langle F_y\rangle &= \mathrm{Tr}(\exp(-i\hat{\mathbf{H}}_{diag}t)\hat{\mathbf{D}}\exp(i\hat{\mathbf{H}}_{diag}t)\hat{\mathbf{D}})\\
&= \frac{1}{4}(1+\sin(2\phi))\cos\left(\left[\frac{\omega_I+\omega_S}{2}-\frac{\sqrt{4(\pi J)^2+(\omega_I-\omega_S)^2}}{2}+\pi J\right]t\right)\\
&+\frac{1}{4}(1+\sin(2\phi))\cos\left(\left[\frac{\omega_I+\omega_S}{2}+\frac{\sqrt{4(\pi J)^2+(\omega_I-\omega_S)^2}}{2}-\pi J\right]t\right)\\
&+\frac{1}{4}(1-\sin(2\phi))\cos\left(\left[\frac{\omega_I+\omega_S}{2}+\frac{\sqrt{4(\pi J)^2+(\omega_I-\omega_S)^2}}{2}+\pi J\right]t\right)\\
&+\frac{1}{4}(1-\sin(2\phi))\cos\left(\left[\frac{\omega_I+\omega_S}{2}-\frac{\sqrt{4(\pi J)^2+(\omega_I-\omega_S)^2}}{2}-\pi J\right]t\right)\end{aligned}$$

[‡‡‡‡]It is not that I do not want to show it here. There are really a lot of calculations here and no space to show them, or rather the space would be wasted with page after page of matrices. They are not difficult to do, just very, very tedious. Fortunately, we have computer software to bear the brunt of the drudgery. The first people to do these calculations before the general availability of computers must have been very dedicated and patient indeed.

Figure 17.2 Spectrum of coupled spins I and S under strong coupling.

We simplify things a bit more by realizing that in the rotating frame of reference the sum of the chemical shift frequencies of spin I and spin S is arbitrary since we are free to choose whatever rotating frame frequency we wish. If we choose it to be exactly half-way between the two then their sum will be zero since one will have positive frequency and the other a negative frequency. With this assumption in hand we can now begin to say something about the appearance of the spectrum. As with the weak-coupling case, there will be four peaks in the spectrum located at frequencies:

1. $-\dfrac{\sqrt{4(\pi J)^2+(\omega_I-\omega_S)^2}}{2}+\pi J$
2. $+\dfrac{\sqrt{4(\pi J)^2+(\omega_I-\omega_S)^2}}{2}-\pi J$
3. $+\dfrac{\sqrt{4(\pi J)^2+(\omega_I-\omega_S)^2}}{2}+\pi J$
4. $-\dfrac{\sqrt{4(\pi J)^2+(\omega_I-\omega_S)^2}}{2}-\pi J$

However, unlike the weak-coupling case, the intensities of these peaks will not all be the same. The first two are modulated by a $1+\sin(2\phi)$ term while the last two are modulated by $1-\sin(2\phi)$ (Figure 17.2). Thus, depending on the value of ϕ, which in turn depends on $\omega_I-\omega_S$, the inner peaks (at frequencies 1 and 2) in the spectrum will have larger intensities than the outer two peaks (at frequencies 3 and 4). In the limit as $\omega_I - \omega_S$ approaches zero the outer peaks will vanish and there will be only one peak in the spectrum.

Further Reading

1. P. Atkins and R. Friedman, *Molecular Quantum Mechanics*, Oxford University Press, London, 4th edn, 2005.
2. P. J. Hore, J. A. Jones and S. Wimperis, *NMR: The Toolkit*, Oxford Science Publications, London, 2009.
3. J. Keeler, *Understanding NMR Spectroscopy*, John Wiley & Sons, Chichester, 2nd edn, 2010.

18 The Product Operator Formalism

A mathematical theory is not to be considered complete until you have made it so clear that you can explain it to the first man whom you meet on the street.

David Hilbert

18.1 The Density Operator Method Revisited

In the last chapter we encountered the density matrix method of analysis of NMR pulse sequences and all its tedious calculations. Our analysis was for a two-spin system ... the thought of applying it to a three-spin system sends shivers up the spine! It is an exceedingly powerful method but also is exceedingly clumsy to use requiring lengthy and error-prone matrix multiplications. Modern computer software makes the method more "user friendly" in the sense that the drudgery of matrix multiplication is lessened and far fewer human errors occur. However, quick "back of the envelope" calculations are all but inaccessible by this method.

It would be nice to keep all of the power of the density matrix analysis procedure but not the tedious nature, or at least, greatly reduce the tedium. This is accomplished in the product operator model. First, however, we point out a pattern that develops when doing density matrix calculations.[†] Recall eqn (17.10) in which a rotation of equilibrium magnetisation about the x-axis is described:

$$\exp(-i\omega_1\hat{\mathbf{I}}_x t)\hat{\mathbf{I}}_z \exp(i\omega_1\hat{\mathbf{I}}_x t) = \hat{\mathbf{I}}_z \cos(\omega_1 t) - \hat{\mathbf{I}}_y \sin(\omega_1 t) \tag{18.1}$$

[†]This will involve more matrix calculations but only briefly, I promise. The "payoff" is coming.

Essential Mathematics for NMR and MRI Spectroscopists
By Keith C. Brown

Published by the Royal Society of Chemistry, www.rsc.org

Let us do the same thing except now we will do a rotation about the y-axis. We need the exponential rotation matrix corresponding to $\hat{\mathbf{I}}_y$ and using the method outlined in Chapter 17 we find this to be:

$$\hat{\mathbf{I}}_y = \frac{1}{2}\begin{bmatrix} 0 & -i \\ i & 0 \end{bmatrix}$$

Application of this rotation matrix to the equilibrium magnetisation gives:

$$\exp(i\omega_I \hat{\mathbf{I}}_y t) = \begin{bmatrix} \cos\left(\dfrac{\omega_I t}{2}\right) & \sin\left(\dfrac{\omega_I t}{2}\right) \\ -\sin\left(\dfrac{\omega_I t}{2}\right) & \cos\left(\dfrac{\omega_I t}{2}\right) \end{bmatrix}$$

There are two other one-spin rotations we can do ... rotations of $\hat{\mathbf{I}}_x$ about $\hat{\mathbf{I}}_z$ and $\hat{\mathbf{I}}_y$ about $\hat{\mathbf{I}}_z$. Rather than do them here we will just state the results and let the reader verify them.

$$\exp(-i\omega_I \hat{\mathbf{I}}_y t)\hat{\mathbf{I}}_z \exp(i\omega_I \hat{\mathbf{I}}_y t)$$

$$= \begin{bmatrix} \cos\left(\dfrac{\omega_I t}{2}\right) & -\sin\left(\dfrac{\omega_I t}{2}\right) \\ \sin\left(\dfrac{\omega_I t}{2}\right) & \cos\left(\dfrac{\omega_I t}{2}\right) \end{bmatrix} \frac{1}{2}\begin{bmatrix} 1 & 0 \\ 0 & -1 \end{bmatrix} \begin{bmatrix} \cos\left(\dfrac{\omega_I t}{2}\right) & \sin\left(\dfrac{\omega_I t}{2}\right) \\ -\sin\left(\dfrac{\omega_I t}{2}\right) & \cos\left(\dfrac{\omega_I t}{2}\right) \end{bmatrix}$$

$$= \frac{1}{2}\begin{bmatrix} \cos^2\left(\dfrac{\omega_I t}{2}\right) - \sin^2\left(\dfrac{\omega_I t}{2}\right) & 2\sin\left(\dfrac{\omega_I t}{2}\right)\cos\left(\dfrac{\omega_I t}{2}\right) \\ 2\sin\left(\dfrac{\omega_I t}{2}\right)\cos\left(\dfrac{\omega_I t}{2}\right) & -\left(\cos^2\left(\dfrac{\omega_I t}{2}\right) - \sin^2\left(\dfrac{\omega_I t}{2}\right)\right) \end{bmatrix}$$

$$= \frac{1}{2}\begin{bmatrix} \cos(\omega_I t) & \sin(\omega_I t) \\ \sin(\omega_I t) & -\cos(\omega_I t) \end{bmatrix}$$

$$= \frac{1}{2}\begin{bmatrix} 1 & 0 \\ 0 & -1 \end{bmatrix}\cos(\omega_I t) + \frac{1}{2}\begin{bmatrix} 0 & 1 \\ 1 & 0 \end{bmatrix}\sin(\omega_I t)$$

$$= \hat{\mathbf{I}}_z \cos(\omega_I t) + \hat{\mathbf{I}}_x \sin(\omega_I t)$$

If we look carefully we see a pattern here. When $\hat{\mathbf{I}}_z$ is rotated about $\hat{\mathbf{I}}_y$, a mixture of $\hat{\mathbf{I}}_z$ and $\hat{\mathbf{I}}_x$ results. The vector-picture (Chapter 12) for this is the rotation of a magnetisation vector from the equilibrium position on the positive z-axis towards the x-axis. Similarly, when $\hat{\mathbf{I}}_z$ is rotated about $\hat{\mathbf{I}}_x$ a mixture of $\hat{\mathbf{I}}_z$ and $-\hat{\mathbf{I}}_y$ results and again this is easily pictured in the vector-model as a rotation of equilibrium magnetisation from the positive z-axis towards the negative y-axis. Evidently, the exponential matrix may be thought of as representing the axis that the magnetisation is being rotated about. What happens to $\hat{\mathbf{I}}_y$ magnetisation under the influence of the $\hat{\mathbf{I}}_z$ rotation matrix? It rotates towards $-\hat{\mathbf{I}}_x$ ... chemical-shift evolution in other words.

As long as we know what rotates where under the influence of which rotation matrix we need *not* explicitly write out the matrices. Those who have come before us have done all of this many times and have come up with a much shortened way of doing the same calculations.

18.2 The Product Operator Method

The product operator formalism is a welcome shortening of the matrix calculation of the last chapter. The amazing thing is that none of the rigour of density matrix calculations is lost. Thus, we represent a rotation of the equilibrium magnetisation from the z-axis towards the negative y-axis very simply by:

$$\hat{\mathbf{I}}_z \xrightarrow{\beta \hat{\mathbf{I}}_x} \hat{\mathbf{I}}_z \cos(\beta) - \hat{\mathbf{I}}_y \sin(\beta)$$

where β is the rotation angle represented by ωt in our matrix equations. We would previously have written it as:

$$\exp(-i\omega_I \hat{\mathbf{I}}_x t)\hat{\mathbf{I}}_z \exp(i\omega_I \hat{\mathbf{I}}_x t) = \hat{\mathbf{I}}_z \cos(\omega_I t) - \hat{\mathbf{I}}_y \sin(\omega_I t)$$

and explicitly done the calculations using matrices. This of course represents the actions of a pulse of phase x on the equilibrium magnetisation. We can change the phase of the pulse:

$$\hat{\mathbf{I}}_z \xrightarrow{\beta \hat{\mathbf{I}}_y} \hat{\mathbf{I}}_z \cos(\beta) + \hat{\mathbf{I}}_x \sin(\beta) \quad (18.2)$$

and for chemical-shift:[‡]

$$\begin{aligned} \hat{\mathbf{I}}_x &\xrightarrow{\omega_I t \hat{\mathbf{I}}_z} \hat{\mathbf{I}}_x \cos(\omega_I t) + \hat{\mathbf{I}}_y \sin(\omega_I t) \\ \hat{\mathbf{I}}_y &\xrightarrow{\omega_I t \hat{\mathbf{I}}_z} \hat{\mathbf{I}}_y \cos(\omega_I t) - \hat{\mathbf{I}}_x \sin(\omega_I t) \end{aligned} \quad (18.3)$$

[‡]I am using radial frequencies here rather than converting frequencies in hertz to radians per second. This will save us a bit of space when writing out terms later on in the chapter. Also, note the "2" in front of the $\hat{\mathbf{I}}_y\hat{\mathbf{S}}_z$ operator. This is the "normalisation" constant and will be explained shortly.

Let us think about this for a moment. These short formulas represent many tedious matrix computations shortened to a succinct, clear symbolic representation of what is occurring. The same pulse equation may be used for one-spin, two-spin and higher-spin systems. For two-spin and higher-spin systems we must include, in addition to pulse and chemical-shift rotations, scalar coupling. This is represented as follows:[§]

$$\hat{\mathbf{I}}_x \xrightarrow{(2\pi Jt)\hat{\mathbf{I}}_z\hat{\mathbf{S}}_z} \hat{\mathbf{I}}_x\cos(2\pi Jt) + 2\hat{\mathbf{I}}_y\hat{\mathbf{S}}_z\sin(2\pi Jt) \tag{18.4}$$

We note another pattern here. In each of these formulas the "initial" operator on the left side evolves towards a "target" operator under the influence of an "evolution" operator. Furthermore, the result is generally a mixture of starting and target operators with the starting operator being modulated by a cosine function and the target operator by a sine function.

$$\hat{\mathbf{O}}_{\text{initial}} \xrightarrow{(\phi)\hat{\mathbf{O}}_{\text{rotation}}} \hat{\mathbf{O}}_{\text{initial}}\cos(\phi) + \hat{\mathbf{O}}_{\text{target}}\sin(\phi) \tag{18.5}$$

Note that the value of the angle "ϕ" in eqn (18.5) may be such that either the sine or cosine term vanishes. This is very often the case when $\pi/2$ or π pulses are used and may also occur when a delay time is used to advantage such that it causes a coupling sine or cosine term to be equal to $\pi/2$ or π.

You may be wondering about the product operator terms on the right side in eqn (18.4). Specifically, you may be asking "why is there a 2 in $2\hat{\mathbf{I}}_y\hat{\mathbf{S}}_z$?". This is to provide "normalisation" for the product operator.[¶] According to Ernst (who originally proposed the method) in *Principles of Nuclear Magnetic Resonance in One and Two Dimensions* (see the Further Reading section), the operators are trace orthogonal (see eqn (2.25)) and their normalisation depends on the number of spins, N, in the product operator. His orthonormal expression is:

$$\text{Tr}(\hat{\mathbf{B}}_a\hat{\mathbf{B}}_b) = \delta_{ab}2^{N-2} \tag{18.6}$$

[§]Note that coupling in the product operator formalism is treated as weak-coupling only.
[¶]This topic is handled very poorly in the literature with most authors simply saying that the "2" is for normalisation purposes and leaving it at that. It certainly is not the kind of normalisation that we have become used to in this text with respect to vectors. This is exactly the kind of thing that makes it so frustrating for a newcomer to the field.

where δ_{ab} is the Kronecker delta of Chapter 3 and the $\hat{\mathbf{B}}$s are product operators. Equation (18.6) indicates that, for all two-spin product operators, the trace of the product of two operators is either zero or one, depending on whether the $\hat{\mathbf{B}}$s are or are not identical. For $\hat{\mathbf{I}}_z\hat{\mathbf{S}}_z$, the corresponding density matrix (from Chapter 17) is:

$$\hat{\mathbf{I}}_z\hat{\mathbf{S}}_z = \frac{1}{4}\begin{bmatrix} 1 & 0 & 0 & 0 \\ 0 & -1 & 0 & 0 \\ 0 & 0 & -1 & 0 \\ 0 & 0 & 0 & 1 \end{bmatrix}$$

When the trace of the product of this matrix with itself is calculated it is necessary to multiply the matrix by 2 in order to satisfy eqn (18.6):

$$2\hat{\mathbf{I}}_z\hat{\mathbf{S}}_z = \frac{1}{2}\begin{bmatrix} 1 & 0 & 0 & 0 \\ 0 & -1 & 0 & 0 \\ 0 & 0 & -1 & 0 \\ 0 & 0 & 0 & 1 \end{bmatrix}$$

$$\frac{1}{2}\begin{bmatrix} 1 & 0 & 0 & 0 \\ 0 & -1 & 0 & 0 \\ 0 & 0 & -1 & 0 \\ 0 & 0 & 0 & 1 \end{bmatrix} \times \frac{1}{2}\begin{bmatrix} 1 & 0 & 0 & 0 \\ 0 & -1 & 0 & 0 \\ 0 & 0 & -1 & 0 \\ 0 & 0 & 0 & 1 \end{bmatrix} = \begin{bmatrix} \frac{1}{4} & 0 & 0 & 0 \\ 0 & \frac{1}{4} & 0 & 0 \\ 0 & 0 & \frac{1}{4} & 0 \\ 0 & 0 & 0 & \frac{1}{4} \end{bmatrix}$$

$$\mathrm{Tr}\begin{bmatrix} \frac{1}{4} & 0 & 0 & 0 \\ 0 & \frac{1}{4} & 0 & 0 \\ 0 & 0 & \frac{1}{4} & 0 \\ 0 & 0 & 0 & \frac{1}{4} \end{bmatrix} = 1 = \delta_{aa}2^{2-2}$$

Most authors explicitly include the normalisation factor in front of the operator when writing product operator equations (although some do not … beware!). Equation (18.12) will give a simple way to calculate what the normalisation constant should be. Note that, according to this formula, the normalisation constant for $\hat{\mathbf{E}}$ is 1/2 and one generally sees the unit product operator written as $\hat{\mathbf{E}}/2$. A quick way to deduce the value of the normalisation constant is to use a number that makes the constant multiplied into the density matrix equal to 1/2.

As you might deduce from our discussion of density matrices, there are 16 two-spin operators including the identity operator. If we were to take things a step further and calculate the operator matrices for a three-spin system we would find a total of 64 operators. These operators have been given names by Sorensen *et al.* (see Table 18.1).‖

One may deduce how these operators evolve under the influence of any of the others. Of course, since some of them commute there will be no evolution. For example, application of the evolution operator $\hat{\mathbf{I}}_z$ to $\hat{\mathbf{S}}_x$ will produce no effect. This may be easily verified using density matrices and matrix rotation operators.** Table 18.2 lists all of the possible two-spin combinations of evolution, starting and target operators.

Table 18.1 Proposed names for one, two and some three-spin product operators.

$\hat{\mathbf{I}}_z$	Longitudinal magnetisation of spin *I*
$\hat{\mathbf{I}}_x$	In-phase *x*-magnetisation of spin *I*
$\hat{\mathbf{I}}_y$	In-phase *y*-magnetisation of spin *I*
$\hat{\mathbf{S}}_z$	Longitudinal magnetisation of spin *S*
$\hat{\mathbf{S}}_x$	In-phase magnetisation of spin *S*
$\hat{\mathbf{S}}_y$	In-phase magnetisation of spin *S*
$2\hat{\mathbf{I}}_x\hat{\mathbf{S}}_z$	*x*-magnetisation of spin *I* antiphase with respect to spin *S*
$2\hat{\mathbf{I}}_y\hat{\mathbf{S}}_z$	*y*-magnetisation of spin *I* antiphase with respect to spin *S*
$2\hat{\mathbf{I}}_z\hat{\mathbf{S}}_x$	*x*-magnetisation of spin *S* antiphase with respect to spin *I*
$2\hat{\mathbf{I}}_z\hat{\mathbf{S}}_y$	*y*-magnetisation of spin *S* antiphase with respect to spin *I*
$2\hat{\mathbf{I}}_x\hat{\mathbf{S}}_x$	Two-spin coherence
$2\hat{\mathbf{I}}_x\hat{\mathbf{S}}_y$	Two-spin coherence
$2\hat{\mathbf{I}}_y\hat{\mathbf{S}}_x$	Two-spin coherence
$2\hat{\mathbf{I}}_y\hat{\mathbf{S}}_y$	Two-spin coherence
$2\hat{\mathbf{I}}_z\hat{\mathbf{S}}_z$	Longitudinal two-spin order
$4\hat{\mathbf{I}}_x\hat{\mathbf{J}}_z\hat{\mathbf{S}}_z$	*x*-magnetisation of spin *I* in antiphase with respect to spins *J* and *S*
$4\hat{\mathbf{I}}_x\hat{\mathbf{J}}_x\hat{\mathbf{S}}_z$	Two-spin coherence of spins *I* and *J* in antiphase with respect to spin *S*
$4\hat{\mathbf{I}}_x\hat{\mathbf{J}}_x\hat{\mathbf{S}}_x$	Three-spin coherence
$4\hat{\mathbf{I}}_z\hat{\mathbf{J}}_z\hat{\mathbf{S}}_z$	Longitudinal three-spin order
$\hat{\mathbf{E}}/2$	Unity operator

‖I include some three-spin operators in Table 18.1 but do not discuss their use until later.

**If you have read Chapter 17 you know why it is too tedious to do here. After all, that is why we are looking at product operators … to get around the tedium of matrix multiplications. You have all of the tools of Chapter 17 to verify this statement for yourself if you want to do so.

Table 18.2 A table of two-spin product operator evolutions. Top row: evolution operators. Left column: "starting" operators. The bulk of the table consists of the "target" operators resulting from the application of the evolution operators acting on the starting operators.

	$\hat{\mathbf{I}}_z$	$\hat{\mathbf{I}}_x$	$\hat{\mathbf{I}}_y$	$\hat{\mathbf{S}}_z$	$\hat{\mathbf{S}}_x$	$\hat{\mathbf{S}}_y$	$2\hat{\mathbf{I}}_z\hat{\mathbf{S}}_z$
$\hat{\mathbf{I}}_z$	$\hat{\mathbf{E}}/2$	$-\hat{\mathbf{I}}_y$	$-\hat{\mathbf{I}}_x$	$\hat{\mathbf{E}}/2$	$\hat{\mathbf{E}}/2$	$\hat{\mathbf{E}}/2$	$\hat{\mathbf{E}}/2$
$\hat{\mathbf{I}}_x$	$\hat{\mathbf{I}}_y$	$\hat{\mathbf{E}}/2$	$-\hat{\mathbf{I}}_z$	$\hat{\mathbf{E}}/2$	$\hat{\mathbf{E}}/2$	$\hat{\mathbf{E}}/2$	$2\hat{\mathbf{I}}_y\hat{\mathbf{S}}_z$
$\hat{\mathbf{I}}_y$	$-\hat{\mathbf{I}}_x$	$\hat{\mathbf{I}}_z$	$\hat{\mathbf{E}}/2$	$\hat{\mathbf{E}}/2$	$\hat{\mathbf{E}}/2$	$\hat{\mathbf{E}}/2$	$-2\hat{\mathbf{I}}_x\hat{\mathbf{S}}_z$
$\hat{\mathbf{S}}_z$	$\hat{\mathbf{E}}/2$	$\hat{\mathbf{E}}/2$	$\hat{\mathbf{E}}/2$	$\hat{\mathbf{E}}/2$	$-\hat{\mathbf{S}}_y$	$\hat{\mathbf{S}}_x$	$\hat{\mathbf{E}}/2$
$\hat{\mathbf{S}}_x$	$\hat{\mathbf{E}}/2$	$\hat{\mathbf{E}}/2$	$\hat{\mathbf{E}}/2$	$\hat{\mathbf{S}}_y$	$\hat{\mathbf{E}}/2$	$-\hat{\mathbf{S}}_z$	$2\hat{\mathbf{I}}_z\hat{\mathbf{S}}_y$
$\hat{\mathbf{S}}_y$	$\hat{\mathbf{E}}/2$	$\hat{\mathbf{E}}/2$	$\hat{\mathbf{E}}/2$	$\hat{\mathbf{E}}/2$	$\hat{\mathbf{S}}_z$	$\hat{\mathbf{E}}/2$	$-2\hat{\mathbf{I}}_z\hat{\mathbf{S}}_x$
$2\hat{\mathbf{I}}_z\hat{\mathbf{S}}_z$	$\hat{\mathbf{E}}/2$	$-2\hat{\mathbf{I}}_y\hat{\mathbf{S}}_z$	$2\hat{\mathbf{I}}_x\hat{\mathbf{S}}_z$	$\hat{\mathbf{E}}/2$	$2\hat{\mathbf{I}}_z\hat{\mathbf{S}}_y$	$2\hat{\mathbf{I}}_z\hat{\mathbf{S}}_x$	$\hat{\mathbf{E}}/2$
$2\hat{\mathbf{I}}_x\hat{\mathbf{S}}_z$	$2\hat{\mathbf{I}}_y\hat{\mathbf{S}}_z$	$\hat{\mathbf{E}}/2$	$-2\hat{\mathbf{I}}_z\hat{\mathbf{S}}_z$	$\hat{\mathbf{E}}/2$	$2\hat{\mathbf{I}}_x\hat{\mathbf{S}}_y$	$2\hat{\mathbf{I}}_x\hat{\mathbf{S}}_x$	$\hat{\mathbf{I}}_y$
$2\hat{\mathbf{I}}_y\hat{\mathbf{S}}_z$	$2\hat{\mathbf{I}}_x\hat{\mathbf{S}}_z$	$2\hat{\mathbf{I}}_z\hat{\mathbf{S}}_z$	$\hat{\mathbf{E}}/2$	$\hat{\mathbf{E}}/2$	$-2\hat{\mathbf{I}}_y\hat{\mathbf{S}}_y$	$2\hat{\mathbf{I}}_y\hat{\mathbf{S}}_x$	$-\hat{\mathbf{I}}_x$
$2\hat{\mathbf{I}}_z\hat{\mathbf{S}}_x$	$\hat{\mathbf{E}}/2$	$-2\hat{\mathbf{I}}_y\hat{\mathbf{S}}_x$	$2\hat{\mathbf{I}}_x\hat{\mathbf{S}}_x$	$2\hat{\mathbf{I}}_z\hat{\mathbf{S}}_y$	$\hat{\mathbf{E}}/2$	$-2\hat{\mathbf{I}}_z\hat{\mathbf{S}}_z$	$\hat{\mathbf{S}}_y$
$2\hat{\mathbf{I}}_z\hat{\mathbf{S}}_y$	$\hat{\mathbf{E}}/2$	$-2\hat{\mathbf{I}}_y\hat{\mathbf{S}}_y$	$2\hat{\mathbf{I}}_x\hat{\mathbf{S}}_y$	$-2\hat{\mathbf{I}}_z\hat{\mathbf{S}}_x$	$2\hat{\mathbf{I}}_z\hat{\mathbf{S}}_z$	$\hat{\mathbf{E}}/2$	$-\hat{\mathbf{S}}_x$
$2\hat{\mathbf{I}}_x\hat{\mathbf{S}}_x$	$2\hat{\mathbf{I}}_y\hat{\mathbf{S}}_x$	$\hat{\mathbf{E}}/2$	$-2\hat{\mathbf{I}}_z\hat{\mathbf{S}}_x$	$2\hat{\mathbf{I}}_x\hat{\mathbf{S}}_y$	$\hat{\mathbf{E}}/2$	$-2\hat{\mathbf{I}}_x\hat{\mathbf{S}}_z$	$\hat{\mathbf{E}}/2$
$2\hat{\mathbf{I}}_x\hat{\mathbf{S}}_y$	$2\hat{\mathbf{I}}_y\hat{\mathbf{S}}_y$	$\hat{\mathbf{E}}/2$	$-2\hat{\mathbf{I}}_z\hat{\mathbf{S}}_y$	$-2\hat{\mathbf{I}}_x\hat{\mathbf{S}}_x$	$2\hat{\mathbf{I}}_x\hat{\mathbf{S}}_z$	$\hat{\mathbf{E}}/2$	$\hat{\mathbf{E}}/2$
$2\hat{\mathbf{I}}_y\hat{\mathbf{S}}_x$	$-2\hat{\mathbf{I}}_x\hat{\mathbf{S}}_x$	$2\hat{\mathbf{I}}_z\hat{\mathbf{S}}_x$	$\hat{\mathbf{E}}/2$	$2\hat{\mathbf{I}}_y\hat{\mathbf{S}}_y$	$\hat{\mathbf{E}}/2$	$-2\hat{\mathbf{I}}_y\hat{\mathbf{S}}_z$	$\hat{\mathbf{E}}/2$
$2\hat{\mathbf{I}}_y\hat{\mathbf{S}}_y$	$-2\hat{\mathbf{I}}_x\hat{\mathbf{S}}_y$	$2\hat{\mathbf{I}}_z\hat{\mathbf{S}}_y$	$\hat{\mathbf{E}}/2$	$-2\hat{\mathbf{I}}_y\hat{\mathbf{S}}_x$	$2\hat{\mathbf{I}}_y\hat{\mathbf{S}}_z$	$\hat{\mathbf{E}}/2$	$\hat{\mathbf{E}}/2$

We are now in a position to analyse many NMR pulse sequences easily and quickly. We may use Table 18.2 to rapidly determine the form of the evolution formula. For example, starting with say, $\hat{\mathbf{S}}_x$ magnetisation, what will be the effect of the coupling operator (the longitudinal two-spin order) on it? The starting operator is $\hat{\mathbf{S}}_x$, the evolution operator is $2\hat{\mathbf{I}}_z\hat{\mathbf{S}}_z$ and from the table we see that the target operator is $2\hat{\mathbf{I}}_z\hat{\mathbf{S}}_y$, antiphase *y*-magnetisation of *S* with respect to *I*.

$$\hat{\mathbf{S}}_x \xrightarrow{(2\pi Jt)\hat{\mathbf{I}}_z\hat{\mathbf{S}}_z} \hat{\mathbf{S}}_x\cos(2\pi Jt) + 2\hat{\mathbf{I}}_z\hat{\mathbf{S}}_y\sin(2\pi Jt)$$

Much has been written on the product operator method and each of the books and articles in the Further Reading section listed at the end of the chapter is a good starting point. Especially useful and simple for those new to product operators is Hore's book. For those who do not mind a bit of a challenge, the original paper by Sorensen *et al.* is very illuminating.

18.3 Multiple Quantum Coherence

Table 18.2 contains some operators that have two transverse plane subscripts such as $2\hat{\mathbf{I}}_x\hat{\mathbf{S}}_x$ or $2\hat{\mathbf{I}}_y\hat{\mathbf{S}}_x$. These operators represent coherences with coherence orders higher than (and lower than) ± 1. See Chapter 19 for more detail on coherence order including a definition. For now, we will accept that there are coherence orders ... $-2, -1, 0, 1, 2$ ... and that the single-spin transverse operators represent coherence orders of ± 1.

Operators such as $2\hat{\mathbf{I}}_x\hat{\mathbf{S}}_x$ may be represented by a combination of raising and lowering operators:

$$\begin{aligned}2\hat{\mathbf{I}}_x\hat{\mathbf{S}}_x &= \frac{1}{2}(\hat{\mathbf{I}}_+ + \hat{\mathbf{I}}_-)(\hat{\mathbf{S}}_+ + \hat{\mathbf{S}}_-) \\ &= \frac{1}{2}(\hat{\mathbf{I}}_+\hat{\mathbf{S}}_+ + \hat{\mathbf{I}}_+\hat{\mathbf{S}}_- + \hat{\mathbf{I}}_-\hat{\mathbf{S}}_+ + \hat{\mathbf{I}}_-\hat{\mathbf{S}}_-)\end{aligned}$$

The first and last terms represent *double quantum coherence* (DQC) while the second and third represent *zero quantum coherence* (ZQC). Similarly, for $2\hat{\mathbf{I}}_y\hat{\mathbf{S}}_y$:[††]

$$\begin{aligned}2\hat{\mathbf{I}}_y\hat{\mathbf{S}}_y &= -\frac{1}{2}(\hat{\mathbf{I}}_+ - \hat{\mathbf{I}}_-)(\hat{\mathbf{S}}_+ - \hat{\mathbf{S}}_-) \\ &= \frac{1}{2}(-\hat{\mathbf{I}}_+\hat{\mathbf{S}}_+ + \hat{\mathbf{I}}_+\hat{\mathbf{S}}_- + \hat{\mathbf{I}}_-\hat{\mathbf{S}}_+ - \hat{\mathbf{I}}_-\hat{\mathbf{S}}_-)\end{aligned}$$

which is much the same as $2\hat{\mathbf{I}}_x\hat{\mathbf{S}}_x$ except for the sign changes. We might represent these two operators in a shorter form by defining some new operators:

$$2\hat{\mathbf{I}}_x\hat{\mathbf{S}}_x = \frac{1}{2}(\mathbf{D\hat{Q}C}_x + \mathbf{Z\hat{Q}C}_x)$$

$$2\hat{\mathbf{I}}_y\hat{\mathbf{S}}_y = \frac{1}{2}(-\mathbf{D\hat{Q}C}_x + \mathbf{Z\hat{Q}C}_x)$$

Subtraction of these two multiple coherence operators will result in pure DQC:

$$2\hat{\mathbf{I}}_x\hat{\mathbf{S}}_x - 2\hat{\mathbf{I}}_y\hat{\mathbf{S}}_y = \mathbf{D\hat{Q}C}_x$$

and addition gives pure ZQC:

$$2\hat{\mathbf{I}}_x\hat{\mathbf{S}}_x + 2\hat{\mathbf{I}}_y\hat{\mathbf{S}}_y = \mathbf{Z\hat{Q}C}_x$$

[††]Do not forget the *i* if you are doing the calculations yourself.

where:

$$\mathbf{D\hat{Q}C}_x \equiv \hat{\mathbf{I}}_+\hat{\mathbf{S}}_+ + \hat{\mathbf{I}}_-\hat{\mathbf{S}}_-$$

$$\mathbf{Z\hat{Q}C}_x \equiv \hat{\mathbf{I}}_+\hat{\mathbf{S}}_- + \hat{\mathbf{I}}_-\hat{\mathbf{S}}_+$$

Similarly:

$$2\hat{\mathbf{I}}_x\hat{\mathbf{S}}_y = \frac{1}{2i}(\mathbf{D\hat{Q}C}_y - \mathbf{Z\hat{Q}C}_y)$$

$$2\hat{\mathbf{I}}_y\hat{\mathbf{S}}_x = \frac{1}{2i}(\mathbf{D\hat{Q}C}_y + \mathbf{Z\hat{Q}C}_y)$$

where:

$$\mathbf{D\hat{Q}C}_y \equiv \hat{\mathbf{I}}_+\hat{\mathbf{S}}_+ - \hat{\mathbf{I}}_-\hat{\mathbf{S}}_-$$

$$\mathbf{Z\hat{Q}C}_y \equiv \hat{\mathbf{I}}_+\hat{\mathbf{S}}_- - \hat{\mathbf{I}}_-\hat{\mathbf{S}}_+$$

Of interest in the next chapter will be the chemical-shift evolution frequencies of these new operators, $\mathbf{D\hat{Q}C}$ and $\mathbf{Z\hat{Q}C}$. We approach that by looking at the product operator evolution of the components of these operators. What we want is:

$$\mathbf{Z\hat{Q}C}_x \xrightarrow{\omega_I t\hat{\mathbf{I}}_z + \omega_S t\hat{\mathbf{S}}_z} ?$$

where we are examining the chemical-shift evolution properties of the zero quantum coherence operator under the influence of both the I and S spin chemical-shift evolution operators. As we saw above, $\mathbf{Z\hat{Q}C}_x$ consists of a mixture of $2\hat{\mathbf{I}}_x\hat{\mathbf{S}}_x$ and $2\hat{\mathbf{I}}_y\hat{\mathbf{S}}_y$:

$$\mathbf{Z\hat{Q}C}_x = 2\hat{\mathbf{I}}_x\hat{\mathbf{S}}_x + 2\hat{\mathbf{I}}_y\hat{\mathbf{S}}_y$$

Using this, we can write:

$$\begin{aligned}2\hat{\mathbf{I}}_x\hat{\mathbf{S}}_x + 2\hat{\mathbf{I}}_y\hat{\mathbf{S}}_y \xrightarrow{\omega_I t\hat{\mathbf{I}}_z + \omega_S t\hat{\mathbf{S}}_z} &[\hat{\mathbf{I}}_x\cos(\omega_I t) + \hat{\mathbf{I}}_y\sin(\omega_I t)][\hat{\mathbf{S}}_x\cos(\omega_S t) + \hat{\mathbf{S}}_y\sin(\omega_S t)]\\ &+[\hat{\mathbf{I}}_x\cos(\omega_I t) - \hat{\mathbf{I}}_y\sin(\omega_I t)][\hat{\mathbf{S}}_x\cos(\omega_S t) - \hat{\mathbf{S}}_y\sin(\omega_S t)]\end{aligned}$$

Expanding and collecting terms:

$$\begin{aligned}2\hat{\mathbf{I}}_x\hat{\mathbf{S}}_x + 2\hat{\mathbf{I}}_y\hat{\mathbf{S}}_y \xrightarrow{\omega_I t\hat{\mathbf{I}}_z + \omega_S t\hat{\mathbf{S}}_z} &(2\hat{\mathbf{I}}_x\hat{\mathbf{S}}_x + 2\hat{\mathbf{I}}_y\hat{\mathbf{S}}_y)[\cos(\omega_I t)\cos(\omega_S t) + \sin(\omega_I t)\sin(\omega_S t)]\\ &+ (2\hat{\mathbf{I}}_y\hat{\mathbf{S}}_x - 2\hat{\mathbf{I}}_x\hat{\mathbf{S}}_y)[\sin(\omega_I t)\cos(\omega_S t) - \cos(\omega_I t)\sin(\omega_S t)]\end{aligned}$$

and using the trigonometric identity (A.10):

$$2\hat{\mathbf{I}}_x\hat{\mathbf{S}}_x + 2\hat{\mathbf{I}}_y\hat{\mathbf{S}}_y \xrightarrow{\omega_I t\hat{\mathbf{I}}_z + \omega_S t\hat{\mathbf{S}}_z} (2\hat{\mathbf{I}}_x\hat{\mathbf{S}}_x + 2\hat{\mathbf{I}}_y\hat{\mathbf{S}}_y)\cos[((\omega_I - \omega_S))t]$$
$$+ (2\hat{\mathbf{I}}_y\hat{\mathbf{S}}_x - 2\hat{\mathbf{I}}_x\hat{\mathbf{S}}_y)\sin[((\omega_I - \omega_S))t]$$

or, in terms of our new operators:

$$\mathbf{Z\hat{Q}C}_x \xrightarrow{\omega_I t\hat{\mathbf{I}}_z + \omega_S t\hat{\mathbf{S}}_z} \mathbf{Z\hat{Q}C}_x \cos([\omega_I - \omega_S]t) + \mathbf{Z\hat{Q}C}_y \sin([\omega_I - \omega_S]t) \quad (18.7)$$

We note immediately that the chemical-shift precession frequency of our ZQC consists of the *difference* in precession frequencies of the two spins involved. We also note the parallel between the form of the evolution of single quantum coherences that we have seen above and that of multiple quantum coherences; that is, the "initial" operator evolves towards a mixture of initial and target operators (eqn (18.5)). The next section in this chapter provides a visualisation of this.

A similar analysis of the chemical-shift evolution of the double quantum coherence operators $\mathbf{D\hat{Q}C}_y$ and $\mathbf{Z\hat{Q}C}_y$ will reveal that their evolution frequencies are the *sum* of the frequencies of the two spins involved. Thus, we would see:

$$\mathbf{D\hat{Q}C}_y \xrightarrow{\omega_I t\hat{\mathbf{I}}_z + \omega_S t\hat{\mathbf{S}}_z} \mathbf{D\hat{Q}C}_y \cos([\omega_I + \omega_S]t) + \mathbf{D\hat{Q}C}_x \sin([\omega_I + \omega_S]t) \quad (18.8)$$

for the evolution of $\mathbf{D\hat{Q}C}_y$.

18.4 Product Operator Rotations

It is very useful to be able to associate some pictures with our product operator rotations. We have already seen something like this in the vector-model, however, with product operators the axes are not Cartesian but rather are represented by the rotation operator, starting operator and target operators in an orthogonal operator space rather than a Euclidean 3D space. Otherwise, the pictures appear Cartesian in nature, are very easy to understand and are helpful in learning how the various kinds of rotations work.

We begin by looking at pulse rotations (Figure 18.1).[‡‡]

[‡‡]I should mention that a very few authors use a left-hand rotation system to portray rotations about the various product operator axes. In these cases the signs will be reversed with respect to what is shown here. Most, however, prefer to use the right-hand rotation and I have used it throughout the text.

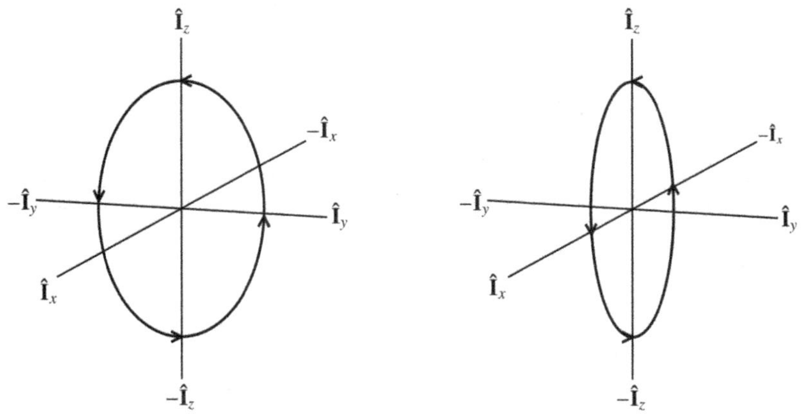

Figure 18.1 Rotation of equilibrium magnetisation about $\hat{I}_x$ and $\hat{I}_y$ *via* application of a $\vec{B}_1$-field pulse.

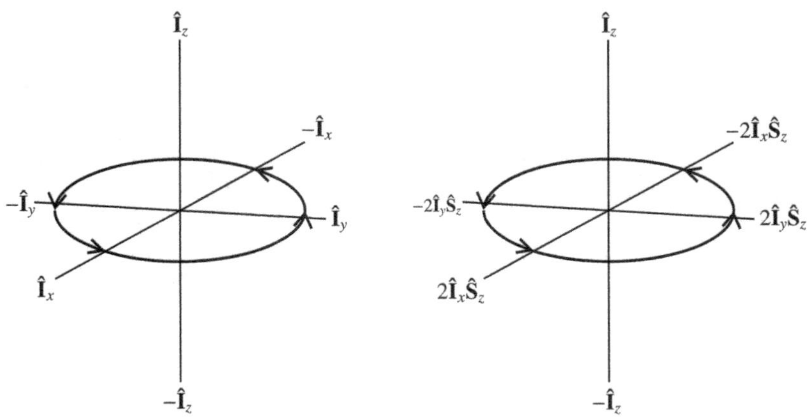

Figure 18.2 Chemical shift evolution of in-phase and antiphase magnetisations *via* the application of rotation operator, $\hat{I}_z$.

We see that the equilibrium magnetisation, $\hat{\mathbf{I}}_z$, follows the path indicated by the above equations for x-phase pulses. We also see that we can potentially rotate $\hat{\mathbf{I}}_z$ from the equilibrium position to the exact opposite position using a π-pulse.

Chemical-shift evolution appears as a rotation about the $\hat{\mathbf{I}}_z$ axis (Figure 18.2).

So far our axes appear to be Cartesian, however, we depart from this nomenclature when considering coupling. The picture remains the same, however, only the axis labels change (Figure 18.3).

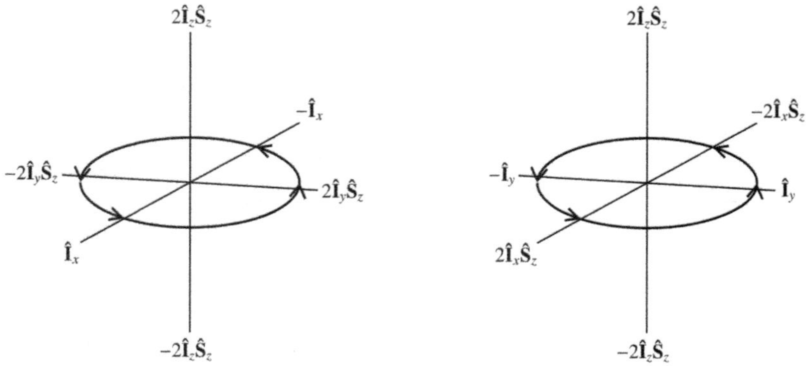

Figure 18.3 Coupling evolution viewed as rotation of in-phase and antiphase magnetisation about $2\hat{I}_z\hat{S}_z$.

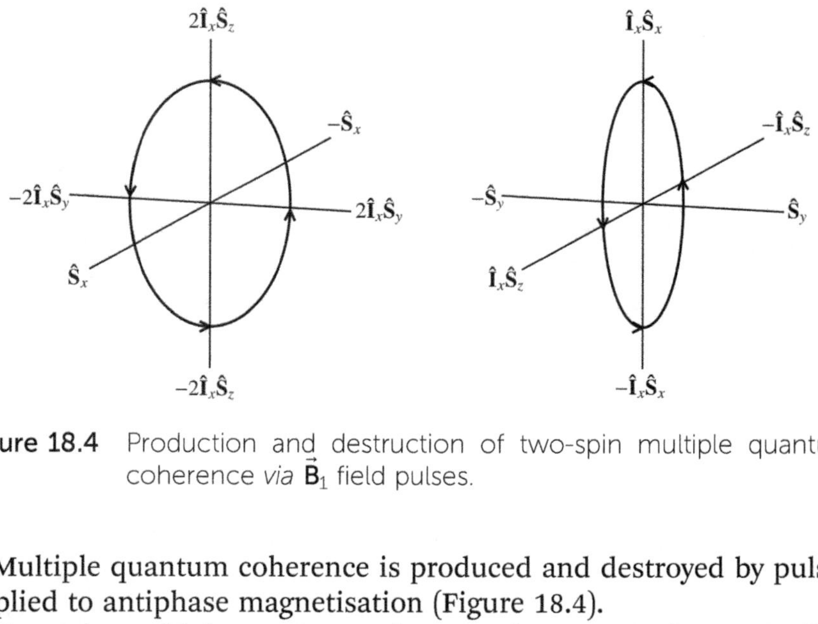

Figure 18.4 Production and destruction of two-spin multiple quantum coherence *via* $\vec{\mathbf{B}}_1$ field pulses.

Multiple quantum coherence is produced and destroyed by pulses applied to antiphase magnetisation (Figure 18.4).

Two-spin multiple quantum coherence does not undergo coupling evolution but does undergo chemical-shift evolution (Figures 18.5 and 18.6).

Note that in all of the figures the *I*-spin operator acts only on *I* spins and the S-spin operator acts only on the *S* spins. It seems intuitive that it should be so but one could also show this using density matrices to prove that *I*-spin density operators commute with S-spin operators. Also, all of these rotations can be deduced from examination of Table 18.2.

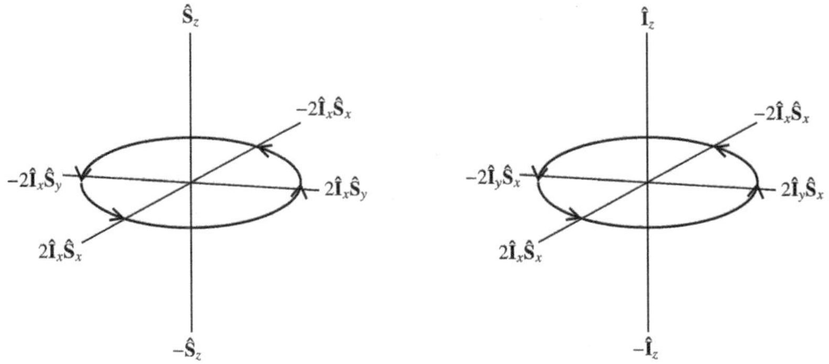

Figure 18.5 Chemical shift evolution of two-spin multiple quantum coherence.

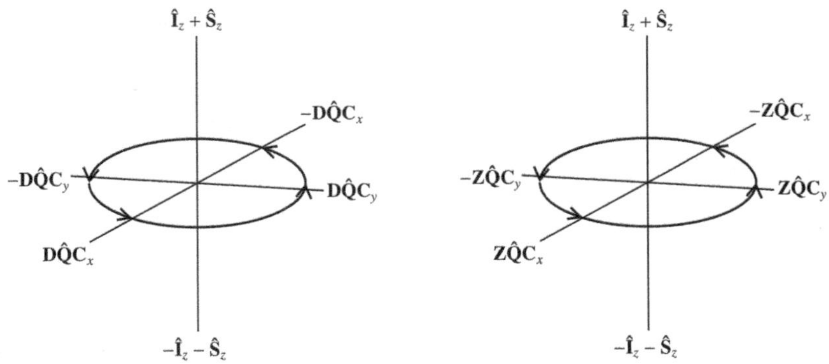

Figure 18.6 Chemical shift evolution of pure $\mathbf{D\hat{Q}C}$ and pure $\mathbf{Z\hat{Q}C}$.

18.5 Pulse Sequence Analysis

When we do modern Fourier transform NMR spectroscopy we usually issue a series of pulses and time delays before observing the result as a spectrum of some type. The simplest example of this is the 1D ^{1}H hydrogen spectrum. This is just a pulse of some angle and phase followed by acquisition of data (Figure 18.7).[§§]

$\hat{\sigma}(0)$ is the initial equilibrium state and $\hat{\sigma}(1)$ represents the state immediately after the pulse. From above, the product operator analysis of this is:

$$\hat{\mathbf{I}}_z \xrightarrow{\beta\hat{\mathbf{I}}_x} \hat{\mathbf{I}}_z \cos(\beta) - \hat{\mathbf{I}}_y \sin(\beta)$$

[§§]We will assume a scalar coupled two-spin system for all analyses in this section. Furthermore, we will assume a heteronuclear system in which spins I and S correspond to different nuclei unless otherwise noted.

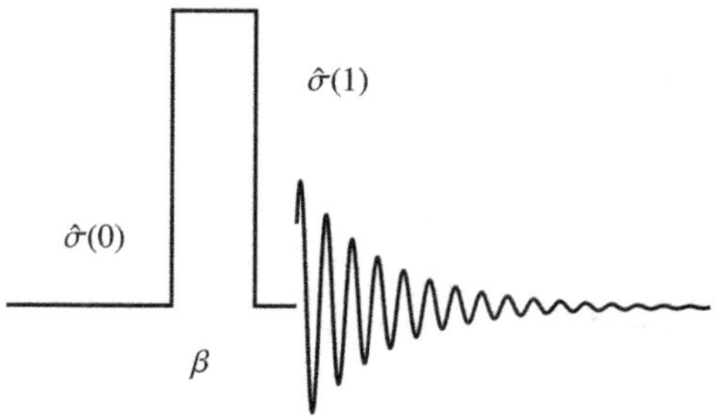

Figure 18.7 A simple pulse-read sequence.

The pulse produces $-\hat{\mathbf{I}}_y$ magnetisation that is then observed during the acquisition period (represented by the decaying sine wave in Figure 18.7). Not all product operators are observable ... strictly speaking, only $\hat{\mathbf{I}}_x$, $\hat{\mathbf{I}}_y$, $\hat{\mathbf{S}}_x$ and $\hat{\mathbf{S}}_y$ magnetisations are observable. However, in a practical sense, since antiphase magnetisation evolves into in-phase magnetisation during acquisition, it too is observable. How do we know that some operators are not observable? Recall that the expectation value may be calculated thus:

$$\langle \mathrm{A} \rangle = \mathrm{Tr}(\hat{\sigma}\hat{\mathbf{A}})$$

where we take the density matrix for the state of interest, $\hat{\sigma}$, and multiply by the corresponding operator $\hat{\mathbf{A}}$ for which we desire information. We then take the trace of the resulting matrix. This operation was (painfully) detailed in Section 17.4. If we do this set of operations for the expectation value of, say $2\hat{\mathbf{I}}_z\hat{\mathbf{S}}_z$, we find that the expectation value of $2\hat{\mathbf{I}}_z\hat{\mathbf{S}}_z$ is zero ... in other words it is unobservable:[¶¶]

$$\begin{aligned}
&\langle I_zS_z\rangle \mathrm{Tr}(\hat{\sigma}(1)2\hat{\mathbf{I}}_z\hat{\mathbf{S}}_z)\\
&= \mathrm{Tr}\left(\frac{1}{2}\begin{bmatrix} 0 & i & i & 0 \\ -i & 0 & 0 & i \\ -i & 0 & 0 & i \\ 0 & -i & -i & 0 \end{bmatrix}\frac{1}{2}\begin{bmatrix} 1 & 0 & 0 & 0 \\ 0 & -1 & 0 & 0 \\ 0 & 0 & -1 & 0 \\ 0 & 0 & 0 & 1 \end{bmatrix}\right)\\
&= \mathrm{Tr}\left(\frac{1}{4}\begin{bmatrix} 0 & i & i & 0 \\ i & 0 & 0 & -i \\ i & 0 & 0 & -i \\ 0 & -i & -i & 0 \end{bmatrix}\right)\\
&= 0
\end{aligned}$$

[¶¶]Most texts simply state that a particular operator is either observable or unobservable ... without explanation.

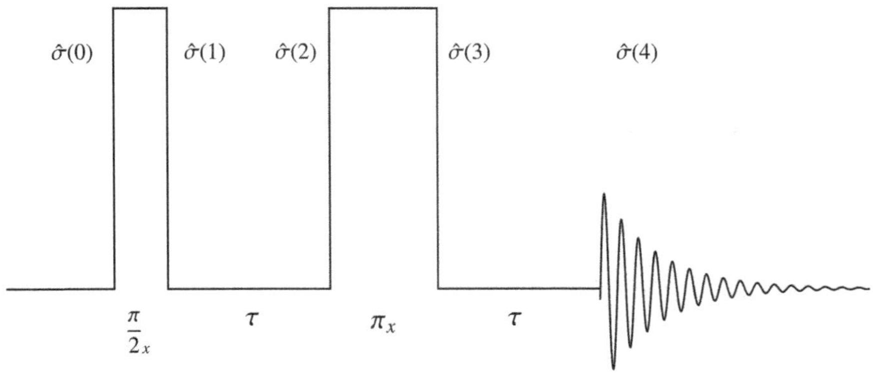

Figure 18.8 The homonuclear spin-echo pulse sequence.

Our next pulse-sequence is the homonuclear spin-echo (Figure 18.8). Here we are pulsing the *I*-spin only. The S-spin is not pulsed but is involved *via* scalar coupling. There are four states to consider here. $\hat{\sigma}(0)$ is, as usual, the equilibrium state and is the starting point for all standard NMR experiments. $\hat{\sigma}(1)$ is the state immediately after the $\pi/2$-pulse, $\hat{\sigma}(2)$ is the state immediately before the π-pulse, $\hat{\sigma}(3)$ is the state after the π-pulse and $\hat{\sigma}(4)$ at the end of the second time delay period, *t*. Note that we have labelled the phases of the pulses in the diagram. The state $\hat{\sigma}(1)$ is calculated as:

$$\hat{\mathbf{I}}_z \xrightarrow{\pi/2\hat{\mathbf{I}}_x} -\hat{\mathbf{I}}_y$$

During the first time delay period *t*, chemical-shift and scalar coupling evolution will occur. These may be calculated sequentially, the order of calculation does not matter since the chemical-shift operator $\hat{\mathbf{I}}_z$ commutes with the coupling operator $2\hat{\mathbf{I}}_z\hat{\mathbf{S}}_z$. How do we know this? Recall the discussion of exponentials of matrices in Chapter 2. There we showed that if two matrices commute then their exponentials commute. Also recall from Chapter 17 that in applying the exponentials of operators to effect multiple rotations in operator space such as chemical-shift and coupling rotations they are applied sequentially:

$$\hat{\sigma}(2) = \exp(-i\omega_I\hat{\mathbf{I}}_z\tau)\ \exp(-i2\pi J\hat{\mathbf{I}}_z\hat{\mathbf{S}}_z\tau)\ \hat{\sigma}(1)\ \exp(i2\pi J\hat{\mathbf{I}}_z\hat{\mathbf{S}}_z\tau)\ \exp(i\omega_I\hat{\mathbf{I}}_z\tau)$$

However, since $\hat{\mathbf{I}}_z$ commutes with the coupling operator $2\hat{\mathbf{I}}_z\hat{\mathbf{S}}_z$ so do their exponentials (see eqn (2.62)) and it does not matter the order in which they are applied:

$$\hat{\sigma}(2) = \exp(-i\pi J2\hat{\mathbf{I}}_z\hat{\mathbf{S}}_z\tau)\ \exp(-i\omega_I\hat{\mathbf{I}}_z\tau)\ \hat{\sigma}(1)\exp(i\omega_I\hat{\mathbf{I}}_z\tau)\ \exp(i\pi J2\hat{\mathbf{I}}_z\hat{\mathbf{S}}_z\tau)\ \exp(i\omega_I\hat{\mathbf{I}}_z\tau)$$

Let us do the chemical-shift calculation first:

$$-\hat{\mathbf{I}}_y \xrightarrow{\omega_I \tau \hat{\mathbf{I}}_z} -\hat{\mathbf{I}}_y \cos(\omega_I \tau) + \hat{\mathbf{I}}_x \sin(\omega_I \tau)$$

We follow this with the scalar coupling calculation on the result of the chemical-shift calculation:

$$\begin{aligned} -\hat{\mathbf{I}}_y \cos(\omega_I \tau) + \hat{\mathbf{I}}_x \sin(\omega_I \tau) &\xrightarrow{(2\pi J t)\hat{\mathbf{I}}_z \hat{\mathbf{S}}_z} \\ &- \hat{\mathbf{I}}_y \cos(\omega_I \tau)\cos(\pi J \tau) + 2\hat{\mathbf{I}}_x \hat{\mathbf{S}}_z \cos(\omega_I \tau)\sin(\pi J \tau) \\ &+ \hat{\mathbf{I}}_x \sin(\omega_I \tau)\cos(\pi J \tau) + 2\hat{\mathbf{I}}_y \hat{\mathbf{S}}_z \sin(\omega_I \tau)\sin(\pi J \tau) \end{aligned}$$

This is the state at the end of the first time-delay. Now we calculate the state $\hat{\sigma}(3)$ immediately after the π-pulse. Recall that the I-spin operators affect only I spins. Furthermore, rotation operator $\hat{\mathbf{I}}_x$ will have no effect on spin-state operator $\hat{\mathbf{I}}_x$.[||] Thus, when we apply an $\hat{\mathbf{I}}_x$ π-pulse only the $\hat{\mathbf{I}}_y$ terms are affected:

$$\begin{gathered} - \hat{\mathbf{I}}_y \cos(\omega_I \tau)\cos(\pi J \tau) + 2\hat{\mathbf{I}}_x \hat{\mathbf{S}}_z \cos(\omega_I \tau)\sin(\pi J \tau) \\ + \hat{\mathbf{I}}_x \sin(\omega_I \tau)\cos(\pi J \tau) + 2\hat{\mathbf{I}}_y \hat{\mathbf{S}}_z \sin(\omega_I \tau)\sin(\pi J \tau) \\ \Downarrow \pi \hat{\mathbf{I}}_x \\ + \hat{\mathbf{I}}_y \cos(\omega_I \tau)\cos(\pi J \tau) + 2\hat{\mathbf{I}}_x \hat{\mathbf{S}}_z \cos(\omega_I \tau)\sin(\pi J \tau) \\ + \hat{\mathbf{I}}_x \sin(\omega_I \tau)\cos(\pi J \tau) - 2\hat{\mathbf{I}}_y \hat{\mathbf{S}}_z \sin(\omega_I \tau)\sin(\pi J \tau) \end{gathered}$$

Next we calculate the state $\hat{\sigma}(4)$ after the second time delay t. As for the first time delay, both chemical-shift and scalar coupling evolution occur. First the chemical-shift evolution:

$$\begin{gathered} + \hat{\mathbf{I}}_y \cos(\omega_I \tau)\cos(\pi J \tau) + 2\hat{\mathbf{I}}_x \hat{\mathbf{S}}_z \cos(\omega_I \tau)\sin(\pi J \tau) \\ + \hat{\mathbf{I}}_x \sin(\omega_I \tau)\cos(\pi J \tau) - 2\hat{\mathbf{I}}_y \hat{\mathbf{S}}_z \sin(\omega_I \tau)\sin(\pi J \tau) \xrightarrow{\omega_I \tau \hat{\mathbf{I}}_z} \\ + \hat{\mathbf{I}}_y \cos(\omega_I \tau)\cos(\pi J \tau)\cos(\omega_I \tau) - \hat{\mathbf{I}}_x \cos(\omega_I \tau)\cos(\pi J \tau)\sin(\omega_I \tau) \\ + 2\hat{\mathbf{I}}_x \hat{\mathbf{S}}_z \cos(\omega_I \tau)\sin(\pi J \tau)\cos(\omega_I \tau) + 2\hat{\mathbf{I}}_y \hat{\mathbf{S}}_z \cos(\omega_I \tau)\sin(\pi J \tau)\sin(\omega_I \tau) \\ + \hat{\mathbf{I}}_x \sin(\omega_I \tau)\cos(\pi J \tau)\cos(\omega_I \tau) + \hat{\mathbf{I}}_y \sin(\omega_I \tau)\cos(\pi J \tau)\sin(\omega_I \tau) \\ - 2\hat{\mathbf{I}}_y \hat{\mathbf{S}}_z \sin(\omega_I \tau)\sin(\pi J \tau)\cos(\omega_I \tau) + 2\hat{\mathbf{I}}_x \hat{\mathbf{S}}_z \sin(\omega_I \tau)\sin(\pi J \tau)\sin(\omega_I \tau) \end{gathered}$$

[||] A moment's thought will convince you that rotation of $\hat{\mathbf{I}}_x$ magnetisation about the $\hat{\mathbf{I}}_x$ -axis results in $\hat{\mathbf{I}}_x$ magnetisation … in other words, no net rotation occurs. This is reflected in Table 18.2.

Note that we have taken each term on the left side of the operation arrow and evolved it corresponding to the rules laid down in Table 18.2. Note also that the sine and cosine terms from the previous operations survive and are included in the current state. Thus, the first term on the left evolves as:

$$+\hat{\mathbf{I}}_y\cos(\omega_I\tau)\cos(\pi J\tau)\xrightarrow{\omega_I\tau\hat{\mathbf{I}}_z}$$

$$+\hat{\mathbf{I}}_y\cos(\omega_I\tau)\cos(\pi J\tau)\cos(\omega_I\tau)-\hat{\mathbf{I}}_x\cos(\omega_I\tau)\cos(\pi J\tau)\sin(\omega_I\tau)$$

and so on. Figure 18.2 is helpful in visualizing this evolution. Lastly, we apply the scalar coupling evolution:

$$\begin{aligned}
&+\hat{\mathbf{I}}_y\cos(\omega_I\tau)\cos(\pi J\tau)\cos(\omega_I\tau)-\hat{\mathbf{I}}_x\cos(\omega_I\tau)\cos(\pi J\tau)\sin(\omega_I\tau)\\
&+2\hat{\mathbf{I}}_x\hat{\mathbf{S}}_z\cos(\omega_I\tau)\sin(\pi J\tau)\cos(\omega_I\tau)+2\hat{\mathbf{I}}_y\hat{\mathbf{S}}_z\cos(\omega_I\tau)\sin(\pi J\tau)\sin(\omega_I\tau)\\
&+\hat{\mathbf{I}}_x\sin(\omega_I\tau)\cos(\pi J\tau)\cos(\omega_I\tau)+\hat{\mathbf{I}}_y\sin(\omega_I\tau)\cos(\pi J\tau)\sin(\omega_I\tau)\\
&-2\hat{\mathbf{I}}_y\hat{\mathbf{S}}_z\sin(\omega_I\tau)\sin(\pi J\tau)\cos(\omega_I\tau)+2\hat{\mathbf{I}}_x\hat{\mathbf{S}}_z\sin(\omega_I\tau)\sin(\pi J\tau)\sin(\omega_I\tau)
\end{aligned}$$

$$\xrightarrow{(\pi Jt)\hat{\mathbf{I}}_z\hat{\mathbf{S}}_z}$$

$$\begin{aligned}
&+\hat{\mathbf{I}}_y\cos(\omega_I\tau)\cos(\pi J\tau)\cos(\omega_I\tau)\cos(\pi J\tau)-2\hat{\mathbf{I}}_x\hat{\mathbf{S}}_z\cos(\omega_I\tau)\cos(\pi J\tau)\cos(\omega_I\tau)\sin(\pi J\tau)\\
&-\hat{\mathbf{I}}_x\cos(\omega_I\tau)\cos(\pi J\tau)\sin(\omega_I\tau)\cos(\pi J\tau)-2\hat{\mathbf{I}}_y\hat{\mathbf{S}}_z\cos(\omega_I\tau)\cos(\pi J\tau)\sin(\omega_I\tau)\sin(\pi J\tau)\\
&+2\hat{\mathbf{I}}_x\hat{\mathbf{S}}_z\cos(\omega_I\tau)\sin(\pi J\tau)\cos(\omega_I\tau)\cos(\pi J\tau)+\hat{\mathbf{I}}_y\cos(\omega_I\tau)\sin(\pi J\tau)\cos(\omega_I\tau)\sin(\pi J\tau)\\
&+2\hat{\mathbf{I}}_y\hat{\mathbf{S}}_z\cos(\omega_I\tau)\sin(\pi J\tau)\sin(\omega_I\tau)\cos(\pi J\tau)-\hat{\mathbf{I}}_x\cos(\omega_I\tau)\sin(\pi J\tau)\sin(\omega_I\tau)\sin(\pi J\tau)\\
&+\hat{\mathbf{I}}_x\sin(\omega_I\tau)\cos(\pi J\tau)\cos(\omega_I\tau)\cos(\pi J\tau)+2\hat{\mathbf{I}}_y\hat{\mathbf{S}}_z\sin(\omega_I\tau)\cos(\pi J\tau)\cos(\omega_I\tau)\sin(\pi J\tau)\\
&+\hat{\mathbf{I}}_y\sin(\omega_I\tau)\cos(\pi J\tau)\sin(\omega_I\tau)\cos(\pi J\tau)-2\hat{\mathbf{I}}_x\hat{\mathbf{S}}_z\sin(\omega_I\tau)\cos(\pi J\tau)\sin(\omega_I\tau)\sin(\pi J\tau)\\
&-2\hat{\mathbf{I}}_y\hat{\mathbf{S}}_z\sin(\omega_I\tau)\sin(\pi J\tau)\cos(\omega_I\tau)\cos(\pi J\tau)+\hat{\mathbf{I}}_x\sin(\omega_I\tau)\sin(\pi J\tau)\cos(\omega_I\tau)\sin(\pi J\tau)\\
&+2\hat{\mathbf{I}}_x\hat{\mathbf{S}}_z\sin(\omega_I\tau)\sin(\pi J\tau)\sin(\omega_I\tau)\cos(\pi J\tau)+\hat{\mathbf{I}}_y\sin(\omega_I\tau)\sin(\pi J\tau)\sin(\omega_I\tau)\sin(\pi J\tau)
\end{aligned}$$

Now, the reader will most definitely be forgiven for thinking that this expression is a handful! However, think what it would have involved to do this by the full-blown density matrix method of Chapter 17. The perceptive reader will realise that it can be dramatically simplified by cancellation of terms and the application of trigonometric identities. For example, all of the terms in $\hat{\mathbf{I}}_x$ add up to zero as do all of the antiphase terms. The only terms left are in $\hat{\mathbf{I}}_y$:

$$\begin{aligned}&\hat{\mathbf{I}}_y \cos(\omega_I \tau)\cos(\pi J \tau)\cos(\omega_I \tau)\cos(\pi J \tau)\\&\quad + \hat{\mathbf{I}}_y \cos(\omega_I \tau)\sin(\pi J \tau)\cos(\omega_I \tau)\sin(\pi J \tau)\\&\quad + \hat{\mathbf{I}}_y \sin(\omega_I \tau)\cos(\pi J \tau)\sin(\omega_I \tau)\cos(\pi J \tau)\\&\quad + \hat{\mathbf{I}}_y \sin(\omega_I \tau)\sin(\pi J \tau)\sin(\omega_I \tau)\sin(\pi J \tau)\end{aligned}$$

Application of eqn (A.7) from Appendix A leaves:

$$+\hat{\mathbf{I}}_y$$

a rather drastic simplification to say the least, in which there is no chemical-shift or scalar coupling modulation. We can use the vector-model of Chapter 12 to show the homonuclear spin-echo sequence pictorially (Figure 18.9).

As it turns out, for the spin-echo pulse-sequence the chemical-shift operation is not necessary. Leaving it out results in exactly the same

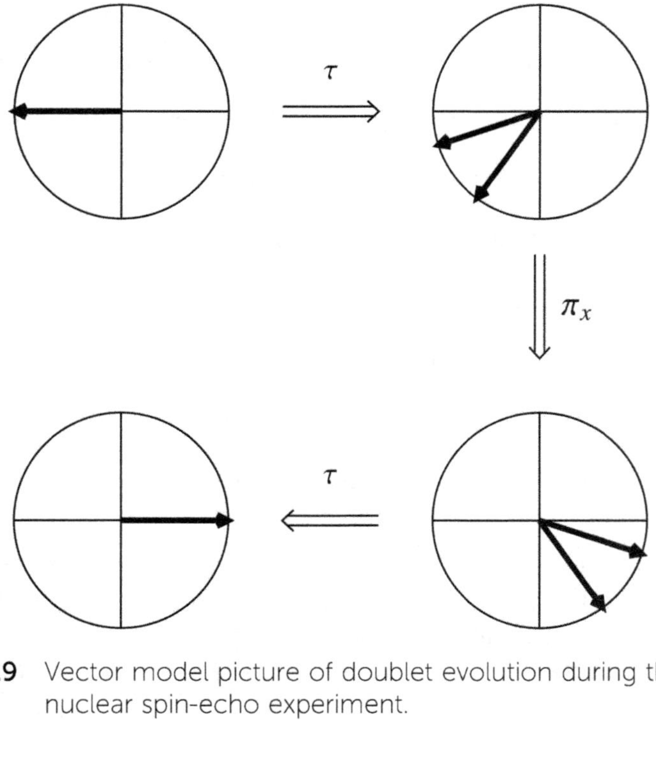

Figure 18.9 Vector model picture of doublet evolution during the homonuclear spin-echo experiment.

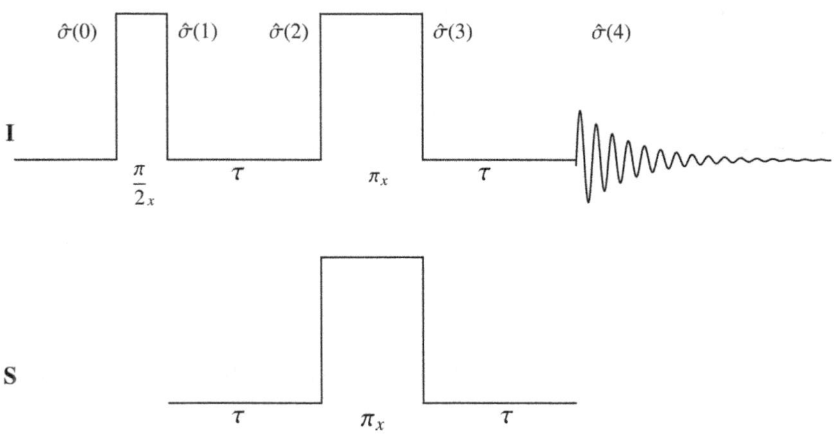

Figure 18.10 The heteronuclear spin-echo pulse sequence.

final state, no matter the value of t. Also, note that the diverging vectors due to scalar coupling come back together after the second delay period. This is known as *refocusing*. The vanishing of chemical-shift can perhaps be seen a bit more clearly in the heteronuclear spin-echo sequence (Figure 18.10).

Here, we are now pulsing both spins ... the π-pulse is applied simultaneously to spins I and S. Analysis as for the homonuclear spin-echo sequence results in a value for $\hat{\sigma}(4)$ of:

$$\hat{\sigma}(4) = +\hat{\mathbf{I}}_y \cos(2\pi J\tau) - 2\hat{\mathbf{I}}_x\hat{\mathbf{S}}_z \sin(2\pi J\tau) \tag{18.9}$$

after simplification. Note that there is no chemical-shift modulation in the result, just coupling. This outcome may be obtained with or without using the chemical-shift product operator. For this reason the spin-echo is very useful as a "building block" sequence to insert into more complex pulse-sequences in order to negate the effects of chemical-shift, one of which is severe phase distortion of the final spectrum.

The INEPT (**I**nsensitive **N**uclei **E**nhancement by **P**olarisation **T**ransfer) experiment analysis is our next project (Figure 18.11).***

We first note the presence of a spin-echo building block incorporating two delay times, each of length τ_1, and a π-pulse. From our

***Interestingly, when this experiment was first proposed, I was used to represent the insensitive nucleus, for example ^{13}C, and S the sensitive nucleus, such as ^{1}H. Somehow, over the years, the two were swapped and now one generally finds that I is used for the sensitive nucleus and S for the insensitive nucleus.

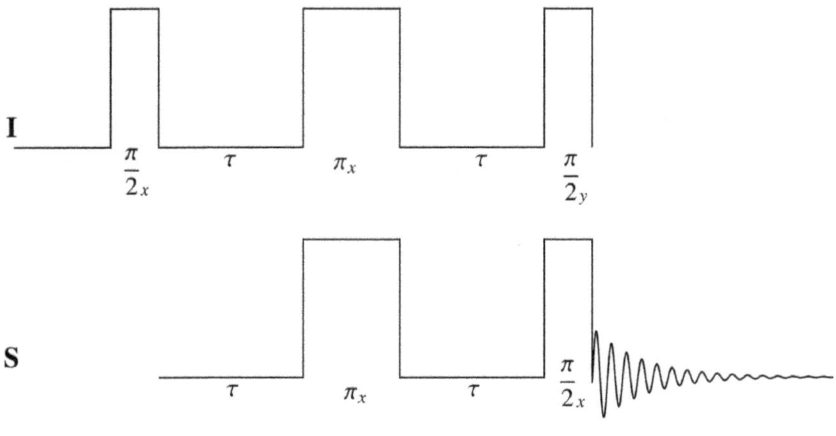

Figure 18.11 The INEPT heteronuclear pulse sequence.

preceding discussion we know that we need not include chemical-shift in this portion of the analysis. Also, happily, the state just prior to the I and S $\pi/2$ pulses at the end is the same as our result for the heteronuclear spin-echo (18.9). We continue from that point and apply the $\pi/2$-pulses on spins I and S:

$$\begin{aligned} &+\hat{\mathbf{I}}_y \cos(\pi J\tau) - 2\hat{\mathbf{I}}_x\hat{\mathbf{S}}_z \sin(\pi J\tau) \xrightarrow{\pi/2\hat{\mathbf{I}}_y} \\ &+\hat{\mathbf{I}}_y \cos(\pi J\tau) + 2\hat{\mathbf{I}}_z\hat{\mathbf{S}}_z \sin(\pi J\tau) \xrightarrow{\pi/2\hat{\mathbf{S}}_x} \\ &+\hat{\mathbf{I}}_y \cos(\pi J\tau) - 2\hat{\mathbf{I}}_z\hat{\mathbf{S}}_y \sin(\pi J\tau) \end{aligned} \tag{18.10}$$

In these two steps we see the polarisation transfer occur. The $\pi/2$-pulse on spin I converts the antiphase magnetisation $-2\hat{\mathbf{I}}_x\hat{\mathbf{S}}_z$ to $2\hat{\mathbf{I}}_z\hat{\mathbf{S}}_z$ and the S-spin $\pi/2$-pulse changes $2\hat{\mathbf{I}}_z\hat{\mathbf{S}}_z$ to $-2\hat{\mathbf{I}}_z\hat{\mathbf{S}}_y$ and we say that polarisation has been transferred from spin I to spin S. Since the spectrum is detected at S-spin frequencies the only detected term will be the antiphase magnetisation. The spectrum for two spins I and S will look like a doublet with one peak up and one peak down … in other words antiphase peaks. We do not usually want this and we will see a modification of the INEPT experiment in the next section that will bring the antiphase peaks into phase.

18.6 Three Spins and Higher

In the last section we considered the product operator analysis of two-spin systems only … I and S. We now expand our horizons somewhat to include three- and four-spin systems, *I*, *J*, *K* and *S* in our nomenclature.[†††] For three spins we have 64 product operators to keep track of and for four spins there are 256 operators! Considering the above analysis of the spin-echo sequence and the number of final terms generated, we might expect that for three and four spins there would probably be many more terms. This is the case and we generally do not often do calculations for three and four-spin systems for just this reason. However, it does occasionally become necessary and we do so here in order to show the utility of the product operator formalism in helping to set up experimental parameters in our NMR spectrometers.

For three spins we have three coupling operators available:

$$2\hat{\mathbf{I}}_z\hat{\mathbf{S}}_z \quad 2\hat{\mathbf{J}}_z\hat{\mathbf{S}}_z \quad 2\hat{\mathbf{K}}_z\hat{\mathbf{S}}_z$$

These will affect things in the expected fashion:

$$\hat{\mathbf{J}}_x \xrightarrow{(2\pi J_{JS}t)2\hat{\mathbf{J}}_z\hat{\mathbf{S}}_z} \hat{\mathbf{J}}_x\cos(\pi J_{JS}t) + 2\hat{\mathbf{J}}_y\hat{\mathbf{S}}_z\sin(\pi J_{JS}t)$$

$$\hat{\mathbf{S}}_y \xrightarrow{(2\pi J_{JS}t)2\hat{\mathbf{J}}_z\hat{\mathbf{S}}_z} \hat{\mathbf{S}}_y\cos(\pi J_{JS}t) - 2\hat{\mathbf{J}}_z\hat{\mathbf{S}}_x\sin(\pi J_{JS}t)$$

$$\hat{\mathbf{I}}_x\hat{\mathbf{J}}_z \xrightarrow{(2\pi J_{IJ}t)2\hat{\mathbf{I}}_z\hat{\mathbf{J}}_z} \hat{\mathbf{I}}_x\hat{\mathbf{J}}_z\cos(\pi J_{IJ}t) + \hat{\mathbf{I}}_y\sin(\pi J_{IJ}t)$$

Note that to keep things straight we are subscripting the *J* coupling constant. We may also encounter three-spin antiphase terms that are rotated by the coupling operator:[‡‡‡]

$$4\hat{\mathbf{I}}_z\hat{\mathbf{J}}_x\hat{\mathbf{S}}_z \xrightarrow{(2\pi J_{JS}t)2\hat{\mathbf{J}}_z\hat{\mathbf{S}}_z} 4\hat{\mathbf{I}}_z\hat{\mathbf{J}}_x\hat{\mathbf{S}}_z\cos(\pi J_{JS}t) + 2\hat{\mathbf{I}}_z\hat{\mathbf{J}}_y\sin(\pi J_{JS}t)$$

To understand how this works look to the rotation operator, in this case $2\hat{\mathbf{J}}_z\hat{\mathbf{S}}_z$. This tells us that the two spins to be concerned about are *J* and *S*. Then look to the initial operator, in this case $4\hat{\mathbf{I}}_z\hat{\mathbf{J}}_x\hat{\mathbf{S}}_z$, and disregard the spin that is not affected by the rotation operator. In this

[†††]Some authors use a numeric subscript notation to keep track of spins. Something such as I_{kx} and I_{lx} might be used to differentiate between spins, homonuclear or heteronuclear. I find this method a bit visually confusing and prefer to use *I*, *J*, *K* and *S* to clearly differentiate between spins. Since we are most often concerned with carbons having from zero to three hydrogens attached I generally use *S* to represent carbon and *I*, *J* and *K* for hydrogen. I do realise the potential for confusion when using *J* but I believe that the context that it is used in will take care of things. Also, note that in the text the spin operator $\hat{\mathbf{J}}$ is in boldface, whereas the coupling constant *J* is not.

[‡‡‡]Note the "4" in the three-spin operators. This is the normalisation constant discussed above. Construction of these three-spin operators from their one-spin matrices results in a constant of 1/8 in front of the matrix. The number needed to convert this to 1/2 is 4, according to the "short-cut" method mentioned above in connection with the normalisation constants.

case we disregard the $\hat{\mathbf{I}}_z$ portion of the operator and imagine that we are looking only at the $\hat{\mathbf{J}}_x\hat{\mathbf{S}}_z$. We then apply the rules for two-spin operators to get $\hat{\mathbf{J}}_y$ and then add $\hat{\mathbf{I}}_z$ back into the operator term to get $2\hat{\mathbf{I}}_z\hat{\mathbf{J}}_y$.

Recall that we found that two-spin multiple quantum coherence (MQC) does not evolve under the influence of the coupling operator. This is not necessarily the case for three and higher spins and we shall have to modify our understanding of this. Let us look at this, using for example, $4\hat{\mathbf{I}}_y\hat{\mathbf{J}}_x\hat{\mathbf{S}}_z$ as our initial state.

$$4\hat{\mathbf{I}}_y\hat{\mathbf{J}}_x\hat{\mathbf{S}}_z \xrightarrow{(2\pi J_{IJ}t)2\hat{\mathbf{I}}_z\hat{\mathbf{J}}_z} 4\hat{\mathbf{I}}_y\hat{\mathbf{J}}_x\hat{\mathbf{S}}_z$$

$$4\hat{\mathbf{I}}_y\hat{\mathbf{J}}_x\hat{\mathbf{S}}_z \xrightarrow{(2\pi J_{IJ}t)2\hat{\mathbf{I}}_z\hat{\mathbf{S}}_z} 4\hat{\mathbf{I}}_y\hat{\mathbf{J}}_x\hat{\mathbf{S}}_z\cos(\pi J_{IS}t) - 2\hat{\mathbf{I}}_x\hat{\mathbf{J}}_x\sin(\pi J_{IS}t)$$

In the first case the coupling operator is $2\hat{\mathbf{I}}_z\hat{\mathbf{J}}_z$ and is operating on the $\hat{\mathbf{I}}_y\hat{\mathbf{J}}_x$ portion of $4\hat{\mathbf{I}}_y\hat{\mathbf{J}}_x\hat{\mathbf{S}}_z$. This being the multiple quantum coherence portion there will be no coupling evolution. In the second case, $2\hat{\mathbf{I}}_z\hat{\mathbf{S}}_z$ is operating on the $\hat{\mathbf{I}}_y\hat{\mathbf{S}}_z$ portion of $4\hat{\mathbf{I}}_y\hat{\mathbf{J}}_x\hat{\mathbf{S}}_z$ and using the rotation rules from Table 18.2, changes it to $-\hat{\mathbf{I}}_x$. So, even though this three-spin multiple quantum coherence evolves under the influence of a coupling operator it is not the MQC portion that undergoes evolution but rather the antiphase portion. A three-spin operator such as $4\hat{\mathbf{I}}_y\hat{\mathbf{J}}_x\hat{\mathbf{S}}_x$ will not undergo coupling evolution at all since there is no antiphase portion to it. A bit more surprising in light of what we have learned concerning the coupling evolution of two-spin MQC is:

$$2\hat{\mathbf{J}}_x\hat{\mathbf{S}}_x \xrightarrow{(2\pi J_{IJ}t)2\hat{\mathbf{I}}_z\hat{\mathbf{J}}_z} 2\hat{\mathbf{J}}_x\hat{\mathbf{S}}_x\cos(\pi J_{IJ}t) + 4\hat{\mathbf{I}}_z\hat{\mathbf{J}}_y\hat{\mathbf{S}}_x\sin(\pi J_{IJ}t)$$

for a three-spin system. MQC between spins *J* and *S* will not evolve under a *JS* coupling operator but *will* do so under an *IJ* or *IS* operator. This is a tricky one to remember!

We will now revisit the INEPT experiment and explore it for different numbers of spins. Our previous analysis for two spins indicated that the result of this pulse sequence is an antiphase doublet. This is not particularly useful, especially since it is a ^{13}C detected experiment and we generally wish to use the decoupler to enhance the signal intensity. Turning on the decoupler at the end of the basic INEPT sequence will cause the antiphase doublet to collapse to zero signal. We would like to modify the INEPT sequence so that we may use the decoupler that will collapse the multiplets into singlets and provide a sensitivity enhancement. Enter the refocused INEPT sequence sometimes called INEPTRD (Figure 18.12).

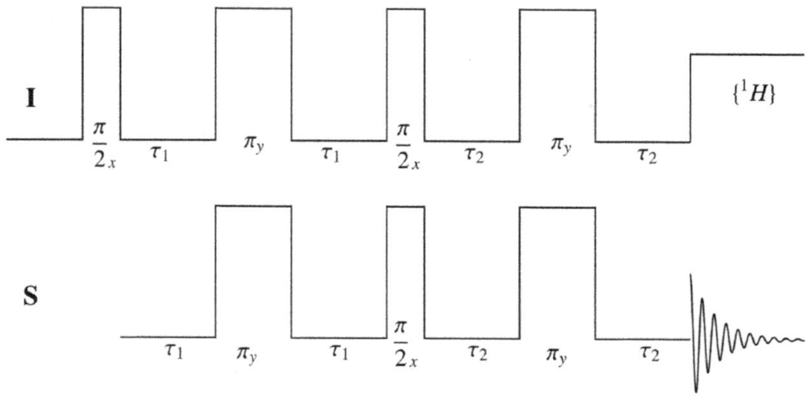

Figure 18.12 The INEPTRD heteronuclear pulse sequence for two spins, *I* and *S*.

A second spin-echo block has been added to the end of the basic INEPT sequence. The purpose of this is to allow the antiphase magnetisation to evolve into in-phase magnetisation while disallowing chemical-shift evolution. The two-spin analysis for this pulse-sequence proceeds from where we left off for the basic INEPT sequence (eqn (18.10)):

$$\hat{\mathbf{I}}_y \cos(2\pi J\tau_1) - 2\hat{\mathbf{I}}_z\hat{\mathbf{S}}_y \sin(2\pi J\tau_1)$$

Application of the second spin-echo using the time-delay τ_2 gives:

$$\hat{\mathbf{I}}_y \cos(2\pi J\tau_1) - 2\hat{\mathbf{I}}_z\hat{\mathbf{S}}_y \sin(2\pi J\tau_1) \xrightarrow{\pi(\hat{\mathbf{I}}_x+\hat{\mathbf{S}}_x)} \xrightarrow{(2\pi J\tau_2)\hat{\mathbf{I}}_z\hat{\mathbf{S}}_z}$$

$$- \hat{\mathbf{I}}_y \cos(2\pi J\tau_1) \cos(2\pi J\tau_2) + 2\hat{\mathbf{I}}_x\hat{\mathbf{S}}_z \cos(2\pi J\tau_1) \sin(2\pi J\tau_2)$$

$$- 2\hat{\mathbf{I}}_z\hat{\mathbf{S}}_y \sin(2\pi J\tau_1) \cos(2\pi J\tau_2) + \hat{\mathbf{S}}_x \sin(2\pi J\tau_1) \sin(2\pi J\tau_2)$$

Since the observed spin is *S*, the only observable portion of this state is:

$$-2\hat{\mathbf{I}}_z\hat{\mathbf{S}}_y \sin(2\pi J\tau_1) \cos(2\pi J\tau_2) + \hat{\mathbf{S}}_x \sin(2\pi J\tau_1) \sin(2\pi J\tau_2)$$

This is better than the basic INEPT experiment but still has some antiphase magnetisation in it that will introduce some phase distortion into the spectrum. The delay time τ_1 is generally set to $1/(4J_{IS})$ that will remove the dependence on $\cos(2\pi J_{IS}\tau_1)$. Setting $\tau_2 = 1/(4J_{IS})$ as well will then result in only the in-phase term $\hat{\mathbf{S}}_x$ surviving:

$$\hat{\mathbf{S}}_x$$

In a real sample all of the J_{IS} couplings will not necessarily be equal and a compromise must be made in which a value of τ_2 is

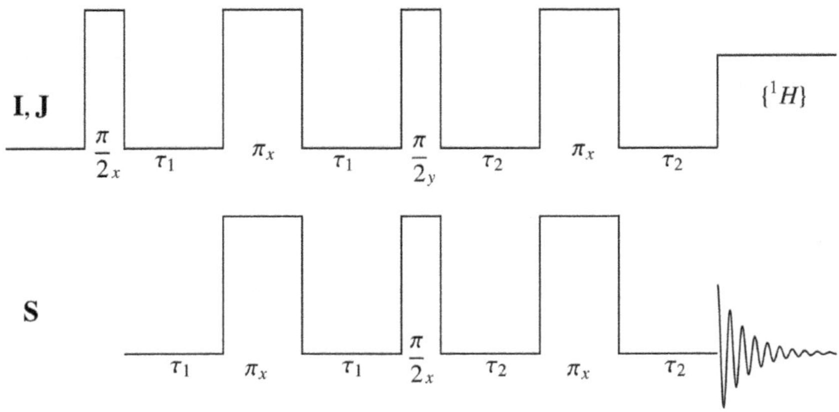

Figure 18.13 The INEPTRD heteronuclear pulse sequence for three spins, *I*, *J* and *S*. Spins *I* and *J* are homonuclear.

chosen to provide the best enhancement of $\hat{\mathbf{S}}_x$. Turning on the *I*-spin decoupler will get rid of any residual $2\hat{\mathbf{I}}_z\hat{\mathbf{S}}_y$ that is not taken care of by the τ_2 setting so that only in-phase $\hat{\mathbf{S}}_x$ magnetisation will be left. We can improve on this further but first we present the results of the product operator analysis of the refocused INEPT for three and four spins.

What will happen in a three-spin system? Let us give it a try using *I*, *J* and *S* spins (Figure 18.13). We will assume that spins *I* and *J* represent spins of the same nucleus (most commonly, hydrogen) and that the J-coupling constant between *I* and *S* and between *J* and *S* are the same. Also, for simplicity we will assume that the J-coupling constant between spins *I* and *J* is zero. We start with spin-state $\hat{\mathbf{I}}_z + \hat{\mathbf{J}}_z + \hat{\mathbf{S}}_z$.

We do first, a $\pi/2$-pulse on spins *I* and *J*:

$$(\hat{\mathbf{I}}_z + \hat{\mathbf{J}}_z) \xrightarrow{\pi/2(\hat{\mathbf{I}}_x + \hat{\mathbf{J}}_x)} -(\hat{\mathbf{I}}_y + \hat{\mathbf{J}}_y)$$

followed by a heteronuclear spin-echo. Recall that for the spin-echo building block we need not include chemical-shift. At the end of the first spin-echo and the $\pi/2$-pulses on *I*, *J* and *S* (which effect the polarisation transfer as we saw earlier) the product operator state will be:

$$\hat{\mathbf{I}}_y \cos(2\pi J\tau_1) + \hat{\mathbf{J}}_y \cos(2\pi J\tau_1) - 2\hat{\mathbf{I}}_z\hat{\mathbf{S}}_y \sin(2\pi J\tau_1) - 2\hat{\mathbf{J}}_z\hat{\mathbf{S}}_y \sin(2\pi J\tau_1)$$

It is often the case in a pulse-sequence that at this stage some of the terms will not contribute to the observed signal at the end of the sequence. Recognising when this will be so comes with hard-won experience and in this case we recognise that the $\hat{\mathbf{I}}_y$ and $\hat{\mathbf{J}}_y$ terms will not contribute any further to observable *S*-magnetisation. Therefore we drop them from further consideration and simplify our current expression to:

$$-2\hat{\mathbf{I}}_z\hat{\mathbf{S}}_y \sin(2\pi J\tau_1) - 2\hat{\mathbf{J}}_z\hat{\mathbf{S}}_y \sin(2\pi J\tau_1)$$

It is only the antiphase magnetisation at this point that will contribute ultimately to the observed signal. From here we launch into the second heteronuclear spin-echo portion of the pulse-sequence. At the end of the spin-echo the state is:[§§§]

$$\begin{aligned} &2\hat{\mathbf{I}}_z\hat{\mathbf{S}}_y \sin(2\pi J\tau_1)\cos(4\pi J\tau_2) + 4\hat{\mathbf{I}}_z\hat{\mathbf{J}}_z\hat{\mathbf{S}}_x \sin(2\pi J\tau_1)\sin(4\pi J\tau_2) \\ &-2\hat{\mathbf{J}}_z\hat{\mathbf{S}}_y \sin(2\pi J\tau_1)\cos(4\pi J\tau_2) + \hat{\mathbf{S}}_x \sin(2\pi J\tau_1)\sin(4\pi J\tau_2) \end{aligned} \tag{18.11}$$

When the decoupler is turned on all of the antiphase magnetisation will collapse and we are left with only:

$$\hat{\mathbf{S}}_x \sin(2\pi J\tau_1)\sin(4\pi J\tau_2)$$

The INEPT+ pulse-sequence adds a $\pi/2$-pulse to the *I* (and *J* and *K*) spins at the end of the INEPTRD sequence. The state at this point is as in eqn (18.11) and the application of the $\pi/2$-pulse changes all of the antiphase magnetisation into multiple quantum coherence which is unobservable, leaving only observable $\hat{\mathbf{S}}_x$ magnetisation.

The COSY 2D shift correlation experiment uses a deceptively simple pulse-sequence (Figure 18.14).

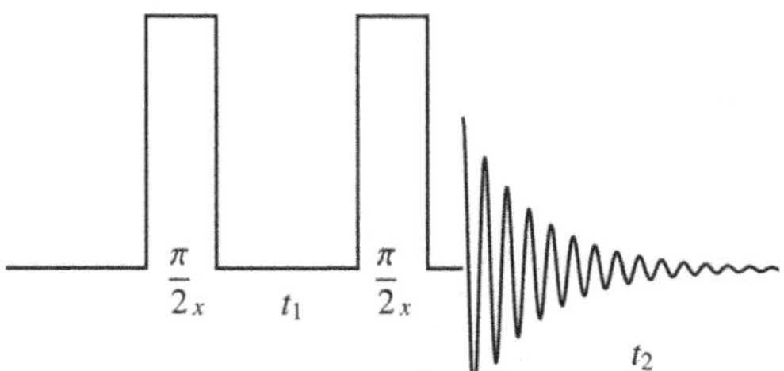

Figure 18.14 The COSY pulse sequence.

[§§§]If you are reading or have read O. Sorensen and R. Ernst's paper on the elimination of distortion in polarisation transfer experiments (see the Further Reading section) you will no doubt notice a difference in the arguments to the trigonometric functions. This is because they use $\tau/2$ to represent the INEPTRD time interval, whereas I have used τ. Otherwise, the results of their analysis are identical to those shown here. By the way, this is an excellent paper to test your knowledge of product operators. Interestingly, it predates their seminal paper introducing the product operator formalism as we now know it.

This short sequence of two $\pi/2$-pulses with a variable time-delay delay between them generates a veritable storm of terms, even for the two-spin case, in the product operator analysis. Following one of the two spins.[¶¶¶] First, the $\pi/2$-pulse:

$$\hat{\mathbf{I}}_z \xrightarrow{\pi/2\hat{\mathbf{I}}_x} -\hat{\mathbf{I}}_y$$

followed by chemical-shift during t_1:

$$-\hat{\mathbf{I}}_y \xrightarrow{\omega_I t_1 \hat{\mathbf{I}}_z} -\hat{\mathbf{I}}_y \cos(\omega_I t_1) + \hat{\mathbf{I}}_x \sin(\omega_I t_1)$$

and coupling during t_1:

$$- \hat{\mathbf{I}}_y \cos(\omega_I t_1) + \hat{\mathbf{I}}_x \sin(\omega_I t_1) \xrightarrow{(\pi J t_1)\hat{\mathbf{I}}_z\hat{\mathbf{J}}_z}$$

$$- \hat{\mathbf{I}}_y \cos(\omega_I t_1) \cos(\pi J t_1) + 2\hat{\mathbf{I}}_x\hat{\mathbf{J}}_z \cos(\omega_I t_1) \sin(\pi J t_1)$$

$$+ \hat{\mathbf{I}}_x \sin(\omega_I t_1) \cos(\pi J t_1) + 2\hat{\mathbf{I}}_y\hat{\mathbf{J}}_z \sin(\omega_I t_1) \sin(\pi J t_1)$$

and the second $\pi/2$-pulse (on *both* nuclei this time since this is a homonuclear experiment):

$$- \hat{\mathbf{I}}_y \cos(\omega_I t_1) \cos(\pi J t_1) + 2\hat{\mathbf{I}}_x\hat{\mathbf{J}}_z \cos(\omega_I t_1) \sin(\pi J t_1)$$

$$+ \hat{\mathbf{I}}_x \sin(\omega_I t_1) \cos(\pi J t_1) + 2\hat{\mathbf{I}}_y\hat{\mathbf{J}}_z \sin(\omega_I t_1) \sin(\pi J t_1)$$

$$\xrightarrow{\pi/2(\hat{\mathbf{I}}_x + \hat{\mathbf{J}}_x)}$$

$$- \hat{\mathbf{I}}_z \cos(\omega_I t_1) \cos(\pi J t_1) - 2\hat{\mathbf{I}}_x\hat{\mathbf{J}}_y \cos(\omega_I t_1) \sin(\pi J t_1)$$

$$+ \hat{\mathbf{I}}_x \sin(\omega_I t_1) \cos(\pi J t_1) - 2\hat{\mathbf{I}}_z\hat{\mathbf{J}}_y \sin(\omega_I t_1) \sin(\pi J t_1)$$

[¶¶¶] I must admit that I have "cheated" here a bit (or actually, a lot) on this particular calculation. Rather than do them by hand I have produced a standalone product operator calculator that does all of the tedious work and that may be freely downloaded. The program is named "wxProdOp" and is currently located at https://sourceforge.net/projects/wxProdOp. One may also do so using symbolic algebra software with add-on packages. See the Further Reading section, specifically, P. Guntert in International Journal of Quantum Chemistry.

Next, chemical-shift during t_2:

$$-\hat{\mathbf{I}}_z\cos(\omega_I t_1)\cos(\pi J t_1) - 2\hat{\mathbf{I}}_x\hat{\mathbf{J}}_y\cos(\omega_I t_1)\sin(\pi J t_1)$$
$$+\hat{\mathbf{I}}_x\sin(\omega_I t_1)\cos(\pi J t_1) - 2\hat{\mathbf{I}}_z\hat{\mathbf{J}}_y\sin(\omega_I t_1)\sin(\pi J t_1)$$
$$\xrightarrow{\omega_I t_2\hat{\mathbf{I}}_z+\omega_J t_2\hat{\mathbf{J}}_z}$$
$$-\hat{\mathbf{I}}_z\cos(\omega_I t_1)\cos(\pi J t_1) - 2\hat{\mathbf{I}}_x\hat{\mathbf{J}}_y\cos(\omega_I t_1)\sin(\pi J t_1)\cos(\omega_I t_2)\cos(\omega_J t_2)$$
$$+2\hat{\mathbf{I}}_x\hat{\mathbf{J}}_x\cos(\omega_I t_1)\sin(\pi J t_1)\sin(\omega_J t_2)\cos(\omega_I t_2)$$
$$-2\hat{\mathbf{I}}_y\hat{\mathbf{J}}_y\cos(\omega_I t_1)\sin(\pi J t_1)\cos(\omega_J t_2)\sin(\omega_I t_2)$$
$$+2\hat{\mathbf{I}}_y\hat{\mathbf{J}}_x\cos(\omega_I t_1)\sin(\pi J t_1)\sin(\omega_J t_2)\sin(\omega_I t_2)$$
$$+\hat{\mathbf{I}}_x\sin(\omega_I t_1)\cos(\pi J t_1)\cos(\omega_I t_2)$$
$$+\hat{\mathbf{I}}_y\sin(\omega_I t_1)\cos(\pi J t_1)\sin(\omega_I t_2)$$
$$-2\hat{\mathbf{I}}_z\hat{\mathbf{J}}_y\sin(\omega_I t_1)\sin(\pi J t_1)\cos(\omega_J t_2)$$
$$+2\hat{\mathbf{I}}_z\hat{\mathbf{J}}_x\sin(\omega_I t_1)\sin(\pi J t_1)\sin(\omega_J t_2)$$

and lastly, coupling during t_2 (remembering that $\hat{\mathbf{I}}_z$ and two-spin MQC do not undergo coupling evolution):

$$\text{previous expression} \xrightarrow{(\pi J t_2)\hat{\mathbf{I}}_z\hat{\mathbf{J}}_z}$$
$$-\hat{\mathbf{I}}_z\cos(\omega_I t_1)\cos(\pi J t_1) - 2\hat{\mathbf{I}}_x\hat{\mathbf{J}}_y\cos(\omega_I t_1)\sin(\pi J t_1)\cos(\omega_I t_2)\cos(\omega_J t_2)$$
$$+2\hat{\mathbf{I}}_x\hat{\mathbf{J}}_x\cos(\omega_I t_1)\sin(\pi J t_1)\sin(\omega_J t_2)\cos(\omega_I t_2)$$
$$-2\hat{\mathbf{I}}_y\hat{\mathbf{J}}_y\cos(\omega_I t_1)\sin(\pi J t_1)\cos(\omega_J t_2)\sin(\omega_I t_2)$$
$$+2\hat{\mathbf{I}}_y\hat{\mathbf{J}}_x\cos(\omega_I t_1)\sin(\pi J t_1)\sin(\omega_J t_2)\sin(\omega_I t_2)$$
$$+\hat{\mathbf{I}}_x\sin(\omega_I t_1)\cos(\pi J t_1)\cos(\omega_I t_2)\cos(\pi J t_2)$$
$$+2\hat{\mathbf{I}}_y\hat{\mathbf{J}}_z\sin(\omega_I t_1)\cos(\pi J t_1)\cos(\omega_I t_2)\sin(\pi J t_2)$$
$$+\hat{\mathbf{I}}_y\sin(\omega_I t_1)\cos(\pi J t_1)\sin(\omega_I t_2)\cos(\pi J t_2)$$
$$-2\hat{\mathbf{I}}_x\hat{\mathbf{J}}_z\sin(\omega_I t_1)\cos(\pi J t_1)\sin(\omega_I t_2)\sin(\pi J t_2)$$
$$-2\hat{\mathbf{I}}_z\hat{\mathbf{J}}_y\sin(\omega_I t_1)\sin(\pi J t_1)\cos(\omega_J t_2)\cos(\pi J t_2)$$
$$+\hat{\mathbf{J}}_x\sin(\omega_I t_1)\sin(\pi J t_1)\cos(\omega_J t_2)\sin(\pi J t_2)$$
$$+2\hat{\mathbf{I}}_z\hat{\mathbf{J}}_x\sin(\omega_I t_1)\sin(\pi J t_1)\sin(\omega_J t_2)\cos(\pi J t_2)$$
$$+\hat{\mathbf{J}}_y\sin(\omega_I t_1)\sin(\pi J t_1)\sin(\omega_J t_2)\sin(\pi J t_2)$$

Of these, the first five are unobservable terms, which leaves:

$$+\,\hat{\mathbf{I}}_x \sin(\omega_I t_1)\cos(\pi J t_1)\cos(\omega_I t_2)\cos(\pi J t_2)$$

$$+\,2\hat{\mathbf{I}}_y\hat{\mathbf{J}}_z \sin(\omega_I t_1)\cos(\pi J t_1)\cos(\omega_I t_2)\sin(\pi J t_2)$$

$$+\,\hat{\mathbf{I}}_y \sin(\omega_I t_1)\cos(\pi J t_1)\sin(\omega_I t_2)\cos(\pi J t_2)$$

$$-\,2\hat{\mathbf{I}}_x\hat{\mathbf{J}}_z \sin(\omega_I t_1)\cos(\pi J t_1)\sin(\omega_I t_2)\sin(\pi J t_2)$$

$$-\,2\hat{\mathbf{I}}_z\hat{\mathbf{J}}_y \sin(\omega_I t_1)\sin(\pi J t_1)\cos(\omega_J t_2)\cos(\pi J t_2)$$

$$+\,\hat{\mathbf{J}}_x \sin(\omega_I t_1)\sin(\pi J t_1)\cos(\omega_J t_2)\sin(\pi J t_2)$$

$$+\,2\hat{\mathbf{I}}_z\hat{\mathbf{J}}_x \sin(\omega_I t_1)\sin(\pi J t_1)\sin(\omega_J t_2)\cos(\pi J t_2)$$

$$+\,\hat{\mathbf{J}}_y \sin(\omega_I t_1)\sin(\pi J t_1)\sin(\omega_J t_2)\sin(\pi J t_2)$$

This is a *lot* of terms for such a simple pulse-sequence! Fortunately, there is another way to look at it that we will explore in the next section.

For this pulse sequence the important parts to look at in our very complex-looking expression are the chemical-shift portions. You will note that the first four terms all are modulated by sine or cosine terms in which only *I*-spin chemical-shift is involved. Since this is a two-dimensional experiment in which each dimension is a chemical-shift dimension this means that these terms will give rise to peaks at or very near the diagonal. The last four terms on the other hand are each modulated by *two* chemical-shift frequencies. These will give rise to peaks in the spectrum that are some distance away from the diagonal and are referred to as *off-diagonal* peaks.

18.7 The Product Operator Tree

Even though the product operator model drastically simplifies calculations of the states of a spin system during a sequence of rf pulses and time delays, we have seen that there can still be a lot of tedious writing of terms, as we saw with the COSY analysis. This is especially evident in longer sequences in which terms multiply at each pulse and time delay. Furthermore, the problem is exacerbated when considering three- and four-spin systems in which many more terms are generated.

Much of the tedium is taken up in the writing of the sine and cosine terms and there is a way to reduce this work by using what we call here

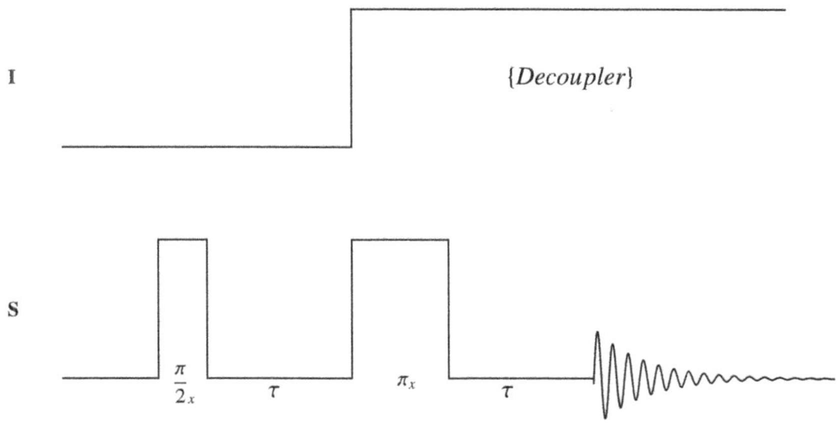

Figure 18.15 The J-modulation pulse sequence.

the operator tree.[||||] Recall that for each application of a rotation operator a cosine and sine term is generated (see eqn (18.5)). Since this is the case every time a rotation operator is applied to a density operator we can leave out the explicit sine and cosine trigonometric terms and draw a tree diagram representation of the rotation. The left branch corresponds to the cosine term and the right branch to the sine term. If the sine or cosine term vanishes as for example a $\pi/2$ or π-pulse then we can simply draw one vertical line. We will use a tree diagram to analyse the J-modulation pulse-sequence (Figure 18.15).

The top of the tree in this case is the first pulse on the S-spin (see Figure 18.16). Of course, applying a pulse rotation operator using a pulse angle of $\pi/2$ results in a cosine value of zero and a sine value of one leaving only the "target" operator. This is followed by a time-delay of $\tau = 1/J$ during which chemical-shift and coupling evolutions take place. As in our previous examples there is a spin-echo block in the J-modulation sequence and we are therefore allowed to disregard chemical-shift evolution during this time period. Since $\tau = 1/J$ the coupling evolution also simplifies to a single term. During the second half of the spin-echo there will be no coupling evolution since ... there is effectively no coupling! The (very simple) product operator tree diagram is shown in Figure 18.16.

The same type of tree diagram for a three-spin system looks as in Figure 18.17.

[||||] I first encountered this in Ray Freeman's *A Handbook of Nuclear Magnetic Resonance* (see the Further Reading section) and have used it ever since for quick "back of the envelope" calculations. Oddly, very few others have mentioned the technique in any publications that I have seen ... which probably says that I am not as well-read as I should be.

$$\mathbf{S}_z$$

$$\frac{\pi}{2}\mathbf{S}_x\ \textit{pulse}\ \downarrow$$

$$-\mathbf{S}_y$$

$$(\pi J \frac{1}{J}) 2\hat{\mathbf{I}}_z\hat{\mathbf{S}}_z\ \textit{coupling}\ \downarrow$$

$$+\mathbf{S}_y$$

$$\pi\mathbf{S}_x\ \textit{pulse}\ \downarrow$$

$$-\mathbf{S}_y$$

$$\mathbf{I}\ \textit{decoupler}\ \downarrow$$

$$-\mathbf{S}_y$$

Figure 18.16 Product operator tree diagram for a heteronuclear two-spin system in the the J-modulation experiment.

$$\mathbf{S}_z$$

$$\frac{\pi}{2}\mathbf{S}_x\ \textit{pulse}\ \downarrow$$

$$-\mathbf{S}_y$$

$$(\pi J \frac{1}{J}) 2\hat{\mathbf{I}}_z\hat{\mathbf{S}}_z\ \textit{coupling}\ \downarrow$$

$$-\mathbf{S}_y$$

$$\pi\mathbf{S}_x\ \textit{pulse}\ \downarrow$$

$$+\mathbf{S}_y$$

$$\mathbf{I}\ \textit{decoupler}\ \downarrow$$

$$+\mathbf{S}_y$$

Figure 18.17 Product operator tree diagram for a heteronuclear three-spin system in the the J-modulation experiment.

We note that the magnetisation for the three-spin system is of opposite sign to that for the two-spin system. This is as expected in view of the vector analysis of the same pulse-sequence in Chapter 12.

A bit more instructive perhaps is the same type of tree diagram for the COSY experiment (Figure 18.18):

This presents us with the same information as our recent COSY analysis but now without all of the cosine and sine terms to obscure

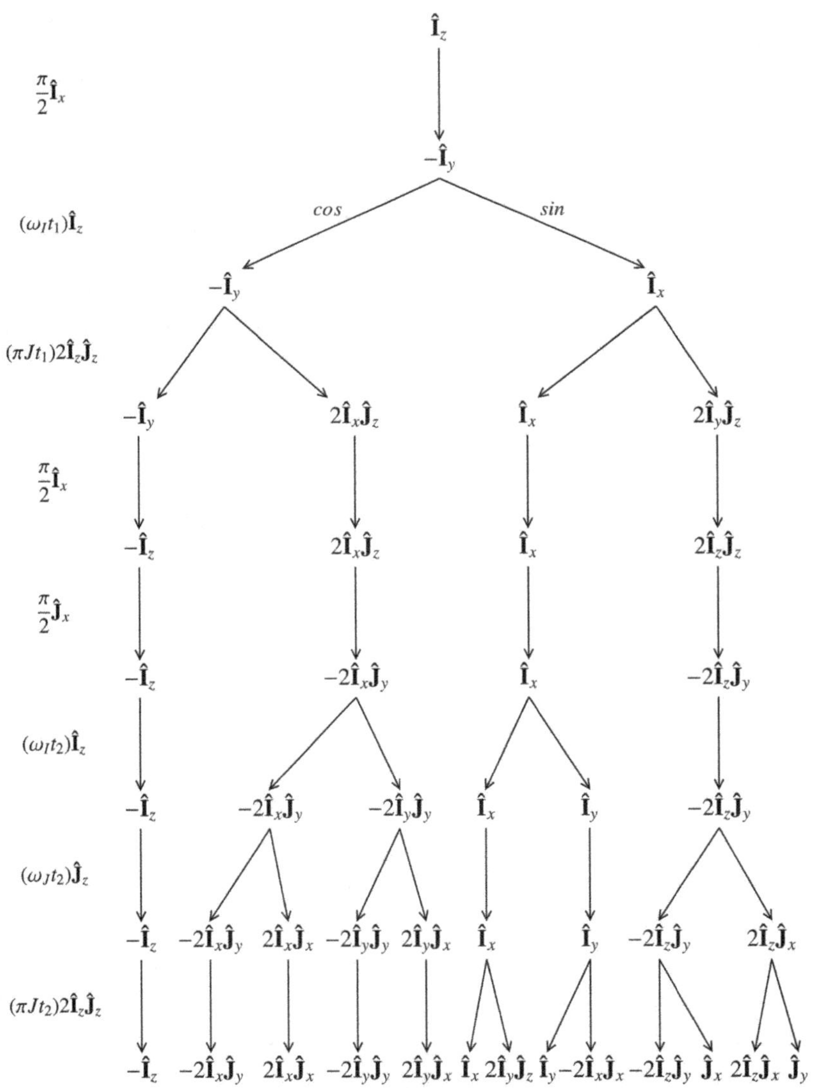

Figure 18.18 Product operator tree diagram for a homonuclear two-spin system in the COSY experiment.

things. One of the major advantages of the tree analysis is that it allows one to see very easily branches of the tree that lead to observable magnetisation and branches that do not. Here, we see that the entire left side of the tree after the first chemical-shift evolution does not lead to observable magnetisation, whereas all of the right side leads to observable magnetisation. Note that in the above diagram the points at which there is a split into sine and cosine terms is shown with the cosine term to the left and the sine term to the right. If the trigonometric terms for any of the operators are required all one needs to do is follow the path from the start to the operator(s) in question and assign trigonometric terms based on the directions in which the tree is traversed.

18.8 Single Transition Product Operators

One may think of the density matrix at a particular instant in time as a linear combination of base operators. In a general sense this may be written as:

$$\hat{\sigma}(t) = \sum_S b_S \hat{\mathbf{B}}_S$$

Indeed, this is embodied in many of the preceding product operator equations such as:

$$\hat{\mathbf{I}}_z \xrightarrow{\beta \hat{\mathbf{I}}_x} \hat{\mathbf{I}}_z \cos(\beta) - \hat{\mathbf{I}}_y \sin(\beta)$$

where, after the application of the x-phase pulse on spin I, the density matrix consists of a mixture of $\hat{\mathbf{I}}_z$ and $\hat{\mathbf{I}}_y$.

The formally proposed**** and generally accepted definition of a spin-1/2 Cartesian product operator is as follows:

$$\hat{\mathbf{B}}_S \equiv 2^{(q-1)} \prod_{k=1}^{N} (\hat{\mathbf{I}}_{k\nu})^{a_{sk}} \tag{18.12}$$

Here, $\hat{\mathbf{I}}_{k\nu}$ is a single-spin operator, q is the number of single-spin operators in the product, N is the total number of $I = 1/2$ nuclei, k is the index of the nucleus, ν is (in the case of Cartesian operators) x, y or z and a_{sk} is equal to one for the q operators in the product and zero

****See Sorensen *et al.* in the Further Reading section. This is the central paper in product operators. There is much to learn from it.

for the rest. The product is the outer product that we explored in Section 17.4. The factor, $2^{(q-1)}$ is the normalisation constant discussed in Section 18.2. Within a set of product operators, individual operators are orthogonal to the others in the sense that the trace of their normal matrix product is zero:

$$\mathrm{Tr}(\hat{\mathbf{B}}_S\hat{\mathbf{B}}_T) = \delta_{ST}2^{(N-2)}$$

as discussed above with respect to eqn (18.6). The definition, eqn (18.12), is a bit convoluted to accept at face value so let us do a couple of examples and see how it works. We will build the two-spin operators for $\hat{\mathbf{I}}_z$ and $2\hat{\mathbf{I}}_x\hat{\mathbf{S}}_y$. Recall that when building the density matrix for $\hat{\mathbf{I}}_z$ we did the outer product of single-spin $\hat{\mathbf{I}}_z$ and $\hat{\mathbf{E}}$. We do that here:

$$\hat{\mathbf{I}}_z = 2^{(1-1)}\hat{\mathbf{I}}_z\hat{\mathbf{E}}$$

where it is understood that the operator on the left is for a two-spin system. Since $q=1$ the term $a_{sk}=1$ for the first operator $\hat{\mathbf{I}}_z$ on the right and 0 for the subsequent ones ... in other words unit operators, represented by $\hat{\mathbf{E}}$. For $2\hat{\mathbf{I}}_x\hat{\mathbf{S}}_y$:

$$2\hat{\mathbf{I}}_x\hat{\mathbf{S}}_y = 2^{(2-1)}\hat{\mathbf{I}}_x\hat{\mathbf{S}}_y$$

This time, $q=2$ so the a_{sk} terms are all equal to 1. $\hat{\mathbf{E}}$ is implied here and usually not written as part of the product. Note that if we were to strictly use the definition, eqn (18.12), we would write the operators with the same symbols but numbered subscripts. If we were to do this it would be:[††††]

$$2\hat{\mathbf{I}}_{1x}\hat{\mathbf{I}}_{2y} = 2^{(2-1)}\hat{\mathbf{I}}_{1x}\hat{\mathbf{I}}_{2y}$$

Cartesian spin-1/2 operators are built this way and, as we have seen in the immediately previous sections, are very useful. However, we need not use operators that are Cartesian ... it is sometimes useful to consider other operator basis sets. In particular, the single-transition basis set is very convenient in some circumstances. We have already seen a couple of these in the ladder operators, introduced and used

[††††]As I have said before in this chapter, I find it visually confusing to look at long strings of numerically subscripted operators and can better follow the flow of things using *I*, *J*, *K* and *S*.

extensively in Chapter 15. The matrix representations of the one-spin ladder operators are, from their definitions in Chapter 15:

$$\hat{\mathbf{I}}_{+} = \hat{\mathbf{I}}_x + i\hat{\mathbf{I}}_y = \frac{1}{2}\begin{bmatrix} 0 & 1 \\ 0 & 0 \end{bmatrix}$$

$$\hat{\mathbf{I}}_{-} = \hat{\mathbf{I}}_x - i\hat{\mathbf{I}}_y = \frac{1}{2}\begin{bmatrix} 0 & 0 \\ 1 & 0 \end{bmatrix}$$

These two operators are constructed from a linear combination of $\hat{\mathbf{I}}_x$ and $\hat{\mathbf{I}}_y$, and we may construct two more using linear combinations of $\hat{\mathbf{I}}_z$ and $\hat{\mathbf{E}}$:

$$\hat{\mathbf{I}}_{\alpha} = \hat{\mathbf{I}}_z + \frac{1}{2}\hat{\mathbf{E}} = \frac{1}{2}\begin{bmatrix} 1 & 0 \\ 0 & 0 \end{bmatrix}$$

$$\hat{\mathbf{I}}_{\beta} = \hat{\mathbf{I}}_z - \frac{1}{2}\hat{\mathbf{E}} = \frac{1}{2}\begin{bmatrix} 0 & 0 \\ 0 & 1 \end{bmatrix}$$

Together, these form a new basis set of operators called the *single transition* operators.

The two new operators, $\hat{\mathbf{I}}_{\alpha}$ and $\hat{\mathbf{I}}_{\beta}$, are called *projection* operators and, as in Chapter 3 where they were introduced, they may be represented as an outer product of vectors:

$$\hat{\mathbf{I}}_{\alpha} = |\alpha\rangle\langle\alpha|$$

$$\hat{\mathbf{I}}_{\beta} = |\beta\rangle\langle\beta|$$

The ladder operators may also be represented in this way:

$$\hat{\mathbf{I}}_{+} = |\alpha\rangle\langle\beta|$$

$$\hat{\mathbf{I}}_{-} = |\beta\rangle\langle\alpha|$$

The effect of these operators on the α and β states is:

$$\hat{\mathbf{I}}_{\alpha}|\alpha\rangle = |\alpha\rangle\langle\alpha|\alpha\rangle = |\alpha\rangle$$

$$\hat{\mathbf{I}}_{\alpha}|\beta\rangle = |\alpha\rangle\langle\alpha|\beta\rangle = 0$$

$$\hat{\mathbf{I}}_{\beta}|\beta\rangle = |\beta\rangle\langle\beta|\beta\rangle = |\beta\rangle$$

$$\hat{\mathbf{I}}_{\beta}|\alpha\rangle = |\beta\rangle\langle\beta|\alpha\rangle = 0$$

$$\hat{\mathbf{I}}_+ \,|\,\alpha\rangle = |\,\alpha\rangle\langle\beta\,|\,\alpha\rangle = 0$$

$$\hat{\mathbf{I}}_+ \,|\,\beta\rangle = |\,\alpha\rangle\langle\beta\,|\,\beta\rangle = |\,\alpha\rangle$$

$$\hat{\mathbf{I}}_- \,|\,\beta\rangle = |\,\beta\rangle\langle\alpha\,|\,\beta\rangle = 0$$

$$\hat{\mathbf{I}}_- \,|\,\alpha\rangle = |\,\beta\rangle\langle\alpha\,|\,\alpha\rangle = |\,\beta\rangle$$

These are all one-spin operators. What about two-spin single transition operators. We construct these in a similar fashion to the Cartesian operators in eqn (18.12):

$$\hat{\mathbf{B}}_S = \prod_{k=1}^{N} (\hat{\mathbf{I}}_{k\nu})$$

where now ν refers to $+$, $-$, α or β and k is the index number of the nucleus. For two spins there are, again, 16 possible combinations:

1. $\hat{\mathbf{I}}_\alpha\hat{\mathbf{S}}_\alpha,\ \hat{\mathbf{I}}_\alpha\hat{\mathbf{S}}_\beta,\ \hat{\mathbf{I}}_\beta\hat{\mathbf{S}}_\alpha,\ \hat{\mathbf{I}}_\beta\hat{\mathbf{S}}_\beta$
2. $\hat{\mathbf{I}}_\alpha\hat{\mathbf{S}}_+,\ \hat{\mathbf{I}}_\alpha\hat{\mathbf{S}}_-,\ \hat{\mathbf{I}}_\beta\hat{\mathbf{S}}_+,\ \hat{\mathbf{I}}_\beta\hat{\mathbf{S}}_-$

   $\hat{\mathbf{I}}_+\hat{\mathbf{S}}_\alpha,\ \hat{\mathbf{I}}_-\hat{\mathbf{S}}_\alpha,\ \hat{\mathbf{I}}_+\hat{\mathbf{S}}_\beta,\ \hat{\mathbf{I}}_-\hat{\mathbf{S}}_\beta$
3. $\hat{\mathbf{I}}_+\hat{\mathbf{S}}_+,\ \hat{\mathbf{I}}_-\hat{\mathbf{S}}_-$
4. $\hat{\mathbf{I}}_+\hat{\mathbf{S}}_-,\ \hat{\mathbf{I}}_-\hat{\mathbf{S}}_+$

Group 1 consists of the population terms that do not involve transitions, group 2 is the single quantum coherence[‡‡‡‡] transition set, group 3 is the double quantum coherence transition set and group 4 is the set of zero quantum coherence transitions.

As above, we may associate outer products of spin wave functions with each single transition operator. Group 1 is simple … just the appropriate wavefunction for the population:

$$1.\ \hat{\mathbf{I}}_\alpha\hat{\mathbf{S}}_\alpha = |\alpha\alpha\rangle\langle\alpha\alpha|,\ \hat{\mathbf{I}}_\alpha\hat{\mathbf{S}}_\beta = |\alpha\beta\rangle\langle\alpha\beta|$$
$$\hat{\mathbf{I}}_\beta\hat{\mathbf{S}}_\alpha = |\beta\alpha\rangle\langle\beta\alpha|,\ \hat{\mathbf{I}}_\beta\hat{\mathbf{S}}_\beta = |\beta\beta\rangle\langle\beta\beta| \tag{18.13}$$

[‡‡‡‡]I will look more closely at the coherence concept in Chapter 19.

Group 2, the single quantum coherence transition operators, represents transitions in which a single quantum number changes or, a single spin of the pair "flips":

$$
\begin{aligned}
&2.\ \hat{\mathbf{I}}_{\alpha}\hat{\mathbf{S}}_{+} = |\alpha\alpha\rangle\langle\alpha\beta|,\ \hat{\mathbf{I}}_{\alpha}\hat{\mathbf{S}}_{-} = |\alpha\alpha\rangle\langle\alpha\beta| \\
&\hat{\mathbf{I}}_{\beta}\hat{\mathbf{S}}_{+} = |\beta\alpha\rangle\langle\beta\beta|,\ \hat{\mathbf{I}}_{\beta}\hat{\mathbf{S}}_{-} = |\beta\beta\rangle\langle\beta\alpha| \\
&\hat{\mathbf{I}}_{+}\hat{\mathbf{S}}_{\alpha} = |\alpha\alpha\rangle\langle\beta\alpha|,\ \hat{\mathbf{I}}_{-}\hat{\mathbf{S}}_{\alpha} = |\beta\alpha\rangle\langle\alpha\alpha| \\
&\hat{\mathbf{I}}_{+}\hat{\mathbf{S}}_{\beta} = |\alpha\beta\rangle\langle\beta\beta|,\ \hat{\mathbf{I}}_{-}\hat{\mathbf{S}}_{\beta} = |\beta\beta\rangle\langle\alpha\beta|
\end{aligned}
\tag{18.14}
$$

And for the last two groups:

$$
\begin{aligned}
&3.\ \hat{\mathbf{I}}_{+}\hat{\mathbf{S}}_{+} = |\alpha\alpha\rangle\langle\beta\beta|,\ \hat{\mathbf{I}}_{-}\hat{\mathbf{S}}_{-} = |\beta\beta\rangle\langle\beta\beta| \\
&4.\ \hat{\mathbf{I}}_{+}\hat{\mathbf{S}}_{-} = |\alpha\beta\rangle\langle\beta\alpha|,\ \hat{\mathbf{I}}_{-}\hat{\mathbf{S}}_{+} = |\beta\alpha\rangle\langle\alpha\beta|
\end{aligned}
\tag{18.15}
$$

Why consider these forms of product operators? They are useful in the description of single transitions as their name implies. For a two-spin system there are 4 energy levels and we can picture the single-quantum transitions between them as in Figure 18.19.

The double and zero quantum transitions appear in Figure 18.20.

It might seem difficult to understand why it is that the transition from, say, the $|\alpha\beta\rangle$ state to the $|\alpha\alpha\rangle$ state is associated with the operator $\hat{\mathbf{I}}_{\alpha}\hat{\mathbf{S}}_{+}$ rather than $\hat{\mathbf{I}}_{\alpha}\hat{\mathbf{S}}_{-}$. We can easily show why this is so:

$$
\begin{aligned}
\hat{\mathbf{I}}_{\alpha}\hat{\mathbf{S}}_{+}|\alpha\beta\rangle &= |\alpha\alpha\rangle\langle\alpha\beta|\alpha\beta\rangle \\
&= |\alpha\alpha\rangle
\end{aligned}
$$

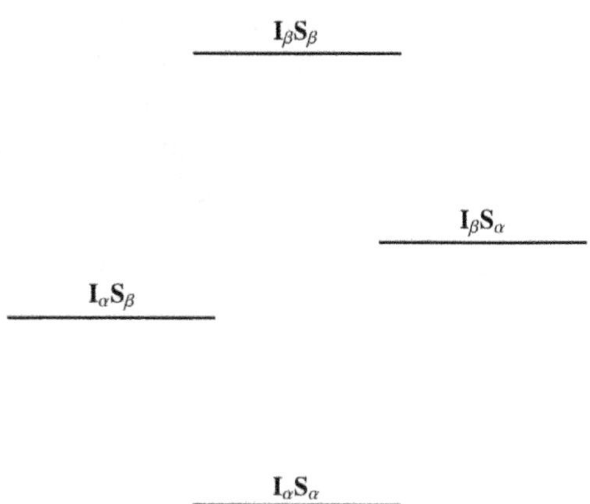

Figure 18.19 Two-spin energy levels and population operators.

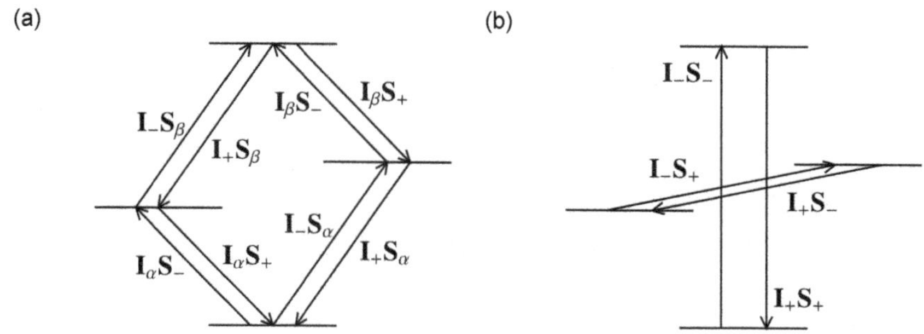

Figure 18.20 (a) Two-spin single quantum transition operators. (b) Two-spin double and zero quantum transition operators.

Thus, $\hat{\mathbf{I}}_\alpha\hat{\mathbf{S}}_+$ takes the state $|\alpha\beta\rangle$ to the state $|\alpha\alpha\rangle$ ... the $\hat{\mathbf{S}}_+$ portion of the operator is raising the angular momentum quantum number by one, from $-1/2$ to $+1/2$ and therefore the state changes from $\alpha\beta$ to $\alpha\alpha$. We will have occasion to use the single-transition operators in later chapters.

Further Reading

1. J. Cavanagh, *Protein NMR Spectroscopy: Principles and Practice*, Academic Press, San Diego, 1996.
2. R. R. Ernst, G. Bodenhausen and A. Wokaun, *Principles of Nuclear Magnetic Resonance in One and Two Dimensions*, Oxford Science Publications, Oxford, 1987.
3. R. Freeman, *A Handbook of Nuclear Magnetic Resonance*, Longman Scientific Technical, London, 1987.
4. P. Guntert, *Int. J. Quantum Chem.*, 2006, **106**, 344–350.
5. P. J. Hore, J. A. Jones and S. Wimperis, *NMR: The Toolkit*, Oxford Science Publications, London, 2009.
6. J. Keeler, *Understanding NMR Spectroscopy*, John Wiley & Sons, Chichester, Second edn., 2010.
7. M. H. Levitt, *Spin Dynamics*, John Wiley & Sons, Chichester, 2001.
8. G. D. Mateescu and A. Valeria, *2D NMR: Density Matrix and Product Operator Treatment*, Prentice Hall, Englewood Cliffs, N.J., 1993.
9. O. Sorensen, G. W. Eich, M. H. Levitt, G. Bodenhausen and R. R. Ernst, *Prog. NMR Spect.*, 1983, **16**, 163–192.
10. O. Sorensen and R. R. Ernst, *J. Magn. Reson.*, 1983, **51**, 477–489.

19 Coherence Transfer

There is no philosophy which is not founded upon knowledge of the phenomena, but to get any profit from this knowledge it is absolutely necessary to be a mathematician.

Daniel Bernoulli

19.1 Phase Cycling

The quadrature detection scheme, you will recall, is based on simultaneously sampling two signals that are 90° out of phase with each other (Figure 19.1). This method allows us to discriminate between signals that are positive and negative with respect to the transmitter frequency.

There are at least two hardware-related problems with this scheme. First, if the detectors are not perfectly matched the output of one may be slightly larger than the other. Looking at Figure 8.11 we can see visually that if the two channels are not perfectly balanced then there will be incomplete cancellation of the left-hand peaks in the figure when Figure 8.11a is added to Figure 11.8b. There will be a "symmetric" artefact in the spectrum referred to as a *quadrature image*. By symmetric, we mean that there will be a signal positioned on the other side of the centre of the spectrum at an equal distance from the centre whose magnitude will depend on the magnitude of the detector imbalance.

The second problem is that the phase difference between the two channels must be exactly 90° or, again, there will be an quadrature image in the spectrum. Essentially, one channel will "see" some of the other channel's signal that will end up being part of the transformed spectrum as a signal in the wrong part of the spectrum.

Essential Mathematics for NMR and MRI Spectroscopists
By Keith C. Brown

Published by the Royal Society of Chemistry, www.rsc.org

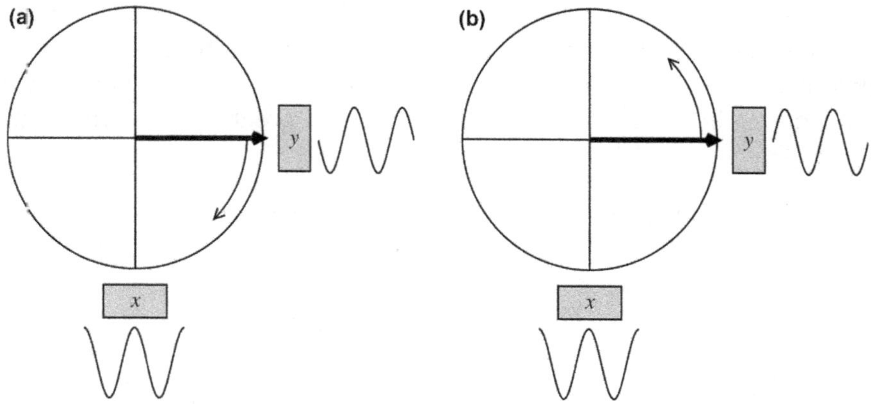

Figure 19.1 Symbolic visualisation of quadrature detectors.

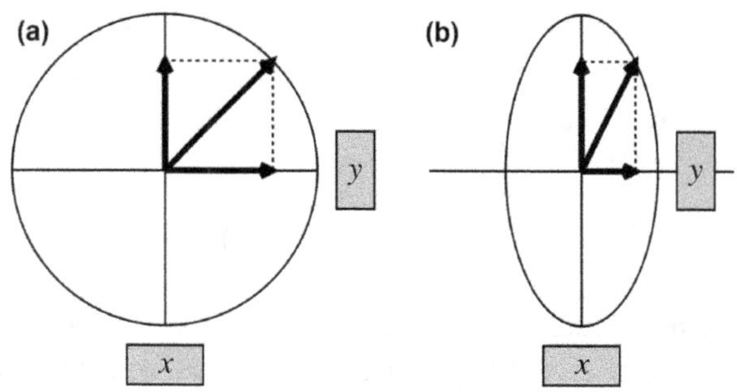

Figure 19.2 (a) Ideally balanced detectors; (b) x-detector less sensitive than y-detector.

In Figure 19.2, diagram a) shows the ideal situation in which both detectors are equally sensitive. Diagram b) shows what the x-detector "sees" if it is less sensitive than the y-detector. The x-component of the rotating vector appears smaller than it actually is. If we were to do nothing and acquire the signal just as shown we would end up with some extra signal in the wrong place in the spectrum.

We can, of course, model this mathematically. Recall eqn (8.47) and (8.48):

$$F(\omega)_x = \frac{1}{2}\left[\frac{\alpha}{\alpha^2+(\omega-\omega_x)^2}+\frac{\alpha}{\alpha^2+(\omega+\omega_x)^2}\right]+\frac{i}{2}\left[\frac{(\omega-\omega_x)}{\alpha^2+(\omega-\omega_x)^2}+\frac{(\omega+\omega_x)}{\alpha^2+(\omega+\omega_x)^2}\right]$$

$$F(\omega)_y = \frac{1}{2}\left[\frac{\alpha}{\alpha^2+(\omega-\omega_x)^2}-\frac{\alpha}{\alpha^2+(\omega+\omega_x)^2}\right]+\frac{i}{2}\left[\frac{(\omega-\omega_x)}{\alpha^2+(\omega-\omega_x)^2}-\frac{(\omega+\omega_x)}{\alpha^2+(\omega+\omega_x)^2}\right]$$

These represent the (balanced) signals from our two channels, x and y. We saw in Chapter 8 that these lead to a single signal in the spectrum. Now let us introduce an imbalance into the x-channel by multiplying by a factor, ϵ which may be either greater or less than one:

$$F(\omega)_x = \frac{\epsilon}{2}\left[\frac{\alpha}{\alpha^2 + (\omega - \omega_x)^2} + \frac{\alpha}{\alpha^2 + (\omega + \omega_x)^2}\right] + i\frac{\epsilon}{2}\left[\frac{(\omega - \omega_x)}{\alpha^2 + (\omega - \omega_x)^2} + \frac{(\omega + \omega_x)}{\alpha^2 + (\omega + \omega_x)^2}\right]$$

As in Chapter 8, we add the real parts of the x and y channels.

$$F(\omega) = \frac{1}{2}\left[\frac{\alpha + \epsilon\alpha}{\alpha^2 + (\omega - \omega_x)^2} + \frac{\alpha - \epsilon\alpha}{\alpha^2 + (\omega + \omega_x)^2}\right]$$

If ϵ is equal to one then this collapses to

$$F(\omega) = \frac{\alpha}{\alpha^2 + (\omega - \omega_x)^2}$$

which is just the spectrum for the perfectly balanced detectors. If we let $\epsilon = 0.8$ a plot of the results of the addition is shown in Figure 19.3.[†]

The quadrature image can be suppressed by cycling the phase of the transmitter. At each pulse the phase is advanced by 90°.

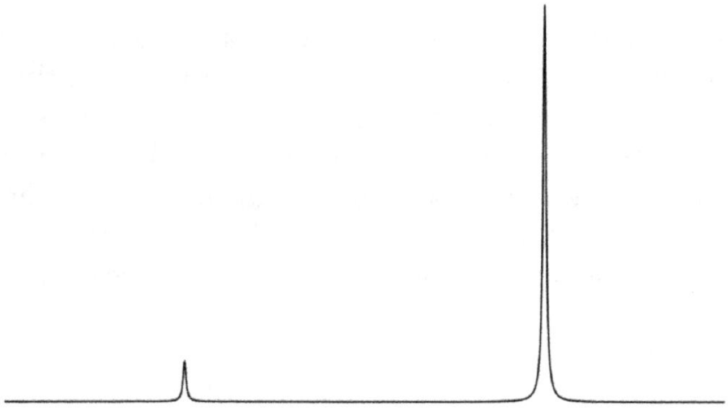

Figure 19.3 Quadrature image of main peak in the spectrum.

[†]This gives us a *really* big quad image. It is only so large here in order to see the effect clearly in the figure. If your spectrometer does this it is time to have it serviced … or replaced.

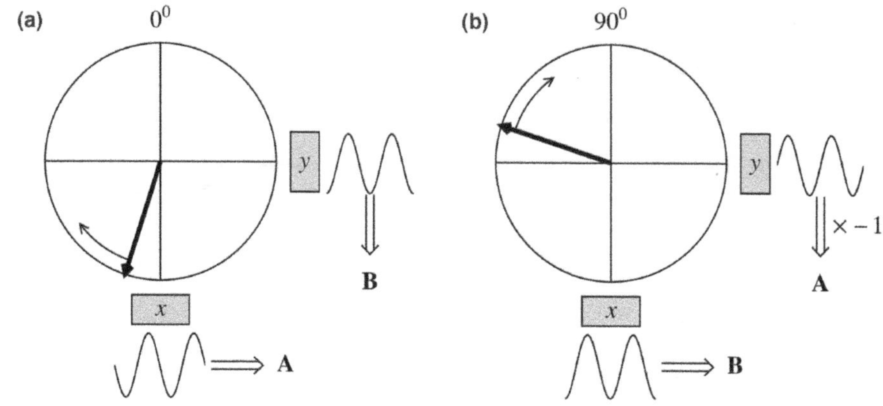

Figure 19.4 Transmitter phase advanced by 90°.

In Figure 19.4 we see that as the phase of the transmitter pulse is advanced 90° the character of the detected signals change. For the 0° pulse the x-detector delivers a sine signal and the y-detector a cosine signal. The x-detector signal is stored in memory location A and the y-detector signal in memory location B. Advancing the transmitter phase at the next pulse, the x-detector now sees a cosine signal and the y-detector an inverted sine signal. The x-detector signal now goes into memory location B, while the inverse of the y-detector signal goes to memory location A. The value of this scheme lies in the fact that after this transmitter phase shift, the signals in memory locations A and B each have a component from the x-detector and the y-detector. If one of the detectors is weaker than the other then this weakness is shared equally between the two memory locations.

Generally, the transmitter phase is advanced through two more 90° steps for a complete, four-step phase cycling scheme, referred to as the CYCLOPS phase cycle (Figure 19.5).

Another problem that occurs is one in which the pulses used in a pulse program are imperfect, especially 180° pulses. By imperfect we mean that the applied pulse does not rotate the magnetisation to the desired angle, either because it was improperly calibrated or because of $\vec{\mathbf{B}}_1$ field inhomogeneity. We can see what happens in this situation using the homonuclear spin echo pulse sequence (Figure 19.6).

As we saw in Chapter 18, for a two-spin system this pulse-sequence results in magnetisation at $\sigma(4)$ that is independent of the chemical-shift of the magnetisation. That analysis assumed a perfect 180° pulse and magnetic field homogeneity.

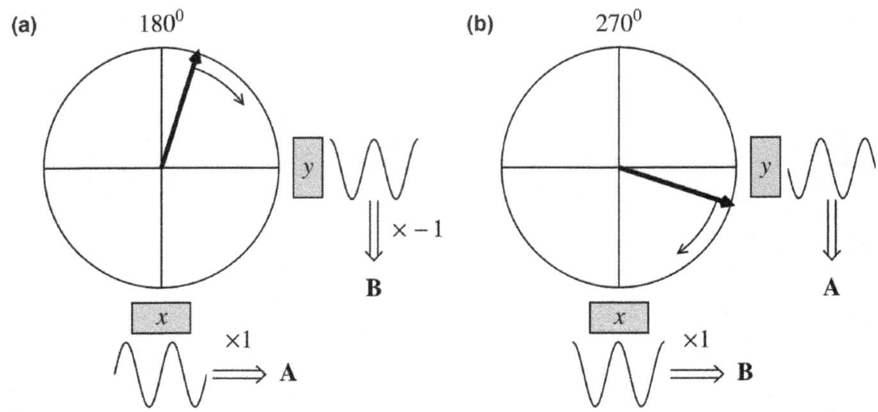

Figure 19.5 Transmitter phase advanced by 180° and 270°.

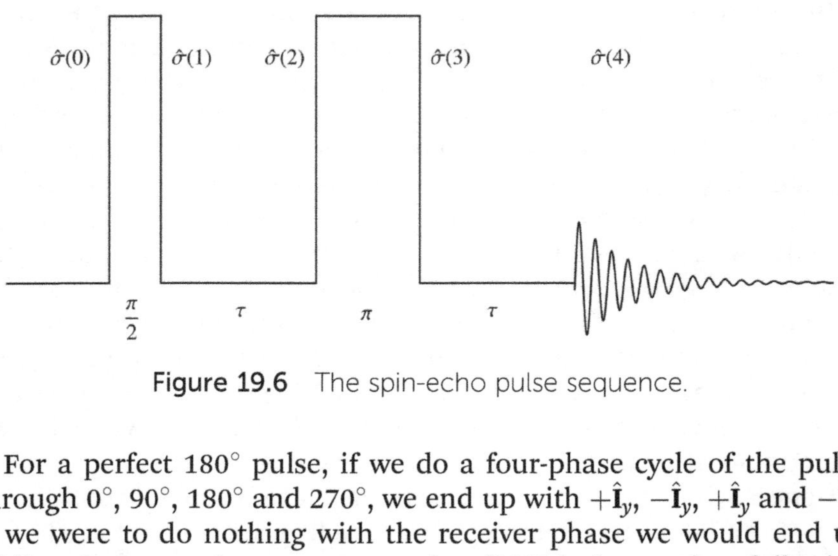

Figure 19.6 The spin-echo pulse sequence.

For a perfect 180° pulse, if we do a four-phase cycle of the pulse through 0°, 90°, 180° and 270°, we end up with $+\hat{\mathbf{I}}_y$, $-\hat{\mathbf{I}}_y$, $+\hat{\mathbf{I}}_y$ and $-\hat{\mathbf{I}}_y$. If we were to do nothing with the receiver phase we would end up adding these together to get zero signal! With the receiver following the phases of the pulses, this will result in the receiver "seeing" $+\hat{\mathbf{I}}_y$, $-\hat{\mathbf{I}}_x$, $-\hat{\mathbf{I}}_y$ and $+\hat{\mathbf{I}}_x$. Again, addition of these will result in zero signal. What are we to do?

Using a pulse phase cycle of 0°, 90°, 180° and 270°, and a receiver phase cycle of 0°, 180°, 0°, 180° will give us $+\hat{\mathbf{I}}_y$, $+\hat{\mathbf{I}}_y$, $+\hat{\mathbf{I}}_y$ and $+\hat{\mathbf{I}}_y$ that will add to give a non-zero signal. Furthermore, each cycle of the receiver sees exactly what the others see. This receiver phase cycle has been in use for many years and is referred to as EXORCYCLE. Its use is not limited to simply following the generated signals so that they sum correctly. It is also useful in suppressing the artefacts that appear as a result of imperfect 180° pulses. What if the 180° pulse is off by a small amount, say, β°? The "180°" pulse is now $180° \pm \beta°$. The first pass through the spin-echo pulse sequence with $(\pi/2)_x$ and π_x pulses and a receiver phase of zero degrees gives us a product operator analysis result at $\sigma(4)$ of (simplified!):‡

$$+\hat{\mathbf{I}}_y \cos^2(\omega_I\tau)\cos^2(\pi J\tau)\cos(\beta) - 2\hat{\mathbf{I}}_x\hat{\mathbf{S}}_z \cos(2\omega_I\tau)\cos(\pi J\tau)\sin(\pi J\tau)\cos(\beta)$$

$$-\hat{\mathbf{I}}_x \cos(\omega_I\tau)\cos(2\pi J\tau)\sin(\omega_I\tau)\cos(\beta) - 2\hat{\mathbf{I}}_y\hat{\mathbf{S}}_z\cos(\omega_I\tau)\sin(\omega_I\tau)\sin(2\pi J\tau)\cos(\beta)$$

$$+2\hat{\mathbf{I}}_x\hat{\mathbf{S}}_z \cos^2(\omega_I\tau)\sin^2(\pi J\tau) + 2\hat{\mathbf{I}}_y\hat{\mathbf{S}}_z \cos(\omega_I\tau)\sin(2\pi J\tau)\sin(\omega_I\tau)$$

$$+\hat{\mathbf{I}}_x \sin(\omega_I\tau)\cos(2\pi J\tau)\cos(\omega_I\tau) + \hat{\mathbf{I}}_y \sin^2(\omega_I\tau)\cos^2(\omega_I\tau)$$

$$+\hat{\mathbf{I}}_y \sin^2(\omega_I\tau)\sin^2(\pi J\tau)\cos(\beta)$$

where we have used eqn (A.9) and (A.10) to simplify $\cos(\pi \pm \beta)$ to $-\cos(\beta)$. It turns out that the $\sin(\pi \pm \beta)$ terms are all associated with unobservable magnetisation and are left out of the above expression. We can clearly see that in this case, instead of a nice simple result, $+\hat{\mathbf{I}}_y$, we have a state that is full of operators that will give all sorts of chemical shift artefacts in the final spectrum. We really do not want this. The EXORCYCLE phase cycling scheme will take care of this problem for us.

If we do three more acquisitions with the same $(\pi/2)_x$-pulse but advancing π-pulse phases, y, $-x$, $-y$, the product operator analyses add up to give $+\hat{\mathbf{I}}_y + \hat{\mathbf{I}}_y \cos(\beta) + +\hat{\mathbf{I}}_y + \hat{\mathbf{I}}_y \cos(\beta)$.§ This is almost what we would get if the π-pulse were perfect with a slight attenuation from the $\cos(\beta)$ term. Furthermore, and more importantly, all of the artefact terms have been removed.

19.2 Coherence Order

Recall from Chapter 17 that the form of the density matrix is:

$$\hat{\sigma} = \begin{bmatrix} c_\alpha c_\alpha^* & c_\alpha c_\beta^* \\ c_\beta c_\alpha^* & c_\beta c_\beta^* \end{bmatrix} \quad \text{or} \quad \begin{bmatrix} \sigma_{\alpha\alpha} & \sigma_{\alpha\beta} \\ \sigma_{\beta\alpha} & \sigma_{\beta\beta} \end{bmatrix} \quad \text{or} \quad \begin{bmatrix} \langle\alpha|\hat{\sigma}|\alpha\rangle & \langle\alpha|\hat{\sigma}|\beta\rangle \\ \langle\beta|\hat{\sigma}|\alpha\rangle & \langle\beta|\hat{\sigma}|\beta\rangle \end{bmatrix}$$

‡You can either go through this by hand or use the wxProdOp product operator calculator at https://sourceforge.net/projects/wxprodop.

§I will leave this exercise to you, the reader. One word of caution … be very careful with signs.

(see eqn (17.4)). One may also write this in terms of the Cartesian spin operators (eqn (17.6)):

$$\hat{\sigma} = a_E\hat{\mathbf{E}} + a_x\hat{\mathbf{I}}_x + a_y\hat{\mathbf{I}}_y + a_z\hat{\mathbf{I}}_z$$

$$= \begin{bmatrix} a_E & 0 \\ 0 & a_E \end{bmatrix} + \begin{bmatrix} 0 & a_x \\ a_x & 0 \end{bmatrix} + \begin{bmatrix} 0 & -ia_y \\ ia_y & 0 \end{bmatrix} + \begin{bmatrix} a_z & 0 \\ 0 & -a_z \end{bmatrix}$$

$$= \begin{bmatrix} a_E + a_z & a_x - ia_y \\ a_x + ia_y & a_E - a_z \end{bmatrix}$$

Yet another way is to go back to the definition of the density matrix and use it (for a one-spin system here) emphasizing the basis-set projection operators of Chapter 17:

$$|\Psi\rangle\langle\Psi| = \begin{bmatrix} c_\alpha|\alpha\rangle \\ c_\beta|\beta\rangle \end{bmatrix} \begin{bmatrix} c_\alpha^*\langle\alpha| & c_\beta^*\langle\beta| \end{bmatrix}$$

$$= c_\alpha c_\alpha^*|\alpha\rangle\langle\alpha| + c_\alpha c_\beta^*|\alpha\rangle\langle\beta| + c_\beta c_\alpha^*|\beta\rangle\langle\alpha| + c_\beta c_\beta^*|\beta\rangle\langle\beta|$$

$$= c_\alpha c_\alpha^* \begin{bmatrix} 1 & 0 \\ 0 & 0 \end{bmatrix} + c_\alpha c_\beta^* \begin{bmatrix} 0 & 1 \\ 0 & 0 \end{bmatrix} + c_\beta c_\alpha^* \begin{bmatrix} 0 & 0 \\ 1 & 0 \end{bmatrix} + c_\beta c_\beta^* \begin{bmatrix} 0 & 0 \\ 0 & 1 \end{bmatrix}$$

$$= \begin{bmatrix} c_\alpha c_\alpha^* & c_\alpha c_\beta^* \\ c_\beta c_\alpha^* & c_\beta c_\beta^* \end{bmatrix}$$

A coherence or coherent superposition of states is defined in terms of the density operator as:

$$\hat{\sigma}_{st} = \langle s|\hat{\sigma}|t\rangle \ (s \neq t)$$

These are just the individual off-diagonal components of the density matrix, as was noted in Chapter 17. Recall also, from Chapter 12 that the operator $\hat{\mathbf{I}}_z$ produces the quantum number associated with the state that it is operating on:

$$\hat{\mathbf{I}}_z|s\rangle = m_s|s\rangle$$

$$\hat{\mathbf{I}}_z|t\rangle = m_t|t\rangle$$

or

$$\hat{\mathbf{I}}_z|\alpha\rangle = \frac{1}{2}|\alpha\rangle$$

$$\hat{\mathbf{I}}_z|\beta\rangle = -\frac{1}{2}|\beta\rangle$$

and that the raising and lowering operators take the current state up or down in characteristic quantum number, respectively:

$$\text{if } m_t = m_s + 1$$

$$\text{then}$$

$$\hat{\mathbf{I}}_+ |s\rangle = |t\rangle$$

$$\hat{\mathbf{I}}_- |t\rangle = |s\rangle$$

The *coherence order* of a coherence, p, is then defined as the *difference* in quantum numbers between states s and t:

$$p_{st} \equiv m_s - m_t$$

Looking closely at the density matrix from above, in the form:

$$\hat{\sigma} = \begin{bmatrix} \langle\alpha|\hat{\sigma}|\alpha\rangle & \langle\alpha|\hat{\sigma}|\beta\rangle \\ \langle\beta|\hat{\sigma}|\alpha\rangle & \langle\beta|\hat{\sigma}|\beta\rangle \end{bmatrix}$$

for the $\langle\alpha|\hat{\sigma}|\alpha\rangle$ term, each state has a characteristic quantum number of 1/2 and we find that the coherence order for this term is therefore zero. Similarly, $\langle\beta|\hat{\sigma}|\beta\rangle$ has a coherence order of $p = 0$. For the off-diagonal terms, $\langle\alpha|\hat{\sigma}|\beta\rangle$ has a coherence order of $p = +1$ and $\langle\beta|\hat{\sigma}|\alpha\rangle$ a coherence order of $p = -1$.

The coherence order of a spin system is changed by the application of pulses. We can see this if we look at a simple pulse experiment:

$$\hat{\mathbf{I}}_z\left(\frac{1}{2}\begin{bmatrix} 1 & 0 \\ 0 & -1 \end{bmatrix}\right) \xrightarrow{\pi/2\hat{\mathbf{I}}_x} \hat{\mathbf{I}}_z \cos\left(\frac{\pi}{2}\right) - \hat{\mathbf{I}}_y \sin\left(\frac{\pi}{2}\right)$$

$$\Downarrow$$

$$= -\hat{\mathbf{I}}_y = \frac{1}{2}\begin{bmatrix} 0 & -i \\ i & 0 \end{bmatrix}$$

A little reflection will reveal that in applying a $\pi/2$-pulse we have equalised the populations of the α and β states of our single spin-1/2. We have also created off-diagonals in the density matrix which represent the coherences. The $(\pi/2)_x$-pulse converts $\hat{\mathbf{I}}_z$ to $\hat{\mathbf{I}}_y$, *i.e.* From a state with coherence order zero to one with ± 1 coherence order. For a single spin-1/2 this is as far as it can go. Note that the creation of coherence does so such that both $+$ and $-$ coherence order terms are created simultaneously. Similarly, $\hat{\mathbf{I}}_x$ consists of a mixture $p = \pm 1$ coherences.

For a two-spin-1/2 system, say, spins I and S:

$$\hat{\sigma} = \begin{bmatrix} \langle \alpha_I\alpha_S|\hat{\sigma}|\alpha_I\alpha_S\rangle & \langle \alpha_I\alpha_S|\hat{\sigma}|\alpha_I\beta_S\rangle & \langle \alpha_I\alpha_S|\hat{\sigma}|\beta_I\alpha_S\rangle & \langle \alpha_I\alpha_S|\hat{\sigma}|\beta_I\beta_S\rangle \\ \langle \alpha_I\beta_S|\hat{\sigma}|\alpha_I\alpha_S\rangle & \langle \alpha_I\beta_S|\hat{\sigma}|\alpha_I\beta_S\rangle & \langle \alpha_I\beta_S|\hat{\sigma}|\beta_I\alpha_S\rangle & \langle \alpha_I\beta_S|\hat{\sigma}|\beta_I\beta_S\rangle \\ \langle \beta_I\alpha_S|\hat{\sigma}|\alpha_I\alpha_S\rangle & \langle \beta_I\alpha_S|\hat{\sigma}|\alpha_I\beta_S\rangle & \langle \beta_I\alpha_S|\hat{\sigma}|\beta_I\alpha_S\rangle & \langle \beta_I\alpha_S|\hat{\sigma}|\beta_I\beta_S\rangle \\ \langle \beta_I\beta_S|\hat{\sigma}|\alpha_I\alpha_S\rangle & \langle \beta_I\beta_S|\hat{\sigma}|\alpha_I\beta_S\rangle & \langle \beta_I\beta_S|\hat{\sigma}|\beta_I\alpha_S\rangle & \langle \beta_I\beta_S|\hat{\sigma}|\beta_I\beta_S\rangle \end{bmatrix}$$

the coherence orders in the density matrix are:

$$\begin{bmatrix} p=0 & p=1 & p=1 & p=2 \\ p=-1 & p=0 & p=0 & p=1 \\ p=-1 & p=0 & p=0 & p=1 \\ p=-2 & p=-1 & p=-1 & p=0 \end{bmatrix} \tag{19.1}$$

where we have used, for example, $\langle \alpha_I\alpha_S|\hat{\sigma}|\alpha_I\beta_S\rangle$ to determine that the coherence order is $p = +1$ by comparing the differences between the characteristic quantum numbers from the bra and ket states as per the definition of coherence order above. The overall coherence order is the sum of the differences between the I spin and the S spin. We note that for the two-spin system there are zero-order quantum coherences, ZQC, single-order quantum coherences, SQC, and double-order quantum coherences, DQC.

Our density matrix is often represented in the literature in terms of coherence orders as:

$$\hat{\sigma} = \sum_{p=-2L}^{p=2L} \hat{\sigma}^p \tag{19.2}$$

where L represents the sum of the spin quantum numbers of the nuclei of interest. So, for two spins-1/2, L would be 1. Thus, $\hat{\sigma}^{-1}$ means that portion of the overall density matrix whose coherence order values are -1. So for our two-spin system:

$$\hat{\sigma}^{-1} = \begin{bmatrix} 0 & 0 & 0 & 0 \\ \langle \alpha_I\beta_S|\hat{\sigma}|\alpha_I\alpha_S\rangle & 0 & 0 & 0 \\ \langle \beta_I\alpha_S|\hat{\sigma}|\alpha_I\alpha_S\rangle & 0 & 0 & 0 \\ 0 & \langle \beta_I\beta_S|\hat{\sigma}|\alpha_I\beta_S\rangle & \langle \beta_I\beta_S|\hat{\sigma}|\beta_I\alpha_S\rangle & 0 \end{bmatrix}$$

where only the matrix components that represent a coherence order change of -1 are included.

A look at a few of our two-spin states derived in Section 17.4 is revealing:

$$\hat{\mathbf{I}}_x = \frac{1}{4}\begin{bmatrix} 0 & 0 & 1 & 0 \\ 0 & 0 & 0 & 1 \\ 1 & 0 & 0 & 0 \\ 0 & 1 & 0 & 0 \end{bmatrix}$$

We see that, as is the case for a single spin, $\hat{\mathbf{I}}_x$ consists of $p = \pm 1$ coherences as does $\hat{\mathbf{S}}_x$ (referring to eqn (19.1)):

$$\hat{\mathbf{S}}_x = \frac{1}{4}\begin{bmatrix} 0 & 1 & 0 & 0 \\ 1 & 0 & 0 & 0 \\ 0 & 0 & 0 & 1 \\ 0 & 0 & 1 & 0 \end{bmatrix}$$

since they each have entries in only $p = \pm 1$ areas of the matrices. In terms of eqn (19.2), $\hat{\mathbf{I}}_x$ would be:

$$\hat{\sigma} = \frac{1}{4}\begin{bmatrix} 0 & 0 & 1 & 0 \\ 0 & 0 & 0 & 1 \\ 0 & 0 & 0 & 0 \\ 0 & 0 & 0 & 0 \end{bmatrix} + \frac{1}{4}\begin{bmatrix} 0 & 0 & 0 & 0 \\ 0 & 0 & 0 & 0 \\ 1 & 0 & 0 & 0 \\ 0 & 1 & 0 & 0 \end{bmatrix}$$

The first matrix contains the contributions from $p = 1$ coherence and the second contains the contributions from $p = -1$.

One of the more interesting coherences is $2\hat{\mathbf{I}}_x\hat{\mathbf{S}}_x$:

$$2\hat{\mathbf{I}}_x\hat{\mathbf{S}}_x = \frac{1}{4}\begin{bmatrix} 0 & 0 & 0 & 1 \\ 0 & 0 & 1 & 0 \\ 0 & 1 & 0 & 0 \\ 1 & 0 & 0 & 0 \end{bmatrix}$$

which we see consists of a mixture of $p = \pm 2$ (DQC) and $p = 0$ (ZQC) coherences according to our coherence order matrix map, eqn (19.1).

The ladder operators from Chapter 15 are unique in that they consist not of mixtures of coherence orders but only *one* coherence order each. We can see this by using their definitions:

$$\hat{\mathbf{I}}_+ = \hat{\mathbf{I}}_x + i\hat{\mathbf{I}}_y$$

$$\hat{\mathbf{I}}_- = \hat{\mathbf{I}}_x - i\hat{\mathbf{I}}_y$$

and writing out their (one-spin) matrices:

$$\hat{\mathbf{I}}_{+}=\frac{1}{2}\begin{bmatrix}0&1\\1&0\end{bmatrix}+\frac{i}{2}\begin{bmatrix}0&-i\\i&0\end{bmatrix}=\begin{bmatrix}0&1\\0&0\end{bmatrix}=|\alpha\rangle\langle\beta|$$

$$\hat{\mathbf{I}}_{-}=\frac{1}{2}\begin{bmatrix}0&1\\1&0\end{bmatrix}-\frac{i}{2}\begin{bmatrix}0&-i\\i&0\end{bmatrix}=\begin{bmatrix}0&0\\1&0\end{bmatrix}=|\beta\rangle\langle\alpha|$$

We see here that $\hat{\mathbf{I}}_{+}$ consists only of coherence order $p=+1$ and $\hat{\mathbf{I}}_{-}$ consists of only $p=-1$ coherence order.

The behaviour of coherences under the influence of phase shifts in the applied pulses is of considerable interest to us as it will allow us to design effective phase cycling procedures. We have been representing the phases of pulses using subscripts x, y, $-x$ and $-y$ but we could just as well say 0°, 90°, 180° and 270°, by (arbitrarily) taking an x-phased pulse to be at 0°. We could apply, for example, a $(\pi/2)_{90}$-pulse to generate $\hat{\mathbf{I}}_x$ magnetisation or we could apply a $\hat{\mathbf{R}}(90)$ rotation operator to a $(\pi/2)_0$-pulse to generate a $(\pi/2)_{90}$ by a 90° rotation about the z-axis which then produces $\hat{\mathbf{I}}_x$ magnetisation:

$$\begin{aligned}\hat{\mathbf{U}}_{\phi}&=\exp(-i\phi\hat{\mathbf{I}}_z)\hat{\mathbf{U}}_0\exp(i\phi\hat{\mathbf{I}}_z)\\ \hat{\sigma}(\pm1)&=\hat{\mathbf{U}}_{\phi}\hat{\sigma}(0)\hat{\mathbf{U}}_{\phi}^{-1}\end{aligned}\qquad(19.3)$$

where U refers to a rotation operator or *propagator* and ±1 refers the coherence orders +1 and −1 generated by a 90° pulse, in this case from equilibrium magnetisation. It is as if we have issued a "z-axis" pulse in conjunction with a $(\pi/2)_0$-pulse. To be clear, we are using $\hat{\mathbf{I}}_z$ here simply as a rotation operator, *not* as a chemical-shift operator.¶ We could generalise this to include any coherence order changes:

$$\hat{\sigma}(l)=\hat{\mathbf{U}}_{\phi}\ \hat{\sigma}(k)\ \hat{\mathbf{U}}_{\phi}^{-1}$$

where our phase adjusted operator, $\hat{\mathbf{U}}_{\phi}$, changes k-order coherence into l-order coherence. Now consider the effect of our rotation operator on a single component of the density matrix, say, $|\alpha\rangle\langle\beta|$:

$$\begin{aligned}&\exp(-i\phi\hat{\mathbf{I}}_z)|\alpha\rangle\langle\beta|\exp(i\phi\hat{\mathbf{I}}_z)\\ &=\exp\left(-i\phi\frac{1}{2}\right)|\alpha\rangle\langle\beta|\exp\left(i\phi\frac{-1}{2}\right)\\ &=\exp(-i\phi1)|\alpha\rangle\langle\beta|\end{aligned}$$

¶Much of the current literature seems to gloss over this little detail so that it appears that a pulse is issued followed by a chemical-shift operation … very confusing at first.

where we have used eqn (3.62) to obtain the second line. Again we can generalise this to include any coherence order:

$$\begin{aligned}
&\exp(-i\phi\hat{\mathbf{I}}_z)|s\rangle\langle t|\exp(i\phi\hat{\mathbf{I}}_z)\\
&\quad=\exp(-i\phi m_s)|s\rangle\langle t|\exp(i\phi m_t)\\
&\quad=\exp(-i\phi(m_s-m_t))|s\rangle\langle t|\\
&\quad=\exp(-i\phi p_{st})|s\rangle\langle t|
\end{aligned}\tag{19.4}$$

We take this a bit further to come up with an important result in a step-by-step fashion. First, we issue a pulse of zero phase:

$$\hat{\mathbf{U}}_0\hat{\sigma}(k)\hat{\mathbf{U}}_0^{-1}=\hat{\sigma}(l)$$

Then we use the idea introduced in eqn (19.3) that a ϕ phase pulse is equivalent to a zero-phase pulse followed by a ϕ radian rotation about the z-axis:

$$\hat{\mathbf{U}}_\phi\hat{\sigma}(k)\hat{\mathbf{U}}_\phi^{-1}=\{\exp(-i\phi\hat{\mathbf{I}}_z)\hat{\mathbf{U}}_0\exp(i\phi\hat{\mathbf{I}}_z)\}\hat{\sigma}(k)\{\exp(-i\phi\hat{\mathbf{I}}_z)\hat{\mathbf{U}}_0\exp(i\phi\hat{\mathbf{I}}_z)\}$$

Here we have replaced the ϕ phase propagator with the zero-phase propagator multiplied by the appropriate exponentials, as in eqn (19.3). We note that the central three terms in this expression can be replaced using eqn (19.4):

$$\hat{\mathbf{U}}_\phi\hat{\sigma}(k)\hat{\mathbf{U}}_\phi^{-1}=\exp(-i\phi\hat{\mathbf{I}}_z)\hat{\mathbf{U}}_0\underbrace{\exp(i\phi\hat{\mathbf{I}}_z)\hat{\sigma}(k)\exp(-i\phi\hat{\mathbf{I}}_z)}_{=\exp(ip_k\phi)\hat{\sigma}(k)}\hat{\mathbf{U}}_0\exp(i\phi\hat{\mathbf{I}}_z)$$

$$\Downarrow$$

$$=\exp(ip_k\phi)\exp(-i\phi\hat{\mathbf{I}}_z)\hat{\mathbf{U}}_0\hat{\sigma}(k)\hat{\mathbf{U}}_0\exp(i\phi\hat{\mathbf{I}}_z)$$

The propagator and density matrix terms in this expression represent the zero-phase pulse on $\hat{\sigma}(k)$ to produce $\hat{\sigma}(l)$:

$$\exp(ip_k\phi)\exp(-i\phi\hat{\mathbf{I}}_z)\underbrace{\hat{\mathbf{U}}_0\hat{\sigma}(k)\hat{\mathbf{U}}_0}_{=\hat{\sigma}(l)}\exp(i\phi\hat{\mathbf{I}}_z)$$

$$\Downarrow$$

$$=\exp(ip_k\phi)\exp(-i\phi\hat{\mathbf{I}}_z)\hat{\sigma}(l)\exp(i\phi\hat{\mathbf{I}}_z)$$

Finally, the last three terms here can be replaced using eqn (19.4) again to arrive at our ultimate goal:

$$\exp(ip_k\phi)\underbrace{\exp(-i\phi\hat{\mathbf{I}}_z)\hat{\sigma}(l)\exp(i\phi\hat{\mathbf{I}}_z)}_{=\exp(-ip_l\phi)\sigma(l)}$$
$$\Downarrow \tag{19.5}$$
$$=\exp(ip_k\phi)\exp(-ip_l\phi)\sigma(l)$$
$$=\exp(-i\phi\Delta p)\sigma(l)$$

What this series of equations is saying is that if we start with $\hat{\sigma}(k)$ and issue a pulse with phase angle ϕ we end up with $\hat{\sigma}(l)$, which is what we would have had if the pulse phase was zero, multiplied by $\exp(-i\phi\Delta p)$. More importantly, it means that different changes in coherence order will experience *different* phase changes when pulse phases change. We shall see below that this can be used to our advantage in selecting the coherences that we want and rejecting those that are of no use to us. We have had to trek through a jungle of equations to get here but final result, eqn (19.5), is wonderfully simple and, as we will see, very powerful.

Eqn (19.4) offers another method to determine the coherence order of various operators, as we did above in our matrix analysis. Let us try to ferret out the coherence order of each of these operators using eqn (19.3), starting with $\hat{\mathbf{I}}_+$. For $\hat{\mathbf{I}}_+$ we write:

$$\hat{\mathbf{I}}_+ \equiv \hat{\mathbf{I}}_x + i\hat{\mathbf{I}}_y \xrightarrow{\phi\hat{\mathbf{I}}_z} [\hat{\mathbf{I}}_x\cos(\phi) + \hat{\mathbf{I}}_y\sin(\phi)] + [i\hat{\mathbf{I}}_y\cos(\phi) - i\hat{\mathbf{I}}_x\sin(\phi)]$$
$$= (\hat{\mathbf{I}}_x + i\hat{\mathbf{I}}_y)\cos(\phi) - (\hat{\mathbf{I}}_x + i\hat{\mathbf{I}}_y)i\sin(\phi)$$
$$= (\hat{\mathbf{I}}_x + i\hat{\mathbf{I}}_y)(\cos(\phi) - i\sin(\phi))$$
$$= (\hat{\mathbf{I}}_x + i\hat{\mathbf{I}}_y)\exp(-i\phi)$$
$$= \hat{\mathbf{I}}_+\exp(-i\phi)$$

(using the imaginary number property $i^2 = -1$) and a similar calculation for $\hat{\mathbf{I}}_-$ gives us:

$$\hat{\mathbf{I}}_- \xrightarrow{\phi\hat{\mathbf{I}}_z} \hat{\mathbf{I}}_-\exp(i\phi)$$

We see from these that the coherence order for $\hat{\mathbf{I}}_+$ is $p = +1$ and for $\hat{\mathbf{I}}_-$ the coherence order is $p = -1$. Since $\hat{\mathbf{I}}_x$ and $\hat{\mathbf{I}}_y$ are equal mixtures of

$\hat{\mathbf{I}}_+$ and $\hat{\mathbf{I}}_-$ (see eqn (15.22)), we can conclude that each of these operators has a mixture of +1 and −1 coherence orders in equal amounts, as we also saw above.

What about two-spin product operators? Let us look at $2\hat{\mathbf{I}}_x\hat{\mathbf{S}}_z$, antiphase magnetisation. Using eqn (15.22) again we write:

$$2\hat{\mathbf{I}}_x\hat{\mathbf{S}}_z = (\hat{\mathbf{I}}_+ + \hat{\mathbf{I}}_-)\,\hat{\mathbf{S}}_z = \hat{\mathbf{I}}_+\hat{\mathbf{S}}_z + \hat{\mathbf{I}}_-\hat{\mathbf{S}}_z$$

The term $\hat{\mathbf{I}}_+\hat{\mathbf{S}}_z$ consists of $\hat{\mathbf{I}}_+$ with coherence order +1 and $\hat{\mathbf{S}}_z$, which has $p=0$. The sum of these two coherence orders is the overall coherence order of the operator. Thus, $\hat{\mathbf{I}}_+\hat{\mathbf{S}}_z$ has a coherence order of +1 and similarly, $\hat{\mathbf{I}}_-\hat{\mathbf{S}}_z$ has an overall coherence order of −1. From this we conclude that $2\hat{\mathbf{I}}_x\hat{\mathbf{S}}_z$, like $\hat{\mathbf{I}}_x$, has an equal mixture of +1 and −1 coherence orders.

A similar procedure for $2\hat{\mathbf{I}}_x\hat{\mathbf{S}}_x$ gives:

$$2\hat{\mathbf{I}}_x\hat{\mathbf{S}}_x = (\hat{\mathbf{I}}_+ + \hat{\mathbf{I}}_-)(\hat{\mathbf{S}}_+ + \hat{\mathbf{S}}_-) = \hat{\mathbf{I}}_+\hat{\mathbf{S}}_+ + \hat{\mathbf{I}}_-\hat{\mathbf{S}}_- + \hat{\mathbf{I}}_-\hat{\mathbf{S}}_+ + \hat{\mathbf{I}}_+\hat{\mathbf{S}}_-$$

The first term, $\hat{\mathbf{I}}_+\hat{\mathbf{S}}_+$, consists of raising operators $\hat{\mathbf{I}}_+$ and $\hat{\mathbf{S}}_+$ each of which has a coherence order of +1. The overall coherence order for this product operator is therefore +2. Similarly for the second, third and fourth product operators the overall coherence orders are −2, 0 and 0, respectively. We say that $2\hat{\mathbf{I}}_x\hat{\mathbf{S}}_x$ contains a mixture of double (DQC) and zero quantum (ZQC) coherence orders in equal amounts that is, again, what we saw in our product operator matrix analysis in Chapter 18.

It is worth noting that one cannot produce double quantum coherence or higher from single quantum coherence. This can only result from a pulse issued on a system that has antiphase magnetisation in it:

$$\hat{\mathbf{I}}_x \xrightarrow{\beta\hat{\mathbf{I}}_y} \hat{\mathbf{I}}_x\cos(\beta) - \hat{\mathbf{I}}_z\sin(\beta)$$

A pulse on SQC simply produces equilibrium magnetisation from which it was created in the first place. On the other hand, if the SQC is allowed to evolve *via* free precession into antiphase magnetisation, then a pulse can then convert the antiphase magnetisation to DQC and ZQC:

$$\hat{\mathbf{I}}_x \xrightarrow{(\pi Jt)\hat{\mathbf{I}}_z\hat{\mathbf{S}}_z} \hat{\mathbf{I}}_x\cos(\pi Jt) + \hat{\mathbf{I}}_y\hat{\mathbf{S}}_z\sin(\pi Jt)$$

$$\hat{\mathbf{I}}_y\hat{\mathbf{S}}_z\sin(\pi Jt) \xrightarrow{\pi/2\hat{\mathbf{S}}_x} -\hat{\mathbf{I}}_y\hat{\mathbf{S}}_y\sin(\pi Jt)$$

19.3 Coherence Transfer Pathways

Let us start by considering the fate of $\hat{\mathbf{I}}_x$ magnetisation immediately after it is generated by a pulse. As we have seen in Chapter 18, under the influence of $\hat{\mathbf{I}}_z$, $\hat{\mathbf{I}}_x$ evolves into a mixture of $\hat{\mathbf{I}}_x$ and $\hat{\mathbf{I}}_y$:

$$\hat{\mathbf{I}}_x \xrightarrow{\omega_0 t \hat{\mathbf{I}}_z} \hat{\mathbf{I}}_x \cos(\omega_0 t) + \hat{\mathbf{I}}_y \sin(\omega_0 t)$$

We have also seen that $\hat{\mathbf{I}}_x$ magnetisation can be thought of as a mixture of $p=+1$ and $p=-1$ coherences represented by $\hat{\mathbf{I}}_+$ and $\hat{\mathbf{I}}_-$. Substituting this into our last equation gives us:

$$\begin{aligned}
&\hat{\mathbf{I}}_x \cos(\omega_0 t) + \hat{\mathbf{I}}_y \sin(\omega_0 t) \\
&= \frac{1}{2}(\hat{\mathbf{I}}_+ + \hat{\mathbf{I}}_-)\cos(\omega_0 t) + \frac{1}{2i}(\hat{\mathbf{I}}_+ - \hat{\mathbf{I}}_-)\sin(\omega_0 t) \\
&= \frac{1}{2}[(\hat{\mathbf{I}}_+ \cos(\omega_0 t) - i\hat{\mathbf{I}}_+ \sin(\omega_0 t)) + (\hat{\mathbf{I}}_- \cos(\omega_0 t) + i\hat{\mathbf{I}}_- \sin(\omega_0 t))] \\
&= \frac{1}{2}[\hat{\mathbf{I}}_+ \exp(-i\omega_0 t) + \hat{\mathbf{I}}_- \exp(i\omega_0 t)]
\end{aligned}$$

The time derivative (see Faraday's Law, Chapter 11) of this is the signal entering the detector (Figure 19.7).

$$\begin{aligned}
S_{\mathrm{fid}} &= \frac{\mathrm{d}}{\mathrm{d}t}\left(\frac{1}{2}[\hat{\mathbf{I}}_+ \exp(-i\omega_0 t) + \hat{\mathbf{I}}_- \exp(i\omega_0 t)]\right) \\
&= \frac{i\omega_0}{2}[\hat{\mathbf{I}}_- \exp(i\omega_0 t) - \hat{\mathbf{I}}_+ \exp(-i\omega_0 t)]
\end{aligned}$$

Note that ω_0 is the laboratory-frame Larmor frequency. Also entering the detector is a reference signal, ω_{ref} that may have a phase angle added to it (which we shall leave out here since it does not affect things directly):

$$S_{\mathrm{ref}} = \cos(\omega_{\mathrm{ref}}\, t)$$

The reference signal is mixed with the raw NMR signal and simply multiplies into it to give an output signal. The quadrature detection scheme splits the raw and reference signals into two paths. In path A, the two are simply mixed and sent on. In path B, the reference signal

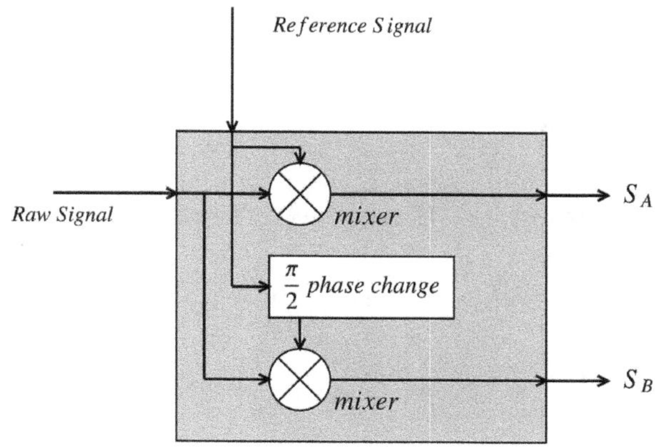

Figure 19.7 The modern quadrature detection system. The raw NMR signal from the probe is mixed with the reference frequency and the difference is output to the digitiser. Quadrature detection is achieved by mixing the raw signal with the unchanged reference frequency for channel *A* and the reference frequency advanced in phase by 90° for channel *B*.

has a phase of $\pi/2$ added to it before mixing occurs. For path *A* we find:

$$\begin{aligned} S_{\text{fid}} \times S_{\text{ref}} &\approx i[\hat{\mathbf{I}}_- \exp(i\omega_0 t) - \hat{\mathbf{I}}_+ \exp(-i\omega_0 t)]\cos(\omega_{\text{ref}} t) \\ &= \frac{i}{2}\hat{\mathbf{I}}_- \exp(i(\omega_0 + \omega_{\text{ref}})t) + \frac{i}{2}\hat{\mathbf{I}}_- \exp(i(\omega_0 - \omega_{\text{ref}})t) \\ &\quad - \frac{i}{2}\hat{\mathbf{I}}_+ \exp(-\,i(\omega_0 + \omega_{\text{ref}})t) - \frac{i}{2}\hat{\mathbf{I}}_+ \exp(-\,i(\omega_0 - \omega_{\text{ref}})t) \end{aligned}$$

where we have used identities (A.18) and (A.19). We have left $\omega_0/2$ out since this is just a multiplicative factor and not important the ultimate point that will be made below. "i", however, will be important to us.

The terms with $\omega_0 + \omega_{\text{ref}}$ are very high-frequency terms that are easily filtered out of the signal (see Chapter 22) so that after filtration the signal on channel *A* is:

$$S_A = \frac{i}{2}\hat{\mathbf{I}}_- \exp(i(\omega_0 - \omega_{\text{ref}})t) - \frac{i}{2}\hat{\mathbf{I}}_+ \exp(-\,i(\omega_0 - \omega_{\text{ref}})t)$$

Now, things begin to get interesting. In channel *B*, the reference signal is phase shifted by $\pi/2$ radians and, using a similar analysis to that for S_A, we see (again using identities (A.18) and (A.19) and using $\cos(\alpha + \pi/2) = -\sin(\alpha)$):

$$
\begin{aligned}
S_{\text{fid}} \times S_{\text{ref}} &\approx i[\hat{\mathbf{I}}_- \exp(i\omega_0 t) - \hat{\mathbf{I}}_+ \exp(-i\omega_0 t)]\cos\left(\omega_{\text{ref}} t + \frac{\pi}{2}\right) \\
&= i[\hat{\mathbf{I}}_- \exp(i\omega_0 t) - \hat{\mathbf{I}}_+ \exp(-i\omega_0 t)]\sin(\omega_{\text{ref}} t) \\
&= -\frac{1}{2}\hat{\mathbf{I}}_- \exp(i(\omega_0 + \omega_{\text{ref}})t) + \frac{1}{2}\hat{\mathbf{I}}_- \exp(i(\omega_0 - \omega_{\text{ref}})t) \\
&\quad -\frac{1}{2}\hat{\mathbf{I}}_+ \exp(-i(\omega_0 + \omega_{\text{ref}})t) + \frac{1}{2}\hat{\mathbf{I}}_+ \exp(-i(\omega_0 - \omega_{\text{ref}})t)
\end{aligned}
$$

and, after filtering out the high-frequency $\omega_0 + \omega_{\text{ref}}$ components:

$$
S_B = \frac{1}{2}\hat{\mathbf{I}}_- \exp(i(\omega_0 - \omega_{\text{ref}})t) + \frac{1}{2}\hat{\mathbf{I}}_+ \exp(-i(\omega_0 - \omega_{\text{ref}})t)
$$

The quadrature detection scheme considers the data collected from channels A and B to be the real and imaginary components respectively of a complex number. Thus, we add them together in this fashion:

$$
\begin{aligned}
S_T &= S_A + iS_B \\
&= \frac{i}{2}\hat{\mathbf{I}}_- \exp(i(\omega_0 - \omega_{\text{ref}})t) - \frac{i}{2}\hat{\mathbf{I}}_+ \exp(-i(\omega_0 - \omega_{\text{ref}})t) \\
&\quad + \frac{1}{2}\hat{\mathbf{I}}_- \exp(i(\omega_0 - \omega_{\text{ref}})t) + \frac{1}{2}\hat{\mathbf{I}}_+ \exp(-i(\omega_0 - \omega_{\text{ref}})t) \\
&= i\hat{\mathbf{I}}_- \exp(i(\omega_0 - \omega_{\text{ref}})t)
\end{aligned}
\tag{19.6}
$$

Note that the contribution of $\hat{\mathbf{I}}_+$ to the detected signal has vanished and that only $\hat{\mathbf{I}}_-$ is left. We could, if we wished, select the $\hat{\mathbf{I}}_+$ component by subtracting channel B rather than adding it or, equivalently, subtracting $\pi/2$ from the reference signal rather than adding it. The convention, however, is to select $\hat{\mathbf{I}}_-$.

The coherence transfer pathway is basically a map of the changes in coherence order during a pulse-sequence. Pulses generally create both positive and negative coherence orders but as was just noted, we are only concerned with coherences of negative order. Furthermore, only coherences of order -1 are observable. Thus, when mapping out coherence orders we discard any paths leading to anything but coherence order -1.

A very simple example is the $\pi/2$ pulse-acquire experiment.

In Figure 19.8 we have explicitly shown both the +1 and −1 coherence orders since, as you will recall, a $\pi/2$ pulse on a spin system at equilibrium creates both of these coherences. Generally, one does not do this for the reasons just given concerning what the final signal is but since we are just beginning we will leave it in this particular figure. Ordinarily, one would show only the paths that ultimately lead to coherence order −1.

A slightly more complex example, the spin-echo, shows that a π-pulse simply changes the signs of coherence orders (Figure 19.9).

You can confirm this for yourself by winding up the full-blown density matrix method and subjecting, say, an $\hat{\mathbf{I}}_x$ matrix to a π_y-pulse.

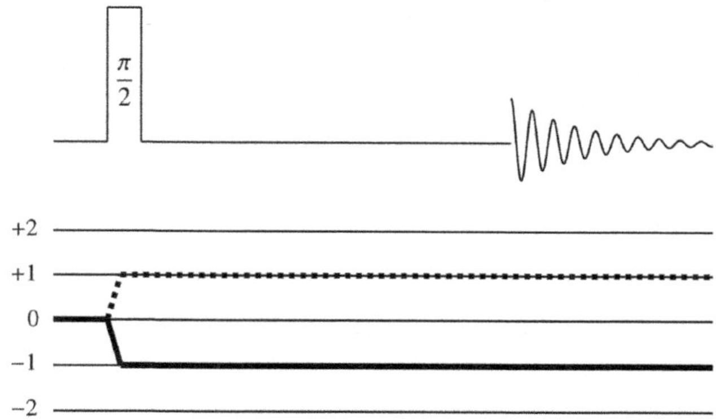

Figure 19.8 Coherence-transfer pathway diagram for a simple pulse-acquire experiment. All possible pathways are shown.

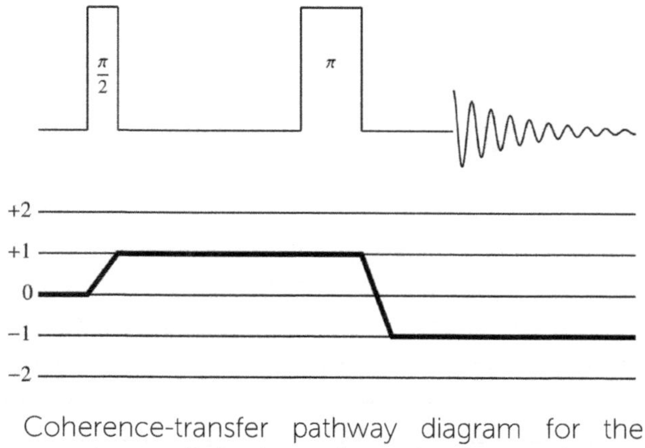

Figure 19.9 Coherence-transfer pathway diagram for the spin-echo experiment. Only the pathway leading to signal acquisition is shown.

In this case the +1 coherence order is converted to −1 coherence order *via* the π_y-pulse. Note that we are now showing *only* the path that leads to −1 coherence order, as is the convention. Note also, that the coherence order(s) do not change between pulses during free-precession evolution periods. Only the application of an rf pulse can change coherence order.

Before we proceed it is worth looking at the coherence order changes, Δp, in these coherence transfer pathway (CTP) diagrams. In Figure 19.8 Δp is simply −1 (following the solid line). In Figure 19.9 Δp is +1 after the first pulse and then −1 after the second pulse. Also, note that the overall change from start to finish in p, in Figure 19.9 is −1 … this is a characteristic of all CTP diagrams. It is Δp and eqn (19.5) that will guide us in the development of phase cycles appropriate for our needs.

Heteronuclear experiments can be represented with two CTP diagrams (Figure 19.10).

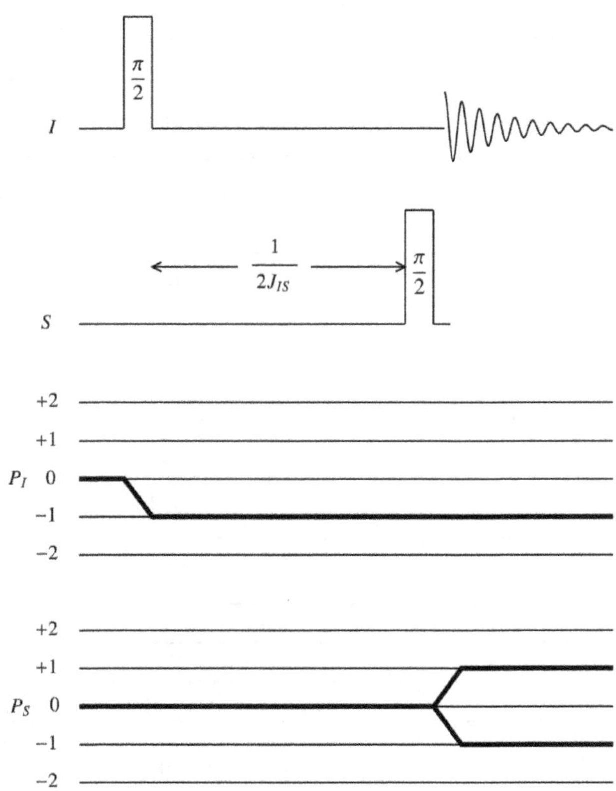

Figure 19.10 A heteronuclear CTP diagram.

This pulse-sequence is a bit unusual in that we are looking for *no* signal from the *I* spin and therefore the CTP diagram is also different from the normal versions.[||] That we will see no signal here is evident when you consider that after the *S*-spin $\pi/2$-pulse the total coherence is +2 and −2 since DQC is generated (do the product operator analysis to confirm this for yourself). Generally, however, if one wants to observe a signal for the *I*-spin then the *S*-spin diagram must end at coherence order zero. We will see this later in the analysis of the HMQC experiment.

19.4 Phase Cycling Revisited

We now have the tools needed to design phase cycles for multiple pulse-sequences so that the desired signals are selected and others are rejected. We will start with the very simplest pulse-sequence, the pulse-acquire experiment (Figure 19.11).

We have seen this already in Section 19.1 in which the pulse phase was increased by 90° for each of a cycle of four scans. With each phase change, the way the data were stored was changed as well. In that section the data were routed to different memory locations with sign changes, depending on the pulse phase. An equivalent way to do this is to follow the pulse phases with the receiver phase. We can see why this should work from our analysis of coherence order changes above.

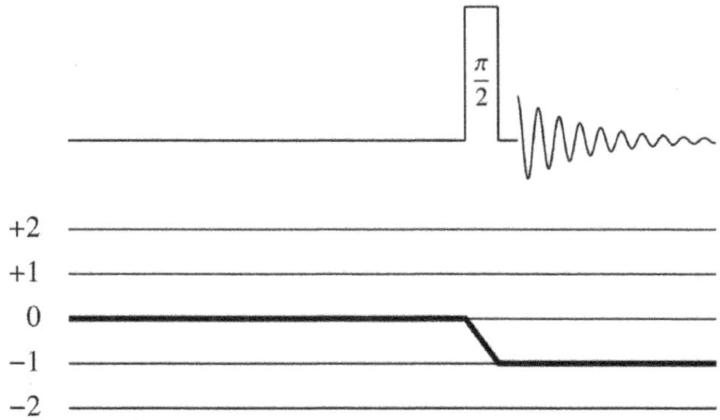

Figure 19.11 CTP diagram for the pulse-acquire experiment. Only the pathway leading to signal acquisition is shown.

[||] I use this pulse-sequence for calibration of ^{13}C and ^{15}N pulse widths by looking for zero signal in ^{1}H when the ^{13}C pulse angle is just $\pi/2$.

We start by setting up a table (Table 19.1) of pulse and receiver phases based on eqn (19.5). Δp we see is −1 from the CTP diagram.

This is the CYCLOPS phase cycle of Section 19.1. At each increment of the pulse phase the receiver is set to follow the resulting phase acquired by the coherences, calculated *via* eqn (19.5). The receiver "sees" the same thing each time. The simple vector picture of this is shown in Figure 19.12.

For the EXORCYCLE phase cycle discussed above we can do a similar analysis. Keeping the first $\pi/2$ pulse phase 0° for simplicity,

Table 19.1 The CYCLOPS phase cycle.

Pulse phase	$-\Delta p\phi$	Receiver phase
0	0	0
90	90	90
180	180	180
270	270	270

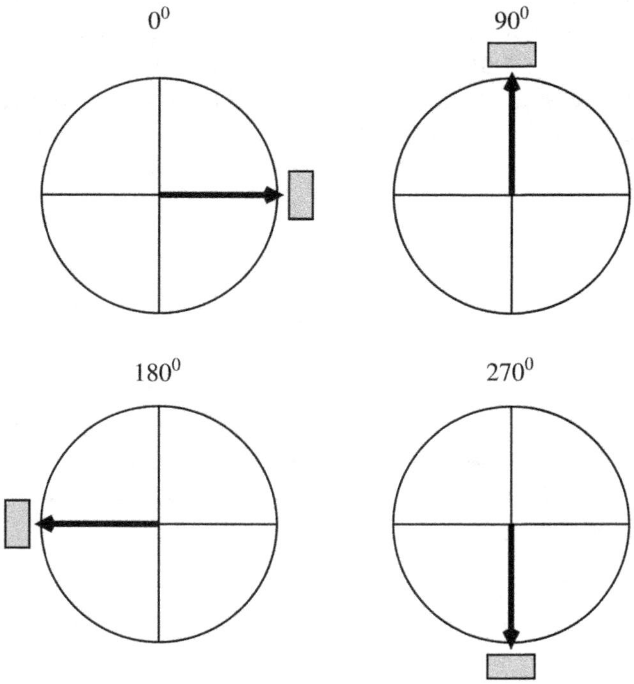

Figure 19.12 Advancing the receiver phase with advanced pulse phase in the pulse-acquire experiment.

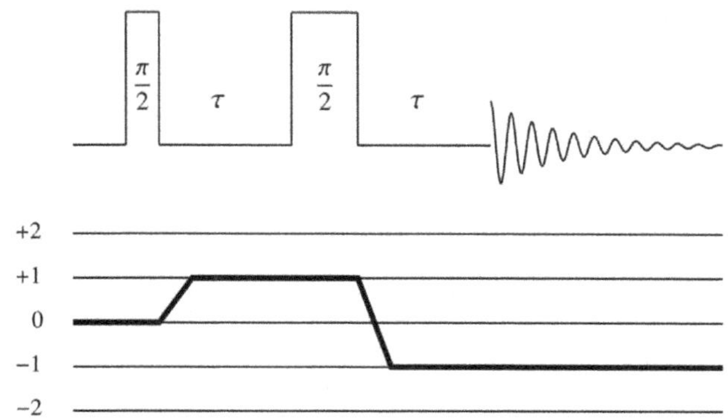

Figure 19.13 CTP diagram for a spin-echo experiment. Only the pathway leading to signal acquisition is shown.

Table 19.2 Abbreviated 4-step EXORCYCLE phase cycle for a spin-echo pulse sequence. Equivalent angles are shown in brackets.

π Pulse phase	$-\Delta p\phi$	Receiver phase
0	0	0
90	180	180
180	360 (0)	0
270	540 (180)	180

if we increment the π-pulse phase by 0°, 90°, 180° and 270° we can deduce the required receiver phase shifts (Figure 19.13 and Table 19.2).

This is as we saw above in our previous analysis of the EXORCYCLE phase cycle. If we were to cycle the $\pi/2$-pulse as well, things become more complicated. First, the phase of the first pulse carries through and is incorporated along with the phase of the second pulse into the coherences. Secondly, in writing the complete phase cycle we keep one pulse phase constant while cycling the other, which means in this case, a sixteen-step cycle (Table 19.3).

The next pulse-sequence and phase cycle scheme that we will consider at is that of the COSY experiment (Figure 19.14). We now have three possibilities for CTP diagrams.

Type *a* is referred to as *N*-type, *b* is *P*-type and *c* is absorption mode COSY.** To select the *a* pathway that goes through coherence order

**I will not go into detail about *N*, *P* and absorption here since Keeler does such an excellent job of it in his book. See the Further Reading section at the end of the chapter.

Table 19.3 Complete 16-step phase cycle for a spin-echo pulse sequence. Angles in brackets are equivalent angles. Δp for the first pulse is +1 and −2 for the second pulse, from the CTP diagram. The phase acquired by the coherences as a result of the application of the $\pi/2$-pulse is added to those acquired as a result of the application of the π-pulse to arrive at the receiver phase.

ϕ_1	$-\Delta p\phi_1$	ϕ_2	$-\Delta p\phi_2$	Receiver phase = $-\Delta p\phi_1$–$\Delta p\phi_2$
0	0	0	0	0
0	0	90	180	180
0	0	180	360 (0)	0
0	0	270	540 (180)	180
90	− 90 (270)	0	0	270
90	− 90 (270)	90	180	90
90	− 90 (270)	180	360 (0)	270
90	− 90 (270)	270	540 (180)	90
180	− 180 (180)	0	0	180
180	− 180 (180)	90	180	0
180	− 180 (180)	180	360 (0)	180
180	− 180 (180)	270	540 (180)	0
270	− 270 (90)	0	0	90
270	− 270 (90)	90	180	270
270	− 270 (90)	180	360 (0)	90
270	− 270 (90)	270	540 (180)	270

change +1 after the first pulse (leaving the phase of the first pulse constant) we can use the phase cycle in Table 19.4.

As always, the receiver phase follows the phase acquired by the coherence. Selection of path *b* that goes through coherence order change −1 as a result of the application of pulse 1 by phase cycling pulse 2 (Table 19.5).

For the absorption mode CTP with symmetrical pathways it is not necessary to phase cycle to pick out a particular path since only these paths are generated (except possibly DQC and ZQC from pulse 2 … but those are not observable).

There are some simplifications that we can apply to phase cycles that will decrease the number of steps. Let us start with a model pulse-sequence and map all of the coherences generated by the pulses (see Figure 19.15).

As noted above, the first pulse produces only ±1 coherence from equilibrium magnetisation. However, after an evolution time t_1 there will have been evolution of scalar coupling so that antiphase

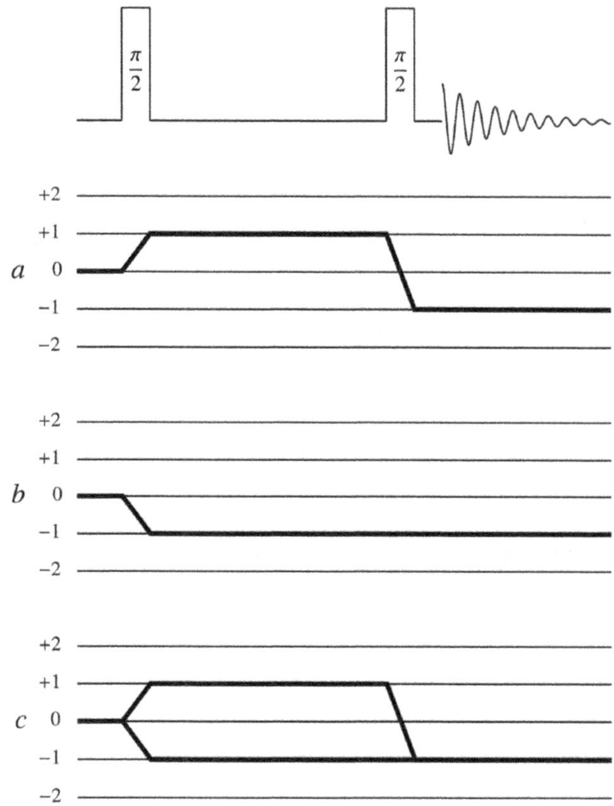

Figure 19.14 Three possible coherence-transfer pathways for the COSY experiment.

Table 19.4 Phase cycle for N-type COSY.

Pulse 2 phase	$-\Delta p_2\phi$	Receiver phase
0	0	0
90	−90 (270)	180
180	−180 (180)	0
270	−270 (90)	180

magnetisation is produced and, *via* chemical-shift evolution, more ± 1 coherence. A second $\pi/2$ pulse converts all of these into ZQC, SQC and DQC. How are we to filter out what we want and discard the rest? Let us suppose that we want only the coherences from the paths shown in Figure 19.16.

We note from this CTP diagram that the first pulse creates ± 1 coherence. The second pulse creates many coherences but we want

Table 19.5 Phase cycle for P-type COSY.

Pulse phase	$-\Delta p\phi$	Receiver phase
0	0	0
90	90	90
180	180	180
270	270	270

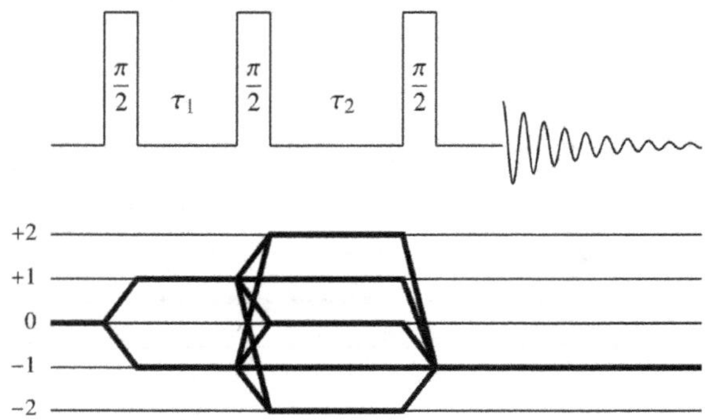

Figure 19.15 A three-pulse sequence and the CTP map of all pathways leading to -1 coherence.

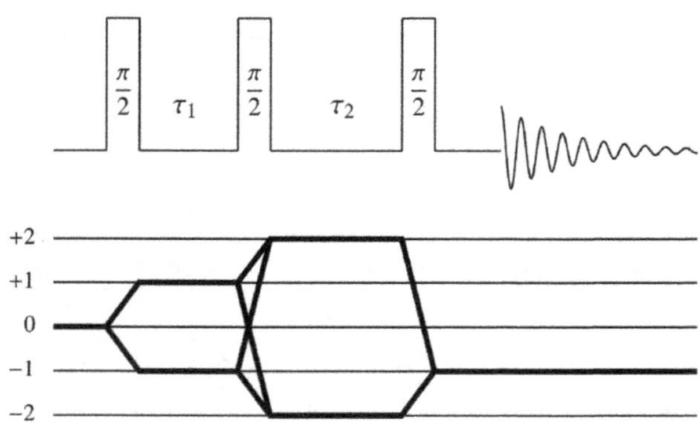

Figure 19.16 A CTP diagram for a three-pulse experiment.

only $\Delta p_2 = \pm 1$ and ± 3. The last pulse also creates multiple coherences but since only -1 coherence is observable and we wish to select only that which has gone through ± 2 we use a phase cycle that looks for $\Delta p_3 = +1$ and -3. To select the upper path with $p = +2$ that is created from the second pulse and that comes from a coherence change of -3 we would use a phase cycle such as that in Table 19.6.

Table 19.6 Phase cycle to select the P-type coherence *via* $\Delta p = -3$ pathway of Figure 19.16.

Pulse 3 phase	$-\Delta p_3 \phi_3$	Receiver phase
0	0	0
90	270	270
180	540 (180)	180
270	810 (90)	90

Table 19.7 Phase cycle to select the P-type coherence *via* $\Delta p = +1$ pathway of Figure 19.16.

Pulse 3 phase	$-\Delta p_3 \phi_3$	Receiver phase
0	0	0
90	−90 (270)	270
180	−180 (180)	180
270	−270 (90)	90

Table 19.8 Phase cycle to reject coherence *via* the $p = +1$ pathway of Figure 19.16.

Pulse 2 phase	$-\Delta p_3 \phi_3$	Receiver phase
0	0	0
90	180	270
180	0	180
270	180	90

A bit of reflection (and a similar table) will reveal that this phase cycle will also select the path corresponding to coherence change of +1 (see Table 19.7).

Each table has the same pulse-3 and receiver phase cycle and indicates that both the $\Delta p_3 = +1$ and $\Delta p_3 = +3$ paths will be selected. Other paths will be suppressed. We can see this, again through the phase cycle tables. For example, any coherence that is at, say, level +1 by the time of the third pulse will have to go through a coherence change $\Delta p_3 = -2$ in order to get through to the receiver (Table 19.8).

Here, we have used the *same* phase cycles as the previous two tables rather than follow the phases in the second column. How this coherence change is suppressed by this phase cycle scheme may be illustrated pictorially *via* a vector diagram (Figure 19.17).

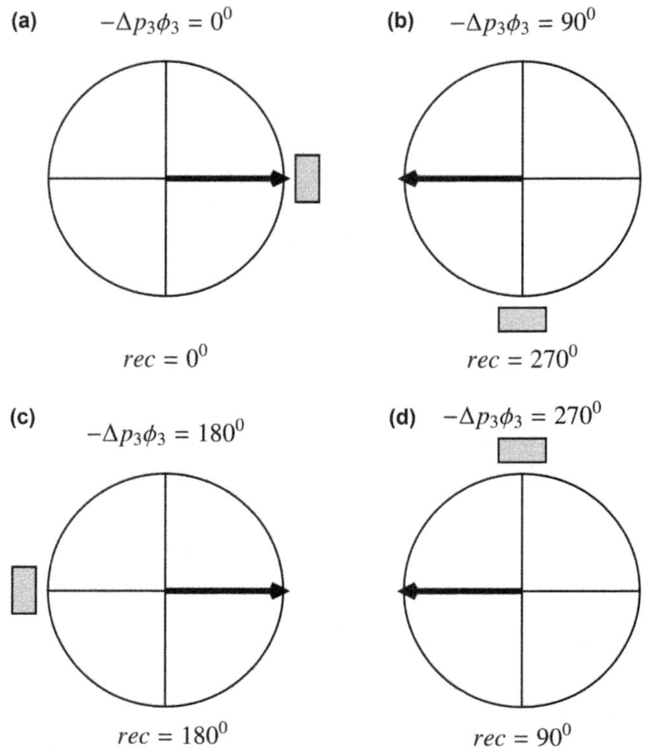

Figure 19.17 A visualisation of the phase cycle designed to reject coherence in Table 19.8. The positions of the vectors represent the acquired phase of the signal after the third pulse of Figure 9.16. The receiver phase is such that (a) and (c) cancel each other, as do (b) and (d). Summing these results in zero signal.

The same will be the case for all but $\Delta p_3 = -3$ and $\Delta p_3 = +1$ coherence change. In fact, if we were to laboriously write out all of the tables for all possible coherence changes we would find an interesting pattern. Coherence level changes that are four integer numbers apart will be let through and all others will be suppressed. Thus, in our case we might write the possible Δps:

$$-4,\ \mathbf{-3},\ -2,\ -1,\ 0,\ \mathbf{+1},\ +2,\ +3,\ 4$$

where we have emphasised the coherence changes that are allowed through the phase cycle filter for our current case. The Δps that are let through depend on the number of scans in the phase cycle that complete a 360° cycle. Commonly this is 4 and, as we see in the current case, the allowed Δps are four numbers apart. If, for some

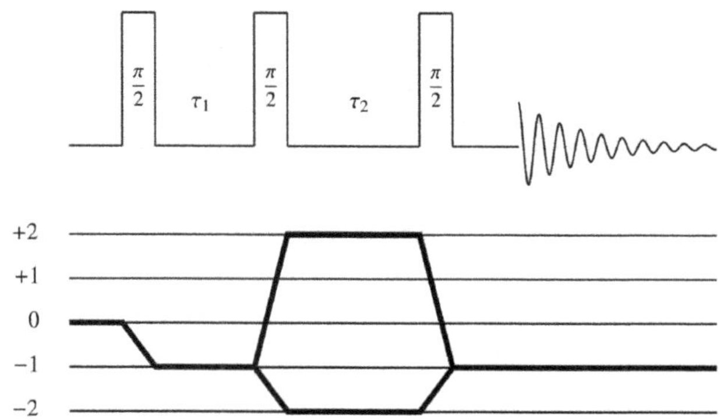

Figure 19.18 CTP diagram to select $p = -1$ and $p = +2$ *via* the $p = -1$ path.

reason we did not want to allow $\Delta p_2 = -1$ through then we could increase the number of scans in the cycle to, say, six and select for $\Delta p_2 = +3$. Since they are six integers apart, $\Delta p_2 = -3$ would also be selected.

Allowing only the $\Delta p_2 = -1$ and $\Delta p_2 = +3$ paths through leaves the CTP diagram for a *P*-type spectrum (Figure 19.18).

In this case we have to select $\Delta p_2 = -1$ and $\Delta p_2 = +3$ so that only the lower coherence of -1 gets through the filter. As with the absorption mode cycle above, we must also cycle the third pulse to select $\Delta p_3 = -3$ and $\Delta p_3 = -1$, also for the reasons given above. The complete phase cycle will have 16 components and we cycle each pulse independently of the other. As we saw above, a cycle that selects for $\Delta p = -1$ will also select $\Delta p = +3$ since they are 4 integers apart and we are using a four-phase cycle. The same can be said for $\Delta p = -3$ and $+1$. A possible phase cycle scheme is shown in Table 19.9.

If we want a *N*-type spectrum we select for $\Delta p_2 = +1$, $\Delta p_2 = -3$, $\Delta p_3 = +1$ and $\Delta p_3 = -3$ *via* the phase cycle shown in Table 19.10.

This table selects the $\Delta p_2 = +1$ path but as we have seen, it also selects the $\Delta p_2 = -3$ path. Our CTP diagram would look like that shown in Figure 19.19.

The HMQC (Heteronuclear Multiple Quantum Coherence) experiment has the pulse-sequence and CTP diagram shown in Figure 19.20.

A possible phase cycle for this is shown in Table 19.11.

In this scheme, only the *S*-spin pulses are cycled in order to select coherence level zero at the end of the pulse-sequence. This is necessary so that detectable -1 coherence is available for detection on the *I*-spin nucleus.

Table 19.9 16-step phase cycle for P-type coherence selection from a three-pulse sequence (Figure 19.18). ϕ_2 and ϕ_3 are the phase angles of pulses two and three, respectively. Angles in brackets are equivalent angles.

ϕ_2	$-\Delta p_2\phi_2$	ϕ_3	$-\Delta p_3\phi_3$ (for $\Delta p_3 = +1$)	Receiver phase = $-\Delta p_2\phi_2$-$\Delta p_3\phi_3$
0	0	0	0	0
0	0	90	−90 (270)	270
0	0	180	−180 (180)	180
0	0	270	−270 (90)	90
90	90	0	0	90
90	90	90	−90 (270)	0
90	90	180	−180 (180)	270
90	90	270	−270 (90)	180
180	180	0	0	180
180	180	90	−90 (270)	90
180	180	180	−180 (180)	0
180	180	270	−270 (90	270
270	270	0	0	270
270	270	90	−90 (270)	180
270	270	180	−180 (180)	90
270	270	270	−270 (90	0

19.5 Pulsed Field Gradients

Phase cycles can be long … *very* long, especially when all pulses in a sequence are cycled. For example, if we wished to implement a simple CYCLOPS for the first pulse of the sequence used to generate Tables 19.9 and 19.10 we would have to do so independently of the other cycles meaning a four-fold increase in the number of cycles in the table. A 64-step cycle, in other words.

The way that we get around a long set of phase cycles is to use pulsed field gradients. Ordinarily, in NMR spectroscopy,[††] we want the external magnetic field to be completely uniform over the bulk of the sample so that chemical-shift precessions are the same everywhere within the sample. A field gradient pulse seems to defy this by purposely changing the strength of the field linearly along the *z*-axis. Of course, chemical-shift frequencies along the *z*-axis will vary if we do this.

[††]MRI spectroscopy, of course, is a different story. Linear field inhomogeneity is essential in that discipline as we shall see in Chapter 22.

Table 19.10 16-step phase cycle table for N-type coherence selection from a three-pulse sequence. ϕ_2 and ϕ_3 are the phase angles of pulses two and three, respectively. Angles in brackets are equivalent angles.

ϕ_2	$-\Delta p_2\phi_2$ $(\Delta p_2 = +1)$	ϕ_3	$-\Delta p_3\phi_3$ $(\Delta p_3 = +1)$	Receiver phase = $-\Delta p_2\phi_2 - \Delta p_3\phi_3$
0	0	0	0	0
0	0	90	− 90 (270)	270
0	0	180	− 180 (180)	180
0	0	270	− 270 (90)	90
90	−90 (270)	0	0	270
90	−90 (270)	90	−90 (270)	180
90	−90 (270)	180	−180 (180)	90
90	−90 (270)	270	−270 (90)	0
180	−180 (180)	0	0	180
180	−180 (180)	90	−90 (270)	90
180	−180 (180)	180	−180 (180)	0
180	−180 (180)	270	−270 (90)	270
270	−270 (90)	0	0	90
270	−270 (90)	90	−90 (270)	0
270	−270 (90)	180	−180 (180)	270
270	−270 (90)	270	−270 (90)	180

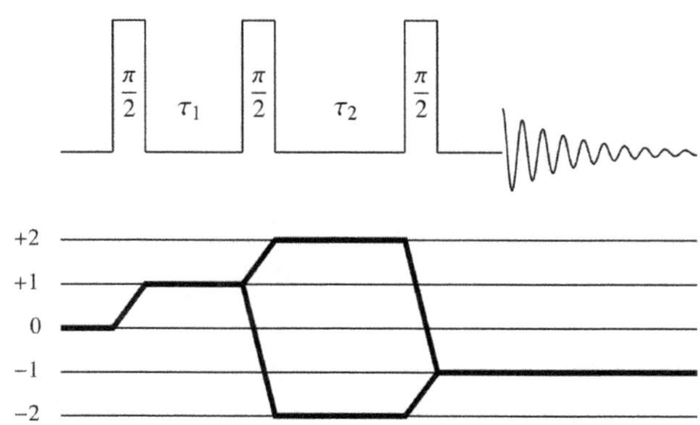

Figure 19.19 CTP diagram to select $p = -1$ and $p = +2$ *via* the $p = +1$ path.

Figure 19.21 shows how this works. We assume that the gradient field strength is zero at $z = 0$ and increases (or possibly decreases) towards $+z$. For situation a) in the figure, the field experienced by the nuclei in the sample is (hopefully) the same throughout the sample and the precession frequency about the external field, B_0, is constant everywhere in the sample resulting in a sharp signal at ω_1. We say that the

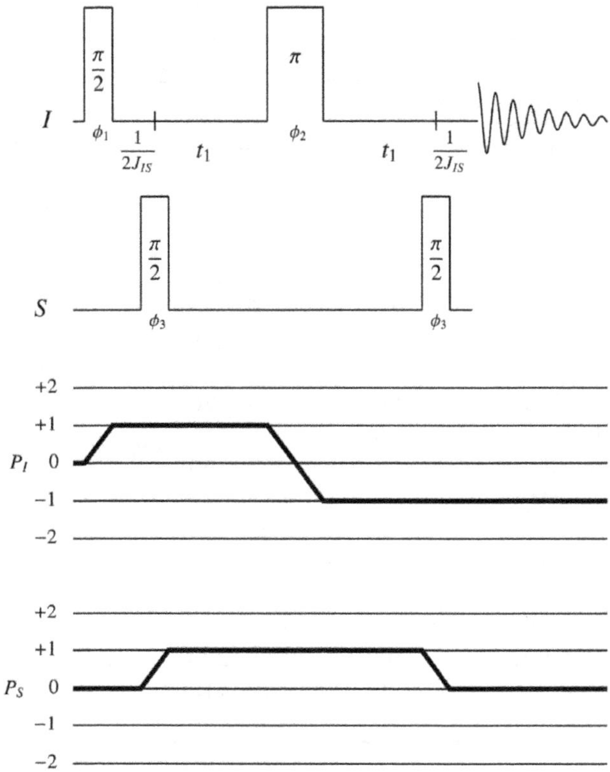

Figure 19.20 CTP diagram for the HMQC experiment.

Table 19.11 A possible phase cycle for the HMQC pulse sequence of Figure 19.20.

ϕ_3	$-\Delta p_3\phi_3$	ϕ_4	$-\Delta p_4\phi_4$ ($\Delta p_3 = +1$)	Receiver phase
0	0	0	0	0
90	− 90 (270)	0	0	270
180	− 180 (180)	0	0	180
270	− 270 (90)	0	0	90
0	0	180	180	180
90	− 90 (270)	180	180	90
180	− 180 (180)	180	180	0
270	− 270(90)	180	180	270

phase acquired by the spins is constant. In situation b) the field experienced by the nuclei is linearly increasing (or decreasing) with the addition of a field gradient, G, between $-z$ and $+z$ meaning that the

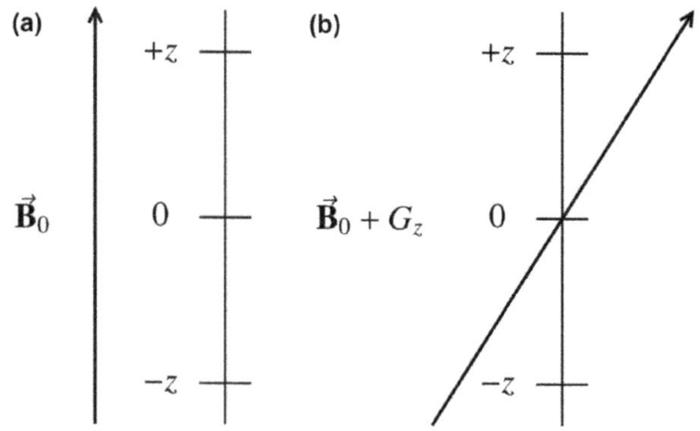

Figure 19.21 Gradient field applied to a sample. (a) no gradient, normal $\vec{\mathbf{B}}_0$ field; (b) gradient applied, $\vec{\mathbf{B}}_0$ field plus gradient field in the z-direction. Note that this does not mean that the direction of the field has changed but rather its intensity changes as a function of z.

nuclear precession frequencies will be changing linearly with the changing field. In this case the *acquired phase* of the spins is spatially dependent … in other words, it depends on where the spin is along the z-axis. The commonly used units for the field gradient are Gauss per centimetre … most NMR probes are capable of producing a maximum field gradient of between 50 and 60 G cm^{-1}. If we multiply by the distance between the bottom and top of the active volume of the sample we get a number that represents the field gradient as a function of position in the sample.

The field in the sample during the time that the gradient is on is:

$$\vec{\mathbf{B}}_z = \vec{\mathbf{B}}_0 + \hat{\mathbf{G}}_z$$

which we can convert to frequencies *via* eqn (11.44):

$$-\gamma\vec{\mathbf{B}}_z = -\gamma\vec{\mathbf{B}}_0 - \gamma\hat{\mathbf{G}}_z$$

$\vec{\mathbf{B}}_z$ and by association, ω_z, is a function of position between $\pm z$ in the sample.

Eqn (19.3) showed us that for a θ-radian rotation about the z-axis (conceptually, a z-pulse!) a coherence will acquire a phase according to:

$$\hat{\sigma}^p \xrightarrow{\theta\hat{\mathbf{I}}_x} \exp(-i\theta p)\hat{\sigma}^p$$

$\hat{\sigma}^p$ is that component of the density matrix corresponding to coherence order p as we saw above in eqn (19.2). With a field gradient we modify this such that θ is spatially dependent:

$$\hat{\sigma}^p \xrightarrow{\theta(z)\hat{\mathbf{I}}_x} \exp(-i\theta(z)p)\hat{\sigma}^p$$

The overall acquired phase is related to the spatially dependent phase and the coherence level:

$$\phi(z) = \theta(z)p = p\omega_G zt = -p\gamma G_z t$$

which is analogous to eqn (19.3).

Figure 19.22 indicates that the application of a field gradient causes the acquisition of many different phases to each of the coherences in the sample, depending on their location along the z-axis within the active area of the sample. If we were to attempt to acquire a signal at this point we would see nothing since the phases have been effectively scrambled. However, we can recover the signal by issuing another field gradient pulse (Figure 19.23).

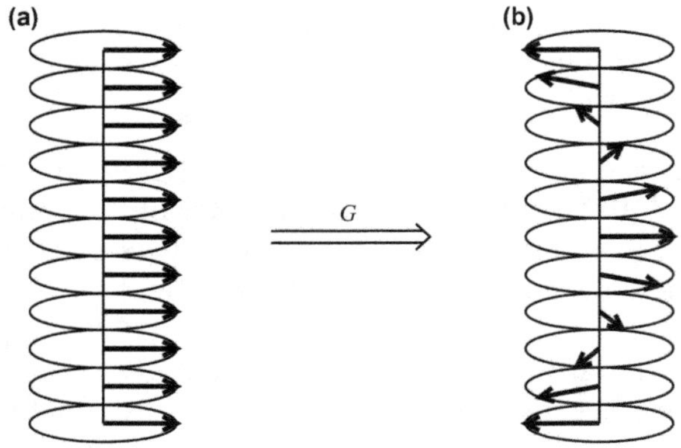

Figure 19.22 (a) no field gradient applied, constant phase acquisition through the bulk of the sample; (b) field gradient applied, spatially dependent acquired phases.

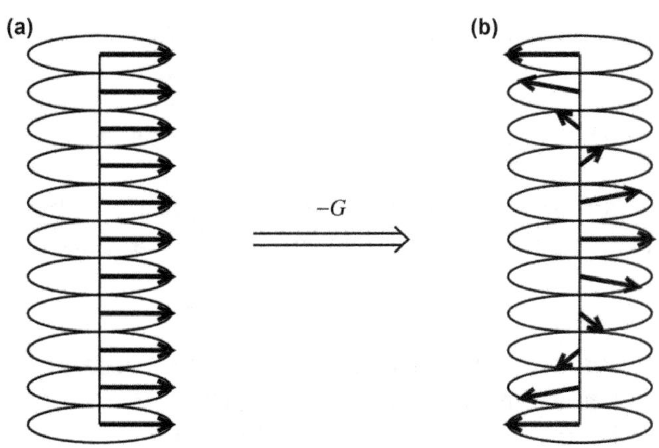

Figure 19.23 Recovery of magnetisation after a second gradient pulse.

This is the key to using field gradient pulses to filter desired coherences from a pulse sequence.

If we issue a gradient pulse and then issue a second gradient pulse we can refocus the original coherences in the sample. This means the overall acquired phase for the coherences is zero:

$$-p_1\gamma G_1 t_1 - p_2\gamma G_2 t_2 = 0$$

Rearrangement of this gives:

$$\frac{G_1 t_1}{G_2 t_2} = -\frac{p_2}{p_1} \tag{19.7}$$

assuming a homonuclear experiment (identical γs). z, of course, disappears since it will always be the same active volume for all gradient pulses. There is actually another parameter to consider and that is the shape factor of the gradient pulse, s. For a rectangular pulse it is normalised to one but for shapes other than this it is necessary to integrate the shape over its time period, t. We use shaped field gradient pulses because rectangular pulses, although workable in theory, cause other problems to occur … most notably undesirable eddy currents. Shaped pulses minimise this effect. A "sine bell" shape is simply the first half of a sine wave and integration over the shape gives:

$$\int_0^{\pi} \sin\left(\frac{\pi t}{\tau}\right) \mathrm{d}t = \left. \frac{-\cos\left(\frac{\pi t}{\tau}\right)}{\frac{\pi}{\tau}} \right|_0^{\pi} = \frac{2\tau}{\pi}$$

The area of a rectangular pulse of height one and the same duration as the sine bell gradient pulse is t and the ratio of the sine bell-shaped pulse area to the rectangular area is:

$$\frac{2\tau}{\pi\tau} = \frac{2}{\pi} \approx 0.64$$

The shape factor for the sine bell pulse is this ratio, $s = 0.64$ (Figure 19.24) and we add it into our current formula:

$$-s_1 p_1 \gamma G_1 t_1 - s_2 p_2 \gamma G_2 t_2 = 0$$

Of course, if we use the same shapes for all of our field gradient pulses then the shape factor cancels out, as do the gyromagnetic ratio for homonuclear experiments and z. We end up using eqn (19.7).

So how do we use field gradient pulses to select coherences? We use eqn (19.7) and a CTP diagram to figure this out. Let us return to the spin echo, shown in Figure 19.25.

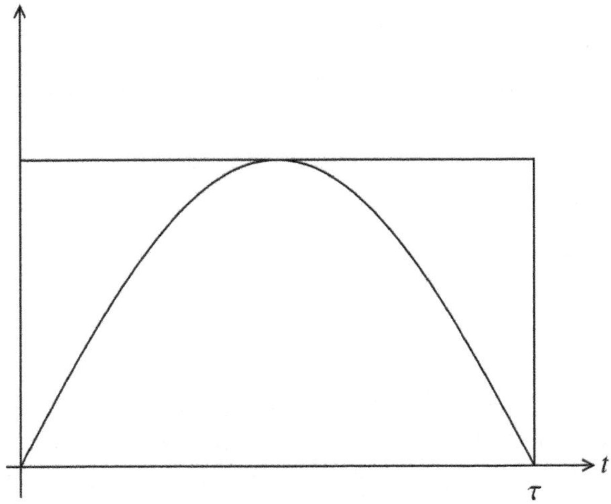

Figure 19.24 Rectangular and sine-bell field gradient pulses each of length τ.

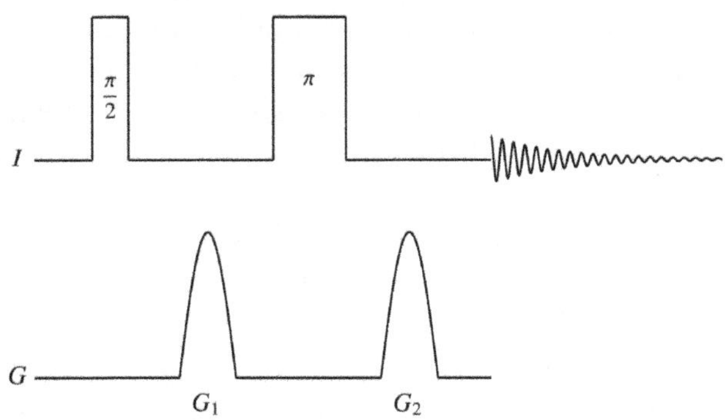

Figure 19.25 Spin-echo pulse sequence with gradient pulses added before and after the 180° refocusing pulse.

In this case there are two equivalent gradient pulses before and after the π-pulse in the spin-echo block. A close look at what happens here will be instructive. The first gradient pulse will "scramble" the coherences in the sample, that is, the coherences will acquire spatially dependent phases as shown in Figure 19.22. Next, the π-pulse changes the signs of all of the coherences as discussed above and finally, the second gradient pulse refocuses the scrambled coherences so that the total phase change is zero:

$$-p_1G_1 - p_2G_2 = 0$$

Note that here we are assuming a homonuclear experiment and equal times and shapes for the gradient pulses. What is the purpose for this? This is useful when the π-pulse is imperfect and creates new magnetisation. We saw in Section 19.1 that an imperfect π-pulse creates a complex set of magnetisation terms and that it is possible to filter out the unwanted magnetisation with the EXORCYCLE phase scheme. The gradient pulses added to the spin-echo do much the same thing. Coherences before the π-pulse survive and are refocused but any new coherences created by the π-pulse are scrambled and do not appear at the end of the spin-echo. The effect is much the same as if the EXORCYCLE scheme had been used. The advantage here, however, is that what took several scans to accomplish in the EXORCYCLE scheme takes only *one* scan here. The experiment is thus shortened or, what is equivalent, the signal-to-noise ratio for an equivalent number of scans is increased.

If a π-pulse is being used to invert z-magnetisation one might use the gradient scheme of Figure 19.26 to help clean things up.

z-magnetisation that enters the pulse sequence is unaffected by the gradient pulses since its coherence order is zero and will therefore acquire zero spatially dependent phase. Any other magnetisation with non-zero coherence order is de-phased and scrambled. It is really only necessary to use one gradient pulse here but using two gradient pulses makes it that much more efficient.

In a similar vein, a gradient pulse inserted between a pair of 90° pulses will select coherence order zero (Figure 19.27).

As with the 180° inversion pulse, any non-zero coherence order magnetisation that exists after the first pulse is scrambled in the time between the two 90° pulses. The surviving zero coherence is then

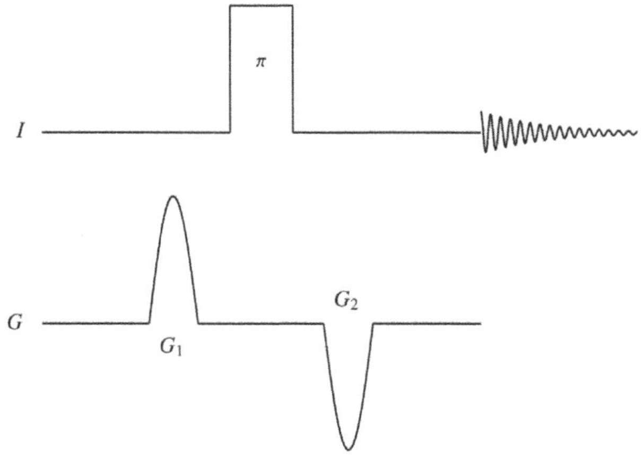

Figure 19.26 A refocusing pulse with gradients of opposite polarity before and after the pulse.

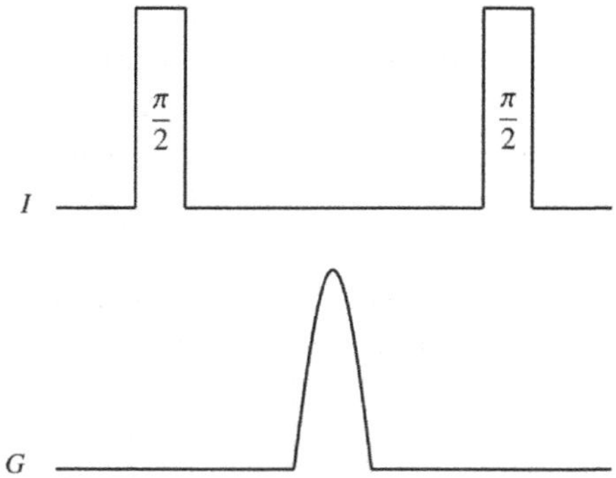

Figure 19.27 The z-filter scheme with two 90° rf pulses bracketing a field gradient.

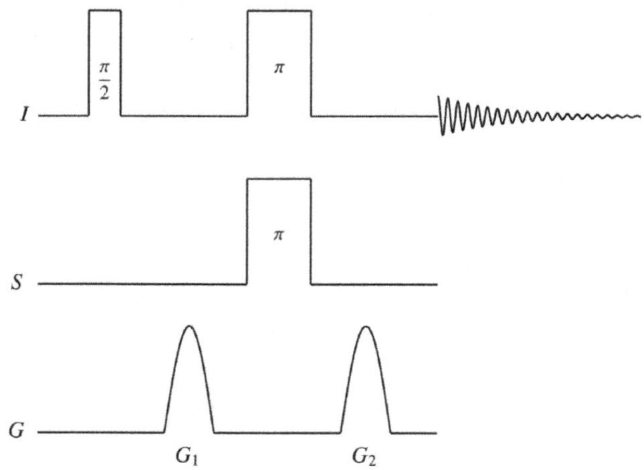

Figure 19.28 Heteronuclear spin-echo pulse sequence with refocusing pulses bracketed by two gradients.

changed to transverse magnetisation by the second pulse. This scheme is referred to as a *z*-filter, for obvious reasons.

Similar considerations apply to heteronuclear pulse sequences. For example, a pulse sequence in which π-pulses are applied simultaneously to both nuclei will benefit from the use of gradient pulses as in the homonuclear case (Figure 19.28).

In this case the *I*-spin magnetisation behaves as discussed above. If the *S*-spin magnetisation is longitudinal (*z*-magnetisation) at the beginning of the pulse block then the first gradient pulse does not affect it and the second scrambles any transverse magnetisation created by an imperfect pulse leaving only inverted *z*-magnetisation. If there is transverse *S*-spin magnetisation it will behave as does the *I*-spin magnetisation.

Although the use of gradient pulses eases the necessity of using long phase cycles to observe the desired signals there are a couple of problems with their use. First, they are generally milliseconds in length ... long enough for significant shift and coupling evolution to take place. For this reason, we generally encounter gradient pulses used within spin-echo blocks in a pulse sequence. Any chemical-shift evolution is then refocused and we minimise phase correction problems in the acquired spectrum. Secondly, gradient pulses are capable of selecting only one coherence at a time, unlike phase cycles. For example, in Figure 19.14 only paths a) and b) would be selectable using gradients. Path c) would not be selectable. Generally, we see in pulse programs a mixture of phase cycling and gradient pulses to achieve the desired result.

Further Reading

1. R. R. Ernst, G. Bodenhausen and A. Wokaun, *Principles of Nuclear Magnetic Resonance in One and Two Dimensions*, Clarendon Press, Oxford, 1988.
2. R. Freeman, *Spin Choreography, Basic Steps in High Resolution NMR*, Oxford University Press, Oxford, 1997.
3. J. Keeler, *Understanding NMR Spectroscopy*, John Wiley & Sons, Chichester, 2nd edn, 2010.
4. M. H. Levitt, *Spin Dynamics*, John Wiley & Sons, Chichester, 2001.
5. T. Parella, *Magn. Reson. Chem.*, 1998, **36**, 467–495.

20 Relaxation and the NOE

Life is good for only two things, discovering mathematics and teaching mathematics.

Simeon Denis Poisson

20.1 Spin Temperature

Consider the equilibrium state of a collection of non-coupled spins-1/2.† In terms of energy levels and the Boltzmann distribution (Section 6.5) there are two levels only and the spins are distributed between them. We can obtain the ratio of the higher energy spins to the lower energy spins using eqn (6.14):

$$\frac{n_1}{n_0} = e^{-\epsilon_1/kT}$$

†Keeler does a very fine job of explaining relaxation in his text. I highly recommend reading it as an introduction to relaxation. It was difficult to know what to write about the topic here that was not already covered by him. My choice was to delve a bit more deeply into the mathematics than him (since this *is* a text involving mathematics). However, in order to avoid being sucked into the quicksand that is relaxation mathematics I will generally only touch on the most fundamental topics. The reader is directed toward the Further Reading section for more in-depth discussions ... and they really *do* go into great depth!

Essential Mathematics for NMR and MRI Spectroscopists
By Keith C. Brown

Published by the Royal Society of Chemistry, www.rsc.org

At room temperature (~300 K) this ratio (for ^{1}H) is:

$$\frac{n_1}{n_0} = \exp\left(\frac{-2\mu B_0}{kT}\right)$$

$$= \exp\left(\frac{2.1411 \times 10^{-27}\mathrm{JT}^{-1} \times 11.75\mathrm{T}}{1.381 \times 10^{-23}\mathrm{JK}^{-1} \times 300\,\mathrm{K}}\right)$$

$$= 0.9999199$$

We expect from this that decreasing the temperature will decrease this ratio. Indeed we see for a temperature of 10 K the ratio drops to 0.99754. Figure 20.1 shows a plot of n_1/n_0 *versus* temperature indicates that the ratio drops precipitously as T approaches absolute zero.

We can envision a situation in which the temperature of a spin-system is 0 K and that in this situation all spins are in the lower energy state. When we do an NMR experiment we frequently apply $\pi/2$ pulses which, in population terms, create a state in which the populations of the lower and upper energy-levels are equal and are said to be saturated. In terms of the Boltzmann equation (eqn (6.15)) this would require that $T = \infty$ or, just as valid, $T = -\infty$. When we issue a π-pulse we invert the spin population. In this situation the temperature is said to be negative! The Boltzmann relation requires this, since for n_1/n_0 to be greater than one T must be negative. In the extreme, when *all* of the spins are in the upper energy level the temperature is −0 K.

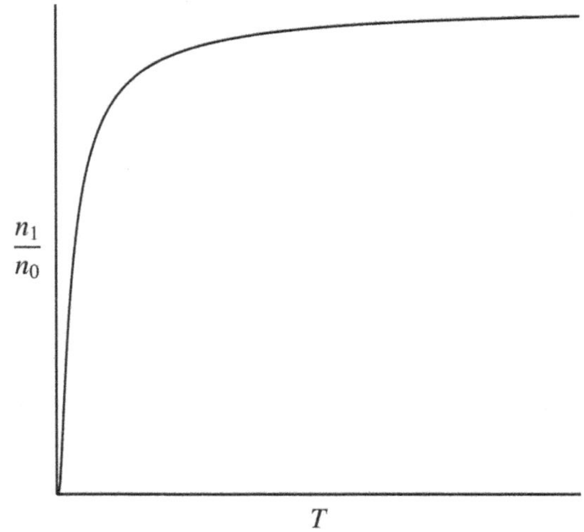

Figure 20.1 Population ratios for a two energy-level spin system as a function of temperature.

The counter-intuitive nature of the idea of negative temperature comes from the assumption that when temperatures increase so does the entropy of a system. This is indeed the case most of the time ... we think primarily of the added energy contributing to temperature increase and secondarily of the increased entropy. However, there is no requirement in thermodynamics that this be so. The derivative of the entropy S with respect to energy E is not necessarily positive:

$$\frac{\partial S}{\partial E} = \pm \frac{1}{T}$$

In some situations, most notably those with a limited number of energy levels, the rate of change of entropy with increasing energy can be negative. For a two-level system the entropy is at a maximum when the populations of each level is the same (or saturated). When the upper-level population increases from this state the entropy decreases, even though the total energy increases. That this must be so can be understood from the statistical mechanics point of view (see Section 6.5). The point where all spins are in the upper level is the minimum entropy state. Think in terms of microstates. There are many, many microstates in a saturated system corresponding to a high entropy but only one for the upper-level state.

Just as counter-intuitive is the fact that when a negative temperature system comes into contact with a positive temperature system energy will flow from the negative temperature system to the positive temperature system. The negative temperature is actually hotter (in the sense that it contains more energy) than the positive temperature!

Since the energy flow for a spin system is mostly between spins we call the temperature associated with a spin state the spin temperature.

20.2 Longitudinal T_1 Relaxation

We first formally encountered the idea of relaxation of the NMR signal in Chapter 12 during the development of the Bloch equations although it was also implicit in the development of eqns (8.47) and (8.48). The concept of two different types of relaxation were introduced, T_1 and T_2 relaxation.

T_1 relaxation that also goes by the names longitudinal relaxation and spin-lattice relaxation can be visualised as the return of the (perturbed) magnetisation back to equilibrium *via* a first-order decay (Figure 20.2):

$$M_0 - M_z = Ae^{-t/T_1} \tag{20.1}$$

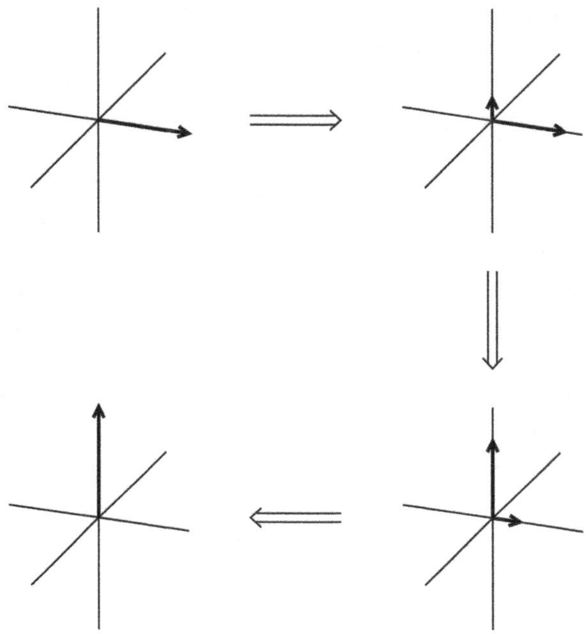

Figure 20.2 Relaxation of transverse magnetisation towards equilibrium (longitudinal) magnetisation.

One may also think in terms of an excited spin system relaxing back to a ground state by releasing energy to the surroundings ... like heat flowing from a hot object to its colder surroundings. This involves the spin temperature idea from above. An excited spin system is at a high spin temperature (unless the spin populations have been inverted) and when it comes into contact with its surroundings (presumably at a lower temperature) energy flows from the spin system to the surroundings or the lattice. The rate of this flow of energy is characterised by a rate constant $R_1 = 1/T_1$. New NMR users may sometimes mistake the T_1 constant for the actual time that it takes for the spin system to relax back to equilibrium when in fact it is a measure of the *rate* of relaxation.

Knowledge of the T_1 relaxation constants for the nuclei of interest in our samples is important for several reasons. First, we generally wish to be able to "scan" our samples multiple times and co-add the results in order to improve the signal-to-noise ratio of the resulting spectrum. However, if we simply pulse the sample continuously without paying attention to the relaxation times of the nuclei we will very likely end up with degraded signal-to-noise in our spectrum. Usually in most pulse programs there is a user-settable relaxation delay time to allow the spin system to relax at least mostly back to equilibrium. How long this delay will be is directly related to the T_1 relaxation constant. In a practical sense, this is probably the primary reason for knowing about T_1.

Measurement of the T_1 constant for a set of spins is generally accomplished by doing a pseudo-2D inversion recovery experiment and analysing the intensities of the resultant peaks.‡ Assuming that the longitudinal relaxation is a first-order process one fits the intensities to an exponential curve and extracts the T_1 constant from the results. It is also possible to estimate the T_1 constant for a peak (or peaks) of interest by using a 1D inversion recovery pulse sequence and setting the interpulse delay such that the peak just nulls out. In eqn (20.1), $A=2M_0$ since at $t=0$, the magnetisation M_z is equal to and the negative of M_0. At the time when M_z passes through zero:

$$\ln(M_0)=\ln(2M_0)-\frac{t_0}{T_1}$$

$$\ln\left(\frac{M_0}{2M_0}\right)=-\frac{t_0}{T_1}$$

$$T_1=\frac{t_0}{\ln(2)}$$

20.3 Transverse T_2 Relaxation

T_2 relaxation (also known as spin-spin relaxation) can be thought of as a loss of phase coherence of the nuclear spin vectors in the sample (Figure 20.3). T_2, as we saw in Chapter 12, is the time constant in a first-order exponential NMR signal relaxation process. Since the NMR

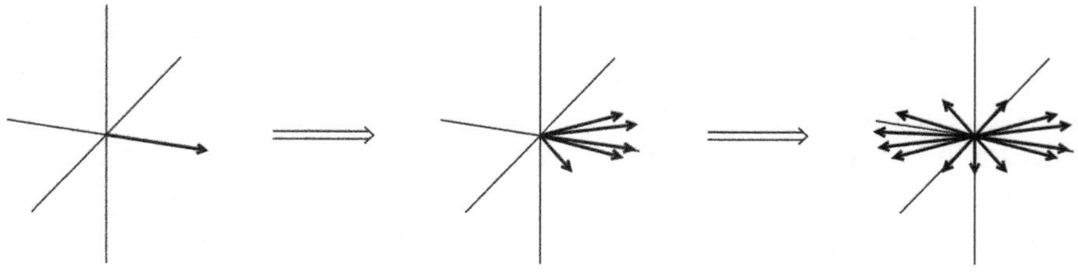

Figure 20.3 Relaxation of transverse magnetisation by loss of phase coherence.

‡A detailed description of the inversion recovery experiment is given in Chapter 21, Section 21.2.

signal embodied in the free-induction decay is detected in the transverse plane its decay is intimately linked with T_2 relaxation.

$$M_x = Ae^{-t/T_2}$$

$$\ln(M_x) = A' - \frac{t}{T_2}$$

$$\frac{1}{M_x}\frac{dM_x}{dt} = -\frac{1}{T_2}$$

$$\frac{dM_x}{dt} = -\frac{M_x}{T_2}$$

In Section 8.5.7 we developed an expression for the real or absorption mode signal:

$$F(\omega) = \frac{\alpha}{\alpha^2 + (\omega - \omega_0)^2}$$

where α is the relaxation rate constant for a first-order process. This is the ubiquitous Lorentzian function. It is of interest to compute the values of the function at the half-height point on both sides of the centre at ω_0. The value of the function at ω_0, the centre of the peak (where $\omega = \omega_0$), is $1/\alpha$. The half-height values are therefore $1/2\alpha$. Setting this equal to the absorption function and solving for ω:

$$\frac{1}{2\alpha} = \frac{\alpha}{\alpha^2 + (\omega - \omega_0)^2}$$

$$\alpha^2 + (\omega - \omega_0)^2 = 2\alpha^2$$

$$(\omega - \omega_0)^2 = \alpha^2$$

$$\omega = \omega_0 \pm \alpha$$

Thus, at $\omega_0 + \alpha$ and $\omega_0 - \alpha$ in the Lorentzian function the value of the function is 1/2 of the full value at ω_0. The distance between these two points is $\omega_0 + \alpha - (\omega_0 - \alpha) = 2\alpha$ and is called the *half-height linewidth*. Furthermore, in Section 12.1 we developed a very similar equation in which the relaxation rate constant was $1/T_2$. Setting $\alpha = 1/T_2$ leads to the conclusion that the half-height linewidth is $2/T_2$ rad s^{-1} or in hertz, $2/\pi T_2$. Note that this is the half-height linewidth in a perfectly homogeneous field. Field inhomogeneity will increase the rate of dephasing that will increase the signal decay rate and consequently increase the linewidth (Figure 20.4).

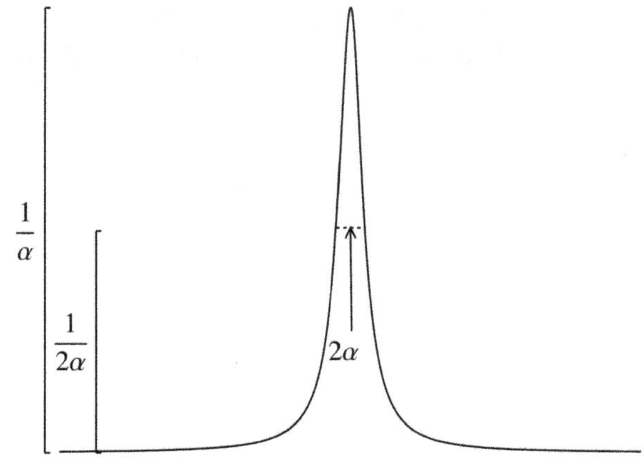

Figure 20.4 The Lorentzian half-height linewidth, 2α.

Since, in the real-world, field homogeneity is never perfect, what we observe in our spectrometers is never the real T_2 but rather T_2^*. This can be viewed as an extra bit added to the relaxation constant:

$$F(\omega) = \frac{\alpha + \beta}{(\alpha + \beta)^2 + (\omega - \omega_0)^2} \tag{20.2}$$

From the above discussion of half-height linewidth, one can see that this results in a linewidth that is larger than the natural one. If we set $\beta = 1/T_H$, where T_H is the transverse relaxation time-constant due to field inhomogeneity, T_2^* is:

$$\frac{1}{T_2^*} = \frac{1}{T_2} + \frac{1}{T_H}$$

T_2^* is usually measured using the CPMG experiment.[§] A quick estimate of T_2^* is to measure the half-height linewidth, as discussed above. However, be careful! Broadening of a linewidth may also be caused by exchange broadening ... essentially a third parameter, δ, added to eqn (20.2). How can you tell if there is exchange broadening? Generally it is temperature-sensitive so changing the temperature of the sample should result in a significant change in the linewidth. Also, an NOE experiment will reveal this. See below for more information about this.

[§]See any competent modern introductory text on NMR spectroscopy for the details of this experiment.

20.4 How Does Relaxation Work?

20.4.1 Equilibrium Magnetisation

When we put our samples into the magnet they tend to reach equilibrium in a short time ... generally seconds. How does this happen? We can rationalise this with a mixed classical/quantum viewpoint. Consider that when the sample first enters the field the spins' magnetic moments are randomly oriented. We have already encountered this picture in Figure 12.1. From the quantum mechanical energy-level viewpoint this corresponds to the situation in which the two levels are equally populated (assuming a single spin-1/2 spin system). At equilibrium we expect that there will be a population difference between the two levels such that the lower energy one will be more populated. In other words some of the spins in the high level will give up some energy to the surroundings and transition to the lower level.

The situation in which the lower level is more populated than the higher level corresponds to there being more spins aligned with the external $\vec{\mathbf{B}}_0$ field than against. The process of relaxation apparently is associated with the net alignment of spins with the external field axis.

It is not strictly correct to think of spins being aligned with or against the field. They are only thus after having been measured. Before measurement they are in a superposition of with and against the field (see eqn (17.1)). Using the vector model this corresponds to the picture as in Figure 12.1b in which more spins are aligned such that they are in the upper hemisphere than in the lower. Summing their z-axis components leads to a net magnetisation as discussed in Chapter 12.

20.4.2 The Spectral Density Function

In dealing with the underlying mechanisms of relaxation we will be looking at molecular motions that are random.¶ Brownian motion is a macroscopic manifestation of the microscopic motions of the many molecules in constant random collisions with the macroscopic particles.

We begin our modelling using eqn (17.19) from the density operator chapter. We want to calculate the rate of change of the population of an energy level *via* the density operator. Also, for simplicity,

¶The key to understanding most of what follows in this section is to keep in mind that $\langle\alpha|\hat{\mathbf{A}}|\beta\rangle$ simply represents a matrix element, $A_{\alpha\beta}$. The mathematical expressions are very complicated looking but the underlying ideas are relatively simple.

we will assume that all states are empty save for one, j. The density matrix is therefore completely populated with zeros except for diagonal element $\langle j|\hat{\sigma}(0)|j\rangle$.[||] A possible density matrix might look like:

$$\begin{bmatrix} 0 & 0 & 0 & 0 \\ 0 & 1 & 0 & 0 \\ 0 & 0 & 0 & 0 \\ 0 & 0 & 0 & 0 \end{bmatrix}$$

We will then be interested in the calculation of the probability of transition from state j to state k.[**] Using eqn (17.14) and our requirements for the density matrix we can say:

$$\langle n|\hat{\sigma}^{\#}(0)|k\rangle = \langle n|\hat{\sigma}(0)|k\rangle = 0$$

and

$$\langle j|\hat{\sigma}^{\#}(0)|j\rangle = \langle j|\hat{\sigma}(0)|j\rangle = 1$$

Also, using eqn (17.21), we can write:

$$\langle j|\hat{\sigma}^{\#}(t)|j\rangle = \langle j|\hat{\sigma}(t)|j\rangle$$

Starting with our interaction notation equation from Chapter 17, we set up the rate of change of state k as:

$$\begin{aligned}\frac{\mathrm{d}}{\mathrm{d}t}\langle k|\hat{\sigma}^{\#}(t)|k\rangle &= \frac{\mathrm{d}}{\mathrm{d}t}\langle k|\hat{\sigma}(t)|k\rangle \\ &= i\langle k|[\hat{\mathbf{H}}^{\#}(t'), \hat{\sigma}^{\#}(0)]|k\rangle + i^2\int_0^t \langle k|[\hat{\mathbf{H}}^{\#}(t), [\hat{\mathbf{H}}^{\#}(t'), \hat{\sigma}^{\#}(0)]]|k\rangle \mathrm{d}t' \\ &= i\langle k|[\hat{\mathbf{H}}^{\#}(t'), \hat{\sigma}^{\#}(0)]|k\rangle - \int_0^t \langle k|[\hat{\mathbf{H}}^{\#}(t), [\hat{\mathbf{H}}^{\#}(t'), \hat{\sigma}^{\#}(0)]]|k\rangle \mathrm{d}t'\end{aligned}$$

All that has been done here is to insert the expression for $\mathrm{d}/\mathrm{d}t\hat{\sigma}^{\#}(t)$ from eqn (17.19) into the first term on the left in the first line. Expanding the multiple commutators using the commutator relations for nested commutators from Appendix A and making use of

[||] It is important to keep these specifications very clearly in mind when following these equations. Otherwise, confusion will reign.

[**] This is modelled on Slichter's exposition that is terrifically difficult to read, mostly because of what he does not say. Hopefully I can make it a bit easier for you.

the closure expression such that we produce a term $\langle m|\hat{\sigma}^{\#}(0)|n\rangle$, (see eqn (3.61)):

$$\frac{\mathrm{d}}{\mathrm{d}t}\langle k|\hat{\sigma}(t)|k\rangle = i\sum_n\left(\langle k|\mathbf{H}^{\#}(t')|n\rangle\underbrace{\langle n|\hat{\sigma}^{\#}(0)|k\rangle}_{\text{(equals 0 for all } n)} - \underbrace{\langle k|\hat{\sigma}^{\#}(0)|n\rangle}_{\text{(equals 0 for all } n)}\langle n|\mathbf{H}^{\#}(t')|k\rangle\right)$$

$$-\int_0^t\left(\left(\langle k|\mathbf{H}^{\#}(t)\mathbf{H}^{\#}(t')\hat{\sigma}^{\#}(0)|k\rangle - \langle k|\mathbf{H}^{\#}(t')\hat{\sigma}^{\#}(0)\mathbf{H}^{\#}(t)|k\rangle\right)\right.$$

$$\left.-\left(\langle k|\mathbf{H}^{\#}(t)\hat{\sigma}^{\#}(0)\mathbf{H}^{\#}(t')|k\rangle + \langle k|\hat{\sigma}^{\#}(0)\mathbf{H}^{\#}(t')\mathbf{H}^{\#}(t)|k\rangle\right)\right)$$

From above, in the first two terms, for all values of n the density matrix elements are zero. To understand this, recall that we have specified that the *only* non-zero value of the density matrix at $t=0$ is $\langle j|\hat{\sigma}^{\#}(0)|j\rangle$; this term never appears in the summation. In the above summation this matrix element never occurs and therefore all of these terms are zero and thus, the first two terms vanish. The third and sixth terms can be written (again using the closure relation to bracket $\sigma^{\#}(0)$ with state vectors):

$$\langle k|\mathbf{H}^{\#}(t)\mathbf{H}^{\#}(t')\hat{\sigma}^{\#}(0)|k\rangle + \langle k|\hat{\sigma}^{\#}(0)\mathbf{H}^{\#}(t')\mathbf{H}^{\#}(t)|k\rangle$$

$$=\sum_n\langle k|\mathbf{H}^{\#}(t)\mathbf{H}^{\#}(t')|n\rangle\underbrace{\langle n|\hat{\sigma}^{\#}(0)|k\rangle}_{\text{(equals 0 for all } n)} + \underbrace{\langle k|\hat{\sigma}^{\#}(0)|n\rangle}_{\text{(equals 0 for all } n)}\langle n|\mathbf{H}^{\#}(t')\mathbf{H}^{\#}(t)|k\rangle$$

As for the first two terms, density matrix elements are zero for all values of n and these terms vanish as well. This leave us with:

$$\frac{\mathrm{d}}{\mathrm{d}t}\langle k|\hat{\sigma}(t)|k\rangle = -\int_0^t\left(-\langle k|\mathbf{H}^{\#}(t')\hat{\sigma}^{\#}(0)\mathbf{H}^{\#}(t)|k\rangle - \langle k|\mathbf{H}^{\#}(t)\hat{\sigma}^{\#}(0)\mathbf{H}^{\#}(t')|k\rangle\right)\mathrm{d}t'$$

and, yet again, we employ the closure "trick" on these terms to enclose the density operator with state vectors:

$$-\sum_n\left(\langle k|\mathbf{H}^{\#}(t')|n\rangle\underbrace{\langle n|\hat{\sigma}^{\#}(0)|n\rangle}_{\text{(equals 1 for } n=j)}\langle n|\mathbf{H}^{\#}(t)|k\rangle\right)$$

$$+\left(\langle k|\mathbf{H}^{\#}(t)|n\rangle\underbrace{\langle n|\hat{\sigma}^{\#}(0)|n\rangle}_{\text{(equals 1 for } n=j)}\langle n|\mathbf{H}^{\#}(t')|k\rangle\right)$$

This time the matrix elements $\langle n|\hat{\sigma}^{\#}(0)|n\rangle$ are zero for all n *except* $n=j$ in which $\langle j|\hat{\sigma}^{\#}(0)|j\rangle=1$ (as was specified above) so the only surviving terms are:

$$-\langle k|\mathbf{H}^{\#}(t')|j\rangle\langle j|\mathbf{H}^{\#}(t)|k\rangle$$

$$-\langle k|\mathbf{H}^{\#}(t)|j\rangle\langle j|\mathbf{H}^{\#}(t')|k\rangle$$

and we have, finally:

$$\frac{\mathrm{d}}{\mathrm{d}t}\langle k|\hat{\sigma}(t)|k\rangle=\int_0^t(\langle k|\mathbf{H}^{\#}(t')|j\rangle\langle j|\mathbf{H}^{\#}(t)|k\rangle + \langle k|\mathbf{H}^{\#}(t)|j\rangle\langle j|\mathbf{H}^{\#}(t')|k\rangle)\mathrm{d}t'$$

Using eqn (17.22), we can convert out of interaction representation:

$$\langle k|\mathbf{H}^{\#}(t)|j\rangle=\exp(i[E_k-E_j]t)\langle k|\mathbf{H}(t)|j\rangle$$

$$\langle k|\mathbf{H}^{\#}(t')|j\rangle=\exp(i[E_k-E_j]t')\langle k|\mathbf{H}(t')|j\rangle$$

$$\langle j|\mathbf{H}^{\#}(t)|k\rangle=\exp(i[E_j-E_k]t)\langle j|\mathbf{H}(t)|k\rangle$$

$$\langle j|\mathbf{H}^{\#}(t')|k\rangle=\exp(i[E_j-E_k]t')\langle j|\mathbf{H}(t')|k\rangle$$

$$\frac{\mathrm{d}}{\mathrm{d}t}\langle k|\hat{\sigma}(t)|k\rangle=\int_0^t\Big(\langle k|\mathbf{H}(t')|j\rangle\langle j|\mathbf{H}(t)|k\rangle\mathrm{e}^{i(E_k-E_j)(t'-t)} + \langle k|\mathbf{H}(t)|j\rangle\langle j|\mathbf{H}(t')|k\rangle\mathrm{e}^{i(E_k-E_j)(t-t')}\Big)\mathrm{d}t'$$

Our equations have been completely general up to this point, $\hat{\mathbf{H}}(t)$ being defined as a general perturbation Hamiltonian for a single-spin system. Random molecular motions are modelled, for NMR purposes, by using a Hamiltonian that describes a molecular property such as position or magnetic field at time t and then again, at some time t' later for each molecule in a large collection of molecules. The random nature of the phenomenon is such that we need to think in terms of average calculations for an ensemble population of spins. With this in mind we convert our transition rate equation to an average calculation and consider our Hamiltonian to be describing the effects of random molecular motions:

$$\frac{\mathrm{d}}{\mathrm{d}t}\overline{\langle k|\hat{\sigma}(t)|k\rangle}=\int_0^t\Big(\overline{\langle k|\mathbf{H}(t')|j\rangle\langle j|\mathbf{H}(t)|k\rangle}\mathrm{e}^{i(E_k-E_j)(t'-t)} + \overline{\langle k|\mathbf{H}(t)|j\rangle\langle j|\mathbf{H}(t')|k\rangle}\mathrm{e}^{i(E_k-E_j)(t-t')}\Big)\mathrm{d}t' \quad (20.3)$$

Since we are considering a stationary-state random process and, recalling Chapter 6 and in particular, the discussion leading up to eqn (6.17) we see that the above equation is similar to a pair of autocorrelation functions. For our purposes here, the autocorrelation function is:

$$G_{kj}(\tau) \equiv \overline{\langle k|\mathbf{H}(\tau)|j\rangle\langle j|\mathbf{H}(0)|k\rangle} \tag{20.4}$$

where we have defined $\tau \equiv t - t'$ and "relocated" the time origin to 0 (in other words, set $t = 0$). As we saw in Chapter 6, $G(\tau)$ is a symmetric even function meaning that $G(\tau) = G(-\tau)$. Substitution into eqn (20.3) gives us:

$$\begin{aligned}\frac{\mathrm{d}}{\mathrm{d}t}\overline{\langle k|\hat{\sigma}(t)|k\rangle} &= \int_0^t \left(G_{kj}(\tau)\mathrm{e}^{-i(E_k-E_j)\tau} + G_{kj}(-\tau)\mathrm{e}^{i(E_k-E_j)\tau}\right)\mathrm{d}\tau \\ &= 2\int_0^t G_{kj}(\tau)\mathrm{e}^{-i(E_k-E_j)\tau}\mathrm{d}\tau \\ &= \int_{-t}^t G_{kj}(\tau)\mathrm{e}^{-i(E_k-E_j)\tau}\mathrm{d}\tau\end{aligned} \tag{20.5}$$

Where we have used eqn (A.56) from Appendix A. Consider $G_{kj}(0)$:

$$G_{kj}(0) = \overline{\langle k|\mathbf{H}(0)|j\rangle^2} \geq 0$$

This says that the average value of the square of $\langle k|\mathbf{H}(0)|j\rangle$ is possibly non-zero. Assuming a non-zero value for $G_{ij}(0)$ and that $\hat{\mathbf{H}}(0)$ and $\hat{\mathbf{H}}(\tau)$ are correlated; that is, $\hat{\mathbf{H}}(0)$ and $\hat{\mathbf{H}}(\tau)$ are approximately equal, then for a small value of τ, $G_{ij}(\tau)$ will be non-zero as well. As τ increases this correlation decreases as well (as we saw for the properties of correlation functions in Chapter 6) and approaches zero for large values of τ giving an even and symmetric function, as noted above (Figure 20.5).

There is a somewhat arbitrarily defined τ-value called the *correlation time* τ_c before which the difference between τ and 0 is considered negligible and beyond which it is considered to be associated with less and less correlation between $\hat{\mathbf{H}}(0)$ and $\hat{\mathbf{H}}(\tau)$. It is frequently defined as the average time required for a rotation of a molecule through one radian. In eqn (20.5), the integration limits are $\pm t$. If $t \gg \tau_c$ then, because $G_{ij}(\tau)$ is less than $G_{ij}(0)$ (see Chapter 6) and decreases as τ increases, the value of the correlation function approaches zero and

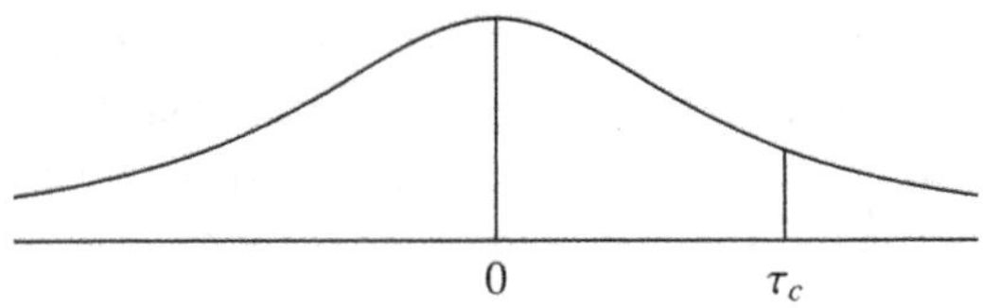

Figure 20.5 The correlation function, $G(\tau)$.

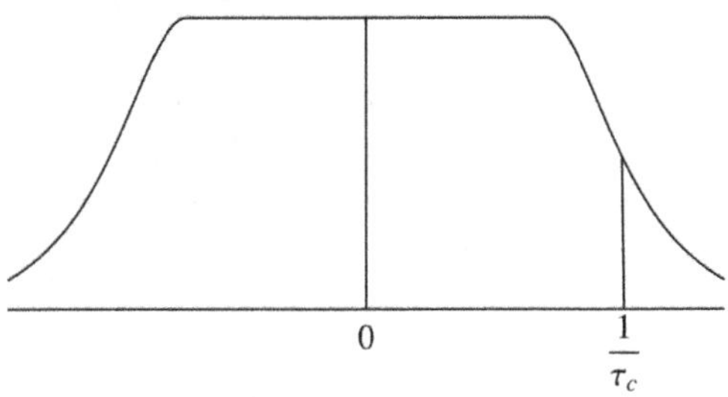

Figure 20.6 The spectral density function, $J(\omega)$.

the limits of integration may be changed to $\pm\infty$ without appreciably changing the result:

$$\frac{\mathrm{d}}{\mathrm{d}t}\overline{\langle k|\hat{\sigma}(t)|k\rangle} = W_{jk} = \int_{-\infty}^{\infty} G_{kj}(\tau)\mathrm{e}^{-i\omega_{kj}\tau}\mathrm{d}\tau \tag{20.6}$$

where we are using W_{jk} to compactly represent the probability of transition per unit time from state j to state k. Note that we have also switched to frequency notation using eqn (13.1). If you have looked at Chapter 8 you will also note that this looks much like a Fourier transform. Indeed, we define a new function called the *spectral density* (Figure 20.6) *via* a transform of this equation:

$$J_{jk}(\omega) \equiv \int_{-\infty}^{\infty} G_{kj}(\tau)\mathrm{e}^{-i\omega_{kj}\tau}\mathrm{d}\tau \tag{20.7}$$

and the inverse relation using the inverse transform:

$$G_{kj}(\tau) = \int_{-\infty}^{\infty} J_{jk}(\omega)\mathrm{e}^{i\omega_{kj}\tau}\mathrm{d}\tau$$

In the simplest cases where the molecule in question is modelled as a sphere the correlation function is well described by an exponential decay:

$$G_{kj}(\tau) = G_{kj}(0)^2\exp\left(-\frac{|\tau|}{\tau_c}\right)$$

Frequently, a *reduced* correlation function is defined:

$$g_{kj}(\tau) \equiv \exp\left(-\frac{|\tau|}{\tau_c}\right)$$

and the correlation function is then:

$$G_{kj}(\tau) = G_{kj}(0)^2\, g_{kj}(\tau)$$

Recalling from Chapter 8 the Fourier transform of an exponential function and specifically, eqn (8.34):

$$\begin{aligned} J(\omega)_{\text{real}} &= G_{kj}(0)^2 \int_{-\infty}^{\infty} \mathrm{e}^{-\tau/\tau_c}\mathrm{e}^{-\omega\tau}\mathrm{d}\tau \\ &= G_{kj}(0)^2 \frac{2/\tau_c}{1/\tau_c^2 + \omega^2} \\ &= G_{kj}(0)^2 \frac{2\tau_c}{1+\omega^2\tau_c^2} \end{aligned} \tag{20.8}$$

The factor of two is there because the result in eqn (8.34) is for a single-sided decay in which the Fourier integral is from zero to ∞. Our current integral is much the same except that it runs from $-\infty$ to $+\infty$, twice the range of the single-sided decay. Hence the factor of two.

An interesting feature of the spectral density function is that it is area remains constant for varying τ_c[††]:

$$\begin{aligned} \text{Area} &= G_{kj}(0)^2 \int_{-\infty}^{\infty} \frac{\tau_c}{1+\tau_c^2\omega^2}\mathrm{d}\omega \\ &= G_{kj}(0)^2 \pi \end{aligned}$$

If we turn things around a bit and keep ω constant at, say, ω_0 and plot $J(\omega_0)$ as a function now of τ_c then we get Figure 20.7.

This indicates that there is a particular value of τ_c for which the spectral density function is a maximum and corresponds to the maximum transition probability. Thus, there is a particular molecular rotation frequency that is better than all others at inducing

[††]The formula for the integration of the Lorentzian function is not easy to find in integration tables but a search for $\int \frac{1}{x^2+a^2}\mathrm{d}x$ gives a result of $\frac{1}{a}\arctan\frac{x}{z} + C$ from which we can easily arrive at the result. I know I said that I would not say "it can be easily shown that…" but this is actually not so difficult if you have some calculus background. I will help you out a bit: $\frac{\tau_c}{1+\tau_c^2\omega^2} = \frac{1/\tau_c}{1/\tau_c+\omega^2} \Rightarrow \frac{a}{a^2+x^2} = a\left(\frac{1}{a^2+x^2}\right)$. You can take it from there or just use a symbolic algebra program.

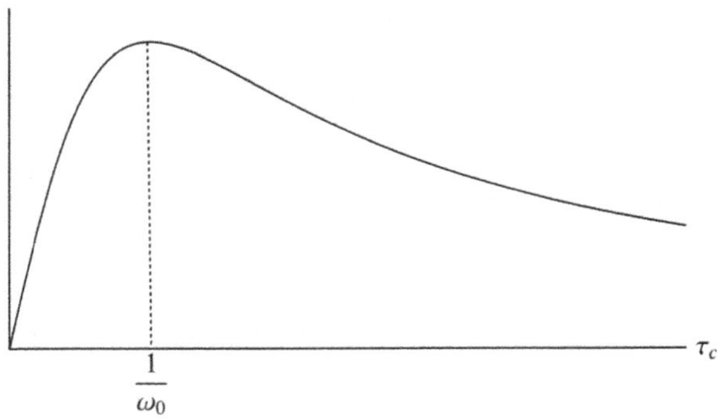

Figure 20.7 $J(\omega)$ as a function of τ_c.

transitions. It also turns out that that frequency is $1/\omega_0$ as can be easily verified by a little differential calculus.[‡‡]

Up to this point, we have been talking about stimulated transitions, that is, transitions for which an external agent was responsible. Of interest is the probability of spontaneous transitions in which a spin goes from a high-energy state to a lower energy state. These may be calculated *via* Einstein's coefficient of *spontaneous emission* (see Atkins and Friedman):

$$A_{fi} = \frac{8\pi h\nu^3}{c_3} B_{fi}$$

where A_{fi} is the coefficient of spontaneous emission and B_{fi} is the coefficient of stimulated absorption. Note that the coefficient is proportional to the *cube* of the transition frequency! Therefore, the higher the transition frequency the higher the probability of emission of a photon and a high-energy to lower-energy transition. The frequencies encountered in NMR spectroscopy are, relatively speaking, very low and the corresponding spontaneous emission probabilities are very, very small. In other words, a spin in a high-energy state will not likely spontaneously transition to a lower-energy state, at least not for a very long time. What is required for the emission process to occur within our lifetimes is stimulated emission. That is, some external mechanism that "pushes" the transition along.

We have already encountered a transition stimulation mechanism in the form of the $\vec{\mathbf{B}}_1$-field excitation pulse. The rf-field stimulates

[‡‡]It *is* easily shown, as a matter of fact, if you have some facility with calculus. However, it is also a bit tedious so you could also approach it using your favourite symbolic algebra software and solve the first derivative of $J(\omega_0) = \frac{\tau_c}{1 + \tau_c^2 \omega_0^2}$

transitions from low-energy to high-energy states and from high-energy to low-energy states. Since, at equilibrium, the low-energy state is more populated than the high-energy state the rate of low-to-high energy transitions is initially larger than those in the opposite direction and the net change is one in which spins are acquiring energy. Eventually, the populations of the high- and low-energy states will be equal and the transition rates between the two will be equal. At this point we say that the spin system is saturated. This is also the situation after a 90° pulse. When the spin populations are equal the spin-temperature for a two-level spin system is infinite, as discussed above.

20.4.3 The Dipole–Dipole Relaxation Mechanism

When we do a simple pulse-acquire NMR experiment we observe relaxation of the spin-system. That is, we see the effect of spins transitioning from high- to low-energy states. However, since the transmitter is off during data acquisition there must be some other mechanism that is providing for stimulated emission. Spontaneous emission, as noted above, is very highly unlikely at the frequencies encountered in NMR spectroscopy. There are two dominant mechanisms in solution spectroscopy, dipole–dipole and chemical-shift anisotropy, in order of importance.

In our consideration of magnetic dipoles in Chapter 11 we found an expression for the energy of a magnetic dipole in the field of another magnetic dipole (see eqn (11.49)):

$$E = \left(\frac{\mu_0}{4\pi}\right)\left(\frac{\vec{\mu}_1 \cdot \vec{\mu}_2}{r^3} - \frac{3(\vec{\mu}_1 \cdot \vec{\mathbf{r}}_2)(\vec{\mu}_2 \cdot \vec{\mathbf{r}}_2)}{r^5}\right)$$

Close inspection indicates that the energy is dependent on the angle between each dipole and the distance vector and on the inverse cube of the distance. As the relative orientation of the magnetic dipoles changes due to molecular rotation so does their energy. Consider two spins in the same rotating molecule. The angle between each dipole remains the same as the molecule rotates but the angle between each dipole and the inter-dipole vector changes. Thus, the first term in the above equation remains constant while the second does not.

As well, the magnetic field flux changes in an oscillatory fashion so that in effect each dipole experiences a $\vec{\mathbf{B}}_1$-field like effect, albeit a small one. Of course, the oscillatory motion must be at or near the transition frequency, ω_0 as shown in Figure 20.8. The other major

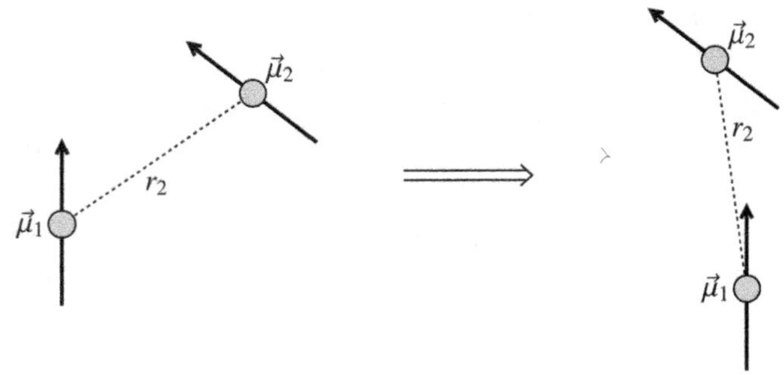

Figure 20.8 Angular relationships between two magnetic dipoles in a rotating molecule.

difference between this and the $\vec{\mathbf{B}}_1$-field is that the $\vec{\mathbf{B}}_1$-field is always oriented in the *x,y*-plane, whereas the oscillatory fields due to magnetic dipole rotation are randomly oriented. The radiation from the $\vec{\mathbf{B}}_1$-field is said to be coherent while that from random rotation is incoherent. This is a major source of stimulated emission in nuclear spin systems.

The local field at a spin is $\vec{\mathbf{B}}_{\text{loc}}$ and consists of the vector sum of the external field, $\vec{\mathbf{B}}_0$, and any fields generated from other sources. In the present case, one of those external sources is that due to dipole–dipole interaction, $\vec{\mathbf{B}}_D$.

$$\vec{\mathbf{B}}_{\text{loc}}(t) = \vec{\mathbf{B}}_0 + \vec{\mathbf{B}}_D(t)$$

$\vec{\mathbf{B}}_0$ is static and $\vec{\mathbf{B}}_D$ is time-dependent. Our current interest lies in $\vec{\mathbf{B}}_D$. We can construct a Hamiltonian from the energy equation given above (and using eqn (15.33)):

$$\hat{\mathbf{H}}_D = \left(\frac{\mu_0 \hbar^2 \gamma_1 \gamma_2}{4\pi r^3}\right)\left(\hat{\mathbf{I}} \cdot \hat{\mathbf{S}} - \frac{3}{r^2}(\hat{\mathbf{I}} \cdot \vec{\mathbf{r}})(\hat{\mathbf{S}} \cdot \vec{\mathbf{r}})\right) \tag{20.9}$$

This may be converted to spherical coordinates§§ by recognizing (using eqns (A.29)) that $\vec{\mathbf{r}}$ expressed in spherical coordinates is:

$$\vec{\mathbf{r}} = (r \sin(\theta) \cos(\phi), r \sin(\theta) \sin(\phi), r \cos(\theta))$$

Since we will be splitting things up and making a proper mess shortly, for clarity let us call the $\hat{\mathbf{I}} \cdot \hat{\mathbf{S}}$ portion of eqn (20.9) the first portion and the part containing $\hat{\mathbf{I}} \cdot \vec{\mathbf{r}}$ the second portion. It does look a little strange at first glance to be computing the dot product of an

§§This following series of calculations is not explicitly shown anywhere in one place in the literature as far as I can see. It is a bit tedious but will be very useful in assessing transition probabilities between energy levels.

operator with another operator, however, if we think of $\hat{\mathbf{I}}$ as a row vector of its component operators ($\hat{\mathbf{I}}_x$, $\hat{\mathbf{I}}_y$, $\hat{\mathbf{I}}_z$) and $\hat{\mathbf{S}}$ as a column vector we may do the dot product of the first portion:

$$\hat{\mathbf{I}}\cdot\hat{\mathbf{S}}=\hat{\mathbf{I}}_x\hat{\mathbf{S}}_x+\hat{\mathbf{I}}_y\hat{\mathbf{S}}_y+\hat{\mathbf{I}}_z\hat{\mathbf{S}}_z \tag{20.10}$$

The dot products of $\hat{\mathbf{I}}$ (again as a row vector) and $\vec{\mathbf{r}}$ (in spherical notation) in the second portion are:

$$\begin{aligned}\frac{(\vec{\mathbf{I}}\cdot\vec{\mathbf{r}})(\vec{\mathbf{S}}\cdot\vec{\mathbf{r}})}{r^2} &= (\hat{\mathbf{I}}_x\sin(\theta)\cos(\phi)+\hat{\mathbf{I}}_y\sin(\theta)\sin(\phi)+\hat{\mathbf{I}}_z\cos(\theta))\\ &\quad\times\ (\hat{\mathbf{S}}_x\sin(\theta)\cos(\phi)+\hat{\mathbf{S}}_y\sin(\theta)\sin(\phi)+\hat{\mathbf{S}}_z\cos(\theta))\\ &= \hat{\mathbf{I}}_x\hat{\mathbf{S}}_x\sin^2(\theta)\cos^2(\phi)+\hat{\mathbf{I}}_x\hat{\mathbf{S}}_y\sin^2(\theta)\sin(\phi)\cos(\phi)\\ &\quad+\hat{\mathbf{I}}_x\hat{\mathbf{S}}_z\sin(\theta)\cos(\phi)\cos(\theta)+\ \hat{\mathbf{I}}_y\hat{\mathbf{S}}_x\sin^2(\theta)\sin(\phi)\cos(\phi)\\ &\quad+\hat{\mathbf{I}}_y\hat{\mathbf{S}}_y\sin^2(\theta)\sin^2(\phi)+\hat{\mathbf{I}}_y\hat{\mathbf{S}}_z\sin(\theta)\sin(\phi)\cos(\theta)\\ &\quad+\hat{\mathbf{I}}_z\hat{\mathbf{S}}_x\sin(\theta)\cos(\phi)\cos(\theta)+\hat{\mathbf{I}}_z\hat{\mathbf{S}}_y\sin(\theta)\sin(\phi)\cos(\theta)+\hat{\mathbf{I}}_z\hat{\mathbf{S}}_z\cos^2(\theta)\end{aligned} \tag{20.11}$$

These may then be converted to ladder operator versions so that we may look at single transitions for a two-spin system. To do so we use:[¶¶]

$$\hat{\mathbf{I}}_x=\frac{\hat{\mathbf{I}}_++\hat{\mathbf{I}}_-}{2}\qquad \hat{\mathbf{I}}_y=\frac{\hat{\mathbf{I}}_+-\hat{\mathbf{I}}_-}{2i}$$

We show an example of how this works for the first term in eqn (20.10):

$$\begin{aligned}\hat{\mathbf{I}}_x\hat{\mathbf{S}}_x&=\frac{1}{2}(\hat{\mathbf{I}}_++\hat{\mathbf{I}}_-)\frac{1}{2}(\hat{\mathbf{S}}_++\hat{\mathbf{S}}_-)\\ &=\frac{1}{4}(\hat{\mathbf{I}}_+\hat{\mathbf{S}}_++\hat{\mathbf{I}}_+\hat{\mathbf{S}}_-+\hat{\mathbf{I}}_-\hat{\mathbf{S}}_++\hat{\mathbf{I}}_-\hat{\mathbf{S}}_-)\end{aligned}$$

For the other two terms:

$$\hat{\mathbf{I}}_y\hat{\mathbf{S}}_y=\frac{1}{4}(-\hat{\mathbf{I}}_+\hat{\mathbf{S}}_++\hat{\mathbf{I}}_+\hat{\mathbf{S}}_-+\hat{\mathbf{I}}_-\hat{\mathbf{S}}_+-\hat{\mathbf{I}}_-\hat{\mathbf{S}}_-)$$

$$\hat{\mathbf{I}}_z\hat{\mathbf{S}}_z=\hat{\mathbf{I}}_z\hat{\mathbf{S}}_z$$

[¶¶]These are from the definitions of $\hat{\mathbf{I}}_+$ and $\hat{\mathbf{I}}_-$. See eqn (15.5).

Thus, the first portion of eqn (20.9) is converted to these ladder operator terms:

$$
\begin{aligned}
&1: \frac{1}{4}(\hat{\mathbf{I}}_+\hat{\mathbf{S}}_+ + \hat{\mathbf{I}}_+\hat{\mathbf{S}}_- + \hat{\mathbf{I}}_-\hat{\mathbf{S}}_+ + \hat{\mathbf{I}}_-\hat{\mathbf{S}}_-)\\
&2: \frac{1}{4}(-\hat{\mathbf{I}}_+\hat{\mathbf{S}}_+ + \hat{\mathbf{I}}_+\hat{\mathbf{S}}_- + \hat{\mathbf{I}}_-\hat{\mathbf{S}}_+ - \hat{\mathbf{I}}_-\hat{\mathbf{S}}_-)\\
&3: \hat{\mathbf{I}}_z\hat{\mathbf{S}}_z
\end{aligned}
\tag{20.12}
$$

We do the same for eqn (20.11):[||]

$$
\begin{aligned}
&1: \hat{\mathbf{I}}_x\hat{\mathbf{S}}_x = \frac{1}{4}(\hat{\mathbf{I}}_+\hat{\mathbf{S}}_+ + \hat{\mathbf{I}}_+\hat{\mathbf{S}}_- + \hat{\mathbf{I}}_-\hat{\mathbf{S}}_+ + \hat{\mathbf{I}}_-\hat{\mathbf{S}}_-)\sin^2(\theta)\cos^2(\phi)\\
&2: \hat{\mathbf{I}}_x\hat{\mathbf{S}}_y = \frac{i}{4}(-\hat{\mathbf{I}}_+\hat{\mathbf{S}}_+ + \hat{\mathbf{I}}_+\hat{\mathbf{S}}_- - \hat{\mathbf{I}}_-\hat{\mathbf{S}}_+ + \hat{\mathbf{I}}_-\hat{\mathbf{S}}_-)\sin^2(\theta)\sin(\phi)\cos(\phi)\\
&3: \hat{\mathbf{I}}_x\hat{\mathbf{S}}_z = \frac{1}{2}(\hat{\mathbf{I}}_+\hat{\mathbf{S}}_z + \hat{\mathbf{I}}_-\hat{\mathbf{S}}_z)\sin(\theta)\cos(\phi)\cos(\theta)\\
&4: \hat{\mathbf{I}}_y\hat{\mathbf{S}}_x = \frac{i}{4}(-\hat{\mathbf{I}}_+\hat{\mathbf{S}}_+ - \hat{\mathbf{I}}_+\hat{\mathbf{S}}_- + \hat{\mathbf{I}}_-\hat{\mathbf{S}}_+ + \hat{\mathbf{I}}_-\hat{\mathbf{S}}_-)\sin^2(\theta)\sin(\phi)\cos(\phi)\\
&5: \hat{\mathbf{I}}_y\hat{\mathbf{S}}_y = \frac{1}{4}(-\hat{\mathbf{I}}_+\hat{\mathbf{S}}_+ + \hat{\mathbf{I}}_+\hat{\mathbf{S}}_- + \hat{\mathbf{I}}_-\hat{\mathbf{S}}_+ - \hat{\mathbf{I}}_-\hat{\mathbf{S}}_-)\sin^2(\theta)\sin^2(\phi)\\
&6: \hat{\mathbf{I}}_y\hat{\mathbf{S}}_z = \frac{i}{2}(-\hat{\mathbf{I}}_+\hat{\mathbf{S}}_z + \hat{\mathbf{I}}_-\hat{\mathbf{S}}_z)\sin(\theta)\sin(\phi)\cos(\phi)\\
&7: \hat{\mathbf{I}}_z\hat{\mathbf{S}}_x = \frac{1}{2}(\hat{\mathbf{I}}_z\hat{\mathbf{S}}_+ + \hat{\mathbf{I}}_z\hat{\mathbf{S}}_-)\sin(\theta)\cos(\phi)\cos(\theta)\\
&8: \hat{\mathbf{I}}_z\hat{\mathbf{S}}_y = \frac{i}{2}(-\hat{\mathbf{I}}_z\hat{\mathbf{S}}_+ + \hat{\mathbf{I}}_z\hat{\mathbf{S}}_-)\sin(\theta)\sin(\phi)\cos(\theta)\\
&9: \hat{\mathbf{I}}_z\hat{\mathbf{S}}_z\cos^2(\theta)
\end{aligned}
\tag{20.13}
$$

Now we substitute these back into eqn (20.11), not forgetting the factor of 3 and the negative sign from the original full Hamiltonian, eqn (20.9). Then we gather like terms together and add in the appropriate terms from eqn (20.10);[***] that is we have separately converted the first portion of eqn (20.9) to give eqn (20.12) and the second

[||] Note that I have left out the trigonometric terms on the left sides of each equation ... too little room!

[***] This really is tedious so I will not do it explicitly here, however, see below regarding the $\hat{\mathbf{B}}$ term. Be careful of signs if you decide to do it for yourself. Do not forget the factor of 3 and the negative sign in eqn (20.9). You will, of course, be applying trigonometric identities and Euler's theorem.

portion of eqn (20.9) to give eqn (20.13). We now recombine them to get our Hamiltonian in spherical coordinates:

$$\begin{aligned}
&\text{let} \quad K = \left(\frac{\mu_0 \hbar \gamma_1 \gamma_2}{4\pi r^3}\right) \\
&\hat{\mathbf{A}} = -K\hat{\mathbf{I}}_z\hat{\mathbf{S}}_z(3\cos^2(\theta) - 1) \\
&\hat{\mathbf{B}} = \frac{1}{4}K(\hat{\mathbf{I}}_+\hat{\mathbf{S}}_- + \hat{\mathbf{I}}_-\hat{\mathbf{S}}_+)(3\cos^2(\theta) - 1) \\
&\hat{\mathbf{C}} = -\frac{3}{2}K(\hat{\mathbf{I}}_z\hat{\mathbf{S}}_+ + \hat{\mathbf{I}}_+\hat{\mathbf{S}}_z)\sin(\theta)\cos(\theta)\exp(-i\phi) \\
&\hat{\mathbf{D}} = -\frac{3}{2}K(\hat{\mathbf{I}}_z\hat{\mathbf{S}}_- + \hat{\mathbf{I}}_-\hat{\mathbf{S}}_z)\sin(\theta)\cos(\theta)\exp(i\phi) \\
&\hat{\mathbf{E}} = -\frac{3}{4}K(\hat{\mathbf{I}}_+\hat{\mathbf{S}}_+)\sin^2(\theta)\exp(-i2\phi) \\
&\hat{\mathbf{F}} = -\frac{3}{4}K(\hat{\mathbf{I}}_-\hat{\mathbf{S}}_-)\sin^2(\theta)\exp(i2\phi) \\
&\mathbf{H}_D = \hat{\mathbf{A}} + \hat{\mathbf{B}} + \hat{\mathbf{C}} + \hat{\mathbf{D}} + \hat{\mathbf{E}} + \hat{\mathbf{F}}
\end{aligned} \tag{20.14}$$

This is the famous (or infamous) dipolar alphabet Hamiltonian associated with dipole-dipole interactions. The constant, K, is called the *dipolar constant*. The $\hat{\mathbf{B}}$ term (not to be confused with the magnetic field $\vec{\mathbf{B}}$!) is especially tricky to extract ... we will look at it here. We wish to combine all $\hat{\mathbf{I}}_+\hat{\mathbf{S}}_-$ and $\hat{\mathbf{I}}_-\hat{\mathbf{S}}_+$ terms in $\hat{\mathbf{B}}$ so we extract them from eqn (20.12) and (20.13). Terms 1 and 2 from eqn (20.12) and terms 1, 2, 4 and 5 from eqn (20.13) all contain these operators. We extract them along with their trigonometric factors:

$$\begin{aligned}
&1: \frac{1}{4}(\hat{\mathbf{I}}_+\hat{\mathbf{S}}_- + \hat{\mathbf{I}}_-\hat{\mathbf{S}}_+) \\
&2: \frac{1}{4}(\hat{\mathbf{I}}_+\hat{\mathbf{S}}_- + \hat{\mathbf{I}}_-\hat{\mathbf{S}}_+) \\
&3: \frac{1}{4}(\hat{\mathbf{I}}_+\hat{\mathbf{S}}_- + \hat{\mathbf{I}}_-\hat{\mathbf{S}}_+)\sin^2(\theta)\cos^2(\phi) \\
&4: \frac{i}{4}(\hat{\mathbf{I}}_+\hat{\mathbf{S}}_- - \hat{\mathbf{I}}_-\hat{\mathbf{S}}_+)\sin^2(\theta)\sin(\phi)\cos(\phi) \\
&5: \frac{i}{4}(-\hat{\mathbf{I}}_+\hat{\mathbf{S}}_- + \hat{\mathbf{I}}_-\hat{\mathbf{S}}_+)\sin^2(\theta)\sin(\phi)\cos(\phi) \\
&6: \frac{1}{4}(\hat{\mathbf{I}}_+\hat{\mathbf{S}}_- + \hat{\mathbf{I}}_-\hat{\mathbf{S}}_+)\sin^2(\theta)\sin^2(\phi)
\end{aligned} \tag{20.15}$$

Now we add them together. Note that the $\hat{\mathbf{I}}_+\hat{\mathbf{S}}_-$ and $\hat{\mathbf{I}}_-\hat{\mathbf{S}}_+$ terms in 4 and 5 in eqn (20.15) cancel due to opposite signs. Note also that, since 3, 4, 5 and 6 are from portion two of eqn (20.9), there is a negative sign and a factor of 3 multiplying into them. Addition of 1, 2, 5 and 6 and gathering of common factors gives:

$$\frac{1}{4}(\hat{\mathbf{I}}_+\hat{\mathbf{S}}_- + \hat{\mathbf{I}}_-\hat{\mathbf{S}}_+)(2 - 3\sin^2(\theta))$$

Now for the "trick":

$$\begin{aligned}&2 - 3\sin^2(\theta)\\ &\quad = 3 - 3\sin^2(\theta) - 1\\ &\quad = 3(1 - \sin^2(\theta)) - 1\\ &\quad = 3\cos^2(\theta) - 1\end{aligned}$$

and we finally end up with our $\hat{\mathbf{B}}$ term looking the way it is presented in the literature:

$$\hat{\mathbf{B}} = \frac{1}{4}(\hat{\mathbf{I}}_+\hat{\mathbf{S}}_- + \hat{\mathbf{I}}_-\hat{\mathbf{S}}_+)(3\cos^2(\theta) - 1)$$

It is convenient to revisit eqn (20.6), the transition probability between energy levels j and k:

$$W_{jk} = \int_{-\infty}^{\infty} G_{kj}(\tau)\mathrm{e}^{i\omega_{kj}(-\tau)}\mathrm{d}\tau$$

which we shall write in its expanded form from which it was derived, using our definition of $G_{jk}(\tau)$:

$$W_{jk} = \int_{-\infty}^{\infty} \overline{\langle k|\hat{\mathbf{H}}_D(\tau)|j\rangle\langle j|\hat{\mathbf{H}}_D(0)|k\rangle}\mathrm{e}^{-i\omega_{kj}\tau}\mathrm{d}\tau$$

and $\hat{\mathbf{H}}_D$ is the dipolar Hamiltonian. It is often the case that the Hamiltonian is split into a time-independent operator part and a time-dependent function part:

$$\hat{\mathbf{H}}_D = \hat{\mathbf{Z}}F(t) \tag{20.16}$$

where $\hat{\mathbf{Z}}$ is the (time-independent!) operator portion and $F(t)$ is the function portion.[†††] We note that the dipolar alphabet Hamiltonian is of this form, or rather is a sum of terms such as this. Furthermore,

[†††]As Neuhaus and Williamson point out (in their Appendix II), many texts use $\hat{\mathbf{A}}$ for the operator portion but this raises the potential for confusion with the $\hat{\mathbf{A}}$ from the dipolar alphabet operator. I will use Neuhaus and Williamson's $\hat{\mathbf{Z}}$, therefore.

each term is easily linked to a specific transition or set of transitions. With this operation in mind we rewrite our probability equation:

$$\begin{aligned} W_{jk} &= \langle k|\hat{\mathbf{H}}_D|j\rangle\langle j|\hat{\mathbf{H}}_D|k\rangle\int_{-\infty}^{\infty}\overline{F(\tau)F(0)}\mathrm{e}^{-i\omega_{kj}\tau}\mathrm{d}\tau \\ &= |Z_{kj}|^2\int_{-\infty}^{\infty}\overline{F(\tau)F(0)}\mathrm{e}^{-i\omega_{kj}\tau}\mathrm{d}\tau \end{aligned} \quad (20.17)$$

With our new way of expressing the Hamiltonian by splitting it into two parts comes the requirement of a redefinition of the correlation function (and the spectral density, ultimately). We will now write it as:

$$G(\tau) = \overline{F(\tau)F(0)}$$

and

$$G(0) = \overline{F(0)^2}$$

At $\tau = 0$, the correlation function is at its maximum. After this, for $\tau > 0$, $G(\tau)$ is decaying towards zero. Very often this decay is modelled as an exponential function, as we saw above in Section 19.4.2:

$$\begin{aligned} G(\tau) &= G(0)g(\tau) \\ &= \overline{F(0)^2}g(\tau) \\ &= \overline{F(0)^2}\exp\left(\frac{|\tau|}{\tau_c}\right) \end{aligned}$$

Inserting into the expression for transition probability:

$$\begin{aligned} W_{jk} &= |Z_{kj}|^2\overline{F(0)^2}\int_{-\infty}^{\infty}g(\tau)\mathrm{e}^{-i\omega_{kj}\tau}\mathrm{d}\tau \\ &= |Z_{kj}|^2\overline{F(0)^2}j(\omega) \end{aligned}$$

and if we use an exponential decay for the correlation function:[‡‡‡]

$$\begin{aligned} W_{jk} &= |Z_{kj}|^2\overline{F(0)^2}j(\omega) \\ &= |Z_{kj}|^2\overline{F(0)^2}\frac{2\tau_c}{1+\omega^2\tau_c^2} \end{aligned} \quad (20.18)$$

[‡‡‡]The factor of 2 in the term representing $j(\omega)$ comes about because we are integrating in eqn (20.17) from $-\infty$ to $+\infty$ since $G(\tau)$ is an even function about zero (see Appendix A). In Chapter 8 when we developed the Fourier transform of the single sided exponential decay we integrated from 0 to $+\infty$ and so, derived a formula that was $\frac{1}{2}$ of this.

In the dipolar alphabet Hamiltonian we have a set of operators, $\hat{\mathbf{Z}}$, and functions, F, in the form of eqn (20.16):

$$\begin{aligned}
\hat{\mathbf{A}} &= -\hat{\mathbf{I}}_z\hat{\mathbf{S}}_z F_0 \\
\hat{\mathbf{B}} &= \frac{1}{4}(\hat{\mathbf{I}}_+\hat{\mathbf{S}}_- + \hat{\mathbf{I}}_-\hat{\mathbf{S}}_+)F_0 \\
\hat{\mathbf{C}} &= -(\hat{\mathbf{I}}_z\hat{\mathbf{S}}_+ + \hat{\mathbf{I}}_+\hat{\mathbf{S}}_z)F_1 \\
\hat{\mathbf{D}} &= -(\hat{\mathbf{I}}_z\hat{\mathbf{S}}_- + \hat{\mathbf{I}}_-\hat{\mathbf{S}}_z)F_1^* \\
\hat{\mathbf{E}} &= -(\hat{\mathbf{I}}_+\hat{\mathbf{S}}_+)F_2 \\
\hat{\mathbf{F}} &= -(\hat{\mathbf{I}}_-\hat{\mathbf{S}}_-)F_2^*
\end{aligned} \tag{20.19}$$

F_0 represents, from eqn (20.14), $K(3\cos^2(\theta) - 1)$, $F_1 = \frac{3}{2}K\sin(\theta) \times \cos(\theta)\exp(-i\phi)$ and $F_2 = \frac{3}{4}K\sin^2(\theta)\exp(-i2\phi)$.

We need to evaluate the $\overline{F(0)^2}$ term in eqn (20.18) before we can proceed. This term applies to the full dipolar Hamiltonian but what we will do is break it up into manageable pieces as above and evaluate the individual terms F_i. Harris gives a definite integral formula for the average value of a function $f(\theta)$ (see the Further Reading section):

$$\overline{f(\theta)} = \frac{1}{2}\int_0^{\pi}\sin(\theta)f(\theta)\mathrm{d}\theta$$

which we can use to evaluate $\overline{F_0^2}$ in eqn (20.19):§§§

$$\begin{aligned}
\overline{F_0^2} &= \frac{1}{2}\int_0^{\pi}\sin(\theta)F_0^2\mathrm{d}\theta \\
&= \frac{1}{2}K^2\int_0^{\pi}\sin(\theta)(3\cos^2(\theta) - 1)^2\mathrm{d}\theta \\
&= \frac{4}{5}K^2
\end{aligned}$$

§§§This is very easy to do with wxMaxima or your favourite symbolic algebra software.

Similarly for F_1 and F_2:

$$\overline{F_1^2} = \overline{(F_1^*)^2} = \frac{1}{2}\left(-\frac{3}{2}\right)^2 \int_0^{\pi} \sin(\theta) F_1^2 \mathrm{d}\theta$$

$$= \frac{1}{2}\left(-\frac{3}{2}\right)^2 K^2 \int_0^{\pi} \sin(\theta)(\sin(\theta)\cos(\theta))^2 \mathrm{d}\theta$$

$$= \frac{2}{15} \times \frac{9}{4} K^2$$

$$= \frac{3}{10} K^2$$

$$\overline{F_2^2} = \overline{(F_2^*)^2} = \frac{1}{2}\left(-\frac{3}{4}\right)^2 \int_0^{\pi} \sin(\theta) F_2^2 \mathrm{d}\theta$$

$$= \frac{1}{2}\left(-\frac{3}{4}\right)^2 K^2 \int_0^{\pi} \sin(\theta)(\sin^2(\theta))^2 \mathrm{d}\theta$$

$$= \frac{8}{15} \times \frac{9}{16} K^2$$

$$= \frac{3}{10} K^2$$

Evaluation of the operator portion of eqn (20.19) involves realizing which transitions we are looking at. Each operator in the dipolar alphabet operates on a particular transition.

Operator $\hat{\mathbf{A}}$ causes no change in spin states; it simply extracts the angular momentum quantum numbers from whichever state it is operating on. $\hat{\mathbf{B}}$ causes "flip-flop" transitions and is associated with W_0 in Figure 20.9b. Operators $\hat{\mathbf{C}}$ and $\hat{\mathbf{D}}$ cause single spin flips and also extract angular momentum quantum numbers and are associated with W_I and

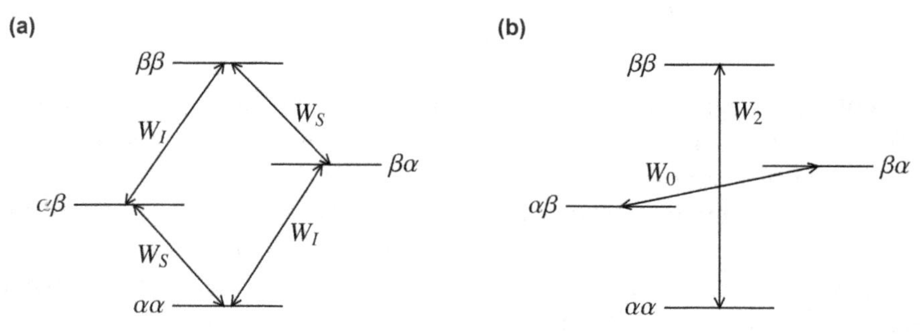

Figure 20.9 (a) Single-spin transitions for a two-spin, IS system; (b) zero- and two-spin transitions for a two-spin, IS system.

W_S in Figure 20.9a. $\hat{\mathbf{E}}$ and $\hat{\mathbf{F}}$ cause two simultaneous spin flips and are associated with W_2 in Figure 20.9b. We saw for example, in Section 18.8, that the operator $\hat{\mathbf{I}}_\alpha\hat{\mathbf{S}}_+$ is involved in the $|\alpha\beta\rangle$ to $|\alpha\alpha\rangle$ transition.

$$\begin{aligned}
&\hat{\mathbf{I}}_\alpha\hat{\mathbf{S}}_+|\alpha\beta\rangle \\
&\quad=\hat{\mathbf{I}}_\alpha|\alpha\rangle\hat{\mathbf{S}}_+|\beta\rangle \\
&\quad=|\alpha\alpha\rangle
\end{aligned}$$

Application of, say, $\hat{\mathbf{I}}_z\hat{\mathbf{S}}_+$ to the $|\alpha\beta\rangle$ state will cause a transition to the $|\alpha\alpha\rangle$ state and extract a quantum number of $+1/2$. $|Z_S|^2$ is then $(1/2)^2$. We then calculate W_S to be:

$$\begin{aligned}
&|Z_S|^2\overline{F_0^2}j(\omega) \\
&\quad=\left(\frac{1}{2}\right)^2\left(\frac{3}{10}K^2\right)j(\omega_S) \\
&\quad=\frac{3}{40}\left(\frac{\mu_0\hbar\gamma_I\gamma_S}{4\pi r^3}\right)^2\frac{2\tau_c}{1+\omega_S\tau_c^2} \\
&\quad=\frac{3}{20}\frac{\mu_0^2\hbar^2\gamma_I^2\gamma_S^2}{16\pi^2 r^6}\frac{\tau_c}{1+\omega_S\tau_c^2}
\end{aligned}$$

Similarly, W_I, W_0 and W_2 are:

$$\begin{aligned}
W_I&=\frac{3}{20}\frac{\mu_0^2\hbar^2\gamma_I^2\gamma_S^2}{16\pi^2 r^6}j(\omega_I)=\frac{3}{20}\frac{\mu_0^2\hbar^2\gamma_I^2\gamma_S^2}{16\pi^2 r^6}\frac{\tau_c}{1+\omega_S\tau_c^2} \\
W_0&=\frac{1}{10}\frac{\mu_0^2\hbar^2\gamma_I^2\gamma_S^2}{16\pi^2 r^6}j(\omega_I-\omega_S)=\frac{1}{10}\frac{\mu_0^2\hbar^2\gamma_I^2\gamma_S^2}{16\pi^2 r^6}\frac{\tau_c}{1+(\omega_I-\omega_S)^2\tau_c^2} \qquad (20.20) \\
W_2&=\frac{3}{5}\frac{\mu_0^2\hbar^2\gamma_I^2\gamma_S^2}{16\pi^2 r^6}j(\omega_I+\omega_S)=\frac{3}{5}\frac{\mu_0^2\hbar^2\gamma_I^2\gamma_S^2}{16\pi^2 r^6}\frac{\tau_c}{1+(\omega_I+\omega_S)^2\tau_c^2}
\end{aligned}$$

Why $(\omega_I-\omega_S)$ and $(\omega_I+\omega_S)$? With reference to Figures 18.20b and 20.9 we see that the W_0 pathway is governed by the $\hat{\mathbf{I}}_+\hat{\mathbf{S}}_-$ and $\hat{\mathbf{I}}_-\hat{\mathbf{S}}_+$ operators whose evolution frequency we saw in eqn (18.6) is the difference in frequency between spins I and S. Similarly, for the W_2 pathway the evolution frequency is the sum of the frequencies of spins I and S.

So, we now have the transition probabilities per second for transitions induced by dipolar interactions as a result of random molecular tumbling in solution. We shall use this in exploration of the Nuclear Overhauser Effect.

Dipolar coupling is usually the most important mechanism of relaxation in solution but not the only one. There is also chemical-shift anisotropy (CSA) induced relaxation in which, again as a result of random molecular tumbling, the size of the induced field varies and provides a relaxation pathway. By induced field, we mean that which is generally considered to be set up as a result of electronic motion in the molecule that itself is induced by the external field. This is the effect that is responsible for the chemical shift.

20.5 The Nuclear Overhauser Enhancement

The nuclear Overhauser effect (NOE) is a direct consequence of dipolar relaxation as we shall see. The analysis is relatively simple, requiring only some familiarity elementary rate calculations such as one might encounter in a course on chemical kinetics. We proceed by considering the analysis of a two-spin system and, with reference to Figures 20.9a and 20.9b, write an equation that describes the rate of change of a population of a level in the figures. For level $\alpha\alpha$ we write:

$$\frac{\mathrm{d}P_{\alpha\alpha}}{\mathrm{d}t} = -(W_I + W_S + W_2)P_{\alpha\alpha} + W_I P_{\beta\alpha} + W_S P_{\alpha\beta} + W_2 P_{\beta\beta} + C$$

$\mathrm{d}P_{\alpha\alpha}/\mathrm{d}t$ is the rate of net increase in the population of the $\alpha\alpha$ level and the first term represents the rate at which population leaves the $\alpha\alpha$ level and the other terms represent the rates at which population enters the $\alpha\alpha$ level. C is a constant to ensure that the population reaches its equilibrium value, $P^0_{\alpha\alpha}$. C is easily determined by setting $\mathrm{d}P_{\alpha\alpha}/\mathrm{d}t = 0$ and solving for C:

$$C = (W_I + W_S + W_2)\,P^0_{\alpha\alpha} - W_I P^0_{\beta\alpha} - W_S P^0_{\alpha\beta} - W_2 P^0_{\beta\beta}$$

We substitute for C and write our rate equation in terms of the difference in population from equilibrium population, $\Delta P_{\alpha\alpha} = P_{\alpha\alpha} - P^0_{\alpha\alpha}$, *etc.*:

$$\begin{aligned}\frac{\mathrm{d}\Delta P_{\alpha\alpha}}{\mathrm{d}t} &= -(W_I + W_S + W_2)P_{\alpha\alpha} + W_I P_{\beta\alpha} + W_S P_{\alpha\beta} + W_2 P_{\beta\beta} \\ &\quad + (W_I + W_S + W_2)P^0_{\alpha\alpha} - W_I P^0_{\beta\alpha} - W_S P^0_{\alpha\beta} - W_2 P^0_{\beta\beta} \\ &= -(W_I + W_S + W_2)(P_{\alpha\alpha} - P^0_{\alpha\alpha}) + W_I(P_{\beta\alpha} - P^0_{\beta\alpha}) + W_S(P_{\alpha\beta} - P^0_{\alpha\beta}) \\ &\quad + W_2(P_{\beta\beta} - P^0_{\beta\beta}) \\ &= -(W_I + W_S + W_2)\Delta P_{\alpha\alpha} + W_I \Delta P_{\beta\alpha} + W_S \Delta P_{\alpha\beta} + W_2 \Delta P_{\beta\beta}\end{aligned}$$

Similarly, for the other three levels:

$$\frac{d\Delta P_{\alpha\beta}}{dt} = -(W_I + W_S + W_2)\Delta P_{\alpha\beta} + W_0\Delta P_{\beta\alpha} + W_I\Delta P_{\beta\beta} + W_S\Delta P_{\alpha\alpha}$$

$$\frac{d\Delta P_{\beta\alpha}}{dt} = -(W_I + W_S + W_2)\Delta P_{\beta\alpha} + W_0\Delta P_{\alpha\beta} + W_I\Delta P_{\alpha\alpha} + W_S\Delta P_{\beta\beta}$$

$$\frac{d\Delta P_{\beta\beta}}{dt} = -(W_I + W_S + W_2)\Delta P_{\beta\beta} + W_I\Delta P_{\alpha\beta} + W_S\Delta P_{\beta\alpha} + W_2\Delta P_{\alpha\alpha}$$

Recalling the definition of the expectation value in eqn (17.3) and the form of the density matrix for $\hat{\mathbf{I}}_z$ and $\hat{\mathbf{S}}_z$, also derived in Chapter 17 we write:[¶¶¶]

$$\begin{aligned}\langle I_z\rangle(t) &= \mathrm{Tr}(\hat{\mathbf{I}}_z\hat{\sigma}(t)) \\ &= \sigma_{11} + \sigma_{22} + \sigma_{33} + \sigma_{44} \\ &= P_{\alpha\alpha} + P_{\alpha\beta} - P_{\beta\alpha} - P_{\beta\beta}\end{aligned} \qquad \begin{aligned}\langle S_z\rangle(t) &= \mathrm{Tr}(\hat{\mathbf{S}}_z\hat{\sigma}(t)) \\ &= \sigma_{11} + \sigma_{22} + \sigma_{33} + \sigma_{44} \\ &= P_{\alpha\alpha} - P_{\alpha\beta} + P_{\beta\alpha} - P_{\beta\beta}\end{aligned}$$

and, again using difference notation:

$$\Delta M_{Iz} = M_{Iz} - M_{Iz}^0$$

where M_{Iz}^0 is to be interpreted as the equilibrium value of the macroscopic magnetisation, M_{Iz} that we first encountered in Chapter 12 with respect to the Bloch equations. For our purposes M_{Iz}^0 may be defined as:[‖‖]

$$M_{Iz}^0 = P_{\alpha\alpha}^0 + P_{\alpha\beta}^0 - P_{\beta\alpha}^0 - P_{\beta\beta}^0$$

Similarly, we define M_{Sz}^0:

$$M_{Sz}^0 = P_{\alpha\alpha}^0 - P_{\alpha\beta}^0 + P_{\beta\alpha}^0 - P_{\beta\beta}^0$$

We note the following identities from the above definitions:

$$M_{Iz} + M_{Sz} = 2(P_{\alpha\alpha} - P_{\beta\beta})$$

$$M_{Iz} - M_{Sz} = 2(P_{\alpha\beta} - P_{\beta\alpha})$$

[¶¶¶] If you are reading Neuhaus and Williamson you will note that I and S are switched with respect to here. It is not that one of us is wrong but rather it is how we have chosen to define the I and S density matrices. See Chapter 17 for more information about this.

[‖‖] The equilibrium magnetisation is actually *proportional* to the populations and there should, by rights, be a constant in the expression but I will leave it out as its inclusion just unnecessarily adds to the complexity of the following equations. Leaving it out will not affect our ultimate results.

$$\Delta M_{Iz} + \Delta M_{Sz} = 2(\Delta P_{\alpha\alpha} - \Delta P_{\beta\beta})$$
$$\Delta M_{Iz} - \Delta M_{Sz} = 2(\Delta P_{\alpha\beta} - \Delta P_{\beta\alpha})$$

We now write:

$$\begin{aligned}\frac{\mathrm{d}\Delta M_{Iz}}{\mathrm{d}t} &= \frac{\mathrm{d}\Delta P_{\alpha\alpha}}{\mathrm{d}t} + \frac{\mathrm{d}\Delta P_{\alpha\beta}}{\mathrm{d}t} - \frac{\mathrm{d}\Delta P_{\beta\alpha}}{\mathrm{d}t} - \frac{\mathrm{d}\Delta P_{\beta\beta}}{\mathrm{d}t} \\ &= -(W_I + W_S + W_2)\Delta P_{\alpha\alpha} + W_I\Delta P_{\beta\alpha} + W_S\Delta P_{\alpha\beta} + W_2\Delta P_{\beta\beta} \\ &\quad -(W_I + W_S + W_2)\Delta P_{\alpha\beta} + W_0\Delta P_{\beta\alpha} + W_I\Delta P_{\beta\beta} + W_S\Delta P_{\alpha\alpha} \\ &\quad +(W_I + W_S + W_2)\Delta P_{\beta\alpha} - W_0\Delta P_{\alpha\beta} - W_I\Delta P_{\alpha\alpha} - W_S\Delta P_{\beta\beta} \\ &\quad +(W_I + W_S + W_2)\Delta P_{\beta\beta} - W_I\Delta P_{\alpha\beta} - W_S\Delta P_{\beta\alpha} - W_2\Delta P_{\alpha\alpha}\end{aligned}$$

After much labour, this will simplify to:****

$$\frac{\mathrm{d}\Delta M_{Iz}}{\mathrm{d}t} = 2(\Delta P_{\beta\beta} - \Delta P_{\alpha\alpha})(W_I + W_2) + 2(\Delta P_{\beta\alpha} - \Delta P_{\alpha\beta})(W_I + W_0)$$

and substituting from our identities above:

$$\begin{aligned}\frac{\mathrm{d}\Delta M_{Iz}}{\mathrm{d}t} &= -(\Delta M_{Iz} + \Delta M_{Sz})(W_I + W_2) - (\Delta M_{Iz} - \Delta M_{Sz})(W_I + W_0) \\ &= -(W_0 + 2W_I + W_2)\Delta M_{Iz} - (W_2 - W_0)\Delta M_{Sz}\end{aligned} \tag{20.21}$$

Similarly:

$$\frac{\mathrm{d}\Delta M_{Sz}}{\mathrm{d}t} = -(W_0 + 2W_S + W_2)\Delta M_{Sz} - (W_2 - W_0)\Delta M_{Iz} \tag{20.22}$$

These are known as the *Solomon equations* and are sometimes written in a shorter form:

$$\begin{aligned}\frac{\mathrm{d}\Delta M_{Iz}}{\mathrm{d}t} &= -\rho_{IS}\Delta M_{Iz} - \sigma_{IS}\Delta M_{Sz} \\ \frac{\mathrm{d}\Delta M_{Sz}}{\mathrm{d}t} &= -\rho_{IS}\Delta M_{Sz} - \sigma_{IS}\Delta M_{Iz}\end{aligned} \tag{20.23}$$

where the ρ_{IS} terms are called the *auto-relaxation* rate constants and the σ_{IS} constant is the *cross-relaxation* rate constant.†††† The

****Try it, either by hand or by using your favourite symbolic algebra software.

††††Do not confuse this σ with that used for the density matrix. This is a simple scalar constant, not a matrix.

interesting portion of these equations is the difference magnetisation multiplied by the σ_{IS} constant. If this is non-zero then the equations say that the rate of change of the *other* spin's *z*-magnetisation is affected by cross-correlation. The cross-relaxation constant includes W_0 and W_2 that we saw in eqn (20.20) are both proportional to $1/r^6$. This means of course that cross-relaxation is very sensitive to the inter-dipole distance and for practical purposes is effective only at short distances.

The NMR signal that we observe is proportional to the difference in populations of the energy levels prior to a read pulse for the spin in question. A read pulse is one that converts unobservable magnetism into observable magnetism. Thus, for spin I say, the signal is proportional to the population differences between the $\alpha\alpha$ and $\beta\alpha$ levels and between the $\alpha\beta$ and $\beta\beta$ levels from their equilibrium population values. We can get a qualitative idea of what the NOE is and how it works by looking at populations under various circumstances. First, for a homonuclear two-spin system at equilibrium we can draw a diagram detailing the equilibrium populations (Figure 20.10a).

The $\alpha\beta$ and $\beta\alpha$ states are very nearly equal in energy and therefore we shall assume that they contain equal numbers at equilibrium. There is, of course, an excess of δ spins in the $\alpha\alpha$ state and the $\beta\beta$ state is deficient by δ spins. We note from the diagram that the population differences are all δ at equilibrium. If we were to apply a read pulse now it would result in a "normal" NMR spectrum with two peaks in it, one at ω_I and one at ω_S. Now suppose that we saturate the S-spin transitions. That is, we mean apply rf-radiation until the populations at each end of the S-transitions are equally populated (Figure 20.10b).

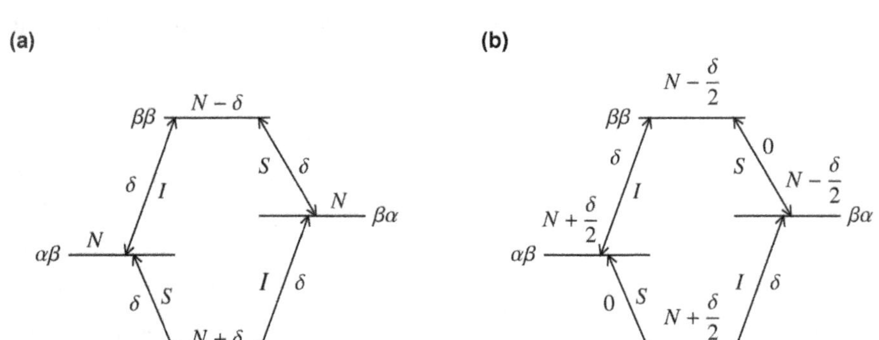

Figure 20.10 (a) Equilibrium spin populations for a two-spin, IS system; (b) Spin populations for saturated spin S.

Since the energy levels at each end of the S transitions are equally populated the population difference between them is now zero. The population difference between the I transition energy levels remains δ. In this situation, if we were to now issue a read pulse there would be a normal intensity peak for the I spin but no peak at all for the S spin.

What would happen if we were to wait for a time before issuing the read pulse? There would be some auto-relaxation *via* the single quantum transition pathways and possibly some cross-relaxation *via* the zero and double quantum transition pathways. It is the latter that is of interest here. For the moment we will neglect auto-relaxation and concentrate on the cross-relaxation pathways. Consider what would be the case if the W_2 pathway is used. The $\beta\beta$ level is depleted and the $\alpha\alpha$ level is enhanced *via* this pathway. In terms of population differences, this means that across the I spin transitions the differences increase. If the W_0 pathway is used then there is a transfer of population from the $\alpha\beta$ to the $\beta\alpha$ energy level (on our diagram at least) and the population differences across the I spin transitions decreases. So, the W_2 pathway leads to an increase in the signal strength and the W_0 pathway leads to a decrease in the signal strength. Collectively, these two observations are known as the nuclear Overhauser effect or more often just referred to as the NOE.

Saturation of one resonance and the observing the effect on another is called steady-state NOE. In this case if, for example, the S-spin resonance is being saturated then the rate of change of the S-spin populations is zero, and from eqn (20.23):

$$\frac{\mathrm{d}\Delta M_{Iz}}{\mathrm{d}t} = 0 = \rho_{IS} M_{Sz}^0 - \sigma_{IS}\Delta M_{Iz}$$

$$\Downarrow$$

$$\frac{M_{Iz} - M_{Iz}^0}{M_{Sz}^0} = \frac{\sigma_{IS}}{\rho_{IS}} = \frac{W_2 - W_0}{W_0 + 2W_I + W_2}$$

Note that M_{Sz} disappears from this since the S-spin transition is saturated. Consideration of the gyromagnetic ratios of I and S leads to:

$$M_{Sz}^0 = \frac{\gamma_S}{\gamma_I} M_{Iz}^0$$
$$\eta_I = \frac{M_{Iz} - M_{Iz}^0}{M_{Sz}^0} = \frac{\gamma_I}{\gamma_S}\frac{\sigma_{IS}}{\rho_{IS}} = \frac{\gamma_I}{\gamma_S}\frac{W_2 - W_0}{W_0 + 2W_I + W_2} \tag{20.24}$$

This is the definition of the NOE enhancement. The first thing we notice is that if the W_2 probability is larger than the W_0 probability then the NOE will be positive. If the W_0 probability is larger than the W_2 probability then the enhancement will be negative. Also, intriguingly, it is possible for these two probabilities to be equal, in which case the NOE enhancement will be zero. Secondly, we see that if the gyromagnetic ratios are different then this may also lead to enhancement of the signal. For example if the I spin is ^{13}C and the S spin is ^{1}H then saturation of the ^{1}H signal will potentially lead to a large enhancement in the ^{13}C signal.

What circumstances might transpire to make the cross-relaxation probabilities equal? To see this we must look into how the correlation time affects the NOE enhancement. Equations (20.20) relate the transition probabilities W_0, W_2, W_I and W_S to the correlation time, τ_c. Substitution of these into eqn (20.24) will give an expression for the NOE enhancement in terms of the correlation time. The expression can be simplified somewhat by making a couple of assumptions. First, we assume a homonuclear spin system so that $\gamma_I/\gamma_S = 1$. This being the case, we can say that $(\omega_I - \omega_S)\,\tau_c \ll 1$. Having made these assumptions we can write:

$$\begin{aligned}\sigma_{IS} &= W_2 - W_0\\ &= \frac{1}{10}K^2\tau_c\left(\frac{6}{1+\omega^2\tau_c^2} - 1\right)\end{aligned}$$

K is the dipolar coupling constant. W_0 collapses to one since it contains the $(\omega_I - \omega_S)\,\tau_c$ term that we are neglecting. Next, ρ_{IS} :

$$\begin{aligned}\rho_{IS} &= W_2 + W_0 + 2W_I\\ &= \frac{1}{10}K^2\tau_c\left(\frac{6}{1+4\omega^2\tau_c^2} + 1 + \frac{3}{1+\omega^2\tau_c^2}\right)\end{aligned}$$

The "4" in the denominator of the W_2 term comes from equating the two frequencies in the $(\omega_I - \omega_S)$ term:

$$\text{let}\quad \omega_I = \omega_S = \omega$$

$$\text{then}\quad (\omega_I + \omega_S)^2 = (2\omega)^2 = 4\omega^2$$

The NOE enhancement is:

$$\eta_I\{S\} = \frac{\sigma_{IS}}{\rho_{IS}} = \frac{\dfrac{6}{1+\omega^2\tau_c^2} - 1}{\dfrac{6}{1+4\omega^2\tau_c^2} + 1 + \dfrac{3}{1+\omega^2\tau_c^2}} = \frac{5+\omega^2\tau_c^2 - 4\omega^4\tau_c^4}{4\omega^4\tau_c^4 + 23\omega^2\tau_c^2 + 10} \quad (20.25)$$

Figure 20.11 shows a plot of this function of the NOE enhancement *versus* $\omega\tau_c$.

The plot shows that the NOE enhancement becomes zero when $\omega\tau_c \approx 1$. Exact calculation gives:

$$5+\omega^2\tau_c^2 - 4\omega^4\tau_c^4 = 0$$

$$\omega\tau_c = \sqrt{\frac{5}{4}} \approx 1.1$$

For a Larmor frequency of 500 MHz or 3.14×10^9 rad s^{-1} (see eqn (7.2)), the correlation time required for this to be the case is approximately 3.18×10^{-10} s. When the correlation time is smaller than this by a couple of orders of magnitude the NOE enhancement is positive. This corresponds to very rapid rotation in solution and is called the *fast motion regime* or the *extreme narrowing regime*. In this case $\omega\tau_c \ll 1$ and the spectral density function (see eqn (20.8)) reduces

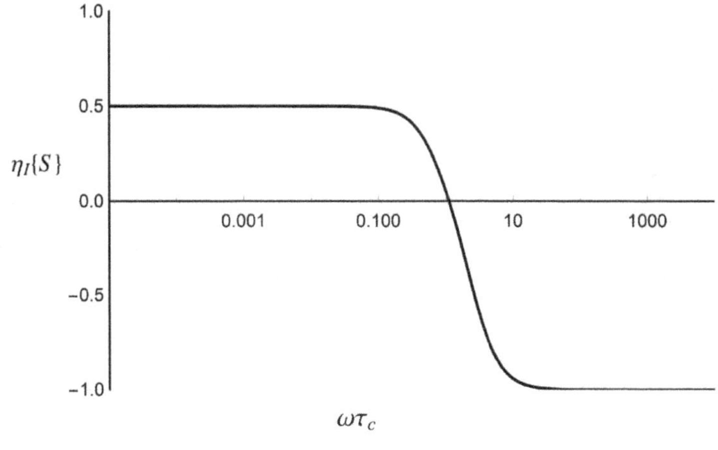

Figure 20.11 Plot of NOE enhancement *vs.* $\omega\tau_c$.

to $j(\omega) = 2\tau_c$ which is independent of Larmor frequency. This also means that under these circumstances, $j(\omega) = j(0) = 2\tau_c$. At the other extreme when the correlation time is a couple of orders of magnitude larger, the NOE enhancement is negative. A large correlation time corresponds to a slow rotation and is called the *slow motion regime* or *spin diffusion regime*. Here $\omega\tau_c \ll 1$ and we can write:

$$j(\omega) = \frac{2\tau_c}{1 + \omega^2\tau_c^2} \approx \frac{2\tau_c}{\omega^2\tau_c^2} = \frac{2}{\omega^2\tau_c}$$

and, since we just saw that $j(0) = 2\tau_c$:

$$j(\omega) = \frac{j(0)}{\omega^2\tau_c^2}$$

The perceptive reader will note that in eqn (20.25) there is no reference to the distance between the dipole moments. The NOE is very often used to determine the proximity of spins to each other and it turns out that this depends on the rate at which these process occur. We have already seen that the distance is incorporated into the dipolar coupling constant in the rate equations for W_I, W_S, W_0 and W_2. Equation (20.25) represents the maximum NOE achievable but the rate at which this is reached depends on $1/r^6$. This is why spins that are closer together can experience a higher NOE than those that are further apart from each other.

There is, to put it mildly, much to learn about relaxation and the nuclear Overhauser effect and we merely lightly touch on the fundamentals here. The reader is directed to the Further Reading section for more (much more) information on these and related topics and in particular, both Claridge and Keeler give very easy-to-follow introductions to the topic.

Further Reading

1. P. Atkins and R. Friedman, *Molecular Quantum Mechanics*, Oxford University Press, London, 4th edn, 2005.
2. T. Claridge, *High-Resolution NMR Techniques in Organic Chemistry*, Elsevier, London, 2009.
3. E. Fukushima and S. B. Roeder, *Experimental Pulsed NMR: A Nuts and Bolts Approach*, Addison-Wesley Publishing Co., Reading, Mass., 1981.

4. R. Harris, *Nuclear Magnetic Resonance Spectroscopy*, Pitman Books Ltd., London, 1983.
5. M. Goldman, *J. Magn. Res.*, 2001, **149**, 160–187.
6. J. Keeler, *Understanding NMR Spectroscopy*, John Wiley & Sons, Chichester, 2nd edn, 2010.
7. M. H. Levitt, *Spin Dynamics*, John Wiley & Sons, Chichester, 2001.
8. J. McConnell, *The Theory of Nuclear Magnetic Relaxation in Liquids*, Cambridge University Press, Cambridge, 1987.
9. D. Neuhaus and M. P. Williamson, *The Nuclear Overhauser Effect in Structural and Conformational Analysis*, John Wiley & Sons, New York, 2nd edn, 2000.
10. N. F. Ramsay, *Phys. Rev.*, 1956, **103**, 20–28.
11. D. Wolf, *Spin-temperature and Nuclear-spin Relaxation in Matter: Basic Principles and Applications*, Oxford University Press, New York, 1979.

21 Data Processing

If a man's wit be wandering, let him study the mathematics.

Francis Bacon

21.1 Fourier Transform Spectroscopy

There are two fundamental ways to obtain our NMR data in the NMR spectrometer. Most modern spectrometers employ *quadrature detection* of the raw NMR signal, however, it is also possible to use only one channel to obtain the data. We will begin by looking at single-channel (often called real) detection as it is the simpler of the two to describe (Figure 21.1).[†]

We will describe the operation of this type of system in more detail in Chapter 22. The point to understand here is that the raw analogue data are converted to discrete digital data ready for Fourier transformation to frequency-domain data. The Fourier transform of single-channel data is frequently called a *real* Fourier transform.

The fundamental data processing operation in modern NMR spectroscopy is the Fourier transform. Since our spectrometers are generally computer controlled, the data are usually digitised and stored in memory as discrete data points. This corresponds directly to the sampled time-domain functions discussed in Chapter 8. Thus, we

[†]In Figure 21.1 you will note the curious designation "audio signal" for the output of the mixer. This is intentional and will be explained in more detail later. Yes, the output frequencies from the mixer in your spectrometer *are* in the audio range and, yes, you can potentially listen to them!

Essential Mathematics for NMR and MRI Spectroscopists
By Keith C. Brown

Published by the Royal Society of Chemistry, www.rsc.org

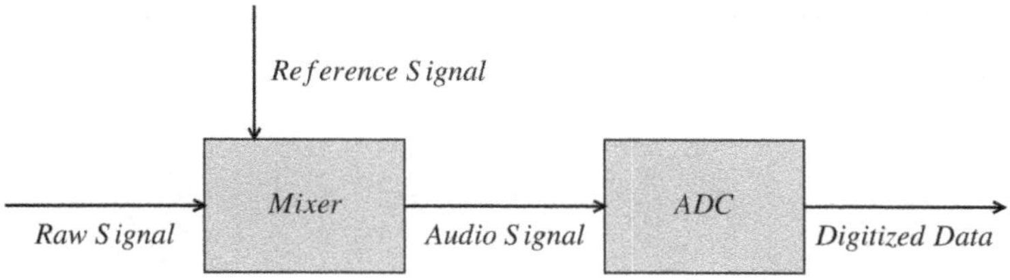

Figure 21.1 Signal flow in a single-channel system.

use the DFT to transform from time-domain data to frequency-domain data:‡

$$F(k) = \sum_{n=0}^{N-1} f(n) e^{-i2\pi kn/N}$$

The N discrete time-domain data points, $f(n)$, are separated by a time Δt and one might reasonably ask "what is the frequency separation of the discrete frequency-domain data resulting from the DFT?". It should be intuitive that the spacing of data points in the frequency-domain should depend on N and on Δt in some fashion.

Let us consider the situation in which during the acquisition time, T, all of the oscillating signals have an integer number of periods. The wave with the largest possible period (and the lowest possible frequency) will have completed exactly one cycle during T. From eqn (7.4) we find that:

$$\nu = \frac{1}{T} \quad \text{or} \quad \omega = \frac{2\pi}{T}$$

In the time-domain, $T = \Delta t N$ or $\Delta t = T/N$.

The data points in the frequency-domain are separated by $\Delta\omega$ (or $\Delta\nu$) and since there are N frequency-domain data points of which $F(N/2)$ is the Nyquist frequency (see Section 8.7) we can say that $\omega_{max} = \Delta\omega\, N/2$, or $\Delta\omega = 2\omega_{max}/N$, in an analogous kind of relationship to that between Δt and T. Regarding the above discussion of the wave with the lowest possible (non-zero) frequency and assuming that the frequency range is from zero to ω_{max}, this frequency must be located at the *second* data point in the frequency-domain. Thus, the frequency

‡Actually, nowadays we generally use the FFT since it is *much* faster than the DFT for large data sets but the FFT is really nothing more than a DFT going at jet-speed and gives exactly the same results as the DFT.

difference between the first and second frequency-domain data points (and all of the other data points) is:

$$\Delta\omega = \frac{2\pi}{T} = \frac{2\pi}{\Delta t N}$$

In regular frequency terms this is:

$$\Delta\nu = \frac{1}{T} = \frac{1}{\Delta t N} \qquad (21.1)$$

We see that there is definitely an inverse relationship between Δt and $\Delta\nu$. ν_{max} is frequently called *SW* (short for **S**weep **W**idth or **S**pectral **W**idth). Δt is frequently called *DW* (for **DW**ell time).

New users of NMR spectrometers are sometimes caught by the relationship between *N* and *SW*. The relationship is:

$$\nu_{\mathrm{max}} = \Delta\nu\frac{N}{2} = \frac{N}{2\Delta t N} = \frac{1}{2\Delta t}$$

$$\text{or} \qquad (21.2)$$

$$SW = \frac{1}{2DW}$$

If one decreases *SW* by a factor of, say, ten then the dwell time will *increase* by a factor of ten. For a given number of data points, *N*, this means that the time required to acquire all *N* data points will also increase by a factor of ten! The total experiment time will change quite drastically. Also, if the dwell time increases then the total acquisition time (*AQ*) must also increase:[§]

$$T = \Delta t N$$

$$\text{or} \qquad (21.3)$$

$$AQ = DW \times TD = \frac{TD}{2SW}$$

Thus, the unwary user increases their acquisition time by a factor of ten and wonders why their experiment time has so drastically increased.[¶]

[§] I am using the Bruker-specific parameters here. See Appendix B for a Bruker-to-Varian translation table.

[¶] Truly, knowledge is power.

So, for the single-channel data acquisition scheme, the spectral width varies from zero hertz up to ν_{max} hertz. Recall from Section 8.9 that the single-channel time-domain data transform to give a spectral data set that is symmetric or anti-symmetric about zero hertz. Since half of the spectral data are redundant we simply throw them out and plot only the other half. In our case we use the negative half. The zero-hertz position on the right side of the spectrum[‖] corresponds to the reference (sometimes called the carrier) frequency applied before digitisation. It is the reference frequency minus the NMR frequency at that point. The thing about this system is that *all* of the NMR frequencies in the sample need to be at or above the reference frequency ... frequencies below the reference frequency ("negative" frequencies) cannot be distinguished from those above using single-channel detection (as we saw in Section 8.9). However, *all* frequencies in the signal will be digitised, either at their correct positions in the spectrum if they fall in the *SW* range or at incorrect positions if they fall outside of the *SW* range. This is called *aliasing*. Analogue filtering of signals outside of the spectral width can reduce but not completely remove these peaks in the spectrum. Digital filtering (see Section 21.5) on the other hand, allows one to set up a spectral width of any desired size and not worry about aliasing of signals outside of the spectral window (Figure 21.2).

In the quadrature scheme, the raw signal is split into two. One portion passes to a phase-sensitive detector** and is mixed with the reference frequency to produce a difference signal that is then passed on for digitisation. The other signal is mixed with a reference frequency that is 90° out of phase with the first one and is, again, passed on for digitisation. There are two types of digitisation to consider, *simultaneous detection* and *sequential detection*. Simultaneous means that the real and imaginary signals are digitised at the same time, whereas sequential means that they are digitised one after the other. We will consider simultaneous digitisation first.

Figure 21.3 implies that there are actually two physical channels involved. This is not usually the case ... generally one signal is obtained from the NMR probe that is then split and one side is mixed with a reference signal to produce the audio frequency signal, while the other is mixed with the same reference signal that has a $\pi/2$ phase

‖In this respect NMR spectroscopy is a bit odd. Normally, one plots increasing frequencies from left to right.

**The phase-sensitive detector is treated in more detail in Section 19.3.

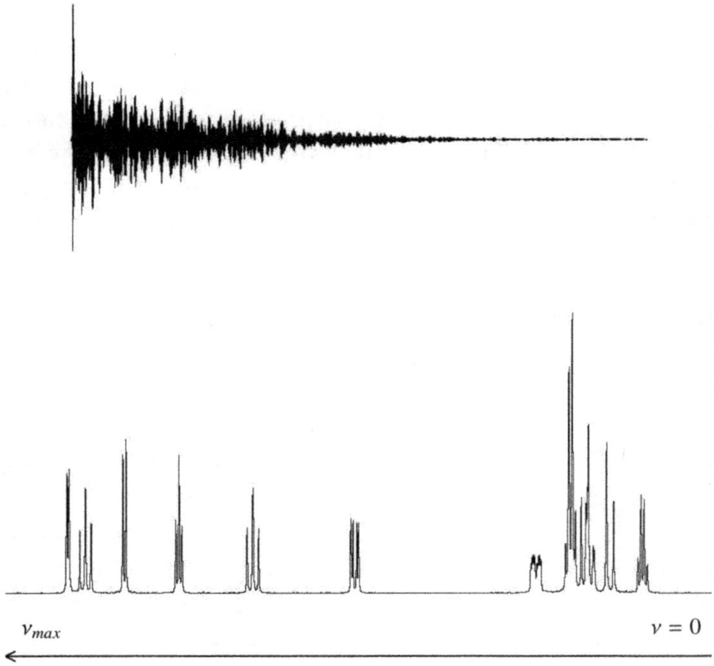

Figure 21.2 AQ, SW and *n* for a single-channel detection scheme.

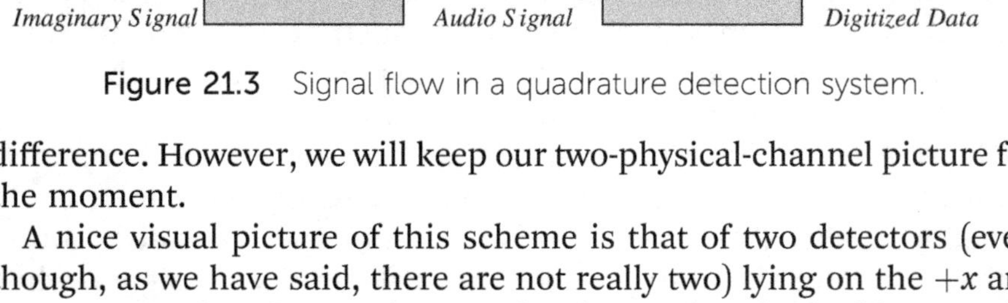

Figure 21.3 Signal flow in a quadrature detection system.

difference. However, we will keep our two-physical-channel picture for the moment.

A nice visual picture of this scheme is that of two detectors (even though, as we have said, there are not really two) lying on the $+x$ and $+y$ axes exactly 90° apart from each other and each "looking" at the rotating magnetisation vector. The "x" detector sees a cosine wave signal (the cosine detector of Section 8.5.7) and the "y" detector sees a sine wave signal (Figure 21.4).

Depending on the sense of rotation of the magnetisation vector, the y-channel signal will be either positive or negative. It is no accident

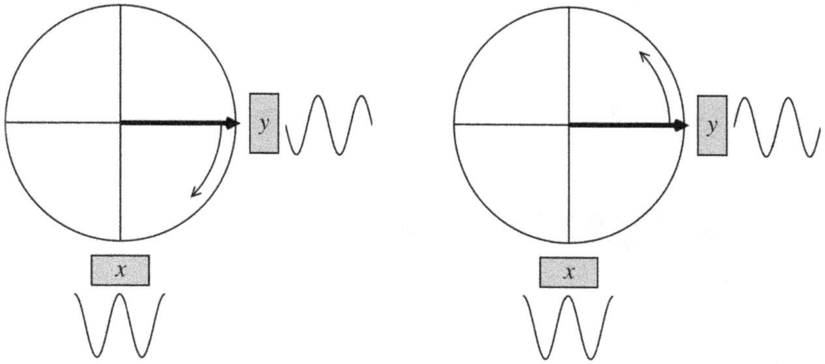

Figure 21.4 Quadrature detection channels.

that the *y*-channel signal appears as a sine wave and the *x*-channel as a cosine wave. Sine and cosine waves are related to each other by a 90° phase difference (see Appendix A):

$$\cos\left(\alpha - \frac{\pi}{2}\right) = \sin(\alpha)$$

$$\sin\left(\alpha - \frac{\pi}{2}\right) = -\cos(\alpha)$$

In the single-channel setup there is only one receiver; let us say for the sake of argument the *x*-channel from Figure 21.4. The receiver will "see" a projection of the precessing magnetisation but will not be able to determine whether it is rotating clockwise or counter-clockwise. In other words, it will not be able to distinguish between frequencies that are positive or negative with respect to the reference frequency. Using the quadrature scheme, we are able to tell whether a magnetisation is precessing clockwise or counter-clockwise, depending on the sign of the *y*-channel signal.

The sampling rate in quadrature simultaneous detection, for a given sweep width, is half that for single-channel detection. Why? The full spectral width is *SW*, as in single-channel detection, but any frequency at up to *SW*/2 will be detected and placed properly into the spectrum. Since, in quadrature detection, we place the reference frequency in the *centre* of the spectrum and allow two channels of acquired data to place the NMR peaks, negative or positive with respect to the reference frequency, the sweep width therefore goes from $-\nu_{max}$ to $+\nu_{max}$ and we need only digitise *half* the sweep width, from the centre to one side. Therefore, the sampling rate need only be half that for single-channel detection in order to obtain a spectrum in the

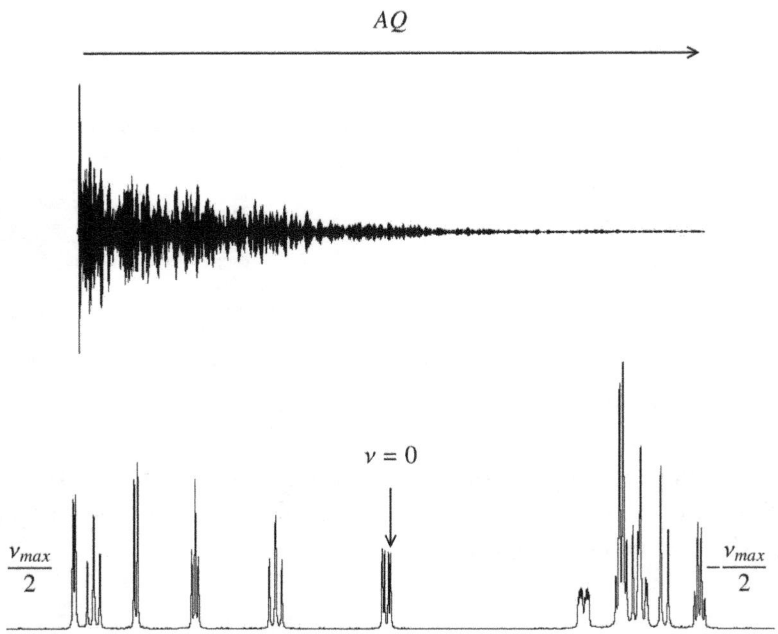

Figure 21.5 AQ, SW and *n* for a quadrature detection scheme.

same amount of time and with the same sweep width as that of a single-channel spectrum (Figure 21.5).

The dwell time in the quadrature scheme will be twice that in the single-channel detection scheme since the sampling frequency is half that of the single-channel method. Therefore, we may relate the sweep width and dwell time:

$$SW = \frac{1}{DW}$$

Since we are sampling *two* data points simultaneously we need only consider half of *TD* when considering the acquisition time:

$$AQ = DW \times \frac{TD}{2} = \frac{TD}{2SW} \tag{21.4}$$

and we see that it is the same as for single-channel detection. In the quadrature detection scheme, *TD* represents the number of real *plus* imaginary data points collected.[††] Since there are two channels in the quadrature detection scheme and the signal from each channel is

[††]Again, I use Bruker-specific language. I apologise to my Varian/Agilent/Jeol colleagues.

offset by $\pi/2$ from the other, we may consider the two signals to each be the components of a complex number.

Quadrature *sequential detection* is an ingenious method of simulating single-channel detection using two channels.[‡‡] Instead of simultaneously digitising each of the two channels they are digitised sequentially with a dwell time between them. The basis of the idea behind this is that we take the sequentially acquired data from the two channels and *pretend* that it is single-channel data which has been acquired with a reference frequency whose phase advances by 90° each time a point is acquired. A bit of reflection will show that this is equivalent to considering that the detectors in Figure 21.4 are used in the order x, y, $-x$, $-y$. A reference frequency phase shift of 0° degrees corresponds to use of the x-detector and 90° the y-detector. 180° and 270° then correspond to the negation of the x and y detectors. The reference frequency is still set at the centre of the sweep width as in simultaneous detection but *appears* to be at one end of the spectrum as in single-channel detection. The advantage of this scheme is that only one digitiser is needed where in the simultaneous scheme two are needed.

Let us use the vector model to examine this. Specifically, we will look at the behaviour of a vector rotating at the rotating frame frequency; zero hertz in other words. In quadrature simultaneous detection this would correspond to a resonance at the centre of the spectral width where the zero hertz point is located. If we were to use quadrature simultaneous detection to observe the rotation of this vector we would see it as motionless within the rotating frame of reference. This is because we would be using the same detectors after each dwell time with no advances in the reference frequency phase. However, alternately using the x and y detectors and changing the sign of every other pair of x and y data, which is equivalent to advancing the reference phase, results in an apparent frequency for the zero-hertz vector.

From Figure 21.6 we see that, using detectors in the order x, y, $-x$, $-y$ results in an apparent frequency for the zero-hertz vector. Furthermore, after four dwell times the vector has concluded one apparent revolution. What is the frequency of the apparent motion of

[‡‡]There is a *serious* paucity of information about this mode of data digitisation in the literature. If your spectrometer is reasonably new then this mode should be available and you should know about it, whether or not you plan to use it. Also, do not confuse this with Time Proportional Phase Increments or TPPI. There is a subtle difference between the two. Quadrature sequential detection refers to the direct detection process, whereas TPPI refers to "sequential detection" in the indirect dimension of an n-dimensional experiment. Knowledge is power.

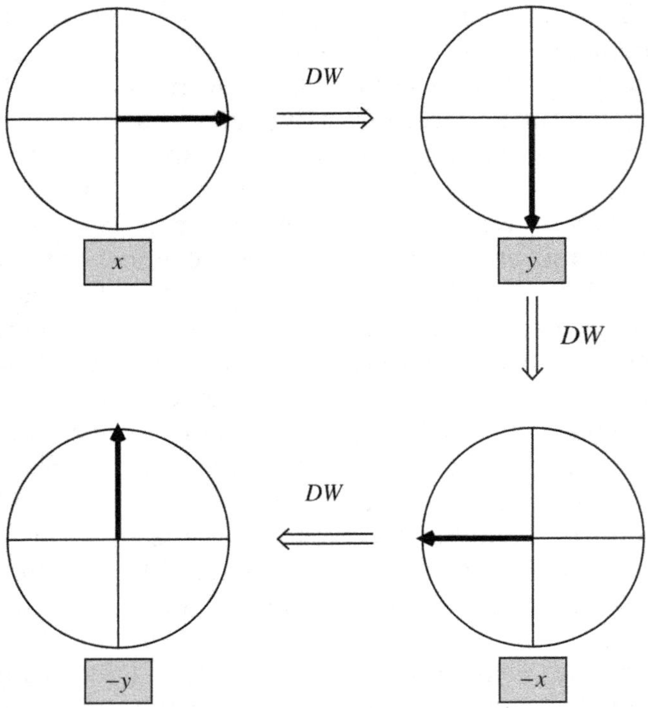

Figure 21.6 Apparent change in 0 Hz precession frequency with advanced receiver phases.

the vector? If the dwell time were, say, 1/8 s then in 1 s the vector would appear to have made two revolutions, *i.e.* $\nu = 2$ Hz. More generally we can say that the apparent frequency of the zero-hertz vector is:

$$\nu_a = \frac{1}{4DW}$$

$$DW = \frac{1}{4\nu_a}$$

We have moved the apparent frequency from zero to ν_a. What happens if we move the apparent frequency all the way to one side of the spectrum (at $1/2\nu_{\text{max}}$)?

$$DW = \frac{1}{4\left(\frac{1}{2}\right)\nu_{\text{max}}} = \frac{1}{2\nu_{\text{max}}} = \frac{1}{2SW}$$

In order to move the zero-hertz frequency to one side of the spectrum we sample at twice the frequency that we would for simultaneous detection or the same frequency that we would for single-channel detection.

21.2 The 2D Fourier Transform

The two-dimensional NMR experiment works by storing a series of free-induction decays (FID), each of which has been acquired with an incremented time-delay in the pulse program. During the incremented time-delay some type of evolution occurs, coupling or chemical-shift perhaps, which is 'encoded' into the FID. Processing pulls out this information and plots it along one of the axes of the two-dimensional plot. The inversion recovery experiment used for calculation of T_1 values gives a nice introduction to the general principles. The pulse-sequence for the inversion recovery experiment is shown in Figure 21.7.

The π-pulse inverts the magnetisation and is followed by a time-delay, t_1 (*not* to be confused with T_1!). This time-delay varies from experiment to experiment, usually going from short to long. Note that the acquisition time has been labelled as t_2. After the t_1-delay a $\pi/2$-pulse rotates the magnetisation into the x,y-plane for acquisition.

Figures 21.8 and 21.9 illustrate what happens to the magnetisation with differing t_1 times. If we do a series of these experiments, saving each t_1 time experiment in a separate FID, we will get a series of spectra in which the intensity of the peaks is such that they start out negative and, as t_1 is incremented, approach zero and become positive. The rate at which this occurs is dependent on the value of T_1 for each peak in the spectrum. The peak intensities may then be fitted to an exponential function and the T_1 value for the peak extracted. We say that the t_1 time period is the evolution period and the t_2 time period is the acquisition period. The T_1 information is then "encoded" in the t_1 period.

The actual two-dimensional NMR experiment is constructed in a very similar fashion. There is a pulse or sequence of pulses (called the

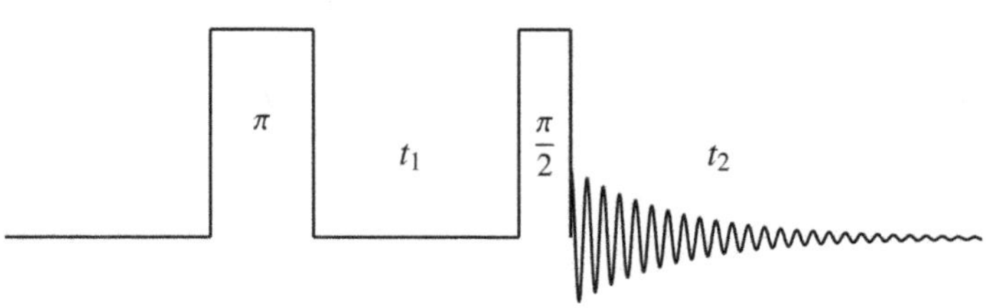

Figure 21.7 The inversion recovery pulse sequence.

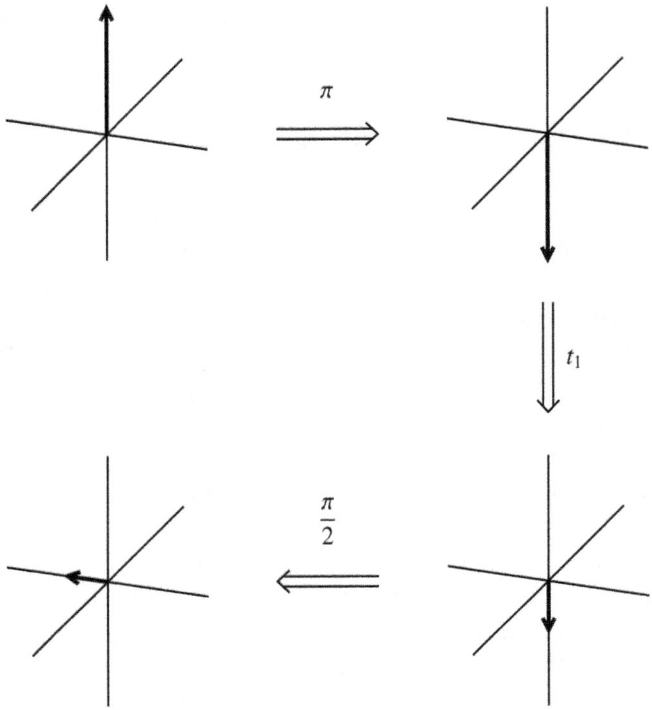

Figure 21.8 Magnetisation response to inversion recovery pulse sequence for short t_1.

preparation time) followed by a variable time, t_1 (called the evolution time) followed possibly by more pulses and delays (the mixing time) and then acquisition (the acquisition time) of the FID during the t_2 time-period. For each value of t_1 an FID is acquired and stored in a serial file so that the ultimate data set is a string of FIDs.

If we transform each of these FIDs we will get a set of spectra much as in Figure 21.10. This corresponds to transforming the t_2 (time) data into f_2 (frequency) data. Then we make a bit of a leap and transform in the t_1 direction to get f_1 data.[§§] What we do here is treat each data point in the t_1 dimension as a data point to be used in a DFT. Thus, data point #1 (for example) in *each* spectrum is considered to be a time-domain data point to be used in the DFT. Therefore, we are doing DFTs in two separate dimensions.

[§§]One of those fundamental concepts that once fully grasped is supremely simple and one that I wish I had thought of.

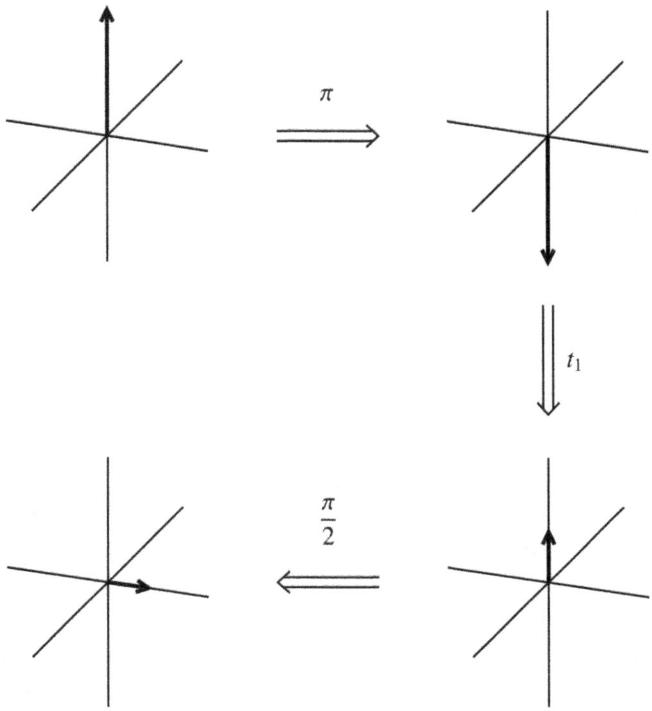

Figure 21.9 Magnetisation response to inversion recovery pulse sequence for long t_1.

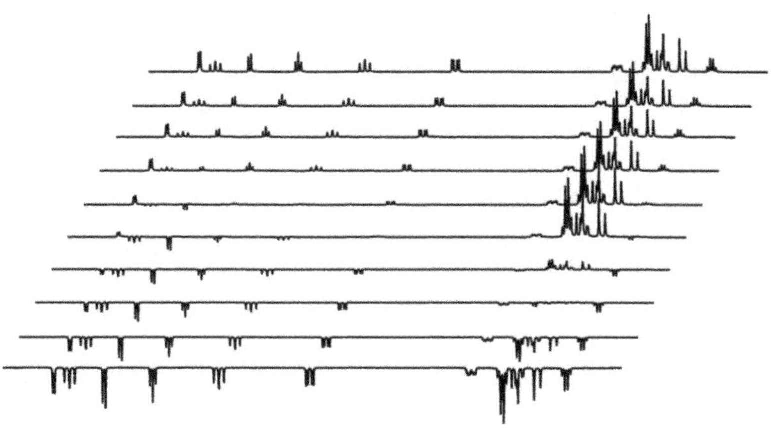

Figure 21.10 Transformed spectra from an inversion recovery experiment.

Formally, the two-dimensional Fourier transform is written as:

$$F(\omega_1, \omega_2) = \int_{-\infty}^{\infty}\int_{-\infty}^{\infty} f(t_1, t_2)\mathrm{e}^{-i2\pi(\omega_1 t_1 + \omega_2 t_2)}\mathrm{d}t_1\mathrm{d}t_2 \qquad (21.5)$$

The transform may also be done in one dimension only, leaving the other in the time-domain:

$$F(\omega_1, t_2) = \int_{-\infty}^{\infty} f(t_1, t_2)\mathrm{e}^{-i2\pi(\omega_1 t_1)}\mathrm{d}t_1$$

21.3 Window Functions

The raw time-domain data or FID obtained in our NMR experiment can be directly Fourier transformed to give a frequency-domain spectrum. More often, however, some pre-transform processing is done using a *window function.*[¶¶] Frequently the application of an exponentially decaying window function is applied to the time-domain data, usually in an attempt to improve the signal-to-noise ratio of the spectrum or to reduce truncation artefacts. Each data point is multiplied by an exponential function of the form:

$$\mathrm{e}^{-LBt}$$

where *LB* is referred to as the line broadening parameter. One can view the effect of this in a couple of ways as was done in Chapter 8. First, it could be seen as an add-on to the free-induction decay analysis of Section 8.5.7:

$$\begin{aligned} f(t) &= U(t)\mathrm{e}^{-\alpha t}\mathrm{e}^{-LBt}\cos(\omega_0 t) \\ &= U(t)\mathrm{e}^{-(\alpha+LB)t}\cos(\omega_0 t) \end{aligned}$$

$$\begin{aligned} F(\omega) = \frac{1}{2}&\left[\frac{\alpha + LB}{(\alpha+LB)^2 + (\omega-\omega_x)^2} + \frac{\alpha + LB}{(\alpha+LB)^2 + (\omega-\omega_x)^2}\right] \\ &+ \frac{i}{2}\left[\frac{(\omega-\omega_x)}{(\alpha+LB)^2 + (\omega-\omega_x)^2} + \frac{(\omega+\omega_x)}{(\alpha+LB)^2 + (\omega+\omega_x)^2}\right] \end{aligned}$$

[¶¶]The term apodisation function is used interchangeably with window function.

This view simply alters the constant in the single-sided exponential decay function, adding in the constant *LB*.

The second and of course, equivalent way of looking at it is as a convolution of the transform of the line-broadening exponential decay function with the transform of the raw FID. As we saw in Section 8.6, this has the effect of smoothing out one function with the other; the Lorentzian lines of the transform of the decaying FID are smoothed even more by the exponential multiplication.

You can see clearly in Figure 21.11 that the application of the line-broadening exponential multiplication leaves the early part of the FID relatively intact, while severely truncating the latter part of the FID, which consists mostly of noise. The effect after Fourier transforming is that the noise level in the spectrum is also reduced and the signal-to-noise ratio is increased. However, as the term "line broadening" suggests, there is a price to pay for using this function ... the lines in the spectrum are broadened by *LB* hertz. Figure 21.12 shows a transformed spectrum with and without application of exponential multiplication.

Exponential multiplication is not the only window function generally available to the NMR spectroscopist. Application of the Gaussian

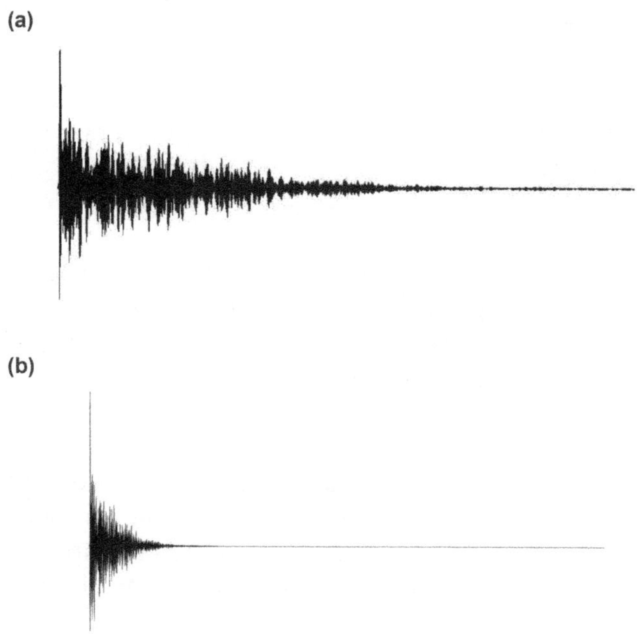

Figure 21.11 (a) Raw FID and (b) FID after exponential multiplication.

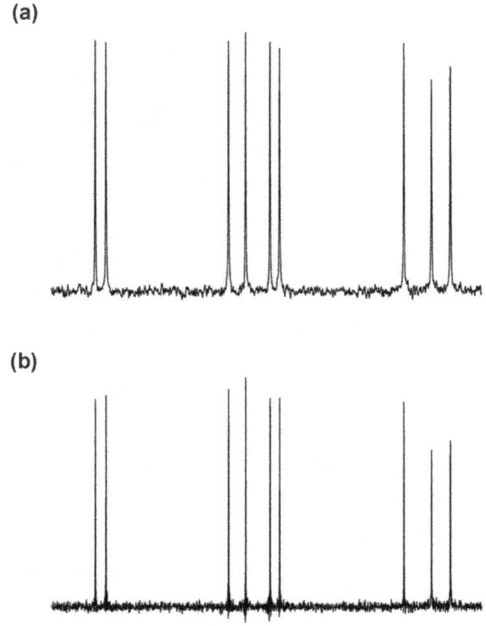

Figure 21.12 (a) 3 Hz exponential multiplication applied prior to Fourier transform; (b) No exponential multiplication applied.

function will enhance resolution at the cost of decreased signal-to-noise.[‖] The form of the function is:

$$e^{-GBt^2}$$

where *GB* is the lineshape parameter. A modified version of this is the shifted Gaussian:

$$e^{-(t-GB)^2}$$

which includes a "shift" parameter, *GB*, which shifts the maximum of the function to some point in time later than zero.

Frequently, one will use a combination of exponential multiplication and shifted Gaussian functions:

$$e^{-LBt}e^{(t-GB)^2}$$

In this case the exponential multiplication is not decaying but rather increasing such that the FID is increased at the beginning and then decays at some time later as the Lorentzian function becomes dominant. The effect in the transformed spectrum is to change the

[‖]Be aware that many publications will refer to a Gaussian resolution enhancement that does *not* use this function but rather the Lorentz-to-Gaussian function.

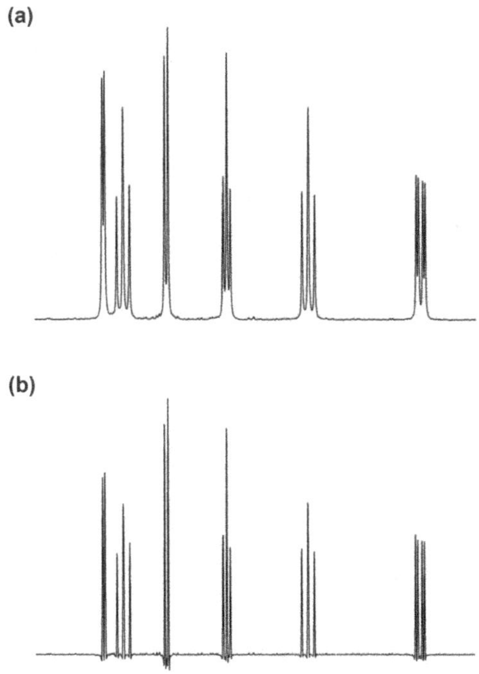

Figure 21.13 (a) Exponential multiplication applied to FID; (b) Lorentz-to-Gaussian multiplication applied to FID.

lineshapes from Lorentz to Gaussian and the price that is paid is an increase in noise since the noise at end of the FID is enhanced. However, since the Gaussian lineshape is thinner than the Lorentzian lineshape there will be an enhancement of resolution in the spectrum with judicious choice of *LB* and *GB* (see Figure 21.13).

Other types of pre-transform window functions are the sine bell function, the shifted sine bell function and the trapezoidal function. Figure 21.14 displays plots of the commonly used window functions.

21.4 Zero Filling

What would happen if we were to add zeros to the end of the acquired data set? Why would we even consider doing this? To see what this does for us, let us consider an amalgam of several of the ideas contained in Chapter 8. Our sampled data will be from the function $\cos(16\pi t) + i\sin(16\pi t)$ containing a single 8 Hz oscillation. Furthermore, the signal will decay exponentially so that we will be modelling

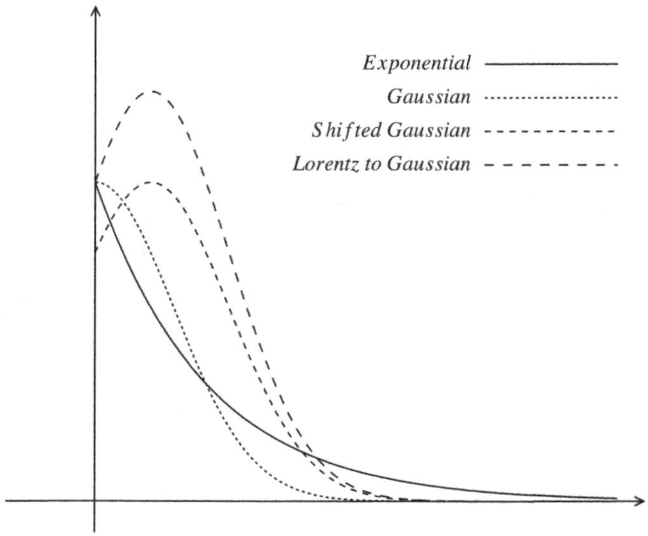

Figure 21.14 Some common window function envelopes.

a free-induction decay signal. In order to clearly see the results of our Fourier transforms we will need to use more than eight data points ... we shall use a 32-point data set.

$$f(t) = (\cos(16\pi t) + i\sin(16\pi t))e^{-\alpha t}$$

The signal that we are modelling consists of a cosine signal plus a sine signal in a quadrature detection scheme. If the signal were eternal (non-decaying in other words) then we would see in the spectrum an infinitely thin spike at a frequency of 8 Hz (see Section 8.5.5). However, since we are modelling a free-induction decay we expect that the observed peak in the spectrum will have finite width (see Section 8.5.7) that will depend on the rate of decay (the value of α). One can consider that this is the convolution of the Fourier transform of $\cos(16\pi t) + i \sin(16\pi t)$ with an exponential decay function.

$$g(t) = \cos(16\pi t) + i\sin(16\pi t) \quad h(t) = e^{-\alpha t}$$

$$f(t) = g(t)h(t)$$

$$F(\omega) = G(\omega)^{*}H(\omega)$$

$$= \frac{\alpha}{\alpha^2 + (\omega - \omega_0)^2} + i\frac{(\omega - \omega_0)}{\alpha^2 + (\omega - \omega_0)^2}$$

We have already seen what this looks like for a continuous function in Figures 8.11 and 8.13. For a discrete 32-point data set it looks like Figure 21.15a.

Note that in Figure 21.15a *all* of the DFT spectrum points are included. That is, 0 Hz is in the centre of the plot with positive frequencies to the right and negative frequencies to the left, as in Section 8.9. Note also that there are no zero-intensity points in the spectrum. The very small, constant-intensity "peaks" are an artefact of the DFT and are inversely proportional to the number of data points.

Let us now add 32 zero points to the end of the last data set to make a total of 64 data points. The Fourier transform will treat this just the same as the 32-point acquired data set but now there are 64 data points going into the transform. Recalling our discussion of the DFT, this means that there will be 64 frequency-domain points generated. Recalling eqn (21.1), since we have changed N from 32 to 64, $\Delta\nu$ will be halved meaning that the *digital resolution* in the frequency domain is doubled. One can see this effect by comparing Figures 21.15a and 21.15b. No new information is added to the spectrum but this technique can be useful to improve the digital resolution of the spectrum.

There is a complication to this, however. Introducing zeros at the end of the data is equivalent to multiplication of the first 32 data points of a 64-point set by 1 and the last 32 by zero. In other words, $\cos(16\pi t) + i\sin(16\pi t)e^{-\alpha t}$ times $\mathrm{Rect}(t/\tau)$ time-shifted to the right by $\tau/2$. Recall

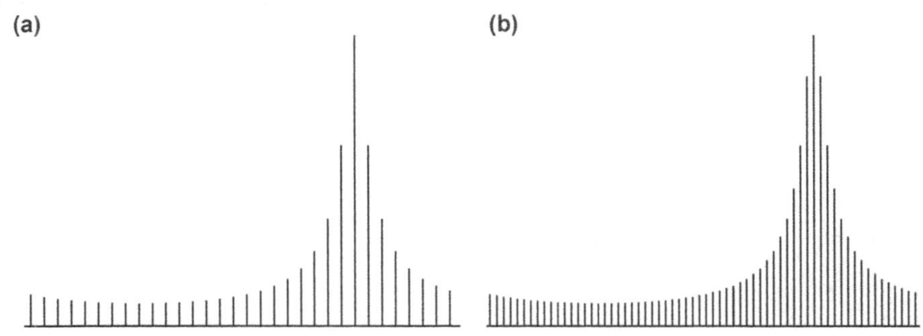

Figure 21.15 (a) The DFT of a 32-point, exponentially decaying, quadrature detected signal; (b) The DFT of a 32-point, zero-filled, exponentially decaying, quadrature detected signal.

from Chapter 8 that a time-shift equates to a multiplication in the frequency-domain by $\exp(i\omega k)$:

$$f\left(t-\frac{\tau}{2}\right)=\mathrm{Rect}\left(\frac{t}{\tau}-\frac{\tau}{2}\right)$$

$$F(\omega)=\tau\frac{\sin\left(\omega\frac{\tau}{2}\right)}{\omega\frac{\tau}{2}}\mathrm{e}^{-i\omega\frac{\tau}{2}}$$

$$=\tau\left[\frac{\cos\left(\omega\frac{\tau}{2}\right)\sin\left(\omega\frac{\tau}{2}\right)}{\omega\frac{\tau}{2}}-i\frac{\sin^2\left(\omega\frac{\tau}{2}\right)}{\omega\frac{\tau}{2}}\right]$$

Convolving this with the previous results.

$$g(t)=\cos(16\pi t)+i\sin(16\pi t)\quad h(t)=\mathrm{e}^{-\alpha t}$$

$$k(t)=\mathrm{Rect}\left(\frac{t}{\tau}-\frac{\tau}{2}\right)$$

$$f(t)=g(t)h(t)k(t)$$

$$F(\omega)=(G(\omega)^{*}H(\omega))^{*}K(\omega)$$

we get a peak at 8 Hz with a line-width due to the exponential decay and "sinc squiggles" that are characteristic of a free-induction decay that has been cut off before substantially decaying to zero. Comparison of a non-cut-off FID and a cut-off FID and the resulting transformed data points are shown in Figures 21.16 and 21.17.

This is one major reason for using window functions, especially the exponential function. If the free-induction decay can be made to

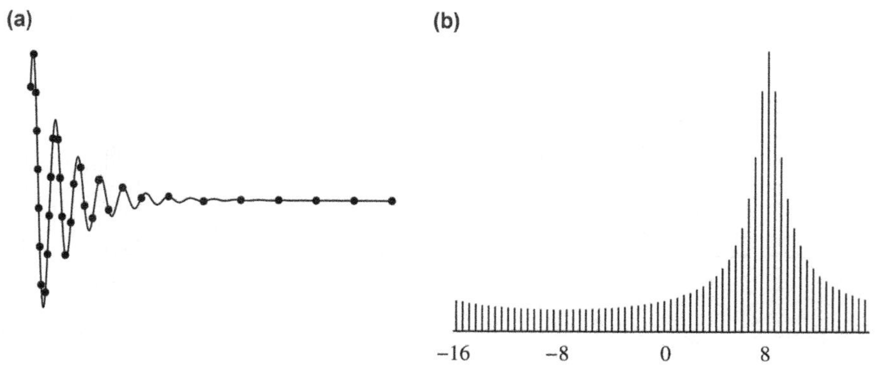

Figure 21.16 (a) FID and sampled data points, decayed to zero before zero-filling; (b) transformed spectrum.

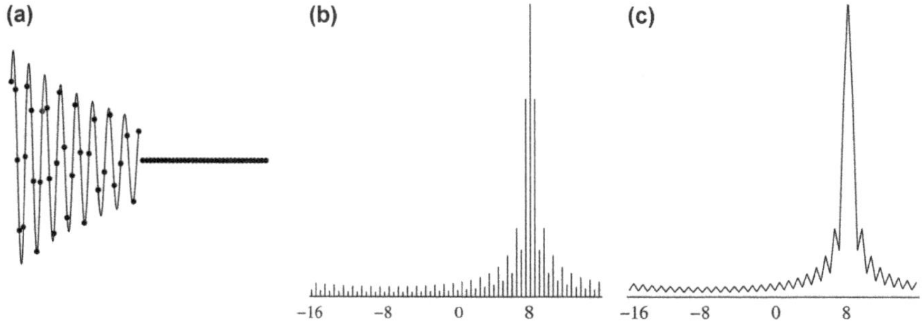

Figure 21.17 (a) FID and sampled data points, not decayed to zero before zero-filling; (b) transformed spectrum; (c) spectral data point joined by lines.

decay mostly to zero before the cut-off then the sinc artefacts can be minimised.

21.5 Digital Filtering

The raw unfiltered spectrum that we obtain is full of noise from various sources; thermal noise and digital noise to name but two. Furthermore, since noise is random in frequency there is noise at *all* frequencies to the left and right of the spectrum. A frequency is a frequency, whether it be noise or legitimate signal and so, noise whose frequency is outside of the spectral width will be digitised along with those within the spectral width. This is the phenomenon of aliasing in which signals outside of the spectral width are digitised and Fourier transformed.*** Aliasing puts the "peaks" at the wrong place in the spectrum. For example, if we are sampling at 32 Hz, the maximum frequency that we can safely digitise and transform is 16 Hz. In this case, 32 Hz is the Nyquist frequency mentioned in Section 8.7. If we sample a signal, for example, $[\cos(40\pi t)+ i\sin(40\pi t)]\exp(-\alpha t)$ where the frequency is 20 Hz then there will be a problem. 20 Hz is beyond the maximum frequency that we can safely digitise. By safely, we mean digitise and Fourier transform such that the signal ends up in the correct spot in the spectrum.

In Figure 21.18 0 Hz is at the centre of the spectrum with positive frequencies to the right of 0 Hz and negative frequencies to the left. In our case of 40 Hz being sampled at 32 Hz the peak ends up at −12 Hz!

***A signal is a signal is a signal and if one outside of the range of the spectral width is present it still appears at the digitiser input and *will* be digitised and transformed.

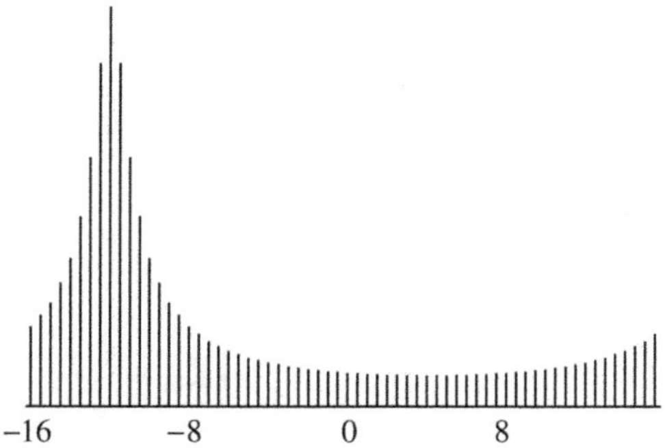

Figure 21.18 Aliased peak at –12 Hz.

Clearly, it would be useful to be able to exclude any frequencies outside of the spectral window, either before digitisation or by clever manipulation in the digital domain after acquisition. We can exclude or at least largely attenuate frequencies outside of the spectral width by using analogue bandpass filters (see Chapter 22) that allow only frequencies within a narrow band to pass through to the digitiser. One problem with these types of filters, however, is that they suffer from "roll off", which means that the frequencies at which signals are cut off are not sharp but rather are gradual cut-offs. Thus, analogue filters, although very helpful, will still let some frequencies through to be aliased into the spectrum. Mostly these frequencies will be noise, the arch enemy of the NMR/MRI spectroscopist. Anything that reduces noise but not the desired signals for the spectroscopist is a Good Thing.

Enter the digital filter. This effectively makes the cut-off frequencies sharp and reduces noise even further than the old-style analogue filters. *Anything*, noise or legitimate signal, outside of the range of the filter will not appear in the spectrum.[†††] The way this works is that the time-domain data are transformed and multiplied with a Rect-style function that is in the frequency-domain rather than the time-domain (Figure 21.19). The result is back-transformed to time-domain data to produce a free-induction decay that has only the specified

[†††]This can be a double-edged sword. On the one hand, you cut out noise from outside of your spectral window and increase your signal-to-noise ratio. On the other hand, any signals outside of the spectral window will not be seen ... they will not even be aliased into the spectrum. Many new users are blissfully unaware of aliasing nowadays since most day-to-day spectra acquired use a digital filter. The user may then miss peaks outside of the spectral window or may encounter aliasing in the F1 dimension in some 2D experiments and not know what they are seeing. Educate yourself because ... knowledge is power.

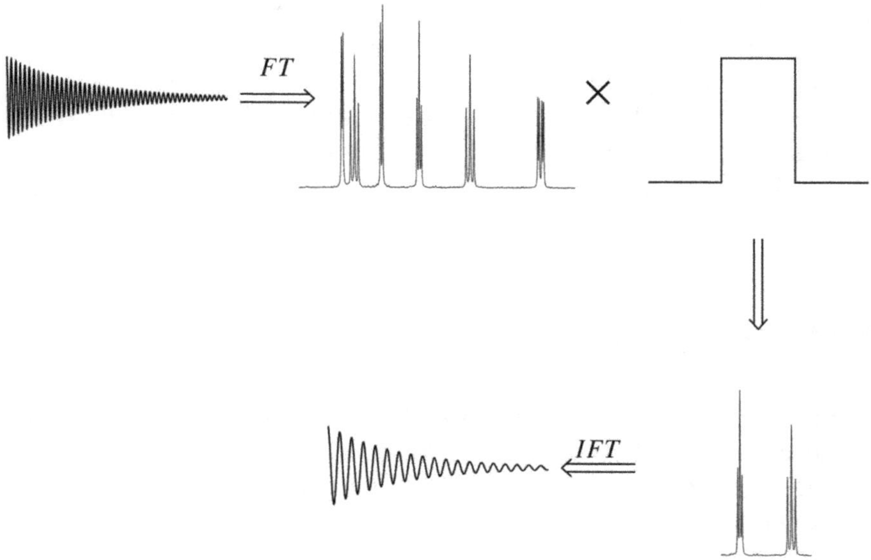

Figure 21.19 Digital filtering scheme.

frequencies in it. Any frequencies outside of the range of the spectral width are, of course, digitised but end up not being present in the final time-domain data. The digital filter does not suffer from the roll-off problem associated with analogue filters.

21.6 Oversampling

Modern NMR spectrometers are generally capable of sampling at rates much higher than at the Nyquist frequency. This higher-than-needed sampling rate is called, appropriately, oversampling. Why do this? The reason is intimately linked with digital filtering. When we oversample we are effectively increasing the spectral width that is being sampled. Recalling eqn (21.2), we can see how this must work. An increase in frequency will mean a decrease in the dwell time, *DW*. This in turn will mean that *SW* will increase. Concurrent with this will be an increase in the number of data points sampled. For example, a ten-times oversampling with ten times *TD* being sampled will sample a spectral width of ten times *SW*. Peaks outside of the spectral window are properly digitised (unless they are ridiculously far out of the window). The data are then convolved with a sinc function whose Fourier transform is a rectangular function, as in Figure 21.19. In this way, any peaks outside of *SW* are filtered out. The data points are then decimated;[‡‡‡] that is, most of them are discarded to slim the total down to *TD* data points.

There are subtleties concerning oversampling that the reader might be interested in. Hoch and Stern (see the Further Reading section at the end of the chapter) discuss this topic in a bit more detail than here.

21.7 Phase Correction

When we acquire our data we generally Fourier-transform followed by phase correction. Before phase correction what we see is usually a mixture of absorption and dispersion spectra. Why phase correct and how does it work?

There are two kinds of phase correction that are frequently used. The first is called zero-order phase correction in which the same amount of correction is applied at every frequency in the spectrum. The second type is 1st-order phase correction and is frequency-dependent; that is, at a particular point in the spectrum (sometimes called the pivot point) there is zero phase correction applied and at point further away from the pivot point more phase correction is added depending on the frequency difference from the point and the pivot point.

The source of zero-order phase error lies in the receivers.[§§§] Ideally, one receiver should detect only the absorption signal and one only the dispersion signal. A simplified diagram of this would be to have the absorption receiver on, say, the *y*-axis and the dispersion receiver on the *x*-axis and then record the signal from the probe as in Figure 21.4. In this case, there is no need for zero-order phase correction as each receiver sees only what it is supposed to see. However, more commonly, the receivers are not located exactly on the axes but rather are offset by a few (or maybe more than a few) degrees as in Figure 21.20).

[‡‡‡]The literal meaning of this term is to leave only every 10th point but generally in our NMR spectrometers this "decimation" factor usually is different from this.

[§§§]There is actually only one receiver in modern instruments and the raw signal is fed into it after being split into two channels, one of which is mixed with a references signal that is 90° out of phase with the other channel. However, it is convenient to think in terms of two receivers.

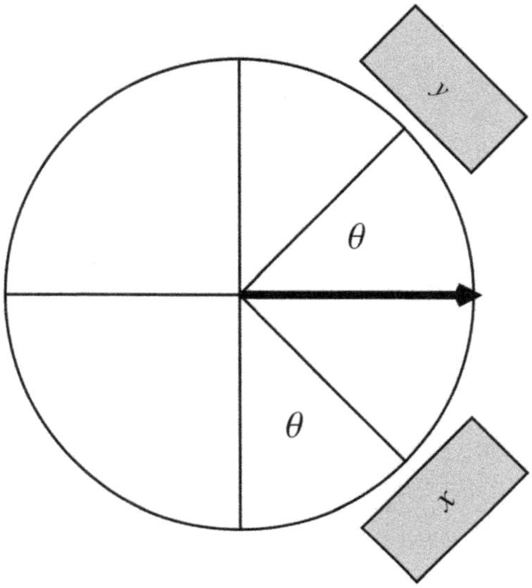

Figure 21.20 Receivers offset from x and y axes by θ.

Here, each receiver is seeing a mixture of absorption and dispersion signals; θ is the phase error. We may model this using:

$$X = A\cos(\theta) - D\sin(\theta)$$

$$Y = A\sin(\theta) + D\cos(\theta)$$

What we actually want is:

$$X = A$$

$$Y = D$$

Transformation of the first to the second corresponds to a simple clockwise vector rotation of θ degrees as encountered in Chapter 3:

$$\begin{bmatrix} X \\ Y \end{bmatrix} = \begin{bmatrix} \cos(\theta) & \sin(\theta) \\ -\sin(\theta) & \cos(\theta) \end{bmatrix} \begin{bmatrix} A\cos(\theta) & -D\sin(\theta) \\ A\sin(\theta) & D\cos(\theta) \end{bmatrix} = \begin{bmatrix} A \\ D \end{bmatrix}$$

This is easily accomplished in software.

1st-order phase correction is needed for a couple of reasons. First, $\vec{\mathbf{B}}_1$ offset causes signals that are not on resonance to arrive in the transverse plane at an offset angle (Figure 21.21a).

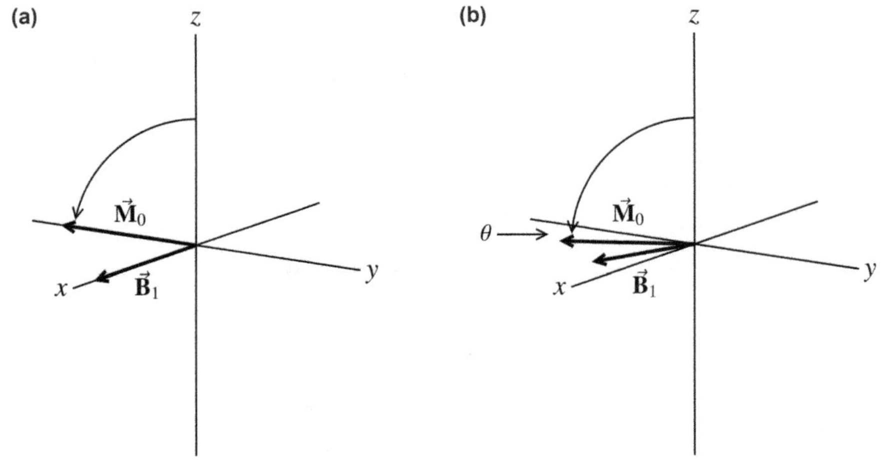

Figure 21.21 (a) On-resonance rotation of the magnetisation vector into the transverse plane; (b) off-resonance rotation of the magnetisation vector into the transverse plane.

Recall from Chapter 12 that the effective field during a pulse is the vector sum of the $\vec{\mathbf{B}}_1$ field and the residual external field, $\vec{\mathbf{B}}_0$. Also, recall that at resonance the external field vanishes leaving only the $\vec{\mathbf{B}}_1$ field lying in the transverse plane. However, for an off-resonance peak the effective field is *not* in the transverse plane. It can be seen in Figure 21.21b that the magnetisation arrives in the plane not on an axis but rather off-axis by θ degrees. The angle θ will depend on how far off-resonance the peak is. Thus, when correcting for this in the NMR software, the correction is applied in a frequency-dependent manner.

The second reason for doing 1st-order phase correction is due to the dead time at the beginning of the FID. In modern spectrometers, the receiver is not turned on immediately after the pulse transmitter is turned off to allow for powerful eddy currents in the probe coils to die away. Not doing so would introduce very powerful signals into the preamplifiers and overload them, possibly to an extent that would prove damaging to them. Therefore, there is a time immediately after the transmitter is turned off and before the receiver is turned on in which there is nothing happening in the hardware … the dead time (*DE* in Bruker language). However, there *is* something happening in the sample … chemical-shift and coupling evolution are occurring. Since there may be resonances at differing frequencies in the sample they will evolve at different rates during the dead time

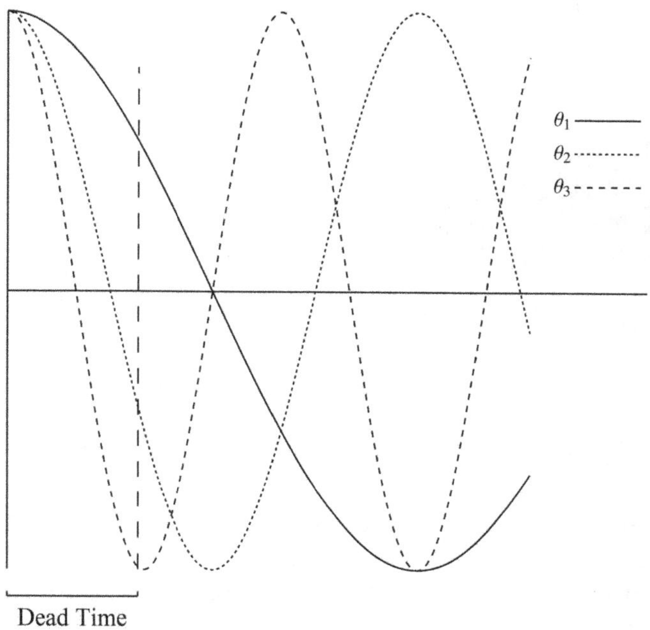

Figure 21.22 Dead time offset of time-domain signals of differing frequencies.

and will be out of phase by a small amount (or a large one if the dead time is long!).

Figure 21.22 shows that during the dead time, signals of different frequencies evolve so as to be out of phase at the end of the dead time by an angle that is frequency-dependent.

21.8 Linear Prediction

Occasionally, for one reason or another, our FID may be "chopped off" at the end of the acquisition; that is, rather than the FID being allowed to decay to zero before stopping acquisition, we cut it off before it has decayed. One sees this very frequently if not always in 2D NMR spectroscopy where the number of data points in the acquired FID is usually quite small and subsequently the FID usually has not decayed fully by the time all of the data points have been filled. If the FID is cut off (see the above discussion of zero filling) then it is probable that one will see sinc "squiggles" on at least the larger peaks in the spectrum (Figure 21.23). These occur at the

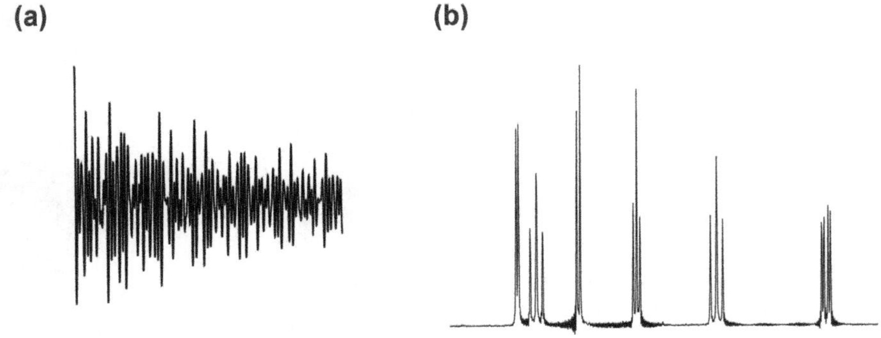

Figure 21.23 (a) truncated FID; (b) transformed spectrum with no exponential multiplication or linear prediction showing sinc-squiggles at the baseline around some peaks.

baseline and appear as an oscillation that dies away from either side of the peak.

The origin of this effect lies in the nature of the Fourier transform. Recall Section 8.5 on useful Fourier pairs and Section 8.6 on convolution. The Fourier transform of the Rect function is a sinc function and the transform of the one-sided exponential decay is a Lorentzian line shape. Convolution of these will insert a sinc squiggle into each peak in a spectrum of peaks. This is precisely what happens with the chopped FID. The "chop" is equivalent to multiplication of the decaying FID by a Rect function. Fourier transformation then produces a spectrum with sinc squiggles at the baseline of the peaks.

It would be nice to avoid this if possible and, as it turns out, it is possible to remove or at least reduce the sinc squiggles. The technique is to model what the FID would have looked like if it had been allowed to decay naturally to zero. This is called forward linear prediction and works by using the acquired data to predict what the behaviour of the FID would have been after the end of data acquisition.

With reference to Section 8.9, we may model the nth data point of a series of data points collected from an exponentially decaying signal as:

$$s(n) = \mathrm{e}^{-\pi n L \Delta t} \mathrm{e}^{-2\pi k n/N} = z^n$$

where

$$z = \mathrm{e}^{-\pi L \Delta t} \mathrm{e}^{-2\pi k/N}$$

Note that z is a constant. With this in mind, the ratio of the nth and $(n-1)$th data points is:

$$\begin{aligned}\frac{S(n)}{S(n-1)} &= \frac{z^n}{z^{(n-1)}} \\ &= \frac{\mathrm{e}^{-\pi nL\Delta t}\mathrm{e}^{-2\pi kn/N}}{\mathrm{e}^{-\pi(n-1)L\Delta t}\mathrm{e}^{-2\pi k(n-1)/N}} \\ &= \mathrm{e}^{-\pi L\Delta t}\mathrm{e}^{-2\pi k/N} \\ &= z\end{aligned}$$

from which we see that:

$$S(n) = S(n-1)z$$

This is a simple linear relationship between one data point and the following data point, implying that it should be possible to predict the values of data points both after the end of data acquisition and before the start of data acquisition. This is exactly the premise of linear prediction, if somewhat simplified, and it allows us to "continue" data collection after it is finished. We predict enough data points to get the signal amplitude low so that sincing is minimised.

One can also reduce or eliminate sincing *via* the exponential multiplication data manipulation. Recall that this technique reduces the signal strength at the end of the FID. If this reduction is large enough then the FID will appear to the Fourier transform software as a simple one-sided exponential decay that is not convolved with a Rect function, eliminating sincing at the bases of the peaks in the spectrum. However, it may be advantageous to perform linear prediction rather than exponential multiplication in some cases, in order to preserve as much resolution as possible (Figure 21.24).

Linear prediction is perhaps most important in multi-dimensional NMR spectroscopy. In 2D NMR the t_1 dimension is generally very small, often consisting only of 256 data points or less. This leads to a truncated interferogram in the second dimension that will display sinc squiggles in the 2D spectrum. Judicious choice of window functions will reduce or eliminate these artefacts, however, the apparent resolution of the 2D peaks will suffer because of the low number of data points. This may be improved considerably by using linear prediction in conjunction with the appropriate window function, as shown in Figure 21.25.

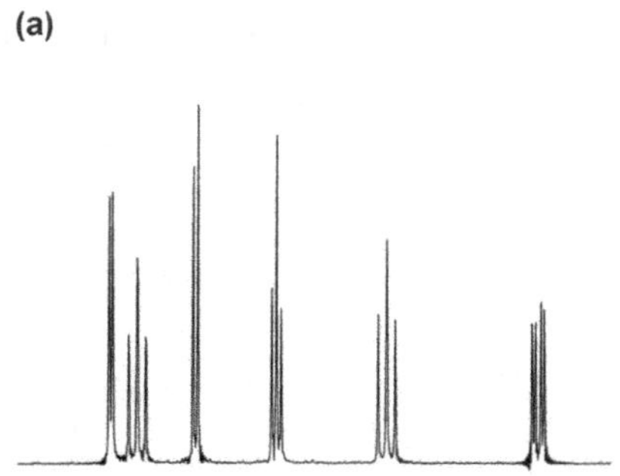

Figure 21.24 (a) FID from Figure 21.23 treated with linear prediction; (b) FID from Figure 21.23 treated with exponential multiplication.

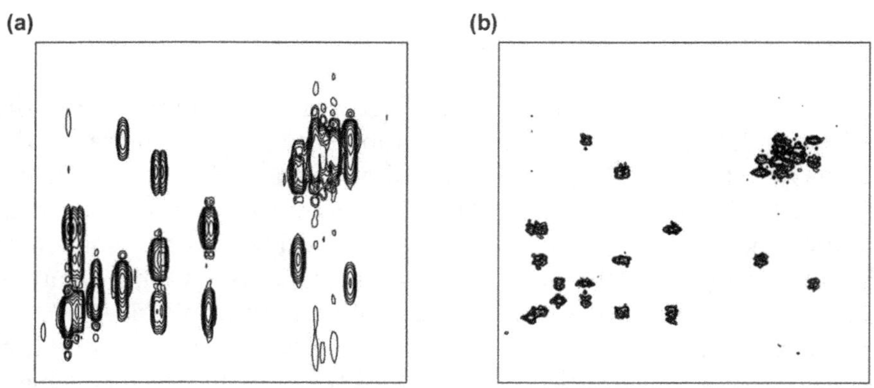

Figure 21.25 (a) COSY spectrum using a QSINE window function in the t_1 domain; (b) COSY spectrum using linear prediction in the t_1 domain.

21.9 Baseline Correction

It is not uncommon to see in our one-dimensional spectra baselines that are not flat but are rather curved or otherwise displaced from being flat. A curved baseline can be fixed by using a software algorithm that subtracts a polynomial function matching the curvature of the baseline from the spectrum. This can be done either manually by the operator or, more commonly, automatically using various built-in baseline correction commands for the software. A rolling baseline may be caused by noise or excessive intensity of the first few data

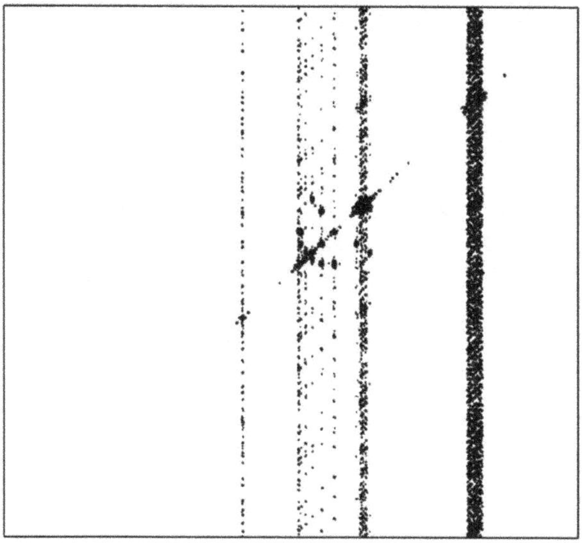

Figure 21.26 COSY spectrum showing t_1 noise.

points, in which case we may use linear prediction in the backwards direction to replace the distorted points by using the normal data points to predict in the backwards direction what the first few points should have been.

The baseline offset problem mentioned in Section 8.9 is not usually a significant problem for us in 1D NMR spectroscopy as modern software can usually compensate for it. We have baseline correction routines that generally will minimise the baseline offset of the DFT as well as flatten the baseline. In 2D spectroscopy, however, it is a problem. These are usually attributed to instrumental instabilities; temperature, $\vec{\mathbf{B}}_1$ field irreproducibility, $\vec{\mathbf{B}}_0$ field instabilities, *etc.*, and the sum of these is called t_1 noise (see Figure 21.26). For example, random changes of millidegrees K in the sample will contribute significantly to the t_1 noise. The noise ridges in the 2D spectrum can be removed or at least appreciably attenuated by a subtractive technique. One defines a region F_2 in which there is t_1 noise but no peaks in the spectrum and uses that region to build a projection that contains only noise peaks. This projection is then subtracted piecewise from bottom to top from the 2D spectrum leaving only (or mostly) the real peaks. Also, for diagonally symmetric 2D experiments one can simply symmetrise the data. However, this procedure can lead to artefacts in the spectrum and one must be aware of this when later interpreting the spectrum.

Further Reading

1. J. Granwehr, *Appl. Magn. Reson.*, 2007, **32**, 113–156.
2. J. C. Hoch and A. S. Stern, *NMR Data Processing*, Wiley-Liss, New York, 1996.

22 Magnetic Resonance Imaging

The merit of painting lies in the exactness of reproduction. Painting is a science and all sciences are based on mathematics. No human inquiry can be a science unless it pursues its path through mathematical exposition and demonstration.

Leonardo Da Vinci

22.1 Selective Pulses

Up to now the only kind of rf pulse that we have considered is the rectangular pulse; that is, a simple on-off pulse. From Chapter 8 we see that the Fourier transform of a rectangular pulse is a un-normalised sinc function or the "Sa" function, as we called it. Changing the length of the rectangular pulse changes the shape of the sinc function (Figure 22.1). The longer the rectangular pulse is the thinner and more compacted the frequency domain sinc function becomes. Consequently, the excitation bandwidth becomes thinner as well.

That the excitation frequency range or bandwidth becomes smaller as we lengthen the pulse leads us to the idea that we might be able to selectively excite some spins in our sample while leaving the others in their equilibrium states. We can indeed do this and in general, we do so using not rectangular pulses but rather shaped pulses. There are many shapes commonly used for various purposes but a very often used one in NMR spectroscopy is the Gaussian-shaped pulse. Using this pulse shape we can control the excitation bandwidth to a very high degree, exciting only the spins of interest and leaving the others alone. Also, interestingly, the Fourier transform of the Gaussian is another Gaussian.

Essential Mathematics for NMR and MRI Spectroscopists
By Keith C. Brown

Published by the Royal Society of Chemistry, www.rsc.org

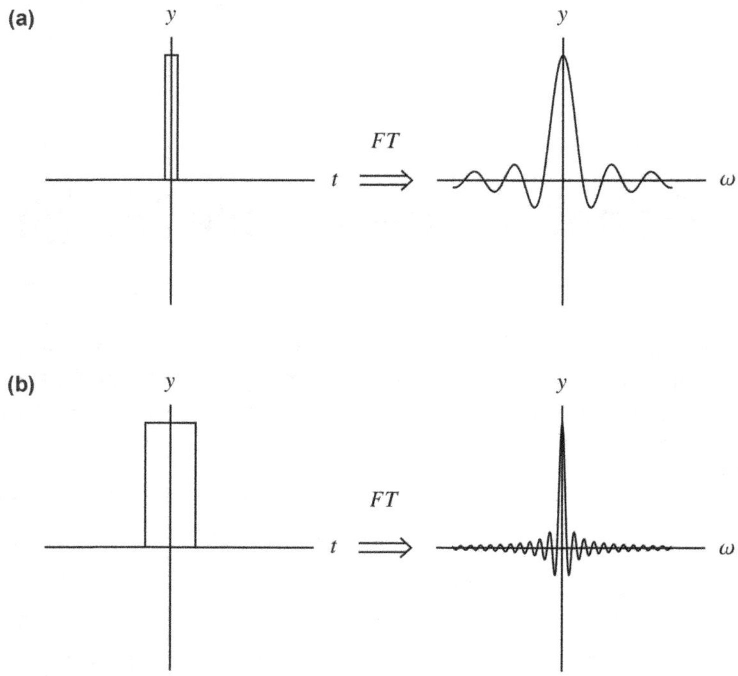

Figure 22.1 (a) A short rectangular pulse transforms to a wide sinc function; (b) A long rectangular pulse transforms to a narrow sinc function.

Figure 22.2 shows that a very high degree of selectivity is possible in our pulse sequences. It is possible to substitute a shaped pulse for a rectangular pulse in a two-dimensional pulse sequence and make it into a selective one-dimensional version of the experiment. For example, the 2D COSY experiment for mapping couplings between protons may be made into a selective 1D experiment by replacing the first rectangular pulse (see Figure 18.14) with a shaped pulse. We then specify which multiplet we wish to irradiate and any coupled multiplets will show in the resulting spectrum, all others not having been excited (Figure 22.3).

The shaped pulse and corresponding bandwidth concepts will be important to us when considering how a magnetic resonance image is obtained.

22.2 Gradients

We saw in Chapter 19 that the use of field gradients can be very useful for filtering out unwanted signals and shortening acquisition times. In MRI spectroscopy field gradients are used, not for filtering, but rather for spatially locating NMR signals.

(a)

(b)

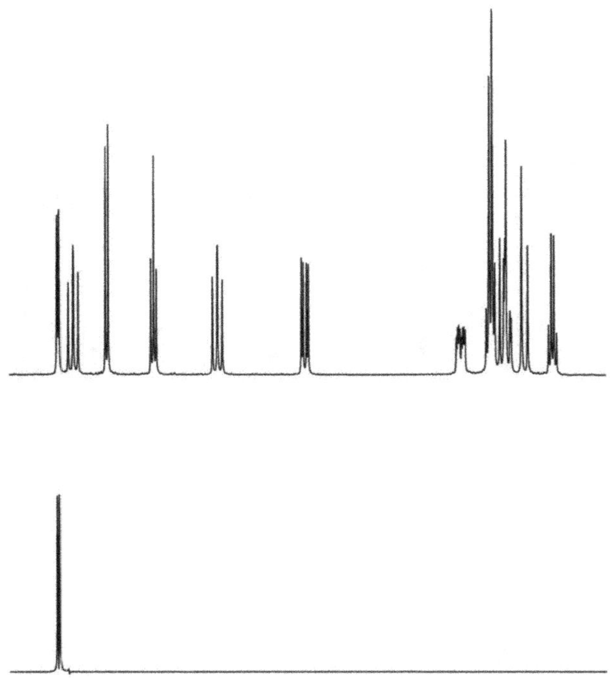

Figure 22.2 (a) ^{1}H spectrum of sucrose octaacetate obtained using a short rectangular pulse; (b) a ^{1}H spectrum of sucrose octa-acetate obtained using a long Gaussian-shaped pulse.

Implicit in the discussion in Chapter 19 was that the area of the gradient pulse is of importance. Recall that, from Figure 19.24, a comparison of the areas of a rectangular pulse and a sine-bell (half-sine wave) pulse showed that the area of the sine bell pulse is 64% that of the rectangular pulse. This was called the shape factor and, as we saw, is important if we use gradient pulses of differing shapes or duration but not if the gradient pulses are identical. In MRI spectroscopy we find that the area under the gradient function *is* of importance in the calculation of phase angles. Thus, if we define a gradient function, $G(t)$ then its area is:

$$A = \int G(t)\mathrm{d}t$$

and if we specify that the duration of the gradient is T s. then we can write a definite integral:

$$A = \int_0^T G(t)\mathrm{d}t$$

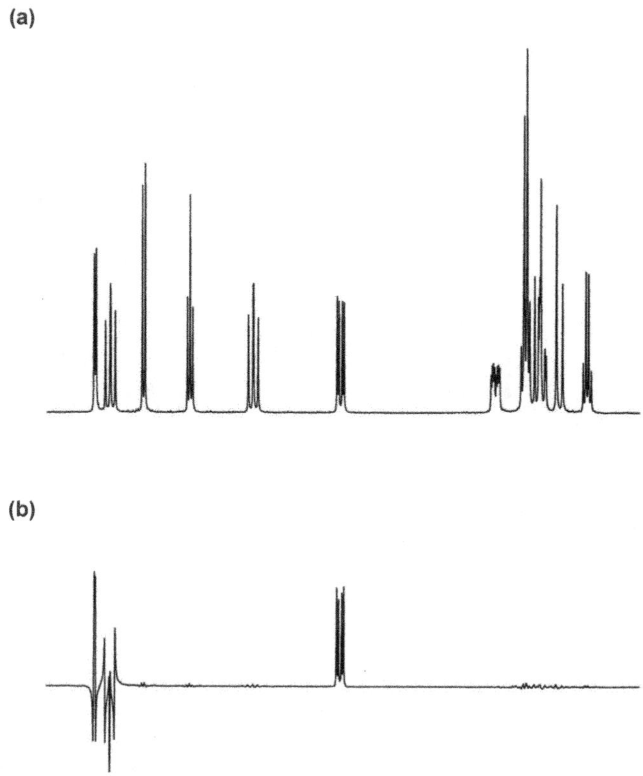

Figure 22.3 (a) Normal 1D ^1H spectrum of sucrose octaacetate; (b) Selective COSY spectrum of sucrose octaacetate.

Also from Chapter 19, we saw that during the application of a gradient pulse, the initially coherent magnetization vectors lose phase coherence during the time that the gradient is applied and acquire a spatially dependent phase (see Figure 19.22). The magnitude of the acquired phase depends on the magnitude of the gradient and on the length of time that the gradient is applied. The Larmor equation tells us the value of the precession frequency of the nucleus in a given magnetic field and with the addition of the field gradient we can write:

$$w_{i,t} = \gamma B_0 + \gamma G(t)x_i$$

Or, in the rotating frame of reference:

$$w_{i,t} = \gamma G(t)x_i \tag{22.1}$$

$w_{i,t}$ is the spatially dependent precession frequency and varies linearly along the field gradient. The spatially dependent acquired phase of the spins during the application of the gradient is:

$$\phi_{i,t} = \int_0^T w_{i,t} \mathrm{d}t = \gamma x_i \int_0^T G(t) \mathrm{d}t$$

For later use we define:

$$k(t) \equiv \gamma \int_0^T G(t) \mathrm{d}t \tag{22.2}$$

and our acquired phase equation becomes:

$$\phi_i = x_i k_t \tag{22.3}$$

22.2.1 The Slice Selection Gradient

In 2D magnetic resonance imaging (MRI) we choose a "slice" through the object to be imaged and then "paint" an image made up of NMR signal intensities distributed through the slice such that a picture of the inside of the object is produced. The problem of course is "how is the slice or region of the object selected?". To approach this we need to review magnetic field gradients and some of the fundamental aspects of the Fourier transform.

Recall from Chapter 19 that we may apply a field gradient through our sample along the *z*-axis such that the magnetic field throughout the sample is not homogeneous but rather changes in a linear fashion along the *z*-axis. In Chapter 19 that axis was the *z*-axis, however, in MRI spectroscopy we use gradients along all three Cartesian axes. In a wide-bore NMR imaging system the *z*-axis is the vertical axis, whereas in a diagnostic medical MRI instrument the *z*-axis is along the horizontal down the centre of the magnet (Figure 22.4).

Application of a field-gradient along the *z*-axis will result in differing precession frequencies along the gradient according to eqn (12.8) that says that for applied field, $\vec{\mathbf{B}}_0$, the precession frequency is ω rad/s or ν Hz. Thus, the field may vary by, for example, 10 mT from one end of a 1-m long object to the other end and since eqn (12.8) relates precession frequency directly with field strength we can write:

$$\Delta\nu = \gamma \Delta B$$

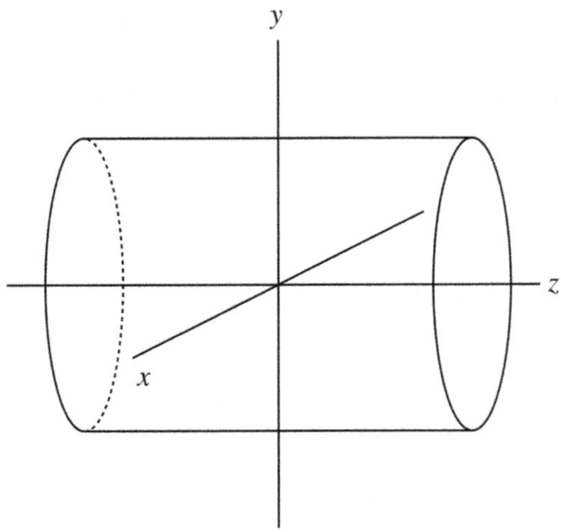

Figure 22.4 Axes in a horizontal MRI magnet. B_0 is parallel to the z-axis.

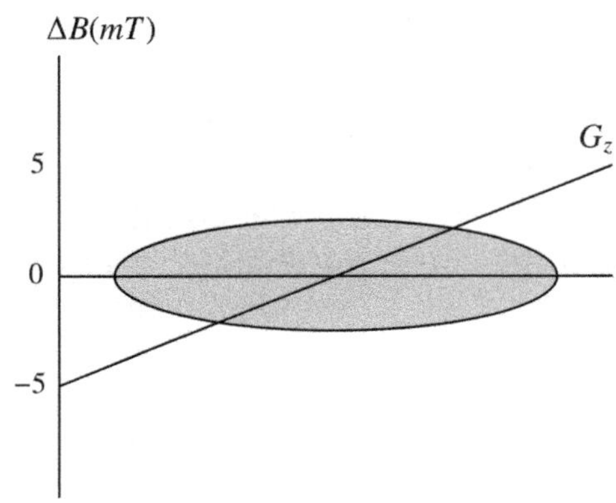

Figure 22.5 Field gradient, G, along the z-axis through an object.

For a z-axis field gradient of 10 mT/m the change in precession frequency over a meter-length object is (42.58 MHz/T)×(1 T/1000 mT)×(10 mT/m) = 0.4258 MHz/m or 4258 Hz/cm (for ^{1}H with $\gamma = 42.58$ MHz/T).

Once the field gradient along the z-axis is set (Figure 22.5) we must then issue an rf pulse such that *only* the nuclei in a specific volume or slice of the object are excited, the rest being unaffected. How are we to do this? Recall from Chapter 8 that the Fourier transform of a square wave or a pulse is a variation of the sinc function and that they form a Fourier pair.

$$\text{Rect}\left(\frac{t}{\tau}\right) \leftrightarrow \tau\text{Sa}\left(\frac{\omega\tau}{2}\right)$$

The rectangular pulse was the time-domain function and the sinc function was in the frequency-domain. Varying the width of the square wave pulse varies the coverage of the sinc function. Recall, also from Chapter 8, that since the rectangular pulse function and the sinc function form a Fourier pair we may use the duality property to write:

$$2\pi\tau \mathrm{Sa}\left(\frac{t\tau}{2}\right) \leftrightarrow \mathrm{Rect}\left(\frac{\omega}{\tau}\right)$$

This says that if we issue a shaped pulse in the form of a sinc function that the frequency range of the transformed Rect function will be sharply limited. We limit the frequency range by selecting the appropriate sinc function values.

We note from Figure 22.6 that the bandwidth of the Fourier transform of the sinc pulse is sharp and limited to only a relatively small set of frequencies. Using this type of pulse in combination with a *z*-axis field gradient allows us to selectively excite only a small number of nuclei with a volume of width Δz through the object.

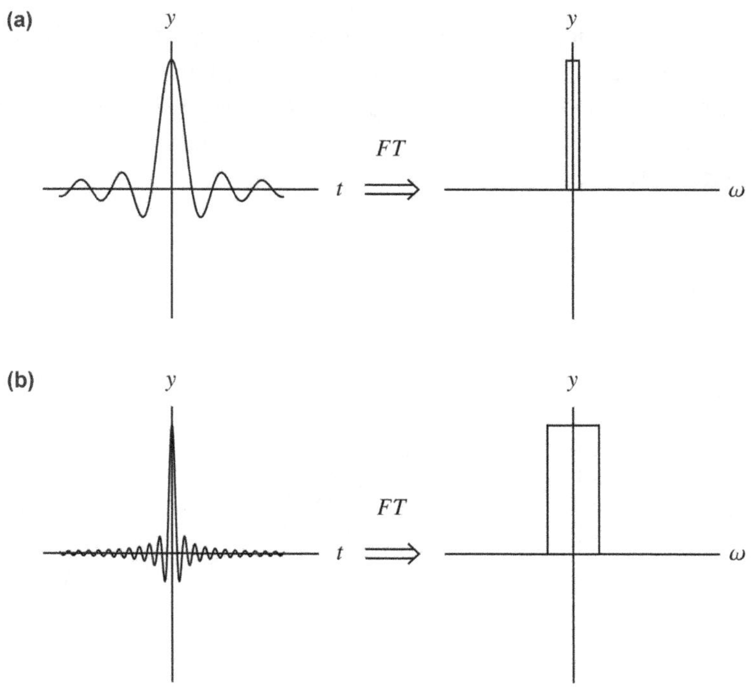

Figure 22.6 (a) A wide sinc function transforms to a narrow rectangular bandwidth; (b) A narrow sinc function transforms to a wide bandwidth.

To see that this is so, consider the 10 mT gradient calculation from above. The calculated gradient per centimeter was 4258 Hz/cm. If our transmitter is set at, for example, the frequency of the external field; that is, at the Larmor frequency corresponding to the $\Delta B = 0$ point in Figure 22.5 and our rectangular bandwidth is 1000 Hz wide then our slice width, Δz, is 1000 Hz×1/4258 Hz/cm = 0.23 cm. Within this region of the object, the precession frequencies of the spins are such that they will be excited by the sinc pulse (Figure 22.7). Outside of this range the spin precession frequencies are either too large or too small to be affected. Thus, we have selected a specific volume of the object that has $\Delta z = 0.23$ cm using a slice select gradient (SSG). We could choose a series of slices in the object by simply stepping through a series of different transmitter frequencies and applying the same field gradient along the *z*-axis.

During the time that the gradient is on, the spins within in it are precessing with differing frequencies *after the sinc pulse is over*, depending on their spatial location in the sample, for exactly the same reason as was shown in Figure 19.22. For further manipulation they need to be brought back to phase coherence, as was done in Figure 19.23 so a negative gradient of the same amplitude but half the time period is issued. It is half since the sinc pulse has substantially rotated the spins into the *x,y*-plane at the half-way point through the positive gradient period at which point they begin to go out of phase due to the positive gradient. A half-period negative gradient brings them back into phase.

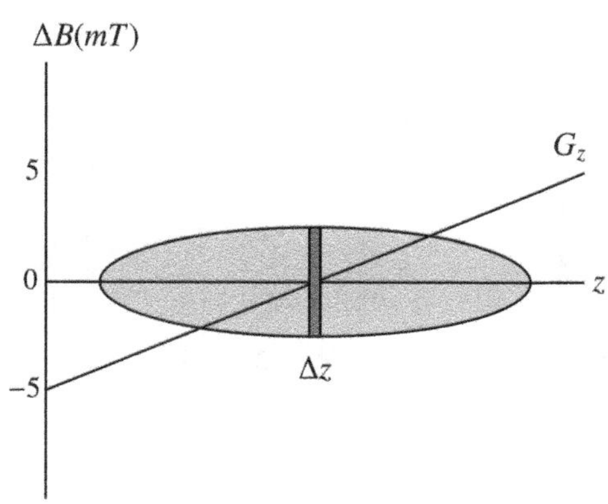

Figure 22.7 Δz slice selection *via* applied *z*-axis gradient and narrow-bandwidth rf pulse. The dark area is the "slice" within which nuclei are excited by the application of a sinc rf pulse.

22.2.2 The Read Gradient

Once the slice of interest is selected and the spins within it are excited we turn our focus to the next of the three dimensions, the frequency encoded dimension that we will take to be the x-axis for simplicity. As with the z-axis, we will be using a gradient; a frequency encode gradient† (Figure 22.8). Application of this gradient will, as with the SSG, cause precession frequencies to differ across the gradient in the x-direction.

The difference here, with respect to the SSG, is that we do *not* issue a negative refocusing gradient; we wish to *retain* the differing spin precession frequencies across the sample in the x-direction. The different precession frequencies correspond to different spatial positions in the x-direction. A visual picture is helpful. Imagine that we have finished the SSG and excited the spins in the selected slice. Our NMR picture at this point is that the magnetisation vector is located in the transverse plane and is rotating at the Larmor precession frequency (see, for example, Figure 18.9). It will be useful to us to picture the object in the x,y-plane as consisting of a matrix of "voxels" (Figure 22.9), much as a modern computer monitor or television screen consists of pixels except that now we are dealing with a three-dimensional volume rather than a two-dimensional surface on a screen.

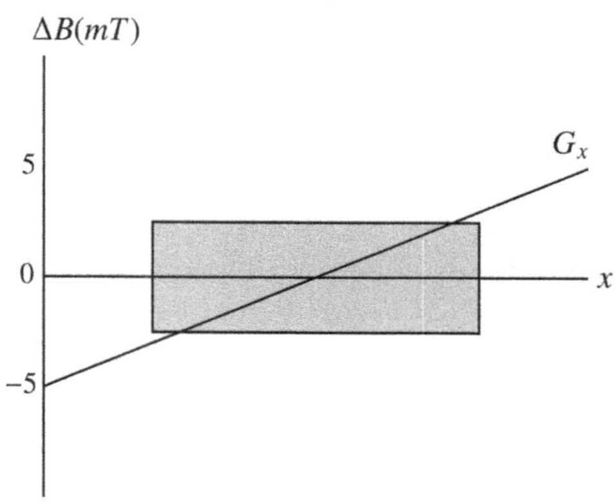

Figure 22.8 Field gradient along the x-axis through the object.

†This nomenclature is a bit misleading for the newcomer to MRI spectroscopy, especially if they have some prior experience in NMR spectroscopy. A more descriptive and often used name for this is the Read gradient, which means that this is the field gradient that is on during the acquisition of data.

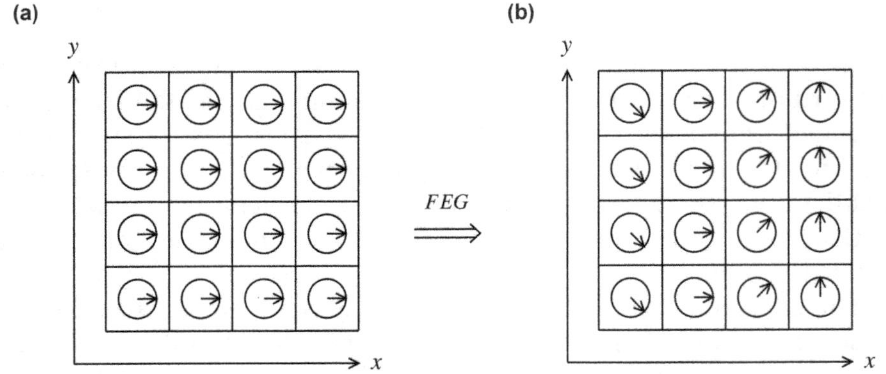

Figure 22.9 (a) Voxels prior to application of the read gradient; (b) Voxels during the read gradient application.

Within each voxel we picture a spin (or a collection of spins) that are rotating in the x,y-plane at the Larmor frequency. Before the read gradient all spins in the sample matrix have the same precession frequency but during the application of the gradient the spin precession frequencies will differ ... in the x-direction.

We note several things from Figure 22.9b. First, we see that during the application of the gradient the spin precession frequencies in the y-direction are all the same. We also observe that, as a result of the differing frequencies, there is an acquired phase in the x-direction and we may write an acquired phase expression for the nuclei affected by the read gradient:

$$\phi_x = xk_x(t) \tag{22.4}$$

using eqn (22.3). Finally, we see that in approximately the middle of the voxel grid there is no phase acquisition and that on one side of this demarcation line phase acquisition is positive and on the other side it is negative.

22.2.3 The Phase Encode Gradient

The third gradient, issued in the y-direction, is the phase encoded gradient (PEG) (Figure 22.10). Its purpose is to tag the spins in each voxel with a phase unique to that voxel in much the same way that the read gradient tags each voxel with a unique phase.

In this case only precession frequencies parallel to the y-axis are affected ... spin precession along the x-axis is unaffected

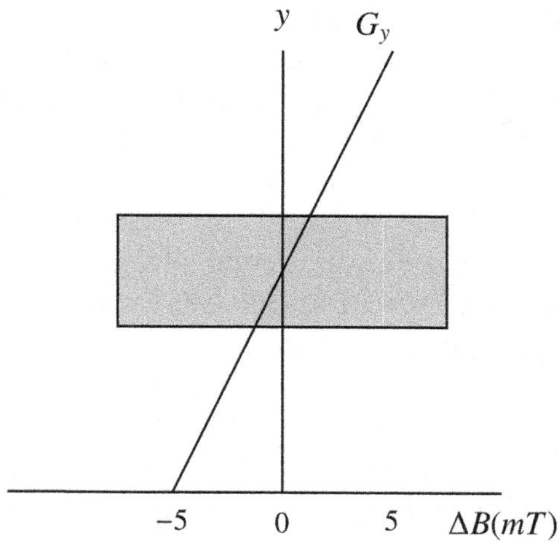

Figure 22.10 Phase encoding gradient along the y-axis through an object.

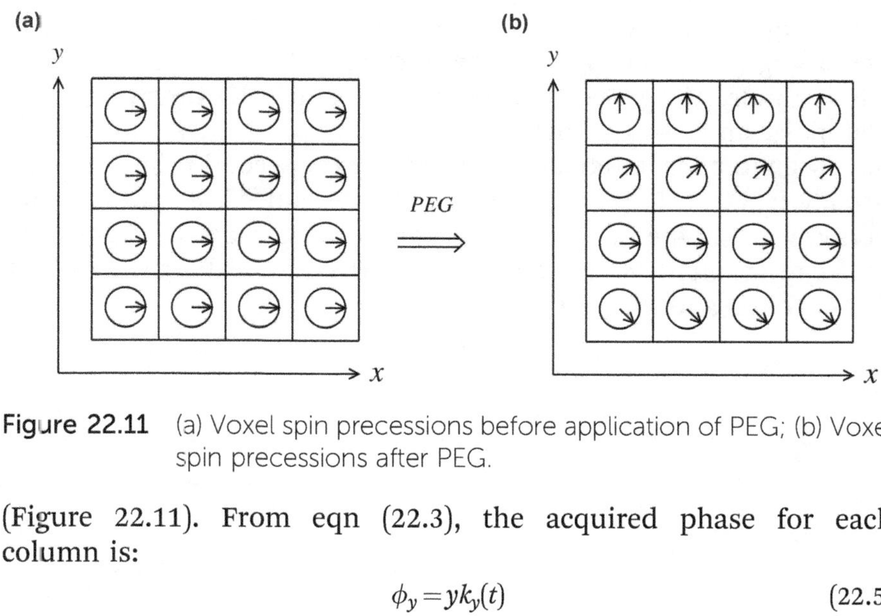

Figure 22.11 (a) Voxel spin precessions before application of PEG; (b) Voxel spin precessions after PEG.

(Figure 22.11). From eqn (22.3), the acquired phase for each column is:

$$\phi_y = yk_y(t) \tag{22.5}$$

As with the read gradient, we note that there is a zero-phase acquisition line in the voxel grid.

22.3 Putting It All Together

Generally, as in NMR spectroscopy, a MRI experiment begins with the issue of a 90° pulse. However, we also wish to observe only a small section or slice through the entire object so we simultaneously issue a slice select gradient (SSG). As discussed above, this will include a refocusing gradient of half the length of the slice select gradient. Also, since we are selecting a slice we use a sinc (or Sa if you wish) pulse in order that the excitation frequency bandwidth is tightly controlled. Figure 22.12 shows a simple MRI pulse sequence using the 90° shaped sinc pulse with a simultaneous slice select gradient.

After the initial slice excitation pulse there follows a phase encode gradient (PEG). In Figure 22.12 it is shown as a series of gradients ranging from positive to negative; that is, a maximum positive sloped gradient to a maximum negative sloped gradient. For each pass through the pulse sequence it will be one particular phase encode gradient. There will be a total of n passes through the pulse sequence and a total of n different PEGs. The value of n is a power of two because of the requirements of the fast Fourier transform and usually ranges from 128 to 512.

In addition to the dephasing caused by the PEG there is also T_2 dephasing (see Chapter 20) beginning immediately after the 90° excitation pulse. A spin-echo sequence consisting of a time period τ, a 180° pulse and a second time period τ will re-phase these spins, as we have seen in previous chapters. The end of the second τ time period is usually labelled T_E (for echo time). At the end of the second time period we may then acquire the signal. During the acquisition time the read gradient (R) is applied along the axis orthogonal to both the PEG axis and the SSG axis.

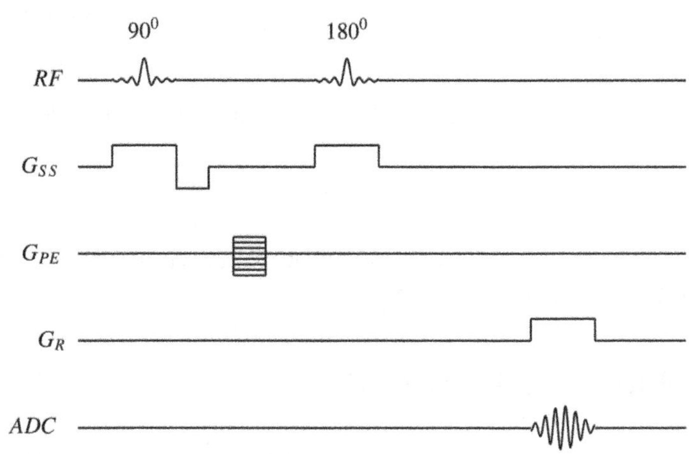

Figure 22.12 A typical spin-echo imaging pulse sequence.

Let us suppose that we are using the x-axis for the read gradient and the y-axis for the phase encode gradient. From above, multiplication of the k_x and k_y values by x or y, respectively will result in a phase angle and that we may then represent the oscillations as in Chapter 7, eqn (7.5):

$$\exp(i\phi_x) = \exp(ik_x x) \quad \text{(read gradient dimension)}$$

$$\exp(i\phi_y) = \exp(ik_y y) \quad \text{(phase encode gradient dimension)}$$

The acquired signal is a composite of all signals from the selected slice. Along, say, the x-axis we have signals of differing phase according to eqn (22.4). Actually, we expect that there might be signals of varying intensity at various positions along the axis ... this is what will ultimately make our image. What we are measuring at each point along the x-axis is the linear density of nuclear spins at points along the axis, $\rho(x)$. Summing up all of the spin densities along the x-axis and multiplying each by its exponential phase factor gives us the overall signal, $S(k)$, that is observed in the MRI receiver:

$$S(k) = \int \rho(x) \exp(ik_x x)\mathrm{d}x \tag{22.6}$$

This is nothing but an inverse Fourier transform from Chapter 8 without the $1/2\pi$ normalisation constant and that has $S(k)$ and $\rho(x)$ as a Fourier pair. To get the spin density as a function of x we just do the forward Fourier transform:

$$\rho(x) = \int S(k) \exp(-ik_x x)\mathrm{d}k_x \tag{22.7}$$

It is a simple step to expand this to a two-dimensional spin density map:

$$\begin{aligned} \rho(x, y) &= \iint S(k_x, k_y)\exp(-ik_x x)\exp(-ik_y y)\mathrm{d}k_x\mathrm{d}k_y \\ &= \iint S(k_x, k_y)\exp(-i[k_x x + k_y y])\mathrm{d}k_x\mathrm{d}k_y \end{aligned} \tag{22.8}$$

The magnetic resonance image calculation is really not different from the NMR 2D experiment calculation except that in MRI spectroscopy the dimensions of the image are position within the object and in NMR spectroscopy the dimensions of the transformed data are in frequency units. The MRI Fourier conjugate variables are x and k and in NMR spectroscopy the Fourier conjugate variables are w and t. Otherwise, their Fourier transform mathematics are identical.

Our defined variable, $k(t)$, is called, in MRI language, k-space. That it represents a space can be seen if we expand our definition (eqn (22.2)) to make it more general:

$$\vec{\mathbf{k}}(t) \equiv \gamma \int_0^T \vec{\mathbf{G}}(t)\mathrm{d}t \tag{22.9}$$

Here, we are considering that the gradient may be in any direction in Cartesian space and we are now representing it as a vector. $\vec{\mathbf{k}}$ is, of course, collinear with $\vec{\mathbf{G}}(t)$ and now being viewed as a vector must inhabit a space, as we saw in Chapter 3. The phase angle is now calculated *via* the inner product of $\vec{\mathbf{r}}$ and $\vec{\mathbf{k}}$:

$$\begin{aligned}\phi &= \gamma \vec{\mathbf{r}} \cdot \int_0^T \vec{\mathbf{G}}(t)\mathrm{d}t \\ &= \vec{\mathbf{r}} \cdot \vec{\mathbf{k}}\end{aligned} \tag{22.10}$$

and is the completely general expression of eqn (22.4) and (22.5) (which assumed the x and y gradients to be exactly aligned with the x and y axes). The Fourier transform is sometimes written using the vector terms for position and k-space:

$$\rho(r) = \int S(k) \exp(-\imath \vec{\mathbf{r}} \cdot \vec{\mathbf{k}})\mathrm{d}^n k$$

where n refers to the number of spatial dimensions (2 and up).

During the read gradient acquisition period the data are being acquired and digitised and as we have seen in previous chapters, it is the precessing spins that induce the acquired signals in the receiver coil. The frequency of the precession is intimately linked with the position of the spin in the imaged object and the received signal is a composite of these frequencies. It is easy to see that this is so in the read gradient dimension (usually along the x-axis) but perhaps not so obvious in the phase encode dimension (usually the y-axis).[‡] To understand that this is so, consider what is happening in the typical imaging experiment. We apply the pulse sequence with a slice select gradient, a phase encode gradient and then a read gradient to acquire

[‡]It is worth noting here that the slice select, phase encode and read gradient axes need not be z, y and x, respectively. Once the slice select axis is chosen the slice plane is determined. Thus, if the slice select axis is the z-axis then the slice plane is the x,y-plane, as is commonly the case and is the one that I am using in this chapter. In MRI language these slices are *axial* or *transverse* slices. One may also choose the x or y axes (as shown in Figure 22.4) as the slice select axis to produce *sagittal* or *coronal* slices respectively. Once the slice select axis is chosen and the slice plane is determined then the phase encode and read axes may be along either of the two remaining axes.

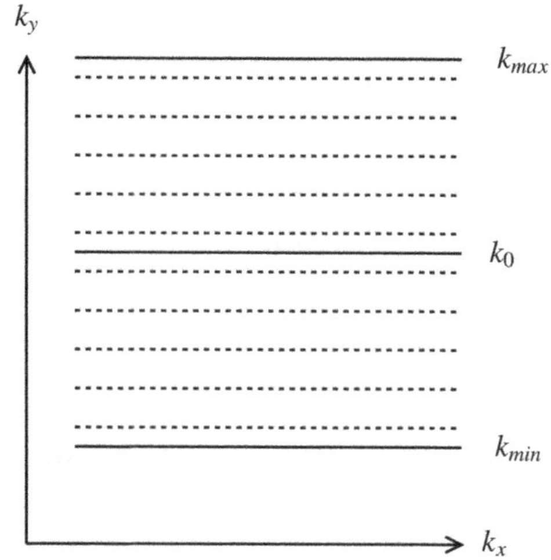

Figure 22.13 Lines in k-space.

a signal. What we have done is to acquire one line in k-space. The sequence is then repeated with a different phase encode gradient; frequently we start with a high positive gradient and change towards a high negative gradient. Recalling eqn (22.9), we see that a different gradient vector produces a different k-space vector or a different line in k-space. For any given voxel along the y-axis there will be a different signal phase since changing the phase encode gradient changes the precession frequency for that voxel. When the experiment is finished there will be n lines in k-space that we arrange in our mind's eye as in Figure 22.13. Fourier transformation of a line corresponds to transformation of the acquired signal for one particular phase encode gradient. If we take the mth data point on each line we will have a set of data in the k_y direction and we will find that for each point there is a different phase corresponding to a different frequency, just as is the case in the k_x direction. Fourier transformation of this line of data points then will give us the spatial positions of spin density in the object.

22.4 The Discrete Fourier Transform

Since the acquired data from the precessing spins in the imaged object are stored in computer memory we must sample, digitise and store discrete values rather than continuous analogue data (see Chapter 21). Consequently, we use the discrete Fourier transform discussed in Chapter 8. As in Chapter 8, the Nyquist theorem applies and, in order to avoid aliasing, we must pay attention to acquisition bandwidth. For the read gradient domain, since there is a finite time, Δt, between sample and digitise events there will be a maximum frequency and we recall eqn (21.4):§

$$AQ = \frac{TD}{2SW}$$

or

$$T_{\mathrm{acqu}} = N\Delta t = \frac{N}{2\Delta\nu}$$

Here, N represents the total number of data points acquired. For quadrature detection, $2\Delta\nu$ is the digitiser bandwidth across the entire field of view (FOV) of interest in the imaged object:

$$2\Delta\nu = \gamma G_x L_x$$

$$\Delta\nu = \frac{\gamma G_x L_x}{2}$$

From eqn (22.10) in the x-dimension:

$$\Delta k_x = \gamma G_x \Delta t$$

and combining these, we have:

$$\Delta k_x = \gamma G_x \Delta t = \frac{\gamma G_x}{2\Delta\nu} = \frac{\gamma G_x}{\gamma G_x L_x} = \frac{1}{L_x} \tag{22.11}$$

This is the Nyquist theorem for k-space in the read direction. In order to adequately map all of the spin density with in an object along the x-axis, Δk_x must be less than or equal to $1/L_x$ where L_x is the length of the object containing excited spins along the x-axis. The Nyquist relation in the phase encode direction is similar:

$$\Delta k_y = \gamma \Delta G_y t_{PEG}$$

$$= \frac{1}{L_y}$$

Here, Δk_y is related to ΔG_y rather than Δt since between each line in k-space G_y is incremented (or decremented). Again, to

§Confusingly, $\Delta\nu$, in the MRI literature, represents the bandwidth rather than the interval between data points in the frequency domain, as it was used in Chapter 21. It is well to note this or misunderstanding will ensue.

avoid aliasing in the PEG dimension, Δk_y must be less than or equal to $1/L_y$.

For N_x data points acquired during T_{acqu}:¶

$$N_x \Delta k_x = \frac{N_x}{L_x} = \frac{1}{\Delta x}$$

$$\text{and} \tag{22.12}$$

$$\Delta k_x \Delta x = \frac{1}{N_x}$$

which is analogous to the frequency-domain equation (21.1). Δx represents the spatial resolution of the image just as $\Delta \nu$ represents the frequency resolution in NMR spectroscopy.

The discrete Fourier transform (DFT) for the read gradient domain (the x-axis, as is common) is much the same as that for the frequency domain DFT of Chapter 8. As we saw in eqn (22.7), spin density and k-space constitute a Fourier pair and therefore we may take our k-space data and transform it to spatially mapped spin density. Recall our definition of the DFT (eqn (8.54)) from Chapter 8:

$$F(j) \equiv \sum_{n=0}^{N-1} f(n) \exp\left(\frac{-i2\pi jn}{N}\right) \quad j = 0, 1, 2, 3 \cdots N-1$$

and the inverse DFT:

$$f(n) = \frac{1}{N} \sum_{k=0}^{N-1} F(j) \exp\left(\frac{i2\pi jn}{N}\right)$$

These are very general expressions; j and n are simply integer numbers and do not refer to what quantities they represent. They represent conjugate variables whose units cancel to give a unitless number when multiplied into each other. We do something very similar in MRI spectroscopy by approximating the Fourier integral (eqn (22.7)) with a summation. In the case of MRI spectroscopy, our conjugate variables are k and x and to insert them into the DFT we would do something like:

$$x_j = j\Delta x$$

$$k_n = n\Delta k$$

¶T_{acqu} is the time that the ADC is collecting data.

And write the DFT as a discrete version of eqn (22.7):

$$\rho(j\Delta x) = \Delta k \sum_{n=0}^{N-1} s(n\Delta k)\exp(-i2\pi j\Delta x n\Delta k)$$

$$\rho(x_j) = \Delta k \sum_{n=0}^{N-1} s(k_n)\exp\left(\frac{-i2\pi jn}{N}\right) \quad \left(\Delta x\Delta k = \frac{1}{N}\right)$$

The factor, Δk, is usually left out of this equation and incorporated into the inverse Fourier transform as part of the normalisation constant. Recall from Chapter 8 that whether the normalisation constant is used in the DFT, the IDFT or both is arbitrary and that we chose there to put it in the inverse transform. Hence, our MRI DFT is:

$$\rho(x_j) = \sum_{n=0}^{N-1} s(k_n)\exp\left(\frac{-i2\pi jn}{N}\right) \tag{22.13}$$

22.5 T_1, T_2 and Spin Density Weighting

The primary goal of the MRI spectroscopist is to attain image contrast in the acquired data; that is, the image should show as clearly as possible the difference between, for example, normal tissue and diseased tissue in the scanned object. To accomplish this, there are three main approaches that are taken; T_1, T_2 and proton density weighting. The pulse sequence that is used and the T_R (time to repeat pulse) and T_E (time to echo) times in the pulse sequence are manipulated so as to emphasise one of these weightings.

We can define a function that measures the contrast of the signal as:

$$\begin{aligned} C_{AB} = {} & \rho(A)[1 - \exp(-T_R/T_{1A})][\exp(-T_E/T_{2A}^*)] \\ & - \rho(B)[1 - \exp(-T_R/T_{1B})][\exp(-T_E/T_{2B}^*)] \end{aligned} \tag{22.14}$$

This is just the signals from two different regions or substances in the object times their respective T_1 and T_2 decay functions, since T_1 and T_2 decay are simultaneously operative. T_R and T_E are the repetition and the echo times, respectively (Figure 22.14).

Recall from Chapter 20 that T_1 refers to longitudinal spin relaxation and that it is frequently modelled as an exponential return of magnetisation to equilibrium, T_1 being the rate constant for the process.

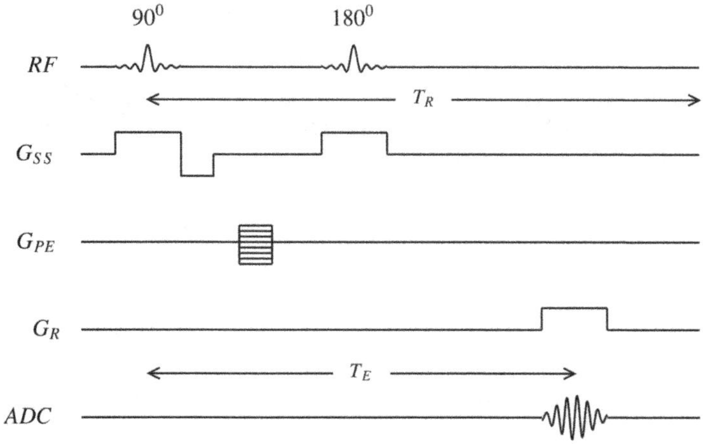

Figure 22.14 A pulse sequence showing T_R and T_E times.

T_2 refers to the loss of coherence in the spins in the transverse plane and is also generally modelled as an exponential decay process. What we see in the decaying FID is actually generally called T_2^* which is a composite of T_2 and T_2', where T_2' is the loss of spin coherence due to magnetic field inhomogeneity. The rate of decay is the sum of the two and we can write:

$$\frac{1}{T_2^*} = \frac{1}{T_2} + \frac{1}{T_2'}$$

The actual value of T_2 can be measured *via* a CPMG spin-echo experiment,‖ which causes the non-coherent spins to refocus (Figure 22.15).

One uses a (NMR) pulse sequence consisting of an initial 90° excitation pulse followed by a series of 180° refocusing pulses. The intensities of the refocused signals are measured and modelled exponentially to extract the T_2 decay parameter.

The basic idea behind T_1/T_2 weighting is that different types of tissues in the imaged object will have different characteristic T_1 and T_2 values. We want to exploit this to enhance the contrast of the image. To T_1-weight the image we use a short T_R and T_E so that there will not be enough time for significant T_2 differences to evolve. The short T_E allows for separation of signals from areas of differing T_1 values (Figure 22.16). We can model this using our contrast function,

‖See any competent book on NMR spectroscopy for a detailed description of the CPMG experiment.

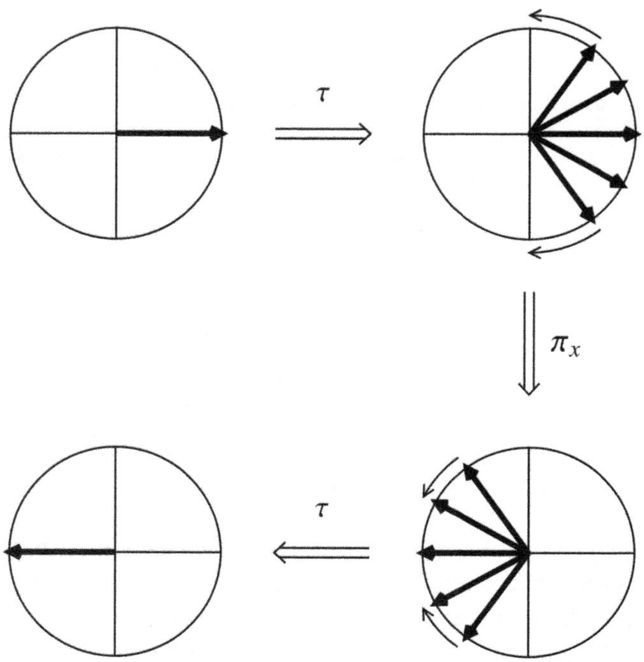

Figure 22.15 T_2 refocusing in a spin-echo pulse sequence.

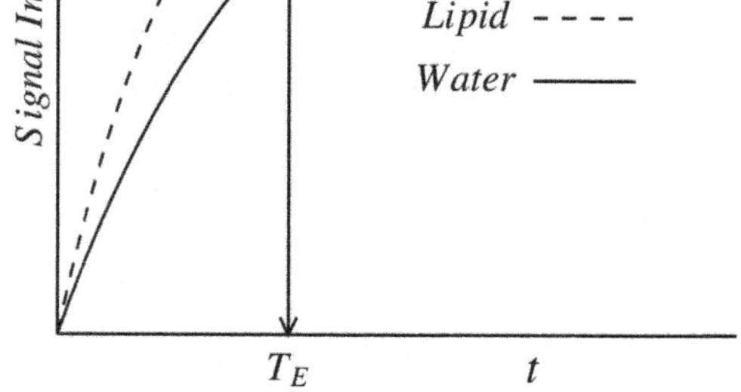

Figure 22.16 T_1-weighted signal intensity difference between lipid and water at short T_E. In an inversion recovery experiment, the material with the shorter T_1 will have the higher signal intensity.

eqn (22.14). If T_R and T_E are short ($T_R \leq T_1$, $T_E \ll T_2^*$) for both A and B then $\exp(-T_E/T_2^*) \approx 1$ and:

$$C_{AB} = \rho(A)\,[1 - \exp(-T_R/T_{1A})] - \rho(B)\,[1 - \exp(-T_R/T_{1B})] \qquad (22.15)$$

A plot of the contrast function (Figure 22.17) shows that there is a point at which the contrast is at a maximum. This suggests that differentiation of the function with respect to T_R and setting the result equal to zero will allow us to calculate the optimum T_R value for the experiment.

$$\frac{dC_{AB}}{dT_R} = \frac{d}{dT_R}[\rho_A[1 - \exp(-T_R/T_{1A})] - \rho_B[1 - \exp(-T_R/T_{1B})]]$$

$$= \frac{d}{dT_R}[\rho_A - \rho(A)\exp(-T_R/T_{1A}) - \rho_B + \exp(-T_R/T_{1B})]$$

$$= \frac{\rho_A}{T_{1A}}\exp(-T_R/T_{1A}) - \frac{\rho_B}{T_{1B}}\exp(-T_R/T_{1B})$$

Setting this equal to zero gives:

$$\frac{\rho_A}{T_{1A}}\exp(-T_R/T_{1A}) = \frac{\rho_B}{T_{1B}}\exp(-T_R/T_{1B})$$

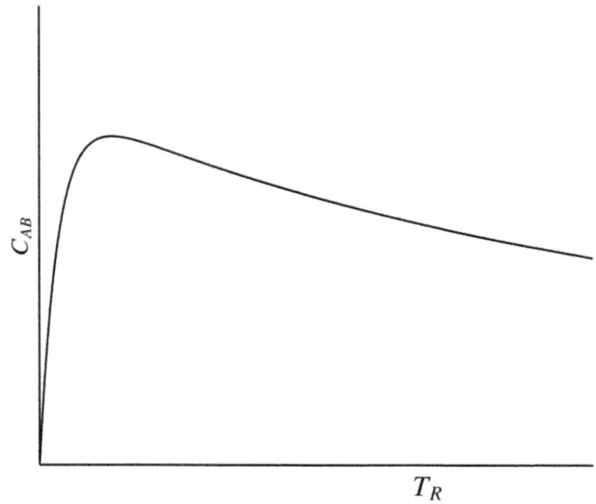

Figure 22.17 Plot of C_{AB} *vs.* T_R for the T_1-weighted contrast function.

and solving for T_R gives the optimum value for T_R in the T_1 weighted experiment:

$$T_{\text{Ropt}} = \frac{T_{1A}T_{1B}\left(\ln\frac{\rho_A}{T_{1A}} - \ln\frac{\rho_B}{T_{1B}}\right)}{T_{1B} - T_{1A}}$$

For a T_2 weighted image we use a long T_E and T_R. This way there has been enough time for T_1 relaxation such that signal intensity differences due to T_1 differences are not as significant. We then look at the intensity differences due to differing T_2s in the tissues (Figure 22.18).

In the T_2 weighted experiment, $T_R \gg T_1$ in which case $\exp(-T_R/T_2^*) \approx 0$ and our contrast function becomes:

$$C_{AB} = \rho(A)\,[\exp(-T_E/T_{2A}^*)] - \rho(B)\,[\exp(-T_E/T_{2B}^*)] \tag{22.16}$$

Again, a plot of eqn (22.16) with T_E as the independent variable shows that there is a value of T_E for which the contrast is maximised (Figure 22.19).

Finally, a spin density weighted signal is just one in which T_1 and T_2 effects have been minimised. Therefore, a short T_E is used to minimise signal intensity differences due to T_2 differences and a long T_R is used to minimise intensity differences due to T_1 differences.

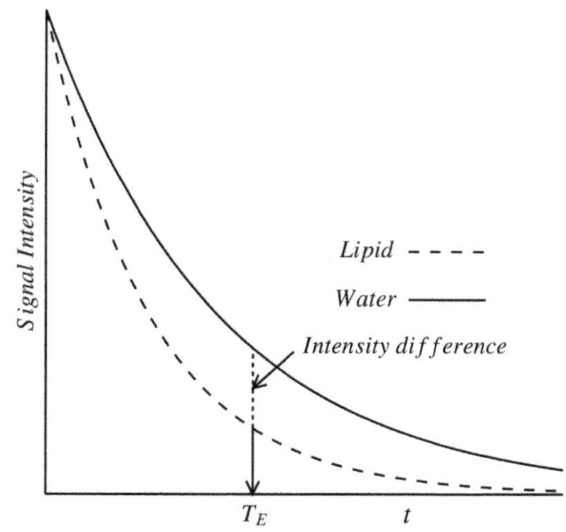

Figure 22.18 T_2 relaxation curves for lipid and water. Judicious choice of T_E gives the largest signal difference between the two. The image is said to be T_2 weighted.

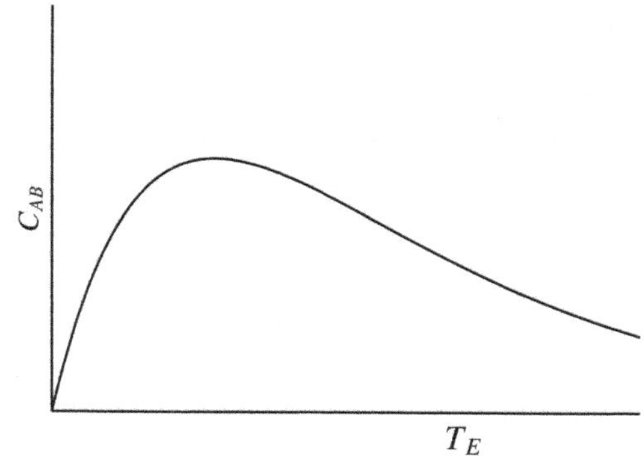

Figure 22.19 Plot of C_{AB} *versus* T_E.

If $T_E \ll T_2^*$ and $T_R \gg T_1$ then $\exp(-T_E/T_2^*) \approx 1$ and $\exp(-T_R/T_1) \approx 0$. Our contrast function collapses to:

$$C_{AB} = \rho(A) - \rho(B) \tag{22.17}$$

In this case, the image contrast depends largely on differences in spin densities throughout the object.

22.6 Pulse Sequences

22.6.1 Inversion Recovery

The inversion recovery pulse sequence is very similar to the one used by NMR spectroscopist except, of course, for the gradients (Figure 22.20). This sequence is useful for emphasising contrast using T_1 differences in the object to be imaged. Recall from Figure 21.10 that, for the inversion recovery experiment, there is a null time for each peak in the spectrum. Each type of spin will have its own characteristic T_1 relaxation constant and we can use this to our advantage. In Figure 21.10 we note that about half-way through the plots some peaks are still slightly negative or zero, while others have progressed more quickly to being strongly positive due to a shorter T_1. We say that the small or zero peaks have been "nulled" and we do not see them in the spectrum. In MRI spectroscopy we can do much the same. If there is some substance that we do not want to contribute to the image we can null it as long as we know its characteristic T_1 value. The null time is calculated from the T_1 value starting with:

$$M_z = M_0(1 - 2\exp(-T_I/T_1))$$

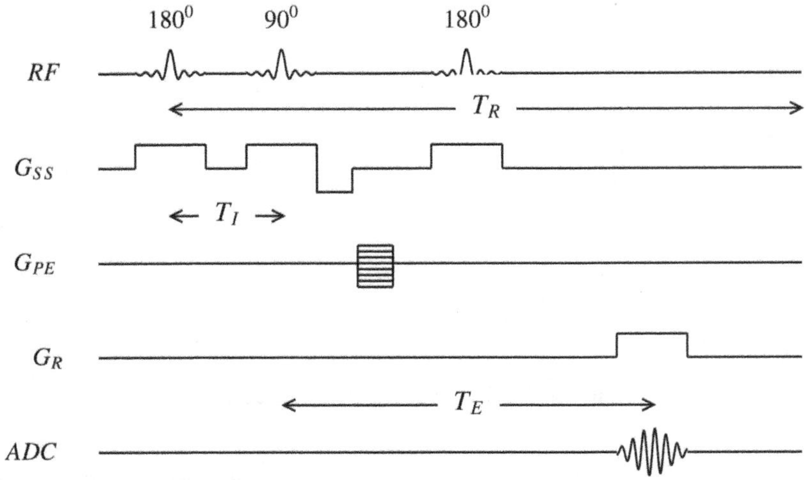

Figure 22.20 The inversion recovery pulse sequence. T_I is the inversion time, T_R is the time-to-repeat the sequence and T_E is the time to echo.

This is for a complete inversion of magnetisation from equilibrium. At the null time, M_z will equal zero:

$$0 = M_0(1 - 2\exp(-T_{\text{null}}/T_1))$$

$$0 = 1 - 2\exp(-T_{\text{null}}/T_1)$$

$$1 = 2\exp(-T_{\text{null}}/T_1)$$

$$\ln(1) = \ln(2) - \frac{T_{\text{null}}}{T_1}$$

$$\ln(2) = \frac{T_{\text{null}}}{T_1}$$

$$T_{\text{null}} = T_1 \ln(2)$$

Thus, prior knowledge of a T_1 value will let us calculate the null time and allow us to cut most of the signal from that material from the image. We could, of course, turn it around and use a measured null time to estimate a T_1 value for some area in the imaged object.

22.6.2 Gradient Echo

The pulse sequence of Figure 22.14 is that of a simple pulse spin echo experiment in which the 180° shaped pulse inverts all of the spins in

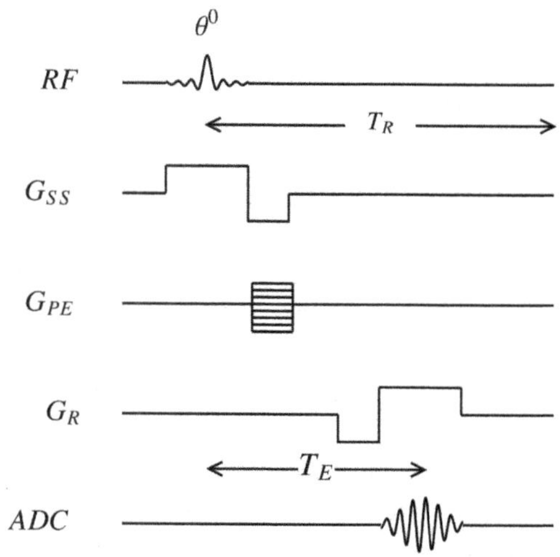

Figure 22.21 Gradient echo pulse sequence.

the transverse plane so that they can refocus at T_E. Refocusing can also be accomplished using gradients only. In Figure 22.21 we see that in the G_R gradient sequence there is a short negative gradient immediately followed by a longer positive gradient. The short negative gradient is called a pre-phase or de-phasing gradient. Its purpose is simply to de-phase the signals along the *x*-axis; rather than a normally decaying FID the pre-phase gradient causes more rapid decay of the signal. The positive portion then re-phases the signal to produce an echo signal at the half-way point through the gradient.

Using eqn (22.10), we can model what is happening in this pulse sequence. The negative G_R gradient causes a differential phase acquisition along the *x*-axis, as we saw above, but since the gradient is negative the acquired phases are opposite to what would occur for a "normal" positive gradient:

$$\phi_{i,\,\text{pre-phase}} = -\gamma x_i \int_0^T G_R(t)\mathrm{d}t$$

After the negative gradient, the positive gradient causes a re-phasing of the spins to occur such that at the time t_{echo} the spins have all re-phased and the echo signal is at its maximum. During this time, phase acquisition is written as:

$$\phi_{i,\,\text{re-phase}} = \gamma x_i \int_0^{t_{\text{echo}}} G_R(t)\mathrm{d}t$$

At time t_{echo} the acquired phase is zero and we can write:

$$\gamma x_i \int_0^{t_{\text{echo}}} G_{\text{R}}(t)\text{d}t = -\gamma x_i \int_0^T G_{\text{R}}(t)\text{d}t$$

What this means is that, since the phase is zero at t_{echo} and negative prior to t_{echo}, for the time period 0 to $T_{\text{acqu}}/2$ (where 0 represents the beginning time of the positive read gradient) we are sampling signals that represent negative values in k-space. At exactly t_{echo} or $T_{\text{acqu}}/2$ the sampled data correspond to $k_x = 0$ and for the second half of the read gradient from $T_{\text{acqu}}/2$ to T_{acqu} we are sampling signals corresponding to positive k_x values.

The advantage of this pulse sequence is that it may be repeated much more rapidly than the simple spin-echo sequence since there is no refocusing rf pulse and associated symmetric de-phase/re-phase time included. Also, the initial rf pulse need not be a full 90° pulse but can be smaller, meaning that recovery of spin magnetisation is more rapid leading to even more rapid repetition of the sequence.

22.6.3 Gradient Spin-Echo

Our first pulse sequences of Figures 22.12 and 22.14 were very simple spin-echo pulse sequences. We generally modify this sequence to include a pre-phasing read gradient in order to have our acquisition cover negative to positive k-space, as was done in the gradient-only echo sequence (Figure 22.22). This time, however, the pre-phase gradient is positive. The application of a 180° pulse after the pre-phasing

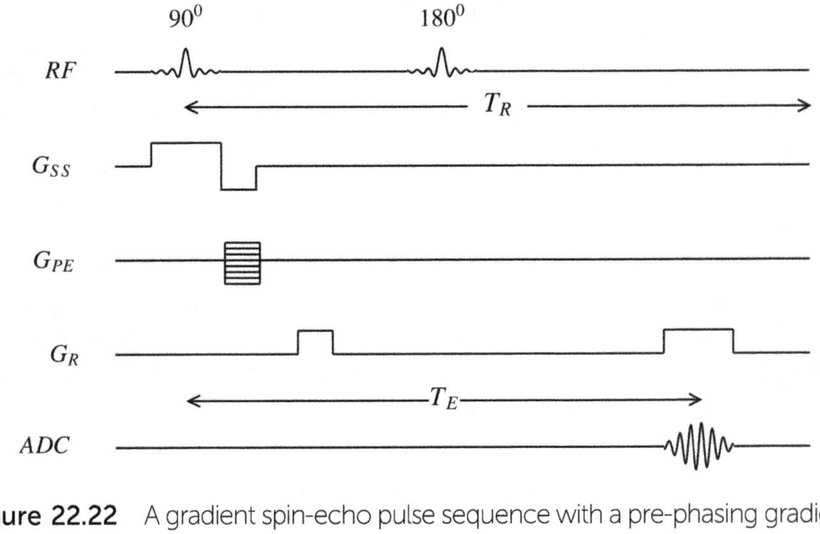

Figure 22.22 A gradient spin-echo pulse sequence with a pre-phasing gradient.

gradient changes the phase signs of the spins so that they are now negative and when the acquisition begins we scan from negative *k*-space through to positive *k*-space, as with the gradient echo sequence above.

One may sometimes use a gradient spin-echo sequence with the pre-phasing read gradient applied after the 180° refocusing pulse. In this case, the pre-phasing gradient will be negative and will perform the same function as in the gradient echo sequence of Figure 22.21; that is, it will set up the spin phases such that when acquisition starts the *k*-space line will be scanned from $-k_{\max}$ to $k_{\max}$.

22.6.4 Fast Spin-Echo

A convenient modification of the gradient spin-echo sequence is to add multiple echoes such that each echo corresponds to a line in *k*-space. In order to accomplish this, the phase encode gradient associated with each line in *k*-space has to be incremented. So, after an echo but before a refocusing rf pulse a negative phase encode gradient is applied with the same magnitude as the previous positive phase encode gradient. This has the effect of refocusing the spins in the *y*-axis. After the refocusing pulse a positive phase encode gradient that is incremented in the appropriate direction as required by *k*-space, is issued and the echo sequence repeats. Each time an echo is acquired the data correspond to the next line in *k*-space.

The utility of this is that, rather than going through an entire pulse sequence for each line in *k*-space, multiple lines are traversed in one sequence making the acquisition of all of the *k*-space data faster. In Figure 22.23 three acquisitions occur, making this particular sequence approximately three times faster than that of Figure 22.22.

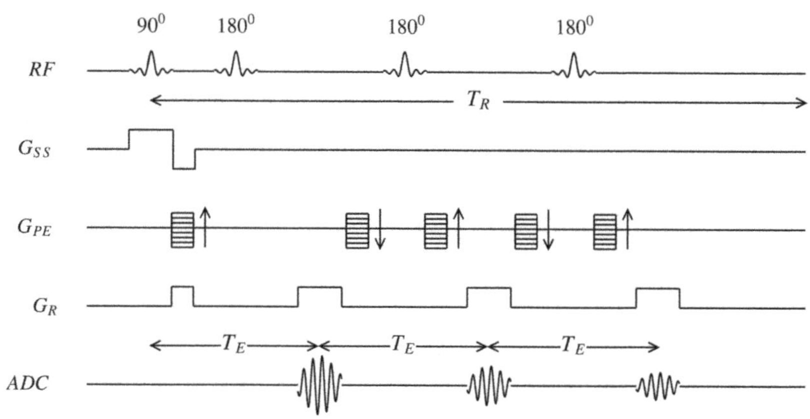

Figure 22.23 Fast gradient spin-echo pulse sequence.

Actually, since the FFT requires a data set whose number is a power of 2 it would be sensible to include four echo acquisitions rather than three.

Further Reading

1. M. A. Bernstein, K. F. King and X. J. Zhou, *Handbook of MRI Pulse Sequences*, Elsevier Amsterdam, 2004.
2. J. M. Boone, J. T. Bushberg, E. M. Leidholdt and J. A. Seibert, *The Essential Physics of Medical Imaging*, LippinCott Williams & Wilkins, Philadelphia, 2002.
3. R. W. Brown, Y. N. Cheng, E. M. Haake, M. R. Thompson and R. Venkatesan, *Magnetic Resonance Imaging*, John Wiley and Sons, 2014.

23 Electronics

It appears to me that if one wants to make progress in mathematics, one should study the masters and not the pupils.

Neils Henrik Abel

23.1 The Modern NMR Spectrometer

The modern NMR spectrometer is truly a marvel of electronic engineering.† Timing of pulses and delays are controlled to very tight tolerances, amplifiers are linear over a wide range of powers and analogue-to-digital conversion is universally employed. Most modern circuit boards used in our spectrometers have surface mounted components that are difficult to repair in the field and usually must be sent to the vendor for repair or replacement. However, it is still possible to repair some things or at least, troubleshoot problems with a bit of electronics background. Also, when contacting the vendor's engineers when there is a problem it is best to have as much detailed information about the issue as possible. Things will go more smoothly and quickly and the engineers will appreciate it. So, instead of saying "there's something wrong with my spectrometer" you should be able to say something like "the 15 volt indicator on our ^{1}H amplifier is not lit".

The starting point in troubleshooting the spectrometer is the block diagram. These may be drawn to any level of complexity. The block diagram in Figure 23.1 is a very general one showing the major

†The material presented here is really very simple and not at the level of say, that of an electrical engineer but I have found that even a simple knowledge of electronics will take you a long way in troubleshooting your spectrometer, which we all have to do at one time or another.

Essential Mathematics for NMR and MRI Spectroscopists
By Keith C. Brown

Published by the Royal Society of Chemistry, www.rsc.org

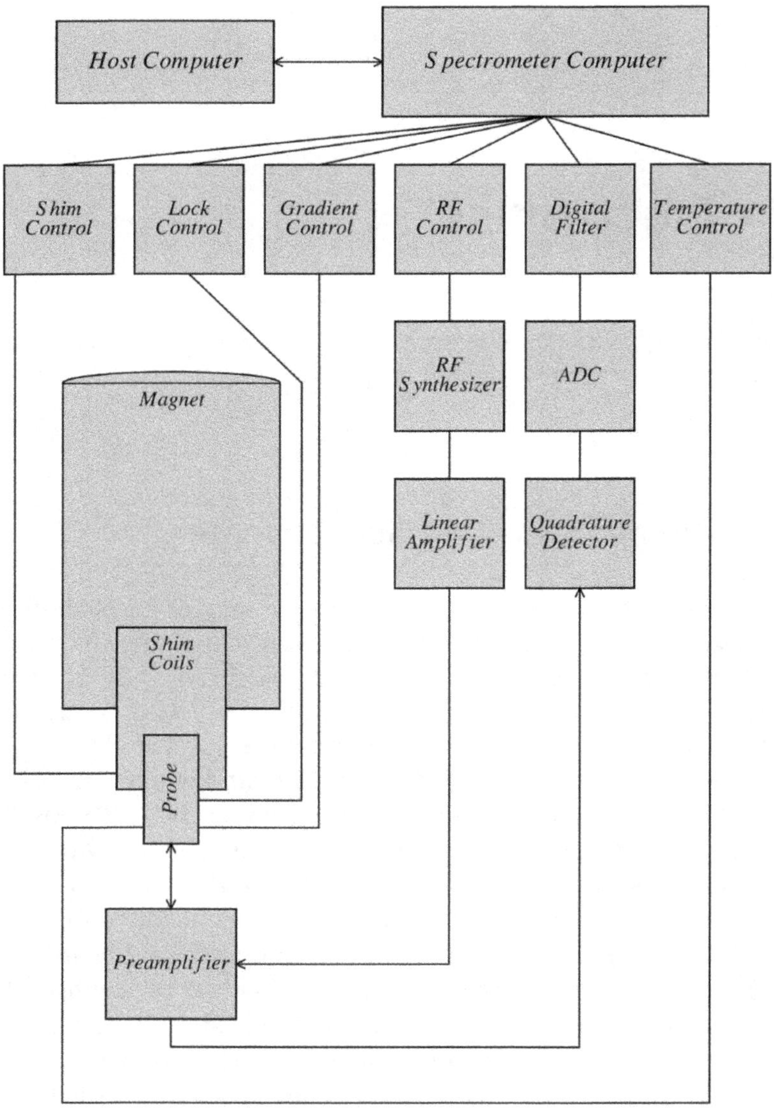

Figure 23.1 Block diagram of an NMR spectrometer.

components of the modern NMR spectrometer and how various signals flow through the system.

Armed with this block diagram you could do a high-level analysis of a problem with your spectrometer. No FID? Is there a magnetic field? A simple test for this is to walk near the magnet with a watch;‡ if the

‡Of course, you might destroy the watch so you should either use a cheap one or find another way to verify that there is a field.

second hand stops there is a field. Are the shim controls working? Look at the appropriate module and see if all of the voltage indicators are lit. If so it is most likely functioning. Is there a pulse being issued? The output port of the power amplifier usually has a pulse indicator lamp tied in to it. If it is not flashing during a pulse then there may be a problem with the amplifier or a module before it. Check that all voltage indicators are lit. Lastly, the problem may be in the receiver section in which case the first thing to do is check that all power indicators are on. Usually, somewhere in the spectrometer system there will be a clue as to what is causing the trouble.

More detailed troubleshooting involves at least a peripheral understanding of electronics. We proceed now to outline some of the simple concepts.

23.2 DC Electronics

DC, of course, stands for *direct current* ... in other words, electrical current flowing through a circuit in one direction only. We frequently encounter DC electronics in circuits that require a battery for power. Possibly the simplest example of this might be the electric torch. In this case the battery current flows out of one pole of the battery, through the light bulb and back to the other pole of the battery.

23.2.1 Ohm's Law and the Resistor

When a battery is connected to a circuit electrical current flows through the circuit from the positive pole of the battery to the negative pole. As the current passes through the circuit it may do some type of work such as light a lamp or power an electric motor. It may also pass through a resistor that, as its name implies, offers resistance to the flow of electricity. On first glance this might seem to be a silly thing to do but it turns out to be very useful in most electronic circuits.

The flow of current through a resistor is indirectly proportional to its resistance value. Thus, the higher the resistance the lower the current. This is encapsulated in Ohm's Law, which is usually stated as:

$$V = IR \tag{23.1}$$

where V is the applied voltage in volts, I is the current in amperes and R is the resistance measured in ohms.

The flow of current in Figure 23.2 is in one direction only and is generally referred to as direct current or simply DC. This will serve as

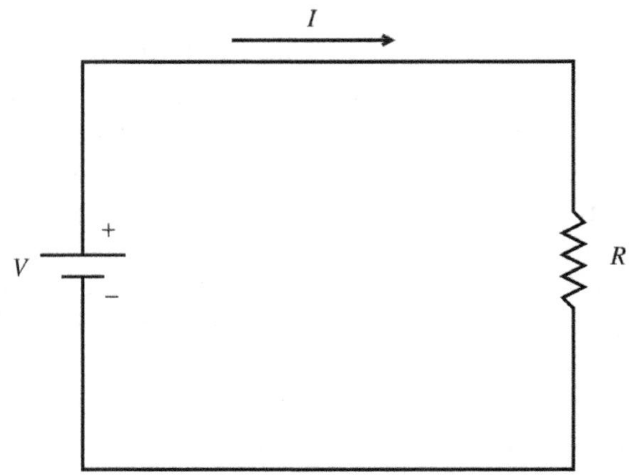

Figure 23.2 A DC voltage applied to a resistance.

our definition of DC. Note that in the figure the current is depicted as flowing from the positive terminal towards the negative terminal. It is the convention in electronics to portray the flow of positive charge … even though we know that it is actually the negatively charged electrons that are in motion. The flow of electrons is of course in the opposite direction to that depicted in Figure 23.2.

That any current flows at all is due to the battery that separates positive and negative charges chemically and sets up an electric field of opposite charge at each pole. When the poles are connected by a wire the mobile electrons in the metal of the wire move towards the positive field and electrons move out of the negative pole's field into the wire producing a current of electricity. With two fields of opposite polarity there will be a potential difference between them that we have called the applied voltage.

What happens if we put two resistors in the circuit, one after the other, instead of one? Logic dictates that the current is determined by the overall resistance so we simply sum the two resistances and use this in Ohm's Law.

$$V = I(R_1 + R_2) \tag{23.2}$$

These resistors are said to be in series with each other (Figure 23.3).

Of course, any number of resistors in series may be summed to give an overall resistance that then may be used in Ohm's Law.

Now, what happens if, instead of being in series, two resistors are put in parallel with each other in the circuit as in Figure 23.4?

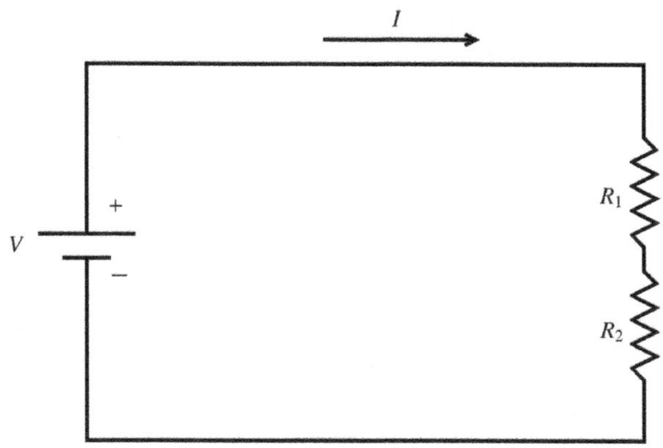

Figure 23.3 Series resistors.

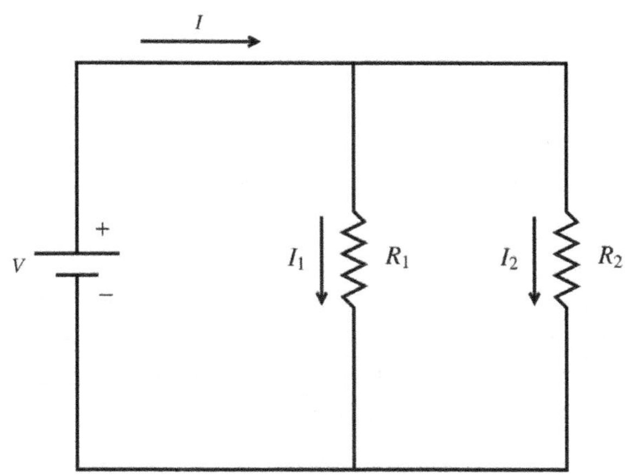

Figure 23.4 Parallel resistors.

In this case, the total current is the sum of the currents flowing through each resistor.

$$I_T = I_1 + I_2$$

$$\frac{V}{R_T} = \frac{V}{R_1} + \frac{V}{R_2}$$

$$= V\frac{R_1 + R_2}{R_1 R_2}$$

We may calculate the value of the total resistance, R_T:

$$\frac{1}{R_T} = \frac{1}{R_1} + \frac{1}{R_2}$$
$$R_T = \frac{1}{\frac{1}{R_1} + \frac{1}{R_2}} \tag{23.3}$$

In both serial and parallel resistive circuits, R_T is the resistance that will replace the series or parallel resistors.

How do we approach a circuit that has *both* series and parallel resistors?

The technique here, in the case of Figure 23.5 is to replace parallel resistors with their equivalent total resistance and then sum that with the series resistors. So, for the series-parallel circuit in Figure 23.5:

$$R_T = R_{\text{parallel}} + R_{\text{series}}$$
$$= \frac{1}{\frac{1}{R_1} + \frac{1}{R_2}} + R_3$$

In Figure 23.6 we see a pair of series resistors that are both in parallel to another resistor. To find the overall resistance of the circuit we first sum the series resistors and then consider their sum to be in parallel with R_1 and do the parallel calculation:

$$R_T = \frac{1}{\frac{1}{R_1} + \frac{1}{R_2 + R_3}}$$

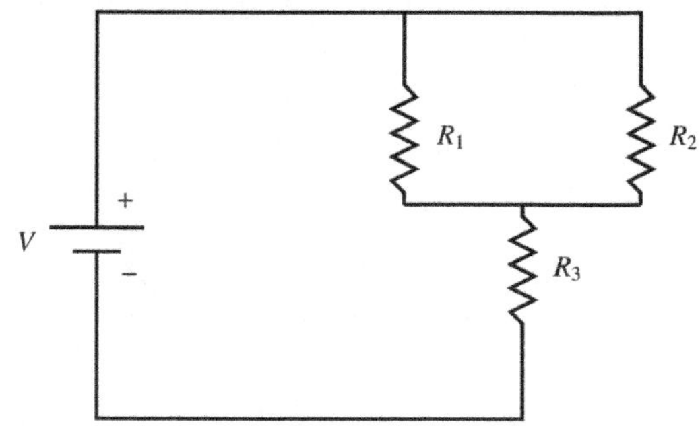

Figure 23.5 Two parallel resistors in series with a third resistor.

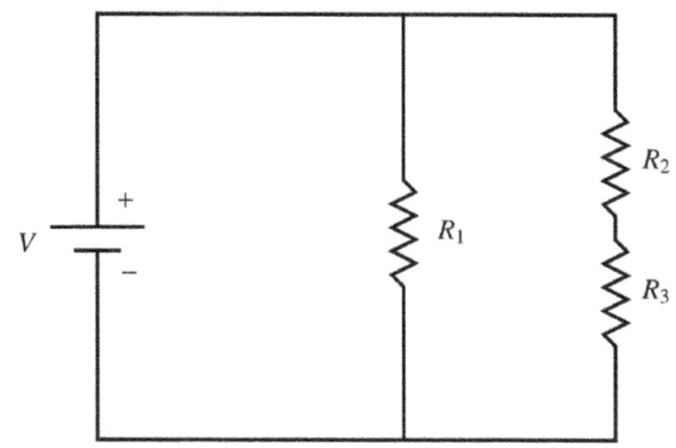

Figure 23.6 Two series resistors in parallel with a third resistor.

One thing that we should perhaps make note of here is that the voltage at any point along a wire is considered to be constant until it encounters a circuit component. This is implied by eqn (23.2):

$$\begin{aligned} V &= I(R_1 + R_2) \\ &= IR_1 + IR_2 \\ &= V_1 + V_2 \end{aligned} \tag{23.4}$$

So, the voltage in the circuit in Figure 23.6, for example, along the positive side from the battery is constant until it encounters the resistors. On the other side of the resistors the voltage has dropped and the voltage drop may be calculated *via* Ohm's Law. Thus, the voltage drop across both R_2 and R_3 in the figure is V, the battery voltage. The voltage drop across R_2 alone is:

$$V_2 = I_2 R_2$$

and for R_3 the voltage drop is:

$$V_3 = I_2 R_3$$

I_2 is just the current flowing through the section of the circuit containing R_2 and R_3. The sum of V_2 and V_3 is, of course, V:

$$V = V_2 + V_3 = I_2(R_2 + R_3)$$

and using this we can calculate the voltage drop across each resistor in terms of V:

$$V_2 = I_2 R_2 = V \frac{R_2}{R_2 + R_3}$$

$$V_3 = I_2 R_3 = V \frac{R_3}{R_2 + R_3}$$

and generally:

$$V_i = IR_i = V\frac{R_i}{\sum R_j} \tag{23.5}$$

A set of resistors connected in series constitutes a *voltage divider*.

A very useful circuit element is the *potentiometer*. This is essentially a resistor with a moveable wiper that can be made to "sample" the voltage drop at any point along the resistor and present it at the wiper contact. It is as if the potentiometer were a voltage divider in which there is an infinite set of resistors in series with individual voltage drops across each that we can measure; in other words, a voltage divider that allows us to select any voltage between zero and the full applied voltage.

In Figure 23.7 the full voltage of the battery is across the potentiometer resistor and a variable voltage V_R is present at the wiper. If the wiper is at the top of the resistor in the figure V_R is the full voltage, V_T. If the wiper is moved to the bottom the voltage becomes zero and any at any position between the two the wiper voltage will be intermediate between the full voltage across the resistor and zero.

What happens if we decrease the total resistance in the circuit towards zero ohms? According to Ohm's Law, eqn (23.1), the current should increase dramatically and ultimately reach infinity, assuming the battery voltage to be constant. Of course, this is not the case in real-world situations. The battery has its own internal resistance (a very low resistance to be sure) that becomes more important as the circuit resistance decreases. Eventually, most of the voltage drop is

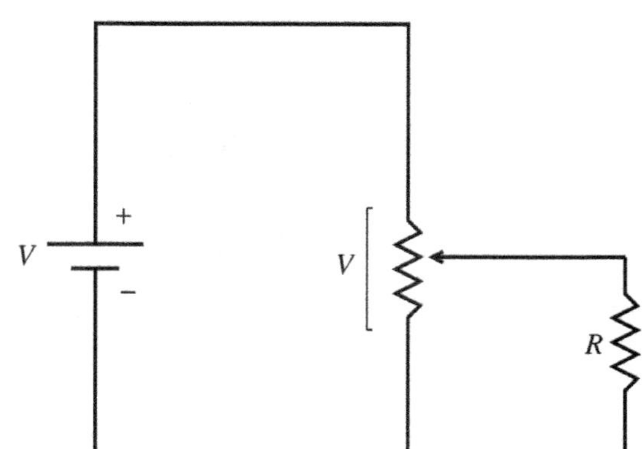

Figure 23.7 Potentiometer voltage across resistor *R*.

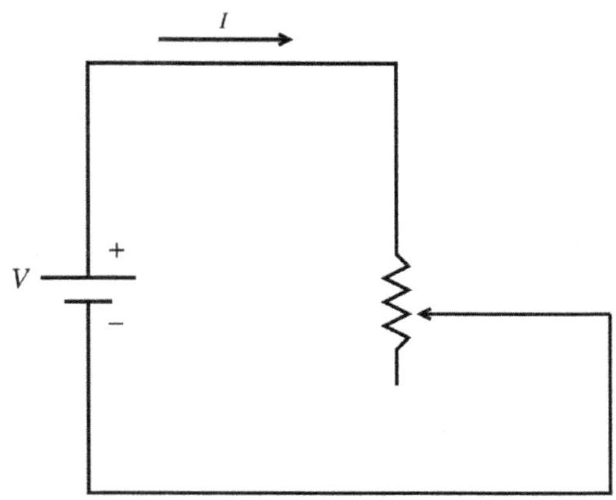

Figure 23.8 A rheostat controlling circuit current.

across the battery's internal resistance that therefore limits the amount of current that can flow out of the battery.

Another use for a variable resistor is as a *rheostat*. This device controls the current flowing through a circuit rather than a voltage.

In Figure 23.8 the rheostat current is at its lowest at the bottom of its travel and highest at the top of the wiper's travel. A typical application for this device is to adjust the current to an incandescent light bulb so that its brightness may be controlled.

It is occasionally of use to define an entity called the *conductance*. To do so we merely take the reciprocal of the resistance:

$$G = \frac{1}{R}$$

From the above analyses one can see that, using the definition of conductance, resistors connected in parallel to each other and viewed as conductances will add in the same way that series-connected resistors do:

$$G_T = \sum G_i$$

Series-connected resistors viewed as conductances will add as parallel-connected resistors do:

$$\frac{1}{G_T} = \frac{1}{\sum G_i}$$

23.2.2 Kirchhoff's Law and Mesh Analysis

We have already encountered Kirchhoff's Law in the form of eqn (23.4). Usually, however, it is stated as:

$$\sum V_i = 0$$

which says that the sum of all of the voltage drops across circuit components around a closed path in a circuit is zero. Thus, in Figure 23.5 for example, travelling in the current direction, the voltage drop across resistors R_1 and R_2 plus the voltage across R_3 plus the voltage from the negative terminal of the battery to the positive terminal all add up to zero. This may seem a trivial statement and it is for the circuits such as we have seen so far. However, for more complex circuits it is very useful. Consider the circuit in Figure 23.9. There are two loop currents shown in the figure and we see that in this model there are two currents flowing in opposite directions through R_1 and R_2. The actual current flows in only one direction and will be the difference between the two, $I_1 - I_2$, along the this leg of the circuit.

We wish to find the voltage between point A and B in Figure 23.9. The matrix method of Chapter 2 is useful here. Traversing the closed paths traced by I_1 and I_2, and using Kirchhoff's Law:

$$\text{for } I_1\text{: } V_{R_1} + V_{R_2} - V_S = 0$$

$$\text{for } I_2\text{: } -V_{R_1} - V_{R_2} + V_{R_3} + V_{R_4} = 0$$

$$\Downarrow$$

$$\text{for } I_1\text{: } (I_1 - I_2)(R_1 + R_2) = 0$$

$$\text{for } I_2\text{: } (I_2 - I_1)(R_1 + R_2) + I_2(R_3 + R_4) = 0$$

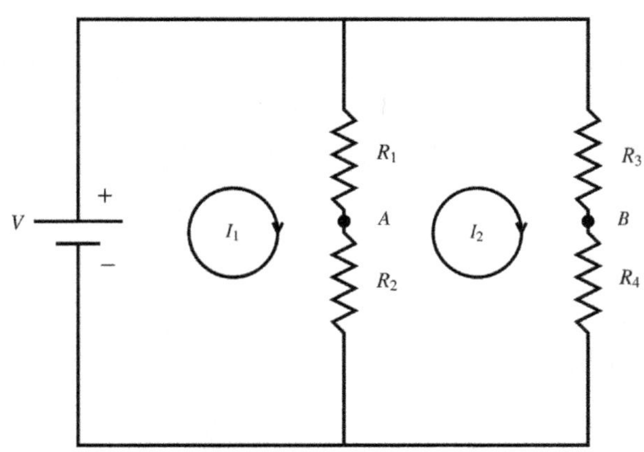

Figure 23.9 Closed-path mesh currents.

and rearranging for the current variables:

$$(R_1+R_2)I_1-(R_1+R_2)I_2=V_S$$
$$-(R_1+R_2)I_1+(R_1+R_2+R_3+R_4)I_2=0$$
$$\Downarrow$$
$$\begin{bmatrix}(R_1+R_2) & -(R_1+R_2)\\ -(R_1+R_2) & (R_1+R_2+R_3+R_4)\end{bmatrix}\begin{bmatrix}I_1\\ I_2\end{bmatrix}=\begin{bmatrix}V_S\\ 0\end{bmatrix}$$

This is a set of three equations in three current variables and, as we saw in Chapter 2, is amenable to solution *via* determinants. Specifically, we can use Cramer's Rule (see eqn (2.41)) to advantage:[§]

$$I_1=\frac{\begin{vmatrix}V_S & -(R_1+R_2)\\ 0 & (R_1+R_2+R_3+R_4)\end{vmatrix}}{\begin{vmatrix}(R_1+R_2) & -(R_1+R_2)\\ -(R_1+R_2) & (R_1+R_2+R_3+R_4)\end{vmatrix}}$$

and

$$I_2=\frac{\begin{vmatrix}(R_1+R_2) & V_S\\ -(R_1+R_2) & 0\end{vmatrix}}{\begin{vmatrix}(R_1+R_2) & -(R_1+R_2)\\ -(R_1+R_2) & (R_1+R_2+R_3+R_4)\end{vmatrix}}$$

As was noted earlier, the actual current flowing through the left leg of the circuit including R_1 and R_2 is $I_L=I_1-I_2$:

$$\begin{aligned}I_L&=I_1-I_2\\ &=\frac{\begin{vmatrix}V_S & -(R_1+R_2)\\ 0 & (R_1+R_2+R_3+R_4)\end{vmatrix}}{\begin{vmatrix}(R_1+R_2) & -(R_1+R_2)\\ -(R_1+R_2) & (R_1+R_2+R_3+R_4)\end{vmatrix}}-I_2=\frac{\begin{vmatrix}(R_1+R_2) & V_S\\ -(R_1+R_2) & 0\end{vmatrix}}{\begin{vmatrix}(R_1+R_2) & -(R_1+R_2)\\ -(R_1+R_2) & (R_1+R_2+R_3+R_4)\end{vmatrix}}\\ &=\frac{V_S}{R_1+R_2}\end{aligned}$$

[§]You probably will suspect by now that I am a fan of symbolic algebra software in general and wxMaxima in particular for doing tedious calculations and this case is no exception. The determinant expansions are slightly complicated and prone to errors for us mere mortals.

The current though the right-hand leg of the circuit is simply I_2:

$$I_2 = \frac{\begin{vmatrix} (R_1+R_2) & V_S \\ -(R_1+R_2) & 0 \end{vmatrix}}{\begin{vmatrix} (R_1+R_2) & -(R_1+R_2) \\ -(R_1+R_2) & (R_1+R_2+R_3+R_4) \end{vmatrix}}$$
$$= \frac{V_S}{R_3+R_4}$$

The voltage or potential between the top of R_1 and point A is calculated using Ohm's Law:

$$V_{R_1} = I_L R_1$$
$$= \frac{V_S}{R_1+R_2} R_1$$

and for the potential between the top of R_3 and point B:

$$V_{R_3} = I_R R_3$$
$$= \frac{V_S}{R_3+R_4} R_3$$

The voltage between points A and B is then the difference between these two:

$$V_{AB} = V_{R_1} - V_{R_3}$$
$$= \frac{V_S}{R_1+R_2} R_1 - \frac{V_S}{R_3+R_4} R_3$$
$$= V_S \left(\frac{R_1R_4 - R_2R_3}{(R_1+R_2)R_4 + (R_1+R_2)R_3} \right)$$

This circuit is actually a special one called the *Wheatstone bridge* circuit. If the ratio of R_1 to R_2 is equal to that of R_3 to R_4 then the voltage between points A and B is zero.

$$\frac{R_1}{R_2} = \frac{R_3}{R_4}$$
$$R_1R_4 = R_2R_3$$
$$R_1R_4 - R_2R_3 = 0$$

Thus, if we set the values of R_1 and R_3 to be equal and if R_2 is an accurate potentiometer then adjustment of the potentiometer until the voltage between points A and B is zero allows us to measure the value of R_4 since at that point in the adjustment of R_2 we see that:

$$R_4 = \frac{R_2 R_3}{R_1} = \frac{R_2 R_1}{R_1} = R_2$$

23.2.3 The Capacitor

A *capacitor* is a circuit device for storing electric charge. It may accept charge from the external circuit and then may let that charge back out into the external circuit, depending on what happens in the external circuit. A capacitor has a capacitance given in Farads (named after the great experimentalist, Michael Faraday) or more often microFarads, which is defined as the amount of charge, Q, that a capacitor will hold for a given applied voltage:

$$C = \frac{Q}{V} \tag{23.6}$$

From this we get:

$$\frac{\mathrm{d}Q}{\mathrm{d}t} = C\frac{\mathrm{d}V}{\mathrm{d}t} = I \tag{23.7}$$

In Figure 23.10 we see a capacitor charging *via* a resistor, R. At first, just as the battery is inserted into the circuit there is no charge on the capacitor but eventually, after current has flowed from the battery through the resistor to the capacitor a charge has built up on the capacitor. The build-up of charge is not, however, instantaneous.

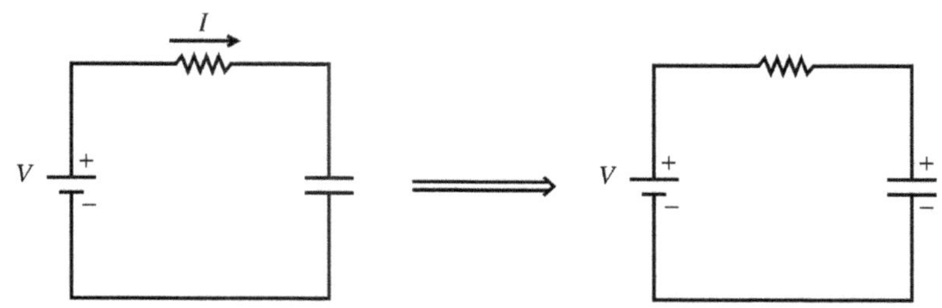

Figure 23.10 Charging a capacitor.

Using the following three relations from eqn (23.1), (23.6) and (11.16):

$$V=\frac{Q}{C},\quad I=\frac{V}{R},\quad I=-\frac{\mathrm{d}Q}{\mathrm{d}t}$$

we can write:

$$I=\frac{Q}{RC}=-\frac{\mathrm{d}Q}{\mathrm{d}t}$$

Why the negative sign in the differential current formula? Recall that in the context of electronics the flow of current refers to the flow of positive charge. We know that it is actually negative charge that flows so we use the negative sign in the expression. This can be confusing to say the least unless it is kept in mind when dealing with electronics.

You will no doubt note that the last equation is a first-order differential equation with a well-defined solution (see eqn (A.49)). The solution to this equation is:

$$\ln Q=-\frac{t}{RC}+K$$

$$\Downarrow$$

$$Q(t)=K'\mathrm{e}^{-t/RC}$$

The value of K' is the value of Q at $t=0$; $Q=CV_T$:

$$Q(t)=CV_T\mathrm{e}^{-t/RC} \tag{23.8}$$

Now, we may calculate the time-dependence of the current in the circuit:

$$\begin{aligned} I(t) &= -\frac{\mathrm{d}Q}{\mathrm{d}t}=-\frac{\mathrm{d}\left(CV_T\,\mathrm{e}^{-t/RC}\right)}{\mathrm{d}t} \\ &= \frac{CV_T}{RC}\mathrm{e}^{-t/RC} \\ &= \frac{V_T}{R}\mathrm{e}^{-t/RC} \end{aligned} \tag{23.9}$$

Equation (23.9) indicates that at $t=0$ the current is V_T/R and that as time proceeds becomes smaller. The value pair, RC, is called the *RC time constant* and the circuit of Figure 23.10 is called an

RC circuit. A dimensional analysis shows us that RC has the units of time:

$$RC \Rightarrow \left(\frac{V}{I}\right)\left(\frac{Q}{V}\right) = \left(\frac{Q}{I}\right) = \left(\frac{Q}{\frac{Q}{t}}\right) = t$$

A larger value of R or of C or both will result in a longer exponential decay of the current. The voltage across the capacitor builds from zero to the full battery voltage in a similar exponential fashion (Figure 23.11).

Of course, as with the resistor, we may connect multiple capacitors in series or parallel with each other. When connected in series, each capacitor contains the same amount of charge, Q. Application of Kirchhoff's Law indicates:

$$\begin{aligned} V_T &= V_{C_1} + V_{C_2} + V_{C_3} \dots \\ \frac{Q}{C_T} &= \frac{Q}{C_1} + \frac{Q}{C_2} + \frac{Q}{C_3} \dots \\ \frac{1}{C_T} &= \frac{1}{C_1} + \frac{1}{C_2} + \frac{1}{C_3} \dots \end{aligned} \tag{23.10}$$

Thus, the total capacitance of series-connected capacitors is calculated in a similar fashion to parallel-connected resistors. For parallel-connected capacitors, the voltage across each capacitor is

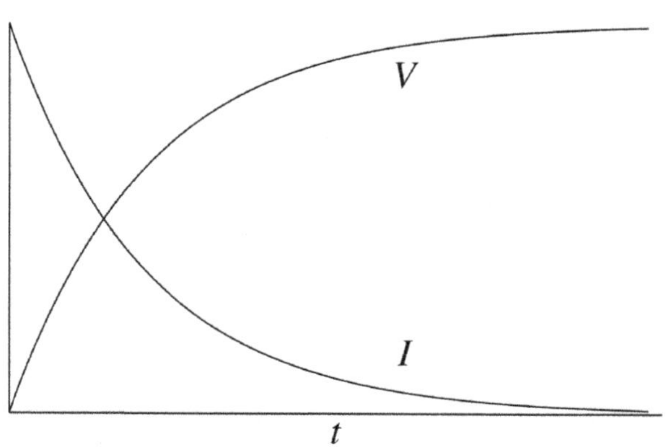

Figure 23.11 Current through and voltage across a capacitor as a function of time.

the same and the total charge is the sum of the charge on each capacitor.

$$
\begin{gathered}
Q_T = Q_1 + Q_2 + Q_3 \ldots \\
\Downarrow \\
C_T V = C_1 V + C_2 V + C_3 V \ldots \\
\Downarrow \\
C_T = C_1 + C_2 + C_3 \ldots
\end{gathered} \tag{23.11}
$$

This time, for parallel-connected capacitors the total capacitance is the sum of all of the capacitors, similar to the calculation for series-connected resistors.

23.2.4 The Inductor

From Chapter 11 we recall Lenz's Law that states that a time-varying magnetic field induces a current in a loop of wire such that the current produces a magnetic field that opposes the changing field. This is expressed in Faraday's Law, eqn (11.30):

$$V = -\frac{\mathrm{d}\Phi}{\mathrm{d}t}$$

which indicates that for a time-varying magnetic field flux the voltage causing the subsequent flow of current in the circuit is such that the current flows so as to oppose the changing field. For N loops this is:

$$V = -N\frac{\mathrm{d}\Phi}{\mathrm{d}t} \tag{23.12}$$

Now, consider a circuit that contains a resistor and a solenoidal coil of N turns (Figure 23.12).

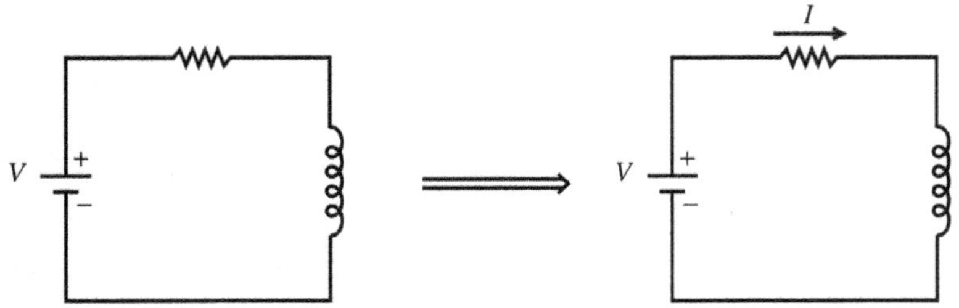

Figure 23.12 Current through an inductor.

When the battery is hooked in to the circuit current will begin to flow and a magnetic field will begin to build. Lenz's Law says that this increasing magnetic field will produce a voltage that will oppose the flow of current from the battery. The current that is flowing in the circuit will be directly proportional to the magnetic field in the coil, as is borne out by eqn (11.23) and the Biot–Savart Law, and by extension, the magnetic flux, Φ. We therefore write:

$$\begin{gathered} V = -N\frac{\mathrm{d}\Phi}{\mathrm{d}t} \\ \Downarrow \\ I \propto \Phi \\ \Downarrow \\ V = L\frac{\mathrm{d}I}{\mathrm{d}t} \end{gathered} \tag{23.13}$$

where L is the constant of proportionality between the current and field flux and is called the *inductance*. A circuit such as this is said to be displaying *self-inductance*; that is, the circuit is inducing a current flow in itself that opposes its main battery-sourced current flow.

In the circuit of Figure 23.12 Kirchhoff's law indicates that the voltages in the circuit are:

$$\begin{aligned} V_T &= V_R + V_L \\ &= IR + L\frac{\mathrm{d}I}{\mathrm{d}t} \end{aligned}$$

Rearranging and integrating (by substitution):

$$\begin{gathered} -\int\frac{\mathrm{d}t}{\tau} = \int\frac{\mathrm{d}I}{I - \frac{V_T}{R}} \quad \left(\tau = \frac{L}{R}\right) \\ \Downarrow \\ -\frac{1}{\tau} + K = \ln\left(I - \frac{V_T}{R}\right) \\ \Downarrow \\ K'\mathrm{e}^{-t/\tau} = I - \frac{V_T}{R} \\ \Downarrow \\ I(t) = \frac{V_T}{R}\left(1 - \mathrm{e}^{-t/\tau}\right) \end{gathered} \tag{23.14}$$

Here, we have substituted $-V_T/R$ for K' since at $t = 0$, I is also zero. In the case of the inductor, the current rises exponentially after the battery is connected, unlike the capacitor circuit that does the opposite. Similarly, the voltage across the inductor falls as time proceeds:

$$\begin{aligned} V_T &= V_R + V_L \\ &\Downarrow \\ V_L &= V_T - V_R \\ &= V_T - I(t)R \\ &= V_T - V_T(1 - \mathrm{e}^{-t/\tau}) \\ &= V_T\mathrm{e}^{-t/\tau} \end{aligned}$$

We have used the current flow in the circuit from Figure 23.12 and Ohm's Law to substitute for V_R. As for the capacitor, the units of t are those of time. A plot of voltage and current *versus* time is shown for the inductor circuit in Figure 23.13.

The astute reader will notice the current and voltage characteristics of inductors with respect to time are opposite to those of the capacitor. As with capacitors, we may ask "How do we treat inductors wired in series and/or parallel". The very definition of inductance (eqn (23.13)) is from the voltage induced by a changing current and we use this to find out how to treat series and parallel inductors. For inductors connected in series each will necessarily experience the

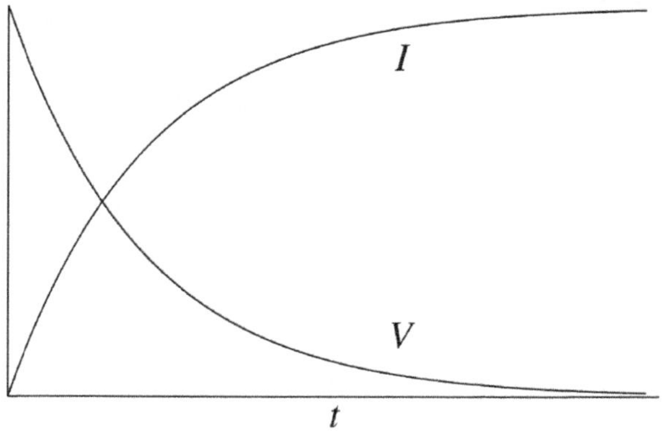

Figure 23.13 Current through and voltage across an inductor as a function of time.

same current changes and will therefore each have their own voltage corresponding to their inductances. The total voltage across all of the inductors will be the sum of their individual voltages:

$$\begin{aligned} V_T &= V_1 + V_2 + V_3 \ \ldots \\ &= L_1 \frac{\mathrm{d}I}{\mathrm{d}t} + L_2 \frac{\mathrm{d}I}{\mathrm{d}t} + L_3 \frac{\mathrm{d}I}{\mathrm{d}t} \cdots \\ &= (L_1 + L_2 + L_3 \ \ldots) \frac{\mathrm{d}I}{\mathrm{d}t} \end{aligned} \tag{23.15}$$

and the total inductance of the series-connected inductors is the sum of their inductances as is the case with resistors.

Parallel-connected inductors will necessarily all have the same voltage across them but will have different currents flowing through them:

$$\begin{aligned} \frac{\mathrm{d}I_T}{\mathrm{d}t} &= \frac{\mathrm{d}I_1}{\mathrm{d}t} + \frac{\mathrm{d}I_2}{\mathrm{d}t} + \frac{\mathrm{d}I_3}{\mathrm{d}t} \cdots \\ &= \frac{V}{L_1} + \frac{V}{L_2} + \frac{V}{L_3} \cdots \\ &= V\left(\frac{1}{L_1} + \frac{1}{L_2} + \frac{1}{L_3} \cdots\right) \\ &= \frac{V}{L_T} \end{aligned} \tag{23.16}$$

23.3 AC Electronics

AC refers to the situation wherein electric current flows alternately in opposite directions within a circuit. This is especially so in our case when dealing with NMR electronics. We regularly deal with oscillating signals from our samples that are translated into high-frequency AC signals in the electronics of the spectrometer. The modern NMR spectrometer does not allow for much "tweaking" of things with one very notable exception . . . the probe. Here, we must tune and match in order to receive the strongest possible signal and it helps to understand why this is needed if we can understand the nature of high-frequency electronics.

23.3.1 Reactance

As we saw in the section on DC circuits the voltage and current characteristics of capacitors and inductors have a time-related behaviour when there is a changing voltage or current, respectively. This is not usually directly important in a DC circuit except when a power source is hooked up or removed ... the DC voltage and current come to an equilibrium value that stays that way until there is a disturbance in the voltage or current. This is not so in AC circuits, which by definition always are changing voltage and current. In the upcoming section on filters we will see a direct application of why this distinction is important to us as NMR/MRI spectroscopists.

Let us look at a capacitor in a simple AC circuit consisting of a source of AC voltage and the capacitor. If the current flowing into the capacitor is varying in a sinusoidal fashion then we can write it as:

$$V_C = V_0 \sin(\omega t)$$

The maximum amplitude of the sine wave voltage is V_0 and its angular frequency is ω. Using this and eqn (23.7):

$$\begin{aligned} I_C &= C\frac{\mathrm{d}V_C}{\mathrm{d}t} \\ &= C\frac{\mathrm{d}(V_0 \sin(\omega t))}{\mathrm{d}t} \\ &= \omega C V_0 \cos(\omega t) \\ &= \omega C V_0 \sin\left(\omega t + \frac{\pi}{2}\right) \\ &= I_0 \sin\left(\omega t + \frac{\pi}{2}\right) \end{aligned}$$

We see here an interesting thing ... there is a $\pi/2$ phase difference between the voltage and current. The voltage lags behind the current across the capacitor by this amount (Figure 23.14).

The ratio V_0/I_0 is called the *reactance* of the capacitor.

$$\frac{V_0}{I_0} = \frac{V_0}{\omega C V_0} = \frac{1}{\omega C} \tag{23.17}$$

The reactance is somewhat analogous to the resistance in the DC circuit. Resistance is a measure of, well, resistance to the flow of current in a DC (and AC) circuit and reactance is much the same. However, we note that the reactance depends not only on the value of

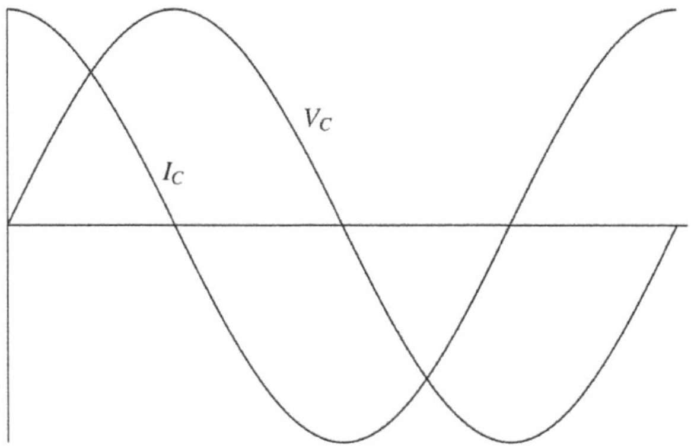

Figure 23.14 AC current in and voltage across a capacitor.

the capacitance but also in the *frequency* of the voltage. Thus, for a given capacitance, a higher frequency current will pass more readily through a capacitor than a lower one because the reactance is lower. For the capacitor the reactance is inversely proportional to the frequency. There is no frequency dependence associated with resistors in AC circuits so their reactance is simply R as in DC circuits. The units for the capacitive reactance are ohms.

Let us look in a similar fashion at an inductor in a simple AC circuit. We shall write for the current through the inductor:

$$I_L = I_0 \sin(\omega t)$$

and from eqn (23.13):

$$\begin{aligned} V_L &= L\frac{\mathrm{d}I_L}{\mathrm{d}t} \\ &= \frac{\mathrm{d}(I_0 \sin(\omega t))}{\mathrm{d}t} \\ &= \omega L I_0 \cos(\omega t) \\ &= \omega L I_0 \sin\left(\omega t + \frac{\pi}{2}\right) \\ &= V_0 \sin\left(\omega t + \frac{\pi}{2}\right) \end{aligned}$$

We see that in the case of the inductor the current lags the voltage by a phase factor of $\pi/2$ (Figure 23.15).

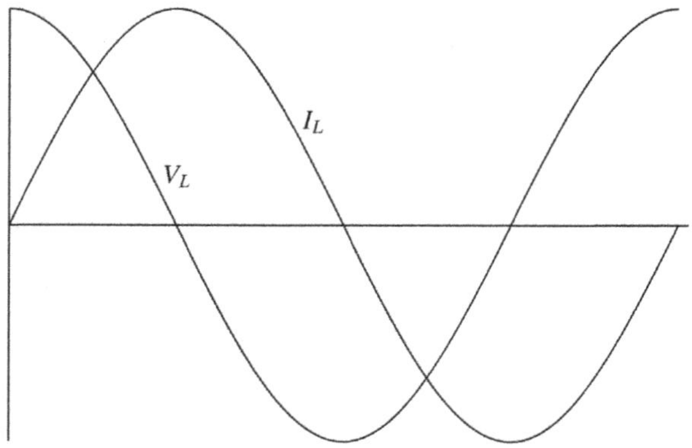

Figure 23.15 AC current in and voltage across an inductor.

The reactance of the inductor is:

$$\frac{V_0}{I_0} = \frac{\omega L I_0}{I_0} = \omega L \tag{23.18}$$

and we see that for an inductor the reactance is directly proportional to the frequency. This means that as the frequency increases the inductor's "resistance" increases. As for the capacitive reactance, the units are ohms.

23.3.2 The Phase Angle and Impedance

In the last section we introduced the idea of the phase angle with respect to a pure capacitive and a pure inductive circuit and in both cases the phase angle was $\pi/2$. We wish to expand somewhat on this idea here. With respect to current and voltage lagging each other, the convention in electronics engineering is to specify the phase angle for what the current does in relation to the voltage. Thus, we would say that in a pure capacitive circuit the current leads the voltage and in a pure inductive circuit the current lags the voltage.

For time-varying voltages *phasor notation* is used to represent the voltage and phase angle. Thus, for a circuit containing a resistor and an inductor with applied voltage $V_0 \sin(\omega t)$ as in Figure 23.16 we may calculate the current through the resistor using this notation. In a resistor there is no lag or lead between current and voltage; they are completely in phase with each other. The applied voltage in the figure has no phase angle associated with it … or rather, has a zero phase angle. The phasor notation for this uses the complex number

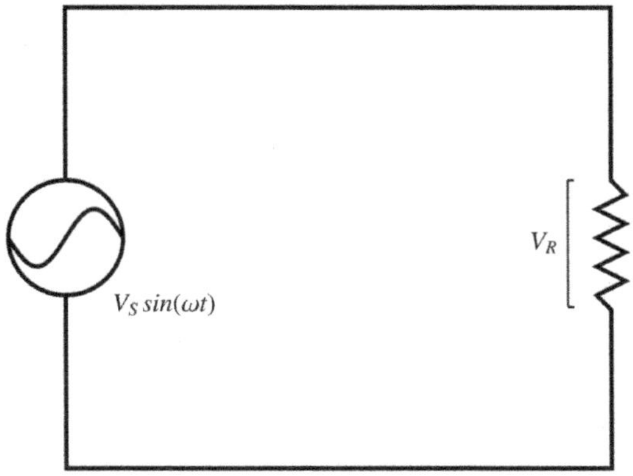

Figure 23.16 Resistor circuit with sinusoidally varying voltage.

ideas of Chapter 2. Thus, the applied voltage in Figure 23.16 is represented as:

$$\mathbf{V} = V_S \mathrm{e}^{\mathrm{i}0}$$

Note that this representation is identical to the polar notation for complex numbers as in eqn (1.7) and simply indicates that the voltage lies along the positive real axis. This will also be the voltage across the resistor and to calculate the current through the resistor we use Ohm's Law:

$$\mathbf{I} = \frac{V_S \mathrm{e}^{i0}}{R\mathrm{e}^{i\phi}} = \frac{V_S}{R}\mathrm{e}^{i(0-\phi)}$$

Since the phase angle of the current through the resistor will be the same as that of the voltage across the resistor, ϕ must also be zero.

$$\mathbf{Z}_R = R\mathrm{e}^{i0} \tag{23.19}$$

$\mathbf{Z}_R$ is the resistive *impedance* in the circuit and as with the voltage in the current example, lies along the positive real axis since its phase angle is zero with R being the magnitude of the number. For AC circuits we use the term impedance to encompass resistive, capacitive and inductive reactance and associated phase angles and we define it here (in polar notation) as:

$$\mathbf{Z}\mathrm{e}^{i(0-\phi)} = \frac{V\mathrm{e}^{i0}}{I\mathrm{e}^{i\phi}} \tag{23.20}$$

We may view this as a generalised version of Ohm's Law that now includes resistors, capacitors and inductors.

Much the same analysis can be done for pure inductive and pure capacitive circuits in which there is only the applied voltage and an inductor or capacitor. Again using $V_0 \sin(\omega t)$ as the applied voltage, for the inductive circuit:

$$\mathbf{I} = \frac{V_0 e^{i0}}{X_L e^{i\phi}} = \frac{V_0}{X_L} e^{i(0-\phi)}$$

We have seen in the preceding analysis of the inductive circuit that the current lags the voltage by $\pi/2$ radians, which means that ϕ must be $\pi/2$. The impedance of the inductor is then:

$$\mathbf{Z}_L = X_L e^{i\pi/2} \tag{23.21}$$

which indicates that it lies along the positive imaginary axis with X_L being the magnitude of the inductive impedance.

For the capacitor:

$$\mathbf{I} = \frac{V_0 e^{i0}}{X_C e^{i\phi}} = \frac{V_0}{X_C} e^{i(0-\phi)}$$

and since the current leads the voltage by $\pi/2$ radians the capacitive impedance is:

$$\mathbf{Z}_C = X_C e^{-i\pi/2} \tag{23.22}$$

and lies along the negative imaginary axis with X_C being the magnitude of the inductive capacitance.

There is a nice visual way of representing these ideas that will be of use to us momentarily. We simply draw a diagram similar to the Argand diagram of Chapter 2 with the inductive reactance along the positive imaginary axis, the resistance along the positive real axis and the capacitive reactance along the negative imaginary axis.

In Figure 23.17 the arrows can be thought of as vectors. ω_L corresponds to a pure inductive circuit, $1/\omega_C$ a pure capacitive circuit and R a pure resistive circuit. The magnitude of the arrow is the reactance and the angle that the arrow makes with the positive real axis is the phase angle between the voltage and the current. We may also take from the figure that the resistance is on the real axis and the capacitance and inductance are on the imaginary axis and we may represent them with complex numbers and use all of the algebra for complex numbers that was developed in Chapter 2. The real portion consists of the resistance and the imaginary portion consists of the inductance and/or the capacitance reactance(s).

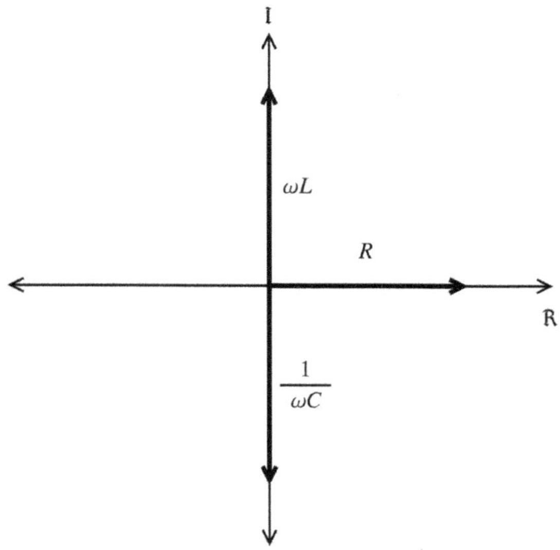

Figure 23.17 Argand-style impedance diagram for pure reactance circuits.

The really fascinating area of AC electrical circuits emerges when we explore combinations of resistors, inductors and capacitors and their associated phase angles and complex representations. In Section 23.2 we discussed direct-current (DC) circuits in which the voltage applied to the circuit is constant. In many electronic circuits, however, the applied voltage is not constant as we have seen but rather, regularly varies in amplitude . . . frequently sinusoidally. The applied voltage may even switch polarity such that it 'pushes' the current through the circuit in the opposite direction. In this case, we say that we are using an AC (alternating current) applied voltage and we use our generalised version of Ohm's Law to do our calculations. In the case of simple DC resistor circuits such as that of Figure 23.4 this becomes Ohm's Law, as we have seen it.

As with DC circuits, Kirchhoff's Law applies and we see in Figure 23.18 that each impedance in a series circuit has its own voltage drop across it.[¶] Also, each will necessarily have the same current flow through it. However, the voltages and currents will not necessarily all have the same phases. Addition of voltages requires an

[¶]Note that I am now using a stylised sinusoidal power source rather than a simple DC battery source to emphasise the AC nature of the circuit . . . more on this later when we look at reading circuit diagrams. Also, I am using a simple rectangle to represent resistors, capacitors and inductors.

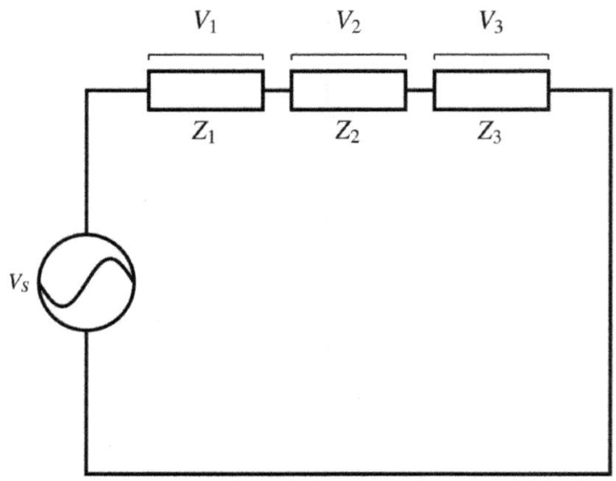

Figure 23.18 Series impedances.

operation equivalent to the addition of complex numbers. In Figure 23.18 we can calculate:

$$
\begin{aligned}
V_S &= V_1 + V_2 + V_3 \\
&= I_S\mathbf{Z}_1 + I_S\mathbf{Z}_2 + I_S\mathbf{Z}_3 \\
&= I_S(\mathbf{Z}_1 + \mathbf{Z}_2 + \mathbf{Z}_3)
\end{aligned}
$$

much the same as in DC circuits (see eqn (23.2)). We can summarise for series-connected impedances:

$$\mathbf{Z}_T = \mathbf{Z}_1 + \mathbf{Z}_2 + \mathbf{Z}_3 \tag{23.23}$$

If we suppose that $\mathbf{Z}_1$, $\mathbf{Z}_2$ and $\mathbf{Z}_3$ are a resistor an inductor and a capacitor, respectively, then using the ideas shown in Figures 23.17 and 23.19 and eqn (23.20), (23.21) and (23.22) we can write the combination of circuit components as:

$$
\begin{aligned}
\mathbf{Z}_T &= R + iX_L - iX_C \\
&= R + i(X_L - X_C) \\
&= Z_T e^{i\phi} \\
Z_T &= \sqrt{R^2 + (X_L - X_C)^2} \\
\phi &= \tan^{-1}\left(\frac{X_L - X_C}{R}\right)
\end{aligned}
\tag{23.24}
$$

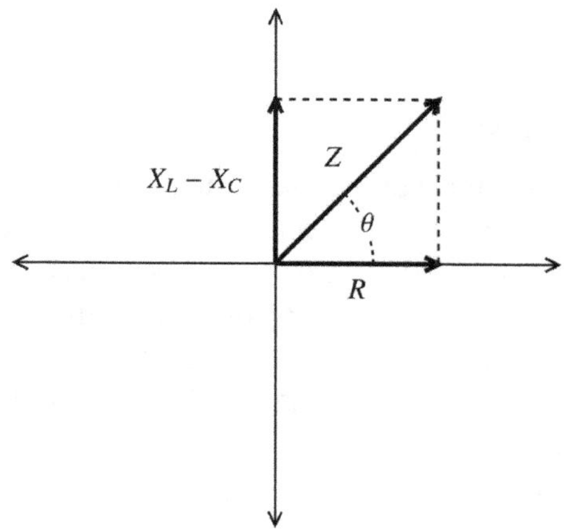

Figure 23.19 Impedance diagram for a RLC series circuit. $X_L > X_C$.

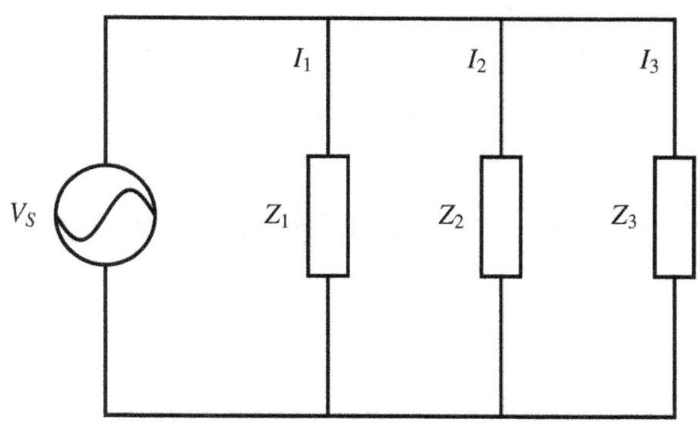

Figure 23.20 Parallel impedances.

Also, just as is the case in DC electronics, we can use the voltage divider principle to our advantage. Equation (23.5) becomes, after substituting impedances for resistances:

$$V_i = IZ_i = V_0 \frac{Z_i}{\sum Z_j} \tag{23.25}$$

In a parallel-impedance circuit the voltage across each impedance is the same but each will have a different current flow through it (Figure 23.20). The corresponding DC equation is (23.5) and the development of eqn (23.25) follows a line similar to that of eqn (23.5).

23.4 AC Circuits

23.4.1 Transformers

In Section 23.2.4 the concept of self-inductance was developed with respect to the inductor. In this situation a circuit containing an inductor has an element in it that opposes the change of current in the circuit. There is another type of inductance called *mutual inductance* that involves two coils of wire in close proximity and a changing current through one of them (Figure 23.21). Our fundamental relationship is eqn (23.12):

$$V_1 = N_1 \frac{\mathrm{d}\Phi_1}{\mathrm{d}t}$$

The change in flux in coil 1 caused by a changing current induces a voltage across the coil, as we have seen. However, the interesting thing is that the changing flux also induces a voltage and a current in the second coil as well.

$$V_2 = N_2 \frac{\mathrm{d}\Phi_2}{\mathrm{d}t}$$

The coils are said to be coupled and we define a coupling coefficient:

$$k = \frac{\Phi_1}{\Phi_2}$$

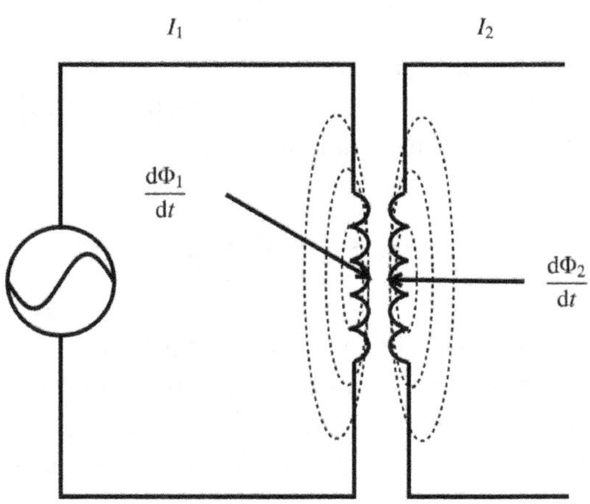

Figure 23.21 Changing magnetic field flux in a transformer.

With this in hand we can modify the equation for V_2:

$$V_2 = kN_2 \frac{d\Phi_1}{dt} = k\frac{N_2}{N_1}V_1$$

In order to get the coupling coefficient as high as possible (the maximum value is, obviously, one) it is often the case that the magnetic field lines are confined to the bulk of an iron core. The flux at coil 2 is almost the same as that at coil 1 and we can say that they are approximately equal. This being the case we can now write:

$$V_2 = \frac{N_2}{N_1}V_1$$

and we see that the voltage across the secondary coil is largely determined by the ratio of turns of wire around the iron core. If $N_1 = N_2$ then the voltage across the secondary coil will be the same as that across the primary coil (Figure 23.22).

This being the case, we can envision building a *step-up transformer* by making the N_2/N_1 ratio larger than one; a ratio of 10:1 for example would convert a primary coil voltage of 10 V to 100 V on the secondary coil. Likewise, we could construct a *step-down transformer* with a N_2/N_1

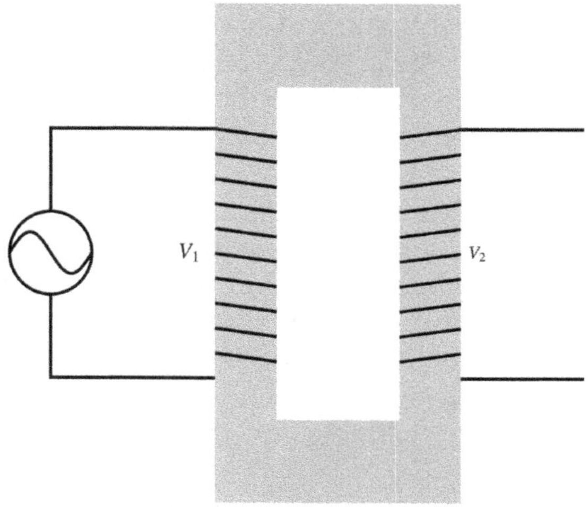

Figure 23.22 The iron-core transformer. The input side on the left is the primary winding and the output side on the right is the secondary winding.

ratio less than one that would convert the primary voltage to a lower secondary voltage. Each of these types of transformers is ubiquitous in our modern society. For the transmission of electricity over long distances voltage is stepped up to several thousand volts and then back down again at the end of the transmission line. Why do this? Recalling eqn (11.12) we may say:

$$E = QV$$

The power that a system uses or supplies is defined as the rate at which energy is expended:

$$P = \frac{E}{t} = \frac{QV}{t} = VI = I^2R \tag{23.26}$$

The first thing that we take from this is that in stepping up the voltage using a transformer we also step down the current by the same factor since we cannot create energy from nowhere ... the product of V and I must remain constant. Secondly, all wires have a small but measurable resistance per unit length that under normal circumstances we just ignore since we usually use very short lengths. However, when transmitting current over long lengths of wire there will inevitably be a loss of energy during the transmission. The power loss we can calculate from eqn (23.26), and is frequently referred to as "*I* squared *R*" loss. To minimise the loss we can do two things; try to lower the resistance of the wire and lower the current. Transmission cables tend to have a large cross-sectional area, which means a low resistance per meter and stepping up the voltage to a large value steps the current down by the same factor. The "*I* squared *R*" loss is thus minimised.

Step-down transformers are in or associated with almost all of our electronics equipment in the home and at work. The voltage at the plug (120 V in North America, 230 V in the UK) is too high for most electronic equipment and must be lowered to a more useable value. One often sees in circuit diagrams step-down transformers; the primary winding (carrying the mains voltage) is usually on the left and the secondary (carrying the output voltage) on the right (Figure 23.23).

It is also possible and indeed, common to have a transformer that steps down to several different voltages (Figure 23.23b). The secondary coil is 'tapped' at various locations along its length and the voltage taken from there.

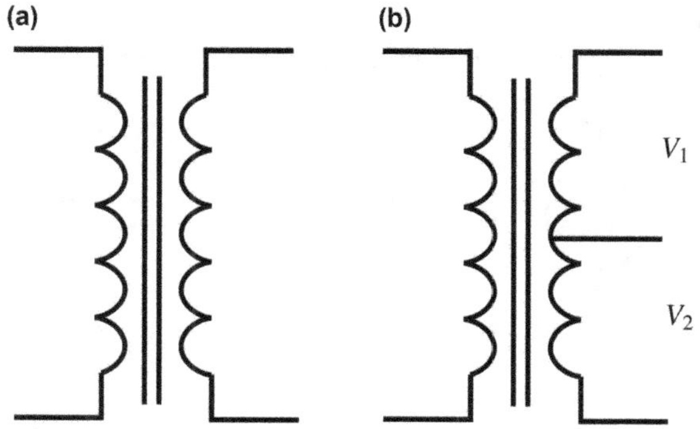

Figure 23.23 (a) Standard iron-core transformer symbol; (b) Multi-tapped iron-core transformer.

23.4.2 Filters

We have not said much about frequency so far except to note that reactances are frequency dependent. The inductive and capacitive reactances from eqn (23.17) and (23.18) are:

$$X_L = \omega L$$

$$X_C = \frac{1}{\omega C}$$

and we see from these definitions that the inductive reactance increases with increasing frequency and the capacitive reactance decreases with increasing frequency. In other words, the resistance that each type of reactance presents to the flow of current is frequency dependent. An inductor would be expected to be relatively difficult to pass current through at high frequencies and a capacitor will pass high-frequency signals more easily than lower-frequency ones. These are very useful characteristics that we will explore a bit more here. Consider the RC circuit of Figure 23.24.

Using the voltage divider eqn (23.25), we can say:

$$\begin{aligned}\frac{V_C}{V_S} &= \frac{X_C e^{-i\frac{\pi}{2}}}{R - iX_C} \\ &= \frac{X_C e^{-i\frac{\pi}{2}}}{\sqrt{R^2 + X_C^2}\exp\left(i \arctan\left(\frac{X_C}{R}\right)\right)} \\ &= \frac{X_C}{\sqrt{R^2 + X_C^2}}\exp\left(i\left(-\frac{\pi}{2} + \arctan\left(\frac{X_C}{R}\right)\right)\right)\end{aligned}$$

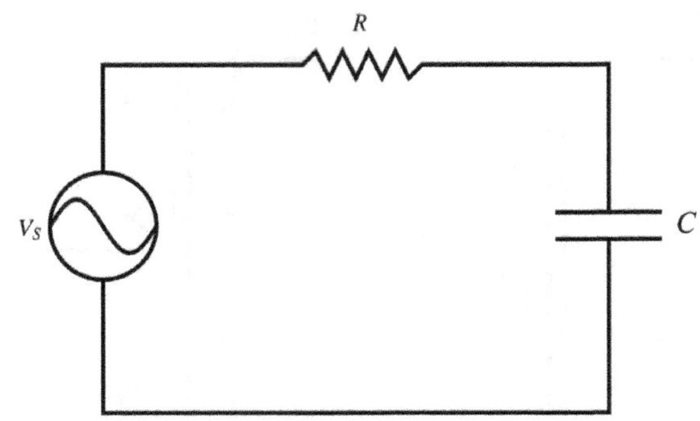

Figure 23.24 An RC filter circuit.

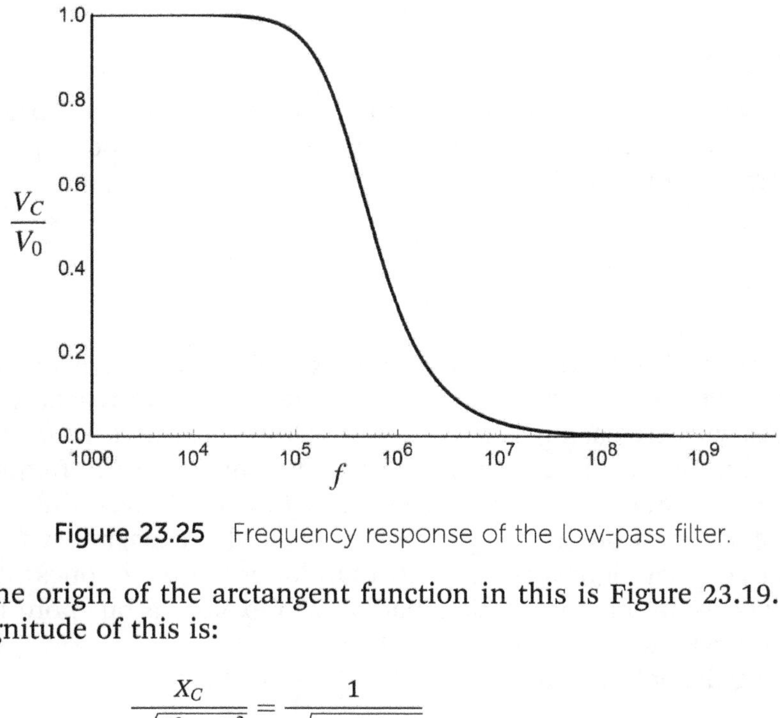

Figure 23.25 Frequency response of the low-pass filter.

The origin of the arctangent function in this is Figure 23.19. The magnitude of this is:

$$\frac{X_C}{\sqrt{R^2+X_C^2}}=\frac{1}{\sqrt{\left(\frac{R}{X_C}\right)^2+1}}$$

$$=\frac{1}{\sqrt{(R\omega C)^2+1}}=\frac{1}{\sqrt{(R2\pi fC)^2+1}}$$

Note that we have changed from radians per second to hertz here as this is the common usage for filters. A plot of the magnitude ratio of V_C/V_0 shows that the magnitude drops off as f increases (Figure 23.25).

Thus, if we were to pass the voltage across the capacitor on to another circuit only lower frequencies would make it through this filter circuit. Accordingly, this is called a *low-pass filter* or sometimes just an LP filter. The frequency response diagram shows the output voltage to input voltage ratio, V_C/V_0 or just $V_{\text{out}}/V_{\text{in}}$. Note that the ratio drops over a wide range of frequencies (the frequency axis is logarithmic) and the area where it is dropping most rapidly is called the shoulder.

What if we pass, instead, the voltage across the resistor on to another circuit? A similar analysis to the above gives us:

$$\frac{R}{\sqrt{R^2+X_C^2}} = \frac{1}{\sqrt{1+\left(\frac{X_C}{R}\right)^2}}$$

$$= \frac{1}{\sqrt{1+\frac{1}{(R\omega C)^2}}} = \frac{1}{\sqrt{1+\frac{1}{(R2\pi fC)^2}}}$$

A plot of this is shown in Figure 23.26.

We see here exactly the opposite behaviour to the low-pass filter. In this case we are looking at the frequency response of a *high-pass filter* or HP filter. Lower-frequency signals either do not pass through or are attenuated.

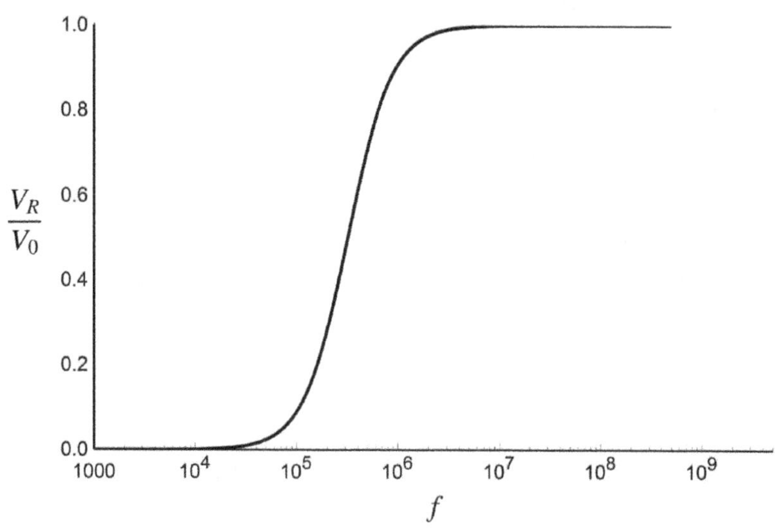

Figure 23.26 Frequency response of the high-pass filter.

How do we specify the pass frequency of a low-pass or high-pass filter circuit? By convention, the critical frequency is defined as that at which $X_C = R$. In both the high-pass and low-pass cases one can see[‖] that when this is so we can say:

$$\frac{V_{\text{out}}}{V_{\text{in}}} = \frac{1}{\sqrt{2}} = 0.707 \quad (23.27)$$

Thus, when the voltage ratio across the capacitor or the resistor is at 0.707 this is the critical frequency of the filter. For the low-pass filter any frequency that is higher than that which results this ratio is said to be attenuated. Just the opposite is the case for the high-pass filter. What is the critical frequency for a given resistor/capacitor combination? Using the definition of capacitive reactance:

$$R = X_C = \frac{1}{\omega C} = \frac{1}{2\pi f C}$$

$$\Downarrow$$

$$f_c = \frac{1}{2\pi RC}$$

So far so good. What if we wanted to pass only a range of frequencies and cut lower and higher ones out? It is possible to construct a *band-pass filter* by simply attaching a high-pass filter to a low-pass filter (Figure 23.27).

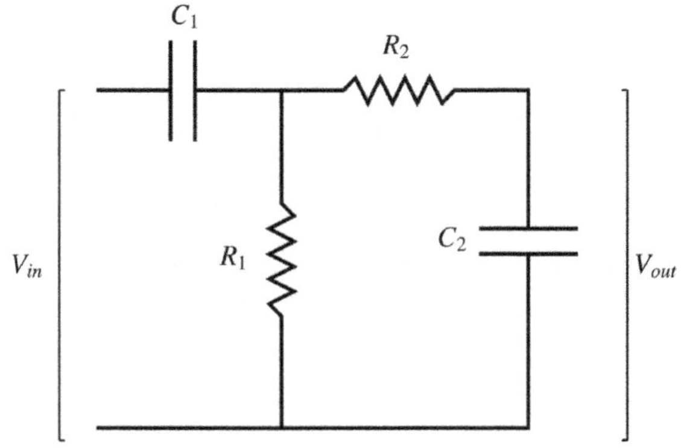

Figure 23.27 A band-pass filter.

[‖]Look at the $V_{\text{out}}/V_{\text{in}}$ magnitude ratio equations for this.

In this figure C_1 and R_1 comprise the high-pass filter and R_2 and C_2 are the low-pass portion. If the critical frequency of the high-pass filter is higher than that of the low-pass filter a band of frequencies is passed through the circuit. The frequency *bandwidth* is defined as the frequency difference between the low-pass critical frequency and the high-pass critical frequency.

In our mind's eye let us break up the operation of the filter into two. The input voltage to the high-pass portion of the circuit we will call V_{in1} and the output of the high-pass portion will be V_{out1}. Similarly for the low-pass half of we define V_{in2} and V_{out2}. From our analysis above we can say for the high-pass portion:

$$V_{out1} = V_{in1}\frac{1}{\sqrt{1+\dfrac{1}{(R_1 2\pi f C_1)^2}}}$$

and for the low-pass portion:

$$V_{out2} = V_{in2}\frac{1}{(R_2 2\pi f C_2)^2+1}$$

However, $V_{in2} = V_{out1}$, so:

$$\begin{aligned} V_{out2} &= V_{out1}\frac{1}{(R_2 2\pi f C_2)^2+1} \\ &= V_{in1}\left(\frac{1}{\sqrt{1+\dfrac{1}{(R_1 2\pi f C_1)^2}}}\right)\left(\frac{1}{(R_2 2\pi f C_2)^2+1}\right) \end{aligned}$$

or:

$$\frac{V_{out2}}{V_{in1}} = \left(\frac{1}{\sqrt{1+\dfrac{1}{(R_1 2\pi f C_1)^2}}}\right)\left(\frac{1}{(R_2 2\pi f C_2)^2+1}\right)$$

A plot of V_{out2}/V_{in1} is shown in Figure 23.28.

Lastly, we will look at the *band-stop filter*. This may also be constructed by using a low-pass and a high-pass filter but this time,

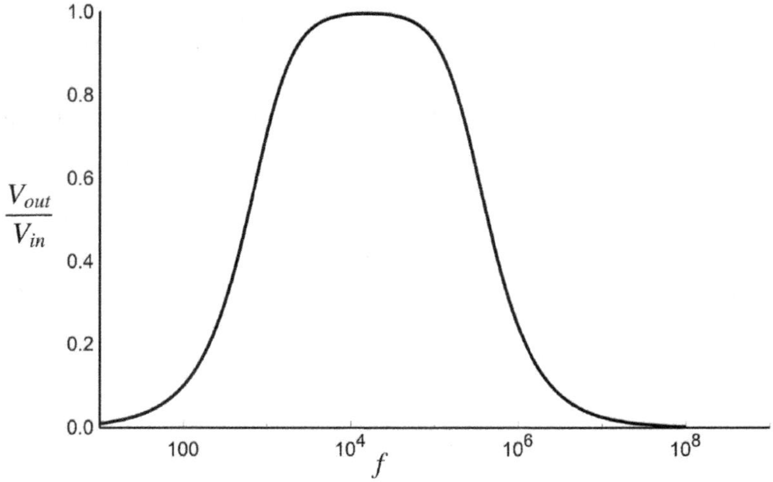

Figure 23.28 Frequency response of a band-pass filter.

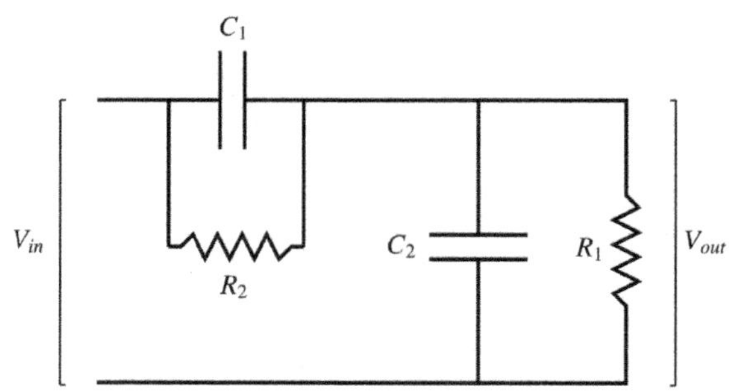

Figure 23.29 A band-stop filter.

rather than connecting them sequentially or in series, we connect them in parallel.

In Figure 23.29 R_1 and C_1 comprise the high-pass portion of the filter and R_2 and C_2 make up the low-pass portion. Each takes the same input voltage, V_{in} and presents its output to the same output line at V_{out}. If the critical frequency of the low-pass filter is lower than that of the high-pass filter then low-frequency signals will pass through with little attenuation. Higher frequencies above the critical frequency for the high-pass filter will also pass relatively unaffected but signals in-between the two will be attenuated. Figure 23.30 illustrates this effect. As with the band-pass filter, the bandwidth is defined as the frequencies that fall in the range between f_{cL} and f_{cH}.

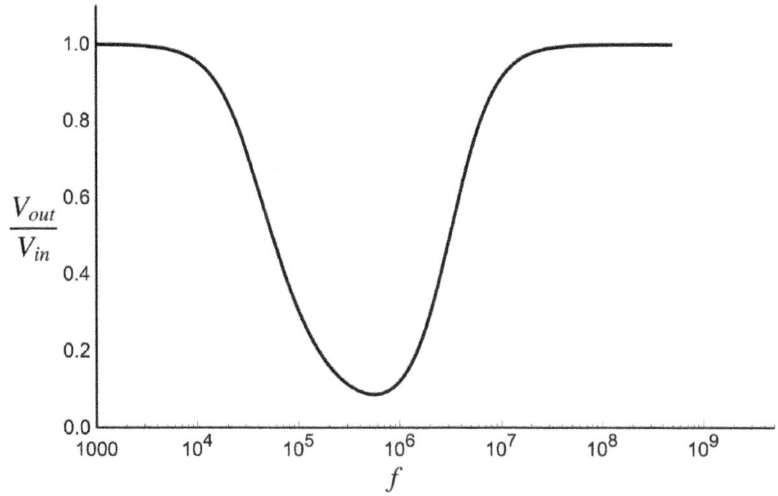

Figure 23.30 Frequency response of the band-stop filter.

For the circuit shown in Figure 23.29 if the values of R_1 and R_2, and of C_1 and C_2 are swapped, f_{cL} and f_{cH} are swapped as well and the circuit becomes a band-pass circuit.

Up until now we have not said anything about the inclusion of the inductor in a circuit. Can we use this circuit component in filter circuits? We can, although it is less common to see this type of filter since capacitors are generally smaller than inductors. Using the by now familiar analysis we can see how this works. Consider the *RL* circuit in Figure 23.31.

The $V_{\text{out}}/V_{\text{in}}$ ratio magnitude is:

$$\frac{X_L}{\sqrt{R^2+X_L^2}} = \frac{1}{\sqrt{\left(\frac{R}{X_L}\right)^2+1}}$$

$$= \frac{1}{\sqrt{\left(\frac{R}{\omega L}\right)^2+1}} = \frac{1}{\sqrt{\left(\frac{R}{2\pi f L}\right)^2+1}}$$

and the frequency-response diagram is shown in Figure 23.32.

As with the *RC* circuit we could construct a *RL* low-pass filter by tapping the voltage across the resistor instead of the inductor.

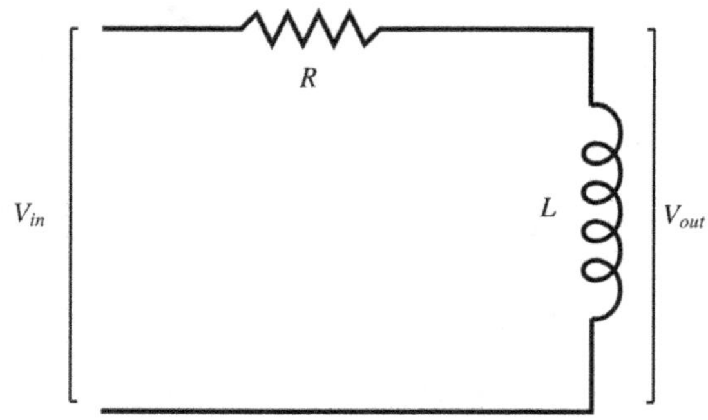

Figure 23.31 A high-pass RL circuit.

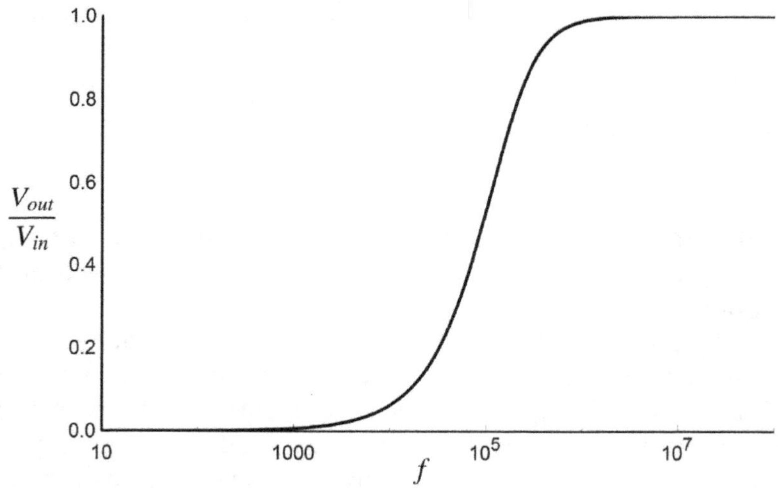

Figure 23.32 Frequency response of the high-pass RL circuit.

The critical frequency in this case is defined as that when $R = X_L$. This being the case, we can write:

$$R = X_L = \omega L = 2\pi f_c L$$
$$\Downarrow$$
$$f_c = \frac{R}{2\pi L}$$

Filters are very commonly used visibly in NMR laboratories at the preamplifier to either stop certain frequencies from passing through or to allow only certain frequencies. Low-pass and high-pass filters are very frequently used in this capacity.

All of the circuits that we have seen so far have included a resistor and either a capacitor or an inductor, except perhaps the very general

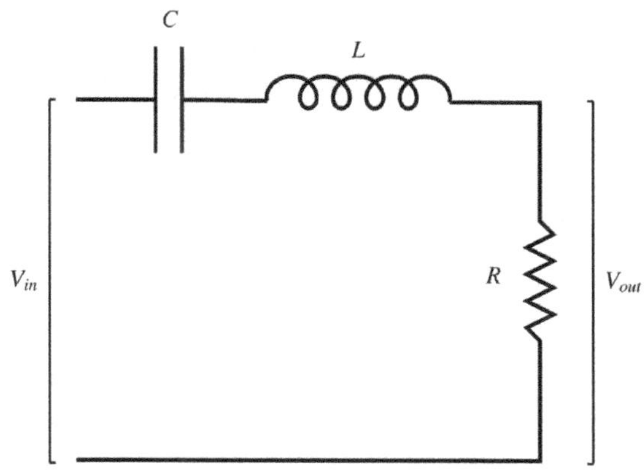

Figure 23.33 Series RLC circuit.

circuit that gave rise to eqn (23.24). We certainly did not look at the frequency characteristics of these types of circuits. We do so now. Consider the series RLC circuit in Figure 23.33.

As for our filter circuit analyses, the V_{out}/V_{in} magnitude ratio from the voltage divider principle is:

$$\frac{R}{\sqrt{R^2 + (X_L - X_C)^2}} = \frac{1}{\sqrt{1 + \left(\frac{2\pi fL}{R} - \frac{1}{2\pi fRC}\right)^2}} \tag{23.28}$$

The first thing to notice is that if $X_L = X_C$ then the ratio is at its maximum point. This being the case, the total impedance of the circuit is:

$$\mathbf{Z}_T = R + i(X_L - X_C) = R$$

We may determine the frequency at which this occurs:

$$\begin{aligned} X_L &= X_C \\ \omega_r L &= \frac{1}{\omega_r C} \\ \omega_r^2 &= \frac{1}{LC} \Rightarrow \omega_r = \frac{1}{\sqrt{LC}} \\ f_r &= \frac{1}{2\pi\sqrt{LC}} \end{aligned} \tag{23.29}$$

A frequency response plot is shown in Figure 23.34.

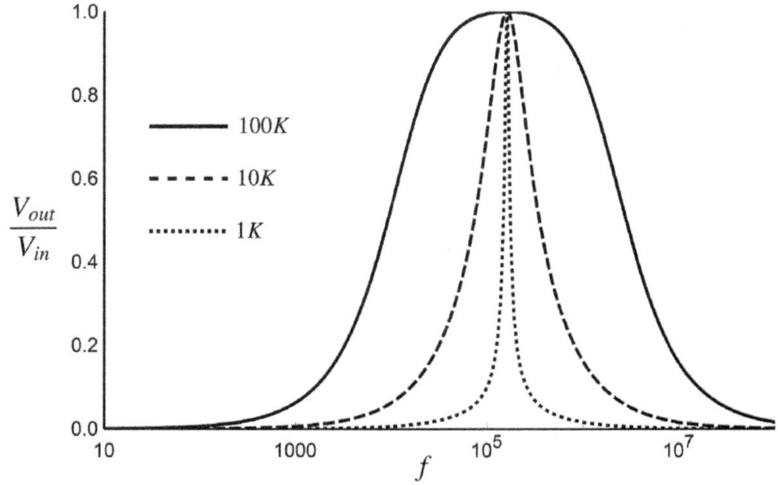

Figure 23.34 Frequency response of the RLC circuit for varying values of R.

Here, we have plotted the frequency response for three different values of R. A large value of R gives us a plot with a wide bandwidth whereas smaller values shrink the bandwidth. The "centre frequency", however, does not change, as expected from the above calculations. One can see that by varying the value of R one can manipulate the range of frequencies allow through the filter. Note also that at the centre frequency all of the input voltage appears across R.

It would seem that this circuit as shown is not much different from the RL and RC bandpass filter circuits, however, let us have a look at what happens to the voltage across the inductor and the capacitor. As with the resistor, the voltage divider ratio of output to input voltage across, say, the capacitor is:

$$\frac{X_C}{\sqrt{R^2 + (X_L - X_C)^2}} = \frac{1}{\sqrt{(2\pi fRC)^2 + ((2\pi f)^2 LC - 1)^2}} \tag{23.30}$$

A plot of $V_{\text{out}}/V_{\text{in}}$ is shown in Figure 23.35.

The voltage across the capacitor at the "centre frequency" is higher than the input voltage; 10 times higher in this case! Something is happening at this frequency that is different from the RC or RL filter circuits. That something is called *resonance* and the RLC circuit is said to be a resonance circuit with a resonant frequency at $f_r = 1/2\sqrt{LC}$. At the resonant frequency, f_r, the $V_{\text{out}}/V_{\text{in}}$ ratio across the capacitor is obtained by substituting f_r for f in eqn (23.30):

$$\frac{V_{\text{out}}}{V_{\text{in}}} = \frac{\sqrt{\frac{L}{C}}}{R} \quad (f = f_r)$$

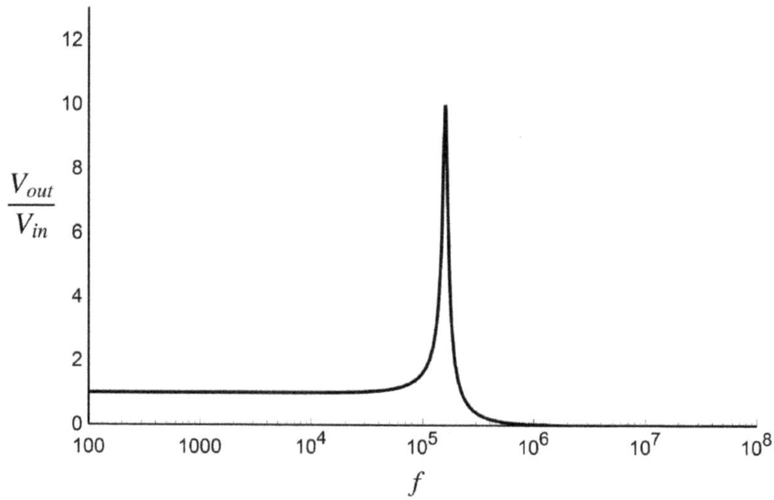

Figure 23.35 V_{out}/V_{in} across capacitor in RLC circuit.

We see from this result that V_{out}/V_{in} is increased if R is decreased and that the ratio can be more than one. The same result is obtained for V_{out}/V_{in} across the inductor. At first glance this would appear to violate Kirchhoff's Law because at the resonant frequency the sum of the magnitudes of the voltages across the inductor, the capacitor and the resistor is generally greater than the input voltage to the circuit. However, for AC circuits we must also include the phase angles in the calculation and when this is done one sees that at resonance the phase angles of the voltages across the inductor and the capacitor are 180° apart. Thus, when added they add to zero leaving only the voltage across the resistor that is exactly equivalent in magnitude and phase to the input voltage.

The bandwidth of the *RLC* filter is defined as those frequencies that pass through the circuit with a V_{out}/V_{in} ratio that is equal to or greater than that which occurs at the critical frequencies. Recall from eqn (23.27) that at the critical point the V_{out}/V_{in} ratio is 0.707 or $1/\sqrt{2}$. Given the shape of the frequency plot we can see that there will be two of these and using the V_{out}/V_{in} ratio for the resistor we can calculate:

$$\frac{R}{\sqrt{R^2 + (X_L - X_C)^2}} = \frac{R}{\sqrt{R^2 + \left(\omega L - \dfrac{1}{\omega C}\right)^2}}$$

At the critical points this ratio is equal to $1/\sqrt{2}$:

$$\frac{R}{\sqrt{R^2+\left(\omega L-\frac{1}{\omega C}\right)^2}}=\frac{1}{\sqrt{2}}$$

$$\Downarrow$$

$$\frac{1}{\sqrt{R^2+\left(\omega L-\frac{1}{\omega C}\right)^2}}=\frac{1}{R\sqrt{2}}$$

The denominators are equal and therefore so must their squares be equal:

$$R^2+\left(\omega L-\frac{1}{\omega C}\right)^2=2R^2$$

$$\Downarrow$$

$$R=\pm\left(\omega L-\frac{1}{\omega C}\right)$$

Taking the positive result:

$$R=\omega L-\frac{1}{\omega C}$$

$$\Downarrow$$

$$\omega RC=\omega^2 LC-1$$

$$\Downarrow$$

$$\omega^2-\omega\frac{R}{L}-\frac{1}{LC}=0$$

Solving for ω:

$$\omega_1=\frac{\frac{R}{L}\pm\sqrt{\left(\frac{R}{L}\right)^2+\frac{4}{LC}}}{2}$$

$$=\frac{R}{2}L\pm\sqrt{\left(\frac{R}{2}L\right)^2+\frac{1}{LC}}$$

Similarly for the negative result:

$$\omega_2 = -\frac{R}{2}L \pm \sqrt{\left(\frac{R}{2}L\right)^2 + \frac{1}{LC}}$$

Using the positive results the bandwidth is then the difference between these two frequencies:

$$B = \omega_1 - \omega_2 = \frac{R}{L} \tag{23.31}$$

From Figure 23.34 and eqn (23.31) we can see that the bandwidth shrinks with decreasing *R*. The "narrowness" is quantified by the *quality factor* as defined by:

$$Q = \frac{\omega_r}{B} = \frac{\omega_r L}{R} = \frac{2\pi f_r L}{R} \tag{23.32}$$

We may do a very similar analysis for a parallel *RLC* circuit (Figure 23.36).

The impedance of the circuit is:

$$\begin{aligned} Z &= \frac{1}{R} + i\omega C + \frac{1}{\omega L} \\ &= \frac{1}{R} + i\omega C + \frac{1}{\omega L} \times \frac{i\omega C}{i\omega C} \\ &= R + i\omega C\left(1 - \frac{1}{\omega^2 LC}\right) \end{aligned}$$

One sees that if $\omega^2 LC = 1$ as in eqn (23.29) then the reactance portion disappears leaving only *R*; another resonance condition.

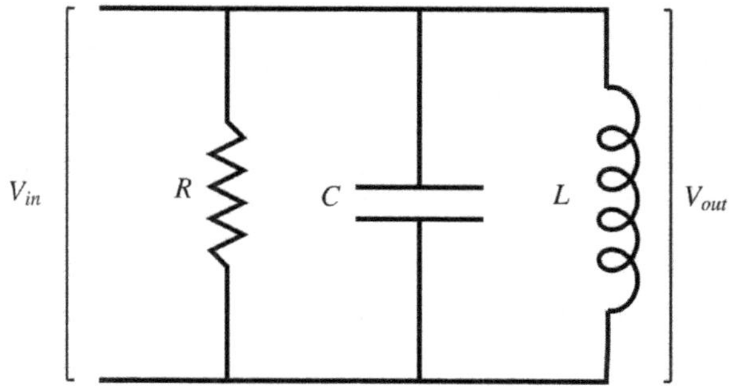

Figure 23.36 A parallel RLC circuit.

23.5 Antenna Theory

At the heart of the NMR spectrometer is a device in the magnet called a probe into which the sample is placed and irradiated with pulses of radio-frequency energy.** The probe is connected to a preamplifier, usually located as close as possible to the magnet to minimise thermal noise since the signals from the nuclei are so weak. At low frequencies we do not have to worry much about the cable connecting things together but we must be very careful at higher frequencies.

The common way to view electronic circuits from a global perspective is to consider them as functional blocks that are interconnected. Each block performs some manipulation on the signal or signals and then passes it on to the next block for subsequent processing. We saw an example of this in the block diagram of an NMR spectrometer in Figure 23.1. Let us consider the implications for the signals as they are passed along through the chain of blocks (Figure 23.37).

Ideally, we want minimal power loss as the signals are transferred from one block to the next. The key to doing this is to consider how the impedances that each block "sees" in the other one affects the transfer of power.

We briefly considered the concept of power in Section 23.4 where we showed that:

$$P = VI = I^2R \tag{23.33}$$

We view things as in Figure 23.37b where Z_A represents the output impedance in block A from Figure 23.37a and Z_B represents the input

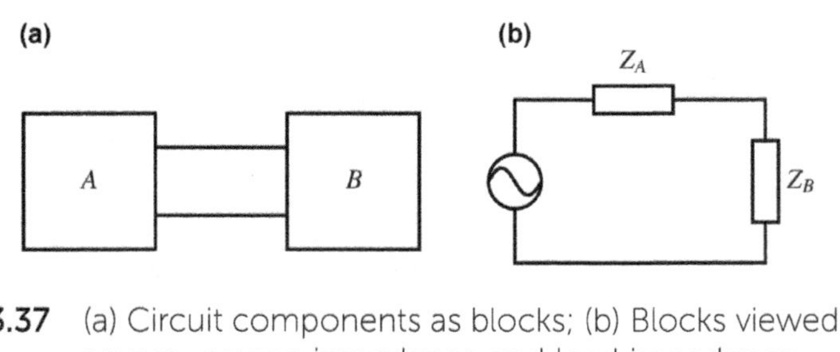

Figure 23.37 (a) Circuit components as blocks; (b) Blocks viewed as signal source, source impedance and load impedance.

**Antenna theory? Whatever does that have to do with NMR spectroscopy? Everything. This is arguably one of the most important practical sections in this chapter.

impedance of block B. Often Z_A is called the source impedance and Z_B the load impedance. What are the optimum values of these impedances for maximum power transfer?

The current flowing in the circuit is:

$$I = \frac{V_{\text{in}}}{\sqrt{(R_A + R_B)^2 + (X_A + X_B)^2}}$$

and from eqn (23.33) the power across Z_B is:

$$P = V_{\text{in}} I = I^2 R_B$$
$$= \frac{V_{\text{in}}^2 R_B}{\sqrt{(R_A + R_B)^2 + (X_A + X_B)^2}}$$

Note that the power is dissipated across the resistance of Z_B. A close look at this will reveal that, keeping R_A and R_B constant for the moment, if $X_A = -X_B$ then the equation becomes:

$$P = \frac{V_{\text{in}}^2 R_B}{(R_A + R_B)^2}$$

Next, we set the derivative of P with respect to R_B, the load resistance, equal to zero to find the maximum power.

$$\frac{dP}{dR_B} = \frac{V_{\text{in}}^2 R_B}{(R_A + R_B)^2}$$
$$= V_{\text{in}}(R_A + R_B)^{-2} - 2V_{\text{in}} R_B (R_A + R_B)^{-3} = 0$$
$$\Downarrow$$
$$(R_A + R_B)^{-2} = 2R_B (R_A + R_B)^{-3}$$
$$\Downarrow$$
$$R_A + R_B = 2R_B$$
$$\Downarrow$$
$$R_A = R_B$$

So, at the condition of maximum power transfer from circuit A to circuit B the resistive portions of their respective impedances must be equal and from above, the reactive portions must be equal and of opposite sign. Putting these together:

$$R_A = R_B \quad X_A = -iX_B$$
$$\Downarrow$$
$$Z_A = Z_B^*$$

indicating that Z_A and Z_B are complex conjugates at maximum power transfer. Adjusting Z_B for maximum power transmission is referred to as *impedance matching*.

When we consider power transmission lines between blocks transferring high-frequency rf energy we must consider them as discrete components in their own right with their own impedances as well as the blocks (Figure 23.38).

Z_0 in Figures 23.38 and 23.39 is referred to as the *characteristic impedance* of the transmission line.

There is another aspect of impedance matching that is worth mentioning. This is the idea of reflected power. When there is a

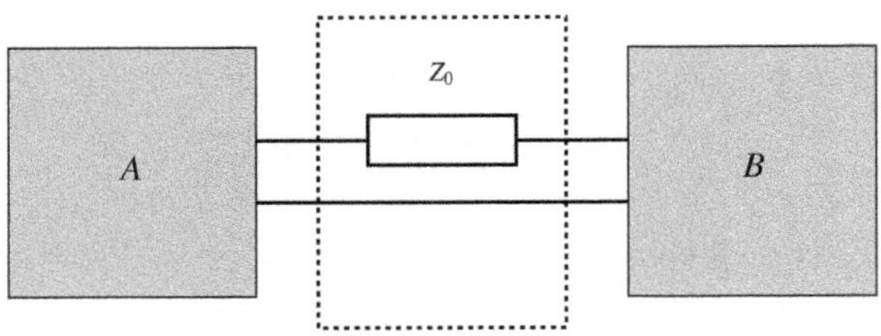

Figure 23.38 RF transmission line between two circuit blocks.

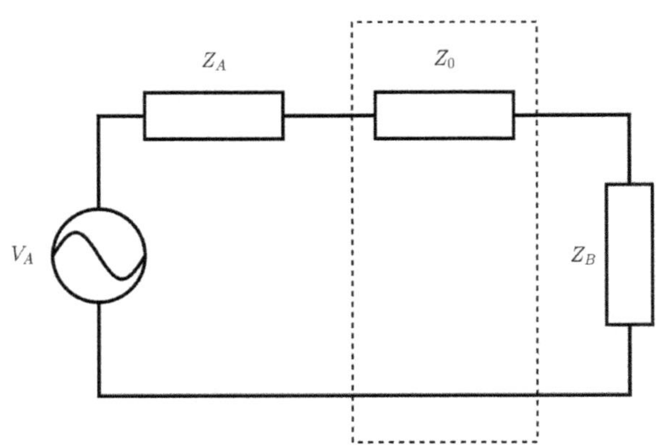

Figure 23.39 Source, transmission line and load impedances.

mismatch of impedances a part of the power sent from A to B is reflected back to A and there are two distinct currents involved, the forward current I_F and the reverse current I_R. Why is there a reflection? When impedances are mismatched the voltage across the load resistance is less than at perfect match, which means the power being delivered to it is lower than the maximum. What happens to the rest of the electrical energy? It is reflected back to the source. There is a close analogue between this and a wave that reaches a new medium and is partially transmitted through the new medium and partially reflected. Light, for example is largely transmitted through glass but a small amount is reflected back from the glass. We can see this from a mathematical point of view by modelling the transmission line.

In Figure 23.40 we show the model of an infinitesimal length of "lossy" transmission line and a "lossless" transmission line in Figure 23.41. By "lossy" is meant that there is power loss due the resistance of the line, both along the line and between the forward and return lines. For short lengths of cable we can safely assume the

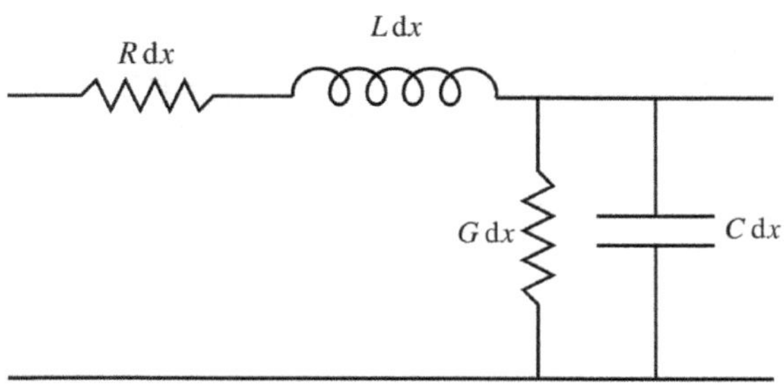

Figure 23.40 "Lossy" impedance model of an infinitesimal length of transmission line.

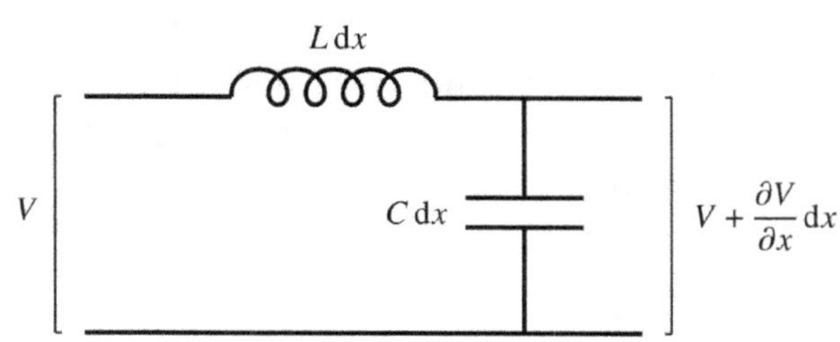

Figure 23.41 Lossless impedance model of an infinitesimal length of transmission line.

resistance of the line to be negligible. From Kirchhoff's voltage Law we write:

$$
\begin{aligned}
V &= V_L + V_C \\
V_C &= V - V_L \\
&\Downarrow \\
V + \frac{\partial V}{\partial x}\partial x &= V - L\frac{\partial I}{\partial t}\partial x \\
&\Downarrow \\
\frac{\partial V}{\partial x} &= -L\frac{\partial I}{\partial t}
\end{aligned}
\tag{23.34}
$$

Similarly we can write, using Kirchhoff's current Law:

$$
\begin{aligned}
I &= I_F + I_C \\
I_F &= I - I_C \\
&\Downarrow \\
I + \frac{\partial I}{\partial x}\partial x &= I - C\frac{\partial V_C}{\partial t}\partial x \\
&= I - C\frac{\partial}{\partial t}\left(V + \frac{\partial V}{\partial x}\partial x\right)\partial x \\
&= I - C\frac{\partial V}{\partial t}\partial x + \frac{\partial V}{\partial x}\partial x^2 \\
&\approx I - C\frac{\partial V}{\partial t}\partial x \\
&\Downarrow \\
\frac{\partial I}{\partial x} &= -C\frac{\partial V}{\partial t}
\end{aligned}
\tag{23.35}
$$

where here I represents the current flowing into this portion of the line and I_F represents the current flowing out of it. These are known as the *telegrapher's equations* and were developed in the late 19th century with the evolution of the telegraph and trans-Atlantic cables. The derivative of each of these with respect to x is:

$$
\begin{aligned}
\frac{\partial^2 V}{\partial x^2} &= -L\frac{\partial^2 I}{\partial t \partial x} \\
\frac{\partial^2 I}{\partial x^2} &= -C\frac{\partial^2 V}{\partial t \partial x}
\end{aligned}
$$

but:

$$\frac{\partial I}{\partial x} = -C\frac{\partial V}{\partial t}$$

$$\frac{\partial V}{\partial x} = -L\frac{\partial I}{\partial t}$$

so:

$$\begin{aligned}\frac{\partial^2 V}{\partial x^2} &= LC\frac{\partial^2 V}{\partial t^2}\\ \frac{\partial^2 I}{\partial x^2} &= LC\frac{\partial^2 I}{\partial t^2}\end{aligned} \tag{23.36}$$

These results should look familiar to you. They are none other than the wave equation of Chapter 7. The solution of the voltage and current equations take the form of eqn (7.18) in which a forward and reverse (*i.e.* reflection) wave are simultaneously travelling along the line.

A fundamental atribute of the transmission line is its characteristic impedance. We must include the line resistance and shunt conductance, G (see Figure 23.40) now as well as the line inductance and shunt capacitance:[††]

$$\begin{aligned}-\frac{\partial V(x,t)}{\partial x} &= RI(x,t) + L\frac{\partial I(x,t)}{\partial t}\\ -\frac{\partial I(x,t)}{\partial x} &= GV(x,t) + C\frac{\partial V(x,t)}{\partial t}\end{aligned} \tag{23.37}$$

Solution of these equations is the same as for the wave equation of Chapter 7 and yields a solution for the time-function:

$$T(t) = \mathrm{e}^{i\omega t}$$

as a possible solution (see eqn (7.15)). With this in mind we write:

$$\begin{aligned}V(x,t) &= V(x)\mathrm{e}^{i\omega t}\\ I(x,t) &= I(x)\mathrm{e}^{i\omega t}\end{aligned} \tag{23.38}$$

[††]The "shunt" capacitance is the equivalent capacitor connecting the upper part of the circuit, containing R and L in Figure 22.40, and the lower part. Sometimes these are called the "hot" and "earths" or "ground" portions of the circuit.

Differentiation gives us:

$$\frac{\partial V(x,t)}{\partial x} = \frac{\mathrm{d}V(x)}{\mathrm{d}x}\mathrm{e}^{i\omega t}$$

$$\frac{\partial V(x,t)}{\partial t} = i\omega V(x)\mathrm{e}^{i\omega t}$$

$$\frac{\partial I(x,t)}{\partial x} = \frac{\mathrm{d}I(x)}{\mathrm{d}x}\mathrm{e}^{i\omega t}$$

$$\frac{\partial I(x,t)}{\partial t} = i\omega I(x)\mathrm{e}^{i\omega t}$$

and substitution into eqn (23.37) gives:

$$-\frac{\mathrm{d}V(x)}{\mathrm{d}x}\mathrm{e}^{i\omega t} = RI(x)\mathrm{e}^{i\omega t} + i\omega LI(x)\mathrm{e}^{i\omega t}$$

$$\begin{aligned}\frac{\mathrm{d}V(x)}{\mathrm{d}x} &= -RI(x) - i\omega LI(x) \\ &= -(R + i\omega L)I(x) \\ &= -ZI(x)\end{aligned} \tag{23.39}$$

and similarly:

$$\frac{\mathrm{d}I(x)}{\mathrm{d}x} = -YV(x)$$

Together these are known as the *Helmholtz equations*. Here, Z is the combination of R and $i\omega L$ and Y is the combination of G and $i\omega C$. These are coupled differential equations (see Appendix A) and have the general solutions:

$$V(x) = V_1\mathrm{e}^{\sqrt{ZY}x} + V_2\mathrm{e}^{-\sqrt{ZY}x}$$

$$I(x) = I_1\mathrm{e}^{\sqrt{ZY}x} + I_2\mathrm{e}^{-\sqrt{ZY}x}$$

Solving for the current, $I(x)$, in eqn (23.39) and substituting from this expression:

$$\begin{aligned}I(x) &= -\frac{1}{Z}\frac{\mathrm{d}V(x)}{\mathrm{d}x} \\ &= \frac{\sqrt{ZY}}{Z}(V_1 e^{\sqrt{ZY}x} + V_2 e^{-\sqrt{ZY}x}) \\ &= \frac{1}{\sqrt{\frac{Z}{Y}}}(V_1 e^{\sqrt{ZY}x} + V_2 e^{-\sqrt{ZY}x}) \\ &= \frac{V_1 e^{\sqrt{ZY}x} + V_2 e^{-\sqrt{ZY}x}}{Z_0} = \frac{V(x)}{Z_0} \\ &\Downarrow \\ Z_0 &= \frac{V(x)}{I(x)}\end{aligned}$$

Thus, Z_0, the characteristic impedance, is the ratio of the forward travelling voltage to the forward travelling current.[‡‡] In this derivation we have used the lossy transmission line model with resistance and conductance included. Our analysis has:

$$Z_0 = \sqrt{\frac{Z}{Y}} = \sqrt{\frac{R + i\omega L}{G + i\omega C}}$$

For a lossless line R and G are each assumed to be zero and we find:

$$Z_0 = \sqrt{\frac{L}{C}}$$

For many short transmission lines this is a reasonable assumption. The *propagation constant* is defined as:

$$\gamma = \sqrt{ZY} = \sqrt{(R + i\omega L)(G + i\omega C)}$$

This is generally a complex number and is frequently written as:

$$\gamma = \alpha + i\beta$$

α is the *attenuation constant* and governs how fast the voltage wave decays as it travels along the line. β is the *phase constant* (since it ends

[‡‡]It is pleasing to note the Ohm's Law form that this takes.

up being part of a phase factor in an exponential term). For $\omega \gg R$ and $\omega \gg G$, γ is:

$$\gamma \approx \sqrt{LC}$$

and $\alpha = 0$, meaning that the transmission line becomes essentially lossless.

Recalling eqn (23.38) the full expressions for the voltage and current are:

$$V(x,t) = (V_1 e^{(\alpha+i\beta)x} + V_2 e^{-(\alpha+i\beta)x}) e^{i\omega t}$$

$$I(x,t) = (I_1 e^{(\alpha+i\beta)x} + I_2 e^{-(\alpha+i\beta)x}) e^{i\omega t}$$

For a high-frequency line these are approximated as:

$$\begin{aligned} V(x,t) &= V_R e^{i(\omega t+i\beta x)} + V_F e^{i(\omega t+i\beta x)} \\ I(x,t) &= I_R e^{i(\omega t+i\beta x)} + I_F e^{i(\omega t+i\beta x)} \end{aligned} \tag{23.40}$$

These are in the D'Alembert form that emphasises the forward and reverse waves of the voltage and current as discussed in Chapter 7 (see eqn (7.18)).

Now let us look a bit more closely at the forward and reverse currents. Differentiating eqn (23.40):

$$\frac{\partial I}{\partial x} = i\beta I_R e^{i(\omega t+i\beta x)} - i\beta I_F e^{i(\omega t+i\beta x)}$$

$$-C\frac{\partial V}{\partial t} = -iC\omega V_R e^{i(\omega t+i\beta x)} - iC\omega V_F e^{i(\omega t+i\beta x)}$$

$$\Downarrow$$

$$\frac{\partial I}{\partial x} = -C\frac{\partial V}{\partial t}$$

The last step is from the telegrapher's eqn (23.35). This being the case we may equate the two expressions:

$$i\beta I_R e^{i(\omega t+i\beta x)} - i\beta I_F e^{i(\omega t+i\beta x)} = -iC\omega V_R e^{i(\omega t+i\beta x)} - iC\omega V_F e^{i(\omega t+i\beta x)}$$

and then break this apart by equating the forward and reverse parts:

$$-i\beta I_F e^{i(\omega t+i\beta x)} = -iC\omega V_F e^{i(\omega t+i\beta x)}$$
$$i\beta I_R e^{i(\omega t+i\beta x)} = -iC\omega V_R e^{i(\omega t+i\beta x)}$$
$$\Downarrow$$
$$\frac{V_F}{I_F} = \frac{\beta}{C} \quad \frac{V_R}{I_R} = -\frac{\beta}{C}$$

Assuming V_F and V_R to have the same signs this result indicates that I_F and I_R have opposite signs. The voltage across the load is the sum of the forward and reverse voltages:

$$V_L = V_F + V_R = Z_L(I_F + I_R)$$

Also, from the discussion on characteristic impedance:

$$I_F = \frac{V_F}{Z_0} \quad I_R = \frac{V_R}{Z_0}$$

and recalling that I_F and I_R have opposite signs:

$$V_F + V_R = Z_L(I_F + I_R) = Z_L\left(\frac{V_F - V_R}{Z_0}\right)$$
$$\Downarrow$$
$$\frac{V_R}{V_F} = \frac{Z_L - Z_0}{Z_L + Z_0}$$

The ratio V_R/V_F is called the *reflection coefficient* and is given the symbol Γ.

$$\Gamma = \frac{V_R}{V_F} = \frac{Z_L - Z_0}{Z_L + Z_0}$$

We see from this that when $Z_L = Z_0$ the reflected voltage is zero and all of the power transmitted to the load is taken up by the load. This is what we want in NMR spectroscopy. The signals from the nuclei in a sample are extremely weak and in order to get the best signal-to-noise ratio we must "match" the probe to the transmission line that it is connected to. The NMR coils in the probe are nothing more than rf antennas and as with any antenna we match it to the line that it is transmitting through.

Of course, we have said nothing about frequencies with respect to impedance matching. As we have seen above, reactances are frequency dependent and will affect the values of the reactances and therefore the power transfer.

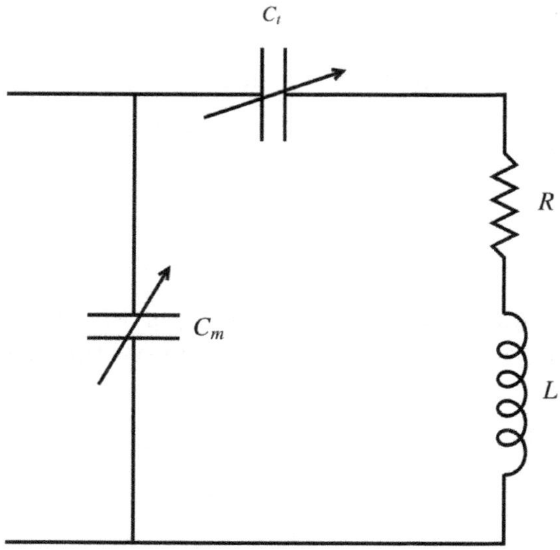

Figure 23.42 NMR probe tuning and matching circuit.

C_t, R and L in Figure 23.42 comprise the series *RLC* circuit of which we spoke above. Actually, this figure represents the electronic components of a typical NMR probe. The inductor is the rf-coil, R is the (small) resistance of the coil and C_t and C_m are the tuning and matching capacitors, respectively. One can think of the coil as an antenna with all of the electrical requirements of any other kind of antenna. In particular, it must be tuned to the appropriate frequency; its characteristic resonant frequency must be at or close to the Larmor frequency of the nucleus of interest. For maximum power transfer it must also be matched to the external components, usually a 50 ohm transmission line that leads to a preamplifier whose function is to immediately amplify the very weak NMR signal.

23.6 Analogue-to-Digital Conversion

We have seen in Chapters 8 and 21 that on modern Fourier transform NMR spectrometers our data consist of discrete data points and that there are consequences of this that we must be aware of. For example, these data points are collected at regular time intervals defined by the dwell time (Figure 23.43). Changing the desired spectral width will change the dwell time and since the two are inversely related an increase in spectral width will result in a decrease in dwell time and an overall decrease in experiment time.

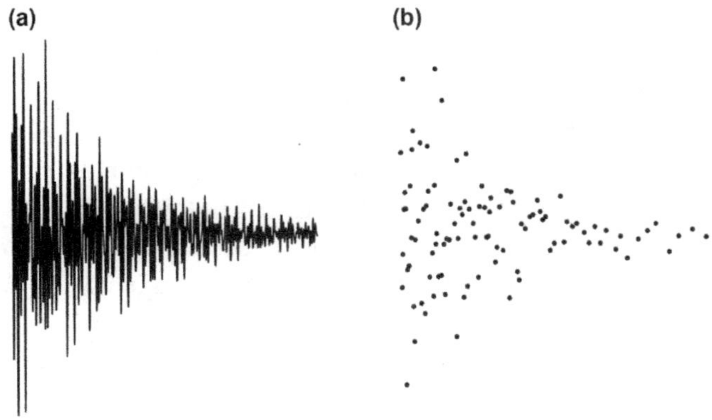

Figure 23.43 (a) Continuous analogue FID; (b) Digitised data points representing the FID.

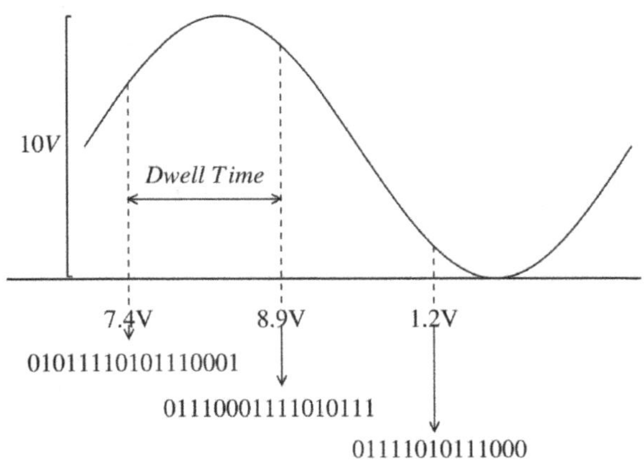

Figure 23.44 Conversion of analogue data to digital data.

The device that is responsible for creating the data points from the analogue data coming from the receiver is the analogue-to-digital converter or ADC. The incoming analogue voltage is recorded in a sample-and-hold module that holds the voltage long enough for the voltage to be digitised and stored. After one dwell time interval the process is repeated. This continues until the total number of data points that the user has requested have been collected.

In Figure 23.44 we see a (cooked up) example of the analogue-to-digital conversion process. Let us suppose for convenience that the maximum input voltage of the 16-bit digitiser is 10 volts. For an analogue voltage that is a maximum of, say +10 V or

10 V peak-to-peak, we see samples taken once every dwell period. The voltage is converted to a binary number amenable to storage in a digital computer for later processing. As we saw earlier in Chapter 8, the digitiser must run at least twice the maximum expected input frequency in order for the digital data to properly represent the analogue data. What happens if the input analogue voltage is greater than 10 V? The digitiser cannot represent voltages more than 10 V and so simply stores the more-than-maximum voltage as 10 V. This is called "clipping" in frequency sampling jargon and refers to the amplitude of the FID being limited by the maximum input voltage of the digitiser. Clipping results in severe baseline distortions that cannot be fixed in software. As a result, one of the things that we must be aware of when collecting our NMR data is how strong the signal is and whether or not it will overload the digitiser.

Since our digitiser outputs 16-bit words it is capable of representing 65 536 different voltages between 0 and 10 volts in 0.000153 V increments and we say that its dynamic range is 65536 to 1. As mentioned above we wish to pay attention to the amplitude of the signal entering the digitiser; if it is too large then clipping occurs. However, we also want to *not* set the signal amplitude (the attenuation, in NMR jargon) too small. Why? If the signal is heavily attenuated then it will be represented by only a few bits in the 16-bit word. If the spectrum contains a mixture of large and small peaks the digitiser may not be able to acquire the small peaks. If the FID contains a very small signal in it, less than 0.000153 V from our example above, then it will not be digitised since 0.000153 V represents one bit.

In the discussion of coherence transfer pathways an expression for the observed magnetisation was put forward in the form of eqn (19.6). We note that this equation involves the difference $(\omega_0 - \omega_{ref})$. Since these are laboratory frame rf frequencies that are relatively close together their difference is a low number that corresponds to the rotating frame frequency. These frequencies are often within the range of 0 to about 3000 rad s^{-1} or 0 to 20 000 Hz. In other words, audio frequencies! It is perfectly possible to hook up an audio amplifier to the analogue signal just prior to digitisation and listen to the decay of the FID (Figure 23.45).[§§]

[§§]This is probably not as easy to do now with modern surface-mount circuit boards being as tightly packed with electronic components. However, this was really quite simple to do in the "old days" when the audio signal was easy to tap in to.

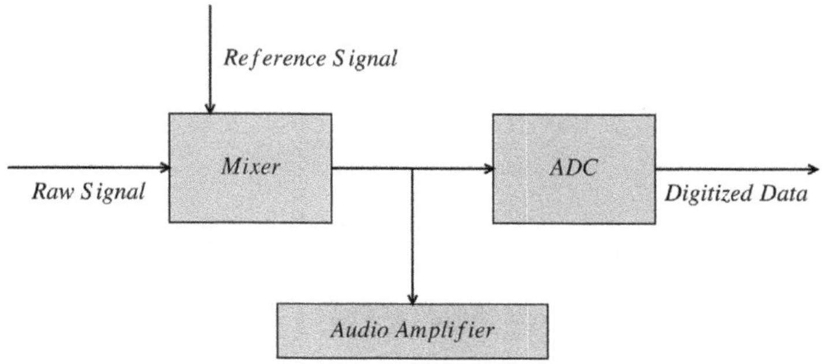

Figure 23.45 Audio circuit to allow one to listen to the FID.

Further Reading

1. R. L. Boylestad, *Introductory Circuit Analysis*, Pearson Prentice Hall, Upper Saddle River, N.J., 2007.
2. E. Fukushima and S. B. Roeder, *Experimental Pulsed NMR: A Nuts and Bolts Approach*, Addison-Wesley Publishing Co., Reading, Mass., 1981.
3. R. G. Lyons, *Understanding Digital Signal Processing*, Prentice Hall, Upper Saddle River, NJ, 2004.

Appendix A Identities, Proofs and Miscellany

A.1 Euler's Formula

$$e^{i\phi} = \cos(\phi) + i\sin(\phi) \tag{A.1}$$

This theorem is proved using the calculus of Maclaurin series. A generalised Maclaurin series is:

$$f(x) = \sum_{n=0}^{\infty} \frac{f^{(n)}(0)}{n!} x^n$$

The Maclaurin series for sin(x) can be obtained using this equation.

$$\begin{aligned}
f(x) &= \sin(x) \Rightarrow f(0) = 0 \\
f'(x) &= \cos(x) \Rightarrow f'(0) = 1 \\
f''(x) &= -\sin(x) \Rightarrow f''(0) = 0 \\
f'''(x) &= -\cos(x) \Rightarrow f'''(0) = -1 \\
&\textit{etc.}
\end{aligned}$$

and generally, $f^{(n)}(0)$ is zero when n is even and alternately 1 and −1 when n is odd. The Maclaurin series for sin(x) is:

$$\sin(x) = x - \frac{x^3}{3!} + \frac{x^5}{5!} - \frac{x^7}{7!} + \cdots = \sum_{n=0}^{\infty} (-1)^n \frac{x^{2n+1}}{(2n+1)!}$$

Essential Mathematics for NMR and MRI Spectroscopists
By Keith C. Brown

Published by the Royal Society of Chemistry, www.rsc.org

Similarly, the Maclaurin series for cos(x) is:

$$\cos(x) = 1 - \frac{x^2}{2!} + \frac{x^4}{4!} - \frac{x^6}{6!} + \cdots = \sum_{n=0}^{\infty} (-1)^n \frac{x^{2n}}{(2n)!}$$

and for e^x:

$$e^x = 1 + x + \frac{x^2}{2!} + \frac{x^3}{3!} + \frac{x^4}{4!} + \cdots = \sum_{n=0}^{\infty} \frac{x^n}{n!}$$

For e^{ix} the series is:

$$\begin{aligned} e^{ix} &= 1 + \frac{(ix)^1}{1!} + \frac{(ix)^2}{2!} + \frac{(ix)^3}{3!} + \frac{(ix)^4}{4!} + \cdots \\ &= 1 + ix - \frac{x^2}{2!} - \frac{ix^3}{3!} + \frac{x^4}{4!} + \frac{ix^5}{5!} + \cdots \\ &= \sum_{n=0}^{\infty} \frac{(ix)^n}{n!} \end{aligned}$$

Looking closely at this we see that e^{ix} is equal to the sum of the Maclaurin series for $i\sin(x)$ and $\cos(x)$:

$$\begin{aligned} e^{ix} &= 1 + ix - \frac{x^2}{2!} - \frac{ix^3}{3!} + \frac{x^4}{4!} + \frac{ix^5}{5!} - \frac{x^6}{6!} \cdots \\ &= 1 - \frac{x^2}{2!} + \frac{x^4}{4!} - \frac{x^6}{6!} + \cdots \\ &\quad + \left(ix - \frac{ix^3}{3!} + \frac{ix^5}{5!} + \cdots\right) \\ &= \cos(x) + i\sin(x) \end{aligned} \tag{A.2}$$

Similarly:

$$e^{-ix} = \cos(x) - i\sin(x) \tag{A.3}$$

It is interesting to note that Isaac Newton was aware of these series representations of sin, cos and e, but not the relationship between them!

There is a very interesting equation that we can write using Euler's Formula and π:

$$e^{i\pi} = \cos(\pi) - i\sin(\pi) = -1$$

or:

$$e^{i\pi} + 1 = 0 \tag{A.4}$$

which is known as Euler's Identity. Eqn (A.4) is said by many to be the most beautiful equation because it links five fundamental constants, 0, 1, i, π and e and because of its simplicity.[†]

A.2 The Binomial Expansion

The binomial expansion theorem is stated as:

$$\begin{aligned}(a+b)^n &= a^n + \binom{n}{1}a^{n-1}b + \binom{n}{2}a^{n-2}b^2 \cdots \binom{n}{k}a^{n-k}b^k \cdots \binom{n}{n-1}ab^{n-1} + b^n \\ &= \sum_{k=0}^{n}\binom{n}{k}a^{n-k}b^k\end{aligned} \tag{A.5}$$

See Chapter 6 for an explanation of the meaning of the $\binom{n}{r}$ term. In order to show that this is so we will use Pascal's identity:

$$\binom{n}{r} = \binom{n-1}{r-1} + \binom{n-1}{r}$$

The proof of this is:

$$\begin{aligned}\binom{n-1}{r-1} + \binom{n-1}{r} &= \frac{(n-1)!}{(r-1)!\,(n-r)!} + \frac{(n-1)!}{r!\,(n-r-1)!} \\ &= (n-1)!\left[\frac{r}{r!\,(n-r)!} + \frac{n-r}{r!\,(n-r)!}\right] \\ &= (n-1)!\left[\frac{n}{r!\,(n-r)!}\right] \\ &= \frac{n!}{r!\,(n-r)!} \\ &= \binom{n}{r}\end{aligned}$$

[†]Some argue that it is not really an equation since there are no variables in it. I think that this is exactly the point. That five fundamental constants that are intimately related in this "equation" is very remarkable.

We use induction to prove the binomial theorem. Imagine that the binomial theorem is satisfied for:

$$(a+b)^n = a^n + \binom{n}{1}a^{n-1}b + \binom{n}{2}a^{n-2}b^2 \cdots \binom{n}{k}a^{n-k}b^k \cdots \binom{n}{n-1}ab^{n-1} + b^n$$

This is easily verified for $n=1, 2, 3$. Now we write:

$$\begin{aligned}
(a+b)^{n+1} &= (a+b)(a+b)^n \\
&= (a+b)\left(a^n + \binom{n}{1}a^{n-1}b + \binom{n}{2}a^{n-2}b^2 \cdots \binom{n}{k}a^{n-k}b^k \cdots \binom{n}{n-1}ab^{n-1} + b^n\right) \\
&= a\left(a^n + \binom{n}{1}a^{n-1}b + \binom{n}{2}a^{n-2}b^2 \cdots \binom{n}{k}a^{n-k}b^k \cdots \binom{n}{n-1}ab^{n-1} + b^n\right) \\
&\quad + b\left(a^n + \binom{n}{1}a^{n-1}b + \binom{n}{2}a^{n-2}b^2 \cdots \binom{n}{k}a^{n-k}b^k \cdots \binom{n}{n-1}ab^{n-1} + b^n\right) \\
&= a^{n+1} + \binom{n}{1}a^n b + \binom{n}{2}a^{n-1}b^2 \cdots \binom{n}{k}a^{n-k+1}b^k \cdots \binom{n}{n-1}a^2b^{n-1} + ab^n \\
&\quad + a^n b + \binom{n}{1}a^{n-1}b^2 + \binom{n}{2}a^{n-2}b^3 \cdots \binom{n}{k}a^{n-k}b^{k+1} \cdots \binom{n}{n-1}ab^n + b^{n+1} \\
&= a^{n+1} + \left[1 + \binom{n}{1}\right]a^n b + \left[\binom{n}{1} + \binom{n}{2}\right]a^{n-1}b^2 \\
&\quad \cdots \left[\binom{n}{k-1} + \binom{n}{k}\right]a^{n-k+1}b^k \cdots \left[\binom{n}{n-1} + 1\right]ab^n + b^{n+1} \\
&= a^{n+1} + \binom{n+1}{1}a^n b + \binom{n+1}{2}a^{n-1}b^2 \cdots \binom{n+1}{k}a^{n-k+1}b^k \cdots \binom{n+1}{n}ab^n + b^{n+1}
\end{aligned}$$

where we have used Pascal's identity to arrive at the last line. If we know that it is true for $n=1, 2, 3$ then it must be so for all n.

We wish to calculate the value of $(1+x)^c$ and in particular, the value of the expression when c is non-integral. The approach to this is to use the binomial theorem that is shown by using, again, the calculus of McLaurin series. We start by defining:

$$f(x) = (1+x)^c$$

where c is non-integral; that is, a fraction. Frequently we encounter $c = 1/2$. Writing the derivatives of the function of increasing order and their values at zero:

$$f(x) = (1+x)^c \Rightarrow f(0) = 1$$

$$f'(x) = c(1+x)^{c-1} \Rightarrow f'(0) = c$$

$$f''(x) = c(c-1)(1+x)^{c-2} \Rightarrow f''(0) = c(c-1)$$

$$f'''(x) = c(c-1)(c-2)(1+x)^{c-3} \Rightarrow f'''(0) = c(c-1)(c-2)$$

etc.

In general:

$$f^n(x) = c(c-1)(c-2)\cdots(c-n+1)(1+x)^{c-n}$$

$$f^n(0) = c(c-1)(c-2)\cdots(c-n+1)$$

and we write the McLaurin series as:

$$\begin{aligned} f(x) &= 1 + \sum_{n=1}^{\infty} \frac{f^n(0)}{n!} x^n \\ &= 1 + \sum_{n=1}^{\infty} \frac{c(c-1)(c-2)\cdots(c-n+1)}{n!} x^n \\ &= 1 + cx + \frac{c(c-1)}{2} x^2 + \frac{c(c-1)(c-2)}{6} x^3 \cdots \\ &= (1+x)^c \end{aligned} \quad \text{(A.6)}$$

This is the binomial expansion formula. For $c = 1/2$ it is:

$$f(x) = 1 - \frac{1}{2}x + \frac{3}{8}x^2 - \frac{5}{16}x^3 \cdots$$

A.3 Trigonometric Identities

A.3.1 The Fundamental Identity

$$\sin^2(\phi)+\cos^2(\phi)=1 \tag{A.7}$$

This is easily shown using a right triangle with length of sides x and y and hypotenuse length r. From this the definitions of sine and cosine are:

$$\sin(\theta)\equiv\frac{x}{r}\ \cos(\theta)\equiv\frac{y}{r}$$

The Pythagorean theorem says that for a right triangle:

$$x^2+y^2=r^2$$

and dividing through by r^2:

$$\frac{x^2}{r^2}+\frac{y^2}{r^2}=1$$

A.3.2 Sine of Sum of Angles

$$\sin(\alpha+\beta)=\sin(\alpha)\,\cos(\beta)+\sin(\beta)\,\cos(\alpha) \tag{A.8}$$

To prove this identity we need to express sine and cosine as exponential functions using Euler's formula:

$$\begin{aligned}
&\cos(\alpha+\beta)+i\sin(\alpha+\beta)=\mathrm{e}^{i(\alpha+\beta)}\\
&=\mathrm{e}^{i\alpha}\mathrm{e}^{i\beta}\\
&=[\cos(\alpha)+i\sin(\alpha)][\cos(\beta)+i\sin(\beta)]\\
&=\cos(\alpha)\cos(\beta)+i\sin(\alpha)\cos(\beta)+i\sin(\beta)\cos(\alpha)-\sin(\alpha)\sin(\beta)\\
&=[\cos(\alpha)\cos(\beta)-\sin(\alpha)\sin(\beta)]+i[\sin(\beta)\cos(\alpha)+\sin(\alpha)\cos(\beta)]
\end{aligned}$$

Equating the real and imaginary parts:

$$\begin{aligned}
\cos(\alpha+\beta)&=\cos(\alpha)\cos(\beta)-\sin(\alpha)\sin(\beta)\\
\sin(\alpha+\beta)&=\sin(\beta)\cos(\alpha)+\sin(\alpha)\cos(\beta)
\end{aligned} \tag{A.9}$$

If angle α is $\pi/2$ radians then eqn (A.9) is the familiar:

$$\cos\left(\frac{\pi}{2}+\beta\right)=-\sin(\beta)$$
$$\sin\left(\frac{\pi}{2}+\beta\right)=\cos(\beta)$$

Similarly:

$$\begin{aligned}\cos(\alpha-\beta)&=\cos(\alpha)\cos(\beta)+\sin(\alpha)\sin(\beta)\\ \sin(\alpha-\beta)&=\sin(\alpha)\cos(\beta)-\sin(\beta)\cos(\alpha)\end{aligned} \tag{A.10}$$

If angle α is $\pi/2$ radians then eqn (A.10) becomes the familiar:

$$\cos\left(\frac{\pi}{2}-\beta\right)=\sin(\beta)$$
$$\sin\left(\frac{\pi}{2}-\beta\right)=-\cos(\beta)$$

A.3.3 Product of Two Sines or Cosines

Using eqn (A.9) and (A.10) we can show that:

$$\begin{aligned}\cos(\alpha)\cos(\beta)&=\frac{\cos(\alpha-\beta)+\cos(\alpha+\beta)}{2}\\ \sin(\alpha)\sin(\beta)&=\frac{\cos(\alpha-\beta)-\cos(\alpha+\beta)}{2}\end{aligned} \tag{A.11}$$

To get the first identity, sum the first lines of eqn (A.9) and (A.10). The second results from the difference between the first lines of eqn (A.9) and (A.10).

Additionally:

$$\begin{aligned}&\sin(\alpha+\beta)+\sin(\alpha-\beta)\\ &\quad=\sin(\alpha)\cos(\beta)+\sin(\beta)\cos(\alpha)+\sin(\alpha)\cos(\beta)-\sin(\beta)\cos(\alpha)\\ &\quad=2\sin(\alpha)\cos(\beta)\end{aligned} \tag{A.12}$$

and

$$\begin{aligned}&\cos(\alpha+\beta)+\cos(\alpha-\beta)\\ &\quad=\cos(\alpha)\cos(\beta)-\sin(\alpha)\sin(\beta)+\cos(\alpha)\cos(\beta)+\sin(\alpha)\sin(\beta)\\ &\quad=2\cos(\alpha)\cos(\beta)\end{aligned} \tag{A.13}$$

A.3.4 Sine and Cosine of 2ϕ

$$\sin(2\phi) = 2\,\sin(\phi)\,\cos(\phi) \tag{A.14}$$

$$\cos(2\phi) = \cos^2(\phi) - \sin^2(\phi) \tag{A.15}$$

To prove eqn (A.14) simply use the preceding sum identity for $\sin(\alpha + \beta)$:

$$\begin{aligned}\sin(2\phi) &= \sin(\phi + \phi)\\ &= \sin(\phi)\cos(\phi) + \sin(\phi)\cos(\phi)\\ &= 2\,\sin(\phi)\cos(\phi)\end{aligned}$$

To prove eqn (A.15) use the preceding sum identity for $\cos(\alpha + \beta)$:

$$\begin{aligned}\cos(2\phi) &= \cos(\phi + \phi)\\ &= \cos(\phi)\cos(\phi) - \sin(\phi)\sin(\phi)\\ &= \cos^2(\phi) - \sin^2(\phi)\end{aligned}$$

A.3.5 Sine and Cosine of a Negative Angle

$$\sin(-\theta) = -\sin(\theta) \tag{A.16}$$

This is easily shown using eqn (A.10). If angle α is equal to zero:

$$\begin{aligned}\sin(0 - \beta) &= \sin(0)\cos(\beta) - \sin(\beta)\cos(0)\\ \sin(-\beta) &= -\sin(\beta)\end{aligned}$$

And for the cosine of a negative angle:

$$\cos(-\phi) = \cos(\phi) \tag{A.17}$$

From eqn (A.9), if angle α is equal to zero:

$$\begin{aligned}\cos(0 - \beta) &= \cos(0)\cos(\beta) + \sin(0)\sin(\beta)\\ \cos(-\beta) &= \cos(\beta)\end{aligned}$$

A.3.6 Cosine and Sine Times a Complex Exponential

We frequently encounter trigonometric functions in the form of the Euler formula. Multiplication by cosine gives several terms:

$$\begin{aligned} &i\exp(i\theta)\cos(\phi) \\ &\quad = i[\cos(\theta) + i\sin(\theta)]\cos(\phi) \\ &\quad = i[\cos(\theta)\cos(\phi) + i\sin(\theta)\cos(\phi)] \end{aligned}$$

Using eqn (A.11) and (A.12) we find that:

$$\begin{aligned} &i[\cos(\theta)\cos(\phi) + i\sin(\theta)\cos(\phi)] \\ &\quad = \frac{i}{2}[\cos(\theta+\phi) + \cos(\theta-\phi) + i\sin(\theta+\phi) + i\sin(\theta-\phi)] \\ &\quad = \frac{i}{2}[\cos(\theta+\phi) + i\sin(\theta+\phi)] + \frac{i}{2}[\cos(\theta-\phi) + i\sin(\theta-\phi)] \\ &\quad = \frac{i}{2}\exp[i(\theta+\phi)] + \frac{i}{2}\exp[i(\theta-\phi)] \end{aligned} \tag{A.18}$$

Similarly:

$$\begin{aligned} &i\exp(-i\theta)\cos(\phi) \\ &\quad = \frac{i}{2}\exp[-i(\theta+\phi)] + \frac{i}{2}\exp[-i(\theta-\phi)] \\ &i\exp(i\theta)\sin(\phi) \\ &\quad = \frac{1}{2}\exp[i(\theta+\phi)] + \frac{1}{2}\exp[i(\theta-\phi)] \\ &i\exp(-i\theta) \\ &\quad = \frac{1}{2}\exp[-i(\theta+\phi)] + \frac{1}{2}\exp[-i(\theta-\phi)] \end{aligned} \tag{A.19}$$

A.3.7 The Law of Cosines

We verify the law of cosines using Figure A.1:

For a triangle with sides of length a, b and c:

$$c^2 = a^2 + b^2 - 2ab\,\cos(\gamma) \tag{A.20}$$

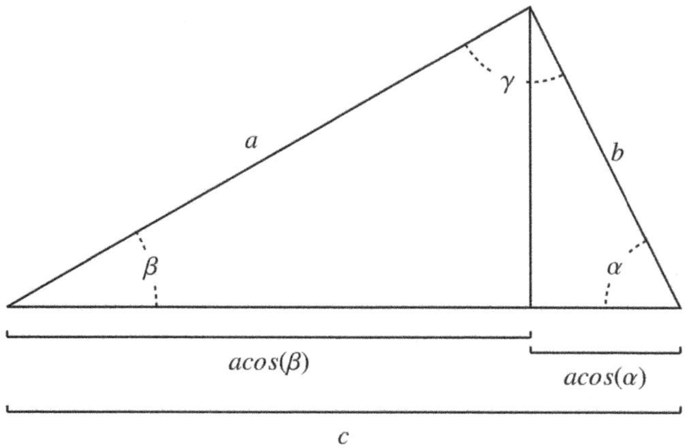

Figure A.1 The law of cosines.

This is a generalisation of the Pythagorean theorem … when $\gamma = 90^o$ it becomes the familiar Pythagorean theorem. The proof is somewhat tedious but not difficult. First, we write for each side:

$$a = b\cos(\gamma) + c\cos(\beta)$$

$$b = a\cos(\gamma) + c\cos(\alpha)$$

$$c = a\cos(\beta) + b\cos(\alpha)$$

You can see how these equations are arrived at from Figure A.1.

These are simultaneous equations so we can use a matrix approach for their solution (see Chapter 2):

$$\begin{aligned} 0\cos(\alpha) \quad & c\cos(\beta) + b\cos(\gamma) = a \\ c\cos(\alpha) \quad & 0\cos(\beta) \quad a\cos(\gamma) = b \\ b\cos(\alpha) \quad & a\cos(\beta) \quad 0\cos(\gamma) = c \end{aligned}$$

or

$$\begin{bmatrix} 0 & c & b \\ c & 0 & a \\ b & a & 0 \end{bmatrix} \begin{bmatrix} \cos(\alpha) \\ \cos(\beta) \\ \cos(\gamma) \end{bmatrix} = \begin{bmatrix} a \\ b \\ c \end{bmatrix}$$

This is in matrix notation and can be written:

$$\hat{\mathbf{A}}\hat{x} = \hat{\mathbf{d}}$$

$$\hat{\mathbf{A}}^{-1}\hat{\mathbf{A}}\hat{\mathbf{x}} = \hat{\mathbf{A}}^{-1}\hat{\mathbf{d}}$$

and therefore

$$\hat{\mathbf{x}} = \hat{\mathbf{A}}^{-1}\hat{\mathbf{d}}$$

$$\hat{\mathbf{x}} = \frac{\operatorname{adj}\hat{\mathbf{A}}}{\det\hat{\mathbf{A}}}\hat{\mathbf{d}}$$

The determinant of $\hat{\mathbf{A}}$ is:

$$\det\hat{\mathbf{A}} = cba + bca = 2abc$$

The adjoint of $\hat{\mathbf{A}}$ is:

$$\begin{bmatrix} -a^2 & ab & ac \\ ab & -b^2 & bc \\ ac & bc & -c^2 \end{bmatrix}$$

so our equation becomes:

$$\begin{bmatrix} \cos(\alpha) \\ \cos(\beta) \\ \cos(\gamma) \end{bmatrix} = \frac{\begin{bmatrix} -a^2 & ab & ac \\ ab & -b^2 & bc \\ ac & bc & -c^2 \end{bmatrix}}{2abc}\begin{bmatrix} a \\ b \\ c \end{bmatrix} = \begin{bmatrix} \dfrac{-a^3 + ab^2 + ac^2}{2abc} \\ \dfrac{a^2b - b^3 + bc^2}{2abc} \\ \dfrac{a^2c + b^2c - c^3}{2abc} \end{bmatrix} = \begin{bmatrix} \dfrac{-a^2 + b^2 + c^2}{2bc} \\ \dfrac{a^2 - b^2 + c^2}{2ac} \\ \dfrac{a^2 + b^2 - c^2}{2ab} \end{bmatrix}$$

From which we write:

$$\begin{aligned} 2bc\cos(\alpha) &= -a^2 + b^2 + c^2 \quad \text{or} \quad a^2 = b^2 + c^2 - 2bc\cos(\alpha) \\ 2ac\cos(\beta) &= a^2 - b^2 + c^2 \quad \text{or} \quad b^2 = a^2 + c^2 - 2ac\cos(\beta) \\ 2ab\cos(\gamma) &= a^2 + b^2 - c^2 \quad \text{or} \quad c^2 = a^2 + b^2 - 2ab\cos(\gamma) \end{aligned} \tag{A.21}$$

which are the law of cosines for each side of the triangle.

A.3.8 Sine and Cosine in Terms of Exponentials

From Euler's law eqn (A.1):

$$\begin{aligned} e^{i\theta} + e^{-i\theta} &= \cos(\theta) + i\sin(\theta) + \cos(\theta) - i\sin(\theta) \\ &= 2\cos(\theta) \\ &\text{or} \\ \cos(\theta) &= \frac{e^{i\theta} + e^{-i\theta}}{2} \end{aligned} \tag{A.22}$$

and

$$\begin{aligned} e^{i\theta} - e^{-i\theta} &= \cos(\theta) + i\sin(\theta) - \cos(\theta) + i\sin(\theta) \\ &= 2i\sin(\theta) \\ &\text{or} \\ \sin(\theta) &= \frac{e^{i\theta} - e^{-i\theta}}{2i} \end{aligned} \tag{A.23}$$

A.4 Direction Cosines

The *direction cosines* are used in various situations involving vectors and are generally expressed as follows:

$$\vec{\mathbf{a}} = \|\vec{\mathbf{a}}\| \; (\cos(\alpha)\vec{\mathbf{u}}_x + \cos(\beta)\vec{\mathbf{u}}_y + \cos(\gamma)\vec{\mathbf{u}}_z)$$

where $\cos(\alpha)$, $\cos(\beta)$ and $\cos(\gamma)$ are the direction cosines and α, β and γ are the angles between the vector $\vec{\mathbf{a}}$ and the positive Cartesian axes (Figure A.2).

We can see by explicitly showing the connections between the direction cosine angles and the vector components what identities are possible. For clarity in Figure A.3, we show only the angle α but it is understood that we also measure angles β and γ as in Figure A.2.

$$\vec{\mathbf{a}} \cdot \vec{\mathbf{u}}_x = \|\vec{\mathbf{a}}\|\cos(\alpha)$$

$$\cos(\alpha) = \frac{a_x}{\|\vec{\mathbf{a}}\|} = \frac{a_x}{\sqrt{a_x^2 + a_y^2 + a_z^2}}$$

$$\cos^2(\alpha) + \cos^2(\beta) + \cos^2(\gamma) = \frac{a_x^2}{\|\vec{\mathbf{a}}\|^2} + \frac{a_y^2}{\|\vec{\mathbf{a}}\|^2} + \frac{a_z^2}{\|\vec{\mathbf{a}}\|^2} = \frac{a_x^2 + a_y^2 + a_z^2}{a_x^2 + a_y^2 + a_z^2} = 1 \tag{A.24}$$

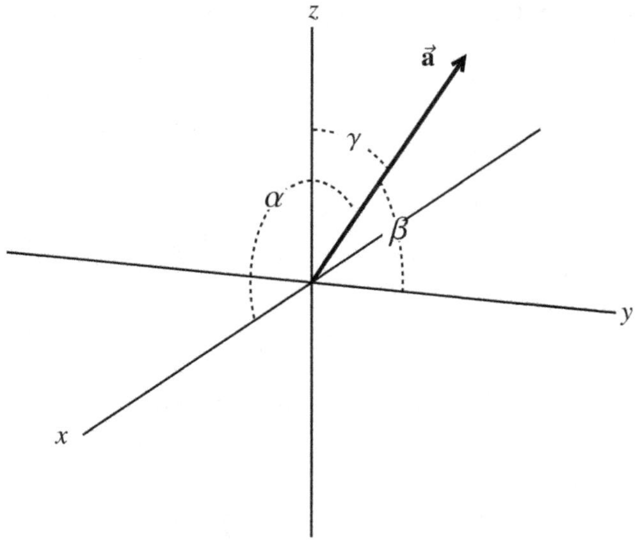

Figure A.2 Direction cosines.

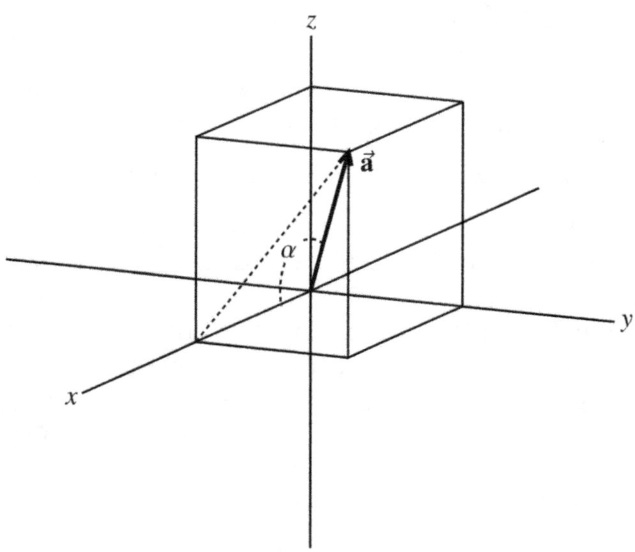

Figure A.3 Direction cosine α.

A.5 The Area of a Parallelogram

Let us consider a labelled diagram of a parallelogram in Figure A.4.

The area of the rectangle is ch and the area of each of the triangles is 1/2 bh. The total area is then:

$$\begin{aligned} A &= c \times h + 2 \times \frac{1}{2} b \times h \\ &= h(c + b) = hd \end{aligned} \tag{A.25}$$

One could also reach this conclusion by a simple geometric argument. Moving the right-hand triangle to the left until its hypotenuse just coincides with that of the left triangle will make a rectangle of length d and height h. The area is then, of course, hd.

Let us look at it from the point of view of a pair of vectors defining the parallelogram (Figure A.5).

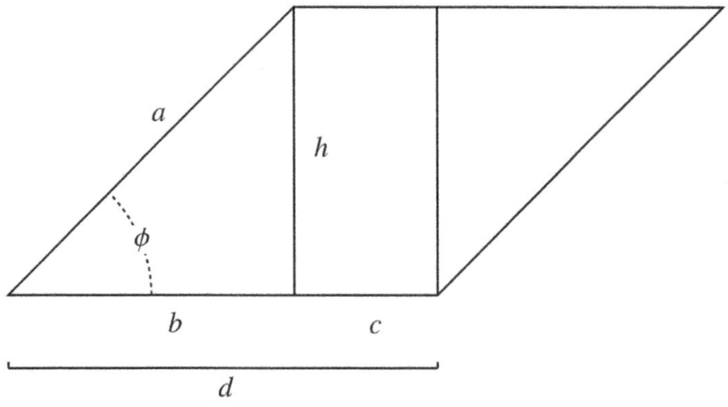

Figure A.4 A parallelogram.

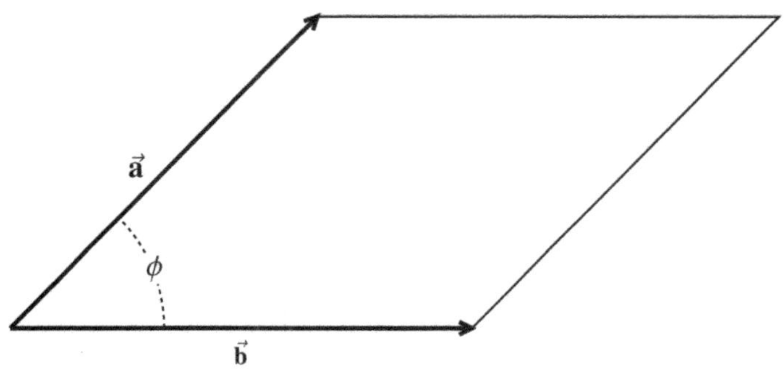

Figure A.5 A parallelogram defined by vectors $\vec{a}$ and $\vec{b}$.

Taking the cross product:

$$\vec{\mathbf{c}} = \vec{\mathbf{a}} \times \vec{\mathbf{b}}$$

Vector $\vec{\mathbf{c}}$ is, of course, orthogonal to the plane defined by vectors $\vec{\mathbf{a}}$ and $\vec{\mathbf{b}}$. However, it is the magnitude of $\vec{\mathbf{c}}$ that we are interested in:

$$\begin{aligned} \|\vec{\mathbf{c}}\| &= \|\vec{\mathbf{a}} \times \vec{\mathbf{b}}\| = \|\vec{\mathbf{a}}\|\,\|\vec{\mathbf{b}}\| \sin(\phi) \\ &= \|\vec{\mathbf{a}}\|\,\|\vec{\mathbf{b}}\| \frac{h}{\|\vec{\mathbf{a}}\|} \\ &= bh \end{aligned}$$

where here, h is again the height of the parallelogram. This result is the same as our previous geometric argument and is the area of the parallelogram defined by vectors $\vec{\mathbf{a}}$ and $\vec{\mathbf{b}}$. Vector $\vec{\mathbf{c}}$ is referred to as the *area vector*.

A.6 The Area of a Trapezoid

This is easily shown, given the parallelogram area just proven. Consider again a parallelogram that has, this time, been cut in half to produce two equal-area trapezoids (Figure A.6).

The bottom or longer base of each trapezoid is of length b and the shorter of length a. We know from the previous discussion that the total area of the two trapezoids is:

$$\text{Area} = (a + b)h$$

and since each trapezoid is of the same area the area of each trapezoid must be:

$$\text{Area} = \frac{(a + b)h}{2} \tag{A.26}$$

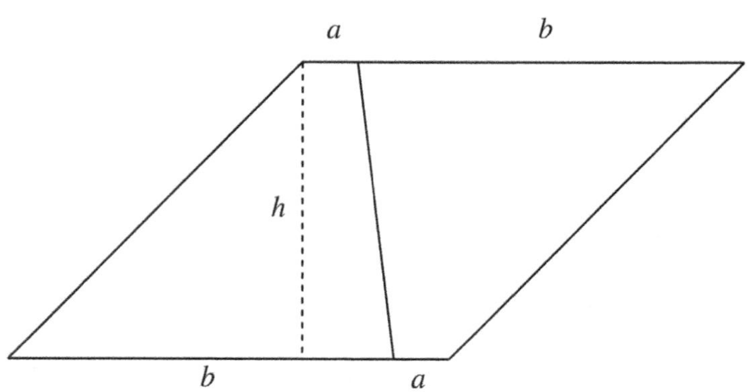

Figure A.6 A trapezoid.

A.7 The Number 'e'

Jacob Bernoulli, in his practical research into interest functions came to recognise a new fundamental number. The letter e, however, was not used until Leonhard Euler began working with it.

To understand where it comes from, consider a compound interest problem. A unit amount of money garners interest at a fractional rate of 1 unit per annum compounded annually. The interest plus the principal after one year is:

$$1 + 1 = 2$$

Now, suppose that we compound the interest biannually and want to know the amount after one year. Since there are now two periods per annum we divide the rate by two. The interest plus principal after the first period is:

$$1 + \frac{1}{2} = 1.5$$

This now becomes the new principal for the second period and we do a similar calculation for the second period:

$$\text{Principal} = 1 + \frac{1}{2} = 1.5$$

$$\text{Interest} = 1.5 \times \frac{1}{2}$$

$$\begin{aligned} \text{Principal} + \text{Interest} &= 1.5 + 1.5 \times \frac{1}{2} \\ &= \left(1 + \frac{1}{2}\right)\left(1 + \frac{1}{2}\right) \\ &= \left(1 + \frac{1}{2}\right)^2 = 1.5^2 = 2.25 \end{aligned}$$

which, by induction, we can generalise for n periods per annum:

$$\left(1 + \frac{1}{n}\right)^n$$

Bernouilli noticed that as n gets larger the calculated interest approaches a particular value. The logical question to ask is "What if

n tends towards infinity?". In mathematical terms we do the limit calculation:

$$\lim_{n\to\infty}\left(1+\frac{1}{n}\right)^n$$

The answer is the irrational number, 2.7182818 ......, which is the number we now call "e". We can show this using the binomial formula (see eqn (A.5)):

$$\begin{aligned}(a+b)^n &= \binom{n}{0}a^n b^0 + \binom{n}{1}a^{n-1}b^1 + \binom{n}{2}a^{n-2}b^2 \cdots \binom{n}{n}a^0 b^n \\ &= \sum_{k=0}^{n}\binom{n}{k}a^{n-k}b^k \\ &= \sum_{k=0}^{n}\frac{n!}{k!\,(n-k)!}a^{n-k}b^k\end{aligned}$$

Applying this to our limit expression:

$$\begin{aligned}\lim_{n\to\infty}\left(1+\frac{1}{n}\right)^n &= \lim_{n\to\infty}\left[\sum_{k}^{n}\binom{n}{k}1^{n-k}\left(\frac{1}{n}\right)^k\right] = \lim_{n\to\infty}\left[\sum_{k}^{n}\frac{n!}{k!\,(n-k)!}\left(\frac{1}{n}\right)^k\right] \\ &= \lim_{n\to\infty}\left[\frac{n!}{0!\,(n-0)!}\left(\frac{1}{n}\right)^0 + \frac{n!}{1!\,(n-1)!}\left(\frac{1}{n}\right)^1 + \frac{n!}{2!\,(n-2)!}\left(\frac{1}{n}\right)^2\right. \\ &\qquad \left.\cdots \frac{n!}{n!\,(n-n)!}\left(\frac{1}{n}\right)^n\right] \\ &= \lim_{n\to\infty}\left[1+1+\frac{n!}{2!\,(n-2)!}\left(\frac{1}{n}\right)^2 + \frac{n!}{3!\,(n-3)!}\left(\frac{1}{n}\right)^3 \cdots \frac{n!}{n!\,(n-n)!}\left(\frac{1}{n}\right)^n\right] \\ &= \lim_{n\to\infty}\left[1+1+\frac{n(n-1)}{2!}\left(\frac{1}{n}\right)^2 + \frac{n(n-1)(n-2)}{3!}\left(\frac{1}{n}\right)^3 \cdots \left(\frac{1}{n}\right)^n\right]\end{aligned}$$

Looking at the third term:

$$\frac{n(n-1)}{2!}\left(\frac{1}{n}\right)^2 = \frac{n(n-1)}{2!\,n^2}$$

we note that as n approaches infinity the portion:

$$\frac{n(n-1)}{n^2}$$

approaches 1. In other words:

$$\lim_{n\to\infty}\frac{n(n-1)}{n^2}=1$$

Also, the same can be done with the following terms:

$$\lim_{n\to\infty}\frac{n(n-1)(n-2)\cdots}{n^k}=1$$

So, keeping this in mind and remembering that we are letting n approach infinity we write:

$$\lim_{n\to\infty}\left(1+\frac{1}{n}\right)^n$$

$$=\lim_{n\to\infty}\left[1+1+\frac{1}{2!}+\frac{1}{3!}+\frac{1}{4!}\cdots\right]$$

This limit is *defined* as e:

$$\mathrm{e}\equiv\lim_{n\to\infty}\left[1+1+\frac{1}{2!}+\frac{1}{3!}+\frac{1}{4!}\cdots\right]$$

Similarly:

$$\mathrm{e}^x\equiv\lim_{n\to\infty}\left[1+x+\frac{x^2}{2!}+\frac{x^3}{3!}+\frac{x^4}{4!}\cdots\right] \quad (A.27)$$

We can use this definition to calculate the value of e to arbitrary precision. Sometimes, however, we are not interested in high precision and want only an estimate of e^x. It is common in these cases to write:

$$\mathrm{e}_x\approx 1+x \quad (A.28)$$

A.8 Alternate Coordinate Systems

Most of us are introduced to an alternate coordinate system in our initial mathematical studies. This is the polar coordinate system (see Figure A.7) consisting of a *radial* portion, r and an *azimuthal* portion θ:

$$x=r\cos(\theta)$$

$$y=r\sin(\theta)$$

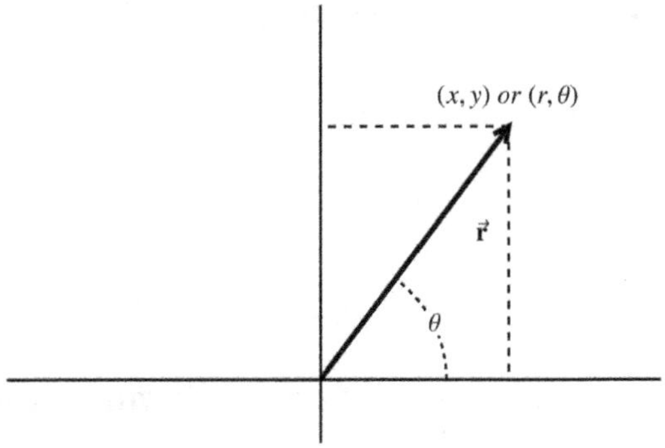

Figure A.7 Polar coordinates.

and from Cartesian to polar:

$$r = \sqrt{x^2 + y^2}$$
$$\theta = \arctan\left(\frac{y}{x}\right)$$

We can envision orthogonal unit vectors in both Cartesian and polar coordinates that obey the rules for such entities as set out in Chapter 4. Thus, the spherical coordinate unit vectors in terms of Cartesian coordinates are:

$$\begin{aligned}\vec{\mathbf{u}}_r &= \frac{\vec{\mathbf{r}}}{\|\vec{\mathbf{r}}\|} = \frac{\vec{\mathbf{r}}}{r}\\ &= x\vec{\mathbf{u}}_x + y\vec{\mathbf{u}}_y\\ &= \cos(\theta)\vec{\mathbf{u}}_x + \sin(\theta)\vec{\mathbf{u}}_y\end{aligned}$$

$$\vec{\mathbf{u}}_\theta \cdot \vec{\mathbf{u}}_r = \begin{bmatrix} a & b \end{bmatrix}\begin{bmatrix}\cos(\theta)\\ \sin(\theta)\end{bmatrix} = a\cos(\theta) + b\sin(\theta)$$

$$a = \frac{-b\sin(\theta)}{\cos(\theta)} \quad a^2 = \frac{b^2\sin^2(\theta)}{\cos^2(\theta)}$$

$$\vec{\mathbf{u}}_\theta \cdot \vec{\mathbf{u}}_\theta = 1 = a^2 + b^2 = \frac{b^2\sin^2(\theta)}{\cos^2(\theta)} + b^2 = \frac{b^2}{\cos^2(\theta)}$$

$$a = -\sin(\theta), \quad b = \cos(\theta)$$

$$\vec{\mathbf{u}}_\theta = -\sin(\theta)\vec{\mathbf{u}}_x + \cos(\theta)\vec{\mathbf{u}}_y$$

This conversion may be written more succinctly:

$$\begin{bmatrix} \vec{\mathbf{u}}_r \\ \vec{\mathbf{u}}_\theta \end{bmatrix} = \begin{bmatrix} \cos(\theta) & \sin(\theta) \\ -\sin(\theta) & \cos(\theta) \end{bmatrix} \begin{bmatrix} \vec{\mathbf{u}}_x \\ \vec{\mathbf{u}}_y \end{bmatrix}$$
$$\Downarrow$$
$$\hat{\mathbf{P}} = \hat{\mathbf{R}}\hat{\mathbf{C}}$$

The reverse transformation from polar to Cartesian unit vectors is simply the inverse of this, as detailed in Chapter 2:

$$\hat{\mathbf{P}} = \hat{\mathbf{R}}\hat{\mathbf{C}}$$
$$\hat{\mathbf{R}}^{-1}\hat{\mathbf{P}} = \hat{\mathbf{R}}^{-1}\hat{\mathbf{R}}\hat{\mathbf{C}}$$
$$\hat{\mathbf{R}}^{-1}\hat{\mathbf{P}} = \hat{\mathbf{C}}$$
$$\Downarrow$$
$$\begin{bmatrix} \vec{\mathbf{u}}_x \\ \vec{\mathbf{u}}_y \end{bmatrix} = \begin{bmatrix} \cos(\theta) & -\sin(\theta) \\ \sin(\theta) & \cos(\theta) \end{bmatrix} \begin{bmatrix} \vec{\mathbf{u}}_r \\ \vec{\mathbf{u}}_\theta \end{bmatrix}$$

If we take the two-dimensional polar coordinate idea and expand it to three dimensions (see Figure A.8) we get *spherical coordinates*, adding a third coordinate, the *elevation* ϕ.

There appears to be no agreement in the literature as to which symbols to use so one occasionally sees ρ in place of r and, ϕ and θ may sometimes be interchanged. Physicists tend to use the

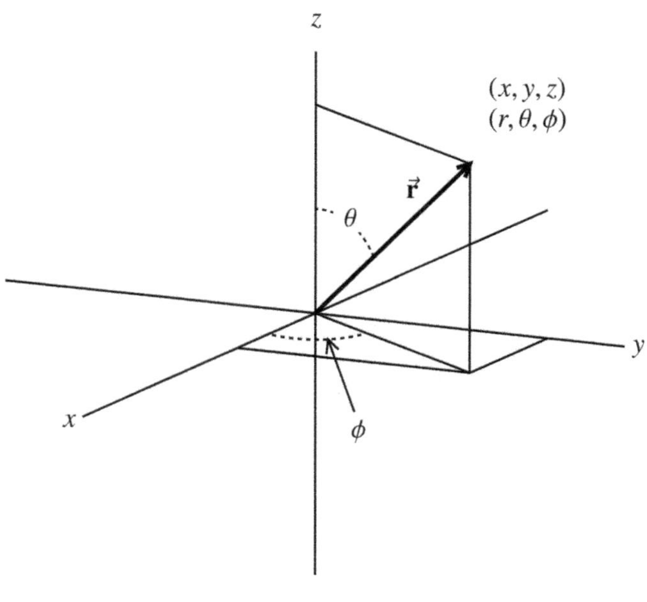

Figure A.8 Spherical coordinates.

definitions we have used in Figure A.8, whereas mathematicians exchange ϕ and θ. We have used r here to label the hypotenuse line. Conversion between spherical and Cartesian coordinates is accomplished in a fashion similar to that of polar coordinates:

$$\begin{aligned}
r &= \sqrt{x^2 + y^2 + z^2} \\
\theta &= \arctan\left(\frac{\sqrt{x^2 + y^2}}{z}\right) \\
\phi &= \arctan\left(\frac{y}{x}\right) \\
x &= r\sin(\theta)\cos(\phi) \\
y &= r\sin(\theta)\sin(\phi) \\
z &= r\cos(\theta)
\end{aligned} \tag{A.29}$$

The transformation between coordinates may be accomplished in much the same way as for polar coordinates. As with polar coordinates, we use the orthogonal properties of the unit vectors to discover the transformation equations from Cartesian to spherical unit vectors:

$$\begin{aligned}
\vec{\mathbf{u}}_r = \frac{\vec{\mathbf{r}}}{r} &= \frac{x}{r}\vec{\mathbf{u}}_x + \frac{y}{r}\vec{\mathbf{u}}_y + \frac{z}{r}\vec{\mathbf{u}}_z \\
&= \sin(\theta)\cos(\phi)\vec{\mathbf{u}}_x + \sin(\theta)\sin(\phi)\vec{\mathbf{u}}_y + \cos(\theta)\vec{\mathbf{u}}_z
\end{aligned}$$

The next conversion needs some explanation. Imagine the Cartesian unit vector, $\vec{\mathbf{u}}_z$, along the z-axis. $\vec{\mathbf{u}}_z \times \vec{\mathbf{u}}_r$ will point in the direction of $\vec{\mathbf{u}}_\phi$ as shown in the Figure A.9. However, $\vec{\mathbf{u}}_\phi$ is not equal to $\vec{\mathbf{u}}_z \times \vec{\mathbf{u}}_r$ since that magnitude of the cross product is dependent on the sine of the angle θ between $\vec{\mathbf{u}}_z$ and $\vec{\mathbf{u}}_r$. Therefore we write:

$$\begin{aligned}
\vec{\mathbf{u}}_\phi \sin(\theta) &= \vec{\mathbf{u}}_z \times \vec{\mathbf{u}}_r \\
\vec{\mathbf{u}}_\phi &= \frac{\vec{\mathbf{u}}_z \times \vec{\mathbf{u}}_r}{\sin(\theta)} \\
&= \frac{1}{\sin(\theta)}\begin{bmatrix} \vec{\mathbf{u}}_x & \vec{\mathbf{u}}_y & \vec{\mathbf{u}}_z \\ 0 & 0 & 1 \\ \sin(\theta)\cos(\phi) & \sin(\theta)\sin(\phi) & \cos(\theta) \end{bmatrix} \\
&= -\sin(\theta)\vec{\mathbf{u}}_x + \cos(\phi)\vec{\mathbf{u}}_y
\end{aligned}$$

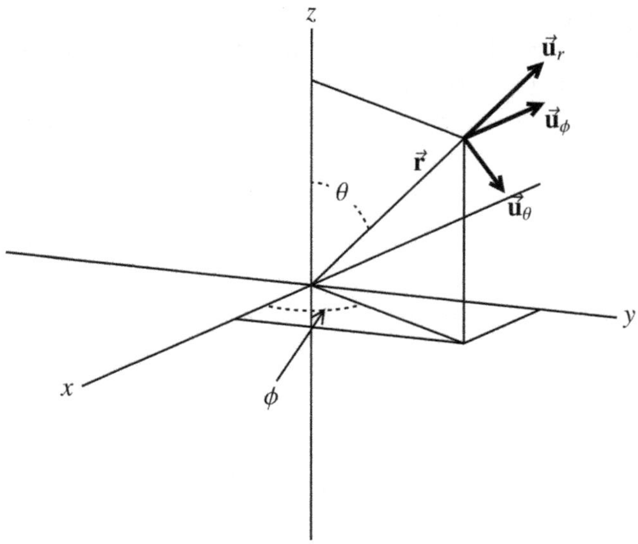

Figure A.9 Spherical coordinate unit vectors.

Last is the unit vector, $\vec{\mathbf{u}}_\theta$. From Figure A.9 we see that:

$$\begin{aligned}\vec{\mathbf{u}}_\theta &= \vec{\mathbf{u}}_\phi \times \vec{\mathbf{u}}_r \\ &= \begin{bmatrix} \vec{\mathbf{u}}_x & \vec{\mathbf{u}}_y & \vec{\mathbf{u}}_z \\ -\sin(\phi) & \cos(\phi) & 0 \\ \sin(\theta)\cos(\phi) & \sin(\theta)\sin(\phi) & \cos(\theta) \end{bmatrix} \\ &= \cos(\theta)\cos(\phi)\vec{\mathbf{u}}_x + \cos(\theta)\sin(\phi)\vec{\mathbf{u}}_y - \sin(\theta)\vec{\mathbf{u}}_z\end{aligned}$$

As with polar coordinates, we can summarise these transformations using matrices:

$$\begin{bmatrix} \vec{\mathbf{u}}_r \\ \vec{\mathbf{u}}_\theta \\ \vec{\mathbf{u}}_\phi \end{bmatrix} = \begin{bmatrix} \sin(\theta)\cos(\phi) & \sin(\theta)\sin(\phi) & \cos(\theta) \\ \cos(\theta)\cos(\phi) & \cos(\theta)\sin(\phi) & -\sin(\theta) \\ -\sin(\phi) & \cos(\phi) & 0 \end{bmatrix} \begin{bmatrix} \vec{\mathbf{u}}_x \\ \vec{\mathbf{u}}_y \\ \vec{\mathbf{u}}_z \end{bmatrix} \tag{A.30}$$

and the inverse transformation:

$$\begin{bmatrix} \vec{\mathbf{u}}_x \\ \vec{\mathbf{u}}_y \\ \vec{\mathbf{u}}_z \end{bmatrix} = \begin{bmatrix} \sin(\theta)\cos(\phi) & \cos(\theta)\cos(\phi) & -\sin(\theta) \\ \sin(\theta)\sin(\phi) & \cos(\theta)\sin(\phi) & \cos(\theta) \\ \cos(\phi) & -\sin(\phi) & 0 \end{bmatrix} \begin{bmatrix} \vec{\mathbf{u}}_r \\ \vec{\mathbf{u}}_\theta \\ \vec{\mathbf{u}}_\phi \end{bmatrix} \tag{A.31}$$

Integral calculus in spherical coordinates requires that we use the appropriate spherical coordinate differentials. We wish to construct a volume element using the appropriate line elements and it is easiest to picture them separately as in Figures A.10–A.12. The radial differential element is the simplest.

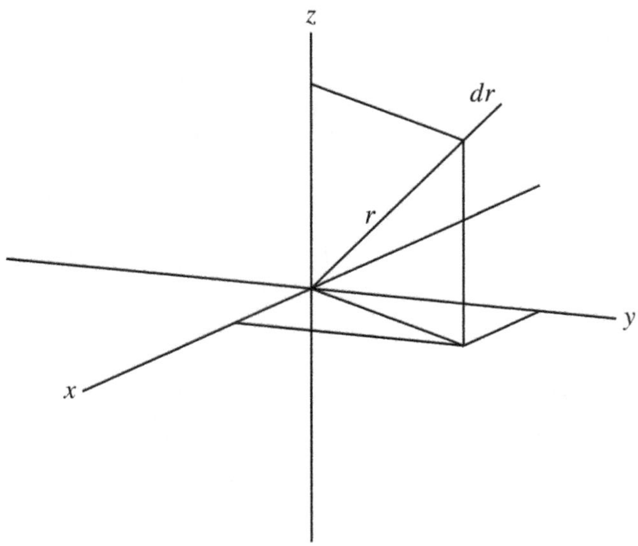

Figure A.10 Spherical coordinates; the radial differential.

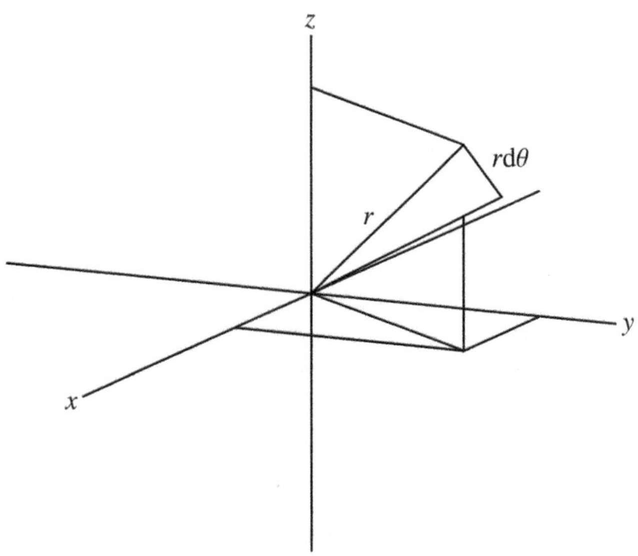

Figure A.11 Spherical coordinates; the azimuthal differential.

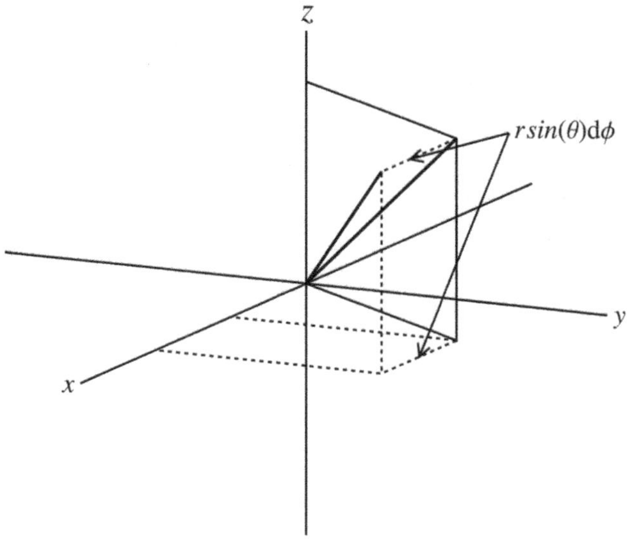

Figure A.12 Spherical coordinates; the elevation differential.

For the azimuthal element we cannot simply use $d\theta$ as the length of this element because it depends on the value of r. We use instead $rd\theta$. Similarly, for the elevation we use $r\sin(\theta)d\phi$. The general displacement differential is then:

$$d\vec{\mathbf{l}} = dr\vec{\mathbf{u}}_r + rd\theta\vec{\mathbf{u}}_\theta + r\sin(\theta)d\phi\vec{\mathbf{u}}_\phi$$

and the complete volume element in spherical coordinates is:

$$\begin{aligned} dV &= dr\, rd\theta r\sin(\theta)d\phi \\ &= r^2\sin(\theta)drd\theta d\phi \end{aligned}$$

When integrating over a full volume the ranges are:

$$r \Rightarrow 0 \text{ to } r$$

$$\theta \Rightarrow 0 \text{ to } \pi$$

$$\phi \Rightarrow 0 \text{ to } 2\pi$$

When integrating over a spherical surface, r is constant so the area element is:

$$\begin{aligned} dA &= rd\theta r\sin(\theta)d\phi \\ &= r^2\sin(\theta)d\theta d\phi \end{aligned}$$

It is sometimes the case that conversion between Cartesian differentials and spherical differentials is necessary. This can be accomplished by taking the partial derivatives of the appropriate functions from eqn (A.29) (see also eqn (4.29)):

$$\mathrm{d}r = \frac{\partial r}{\partial x}\mathrm{d}x + \frac{\partial r}{\partial y}\mathrm{d}y + \frac{\partial r}{\partial z}\mathrm{d}z$$

$$\mathrm{d}r = \frac{x}{r}\mathrm{d}x + \frac{y}{r}\mathrm{d}y + \frac{z}{r}\mathrm{d}z \tag{A.32}$$

and we have

$$\frac{\partial r}{\partial x} = \frac{x}{r} = \sin(\theta)\cos(\phi) \quad \frac{\partial r}{\partial y} = \frac{y}{r} = \sin(\theta)\sin(\phi) \quad \frac{\partial r}{\partial y} = \frac{z}{r} = \cos(\theta)$$

Similarly, we may write formulas to convert from the Cartesian differentials to the spherical differentials, $\mathrm{d}\theta$ and $\mathrm{d}\phi$ (again by differentiating the appropriate functions in eqn (A.29)):

$$\mathrm{d}\theta = \frac{\partial \theta}{\partial x}\mathrm{d}x + \frac{\partial \theta}{\partial y}\mathrm{d}y + \frac{\partial \theta}{\partial z}\mathrm{d}z$$

$$\mathrm{d}\theta = \frac{1}{\sqrt{x^2+y^2}}\frac{xz}{r^2}\mathrm{d}x + \frac{1}{\sqrt{x^2+y^2}}\frac{yz}{r^2}\mathrm{d}y - \frac{\sqrt{x^2+y^2}}{r^2}\mathrm{d}z$$

$$= \frac{\cos(\theta)\cos(\phi)}{r}\mathrm{d}x + \frac{\cos(\theta)\sin(\phi)}{r}\mathrm{d}y - \frac{\sin(\theta)}{r}\mathrm{d}z \tag{A.33}$$

and

$$\frac{\partial \theta}{\partial x} = \frac{\cos(\theta)\cos(\phi)}{r} \quad \frac{\partial \theta}{\partial y} = \frac{\cos(\theta)\sin(\phi)}{r} \quad \frac{\partial \theta}{\partial z} = -\frac{\sin(\theta)}{r}$$

$$\mathrm{d}\phi = \frac{\partial \phi}{\partial x}\mathrm{d}x + \frac{\partial \phi}{\partial y}\mathrm{d}y + \frac{\partial \phi}{\partial z}\mathrm{d}z$$

$$\mathrm{d}\phi = -\frac{y}{x^2+y^2}\mathrm{d}x + \frac{x}{x^2+y^2}\mathrm{d}y$$

$$= -\frac{\sin(\phi)}{r\sin(\theta)}\mathrm{d}x + \frac{\cos(\phi)}{r\sin(\theta)}\cos(\phi)\mathrm{d}y \tag{A.34}$$

and

$$\frac{\partial \phi}{\partial x} = -\frac{\sin(\phi)}{r\sin(\theta)} \quad \frac{\partial \phi}{\partial y} = \frac{\cos(\phi)}{r\sin(\theta)} \quad \frac{\partial \phi}{\partial z} = 0$$

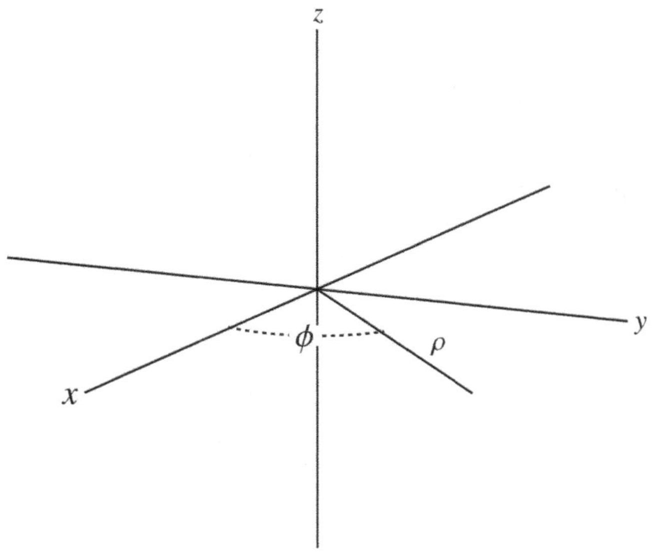

Figure A.13 Cylindrical coordinates, z, ϕ and ρ.

Another coordinate system that is of use is cylindrical coordinates (Figure A.13). Here, we envision a cylinder along the central axis of which we put the z-axis.

There are, of course, three coordinates to worry about, ρ, ϕ and z. The transformations between Cartesian and cylindrical coordinates are:

$$\rho = \sqrt{x^2 + y^2}$$
$$\phi = \arccos\left(\frac{z}{\sqrt{x^2 + y^2}}\right)$$
$$z = z$$
$$x = r\cos(\phi)$$
$$y = r\sin(\phi)$$
$$z = z$$

Conversion between Cartesian and cylindrical unit vectors is accomplished in the same way as was done above for spherical coordinates.

A.9 The Fundamental Theorem of Calculus

The fundamental theorem of calculus is as important to mathematics as gasoline is to a car. It is the foundation upon which everything is built so understanding it is of vital importance. There are two parts to the general fundamental theorem in the calculus. We present them here.

Preliminary to the fundamental theorem, we briefly discuss the mean value theorem of integral calculus. Consider a function $f(x)$ that we integrate between a and b. The mean value theorem says that there is some number c between a and b such that the following is true:

$$\text{for} \quad a \leq c \leq b$$

$$f(c) = \frac{1}{b-a}\int_a^b f(x)\mathrm{d}x$$

We can see what this means with a very simple picture. Imagine that $f(x) = x$ and that we are integrating between 0 and 4 (Figure A.14). Essentially we are finding the area of a triangle of base 4 and height 4. The definite integral is:

$$\int_0^4 x\mathrm{d}x = \left.\frac{x^2}{2}\right|_0^4 = \frac{4^2}{2} - 0 = 8$$

Now, the mean value theorem says that there is a value of x between 0 and 4 at which

$$f(c) \times (4-0) = 8.$$

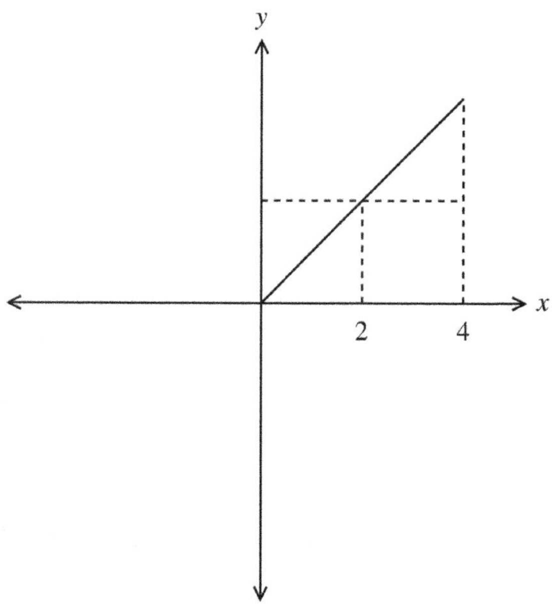

Figure A.14 $f(x) = x$.

In this case the value of x that is needed is 2 such that $f(2)=2$ and the area of a rectangle of width 4–0 and height $f(2)=2$ is 8. From this we can say:

$$f(c)(b-a)=\int_a^b f(x)\mathrm{d}x$$

or

$$f(c)=\frac{1}{b-a}\int_a^b f(x)\mathrm{d}x$$

The first part of the fundamental theorem of calculus is:

$$\begin{aligned}
&\text{for} \quad a \le c \le b \\
&\text{let} \quad G(x)=\int_a^x f(t)\mathrm{d}t \\
&\frac{\mathrm{d}G}{\mathrm{d}x}=G'=\frac{d}{\mathrm{d}x}\int_a^x f(t)\mathrm{d}t=f(x)
\end{aligned} \tag{A.35}$$

We show that this is so by first invoking the definition of the derivative:

$$\begin{aligned}
G'(x) &= \lim_{h\to 0}\frac{G(x+h)-G(x)}{h} \\
&= \lim_{h\to 0}\frac{1}{h}\left[\int_a^{x+h} f(t)\mathrm{d}t-\int_a^x f(t)\mathrm{d}t\right] \\
&= \lim_{h\to 0}\frac{1}{h}\int_x^{x+h} f(t)\mathrm{d}t
\end{aligned}$$

Geometrically, this is saying that the area under the function between a and $x+h$ minus the area under the function between a and x is just the area between x and $x+h$. If h is constrained such that $x+h$ is between a and b then the mean value theorem implies that there is a number c between x and $x+h$ such that:

$$f(c)=\frac{1}{h}\lim_{h\to 0}\int_x^{x+h} f(t)\mathrm{d}t$$

and:

$$G'(x)=\frac{1}{h}\lim_{h\to 0}\int_x^{x+h} f(t)\mathrm{d}t=\lim_{h\to 0}f(c)$$

Since c is between x and $x+h$, the limit as h approaches zero must be $f(x)$:

$$G'(x) = \lim_{h\to 0} f(c) = f(x)$$

$G(x)$ is said to be an *antiderivative* of $f(x)$ and is not unique in this sense. We could as well have a function $F(x)$ whose derivative is also equal to $f(x)$. $G(x)$ and $F(x)$ would then have to differ by a constant:

$$G(x) = F(x) + K$$

The second part of the theorem is stated, assuming $G(x)$ is an antiderivative of $f(x)$:

$$\int_a^b f(x)\mathrm{d}x = G(b) - G(a)$$

We show this by defining $F(x)$ as:

$$F(x) = \int_a^b f(t)\mathrm{d}t$$

and then, from the first part of the theorem above:

$$F'(x) = f(x)$$

which makes $F(x)$ an antiderivatative of $f(x)$ … but so is $G(x)$. So:

$$G(x) = F(x) + K$$

Now, from the first part of the theorem in eqn (A.35), $G(a) = \int_a^a f(t)\mathrm{d}t = 0$, which means that:

$$G(a) = F(a) + K = 0 + K = K$$

and

$$\begin{aligned} G(b) - G(a) &= F(b) + K - K \\ &= G(b) \\ &= \int_a^b f(t)\mathrm{d}t \end{aligned}$$

The significance of this part of the theorem is that it shows the intimate link between differentiation and integration. We can highlight this link more graphically by writing the theorem as:

$$\int_a^b G'(x)\mathrm{d}x = G(b) - G(a)$$

A.10 Some Commonly Used Derivatives and Their Properties

$$\frac{\mathrm{d}}{\mathrm{d}x}(x)=1 \quad \frac{\mathrm{d}}{\mathrm{d}x}\sin(ax)=a\cos(ax)$$

$$\frac{\mathrm{d}}{\mathrm{d}x}\cos(ax)=-a\sin(ax)$$

$$\frac{\mathrm{d}}{\mathrm{d}x}\arcsin(x)=\frac{1}{\sqrt{1+x^2}} \quad \frac{\mathrm{d}}{\mathrm{d}x}\arccos(x)=\frac{-1}{\sqrt{1+x^2}}$$

$$\frac{\mathrm{d}}{\mathrm{d}x}\arctan(x)=\frac{1}{1+x^2} \tag{A.36}$$

$$\frac{\mathrm{d}}{\mathrm{d}x}\mathrm{e}^x=\mathrm{e}^x \quad \frac{\mathrm{d}}{\mathrm{d}x}\ln x=\frac{1}{x}$$

$$\frac{\mathrm{d}}{\mathrm{d}x}kf(x)=k\frac{\mathrm{d}}{\mathrm{d}x}f(x) \quad \frac{\mathrm{d}}{\mathrm{d}x}k=0$$

$$\frac{\mathrm{d}}{\mathrm{d}x}\left(f(x)\pm g(x)\right)=\frac{\mathrm{d}}{\mathrm{d}x}f(x)\pm\frac{\mathrm{d}}{\mathrm{d}x}g(x)$$

The Product Rule:

$$\frac{\mathrm{d}}{\mathrm{d}x}f(x)g(x)=f'(x)g(x)+f(x)g'(x) \tag{A.37}$$

The Quotient Rule:

$$\frac{\mathrm{d}}{\mathrm{d}x}\frac{f(x)}{g(x)}=\frac{g(x)f'(x)-f(x)g'(x)}{g(x)^2} \tag{A.38}$$

The Chain Rule:

$$\frac{\mathrm{d}}{\mathrm{d}x}f(g(x))=g'(x)\,f'(g(x))$$

(for single variable functions)

$$\frac{\mathrm{d}}{\mathrm{d}t}f(x(t),y(t))=\frac{\partial f}{\partial x}\frac{\mathrm{d}x}{\mathrm{d}t}+\frac{\partial f}{\partial y}\frac{\mathrm{d}y}{\mathrm{d}t} \tag{A.39}$$

(for multivariable functions)

The Power Rule:

$$\frac{\mathrm{d}}{\mathrm{d}x}x^n = nx^{n-1} \tag{A.40}$$

Exponential Differentiation:

$$\frac{\mathrm{d}}{\mathrm{d}x}\mathrm{e}^{ax} = a\mathrm{e}^{ax} \tag{A.41}$$

A.11 Partial Derivatives

Frequently, the value of a function will depend on two or more variables:

$$y = f(x_1, x_2 \ldots)$$

Which domain variable do we use to differentiate the function? This will depend on which one you want to find the rate of change in. Thus, when we want to differentiate with respect to x_1 we use that variable and treat all the others as constants:

$$f'_{x1} = \frac{\partial y}{\partial x_1}$$

and the result is termed a *partial derivative* since we have only differentiated with respect to one variable. Note the change in notation. In order to emphasise that we are differentiating a multi-variable function we use the symbol "∂" instead of "d".

Example:

$$y = 3x_1 + 5x_2$$

$$\frac{\partial y}{\partial x_1} = 3 + 5x_2$$

$$\frac{\partial y}{\partial x_2} = 3x_1 + 5$$

Things become more interesting when the second derivative is taken. The second derivative can be in the same variable as the first or another of the possible variables. So, for a function of two variables, there are four possibilities:

$$\frac{\partial}{\partial x_1}\left(\frac{\partial y}{\partial x_1}\right)=\frac{\partial^2 y}{\partial x_1^2}$$

$$\frac{\partial}{\partial x_2}\left(\frac{\partial y}{\partial x_2}\right)=\frac{\partial^2 y}{\partial x_2^2}$$

$$\frac{\partial}{\partial x_1}\left(\frac{\partial y}{\partial x_2}\right)=\frac{\partial^2 y}{\partial x_1 \partial x_2}$$

$$\frac{\partial}{\partial x_2}\left(\frac{\partial y}{\partial x_1}\right)=\frac{\partial^2 y}{\partial x_2 \partial x_1}$$

Most of the time, the last two partial derivatives are equal:

$$\frac{\partial^2 y}{\partial x_1 \partial x_2}=\frac{\partial^2 y}{\partial x_2 \partial x_1} \tag{A.42}$$

We can see that this is so in our example:

$$\frac{\partial^2 y}{\partial x_1 \partial x_2}=3+5=8$$

$$\frac{\partial^2 y}{\partial x_2 \partial x_1}=5+3=8$$

A.12 Parametric Equations

The general coordinate values of lines may be represented by equations, the simplest being those of straight lines. Thus, we learn in our introductory algebra courses that the general straight line equation is:

$$y=mx+b$$

There is, however, another way to represent this information. We relate each of the variables *separately* to a common parameter in a set of *parametric equations*. For example, the standard equation of a unit circle is:

$$x^2+y^2=1$$

but we may also represent this parametrically as:

$$x=\cos(t) \quad y=\sin(t)$$

where t is the parameter (an angle in this case).

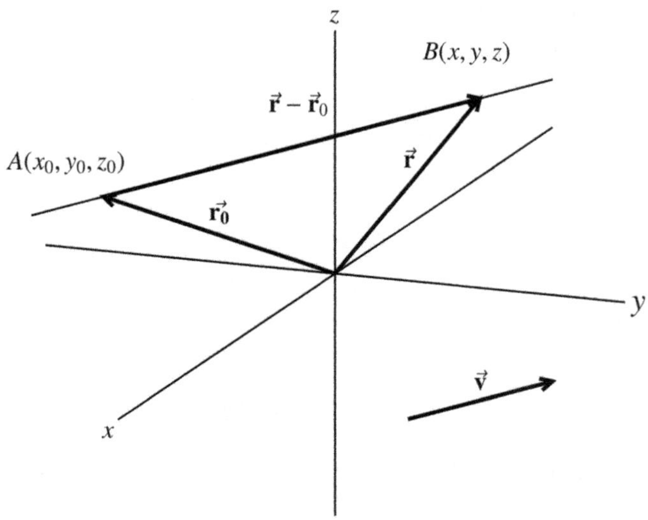

Figure A.15 Parametric equations.

A straight line is easily represented parametrically. Let us start with a straight line and a vector $\vec{\mathbf{v}}$ that is parallel to it (Figure A.15).

From our vector algebra (see Chapter 3) we can say that, since the vector $\vec{\mathbf{r}} - \vec{\mathbf{r}}_0$ is parallel to $\vec{\mathbf{v}}$:

$$\vec{\mathbf{r}} - \vec{\mathbf{r}}_0 = t\vec{\mathbf{v}}$$

or

$$\vec{\mathbf{r}} = \vec{\mathbf{r}}_0 + t\vec{\mathbf{v}}$$

where t is our parameter. This is the parametric vector equation. The components of each of these are:

$$\vec{\mathbf{r}} = x\vec{\mathbf{u}}_x + y\vec{\mathbf{u}}_y + z\vec{\mathbf{u}}_z$$

$$\vec{\mathbf{r}}_0 = x_0\vec{\mathbf{u}}_x + y_0\vec{\mathbf{u}}_y + z_0\vec{\mathbf{u}}_z$$

$$\vec{\mathbf{v}} = v_1\vec{\mathbf{u}}_x + v_2\vec{\mathbf{u}}_y + v_3\vec{\mathbf{u}}_z$$

and in terms of these vector components:

$$x = x_0 + tv_1$$

$$y = y_0 + tv_2$$

$$z = z_0 + tv_3$$

These are our straight-line parametric equations.

A.13 Some Commonly Used Integrals and Their Properties

$$\int k\mathrm{d}x = kx + c \quad \int x^n \mathrm{d}x = \frac{1}{n+1}x^{n+1} + c$$
$$\int \cos(ax)\mathrm{d}x = \frac{\sin(ax)}{a} + c$$
$$\int \frac{1}{x}\mathrm{d}x = \ln|x| + c \quad \int \sin(ax)\mathrm{d}x = \frac{-\cos(ax)}{a} + c \tag{A.43}$$
$$\int \mathrm{e}^x \mathrm{d}x = \mathrm{e}^x + c$$

$$\int f(x) \pm g(x)\mathrm{d}x = \int f(x)\mathrm{d}x \pm \int g(x)\mathrm{d}x$$
$$\int kf(x)\mathrm{d}x = k\int f(x)\mathrm{d}x \tag{A.44}$$

A.14 Integration by Substitution

How do we integrate the following function? We use the method of substitution of variables:

$$\int \frac{\mathrm{d}x}{x-a}$$

Let $u = x - a$ and then $\mathrm{d}u = \mathrm{d}x - \mathrm{d}a = \mathrm{d}x$.

$$\int \frac{\mathrm{d}x}{x-a}$$
$$\Downarrow$$
$$\int \frac{\mathrm{d}u}{u} \tag{A.45}$$
$$= \ln u + c$$
$$= \ln(x-a) + c$$

A.15 Integration by Parts

$$\begin{aligned}\frac{\mathrm{d}}{\mathrm{d}x}f(x)g(x) &= \frac{\mathrm{d}f(x)}{\mathrm{d}x}g(x) + f(x)\frac{\mathrm{d}g(x)}{\mathrm{d}x}\\ &= f'(x)g(x) + f(x)g'(x)\\ &\Downarrow\\ f(x)g(x) &= \int f'(x)g(x)\mathrm{d}x + \int f(x)g'(x)\mathrm{d}x\\ &\Downarrow\\ \int f(x)g'(x)\mathrm{d}x &= f(x)g(x) - \int f'(x)g(x)\mathrm{d}x\end{aligned} \tag{A.46}$$

The following integrals are of use in Fourier transforms and are arrived at by integration by parts ... twice. First:

$$\int_0^\infty \mathrm{e}^{-ax}\cos(bx)\mathrm{d}x$$

$$\text{let}\quad f(x) = \cos(bx)\ g'(x) = \mathrm{e}^{-ax}$$

$$f'(x) = -b\sin(bx)\ g(x) = -\frac{\mathrm{e}^{-ax}}{a}$$

$$\int f(x)g'(x)\mathrm{d}x = f(x)g(x) - \int f'(x)g(x)\mathrm{d}x$$

$$\Downarrow$$

$$\begin{aligned}&1.\ \int_0^\infty \mathrm{e}^{-ax}\cos(bx)\mathrm{d}x = -\frac{\mathrm{e}^{-ax}}{a}\cos(bx)\bigg|_0^\infty - \frac{b}{a}\int_0^\infty \mathrm{e}^{-ax}\sin(bx)\mathrm{d}x\\ &2.\ \frac{1}{a} - \frac{b}{a}\left[-\frac{\mathrm{e}^{-ax}}{a}\sin(bx)\bigg|_0^\infty + \frac{b}{a}\int_0^\infty \mathrm{e}^{-ax}\cos(bx)\mathrm{d}x\right]\\ &= \frac{1}{a} - \frac{b^2}{a^2}\int_0^\infty \mathrm{e}^{-ax}\cos(bx)\mathrm{d}x\\ &\left(1 + \frac{b^2}{a^2}\right)\int_0^\infty \mathrm{e}^{-ax}\cos(bx)\mathrm{d}x = \frac{1}{a}\\ &\int_0^\infty \mathrm{e}^{-ax}\cos(bx)\mathrm{d}x = \frac{1}{a^2 + b^2}\end{aligned} \tag{A.47}$$

and secondly:

$$\int_0^\infty \mathrm{e}^{-ax}\sin(bx)\mathrm{d}x$$

$$\text{let} \quad f(x)=\sin(bx)\ g'(x)=\mathrm{e}^{-ax}$$

$$f'(x)=b\cos(bx)\ g(x)=-\frac{\mathrm{e}^{-ax}}{a}$$

$$\int f(x)g'(x)\mathrm{d}x=f(x)g(x)-\int f'(x)g(x)\mathrm{d}x$$

$$\Downarrow$$

$$1.\ \int_0^\infty \mathrm{e}^{-ax}\sin(bx)\mathrm{d}x=-\frac{\mathrm{e}^{-ax}}{a}\sin(bx)\Big|_0^\infty+\frac{b}{a}\int_0^\infty \mathrm{e}^{-ax}\cos(bx)\mathrm{d}x$$

$$2.\ \frac{b}{a}\left[-\frac{\mathrm{e}^{-ax}}{a}\cos(bx)\Big|_0^\infty+\frac{b}{a}\int_0^\infty \mathrm{e}^{-ax}\sin(bx)\mathrm{d}x\right]$$

$$=\frac{b}{a^2}-\frac{b^2}{a^2}\int_0^\infty \mathrm{e}^{-ax}\sin(bx)\mathrm{d}x \tag{A.48}$$

$$\left(1+\frac{b^2}{a^2}\right)\int_0^\infty \mathrm{e}^{-ax}\sin(bx)\mathrm{d}x=\frac{b}{a^2}$$

$$\int_0^\infty \mathrm{e}^{-ax}\sin(bx)\mathrm{d}x=\frac{b}{a^2+b^2}$$

In each case we have done two integrations by parts; the first in line 1 and the second in line 2 on the integral in line 1. The result is then obtained by recombining the final integral with the original integral by simple algebra.

A.16 The Trapezoid Rule for Estimating Definite Integrals

Occasionally, we need to estimate the value of a definite integral, especially when we are doing so using a digital computer. One method of doing this is to use the trapezoidal rule. We imagine a continuous function, $f(x)$ for which we want the definite integral between x_0 and x_n. We divide the area under the portion of the

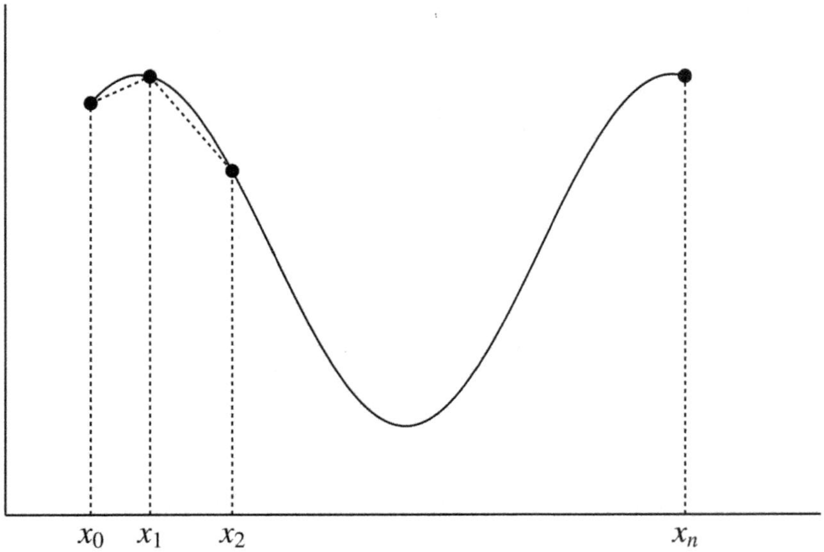

Figure A.16 Secant segments along a curve.

function of interest into n equal-width segments and construct a trapezoid in each segment as in Figure A.16.

Assuming that the width of each trapezoid is the same ($h = x_{k+1} - x_k$) and using eqn (A.26), the area of the trapezoid between x_k and x_{k+1} is:

$$A_n = \frac{h}{2}(f(x_n) + f(x_{n+1}))$$

The total area between x_0 and x_n is:

$$\begin{aligned} A_T &= \frac{h}{2}\sum_{k=0}^{n-1}(f(x_k) + f(x_{k+1})) \\ &= \frac{h}{2}[f(0) + 2f(1) + 2f(2) \cdots 2f(n-1) + f(n)] \end{aligned}$$

This is our approximation of the area between x_0 and x_n under the function. In other words, the approximation of the definite integral between x_0 and x_n:

$$\int_0^{x_n} f(x)\mathrm{d}x \approx \frac{h}{2}\sum_{k=0}^{n-1}(f(x_k) + f(x_{k+1})) \tag{A.49}$$

A.17 Solving Differential Equations

The simplest linear differential equation is the first-order linear differential equation:

$$\frac{\mathrm{d}f(t)}{\mathrm{d}t} = Af(t)$$

or

$$\frac{\mathrm{d}f(t)}{\mathrm{d}t} - Af(t) = 0$$

In the second form it is referred to as a *homogeneous linear differential equation*. Solution of the equation is easily done using a bit of integral calculus:

$$\int \frac{1}{f(t)}\mathrm{d}(f(t)) = A\int f(t) = 0$$

$$\ln f(t) = At + C$$

$$f(t) = \mathrm{e}^{At+C} = \mathrm{e}^{At}\mathrm{e}^{C}$$

Then, at $t = 0$:

$$f(0) = \mathrm{e}^{A\times 0}\mathrm{e}^{C} = \mathrm{e}^{C}$$

so:

$$f(t) = f(0)\mathrm{e}^{At} \tag{A.50}$$

A step up in complexity is the second-order differential equation:

$$\frac{\mathrm{d}^2 f(t)}{\mathrm{d}t^2} + b\frac{\mathrm{d}f(t)}{\mathrm{d}t} + cf(t) = 0$$

or (A.51)

$$\ddot{f}(t) + b\dot{f}(t) + cf(t) = 0$$

We are assuming a coefficient of one for the first term to simplify things later. A second-order ordinary differential equation (ODE) can always be made to be so. Given the solution to the first-order differential equation, we suspect that the solution here will be similar. Our "guess" at a solution is:

$$f(t) = \mathrm{e}^{mt+A}$$

This being the case we then can write:

$$\dot{f}(t) = me^{mt+A}$$
$$\ddot{f}(t) = m^2 e^{mt+A}$$

and:

$$\begin{aligned}\ddot{f}(t) + b\dot{f}(t) + cf(t) &= m^2 e^{mt+A} + bme^{mt+A} + ce^{mt+A} = 0\\ &= e^{mt+A}(m^2 + bm + c)\\ e^{mt+A} &\neq 0 \text{ so}\\ m^2 + bm + c &= 0\end{aligned}$$

This is called the *characteristic equation* and we can easily solve it using the quadratic formula:

$$m = \frac{-b \pm \sqrt{b^2 - 4c}}{2}$$

or

$$m = -\frac{b}{2} \pm \frac{\sqrt{b^2 - 4c}}{2} = r \pm s$$

We see that there are *two* possible solutions:

$$m_1 = r + s \; m_2 = r - s$$

so for $f(t)$ we can write either:

$$f(t) = e^{(r+s)t+A} \text{ or } f(t) = e^{(r-s)t+A}$$

In fact, these are linearly independent solutions since there is no factor that we can multiply one by to get the other. If this is the case then the sum of these solutions is also a solution, similar to the case with vectors (see Chapter 3, eqn (3.2)). Therefore, we can write a completely general *superposition* solution:

$$\begin{aligned}f(t) &= C_1 e^{(r+s)t+A} + C_2 e^{(r-s)t+A}\\ &= e^{rt+A}(C_1 e^{st} + C_2 e^{-st})\end{aligned}$$

We can take this a bit further by looking at the possibilities for the value of s.

Case 1:

$$\text{if } b^2 > 4c$$

$$\text{then}$$

$$\sqrt{b^2 - 4c} \text{ is real}$$

This being so, the solution is the one written above.

Case 2:

$$\text{if } b^2 = 4c$$

$$\text{then}$$

$$\sqrt{b^2 - 4c} = 0$$

$$\text{and}$$

$$s = 0$$

and the solution becomes:

$$\begin{aligned} f(t) &= C_1 e^{(r+0)t+A} + C_2 e^{(r-0)t+A} \\ &= e^{rt+A}(C_1 + C_2) \\ &= K e^{rt+A} \end{aligned}$$

Case 3:

$$\text{if } b^2 < 4c$$

$$\text{then}$$

$$\sqrt{b^2 - 4c} = i\sqrt{|b^2 - 4c|}$$

and our general solution becomes:

$$\begin{aligned} f(t) &= C_1 e^{(r+is)t+A} + C_2 e^{(r-is)t+A} \\ &= e^{rt+A}\left(C_1 e^{ist} + C_2 e^{-ist}\right) \\ &= e^{rt+A}[C_1(\cos(st) + i\sin(st)) + C_2(\cos(st) - i\sin(st))] \\ &= e^{rt+A}[(C_1 + C_2)\cos(st) + i(C_1 - C_2)\sin(st)] \\ &= e^{rt+A}[K_1 \cos(st) + iK_2 \sin(st)] \end{aligned}$$

A special case that occurs frequently enough to warrant attention is the second-order differential equation of the form:

$$\frac{d^2f(t)}{dt^2} + cf(t) = 0$$

with b from eqn (A.51) equal to zero. For example, the stationary wave equation of Chapter 7 is of this form as is the famous Schrodinger equation. In this case:

$$b = 0 \Rightarrow r = 0$$

and

$$s = i\frac{\sqrt{4c}}{2} = \pm i\sqrt{c}$$

and the general solution is:

$$\begin{aligned} f(t) &= e^A\left[K_1\cos(\sqrt{c}t) + K_2\sin(\sqrt{c}t)\right] \\ &= C_1e^{A+i\sqrt{c}t} + C_2e^{A-i\sqrt{c}t} \end{aligned} \tag{A.52}$$

A special case of a first-order differential equation is one in which two functions are *coupled*:

$$\frac{df(t)}{dt} = \dot{f}(t) = +Ag(t) \quad \text{or} \quad \dot{f}(t) - Ag(t) = 0$$

$$\frac{dg(t)}{dt} = \dot{g}(t) = +Af(t) \quad \text{or} \quad \dot{g}(t) - Af(t) = 0$$

Thus, the derivative of one function yields the other function times a constant. How do we go about solving this system of equations? The technique here is to take the derivative of each side:

$$\ddot{f}(t) - A\dot{g}(t) = 0 \quad \text{or} \quad \ddot{g}(t) - A\dot{f}(t) = 0$$

Let us concentrate on the first equation:

$$\ddot{f}(t) - A\dot{g}(t) = 0$$

or

$$\ddot{f}(t) - A^2f(t) = 0$$

where we have substituted $Af(t)$ for $\dot{g}(t)$ from above. This is in the form from the above special case second-order differential equation (and the time factor of wave equation of Chapter 7) and will have a similar solution:

$$f(t)=K_1\cos(At)+iK_2\sin(At)$$

To solve for $g(t)$ we first specify $f(0)$:

$$f(0)=K_1\cos(A\cdot 0)+iK_2\sin(A\cdot 0)=K_1$$

and

$$f(t)=f(0)\cos(At)+iK_2\sin(At)$$

$$\dot{g}(t)=+Af(t)=+A[f(0)\cos(At)+iK_2\sin(At)]$$

and integrating

$$\int\dot{g}(t)\mathrm{d}t=+\int Af(0)\cos(At)\mathrm{d}t+\int AiK_2\sin(At)\mathrm{d}t$$

$$g(t)=+f(0)\int A\cos(At)\mathrm{d}t+iK_2\int A\sin(At)\mathrm{d}t=0$$

$$=+f(0)\sin(At)-iK_2\cos(At)$$

and finally specify $g(0)$:

$$g(0)=+f(0)\sin(A\cdot 0)-iK_2\cos(A\cdot 0)$$

$$=-iK_2$$

and

$$g(t)=+f(0)\sin(At)-g(0)\cos(At)$$

$$=-g(0)\cos(At)+f(0)\sin(At)$$

Since we now know the that $g(0)=-iK_2$ we can write for $f(t)$:

$$f(t)=f(0)\cos(At)-g(0)\sin(At)$$

A.18 Commutator Algebra

The commutator of two operators is defined as:

$$[\hat{\mathbf{A}},\hat{\mathbf{B}}]=\hat{\mathbf{A}}\hat{\mathbf{B}}-\hat{\mathbf{B}}\hat{\mathbf{A}}$$

and the anticommutator as:

$$\{\hat{\mathbf{A}},\hat{\mathbf{B}}\} = \hat{\mathbf{A}}\hat{\mathbf{B}} + \hat{\mathbf{B}}\hat{\mathbf{A}}$$

Reversing the order of the operators gives:

$$[\hat{\mathbf{B}},\hat{\mathbf{A}}] = \hat{\mathbf{B}}\hat{\mathbf{A}} - \hat{\mathbf{A}}\hat{\mathbf{B}} = -(\hat{\mathbf{A}}\hat{\mathbf{B}} - \hat{\mathbf{B}}\hat{\mathbf{A}}) = -[\hat{\mathbf{A}},\hat{\mathbf{B}}]$$
$$[\hat{\mathbf{A}},\hat{\mathbf{B}}] = -[\hat{\mathbf{B}},\hat{\mathbf{A}}]$$

The value of the commutator may or may not be zero. It will depend on the operators and whether or not they have common eigenfunctions. For example, using $\hat{\mathbf{x}}$ and $\hat{\mathbf{p}}$, the position and momentum operators:

$$[\hat{\mathbf{x}},\hat{\mathbf{p}}]|\Psi\rangle = (\hat{\mathbf{x}}\hat{\mathbf{p}} - \hat{\mathbf{p}}\hat{\mathbf{x}})|\Psi\rangle = (x)\left(ih\frac{\partial}{\partial x}\right)|\Psi\rangle - \left(ih\frac{\partial}{\partial x}\right)(x)|\Psi\rangle$$
$$= ih\left(x\frac{\partial|\Psi\rangle}{\partial x} - \left(\frac{\partial}{\partial x}\right)(x)|\Psi\rangle\right) \text{ (using the product rule)}$$
$$= ih|\Psi\rangle$$

and

$$[\hat{\mathbf{x}},\hat{\mathbf{p}}] = ih$$

Thus, $\hat{\mathbf{x}}$ and $\hat{\mathbf{p}}$ do not commute.

Suppose that $\hat{\mathbf{A}} = \hat{\mathbf{C}} + \hat{\mathbf{D}}$. We can then write:

$$[\hat{\mathbf{A}},\hat{\mathbf{B}}] = [(\hat{\mathbf{C}} + \hat{\mathbf{D}}),\hat{\mathbf{B}}]$$
$$= \hat{\mathbf{C}}\hat{\mathbf{B}} + \hat{\mathbf{D}}\hat{\mathbf{B}} - \hat{\mathbf{B}}\hat{\mathbf{C}} - \hat{\mathbf{B}}\hat{\mathbf{D}}$$
$$= [\hat{\mathbf{C}},\hat{\mathbf{B}}] + [\hat{\mathbf{D}},\hat{\mathbf{B}}]$$

Likewise, if $\hat{\mathbf{B}} = \hat{\mathbf{C}} + \hat{\mathbf{D}}$:

$$[\hat{\mathbf{A}},\hat{\mathbf{B}}] = [\hat{\mathbf{A}},(\hat{\mathbf{C}} + \hat{\mathbf{D}})]$$
$$= \hat{\mathbf{A}}\hat{\mathbf{C}} + \hat{\mathbf{A}}\hat{\mathbf{D}} - \hat{\mathbf{C}}\hat{\mathbf{A}} - \hat{\mathbf{D}}\hat{\mathbf{A}}$$
$$= [\hat{\mathbf{A}},\hat{\mathbf{C}}] + [\hat{\mathbf{A}},\hat{\mathbf{D}}]$$

An operator always commutes with itself:

$$[\hat{\mathbf{A}},\hat{\mathbf{A}}] = \hat{\mathbf{A}}\hat{\mathbf{A}} - \hat{\mathbf{A}}\hat{\mathbf{A}} = \hat{\mathbf{0}}$$

and with its inverse:

$$[\hat{\mathbf{A}},\hat{\mathbf{A}}^{-1}] = \hat{\mathbf{A}}\hat{\mathbf{A}}^{-1} - \hat{\mathbf{A}}^{-1}\hat{\mathbf{A}} = \hat{\mathbf{1}} - \hat{\mathbf{1}} = \hat{\mathbf{0}}$$

If the function that $\hat{\mathbf{A}}$ and $\hat{\mathbf{B}}$ operate on is an eigenfunction of both operators, the commutator will be equal to zero:

$$\begin{aligned}[\hat{\mathbf{A}},\hat{\mathbf{B}}]|\Psi\rangle &= \hat{\mathbf{A}}\hat{\mathbf{B}}|\Psi\rangle - \hat{\mathbf{B}}\hat{\mathbf{A}}|\Psi\rangle \\ &= ab|\Psi\rangle - ba|\Psi\rangle \\ &= (ab - ba)|\Psi\rangle \\ &= 0\end{aligned}$$

Nested commutators evaluate as:

$$\begin{aligned}&\left[\hat{\mathbf{A}},\left[\hat{\mathbf{B}},\hat{\mathbf{C}}\right]\right] \\ &= \left[\hat{\mathbf{A}},\left(\hat{\mathbf{B}}\hat{\mathbf{C}}-\hat{\mathbf{C}}\hat{\mathbf{B}}\right)\right] \\ &= \hat{\mathbf{A}}\hat{\mathbf{B}}\hat{\mathbf{C}} - \hat{\mathbf{B}}\hat{\mathbf{C}}\hat{\mathbf{A}} - \hat{\mathbf{A}}\hat{\mathbf{C}}\hat{\mathbf{B}} + \hat{\mathbf{C}}\hat{\mathbf{B}}\hat{\mathbf{A}}\end{aligned}$$

$$\begin{aligned}&\left[\left[\hat{\mathbf{B}},\hat{\mathbf{C}}\right],\hat{\mathbf{A}}\right] \\ &= \left[\left(\hat{\mathbf{B}}\hat{\mathbf{C}}-\hat{\mathbf{C}}\hat{\mathbf{B}}\right),\hat{\mathbf{A}}\right] \\ &\quad \hat{\mathbf{B}}\hat{\mathbf{C}}\hat{\mathbf{A}} - \hat{\mathbf{A}}\hat{\mathbf{B}}\hat{\mathbf{C}} - \hat{\mathbf{C}}\hat{\mathbf{B}}\hat{\mathbf{A}} + \hat{\mathbf{A}}\hat{\mathbf{C}}\hat{\mathbf{B}}\end{aligned}$$

A.19 L'Hopital's Rule

The proof of L'Hopital's rule requires the use of Rolle's theorem and the Cauchy mean value theorem. First, Rolle's theorem:

For a function f that is continuous and differentiable on the interval $[a,b]$ and for which $f(a)=f(b)$ there exists at least one number $a<c<b$ such that $f'(c)=0$.

There are two possibilities for the function f. The first is that it is a constant function on the interval $[a,b]$, in which case $f'(x)=0$ for all x in $[a,b]$ and we may choose any c on the interval. The second case is for f not constant on $[a,b]$. Since $f(a)=f(b)$ there must exist at least one maximum or minimum in $[a,b]$. For the case of the maximum, since $f(c)>f(x)$ for all x in $[a,b]$ and since f is continuous and differentiable on $[a,b]$, c is therefore a critical point for which $f'(c)=0$, as shown in Figure A.17.

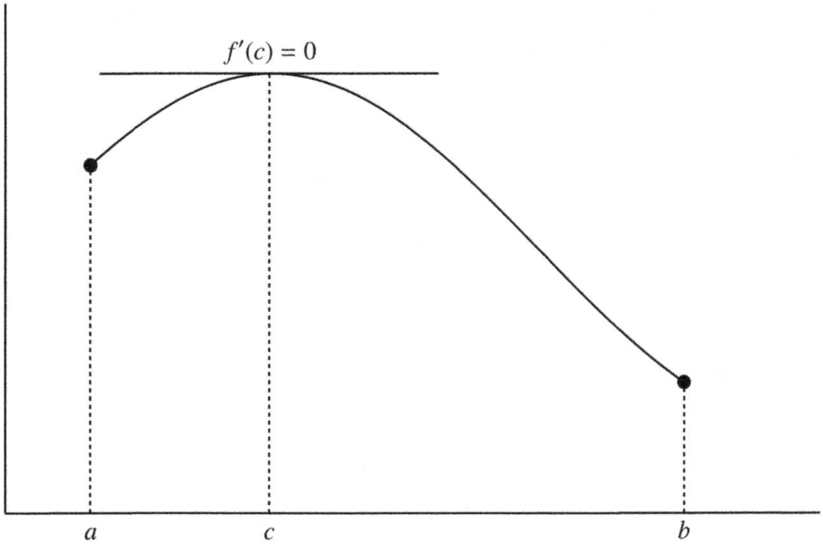

Figure A.17 Maximum point on $f(x)$.

The Cauchy mean value theorem is stated as:

For functions f and g that are each continuous and differentiable on the interval $[a,b]$ and for which g' is non-zero on $[a,b]$ there exists a number $a<c<b$ such that:

$$\frac{f(b)-f(a)}{g(b)-g(a)}=\frac{f'(c)}{g'(c)}$$

The proof of this invokes Rolle's theorem. First, a new function h is defined on the interval $[a,b]$:

$$h(x)=[f(x)-f(a)]\,[g(b)-g(a)]-[f(a)-f(a)]\,[g(x)-g(a)]$$

From this definition we see that $h(a)=0$ and that $h(b)=0$ and therefore $h(a)=h(b)$. Rolle's theorem then implies that:

$$h'(c)=0=f'(c)\,[g(b)-g(a)]-[f(b)-f(a)]\,g'(c)$$

and:

$$f'(c)[g(b)-g(a)]=[f(b)-f(a)]g'(c)$$

$$\frac{f(b)-f(a)}{g(b)-g(a)}=\frac{f'(c)}{g'(c)}$$

Now for L'Hopital's rule. Briefly stated, it is:

$$\text{if } \lim_{x\to a} f(x)=0 \quad \text{and} \quad \lim_{x\to a} g(x)=0$$

and

$$\lim_{x\to a}\frac{f'(x)}{g'(x)}=K \quad (g'(x)\neq 0)$$

then

$$\lim_{x\to a}\frac{f(x)}{g(x)}=K$$

Using the Cauchy mean value theorem on the interval $[x,a]$ we write:

$$\frac{f'(c)}{g'(c)}=\frac{f(x)-f(a)}{g(x)-g(a)}=\frac{f(x)}{g(x)}$$

since $f(a)=g(a)=0$. As stated in the postulate, $\lim_{x\to a}\frac{f'(x)}{g'(x)}=K$, so we may write:

$$\lim_{x\to a}\frac{f'(x)}{g'(x)}=\lim_{x\to a}\frac{f(x)}{g(x)}=K$$

A.20 Even and Odd Functions

Odd functions have the property:

$$f(t)=-f(-t) \tag{A.53}$$

Even functions have the property:

$$f(t)=f(-t) \tag{A.54}$$

For example, sine is an odd function and is *anti-symmetric* about the y-axis and cosine is even and is *symmetric* about the y-axis. This is very easily seen in the Figure A.18a and b.

$$\begin{aligned} \sin(x) &= -\sin(-x) \\ \cos(x) &= \cos(-x) \end{aligned} \tag{A.55}$$

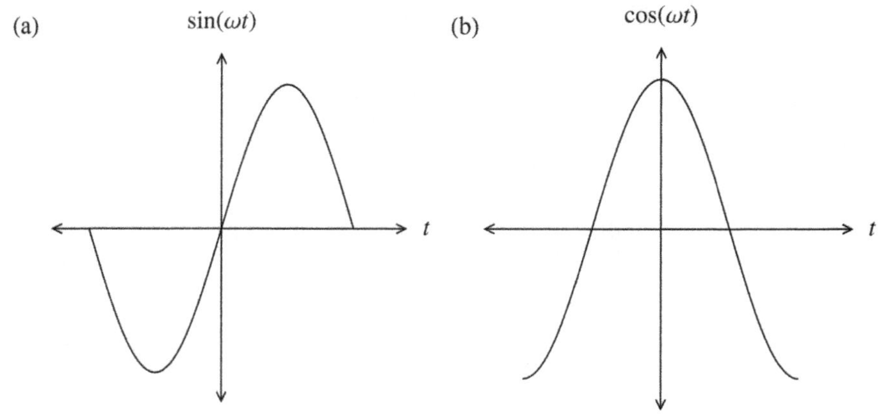

Figure A.18 (a) $f(t) = \sin(\omega t)$, an asymmetric function; (b) $f(t) = \cos(\omega t)$, a symmetric function.

In general, each of these types of functions has some useful properties:

$$\begin{aligned} &\int_{-a}^{a} f(x)\mathrm{d}x = 2\int_{0}^{a} f(x)\mathrm{d}x \quad \text{(even functions)} \\ &\int_{-a}^{a} f(x)\mathrm{d}x = 0 \quad \text{(odd functions)} \end{aligned} \tag{A.56}$$

$$\begin{aligned} &\text{if} \quad \sum_{n=0}^{a} f_n(nx) = \sum_{n=-a}^{0} f_n(-nx) \\ &\text{then} \quad \sum_{n=-a}^{a} f_n(nx) = f(0) + 2\sum_{n=1}^{a} f_n(nx) \end{aligned} \tag{A.57}$$

$$\begin{aligned} &\text{if} \quad \sum_{n=1}^{\infty} f_n(nx) = \sum_{n=1}^{\infty} f_n(-nx) \\ &\text{then} \quad \sum_{n=1}^{\infty} f_n(nx) = \sum_{n=-\infty}^{-1} f_n(nx) \end{aligned} \tag{A.58}$$

$$\begin{aligned} &\text{if} \quad \sum_{n=1}^{\infty} f_n(nx) = \sum_{n=1}^{\infty} -f_n(-nx) \\ &\text{then} \quad \sum_{n=1}^{\infty} f_n(nx) = -\sum_{n=-\infty}^{-1} f_n(nx) \end{aligned} \tag{A.59}$$

Referring to Figure A.18, we can see that for sine and cosine:

$$\sin(\omega t) + \sin(-\omega t) = 0$$

$$\cos(\omega t) + \cos(-\omega t) = 2\cos(\omega t)$$

This also follows from the definitions of even and odd functions, eqn (A.53) and (A.54):

$$\begin{aligned} f(t)_{\text{odd}} + f(-t)_{\text{odd}} &= 0 \\ f(t)_{\text{odd}} - f(-t)_{\text{odd}} &= 2f(t)_{\text{odd}} \\ f(t)_{\text{even}} + f(-t)_{\text{even}} &= 2f(t)_{\text{even}} \\ f(t)_{\text{even}} - f(-t)_{\text{even}} &= 0 \end{aligned} \tag{A.60}$$

Now, suppose we have a periodic function that is neither even nor odd. Is it reasonable to presume that it can be broken down into the sum of even and odd parts? Let us try a very simple linear combination:

$$g(t) = f(t)_{\text{even}} + f(t)_{\text{odd}}$$

Given this we write (using eqn (A.60)):

$$\begin{aligned} g(t) + g(-t) &= \left[f(t)_{\text{even}} + f(-t)_{\text{even}}\right] + \left[f(t)_{\text{odd}} + f(-t)_{\text{odd}}\right] \\ &= \left[f(t)_{\text{even}} + f(t)_{\text{even}}\right] + 0 \\ &= 2f(t)_{\text{even}} \\ f(t)_{\text{even}} &= \frac{1}{2}[g(t) + g(-t)] \end{aligned} \tag{A.61}$$

also:

$$\begin{aligned} g(t) - g(-t) &= \left[f(t)_{\text{even}} - f(-t)_{\text{even}}\right] + \left[f(t)_{\text{odd}} - f(-t)_{\text{odd}}\right] \\ &= 0 + \left[f(t)_{\text{odd}} - f(-t)_{\text{odd}}\right] \\ &= 2f(t)_{\text{odd}} \\ f(t)_{\text{odd}} &= \frac{1}{2}[g(t) - g(-t)] \end{aligned} \tag{A.62}$$

Finally, the products of even and odd functions are of interest:

$$\begin{aligned} f(t)_{\text{odd}} \times g(t)_{\text{odd}} &= (f(t)g(t))_{\text{even}} \\ f(t)_{\text{even}} \times g(t)_{\text{odd}} &= (f(t)g(t))_{\text{odd}} \\ f_{(t)\text{even}} \times g(t)_{\text{even}} &= (f(t)g(t))_{\text{even}} \end{aligned} \tag{A.63}$$

These are easily shown to be so using the definitions of even and odd functions, eqn (A.53) and (A.54):

$$\text{let}\quad f(t) = f(-t) \quad \text{and} \quad g(t) = -g(-t)$$

$$\text{then}\quad f(t)g(t) = h(t) = -f(-t)g(-t) = -h(-t)$$

and we see that an even function times an odd function produces an odd function. The reader may confirm the other equations in eqn (A.63).

A.21 Group Theory

The basic idea behind group theory is encapsulated in a very short set of simple rules. Consider an equation:

$$a \cdot b = c \tag{A.64}$$

From the group theory point of view, we make no assumptions about what a, b and c are nor do we specify what the binary operation, $\cdot$, is. There are four main rules that govern these object's eligibility for group membership.

1. Closure: if a and b are group members then $a \cdot b$ is also a member.
2. Associativity: if a, b and c are group members then $(a \cdot b) \cdot c = a \cdot (b \cdot c)$.
3. Identity: There is a member, e, of the group such that $a \cdot e = e \cdot a = a$.
4. Inverse: For all members, a, of the group there is a corresponding member a^{-1} such that $a \cdot a^{-1} = e$ and $a^{-1} \cdot a = e$.

A.21.1 Closure

We will consider examples of mathematical objects that are group members under the closure rule and some that are not. First, the set of integers is closed under addition, subtraction and multiplication. In this case, the operator, $\cdot$, is "+" or "×" and a, b and c are integers. They are not, however, closed under division. One may take any two integers and add, subtract or multiply them together to produce another number that is also an integer. Division on the other hand may produce a non-integer result and therefore does not satisfy the closure rule in general.

$$2 + 6 = 8 \quad \text{(closed)}$$

$$4 - 9 = -5 \quad \text{(closed)}$$

$$3 \times 7 = 21 \quad \text{(closed)}$$

$$5/2 = 2.5 \quad \text{(not closed)}$$

Three-dimensional vectors are closed under addition, subtraction and cross product multiplication but not dot product multiplication since the result of the dot product is a scalar rather than a vector. Square matrices are closed under addition, subtraction and "normal" multiplication but not under outer-product multiplication since this operation changes the dimensions of the matrices. Vector matrices subjected to either inner or outer product multiplication are not closed since they produce new objects that are not members of the set of vectors.

$$\vec{\mathbf{a}} + \vec{\mathbf{b}} = \vec{\mathbf{c}} \quad \text{(closed)}$$

$$\vec{\mathbf{a}} - \vec{\mathbf{b}} = \vec{\mathbf{d}} \quad \text{(closed)}$$

$$\vec{\mathbf{a}} \times \vec{\mathbf{b}} = \vec{\mathbf{g}} \quad \text{(closed)}$$

$$\vec{\mathbf{a}} \cdot \vec{\mathbf{b}} = \lambda \quad \text{(not closed)}$$

$$\vec{\mathbf{a}} \otimes \vec{\mathbf{b}} = \hat{\mathbf{A}} \quad \text{(not closed)}$$

Non-conjugate complex numbers under addition and multiplication are closed.

$$(a + ib) + (c + id) = (a + c) + i(b + d) \quad \text{(closed)}$$

$$(a + ib) \times (c + id) = (ac - bd) + i(ad + bc) \quad \text{(closed)}$$

Conjugate complex numbers are not closed under addition.

$$(a + ib) + (a - ib) = 2a \quad \text{(not closed)}$$

$$(a + ib) \times (a - ib) = a^2 + b^2 \quad \text{(not closed)}$$

A.21.2 Associativity

The addition and multiplication of integers and real numbers is associative. One may add or multiply three numbers in arbitrary order to arrive at the same result. Subtraction and division of these objects is not associative.

$$(3+4)+6=3+(4+6) \text{ (associative)}$$

$$(4.3+7.1)+1.35=4.3+(7.1+1.35) \text{ (associative)}$$

$$(5\times 2)\times 6=5\times(2\times 6) \text{ (associative)}$$

$$(3-4)-6\neq 3-(4-6) \text{ (non-associative)}$$

Matrices are associative under addition but not subtraction or multiplication.

$$\left(\begin{bmatrix}1 & 2\\3 & 4\end{bmatrix}+\begin{bmatrix}5 & 6\\7 & 8\end{bmatrix}\right)+\begin{bmatrix}7 & 6\\5 & 4\end{bmatrix}=\begin{bmatrix}1 & 2\\3 & 4\end{bmatrix}+\left(\begin{bmatrix}5 & 6\\7 & 8\end{bmatrix}+\begin{bmatrix}7 & 6\\5 & 4\end{bmatrix}\right) \text{ (associative)}$$

$$\left(\begin{bmatrix}1 & 2\\3 & 4\end{bmatrix}-\begin{bmatrix}5 & 6\\7 & 8\end{bmatrix}\right)-\begin{bmatrix}7 & 6\\5 & 4\end{bmatrix}\neq\begin{bmatrix}1 & 2\\3 & 4\end{bmatrix}-\left(\begin{bmatrix}5 & 6\\7 & 8\end{bmatrix}-\begin{bmatrix}7 & 6\\5 & 4\end{bmatrix}\right) \text{ (non-associative)}$$

$$\left(\begin{bmatrix}1 & 2\\3 & 4\end{bmatrix}\times\begin{bmatrix}5 & 6\\7 & 8\end{bmatrix}\right)\times\begin{bmatrix}7 & 6\\5 & 4\end{bmatrix}\neq\begin{bmatrix}1 & 2\\3 & 4\end{bmatrix}\times\left(\begin{bmatrix}5 & 6\\7 & 8\end{bmatrix}\times\begin{bmatrix}7 & 6\\5 & 4\end{bmatrix}\right) \text{ (non-associative)}$$

Vectors are associative under addition but not under subtraction or cross product multiplication. Associativity for the dot product is not defined.

$$(\vec{\mathbf{a}}+\vec{\mathbf{b}})+\vec{\mathbf{c}}=\vec{\mathbf{a}}+(\vec{\mathbf{b}}+\vec{\mathbf{c}}) \text{ (associative)}$$

$$(\vec{\mathbf{a}}-\vec{\mathbf{b}})-\vec{\mathbf{c}}\neq\vec{\mathbf{a}}-(\vec{\mathbf{b}}-\vec{\mathbf{c}}) \text{ (non-associative)}$$

$$(\vec{\mathbf{a}}\times\vec{\mathbf{b}})\times\vec{\mathbf{c}}\neq\vec{\mathbf{a}}\times(\vec{\mathbf{b}}\times\vec{\mathbf{c}}) \text{ (non-associative)}$$

Complex numbers are associative under addition and multiplication but not subtraction.

$$((a+ib)+(c+id))+(e+if)=(a+ib)+((c+id)+(e+if)) \text{ (associative)}$$

$$((a+ib)-(c+id))-(e+if)\neq(a+ib)-((c+id)-(e+if)) \text{ (non-associative)}$$

$$((a+ib)\times(c+id))\times(e+if)=(a+ib)\times((c+id)\times(e+if)) \text{ (associative)}$$

A.21.3 Identity

The integers have the identity member 0 for addition and 1 for multiplication.

$$a+0=0+a=a \quad \text{(additive identity)}$$

$$a\times 1=1\times a=a \quad \text{(multiplicative identity)}$$

The zero matrix is the identity member under addition for matrices and the unity matrix for "normal" multiplication.

$$\begin{bmatrix} a & b \\ c & d \end{bmatrix}+\begin{bmatrix} 0 & 0 \\ 0 & 0 \end{bmatrix}=\begin{bmatrix} 0 & 0 \\ 0 & 0 \end{bmatrix}+\begin{bmatrix} a & b \\ c & d \end{bmatrix}=\begin{bmatrix} a & b \\ c & d \end{bmatrix} \quad \text{(additive identity)}$$

$$\begin{bmatrix} a & b \\ c & d \end{bmatrix}\times\begin{bmatrix} 1 & 0 \\ 0 & 1 \end{bmatrix}=\begin{bmatrix} 1 & 0 \\ 0 & 1 \end{bmatrix}\times\begin{bmatrix} a & b \\ c & d \end{bmatrix}=\begin{bmatrix} a & b \\ c & d \end{bmatrix} \quad \text{(multiplicative identity)}$$

The zero vector is the identity member for vectors under addition. There is no identity member for multiplication since the dot product produces a scalar and the cross product produces a new vector orthogonal to the plane of the two product vectors.

$$\vec{\mathbf{a}}+\vec{\mathbf{0}}=\vec{\mathbf{0}}+\vec{\mathbf{a}}=\vec{\mathbf{a}} \text{ (additive identity)}$$

For complex numbers the identity member under addition is $0+0i$.

$$(a+bi)+(0+0i)=(0+0i)+(a+bi)=(a+bi) \text{ (additive identity)}$$

There is no multiplicative identity member for complex numbers.

A.21.4 Inverse

For integers and real numbers the inverse under addition is the negative of the number.

$$a+a^{-1}=0 \text{ (additive inverse)}$$

There is no multiplicative inverse for the integers but for the reals it is $1/a$.

$$a\times a^{-1}=1 \text{ (multiplicative inverse)}$$

The additive inverse for matrices is very similar to that of the integers.

$$\begin{bmatrix} 1 & 2 \\ 3 & 4 \end{bmatrix} + \begin{bmatrix} -1 & -2 \\ -3 & -4 \end{bmatrix} = \begin{bmatrix} 0 & 0 \\ 0 & 0 \end{bmatrix} \quad \text{(additive inverse)}$$

The non-singular matrices have a multiplicative inverse.

$$\begin{bmatrix} 1 & 2 \\ 3 & 4 \end{bmatrix} \times \begin{bmatrix} -2 & 1 \\ \frac{3}{2} & -\frac{1}{2} \end{bmatrix} = \begin{bmatrix} 1 & 0 \\ 0 & 1 \end{bmatrix} \quad \text{(multiplicative inverse)}$$

Complex numbers have an additive and a multiplicative inverse.

$$(a + ib) + (-a - ib) = 0 + 0i \quad \text{(additive inverse)}$$

$$(a + ib) \times \left(\frac{a}{a^2 + b^2} + i\frac{b}{a^2 + b^2} \right) = 1 \quad \text{(multiplicative inverse)}$$

Appendix B A Bruker to Varian Translation Table

Unfortunately, during the writing of this text, Varian (or rather, Agilent) became non-existent. However, there may still occasionally be a need to translate between the two, perhaps when reading literature descriptions of pulse sequences. We provide a translation table here (Table B.1).

Table B.1 Bruker to Varian translation table.

Parameter	Bruker	Varian
Spectral width	SW	Sw
Offset frequency	O1	Tof
Center decoupling frequency	O2	Dof
Relaxation delay	D1	D1
Acquisition time	AQ	At
Number of scans	NS	Nt
Dummy scans	DS	Ss
Receiver gain	RG	Gain
Number of data points	TD	Np
Temperature	TE	Temp
Fourier transform	FT	Ft
Line broadening	LB	Lb
Exponential window function and Fourier transform	EF	Wft
Automatic phase correction	APK	Aph
Zero-order phase correction parameter	PH0	Rp
1st-order phase correction parameter	PH1	Lp

Essential Mathematics for NMR and MRI Spectroscopists
By Keith C. Brown

Published by the Royal Society of Chemistry, www.rsc.org

Subject Index

Charts and tables are in **bold**.